Super-Intense Laser-Atom Physics IV

NATO ASI Series

Advanced Science Institutes Series

A Series presenting the results of activities sponsored by the NATO Science Committee, which aims at the dissemination of advanced scientific and technological knowledge, with a view to strengthening links between scientific communities.

The Series is published by an international board of publishers in conjunction with the NATO Scientific Affairs Division

A	**Life Sciences**	Plenum Publishing Corporation
B	**Physics**	London and New York
C	**Mathematical and Physical Sciences**	Kluwer Academic Publishers
D	**Behavioural and Social Sciences**	Dordrecht, Boston and London
E	**Applied Sciences**	
F	**Computer and Systems Sciences**	Springer-Verlag
G	**Ecological Sciences**	Berlin, Heidelberg, New York, London,
H	**Cell Biology**	Paris and Tokyo
I	**Global Environmental Change**	

PARTNERSHIP SUB-SERIES

1.	**Disarmament Technologies**	Kluwer Academic Publishers
2.	**Environment**	Springer-Verlag / Kluwer Academic Publishers
3.	**High Technology**	Kluwer Academic Publishers
4.	**Science and Technology Policy**	Kluwer Academic Publishers
5.	**Computer Networking**	Kluwer Academic Publishers

The Partnership Sub-Series incorporates activities undertaken in collaboration with NATO's Cooperation Partners, the countries of the CIS and Central and Eastern Europe, in Priority Areas of concern to those countries.

NATO-PCO-DATA BASE

The electronic index to the NATO ASI Series provides full bibliographical references (with keywords and/or abstracts) to more than 50000 contributions from international scientists published in all sections of the NATO ASI Series.
Access to the NATO-PCO-DATA BASE is possible in two ways:

– via online FILE 128 (NATO-PCO-DATA BASE) hosted by ESRIN,
Via Galileo Galilei, I-00044 Frascati, Italy.

– via CD-ROM "NATO-PCO-DATA BASE" with user-friendly retrieval software in English, French and German (© WTV GmbH and DATAWARE Technologies Inc. 1989).

The CD-ROM can be ordered through any member of the Board of Publishers or through NATO-PCO, Overijse, Belgium.

Series 3: High Technology – Vol. 13

Super-Intense
Laser-Atom Physics IV

edited by

H. G. Muller

FOM-Institute for Atomic and Molecular Physics,
Amsterdam, The Netherlands

and

M. V. Fedorov

General Physics Institute,
Academy of Sciences,
Moscow, Russia

Kluwer Academic Publishers

Dordrecht / Boston / London

Published in cooperation with NATO Scientific Affairs Division

Proceedings of the NATO Advanced Research Workshop on
Super-Intense Laser-Atom Physics (SILAP IV)
Moscow, Russia
August 5–9, 1995

A C.I.P. Catalogue record for this book is available from the Library of Congress.

ISBN-13: 978-94-010-6601-3   e-ISBN-13: 978-94-009-0261-9
DOI: 10.1007/978-94-009-0261-9

Published by Kluwer Academic Publishers,
P.O. Box 17, 3300 AA Dordrecht, The Netherlands.

Kluwer Academic Publishers incorporates the publishing programmes of
D. Reidel, Martinus Nijhoff, Dr W. Junk and MTP Press.

Sold and distributed in the U.S.A. and Canada
by Kluwer Academic Publishers,
101 Philip Drive, Norwell, MA 02061, U.S.A.

In all other countries, sold and distributed
by Kluwer Academic Publishers Group,
P.O. Box 322, 3300 AH Dordrecht, The Netherlands.

Softcover reprint of the hardcover 1st edition 1996

Table of contents

Molecules

CHAPTER II - MULTI-ELECTRON ATOMS

Correlation effects

Coherence transfer

Multiphoton multiple ionization

CHAPTER III - HARD RADIATION QUANTA

X-rays

Two-color processes

Compton scattering

High-harmonic generation

CHAPTER IV - COHERENCE AND INTERFERENCE

Non-linear light propagation

Coherence, interference and wavepackets

PREFACE

Atoms in strong radiation fields are interesting objects for study, and the research field that concerns itself with this study is a comparatively young one. For a long period after the discovery of the photoelectric effect, it was not possible to generate electromagnetic fields that did more than perturb the atom only slightly, and (first-order) perturbation theory could perfectly explain what was going on at those low intensities. The development of the pulsed laser has changed this state of affairs in a rather dramatic way, and fields can be applied that really have a large, or even dominant influence on atomic structure. In the latter case, we speak of *super-intense* fields.

Since the interaction between atoms and electromagnetic waves is characterized by many parameters other than the light intensity, such as frequency, ionization potential, orbit time, etc., it is actually quite difficult to define what is exactly meant by the term 'super-intense'. Obviously the term does not have an absolute meaning, and intensity should always be viewed in relation to other properties of the system. An atom in a radiation field can thus best be described in terms of various ratios of the quantities involved. The nature of the system sometimes drastically changes if the value of one of these parameters exceeds a certain critical value, and the new regime could be called super-intense with respect to that parameter.

One definition calls an electromagnetic field super-intense, when the force exerted by it on an atomic (or molecular) electron is larger than the force that binds the electron to the atom (or molecule). For the archetypal atom, hydrogen, in its ground state, this would only occur above an intensity of $3\,5\ 10^{16}$ W/cm^2, often (but unjustly) called the atomic unit of intensity. Laser technology has progressed to a state that such intensities can be routinely achieved, even with kHz rep-rates.

But just as important as the ability to create such intensities, is the ability to create them in pulses with sufficiently fast rise times. A super-intense laser pulse would be of little use, if none of the target atoms could be exposed to it because they would all have decayed at much lower intensities during the rising edge of the pulse. This problem, sometimes referred to as the Lambropoulos curse, has plagued high-intensity laser research from the very beginning, and explains the drive towards ever shorter pulses. It is especially nasty in the super-intense regime, since (by definition) even frequencies approaching DC will be able to ionize the atoms by simply pulling away their electrons. thus the commonly applied trick for having the target atoms survive the pulse wings, studying very-high-order processes with low-energy photons, is no longer a viable solution.

In the super-intense regime, nothing can stop the electron from reacting to the electromagnetic force. That there is any atomic structure at all at such intensities, is due to the fact that light is purely an AC-phenomenon, and inertia of the electron limits its effects. Although the electron will be forced into an oscillatory motion known as *quiver motion*, this does not prevent it from hanging around the nucleus for an extended period of time before it acquires enough linear velocity ('drift motion', in SILAP jargon) to escape the atomic potential well permanently.

With the definition of super-intense given above, sometimes super-intense fields can be comparatively weak on an absolute scale, if the system under study is sufficiently weakly bound. From noble gases through alkali atoms, excited states, negative ions, to high Rydberg states, the binding gets progressively weaker. For example, a Rydberg state with principal quantum number n around 30 has the electron dwelling about 10^3 au from the nucleus, experiencing a Coulomb force of 10^{-6}, so that fields 10^{12} times smaller than the huge intensity quoted above (i.e. several kW/cm^2) are already super intense. Studying such systems is thus a good way to beat technological limitations on light-intensities, frequencies and pulse durations.

Several other definitions for super-intense can be employed as well. A famous parameter for describing atom-light interaction is the Keldysh parameter γ, giving the intensity (in terms of the quiver energy it produces) in relation to the ionization potential ($\gamma^2 = IP/U_p$). If the intensity gets large ($\gamma < 1$), the situation changes qualitatively, and the atom ionizes by means of barrier tunneling rather than multiphoton ionization. Seen in this light, tunneling could be considered a super-intense phenomenon.

As a measure of intensity one could also use the coupling between states (in terms of a matrix element), and consider this in relation to unity, the binding energy, the level spacing or the photon energy. For ground-state atoms, definitions of super-intense based on these parameters more or less coincide, but for excited states they might diverge from each other as the state approaches the continuum. For states in the continuum, i.e. free-free scattering in the presence of a laser field, critical intensities occur when the maximum quiver amplitude equals the average (drift) velocity of the electron, or when the ponderomotive energy equals the photon energy.

In almost all cases, quiver motion is the dominant feature of all atomic physics in the super-intense regime. Nevertheless, this can result in a very wide variety of phenomena, such as ionization, emission of radiation, exchange of excess photons with the field, double ionization, etcetera. All these aspects fall within the interest sphere of, and were indeed addressed at the series of 'Super-Intense Laser-Atom Physics' conferences. In the following chapters an attempt is made to logically group the various contributions from the attendants of the fourth conference in this series, (SILAP IV), which took place in August 1995 as a part of the LPHYS'95 event held on a boat cruising the river Volga near Moscow. The meeting was a very successful, attracting 89 scientists from 17 countries.

SILAP IV was organized as a NATO 'Advanced Research Workshop', and as both editors of this volume, and organizers of the conference, we gratefully acknowledge the support received from the NATO Science Committee, as well as that from several other funding agencies, such as the U.S. Army, the Russian Fund for Basic Research and the Laser Physics journal. Many citizens of Eastern Europe received individual support from Soros' International Science Foundation, enabling them to attend this conference and contribute to this volume. In addition we want to express our thanks to the General Physics Institute and the FOM-Institute for Atomic and Molecular Physics for the effort they put into organizing the meeting and preparing this volume, especially to Louise Roos, Ernie Lammers and Liudmila Kuzmina. We hope this compilation will show that these efforts and the generosity of the sponsors has not been wasted, and that, apart from reporting on the conference, the book will make an entertaining read and provide a good overview of the state of the art in atoms in strong radiation fields.

H.G. Muller   M.V. Fedorov

GENERAL INTRODUCTION TO THE CHAPTERS

1. Strong-field photoionization

1.1 STABILIZATION

One of the surprising aspects of the super-intense regime, is that atomic structure is present at all. Under some conditions, ionization becomes even less rapid with increasing intensity. This stabilization of atoms in super-intense conditions, has been hotly debated during the past decade, but recently the remains of this discussion have crystallized into something like a consensus. Several different types of stabilization, each with its own phenomenology and mechanism, are distinguished now. *Transient stabilization* is the temporary suppression of ionization due to the fact that it takes time for the electron to move to the nucleus in order to be kicked out of its orbit by it. *Interference stabilization* is the effect that the probability to be kicked out on such an encounter decreases with intensity. In *adiabatic stabilization*, sufficiently hard encounters get less frequent because the electron tends to avoid the nucleus altogether and gets trapped in a greatly expanded wave function. Finally, all kinds of frequency-specific effects can cause *resonance stabilization* of the atom.

Convincing demonstration experiments have been performed for transient stabilization and adiabatic stabilization, the latter described in this chapter. The situation has now progressed to the point where actual applications of the effect are contemplated, and the contribution on the use of stabilization as part of an x-ray laser scheme with enhanced efficiency is an example of this. Also included is an imaginative contribution on the stabilization of circular Rydberg states in orbits similar to those of Trojan asteroids, that might defy the established classification of stabilization, and therefore be the first example of an entirely new class of stabilization mechanisms.

The large role that classical mechanics plays in the explanations and theories put forward in this chapter is amazing, and suggests that stabilization is not exclusively a quantum effect. The classical counterparts of atomic systems in fields attract much interest by themselves in an unrelated (at least that's what we thought!) field of physics, the study of chaos. Two papers in this chapter make the connection.

1.2 STRONG-FIELD ATI

The discovery of Above-Threshold Ionization, (also referred to as Excess-Photon Ionization), now nearly two decades ago, marked the beginning of the research on super-intense laser-atom interaction. The absorption of excess photons accompanying multi-photon ionization, which defines this effect, was the first manifestation of the unrestricted exchange of photons with the field that characterizes the super-intense regime.

'Old fashioned' ATI still has plenty of surprises in store, and still is an active subject of research. One of the continuing points of interest is the study of ATI in atomic hydrogen, an experimentally difficult undertaking, but very rewarding, since it is the only case for which theory is on nearly exact grounds. That even here experiments can come up with surprising results, which must have dynamical origins hitherto neglected in theory (the latter often being based on adiabatic assumptions), was shown by the recent discover of an unexpected ATI series in ionization of atomic H with fs-pulses.

One of the major discoveries the past years was made possible by the development of kHz laser systems, and the resulting high dynamic range in count rate that could be probed. This revealed an irregular decrease of the magnitude of ATI peaks with order, sometimes referred to as the ATI 'plateau' (in analogy with High-Harmonic Generation).

The detailed study of the high-energy tail of the ATI spectrum points to more difference than analogy, though, for the sharp drop-off with order that delimits the plateau on the high-energy side in HHG seems to be absent in ATI spectra. The established explanation of ATI in terms of quiver motion, held that the difference in energy of the photoelectrons in the various ATI peaks was the result of the different time of ejection (classically speaking) of these electrons with respect to the zero crossings of the light field. These different electrons would then start their free life with different amounts of acceleration. The maximum energy would occur for electrons born near a zero crossing, and is 3 times the ponderomotive energy U_p.

In fact the ATI spectra extend to surprisingly high order, much farther than the $3U_p$ expected from quiver dynamics. It is thus likely that rescattering of the photo-electron by the parent ion plays a significant role, and this rescattering seems to be accomplished (for reasons not yet completely understood) by peculiar structures in the angular distribution, called scattering jets. High-order ATI is one of the front ends of this research field, posing many questions that are still open, as will be discussed in this chapter.

1.3 MOLECULES

In the super-intense regime molecules are much destabilized, because the quiver motion tends to disperse the electron clouds, and thus exposes the charge of the bare nuclei. Due to the large inertia of nuclei compared to electrons, the behavior of molecules in short laser pulses is strongly non-adiabatic. This means that in many experiments, molecules do not relax to the much more weakly bound equilibrium structure they might have in the field, but in stead find themselves after removal of several electrons so high on the repulsive part of their potential curves, that they fall apart.

This molecular dissociation by Coulomb explosion is an interesting subject for research. Most important questions are how exactly the fragmentation takes place, and what the order of the various ionization and dissociation events leading to the observed final products is. Correlation and coincidence measurements of fragments of different energy is a powerful tool to figure out the answer to such questions.

In slightly less violent environments, ionization does not occur immediately, but resonant excitation of the molecules taking place at certain internuclear distances completely reshapes the nuclear potential curves, an effect called bond softening. Electronic transition rates are usually so high that adiabatic passage from one electronic state to the other (accompanied by absorption or emission of photons) takes place at such points with high probability or even near certainty. This photon exchange affects the kinetic energy at which the fragments appear by (a multiple of) the photon energy, an effect called Above-Threshold Dissociation because of its phenomenological likeness to Above-Threshold Ionization.

xiii

2. Many-body effects in strong fields

2.1 ELECTRON CORRELATION

Most processes in single-electron systems are by now well understood, and it seems only natural that the center of gravity of the research effort is shifting towards the next challenge, the interplay of correlation and super-intense effects. The obvious model systems here are two-electron atoms, and especially theorists are undertaking heroic efforts to come to an understanding of those systems.

Correlation effects can manifest themselves in various ways, from shifts in energy levels due to screening to processes such as autoionization, in which transfer between electrons of substantial fractions of their energy takes place. This interaction between electrons can be very important, and in the super-intense regime it is qualitatively different from that in unperturbed atoms: In a multi-electron atom, not all electrons might experience intensity in the same way. Therefore weakly bound electrons can be in the super-intense regime and perform wild quiver motions, while more-tightly bound electrons of inner shells are still tied to the nucleus. Interactions between electrons of these different classes can be very violent, leading to unusual effects.

Even interactions between electrons that are all in the super-intense regime can lead to funny effects. Since all these electrons quiver in phase under the effect of the external radiation field, their mutual interaction is affected differently from that with the stationary nucleus, leading to a complete change of atomic shell structure, and to the existence of multiply charged negative ions.

2.2 COHERENCE TRANSFER

One special case of correlation, known as coherence transfer, is that between an inner (or core) electron, and an outer electron in an excited state. If electrons were independent particles, the energy of the core transition would be completely independent of the state the outer electron was in, or even on the fact if it was present at all. But correlation does make the core transition shift, making it possible, for instance, to resonantly drive the core transition in the ion, and not in the atom.

Different games can be played in such systems, often leading to structure in the electron spectrum due to the huge Autler-Townes splitting of the resonant transition. Theoretically, even the possibility for 'zero-photon ionization' exists, when the Rabi frequency of the core transition becomes larger than the binding energy of the outer electron, and the oscillating field generated by the core ejects the outer electron. If after the pulse the core electron ends up in the same state as it started, no net photon absorption has taken place in this ionization event.

2.3 DIRECT MULTIPLE IONIZATION

Multiphoton multiple ionization is at the moment one of the hottest topics in super-intense laser-atom physics. In contrast to multiple ionization with a single photon, which is quite common and occurs (due to electron correlation) in significant amounts (several percent) as soon as the photon energy exceeds the sum of the applicable ionization energies, the corresponding multiphoton process has proved extremely elusive. The reason for this is that multiple ionization has to compete with single ionization, which in the multiphoton case always can occur by a lower-order process.

This does not mean that multiple ionization can not occur, just that it is extremely likely to occur through a stepwise, or sequential process, rather than in a direct event. This is just a manifestation of a the quite general Lambropoulos curse, mentioned in the preface. Higher-order processes usually are completely masked by lower-order ones, since these latter run to completion in the wings of the laser pulse, before the intensity gets high enough to drive the higher-order process in measurable amounts. Once the intensity does get high enough, the initial state (in this case a neutral atom) will be completely depleted, and there will be no targets left for the higher-order effect to occur in.

So even if theoretical calculations would show that above a certain intensity, direct ejection of several electrons from the neutral atom were a significant, or even dominant occurrence, this effect will not be observable if the rising edge of the pulse leading to this intensity is not fast enough to make any neutrals survive to that intensity. Although several tricks can be imagined to enhance the probability for the neutral to survive up to the intensity required for double electron ejection, (e.g. lowering this intensity at which it occurs by using a doubly excited state as a resonance, and using the shortest pulses that can be technologically produced), such attempts have met with little success so far.

Nevertheless, reports of direct double ionization have been around since the early days of MPI, based on the yield of doubly charged ions as a function of intensity. Analyzing these yields in terms of rate equations, the comparative excess of ions at low intensities suggested there was another mechanism than ionization of ground-state *ions* operative when neutrals were still present. This could very well be direct double ionization.

The main problem is, that it could be many other things as well, such as sequential ionization to an excited ion state, followed by subsequent rapid ionization of that excited state. From measuring the charge/mass ratio of ions such excited states can not be distinguished. Alternative explanations based on resonances in the second ionization step at certain (low) intensities are conceivable as well.

The effect of the doubly-charged-ion excess has recently been observed more clearly than ever before in double ionization of He with the Ti:Sapph laser (described in the Di-Mauro paper in the ATI section.) In this system the alternative explanations are unlikely, and it is believed that we are dealing with the direct process here. A strange thing about the kinetics is that the direct double ejection at low intensities seems to have a higher rate than the second ionization step, although the latter is necessarily of lower order!

The *self-knockout* process, where the photoelectron from the first ionization step recollides with its parent ion after having accumulated enough energy (in the course of its quiver motion) to knock out the second electron, was the first mechanism to be forwarded that could explain this. But a quantitative evaluation of this mechanism suffers from discrepancies, since the amount of excess it predicts falls short of that measured by more than an order of magnitude.

The origin of the double-ion excess thus is hotly debated. The self-knockout mechanism is supported by the polarization dependence of the excess, while the absolute levels, and the fact that some excess persists below the theoretical threshold for the self-knockout process, points to a different mechanism. Apart from discussing these problems in detail, there is also a contribution in this chapter that reports experimental observation of what might be direct triple ionization.

3. Hard Radiation Quanta

3.1 HIGH-HARMONIC GENERATION

Ever since its discovery in 1988, the generation of high harmonics belongs to the main stream of research in high-intensity laser physics. Before that time, generation of low harmonics (3rd or 5th) from laser pulses focused in gas jets was well known, and had even evolved to a standard way for generating coherent radiation in the VUV. Main disadvantage of the technique was that the efficiency was very low, and dropped very rapidly with harmonic order, in accordance with perturbative predictions.

The obvious way for generating short-wavelength radiation therefore seemed starting from as small an optical wavelength as possible, if necessary by using frequency doubling or mixing in non-linear crystals, so that the order of the process in the gas jet could be kept as low as possible. How wrong we were!

The exciting, and initially completely baffling aspect of the high harmonics was that the production efficiency did not continue to drop off rapidly, but stayed approximately constant for a wide range of harmonics, from about 15eV to (in some cases) above 100eV. Above this 'plateau' of constant efficiency, the intensity dropped very quickly to an unobservable level. This behavior violates the basic premises of perturbation theory, and, even more surprising, low-frequency driving lasers turned out to generate much wider plateaus than high-frequency ones. The yield of a harmonic around 50eV from a YAG laser was thus much larger than that from an excimer of the same intensity, although for the YAG the process required 45 photons, compared to 9 for the excimer!

Current understanding of high-harmonic generation, explains all this nicely, and indeed makes it obvious (in hindsight!) why this should be so. The key can be found in the very strong coupling between nearby continuum states, which is truly a super-intense effect: a free electron is completely at the mercy of the driving laser field, without any potential well confining its quiver motion. This means that once an electron reaches the continuum in a strong electromagnetic field, the unrestricted quiver motion makes the electron energy swing by a large amount during each optical cycle, an amount completely determined by the dynamics of the free electron.

In intense low-frequency fields, this energy can easily exceed the photon energy, so that each optical cycle many photons must be exchanged with the field. The maximum energy an electron so acquires is inversely proportional to the light frequency squared, leading to exchange of larger energy in the low frequency case, even though the number of photons required is much higher. But there is no rate limitation on the absorption of the required number of photons, which must occur with certainty within one quarter of an optical cycle (after which deceleration occurs, and the photons are re-emitted).

This complete saturation of free-free transitions, known already from non-perturbative ATI spectra, also underlies the HHG process, because once all the required energy is concentrated in a single particle (the fast continuum electron) is becomes comparatively easy to radiate it as a single hard photon. For quantum mechanical reasons, the dominant process is recapture of the free electron into its original state: there it can interfere constructively with the part of the wave function that was not yet ionized. For a high-energy photon to be emitted in this recapture, the electron must both be in the right place (near its parent ion), and have the right energy. Since all recollision energy up to a maximum of 3.17 (the Kulander constant) times U_p occur with approximately equal likelihood (on a log scale!), this explains nicely both the existence of the plateau, and the energy to which it extends (IP+3.1U_p).

Thus, at this stage, the basic mechanism of HHG seems well understood, and relies on a trickling amount of ionization, acceleration of the photoelectron as part of the quiver motion, recollision with the parent ion and recapture in the original state in a Bremsstrahlung event. At the moment, there is no experimental evidence at odds with this theoretical description. On the contrary, the dependence of HHG on the ellipticity of the polarization (which prevents the recollision) strongly supports this mechanism.

Armed with this understanding, research on HHG now branches in several directions. One line is the optimization of the HHG process, in order to boost the efficiency, and make it a really useful source of VUV/XUV radiation. To this end, phase-matching during propagation of the involved beams needs to be addressed. Another line is directed at extending the plateau to even shorter wavelength, by using ions as the generating medium. In addition, the use of polychromatic radiation can tailor recollision energies in order to achieve the desired effect with larger probability.

3.2 PROCESSES INVOLVING X-RAYS

The role of x-rays in high-intensity light-matter interaction continuous to increase. While in the early days infrared lasers and electron spectra were the name of the game, the trend definitely seems to go in the direction of higher energy processes, also including higher-energy photons.

X-rays can come into the picture in several places. They can for instance be generated in collisions of atomic constituents accelerated to high energy by the super-intense field. HHG is a by now well established example of this. But processes other than coherent transition of photoelectrons back to the initial state can lead to production of x-rays as well, and play an important role in systems where collision with particles other than the parent ion is important. Ionization of clusters is an example of this, since there the probability for collision of the photoelectrons with other atoms of the cluster can be very large.

On the other hand, x-rays can be used as incoming radiation. This can serve several purposes that are difficult to accomplish with optical radiation, namely ionization of inner shells (without affecting overlying shells), and generation of high-energy photo-electrons. Although x-rays can not be made super-intense themselves at the moment, they can be used in combination with optical radiation that is super intense, to access parts of parameter space that would otherwise be inaccessible.

The contribution on Laser-Assisted Auger Decay in this chapter presents an example of this, where the high energy of an Auger electron is exploited to study super-intense couplings high up in the continuum. Couplings tend to be stronger there, because quiver motion modulates the velocity, and the same velocity modulation results in a much larger energy modulation if the average energy is large.

Very similar effects occur when two-color ionization with x-rays and super-intense optical radiation takes place (the laser-assisted photo-electric effect). Especially interesting here is that the frequencies of x-ray and optical radiation can be made commensurate (as they would automatically be if the x-rays were produced by HHG), leading to interference effects between the various paths to the same final state.

3.3 RADIATION FROM FREE ELECTRONS

All cases of interaction of a completely free, non-relativistic electron with a plane-wave radiation field can be transformed to the same problem by appropriate scaling of time and distances. Galilean or Lorentz invariance furthermore ensure that the average velocity of the electron can be transformed away as well, provided of course that the radiation field is transformed (e.g. Doppler shifted) to how it looks in the electron rest frame. Thus, for a completely free electron in a plane-wave radiation field, the only critical intensity, above which a qualitative change in behavior occurs, is when the free electron is accelerated to relativistic quiver velocities.

This is not to say that the problem of a free electron in a field of lower intensity is not important, or that the light does not have any effect on a free electron. On the contrary, in many specific problems of free-electron-light interactions, critical intensities separating weak- and strong-field regimes appear to be very low, much lower than the intensity at which quiver motion achieves the relativistic level. Some well-known examples are the Kapitza-Dirac effect, Multiphoton stimulated Bremsstrahlung etc.

One of the most important effects of an electromagnetic field on free electrons is the ponderomotive force, and its importance was already recognized in the early days of ATI research. All experiments to date have been performed under conditions where the laser focus is very large compared to the quiver motion. After all, for a non-relativistic electron, the latter is of necessity several orders smaller than the wavelength of the light, while the focus usually is at least an order larger. The quiver motion of the free electron is therefore nearly the same as that in a homogeneous field, which also means that the electron will approximately have the quiver energy corresponding to the local intensity. At intensities where the quiver motion becomes relativistic, magnetic effects become comparable to electric ones, and the dipole approximation breaks down. Present-day lasers can achieve intensities that make this happen, and although it is impossible to expose atoms to such intensities, free electrons survive up to any intensity.

Even for the interpretation of ionization experiments, it is important to know how free electrons behave, and how the focusing and pulse envelope distort the measured electron spectra and angular distributions. But the free electron can act as a scatterer of electromagnetic radiation as well. The motion of free electrons and in a laser field has been studied both theoretically and experimentally, and from these studies the concept of ponderomotive force has become well established: the position-dependent quiver energy acts as a potential energy as far as the equation of motion of the period-averaged position of the electron is concerned. How the effects of this manifest themselves as electrons from an ionization experiment leave the focus depends on the details of temporal and spatial structure of the focus, and on how the process that produces the electrons reacts to the intensity.

When an electron moves in an inhomogeneous light field, ponderomotive forces change both electron energy and momentum. The recoil of this is absorbed by the laser pulse as a whole. If the density of electrons is large (e.g. atmospheric density), the recoil can be large enough to produce observable changes in the frequency spectrum and angular distribution of the light (red/blue shifts and self-focusing/defocusing)

Although the average velocity of the electron can be transformed away and introduces no new physics, backscattering from a relativistic beam of electrons is experimentally important because the large Doppler shifts it involves can be used to generate gamma rays. Lorentz contraction makes the volume-density of photons seen by the electrons in the beam larger as well, so that relativistic electrons colliding head-on with a laser pulse really experience incredible intensities, at which perturbative Quantum Electro-Dynamics breaks down.

4. Coherence

This chapter contains a number of contributions that are rather diverse in nature, and sometimes fall a little bit outside the main stream of super-intense laser-atom physics. Nevertheless, the term coherence seems to apply to all of them, in one way or another. Propagation of light in media is often dominated by driven oscillations of the polarization (the 'coherences', in terms of the Bloch equations) in those media. Other papers address the effects of the phase relation between various laser pulses, and the effects of coherences in inversionless lasers. Since the papers have otherwise little in common, a general introduction to them all makes little sense, and the papers introduce themselves.

4. 1 PROPAGATION

Propagation of intense laser beams through non-linear media is a topic that is interesting in its own right, but also important to atomic physics from a practical point of view. In processes that emit radiation, like HHG, the single-atom response is extremely weak, and can only be measured because a large number of atoms in the sample under study act coherently. For this to be realized, it is important that the volume in which the intensity is realized to drive the process is large enough, and that all atoms in that volume radiate in phase with the radiation produced by the others (phase matching).

These aspects are especially important for processes that need super-intense driving fields, because with present laser technology such intensities can only be reached by focusing as tightly as possible. This makes the volume with the required intensity very small, and spoils the phase matching. That the latter is true is easily understood by realizing that phase matching actually means momentum conservation of the photons involved in the process, and the large spread in angles of incoming photons in a strongly focused beam tends to produce a comparative deficit of momentum when these photons are absorbed in the course of a non-linear process.

Confining a laser beam in a small, long channel could provide a solution to all these problems. This channeling can occur through a prepared index gradient in a medium, but also through changes of the refractive index caused by the laser beam to be propagated itself (so called self-channeling). Non-linear effects that change the index can be due to near-resonant (possibly multiphoton) transitions in a gaseous medium, or non-linearities in the response of free electrons in a plasma to the radiation field due to relativistic effects in the quiver motion. This chapter contains papers on all these cases.

THE VARIOUS CAUSES OF STABILIZATION

H.G. MULLER
FOM-Institute for Atomic and Molecular Physics
Kruislaan 407, 1098 SJ Amsterdam, Netherlands

1. Classification of Stabilization

One of the most remarkable effects seen in the super-intense regime is that there exists anything like an atom at all, rather than everything decaying into a completely ionized plasma of individual electrons and nuclei. But there are many indications now, that well-defined atomic structures can survive the onslaught of a super-intense light pulse. This unexpected course of events can have several causes, but can be traced back mainly to the fact that an electromagnetic field is an AC phenomenon, and thus on the average has no tendency to displace or accelerate electrons with respect to the nucleus.

As simple models as well as elaborate calculations revealed this unexpected stability of atomic structure, the term stabilization was coined as a generic name for these effects. The common denominator in all cases was that the situation was more stable (i.e. the lifetime of the state was longer or the total amount of ionization smaller) than an imaginary reference situation derived from extrapolation from the behavior of the same system under other conditions. These 'other conditions' could, for instance, be lower irradiance, longer or shorter pulse durations, etcetera.

Not surprisingly, the question if stabilization occurs depends on what conditions one extrapolates from, and the physical causes underlying stabilization are equally diverse. Over the years a variety of effects have been presented as stabilization this way, and they can roughly be grouped into four categories, most commonly referred to as adiabatic, transient (or dynamic), interference, and resonance stabilization. In some models, various types of stabilization can be active at the same time.

1.1 TRANSIENT STABILIZATION

In transient stabilization, the atom is more stable (more resistant to ionization) when subjected to short laser pulses than expected on the basis of Fermi's golden rule (which essentially gives the ionization rate under CW conditions). This effect is intimately connected with the concept of electronic wave packets, and plays a role when one state out of a manifold of states, all lying within the bandwidth of the laser, is being ionized. This is usually the case when a Rydberg state is ionized with a sub-picosecond laser pulse, and a very convincing experimental demonstration of the effect has been given for exactly this system [1]. Like so many things in quantum mechanics, the effect can be explained in two ways, in the energy domain or in the time domain. The descriptions are

1

H. G. Muller and M. V. Fedorov (eds.), Super-Intense Laser-Atom Physics IV, 1–10.
© 1996 *Kluwer Academic Publishers.*

completely complementary, but one or the other might be more appealing depending on if you're a mathematician or an engineer.

In terms of energy levels, stabilization comes about due to two-photon Λ-transitions between the initial state and other (initially empty) states of the manifold. If all the states of the manifold couple to the same continuum, the manifold can be described in terms of a new basis set (consisting of linear combinations of the original manifold states), only one of which couples to the continuum. All other basis states orthogonal to this one do not couple to the continuum at all, and therefore can not be directly ionized. This can be achieved for any value of the couplings of the manifold states to the continuum: the ionizing state is the linear combination with coefficients proportional to the couplings. Any state orthogonal to it has vanishing coupling to the continuum, because the ionization from all its component manifold states interferes destructively.

The initial manifold state that was populated now is described as a linear combination of all new basis states, and most of these basis states are impervious to ionization. Only the population that initially projected onto the single ionizing basis state can be rapidly depleted, leading to a change of the contribution of the initially empty manifold states in proportion to their coupling to the continuum. In terms of the original manifold states this population appearing in the neighboring states is described as the result of a two-photon Λ-transition through a continuum intermediate state.

If there was an energy splitting between the states in the manifold, such as is the case in a Rydberg series, the new basis states are not eigenfunctions of the atomic Hamiltonian, but are coupled to each other by it. As a result, in time population will leak from the non-ionizing basis states to the ionizing one, (with a rate that is independent of the light intensity), and once it arrives there, it can be ionized again. This is the reason why the stabilization is transient, and on a longer time scale total ionization will occur.

In the time domain things can be easily visualized when we mentally picture a manifold of Rydberg states. The coupling of such Rydberg states to a continuum state one photon energy up, scales with the principal quantum number n as $n^{-3/2}$. Thus a group of neighboring Rydberg states with high n nearly all have the same coupling to the continuum, and the ionizing basis state discussed above is simply their sum. Since all Rydberg wave functions look similar near the origin, the wave function of this sum is large there, but rapidly vanishes at larger distances, where the different Rydberg wave functions get dephased. This makes it clear that the ionizing population is that near the origin, and that all population orthogonal to it (i.e. far away from the origin) can not ionize.

Transient ionization is now explained by complete saturation of ionization near the origin, depleting the wave function and creating a *probability hole*, or anti wave packet, there. No matter how high the light intensity, a short laser pulse can never do better than that. To ionize any further, population has to flow in from the outskirts of the Rydberg wave function. The time it takes for all population to reach the origin is by definition the orbit time, so complete ionization can only be achieved with pulses longer than this.

This picture of transient stabilization is of course an idealized one, since electron density far away from the origin is not absolutely un-ionizable. The entire volume within one quiver amplitude from the nucleus can suffer ionization within an optical cycle, and this region can grow without bound if the intensity is cranked up. It is instructive to check how this effect shows up in the essential states picture, because coupling of a single series of levels to one continuum always leads to a *single* non-ionizing level, no matter how strong the coupling. In real life there is more than one continuum, though. In the length or velocity gauge the *direct* coupling of the Rydberg series with continua of other angular momentum ($\Delta\ell > 1$) strictly vanishes, but these continua are coupled *with each other* in very singular ways, making it impossible to separate the directly coupling continuum from higher ones.

In the acceleration gauge, the singular couplings disappear, (in other words, this gauge diagonalizes the singular part of the continuum-continuum interactions) but there the bound states each couple *directly* to higher continua, through higher-order couplings caused by the Taylor-series expansion of the displaced atomic potential, $V(r+\alpha(t))$. These higher-order couplings become important as the intensity is increased, and the number of ionizing linear combinations of bound states is in general equal to the number of continua to which the series is coupled: not only the innermost part of the wave function suffers ionization, but neighboring parts start to do so as well.

1.2 ADIABATIC STABILIZATION

In contrast to transient stabilization, adiabatic stabilization does not require a manifold of states, but can occur for a single isolated level. Since time-dependence in a quantum-mechanical description can only come about by the interference of several stationary states, no time-dependent dynamics is involved, and the stabilization is a *steady-state* property of the ionization rate. It is also defined in a completely different way, namely with respect to the ionization rate extrapolated from *lower intensities*, specifically the perturbative rate as given by Fermi's Golden Rule.

Adiabatic stabilization manifests itself as a dramatic lowering of the (stationary) ionization rate above a certain intensity, to such an extent that the atomic lifetime even starts to increase with intensity. The lifetime vs intensity plot thus exhibits a minimum, which has become known as Death-Valley [2], since, for ground state atoms, the minimum lifetime is extremely short. This makes it virtually impossible to follow the lifetime curve by smoothly turning on the light, since all atoms would decay during the period in which the intensity crosses the Death-Valley region.

The difficulty of crossing Death-Valley in itself does not make adiabatic stabilization a meaningless concept, for one could always imagine an experiment in which the stabilized states are formed after the light intensity is sufficiently far above that near the lifetime minimum, for instance by recombination. But fortunately, it was predicted that the minimum lifetime for Rydberg states can be much longer than that of ground states, and Death Valley can be crossed in this case before all atoms are ionized. This has made experimental demonstration of the effect possible in such Rydberg systems.

In order to get an idea which states are suitable, the cause of adiabatic stabilization should be considered. This cause is most obvious if the atom is described in the Kramers-Henneberger coordinate frame, which oscillates in the way a free electron would. In that frame, the electromagnetic force on the electron is exactly canceled by the inertial forces, and in the absence of an atomic potential electrons would indeed feel no acceleration with respect to this frame!) whatsoever. The atomic nucleus, however, would be stationary in the lab frame, and thus perform the reverse quiver motion in the KH-frame. The temporal variation of the force exerted on the electron by this quivering nucleus then must be responsible for non-periodic effects of the radiation field, such as ionization.

For an electron located at some distance from the average nuclear position, it is now obvious that a saturation mechanism is operative: the maximum force the electron feels during an optical cycle is determined by the distance of closest approach, and once the amplitude α_o of the nuclear excursions gets larger than this distance, increasing α_o only makes that this peak force is felt during a *shorter* time interval. In fact both the time-average and the oscillating Coulomb force on the electron decrease with α_o, the latter leading to a reduced ionization rate even for an electron located near its original position, and the former enhancing this effect by causing the wave function to expand, and the electron to move to a position where the oscillating force is even less.

From this point of view, it is clear that stabilization is to be expected as soon as the quiver amplitude becomes larger than the average size of the original wave function. In the lab frame, the force on the electron seemed to grow without bounds, but apparently all this growing force does is making the electron quiver, while ionization depends on more subtle aspects of the light-matter interaction!

One point should be kept in mind, though: the description in the KH-frame only makes sense for electrons that indeed quiver like free electrons, or in other words, that are approximately at rest in the KH-frame. Therefore the frequency of the light should always be fast compared to the time scale of the electronic motion in the atomic system. Quantum-mechanically, the latter translates to the photon energy being much higher than the typical level spacing of the system. For ground-state atoms this level spacing is approximately the ionization potential, but in Rydberg states it can be arbitrarily smaller.

1.3 INTERFERENCE STABILIZATION

Interference stabilization is much less understood than the other types of stabilization. It was discovered in a rather abstract model for ionization out of a Rydberg state [3]. Phenomenologically, it is very much like transient stabilization, and in the original treatment, with a sudden switch-on of the light field, both transient and interference stabilization play a role. Due to this it was initially not obvious at all that interference stabilization was a mechanism of stabilization in its own right.

The basic idea of interference stabilization is that when a number of levels each couple to the same continuum, the strength of this coupling (= light intensity) can be made so high that the lifetime-broadened states start to overlap. Ionization out of the states than populates identical final states in the continuum, and all this ionization interferes, sometimes destructively. This then can causes a reduction of the ionization rate for a particular superposition of the levels. Described in the basis of the original states, population ionized from one state would be immediately recaptured in the other (by Λ-transitions), leading to trapping of population in the series of bound states.

All this is very similar to transient ionization, except that the overlap of the ionization profiles in the continuum is now caused by *lifetime broadening*, rather than the large laser bandwidth of a short pulse. An important difference, though, is that the effect is not transient at all: the coupling to the continuum apparently locks the phases of the population in the bound states, so that the interference of their respective ionizations remains destructive, and states with a long lifetime result.

Analysis of interference stabilization of a Rydberg series similar to that of transient ionization shows that also here ionization takes place near the origin only, since all couplings with the continuum are of the same sign. At low intensities, the boundary condition here is such that most of the infalling wave function is elastically reflected back into the same orbit it came from. At higher intensities, losses due to ionization occur, and the reflection coefficient with which the boundary condition at the origin can be described decreases from nearly one to a lower value. Eventually it reaches zero, and the infalling electron is knocked into the continuum with certainty when it hits the origin, due to the wildly oscillating nucleus (in the KH-frame) it encounters there.

Interference stabilization now occurs when the reflection coefficient decreases still further, to negative values. The amount of elastic reflection (and with it the probability to stay in a bound state) then again starts to increase, but it has *the opposite sign*. A side effect of this is that the boundary condition at the origin changes from an antinode to a node, which introduces a quantum defect of ½ to the Rydberg series, causing the energy of the stabilized states to lie exactly in between the old levels.

This situation is completely analogous to acoustic resonances in an organ pipe that is closed off at one end by a variable aperture: gradually opening the aperture first causes the resonances (with an odd number of quarter waves in the pipe) to broaden due to acoustic losses, until perfect impedance matching between the pipe and the outside world completely destroys the resonance structure. Opening the aperture still further now turns the pipe into an open-ended one, with resonances for an even number of quarter waves in the pipe, that can grow very narrow if the aperture is opened completely.

The aspect that is least understood in interference stabilization, is what limits the 'overshoot' of the reflection coefficient: can a wildly vibrating nucleus really cause perfectly elastic reflection?

1.4 RESONANCE STABILIZATION

Resonances, and their counterparts window-resonances, can often cause unexpected behavior of ionization rates, especially if the exact frequency at which they occur shifts with light intensity. Observed at a light frequency close to the resonance, the effect of such shifts can be large, and it is often difficult to reconstruct what is going on without having the total picture of ionization rate as a function of both frequency and intensity.

Some aspects of stabilization can sometimes be mimicked by resonances, for instance a decrease of the ionization rate with increasing intensity if this increasing intensity shifts the resonance out of the way so that it loses its influence [4]. A negative slope in the rate vs intensity curve is thus by no means unique. But usually the region over which such a slope exists is very narrow, and the effect is interpreted in a more straightforward way by considering the resonance as an event that temporarily *destabilized* the atom by enhancing the ionization.

A window resonance (a zero in the ionization cross section) actually does decrease the ionization rate when increasing intensity Stark shifts it into resonance [5]. In this case the stabilization also occurs in a limited intensity range, because the ionization rate will start to increase again once the resonance has been passed. But window resonances sometimes can be very broad, especially in the case of Cooper minima, and their stabilizing effects might extend over a large intensity range.

All resonance stabilization has in common that the effect is specific for the photon energy used. Another such effect is channel closure [6], where an ionization process with a certain number of photons becomes energetically impossible due to Stark shift of the continuum limit. In that case ionization can still occur with one more photon, but this process might have a much lower rate than the lower-order channel, so that the closure of the latter causes a significant drop in ionization rate.

2. Validity Range of High-Frequency Theory

The conditions for the validity of high-frequency theory are that the photon energy ω must be much larger than that of the spacing ΔE of the levels involved. But this condition applies to the energy spacing inside the field, which is itself a function of the light intensity. Since binding energies and accompanying level spacings go down with intensity, requiring that the photon energy is large compared to the unperturbed level spacing is an obvious way to ensure that high-frequency theory is valid at any intensity.

Since level spacings go down with intensity, there is also a *non-obvious* region of parameter space where high-frequency theory is valid. On the assumption that it is, a scaling law for the binding of electrons on the average nuclear charge distribution can be derived [7]. For linearly polarized monochromatic light, it turns out that the electrons preferably bind to the endpoints of this charge distribution, with a binding energy (and spacing) that depends on the quiver amplitude as $W \cdot \alpha_0^{-2/3}$.

Realizing that $\alpha_0 = I^{1/2}/\omega^2$, the high-frequency condition then reads

$$\Delta E = \Delta W \cdot (I^{1/2}/\omega^2)^{-2/3} = \Delta W \cdot I^{-1/3} \omega^{4/3} \ll \omega,$$

or

$$I/\omega \gg \Delta W^3. \tag{1}$$

If this condition is fulfilled, the initial assumption used for the calculation of ΔE proves valid, and the reasoning leading to (1) therefore consistent.

One peculiar aspect of (1) is that the high-frequency condition can be satisfied at any intensity, provided that the frequency is *low* enough! The physical picture is that at low frequencies, quiver amplitudes get enormous, and such a wildly quivering electron will

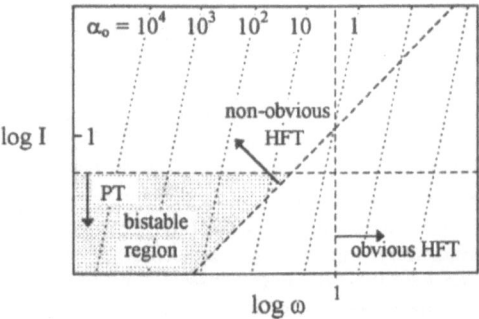

Figure 1. Validity regions of high-frequency and perturbation theory.

on the average be so far away from the nucleus, that they are hardly bound at all, and thus find their quiver motion hardly perturbed by the presence of the nucleus.

This picture is completely different from that painted by perturbation theory, which shows that at low intensities the atom is hardly perturbed at all, and has its electrons nicely sticking to the nucleus. This binding is so strong that the field is not able to set the electron quivering, so that this picture is completely self-consistent too.

It thus seems that atoms become bi-stable entities in the non-obvious region, where (1) is fulfilled for small I. Depending on their history, they can settle either into a quivering mode (the eigenstates in the KH-potential), or a tightly bound mode (the unperturbed atomic eigenstates). Once it is in one of these modes, it will stay there, unless conditions are changed to get out of the bi-stability region. Increasing the intensity into the super-intense regime will always force the system into the quivering mode.

This bistable behavior has been seen in calculations a long time ago already [8], in the form of the appearance of a second ground state 1s'. Adiabatically following this state with decreasing intensity showed that it evolved into a 'shadow pole'. But it has been shown now that it are these shadow states that actually converge towards the KH-eigenstates as the frequency is increased [9].

3. Reverse Engineering of Essential-State Models.

Essential-state models are very convenient, because they are amenable to analytic solution, and can provide much insight that way. They are tricky as well, since it is sometimes very tempting to leave out states that happen to be essential in some non-obvious way. Because of their abstract nature, assumptions on couplings between the states might be completely unrealistic, without anyone noticing. To guard against this, it is always helpful to try to 'reverse engineer' an essential-states model, to find real systems that would lead to the couplings used.

Attempting this immediately provides an important insight: most essential state models apply to the acceleration gauge only! In the other gauges, couplings between continua are extremely singular (diverging as $1/\Delta E^2$ in the length gauge, and as pv $1/\Delta E$ in the velocity gauge). Such couplings can never be neglected, no matter how low the intensity. Any essential-state model that only takes a single continuum into account implicitly makes the assumption that the coupling of this continuum with higher ones is smooth, and thus must have been formulated in the acceleration gauge.

As an interesting aside, there have been attempts to include singular couplings of the type given above between multiple continua in essential state models. [10] Since such couplings are highly non-separable, these models require special tricks to solve them, like changing to a different basis which magically makes the singular couplings disappear. Of course no magic is involved, but the trick is actually the gauge transform to the acceleration gauge formulated in terms of the essential states!

The simplest form of interference stabilization considers a series of bound levels coupled to a single continuum. No degeneracy because of angular momentum is considered, so the model must be one-dimensional. Atomic potentials are such that Rydberg electrons move in a potential well that is very smooth at the outer turning point, but singular near the origin. Making this well quiver (in the KH-frame) has no effect where the potential is smooth, but can lead to ionization in the origin, where the original boundary condition now quivers. The lifetime of the bound states is thus completely determined by the probability for elastic scattering off this quivering boundary condition.

Let us, as a model, analyze reflection from a vibrating hard wall. The boundary condition for a hard wall is $\Psi(0)=0$, and if the wall is vibrating due to the quiver motion this becomes $\Psi(\alpha_o \cos \omega t) = 0$. If the wave function is sufficiently smooth near the boundary, it can be approximated by its Taylor series, and to first order the condition becomes

$$\Psi(0) + \alpha_0 \cos \omega t \, \Psi'(0) = 0. \tag{2}$$

Writing Ψ as the sum of an incoming wave e^{-ikr}, and outgoing elastically reflected (Re^{ikr}) and excited ($Te^{ik'r}e^{-i\omega t}$) waves, applying condition (2) for each frequency component gives

$$R = ((\alpha_0 k/2) (\alpha_0 k'/2) - 1) / ((\alpha_0 k/2) (\alpha_0 k'/2) + 1). \tag{3}$$

This shows exactly the expected sign reversal as $\alpha_o \to \infty$, and thus reflection from a hard wall seems to exhibit interference stabilization. Unfortunately, stabilization occurs only for $\alpha_o k' > 2$, ($k' > k$) and the first-order expansion (2) of the plane waves loses its validity long before that. Replacing (2) with a higher-order expansion, introducing higher powers of $\cos \omega t$, and thus multiphoton couplings to higher continua like $Ae^{ik''r}e^{-2i\omega t}$, does not help: the new expression for R features stabilization at larger $\alpha_o k$, and enlarging the validity range of (1) seems to automatically push Death-Valley out of this range.

This Zeno-like state of affairs suggests that (for the hard wall, but also in a similar treatment of a δ-function well) the rising lifetime in interference stabilization is an artifact of neglecting the direct coupling to higher (ATI) continua that exists in the acceleration gauge. As soon as the continua that *are* taken into account no longer make the dominant contribution to the ATI spectrum, the model shows stabilization, while in reality the missing ionization rate would simply show up in the neglected channels. But it is still only a matter of conjecture to suppose that this would hold for any potential, in which case it would be better to speak about 'Zeno stabilization' instead.

4. The Adiabatic-Stabilization Experiment

Adiabatic stabilization can occur if the photon frequency is high compared to the orbit time. An additional requirement is that the switch-on of the laser pulse is indeed adiabatic, i.e. the unperturbed ground state gets the opportunity to adapt itself to the deformed ground state in the KH-frame. Even if this latter condition is not met, the KH-states are still the most natural states to describe the system, but the initial state will have to be projected onto these states, leading to excitation of many of them. This has been seen in numerical simulations: the most popular turn-on in such calculations is the 5.25-cycle linear slope, and apparently this is too fast for adiabatic following of the ground state.

Non-adiabatic effects can also be the reason that the stabilization seems to go away at even higher intensities: Scaling up the intensity of a fixed pulse shape, leads to an ever increasing intensity slope in the turn-on, and eventually to breakdown of the adiabatic assumption. This means the population can end up in any KH-state, including continuum states. An initial ground state is much more compact than the KH-states, and as a result the electron has a much too large momentum (as well as too large a repulsion with other electrons in a multielectron atom) to be bound by the shallow potential well of the time-

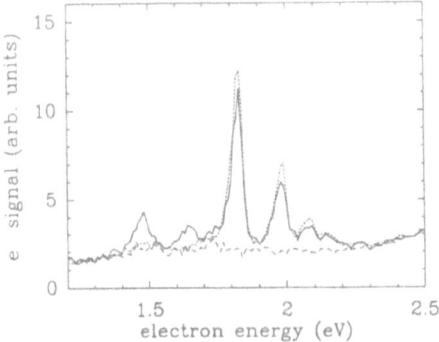

Figure 2. Electron spectra from ionization of *ng*-states in Ne with 620nm, 100fs pulses, followed by a 532nm, 5ns pulse to probe remaining *ng*-population (drawn line). The dotted spectrum, taken in absence of the fs-pulse, shows the initial populations. Above $5 \cdot 10^{13}$ W/cm^2 there is no change in the drawn spectrum, not even upon tripling intensity.

averaged potential. Therefore sudden switch-on usually leads to instant and complete ionization, even when frequency and intensity would favor stabilization [11].

In addition to the conditions mentioned above, in a demonstration experiment, a system should be chosen that does not exhibit competing processes, such as ionization of core electrons in a multielectron atom (for which the high-frequency condition is not satisfied, and that therefore do not stabilize).

A technically important aspect in stabilization experiments, is the role of the volume effect. One can get away with simply focusing a laser in studying processes with a very-high order of non-linearity, since there almost all signal will be caused by those part of the focus in which the intensity is within 20% from the peak intensity. The outer fringes of the focus are much larger, but the ionization rate at the intensity there is completely negligible. In stabilization, exactly the opposite is true, and the 0.1% focal volume in which the intensity is high enough to enter the stabilized regime would contribute even less than 0.1% to the signal. It would be next to impossible to measure the detailed behavior of this small contribution.

Two solutions to this problem are possible, using a top-hat spatial intensity profile by special imaging techniques, or taking the focal fringes for granted but keeping them free of target atoms so they can't contribute to the signal. In the experiments at FOM we opted for the latter, since this was easy to achieve: the experiments were performed on Rydberg states, and preparing the Rydberg states with a high-order process by a (three times) more-tightly focused laser than the one ionizing (or stabilizing) the Rydberg states the focal fringes are virtually free of target atoms. This makes this experiment one of the very few where no averaging over a spatial intensity distribution took place.

A disadvantage of this method, is that a very large volume (about a hundred times the size of the cloud of target atoms) filled with ground-state atoms is also illuminated by the ionizing laser. Any background ionization from this cloud would quickly swamp the ionization signal from the Rydberg atoms, especially since the excitation mechanism of the Rydberg states is not very efficient, making the density of Rydberg targets only 1% to 0.1% of that of ground states. This forced us to do the experiments on atoms with a ground state that is hard to ionize. Neon was preferred over helium because angular-momentum selection rules allow excitation of the same Rydberg state with one more photon, making it easier to deliver the huge excitation energy.

The experiment offered a very large degree of control, exposing atoms to pulses with a well-known peak intensity (not volume averaged, using Stark shift of a ground state-Rydberg transition to calibrate intensity), and measuring both ionization from and populations remaining in the ng-series, for various orientations of the (linear) polariz-ation. The results indicate stabilization above $5 \cdot 10^{13}$ W/cm^2, with about 70% of the atoms surviving the pulse in the initial state.

The strongest basis for the claim that we are dealing with adiabatic stabilization here, is the observed electron spectrum. The involvement of channel closure is ruled out by the fact that the peaks do not shift anywhere near the threshold (in fact, there is no observable shift at all before ionization shuts off). The absence of peaks at positions corresponding to odd half-integer principal quantum numbers rules out interference stabilization.

It is an interesting question how diabatic stabilization can occur for a 5g-state ($\Psi = r^5 e^{-0.2r}$, peaking at r=25) with a quiver amplitude $\sqrt{I}/\omega^2$ of about 7. Definitely such a low value of α_o is not enough to cause dichotomy. It is a popular misconception, though, that dichotomy and stabilization have anything to do with each other. As shown in this book by Mantica, one can occur without the other. But it is true that the oscillating force mostly seen by the electron (at r=25) in the KH-frame does not saturate at such low α_o. However, *local* ionization rates have been shown to drop extremely rapidly with distance from the nucleus, at least as rapidly as r^{-8}, [12], and this weighting causes the ionization from these regions to be negligible. For a 5g-state most ionization occurs much closer to the nucleus ($r^{-8}\Psi^2 = r^2 e^{-0.4r}$ peaks at r=5) than where the bulk of the wave-function is, and it is sufficient that ionization from this region saturates, which it does for $\alpha_o > 5$.

5. Acknowledgement

This work is part of the research programme of the Foundation for Research on Matter (FOM), which is subsidized by the Netherlands Organization for the Advancement of Research (NWO).

6. References

1. Hoogenraad, J., Vrijen, R.B. and Noordam, L.D. (1994) *Phys. Rev. A* **50**, 4133
2. Pont, M. and Gavrila, M. (1990) *Phys. Rev. Lett.* **65**, 2362
3. Fedorov, M.V. and Movsesian, A.M. (1988) *J. Phys. B* **21**, L155
4. Gontier, Y., Rahman, N.K. and Trahin, M. (1975) *Phys. Rev. Lett.* **34**, 779
5. Dimou, L. and Faisal, F.H.M. (1992) *Phys. Lett. A* **177**, 211
6. Reiss, H.R. (1980) *Phys. Rev. A* **22**, 1786
7. Gavrila, M. (1992) in M. Gavrila (ed.), *Atoms in Intense Fields*, Academic Press, Boston, pp. 435
8. Dorr, M., Potvliege, R.M. and Shakeshaft, R. (1990) *Phys. Rev. Lett.* **64**, 2003
9. Fearnside, A.S., Potvliege, R.M. and Shakeshaft, R. (1995) *Phys. Rev. A* **51**, 1471
10. Rzażewski, K., Li Wang and Haus, J.W. (1990) *J. Opt. Soc. Am. B* **7**, 481
11. Geltmann, S. (1995) *Chem. Phys. Lett.* **237**, 286
12. Muller, H.G. and Van Linden van den Heuvell, H.B. (1993) *Laser Phys.* **3**, 694

INTERFERENCE STABILIZATION

M.V. FEDOROV
General Physics Institute, Russian Academy of Sciences,
38 Vavilov St., 117942 Moscow, Russia

Interference stabilization of atoms is discussed. Raman-type transitions to Rydberg states with higher values of the electron orbital momentum ℓ are taken into account in numerical calculations. These transitions are shown to change qualitatively theoretical predictions concerning the dependence of the time of ionization t_i on the light field-strength amplitude $\mathcal{E}_0$: in a strong field, in a rather large interval of $\mathcal{E}_0$, the time of ionization is shown to be stabilized on the level of the order of the classical Kepler period t_K, i.e., to be approximately constant rather than being a growing or falling function of $\mathcal{E}_0$. The results derived are shown to be in a rather good agreement with the existing experiments.

1. Introduction

Strong-field stabilization of atoms is one of the most striking theoretical predictions of the last few years in the field of super-intense laser-atom physics. This is a counterintuitive prediction according to which, in a strong laser field, the ionization rate becomes a decreasing function of the light field-strength amplitude $\mathcal{E}_0$. There are several models predicting and describing field-induced stabilization. They differ from each other both in physics and in their assumed applicability conditions. Interference stabilization of Rydberg atoms [1, 2] is one of these models.

Let us assume that, in the simplest case, initially (at t=0), an atom is excited to some Rydberg level $E_n = -1/2n^2$ with a large principal quantum number n >> 1. (Atomic system of units is used here and below and, hence, $\hbar = m = 1$, e = -1, and c = 137 where ė and m are the electron charge and mass and c is the speed of light). Let such an excited atom be ionized by a light field of a frequency ω such that $\omega > |E_n|$. If the ionizing light field is weak, the corresponding rate of ionization (or ionization width of the level E_n) is determined by the well-known Fermi golden rule

$$\Gamma_i = 2\pi \left| V_{nE} \right|^2 \propto \mathcal{E}_0^2 \qquad (1)$$

11

H. G. Muller and M. V. Fedorov (eds.), Super-Intense Laser-Atom Physics IV, 11–21.
© 1996 Kluwer Academic Publishers.

where V_{nE} denotes the bound-free matrix element of the interaction operator $\hat{V} = -\frac{1}{2}\mathbf{d}\mathcal{E}_0$, $V_{nE} = \langle \psi_n | \hat{V} | \psi_E \rangle$; $\mathbf{d}$ is the atomic dipole moment, $\mathbf{d} = -\mathbf{r}$, and $\mathbf{r}$ is the electron position vector. The weak-field rate of ionization is a growing function, whereas the weak-field time of ionization $t_i(\mathcal{E}_0) = 1/\Gamma_i(\mathcal{E}_0)$ is a falling function of the field-strength amplitude $\mathcal{E}_0$. Eq. (1) is valid as long as the weak-field ionization width Γ_i is smaller than the spacing between neighboring Rydberg levels, $\Gamma_i \ll 1/n^3$. Under this condition, only the originally populated Rydberg level is involved in the process of photoionization. In the opposite case, $\Gamma_i > 1/n^3$, the field is strong, and the process of photoionization can be accompanied by rather efficient Raman-type transitions to neighboring Rydberg levels $E_{n'}$ with $n' = n \pm 1$, $n \pm 2$, ... Subsequent transitions to the continuum from these levels interfere with and partially cancel each other giving rise to suppression of ionization. As a result, in accordance with the theory of interference stabilization [1, 2], the strong-field ionization rate $\Gamma(\mathcal{E}_0)$ becomes a falling function, and the strong-field time of ionization $t_i(\mathcal{E}_0) = 1/\Gamma_i(\mathcal{E}_0)$ becomes a growing function of the field-strength amplitude $\mathcal{E}_0$. Such a behavior of the function $t_i(\mathcal{E}_0)$ is shown by the chain curve in *Figure* 1,

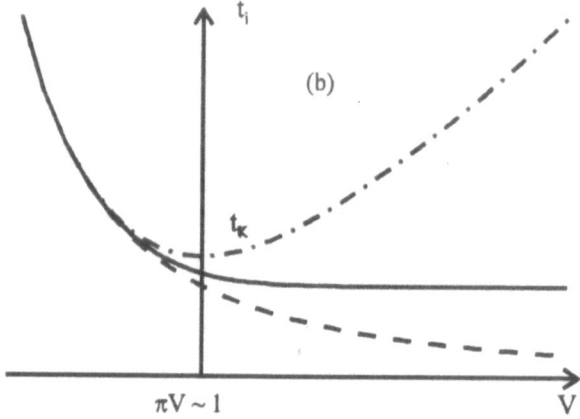

Figure 1. Time of ionization vs the field-parameter V (theoretical predictions).

where V denotes the main field-parameter calculated in the quasiclassical approximation [3, 4]. In this approximation, for S-P transitions, the parameter V is given by

$$V = 0.12\, \mathcal{E}_0\, \omega^{-5/3} \tag{2}$$

In terms of V the bound-free matrix element V_{nE} is equal to $Vn^{-3/2}$. The chain curve of *Figure* 1 has a minimum at $\pi V \approx 1$ and $t_{i\,min} \approx t_K = 2\pi n^3$.

The broken curve of *Figure* 1 shows the weak-field time of ionization extrapolated to the region of a strong field. The solid curve of *Figure* 1 shows qualitatively the behavior of the function $t_i(V)$ predicted by an alternative quasiclassical theory [5]. In accordance with this version of the theory, the main bound-continuum transitions occur in a region of relatively small electron-nucleus distances r, $r \sim r_q = \omega^{-2/3} \ll r_{max} = 2n^2$, where r_{max} is the size of the Rydberg orbit. Even if these transitions are rather fast and efficient, it takes a time of the order of the classical Kepler period t_K for all parts of the electron wave function to reach the region $r \sim r_q$, and this is why the time of ionization is of the order of t_K. For this reason, in accordance with the above-mentioned quasiclasical theory [5], the strong-field time of ionization is expected to be stabilized on the level of the order of the Kepler period t_K, $t_i \approx$ const. $\sim t_K$, rather than to be a growing or decreasing function of ε_0 or V. Unfortunately, earlier [5] this conclusion was justified only by such rather qualitative arguments. One of the goals of this talk is to present quantitative arguments to make a reasonable choice between the two strong-field predictions characterized by the chain [1] and solid [5] curves of *Figure* 1.

Another goal of the present consideration consists of analysis of the behavior of Rydberg complex quasienergies in dependence on the field-strength amplitude ε_0 or the field-parameter V. In accordance with the earlier theoretical predictions, in the strong field ($\pi V > 1$), the field-free Rydberg energies E_n are replaced by the strong-field quasienergies $E^{(n)} = \frac{1}{2} (E_n + E_{n+1})$, and these strong-field quasienergy levels are narrow. However, the simple analytical theory hardly can establish any one-to-one correspondence between E_n and $E^{(n)}$. Analysis of this correspondence and, in a more general sense, analysis of features of the field-driven Rydberg atom both in strong ($\pi V > 1$) and intermediate ($\pi V \sim 1$) fields are the goals of the present research.

Finally, all the earlier presented analytical investigations were based on some approximations. Not all of them are and can be justified well enough. Numerical solutions can be used to investigate sensitivity of the theory to such approximations, and this is also one of the goals of the present consideration.

2. A model for numerical calculations

Let $A_{n,\ell}(t)$ be the probability amplitudes to find an atom in Rydberg states $\psi_{n,\ell}(r)$ with principal and orbital quantum numbers n and ℓ. Equations for the functions $A_{n,\ell}(t)$ can be found directly from the Schrödinger equation with the help of the procedure known as the adiabatic elimination of the continuum [6]. Then, the probability amplitudes $A_{n,\ell}(t)$ appear to be coupled with each other by the so-called complex ac Stark matrix **Q**

$$i \, \dot{A}_{n,\ell}(t) - E_n \, A_{n,\ell}(t) - \sum_{n',\ell'} Q_{n,\ell;n',\ell'} \, A_{n',\ell'}(t) = 0 \qquad (3)$$

Calculated in the second order of the perturbation theory, the real and imaginary parts of the ac Stark matrix Q are determined by the well-known polarizability

$(\alpha_{n,\ell;\,n',\ell'})$ and ionization broadening $(\Gamma_{n,\ell;\,n',\ell'})$ tensors:

$$Q_{n,\ell;\,n',\ell'} = -\frac{1}{4} \, \alpha_{n,\ell;\,n',\ell'} \, \mathcal{E}_0^2 + \frac{i}{2} \, \Gamma_{n,\ell;\,n',\ell'} \qquad (4)$$

In this case, the only nonvanishing elements of the matrix Q correspond to ℓ' $= \ell \pm 2$, where ℓ, $\ell' \geq 0$.

Let us search for the quasienergy solutions of Eqs. (3) for which

$$A_{n,\ell}(t) = C_{n,\ell} \exp(-i\gamma t) \qquad (5)$$

where γ is the complex quasienergy (CQE) and $C_{n,\ell} = \text{const}$. Now, Eqs. (3) take the form

$$\sum_{n',\ell'} Q_{n,\ell;n',\ell'} \, C_{n',\ell'} + E_n \, C_{n',\ell'} = \gamma \, C_{n,\ell} \qquad (6)$$

These equations reduce the original problem to the eigenvalue-eigenvector problem which is very convenient for numerical solution. Let the solutions of this problem yield eigenvalues γ_i and eigenvectors $\{C^{(i)}\}$ which form the set of fundamental solutions of Eqs. (3). Here and below, a symbol in curly braces denotes a column. E.g., the elements of a column $\{C\}$ are $\{C_{n0}\}$, $\{C_{n2}\}$, $\{C_{n4}\}$, where, again, for any given ℓ, $\{C_{n\ell}\}$ are the columns with elements $C_{1\ell}$, $C_{2\ell}$, $C_{3\ell}$, ...

Solutions of the above-described eigenvalue problem can be used to solve the original initial-value problem. The general solution of Eqs. (3) can be presented as a superposition of fundamental solutions

$$\{A(t)\} = \sum_i a_i \, \{C^{(i)}\} \exp(-i\gamma_i t) \qquad (7)$$

where a_i are some new unknown constants. They can be found from the initial conditions $A_{n,\ell}(0) = \delta_{n,n_0} \, \delta_{\ell,0}$, where, now, n_0 denotes the principal quantum number of the initially populated Rydberg level and the initial orbital momentum is assumed to be equal zero. These initial conditions yield the set of equations for the unknown constants a_i:

$$\sum_i a_i C^{(i)}_{n,\ell} = \delta_{n,n_0} \delta_{\ell,0} \tag{8}$$

Again, this set of linear inhomogeneus algebraic equations is very convenient for its numerical solution.

By assuming that Eqs. (6) and (8) are solved, one can use their solution in the form of Eq. (7) to find the total $[w_{res}(t)]$ and partial $[w^{(\ell)}_{res}(t)$ and $w^{(n,\,\ell)}_{res}(t)]$ residual probabilities to find an atom in any or in some specific bound states. By assuming that $w_{res}(t_i) = \frac{1}{2}$, one can find the time of ionization t_i in its dependence on the field-strength amplitude $\mathcal{E}_0$ and on any other parameters of the problem under consideration.

In order to carry out explicitly all the above-described calculations, one has to specify further the ac Stark matrix $\mathbf{Q}$. Moreover, for numerical calculations, inevitably, this matrix has to be truncated in a reasonable way. Let us assume that, in dependence on the principal quantum number n, the ac Stark matrix has a size $2\Delta n$, so that $n_0 - \Delta n \leq n < n_0 + \Delta n$. Let $\ell_{max} = 2s$ be the maximal orbital momentum taken into account in the ac Stark matrix, $0 \leq \ell \leq 2s$, with ℓ being an even integer and s being an integer smaller than (or equal to) $\frac{1}{2}(n-1)$. Hence, in the general case, the total size of the ac Stark matrix taken into account is equal to $2\Delta n(s+1)$. In accordance with the description given above, let us assume that the elements of the ac Stark matrix have the form

$$Q_{n,\ell n',\ell'} = -\pi (i + \alpha) \frac{V^2}{(nn')^{3/2}} [\beta_{\ell+2} \delta_{\ell',\ell+2} + \beta_\ell \delta_{\ell',\ell-2} + \widetilde{\beta}_\ell \delta_{\ell',\ell}] \tag{9}$$

where V (2) and α, β_ℓ, and $\widetilde{\beta}_\ell$ are some constants. The constants $\widetilde{\beta}_\ell$ and β_ℓ characterize, correspondingly, the ac Stark coupling between Rydberg levels with the same orbital momentum ($\ell' = \ell$) and with orbital momenta ℓ and ℓ' differing by 2 ($\ell' = \ell \pm 2$) The constant α characterizes the ac Stark shift and mixing of Rydberg levels. More rigorously, the real part of the ac Stark matrix $\mathbf{Q}$ is assumed to consist of two parts: the diagonal part determining the common shift of all Rydberg levels by the free-electron quiver energy $(\mathcal{E}_0/2\omega)^2$ plus a relatively small correction of the order of ionization broadening [7]. Though much smaller than $(\mathcal{E}_0/2\omega)^2$, this correction may be comparable with the spacing between neighboring Rydberg levels $1/n^3$ and, if so, it has to be taken into account on equal terms with the ionization broadening. As usual, ionization broadening is characterized by the imaginary unit, whereas the above-described additional ac Stark shift is characterized by the real constant α in the brackets on the right-hand side of Eq. (9).

Unfortunately, any specific information about the constants α, β_ℓ and $\widetilde{\beta}_\ell$ is hardly available. For this reason, we can make calculations for a series of reasonably chosen values of these parameters to investigate the dependence of the results on various physical assumptions about features of the system under consideration. The results of calculations are described in the following section.

3. The results of calculations

3.1. QUASIENERGIES

The first group of calculations concerns eigenvalues of the matrix Q, i.e., complex quesienergies $\gamma_i = \gamma'_i + i \gamma''_i$. A typical result is shown in *Figure* 2 for the

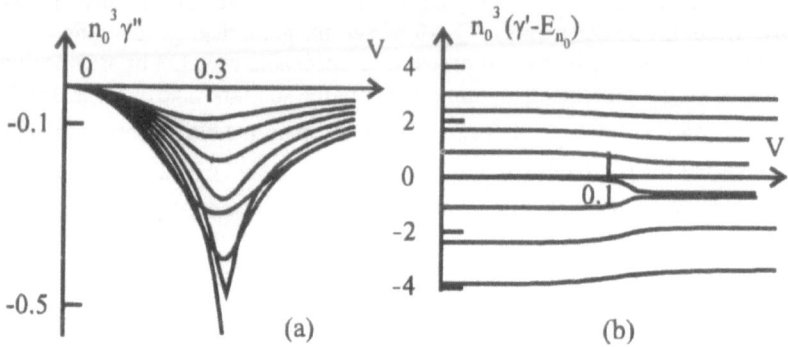

Figure 2. Imaginary (a) and real (b) parts of quasienergies vs the field-parameter V.

case $n_0 = 20$, $\Delta n = 4$, $\alpha = s = \beta_\ell = 0$ and $\widetilde{\beta}_\ell = \delta_{\ell,0}$. In contrast with the predictions of the oversimplified analytical theory [1, 2], all γ''_i (*Figure* 2a) differ from each other, they are finite at any V, and their absolute values do not exceed n_0^{-3}. This result is very important for understanding why the shortest achievable time of ionization remains always of the order of the Kepler period $t_K = 2\pi n_0^3$. Calculations of γ''_i for other values of the ac Stark-parameter α give qualitatively similar results. The curves of *Figure* 2b establish one-t-one correspondence between the weak-field (E_n) and strong-field ($E^{(n)}$) quasienerfies. Additional features of quasienergies and their dependence on the ac Stark shift and mixing parameter α are described in the paper [8].

3.2. EXCITATION OF HIGHER-ℓ STATES

To analyze the role and efficiency of Raman-type transitions to Rydberg states with higher values of the electron angular momentum, we solved the eigenvalue and initial-value problems in the case of a larger-size ac Stark matrix Q with $s \neq 0$. Some results of such calculations are presented in *Figure* 3 for $s = 4$ ($\ell_{max} = 8$), $V = 1.2$, $n_0 = 40$, and $\Delta n = 5$. The parameters β_ℓ and $\widetilde{\beta}_\ell$ were chosen to be equal to: $\{\beta_\ell\} = \{1,1,1,1\}$ and $\{\widetilde{\beta}_\ell\} = \{1,2,2,2,2\}$. This choice of the constants β_ℓ and $\widetilde{\beta}_\ell$ is based on the assumption about equal values of Rydberg-continuum matrix elements for transitions from any Rydberg states with arbitrary initial orbital momentum. Probably, this choice exaggerates the efficiency of transitions in the region of very large ℓ. In reality, both the Rydberg-continuum matrix elements and the constants β_ℓ and $\widetilde{\beta}_\ell$ have to fall smoothly with the growing ℓ. *Figure* 3 shows the ℓ-dependence of the partial relative residual probabilities $w_{res}^{(\ell)}(t) / \sum_\ell w_{res}^{(\ell)}(t)$ to find an atom in Rydberg states with any n and with an arbitrary but given value of ℓ. This calculation shows

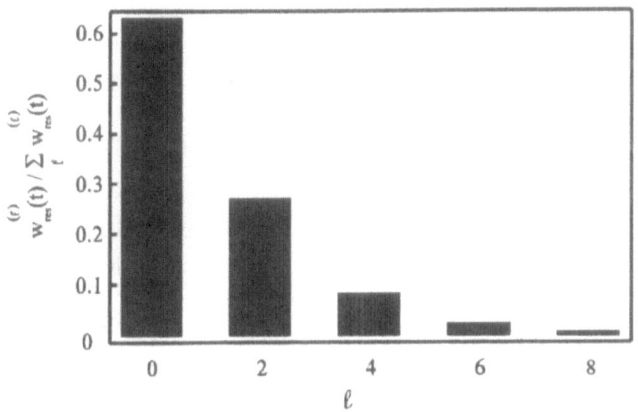

Figure 3. Distribution over ℓ of partial relative probabilities $w_{res}^{(\ell)}(t) / \sum_\ell w_{res}^{(\ell)}(t)$ at $t/t_K = 2.5$.

that, even in a sufficiently strong field, excitation of Rydberg states with high ℓ is not very efficient. However, as it will be shown in subsection 3.4, even these not too efficient transitions affect strongly the strong-fied ionization and stabilization regimes.

3.3. TIME OF IONIZATION

In accordance with the definition of the previous section, the time of ionization t_i is determined as the time when the total residual probability to find an atom in any bound states $w_{res}(t)$ falls to the level of ½. Typical results of calculations of t_i are presented in *Figure* 4. The broken curve corresponds to the case when only Rydberg states with $\ell = 0$ are taken into account ($n_0 = 40$, $\Delta n = 5$, $\alpha = s = \beta_\ell = 0$, and

18

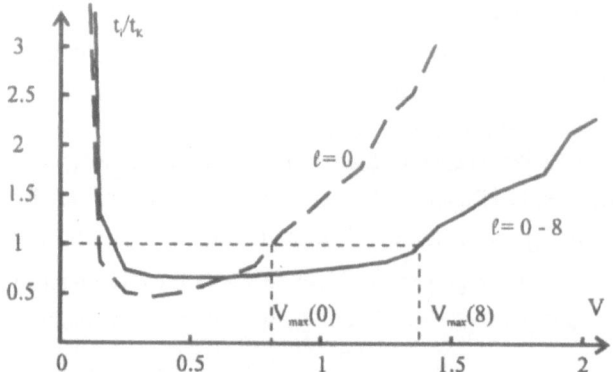

Figure 4. Time of ionization vs the field-parameter V in two
cases corresponding to different numbers of ℓ-states taken into
account in the calculations.

$\widetilde{\beta}_\ell = \delta_{\ell\,0}$); the solid curve describes the results of calculations with all such states and transitions taken into account, for ℓ ranging from 0 to 8 (all the same parameters but $\{\beta_\ell\} = \{1,1,1,1\}$ and $\{\widetilde{\beta}_\ell\} = \{1,2,2,2,2\}$). The main feature of the solid consists of a rather large region of V where the time of ionization t_i remains more or less constant, $t_i \sim t_K$. This region (that may be referred to as the "death-plateau" region) appears due to Raman-type transitions to Rydberg states with higher values of the electron orbital momentum ℓ. The "death-plateau" region grows with an increasing number (ℓ_{max}) of ℓ-states taken into account in the calculations. Let V_{max} be the largest of the two values of the field-parameter V at which $t_i(V) = t_K$. The definition is illustrated in *Figure* 4 for two cases: $\ell_{max} = 0$ and $\ell_{max} = 8$. The numerically found dependence $V_{max}(\ell_{max})$ can be approximated rather well by a logarithmic function

$$V_{max} = 0.84 + 0.26 \times \ln(\ell_{max} + 1) \tag{10}$$

For $\ell_{max} \approx n^{2/3}$ and, e.g., for n =30 this yields $V_{max} \approx 6$. Hence, in this case, the "death-plateau" region is expected to range from $V_{min} \approx 0.34$ to $V_{max} \approx 6$. This is a rather large interval that corresponds to variation of the light intensity $I \propto V^2$ from I_{min} to I_{max} where $I_{max}/I_{min} = (V_{max}/V_{min})^2 \approx (6/0.34)^2 \approx 300$.

3.4. YIELD OF ELECTRONS

The time of ionization in its dependence on the field-parameter V (*Figure* 4) gives a very clear indication of the field-induced stabilization and is very useful for clarifying differences between various regimes of strong-field ionization. However, usually, in experiments, one measures the electron yield or the residual probability to find an atom in bound states rather than the time of ionization. Moreover, very often these characteristics of the process of ionization are measured in their dependence on some parameters different form the field-parameter V. For example, one measures often the dependence of the electron yield on the laser fluence for a given pulse duration, or on the pulse duration for a given fluence [9]. Such dependences can also be calculated in the framework of our model, and the most interesting result is shown in *Figure* 5. The curves of *Figure* 5 show the dependence of the total ionization probability on the dimensionless fluence F= $V^2(\tau/t_K)$ where τ is the pulse duration.

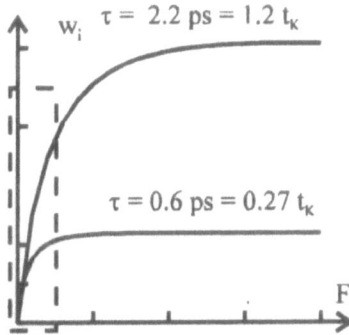

Figure 5. Probability of ionization vs the dimensionless fluence F.

In the experiment [9], atoms of Ba were excited initially to the Rydberg state with n =27 and ℓ= 0, and then they were ionized by a strong light filed. In this case, the Kepler period was equal to t_K = 2.2 ps, whereas the pulse duration τ was equal to either 0.6 ps or to 2.2 ps, and the same values were used in our calculations. Some other parameters used in calculations were: n_0 = 27, α = 0, Δn = 4, s = 4 (ℓ_{max} = 8), $\{\beta_\ell\}$ = {1,1,1,1}, and { $\tilde{\beta}_\ell$ } = {1,2,2,2}. The first main conclusion following from the curves of *Figure* 5 consist of the statement that the level of stabilization depends strongly on the pulse duration τ as compared to the Kepler period t_K: stabilization is much more pronounced in the case $\tau < t_K$ than in the opposite case, $\tau > t_K$. Second, the curves of *Figure* 5 show that, at high fluence, the electron yield saturates ($w_i(F) \approx$ const.) on a level smaller than one. This effect is related to the Raman type transitions to Rydberg states with higher values of the electron orbital momentum. In the case when these transitions are ignored (s = 0), the calculations give curves $w_i(F)$ that have maxima and fall in the large-F limit. Saturation of the curves $w_i(F)$ has the same origin as the above-described "death plateau" arising in the dependence of the time of ionization t_i on the field-parameter V (*Figure* 4). Third, the curves of *Figure* 5 agree (at least qualitatively) with the results of experiment [9]. Some details of the experiment- theory comparison and discussion of alternative theoretical models will be given in the following section. Some other results and details of calculations in the framework of the above-suggested model are described elsewhere [8].

4. Discussion

Probably, the main result of the above-presented consideration consists of the demonstration that the "death-plateau" behavior of the time of ionization, $t_i(V) \approx$ const., can arise due to Raman-type transitions to Rydberg states with higher values of the electron angular momentum ℓ. Though even relative populations in these states remain always rather small (*Figure* 3), they are important enough to change qualitatively predictions concerning the picture of the field-induced interference stabilization. In accordance with the results of the present investigation, as a whole, the time of ionization t_i in its dependence on the field-parameter V first falls in accordance with the predictions of the perturbation theory and then achieves a level about a half-Kepler period and remains more or less constant in a rather large interval of V and light intensities I. These results confirm the idea formulated earlier [5] on the basis of absolutely different approach and concepts .

Another important conclusion of the presented theory consists of its qualitative (if not quantitative) coincidence with experimental results [9]. However, interpretation of the experiment [9] given by its authors differs strongly from our interpretation. The model used for calculations in the paper [9] coincides with our model in which, however, all the transitions to Rydberg states with higher values of the electron angular momentum have to be dropped, i.e., in our notations, it coincides with our model with $s = \beta_\ell = 0$, and $\widetilde{\beta}_\ell = \delta_{\ell,0}$. Evidently, in this case, we would not get any "death-plateau" behavior for the function $t_i(V)$ or saturation of the ionization probability in its dependence on the light fluence [$w_i(F)$]. On the other hand, two other factors, taken into account in the paper [9] and missing in the present theory, are the consideration of pulses with smooth Gaussian envelopes and averaging of the results over distribution of the field in the focal region. In principle, these factors can give rise to the results similar to those described above and explained here by transitions to higher-ℓ states. However, unfortunately, both theoretical calculations and experimental measurements of the paper [9] are restricted by a rather small region of parameters that is not sufficient for clear differentiation between the "death-plateau" and "death-valley" behavior, between the curves $w_i(F)$ with saturation and with maximum, etc. In *Figure* 5 the region of parameters corresponding to the measurements and calculations of the paper [9] is indicated by a box with broken-line boundaries. Continuation of both calculations in the model of the paper [9] and the corresponding measurements to the region of larger values of the light fluence F would be very interesting for comparison of our models, their results, and experimental data.

Another part of interpretation given in the paper [9] is based on the use of ideas about antiwave packets, a hole that can be burned-off in the spatial distribution of the electron density by a strong short light pulse, etc. These ideas are considered as something different from (if not opposite to) the model of interference stabilization. In our opinion, such a contraposition is a rather strong exaggeration. This is evident already from the comparison of mathematical models given above. Ideologically, they are identical though of course, as mentioned above, there are some factors taken or not taken into account in each of these models. Moreover, contrary to the opinion formulated in the paper [9], the model of interference stabilization is valid equally well to both short and long pulses. Independently of the pulse duration, there is only one parameter (πV) that determines the threshold of stabilization by the condition $\pi V \approx 1$. In our opinion, there are several approaches to the description of the same phenomenon, and they are complementary to each other.

5. Acknowledgment

The author acknowledges the support of the International Science Foundation (grant # M6I000) and of the Russian Fund of Basic Research (grant # 93-02-15001).

6. References

1. Fedorov, M.V. and Movsesian, A.M. (1988) Field-induced effects of narrowing of photoelectron spectra and stabilization of Rydberg atoms, *Journ. Phys.* B, **21**, L155-L158.
2. Fedorov, M.V. (1993) Interference stabilization of Rydberg atoms in a strong ionizing field, *Las. Phys.*, **3**, 219-240.
3. Delone, N.B., Goreslavsky, S.P., and Krainov, V.P. (1982) Quasiclassical Coulomb dipole matrix elements, *J.Phys.* B, **15**, L421-L423.
4. Adams, M.S., Fedorov, M.V., Krainov, V.P., and Meyerhofer, D.D. (1995) Comparison of quasiclassical and exact dipole moments for bound-free transitions in hydrogen, *Phys. Rev.* A, in press.
5. Fedorov, M.V. (1994) Quasiclassical atomic electron in a strong light field, *J. Phys.* B, **27**, 4145-4167.
6. Fedorov, M.V. (1997) Resonance ionization of atoms and switching-on of the interaction, *J. Phys.* B, **10**, 2573-2582.
7. Fedorov, M.V. and Movsesian, A.M. (1989) Ac Stark effect and trapping of population on Rydberg levels in a strong ionizing field, *JOSA* B, **6**, 1504-1512.
8. Fedorov, M.V., Tehranchi, M.-M., and Fedorov, S.M. (1995) Interference stabilization of atoms: numerical calculations and physical models, *J.Phys. B*, submitted.
9. Hoogenraad, J., Vrijen, R., and Noordam, L. (1994) Ionization suppression of Rydberg atoms by short laser pulses, *Phys. Rev.* A, **50**, 4133-4138.

COHERENTLY CONTROLLED RYDBERG ATOMS AS GAIN MEDIUM FOR AN X-RAY LASER

R. B. VRIJEN, M. VAN INGEN, AND L. D. NOORDAM
FOM-Institute for Atomic and Molecular Physics
Kruislaan 407, 1098 SJ Amsterdam, The Netherlands

Abstract. The application of atomic Rydberg states as the initial medium in an x-ray laser based on optical field ionization is discussed. After laser preparation of the Rydberg state a second, intense short laser pulse ionizes the core electrons. The Rydberg electron can survive this laser pulse with high probability. Thus the ions that are produced still have one bound electron in a highly excited Rydberg state. The problem of electron-ion recombination, which is required to produce ionic Rydberg states in any recombination x-ray laser scheme, and which is generally slow due to high electron temperatures, is thus circumvented. Calculations on coherently excited lithium Rydberg states are presented which show the evolution of the electron wavepacket after the ionization of the inner electrons. The stability of the Rydberg electron is discussed and experimental evidence for this stability obtained on the photoionization of barium Rydberg states is presented. Finally the feasibility of implementing the proposed scheme under high-density plasma conditions is discussed.

1. Introduction

Super-intense laser fields, which have the power to strip a large number of electrons from atoms or molecules, are now considered by several research groups to be a suitable pump for lasers that emit soft x-rays. A short super-intense laser pulse creates a highly ionized plasma through optical-field ionization of the target atoms[1, 2, 3]. The photoelectrons in the plasma are recaptured into high-lying Rydberg states through electron-ion recombination and subsequently decay to low-lying states due to collisional and radiative transitions. As the ensemble of electrons cascades down, a tran-

H. G. Muller and M. V. Fedorov (eds.), Super-Intense Laser-Atom Physics IV, 23–35.

24

sient inversion may exist in the population of low lying states. During this period of inversion, which typically lasts a few picoseconds, gain may be obtained from the system[4, 5]. Due to the high charge state of the ions, the low-lying transitions can yield radiation in the soft x-ray region, with wavelengths well below 50 nm.

A general problem of the described pumping scheme is the low rate at which recombination of electrons from the plasma into bound Rydberg states takes place. This recombination has to take place on timescales short compared to the lifetime of the upper state of the desired laser transition in order to achieve inversion. Efficient electron-ion recombination requires a low electron temperature. However, the electrons which are ejected by optical-field ionization can have very high energies[6], thereby frustrating efficient recombination.

Nevertheless, Nagata et al.[7] have observed gain in a lithium plasma after completely stripping the lithium atoms with a laser pulse focused to an intensity of 10^{17} W/cm^2. They observed an exponential dependence of the emitted radiation on plasma length for the $3p \rightarrow 1s$ (11.4 nm) and the $2p \rightarrow 1s$ (13.5 nm) transitions in hydrogenic lithium.

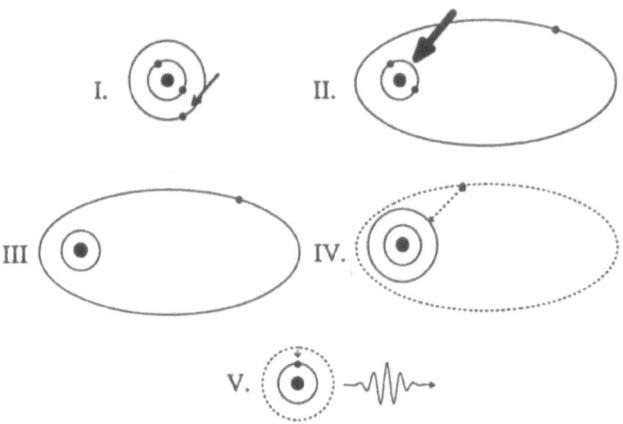

Figure 1. Schematic representation of the proposed laser scheme: I→II. A lithium Rydberg state is created with a relatively weak laser pulse. II→III. A short, intense ($I \approx 10^{17}$ W/cm^2) laser pulse ionizes the two inner electrons. III. An ionic Rydberg state remains since the outer electron is not ionized by the short laser pulse. IV. The Rydberg electron is de-excited to a lower lying state, either by collisional de-excitation, or in a controlled manner by a third laser pulse. V. X-ray lasing occurs on one of the low-lying transitions.

In the experiment special care was taken to provide a bath of low-

energy electrons to enhance the recombination rate. Just before the super-
intense laser pulse, a prepulse, with a much lower intensity, singly ionized
the lithium atoms. The electrons emitted upon this single ionization were
estimated to have a relatively low temperature of 1.5 eV. It is this initial
ensemble of cold electrons that was considered to contribute mainly to the
recombination and the subsequent inversion to yield the observed lasing.

In this paper a recently proposed [8] alternative x-ray laser scheme is
discussed in detail. The basic scheme is schematically depicted in figure 1.
The new scheme exploits the idea of preparing the target before stripping it
completely. The initial target in this scheme is not in the electronic ground
state but in a highly excited Rydberg state, or in a more advanced scheme,
a coherent superposition of Rydberg states. As will be discussed the intense laser pulse that strips the core electrons has a very high probability
not to ionize the Rydberg electron. The Rydberg electron can survive due
to various mechanisms, collectively referred to as stabilization against ionization. After the removal of the inner electrons by the short laser pulse no
recombination is required: a population of highly charged Rydberg ions is
instantly present. This collection of Rydberg ions is then available for the
cascade down to the upper state of a lasing transition in order to achieve inversion. However, opposed to the wide distribution of Rydberg states that
are populated by electron-ion recombination, the Rydberg ions that are
produced after initial preparation of a Rydberg target are in a relatively
small set of states. More specifically, the sudden removal of core electrons
will hardly affect the angular momentum of the Rydberg electron. We can
exploit the well defined state of the Rydberg electron by using a third laser
pulse to stimulate the Rydberg electron down to a lower lying state in the
ion, thereby enhancing the rate at which inversion is created.

Calculations are presented on the evolution of the Rydberg electron
after removal of the inner electrons. The calculations are concentrated on
lithium Rydberg states and hydrogenic-lithium Rydberg ions. The time-
dependent probability to stimulate the Rydberg electron down to a lower
lying state is calculated. Furthermore we will discuss the stability of the
Rydberg electron against photoionization and show experimental evidence
for this stability which was obtained from experiments on barium. Finally
we will discuss the feasibility of implementing the proposed scheme in an
x-ray laser.

2. Coherent manipulation of the Rydberg electron

The gaseous target is prepared in a highly excited state before the inner
electrons are ionized. This state can be chosen to be either a single Rydberg
eigenstate, or a wavepacket. Ionization of the Rydberg electron only takes

place near the nucleus[9], as will be discussed in more detail in the next section. The stability of the outer electron against ionization is based on the large fraction of the Rydberg wavefunction that is far from the nucleus. Using a wavepacket has the big advantage over using a single eigenstate that, at well defined times, all the wavefunction is localized far from the core so that ionization is completely inhibited. The drawback is that yet another laser with a pulse duration short with respect to the Kepler orbit time ($\tau_p < \tau_n$) is needed. The procedure would then be the following. First a wavepacket is launched in the neutral atom. The angular momentum can be selected by the exciting laser and for the present calculations d states are used. It is assumed that there is no change of the angular momentum of the wavepacket due to external forces, neither during its free evolution, nor upon ionization of the inner electrons. Since the overlap for excitation from the ground state mainly originates from the core region, the wavepacket starts near the core. The wavefunction $\psi(r, t)$ is described by (atomic units are used throughout)

$$\psi(r,t) = \sum_n e^{-iE_n t} a_n \phi_n(r) \tag{1}$$

with $\phi_n(r)$ the spatial part of the Rydberg eigenfunctions with principal quantum number n. The present calculations concentrate on a lithium target but are not restricted to this atom. The lithium Rydberg d states are described with hydrogenic nd states since the quantum defect of lithium d states is negligible ($\delta_d = 0.02$). The coefficients a_n are determined by the spectral content of the laser that excites the wavepacket and E_n is the energy of the Rydberg states given by $E_n = -1/(2n^2)$. After excitation the wavepacket oscillates between its inner and outer turning point with a roundtrip time $2\pi \langle n \rangle^3$. When the wavepacket is far from the core, at $\tau_1 = \pi \langle n \rangle^3$ after excitation, a second, very intense laser pulse is used to eject the inner electrons. Under the assumption that the ionization is fast compared to the evolution of the wavefunction, a situation which can easily be achieved experimentally, the wavefunction does not change during the removal of the inner electrons. However, to describe its further evolution the wavefunction is projected on the new basis of ionic Rydberg states $\phi_{n'}(r)$

$$\psi(r, t > \tau_1) = \sum_{n'} e^{-iE_{n'}(t-\tau_1)} a_{n'} \phi_{n'}(r) \tag{2}$$

with the coefficients $a_{n'}$

$$a_{n'} = \sum_n a_n e^{-iE_n \tau_1} \langle \phi_{n'}(r) | \phi_n(r) \rangle \tag{3}$$

with n' the principal quantum number of the *ionic* Rydberg states with energies $E_{n'} = -Z'^2/(2n'^2)$. Z' is the new core charge, with $Z' = 3$ in the case of lithium when both inner electrons are emitted.

The maximum probability to stimulate the Rydberg electron down to a lower lying state $\chi(r)$ occurs when the ionic wavepacket returns to the core. This probability $P_{stim}(t)$ is determined (in the dipole approximation) by

$$P_{stim}(t) \propto |\langle \chi(r)|r|\psi(r,t)\rangle|^2 \qquad (4)$$

We have calculated the evolution of a lithium Rydberg wavepacket explicitly, using equations 1-4. For comparison with future experiments, we have chosen to take $\phi_{n=20}$ as the central state in the wavepacket. The core charge $Z = 1$ is changed to $Z' = 3$ as the wavepacket reaches its outer turning point, simulating the removal of both core electrons.

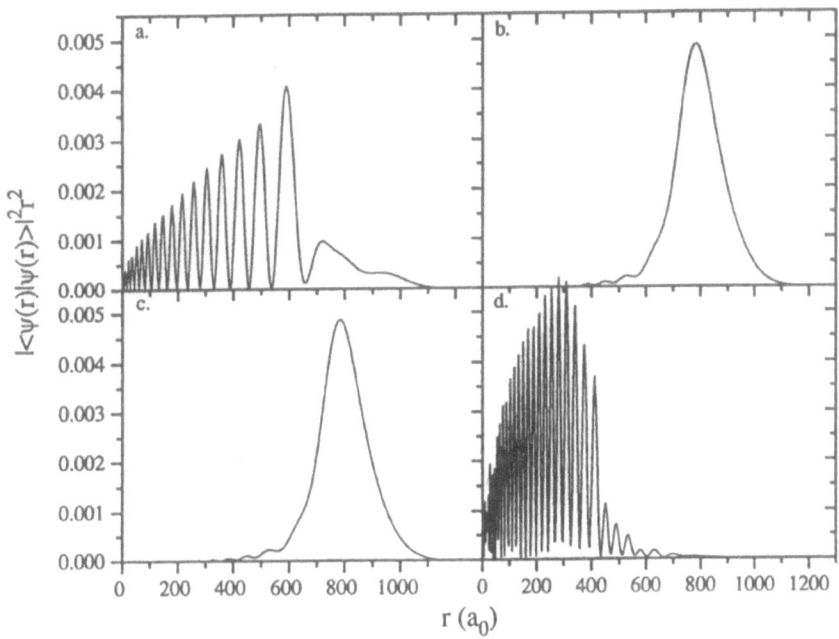

Figure 2. A wavepacket of Rydberg states in lithium, of which the inner electrons are ionized when the wavepacket is at its outer turning point. The amplitudes of the different Rydberg states constituting the wavepacket are described by a gaussian around $n = 20$ with a width of approximately 3 states. The wavepacket is shown at various times: a. $t = 0$, b. $t = \pi\langle n\rangle^3 = 25*10^3$ (a.u.), c. $t = 25*10^3$ (a.u.) directly after the core ionization, and d. $t = 39.4*10^3$ (a.u) when the ionic wavepacket returns to the core.

In figure 2 $\psi(r,t)$ has been plotted at several times: Figure 2a shows the initial wavepacket localized near the core. The wavepacket consists of approximately 3 states, populated with a gaussian distribution in amplitude around $n = 20$.

We have chosen to simulate a dispersed wavepacket that can be excited with a chirped laser pulse, in which the different frequency components do not have the same phase. The phase relation between the different frequencies is such that the atomic wavepacket would be maximally localized if it returned to the core after one roundtrip. As it turns out, the ionic wavepacket also shows a high degree of localization if the wavepacket is excited in this way. The highly localized wavepacket near the core ensures a maximum probability to stimulate the electron to a lower lying state. Figure 2b shows the same wavepacket at $t = \tau\langle n\rangle^3$, when it is at the outer turning point. Figure 2c shows the same packet shortly after the removal of the inner electrons indicating that we have used a sufficient number of ionic Rydberg states to expand the wavefunction, and figure 2d shows the return to the core of the ionic wavepacket.

An estimate for the central n' in the population distribution of the ionic Rydberg states can be made easily. In the sudden approximation, where the wavefunction does not change upon the ejection of the two inner electrons $\langle r\rangle$ is preserved. Since $\langle r\rangle = \frac{3n^2}{2Z}$ we find for the central n' in the ionic wavepacket $n' \approx \sqrt{Z'/Z}\,n = \sqrt{3}n$, which gives $n' \approx 35$ for $n = 20$.

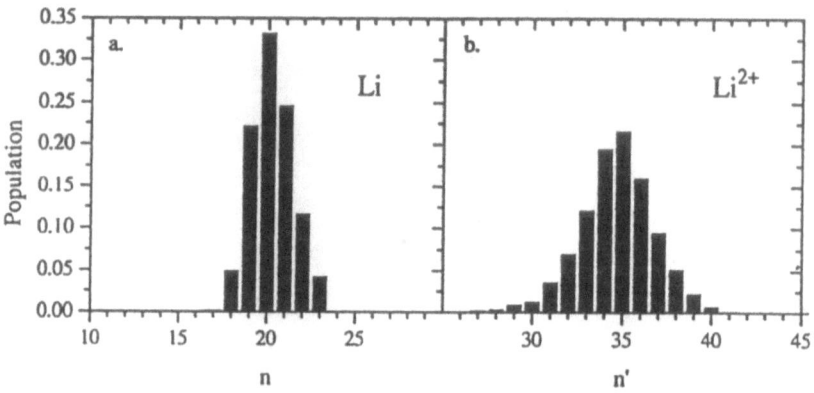

Figure 3. The distribution of the population over a. the atomic lithium and b. the ionic lithium (2+) Rydberg states.

The distribution of the Rydberg population is plotted in figure 3 for the wavepacket shown in figure 2. Figure 3a shows the population of the atomic Rydberg states in the lithium atom (corresponding to the wavepacket in figure 2a and b) and figure 3b shows the population of the Rydberg states in the lithium (2+) ion (corresponding to the wavepacket in figure 2c and d). Assuming $n' = \sqrt{3}n$ the roundtrip time of the ionic Rydberg wavepacket is $\tau_{n'} = 2\pi\langle n'\rangle^3/Z'^2 = \tau_n/\sqrt{3}$, which is in good agreement with the observed recurrence time in figure 2d.

The probability to stimulate the electron down to a lower lying state peaks as the electron wavepacket returns to the core. This probability is calculated as a function of time for a transition to the $5p$ state. The $5p$ state is chosen because it can still be reached with a practical optical wavelength of 259 nm. After population, the $5p$ state can either act as the upper level for a lasing transition, in which case direct inversion is achieved, or further deexcitation can take place to lower-lying upper laser levels, as will be discussed in section 4.

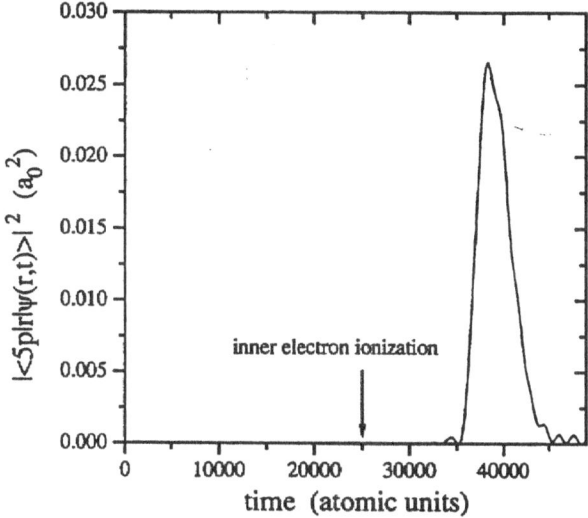

Figure 4. $P_{stim}(t) \propto |\langle\chi(r)|r|\psi(r,t)\rangle|^2$ is shown as a function of time for the wavepacket shown in figure 2. The ionic wavepacket returns to the core at $t = \frac{1}{2}\tau_n + \frac{1}{2}\tau_{n'} = 39.4*10^3$ atomic units, which is where the probability to deexcite the electron peaks.

Figure 4 shows the time-dependence of the probability to stimulate the Rydberg electron down to the $5p$ state. Of course, if the electron returns to the core, not only the probability to stimulate it down peaks, but also the probability to ionize it. However, the cross section to ionize a Rydberg state generally is much lower than the probability to make a transition to a lower lying bound state so we neglect the ionization here.

3. Stability of the Rydberg electron

3.1. TRANSIENT STABILIZATION

Rydberg atoms generally have a small cross section for ionization. This is due to the small overlap of the extended Rydberg electron wavefunction with the core. Energy and momentum conservation require that the electron interacts with the core when the atom is ionized: a free electron can not absorb photons. In addition to having a small cross section, Rydberg electrons are also known to stabilize against ionization under certain conditions. Stabilization against ionization is the general phenomenon that the ionization rate is not proportional to the intensity and the electron can survive much higher intensities than would be expected on basis of the perturbative cross section for ionization of the atom. Various types of stabilization against ionization have been predicted and measured [10, 11, 12, 13, 14]. In the presently proposed scheme we concentrate on a phenomenon called transient stabilization which occurs when a Rydberg atom is ionized with short-pulsed radiation. In the case of transient stabilization the ionization yield at high intensities is not limited by the intensity of the light but by the duration of the ionizing pulse. The name originates from the fact that the residual wavefunction may very well be ionized by longer pulses. For laser pulses shorter than the classical Kepler orbit time (τ_n) of the electron complete ionization is inhibited because ionization is limited to the core region. The fraction of the wavefunction that passes the core during a pulse of length τ_p can be estimated by the fraction τ_p/τ_n. The fraction of the wavefunction that does not pass the core region during the pulse can not be ionized and will survive the laser pulse, irrespective of its intensity.

The predictions of this simple picture are confirmed by quantum mechanical calculations [13, 14]. As a result of the short laser pulse a hole is drilled in the wavefunction near the core: an anti-wavepacket is created. This non-stationary state is described by a coherent superposition of the initial Rydberg state and the neighbouring Rydberg states[15]. These adjacent Rydberg states are populated by a stimulated Raman process via the continuum[13]. The bandwidth necessary for this non-resonant Raman transition is present due to the short duration of the pulse. The requirement for transient stabilization, that the pulse duration is short with respect to the roundtrip time, can not be fulfilled for ground-state electrons ($\tau_n < 10^{-16}$ s) and they will be ionized by the short laser pulse. In a recent experiment we have indeed observed [14] that short pulse irradiation of Rydberg atoms produces only a limited fraction of ions as given by τ_p/τ_n, irrespective of the intensity. In earlier studies [16, 17] the inner-electron ionization of Rydberg atoms was already observed and more detailed data is presented below.

3.2. EXPERIMENTAL OBSERVATION OF INNER-ELECTRON IONIZATION OF BARIUM RYDBERG STATES BY AN ULTRASHORT LASER PULSE

In this section we present experimental evidence for the proposed mechanism of inner-electron ionization of a Rydberg atom by an ultrashort intense laser pulse. It is shown that the outer electron can survive such a laser pulse and remain bound to the atom. We studied the inner electron ionization of barium Rydberg states, in which Z changes from 1 to 2:

$$\text{Ba}(n, l) + 5\hbar\omega \rightarrow \text{Ba}^+(n', l') + e^- \tag{5}$$

Recently Stapelfeldt *et al.*[17] have demonstrated that such a process indeed occurs for optical pulses shorter than the Kepler orbit time of the Rydberg electron: $\tau_p < \tau_n$. Jones *et al.*[16] observed similar behavior when the Rydberg atom was placed in an electric field and the pulse duration was short with respect to the angular beating time (associated with the spacing of the k-states) of the Rydberg electron: $\tau_p < \tau_k$.

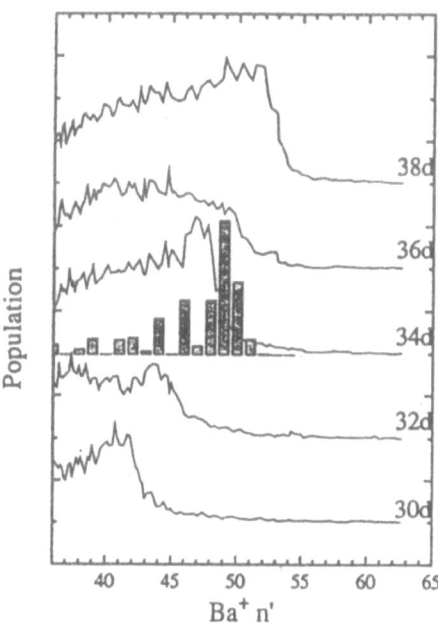

Figure 5. Distribution of population over ionic Ba$^+$ Rydberg states after ionization of the core electron. The distributions are measured with state selective electric field ionization. States below $n' = 37d$ could not be detected. The label on each curve indicates the initial state of the Ba Rydberg electron before core ionization. The histogram shows the expected distribution over the ionic Rydberg states as expected for the hydrogenic 34d.

As described in section 2 the population of ionic Rydberg states after core ionization can be easily calculated if the ejection of the inner electron is fast with respect to the classical roundtrip time of the Rydberg electron, as is the case in the experiment we performed. In this sudden approximation the initial Rydberg state has to be projected on the new basis of Rydberg eigenstates of the Ba$^+$-ion. The probability to populate a Ba$^+(n', l')$ state from an initial $Ba(n, l)$ state is then given by $\langle nl|n'l'\rangle^2$ which is zero for $l \neq l'$. The expected distribution of Rydberg states from this projection peaks around $n' = \sqrt{2}n$.

Under experimental conditions similar to those described by Stapelfeldt et al.[17], we measured the actual distribution of the population over the $Ba^+(n', l')$ Rydberg states after a selected $Ba(n, l)$ initial state had been exposed to a short intense laser pulse. The experiment has been described in detail elsewhere [8]. After inner electron ionization of $Ba(6snd)$ ($25 < n < 38$) states with 200-fs laser pulses at 620 nm, the distribution of population over ionic Rydberg states was measured by means of state-selective electric-field ionization. The experiment was set up such that only ionic and not neutral Rydberg states contributed to the signal. Figure 5 displays the experimental results. The various curves show the detected population using state-selective field ionization as a function of final effective quantum number n' of the Ba^{*+} state that was formed. The five curves show the results for five different initial Rydberg (nd) states. The data demonstrate that it is indeed possible for Rydberg electrons to survive a pulse that is intense enough to ionize an inner electron. The histogram represents the expected population of the ionic Rydberg states if the initial state is the hydrogenic 34d state. Despite the big quantum defect of the barium Rydberg d states, the experimental trend is well reproduced by the hydrogen model.

An important number for an efficient x-ray laser is the ratio between the number of Rydberg electrons that remain bound while the inner electron is removed and the number of atoms of which the outer electron is also ionized. This number should be as high as possible to achieve a high density of Rydberg ions available for x-ray amplification. This ratio of $Ba^+(n, l)/Ba^{++}$ is available from earlier experiments aimed at investigating the stability of the outer electron against photoionization with pulses that are shorter than the Rydberg roundtrip time [14]. With pulses of 200 fs we found a survival probability of 75% for the $Ba(6s27d)$ state, which has a roundtrip time of $\tau_n = 2.2$ ps. The pulse length was limited to 200 fs due to the experimental configuration, but shorter pulses should yield even higher survival probabilities. Furthermore, these experiments have been performed on stationary states instead of wavepackets. This means that there is a constant finite fraction of the wavefunction near the core. A wavepacket allows ionization of the inner electrons at times at which the Rydberg electron has zero probability to be in the core region, yielding higher survival probabilities.

4. Implementation in an x-ray laser

In order to achieve an appreciable gain, both an appreciable inversion and a high density of Rydberg ions are needed. An actual x-ray laser requires a high density plasma as a gain medium. Neither the calculations nor the experiments presented in this paper have been performed under high-density plasma conditions. In the high electric and magnetic fields present in a plasma, no excited Rydberg state will retain its original character for longer than a few picoseconds. Due to collisional (de)excitation the population will spread out over a large number of states. In order to overcome this decay of the Rydberg state, the wavepacket should be prepared in the state with the lowest possible principal quantum number n. Lower n states both have a longer lifetime against collisional (de)excitation because their binding energy and the spacing between neighboring levels is larger, and their roundtrip time is shorter so that the collisional destruction can be beaten more easily because the atom does not stay as long in the highly excited state, using the scheme described above. The lowest possible state is determined by the criterion for transient stabilization, the essential ingredient in the scheme: the roundtrip time of the lowest possible state should be longer than the shortest available laser pulse. Based on the arguments above, a favorable state would be $n = 8$ with a roundtrip time of 78 fs, which is still appreciably longer than the current record for an intense short laser pulse of approximately 20 fs [18, 19].

When regarding the influence of the plasma fields on the Rydberg population one should bear in mind that the entire proposed scheme to obtain inversion can be completed within a window of approximately 100 fs. The different steps and the corresponding timing are listed below.

In this paper we have mainly discussed the preparation of the Rydberg electron in a wavepacket. This procedure requires three laser pulses, all shorter than the roundtrip time of the Rydberg electron: one to excite the wavepacket, one to ionize the inner electrons and one to stimulate the electron down. A slightly less complicated scheme would involve the excitation of a stationary state. This could be done with CW-diode lasers [20, 8], ensuring a constant population in the Rydberg state of choice. Two different short laser pulses would then ionize the inner electrons and stimulate the Rydberg electrons to a lower lying state. In that case a finite fraction of the Rydberg electrons would be ionized upon inner electron ionization and the probability to stimulate it down would not be so much peaked at one particular time, but would have a rather constant, lower value.

The role of the lower state to which the Rydberg electron is stimulated down also depends on the plasma conditions. Any free atom in a p state decays with largest probability to the $1s$ ground state, which would be favorable for x-ray lasing. The rate at which this happens increases with decreasing n. However, it may be the case that the lowest n which can still be reached with a practical wavelength, for example the $5p$ state as described in section 2, still has a radiative decay rate to the $1s$ state, which is lower than the collisional deexcitation rate. In that case the last part of the scheme will again be dominated by collisional population of the upper lasing level, which would then be an even lower lying state. The big advantage of the proposed scheme is then that the source for population now lies relatively close, instead of in the continuum, thereby enhancing the rate at which the upper laser level is populated.

TABLE 1. Timetable for the subsequent steps in the proposed scheme to obtain inversion on low lying states in lithium.

The 100 fs recipe for a lithium x-ray laser	
Time (fs)	**Operation**
$-\infty - 0$	Diode laser excitation $2s \rightarrow 2p$ at 671 nm
0 - 30	Rydberg wave packet excitation of $n \approx 8$, $2p \rightarrow 8d$ at 372 nm
30 - 40	Wait for wavepacket to reach the outer turning point
40 -70	Inner electron ionization at 10^{17} W/cm^2
70 - 95	Wait for ionic wavepacket to return to the core
100 -130	Stimulated emission: $Li^{2+}(n = 14d) \rightarrow Li^{2+}(n = 5p)$ at 290 nm

5. Conclusions

Calculations have been presented on the evolution of a Rydberg electronic wavepacket, when used in a new scheme for an x-ray laser pumped by optical-field ionization. The new scheme involves three short-pulsed lasers: one to excite the Rydberg wavepacket, one very intense laser pulse to ionize the inner electrons when the wavepacket is far from the core, and one to stimulate the population down to the upper lasing level. Quantum mechanical calculations show that the probability to stimulate the electrons down peaks when the ionic Rydberg wavepacket returns to the nucleus.

Experiments on barium Rydberg atoms show evidence for the stability of the Rydberg electron against ionization with a very intense laser pulse and for the mechanism of inner electron ionization.

6. Acknowledgements

The authors like to acknowledge fruitful discussions with H. G. Muller, H. B. van Linden van den Heuvell, J. H. Hoogenraad and J. G. Story. We also thank S. Hünsche for carefully reading the manuscript. The work described in this paper is part of the research program of the Stichting Fundämenteel Onderzoek van de Materie (Foundation for Fundamental Research on Matter) and was made possible by the financial support from the Nederlandse Organisatie voor Wetenschappelijk Onderzoek (Netherlands Organization for the Advancement of Research).

References

1. J. Peyraud and N. Peyraud, J. Appl. Phys. **43**, 2993 (1972).
2. N. H. Burnett and P. B. Corkum, J. Opt. Soc. Am. B **6**, 1195 (1989).
3. P. Amendt, D. C. Eder, and S. C. Wilks, Phys. Rev. Lett. **66**, 2589 (1991).
4. W. W. Jones and A. W. Ali, Appl. Phys. Lett. **26**, 450 (1975).
5. W. W. Jones and A. W. Ali, J. Appl. Phys. **48**, 3118 (1977).
6. T. D. Donnelly, R. W. Lee, and R. W. Falcone, Phys. Rev. A. **51**, R2691 (1995).
7. Y. Nagata, K. Midorikawa, S. Kubodera, M. Obara, H. Tashiro, and K. Toyoda, Phys. Rev. Lett. **71**, 3774 (1993).
8. R. B. Vrijen and L. D. Noordam, accepted for publication in J. Opt. Soc. Am. B, December 1995.
9. A. Giusti-Suzor and P. Zoller, Phys. Rev. A **36**, 5178 (1987).
10. R. J. Vos and M. Gavrila, Phys. Rev. Lett. **68**, 170 (1992).
11. M. Pont and M. Gavrila, Phys. Rev. Lett. **65**, 2362 (1990).
12. M. P. de Boer, J. H. Hoogenraad, R. B. Vrijen, L. D. Noordam, and H. G. Muller, Phys. Rev. Lett. **71**, 3263 (1993).
13. J. Parker and J. C. R. Stroud, Phys. Rev. A **41**, 1602 (1990).
14. J. H. Hoogenraad, R. B. Vrijen, and L. D. Noordam, Phys. Rev. A **50**, 4133 (1994).
15. L. D. Noordam, H. Stapelfeldt, D. I. Duncan, and T. F. Gallagher, Phys. Rev. Lett. **68**, 1496 (1992).
16. R. R. Jones and P. H. Bucksbaum, Phys. Rev. Lett. **67**, 3215 (1992).
17. H. Stapelfeldt, D. G. Papaioannou, L. D. Noordam, and T. F. Gallagher, Phys. Rev. Lett. **67**, 3223 (1991).
18. J. P. Zhou, C. P. Huangand, C. Shi, H. C. Kapteyn, and M. M. Murnane, Opt. Letters. **19**, 126 (1994).
19. C. P. J. Barty, C. L. Gordon III, and B. E. Lemoff, Opt. Letters. **19**, 1442 (1994).
20. Chun-ho lu, G. D. Stevens and H. Metcalf, Appl. Opt. **34**, 2640 (1995)

ON ADIABATIC GENERATION OF TROJAN WAVE PACKETS*

MACIEJ KALINSKI AND J. H. EBERLY

Department of Physics and Astronomy
Univesity of Rochester
Rochester, NY 14627

1. Introduction

A hydrogen atom in a circularly polarized electromagnetic field is a simple quantum system that has been the subject of both experimental and theoretical studies for quite a long time. However one very interesting property of this system has not been noticed until recently. In 1989 Klar [1] had shown that two analytical solutions of the equations of motion exist for this system in the form of circular orbits, one always unstable and one stable for some values of the external field.

The presence of the stable solution allowed us to construct a quantum mechanical rotating-frame Hamiltonian [2] that can generate a locally harmonic energy spectrum. The eigenstates of this Hamiltonian have the form of shape invariant probability clouds that move around the circular orbit.

We have shown [2] that the formal mathematical equivalence between this and the restricted three body problem makes it possible to consider these states as microscopic probability asteroids. The trapping mechanism is essentially the same as in the vicinity of the stable Lagrangian points [3] that are responsible for the location of the Trojan asteroids in the Sun-Jupiter subsystem of the Solar system. Because of this connection we called these states Trojan wavepackets.

The peculiarity of these states lies not only in their connection with celestial mechanics but also in the fact that they are the solutions of the Schrödinger equation that have a stable and shape invariant character. Note that one needs to supply a nonlinearity in the Schrödinger equation to obtain similiar solutions (gaussons) for the free electron [4]. These states also provide an example of the stabilizing effect of quantum mechanics since the classical analysis of the phase space properties gives an estimate of the angular momenta for which such packets exist that is too pessimistic in light of their actual quantum mechanical behavior [5, 6, 7].

H. G. Muller and M. V. Fedorov (eds.), Super-Intense Laser-Atom Physics IV, 37–43.
© *1996 Kluwer Academic Publishers.*

The simplest possible state, in the form of a Gaussian packet, has been shown by us to exist by direct numerical integration of the time and field dependent Schrödinger equation [2, 8]. The lifetime of this state, which we predicted to be infinite within the harmonic approximation has been confirmed to be enormously long by monitoring its ionization [8].

In the following we demonstrate that the harmonic approximation formally predicts that these Gaussian-like Trojan packets are adiabatically connected with those circular Rydberg states with orbital frequency equal to the frequency of the circularly polarized field. We confirm this prediction by identification of Trojan states on the energy diagrams obtained numerically. We propose an experimental method of generation of a Trojan packet by adiabatic switching of the external field. Various techniques of circular state preparation are known [9, 10, 11, 12, 13, 14].

2. Gaussian-like Trojan wavepackets and circular states

The quantum-mechanical Hamiltonian of the hydrogen atom in the presence of a circularly polarized field is given (in the co-rotating frame) by

$$H = \frac{\mathbf{p}^2}{2} - \frac{1}{r} + \mathcal{E}x - \omega L_z, \tag{1}$$

where $\mathcal{E}$ and ω are the field amplitude and frequency. The existence of classical, stable, circular electron trajectories implies [2, 15] that this Hamiltonian can be be approximated by the following harmonic Hamiltonian

$$H_L = E_0 + \omega_+(a_+^\dagger a_+ + \frac{1}{2}) - \omega_-(a_-^\dagger a_- + \frac{1}{2}) + \omega_0(a_0^\dagger a_0 + \frac{1}{2}) \tag{2}$$

with the frequencies $\omega_+, \omega_-, \omega_0$ dependent on the electric field in an implicit way

$$\omega_+ = \omega\sqrt{2 - q + \sqrt{9q^2 - 8q}}\bigg/\sqrt{2},$$

$$\omega_- = \omega\sqrt{2 - q - \sqrt{9q^2 - 8q}}\bigg/\sqrt{2},$$

$$\omega_0 = \omega\sqrt{q}, \tag{3}$$

where $q = 1/\omega^2 r_c^3$, $\mathcal{E} = (1 - q)q^{1/3}\omega^{4/3}$ and r_c is the radius of the stable classical orbit. The energy shift E_0 is equal to the classical electron energy in the rotating frame. This approximation is formally valid when $8/9 < q < 1$, which guarantees the stability of the classical orbit and therefore keeps the frequencies real.

The simplest possible eigenfunction of the Hamiltonian (2) written in cylindrical coordinates is

$$\Psi(r, \phi, z) = Ne^{il_0\phi}e^{-\frac{\omega}{2}[A(r_c\phi)^2 + B(r - r_c)^2 + 2i(C-1)(r - r_c)r_c\phi]}$$
$$\times e^{-\frac{\omega}{2}Dz^2}, \tag{4}$$

where l_0 is the classical angular momentum of the electron in atomic units. The coefficients A, B, C and D can be found analytically as functions of the parameter q [2, 15] and in the limit $\mathcal{E} \to 0$ ($q \to 1$) one obtains

$$\Psi(r, \phi, z)_{\mathcal{E}=0} = Ne^{il\phi}e^{-\frac{\omega}{2}[(r_0\phi)^2+(r-r_0)^2+z^2]}, \tag{5}$$

where $r_0 = l^2$. This function can be identified as the asymptotic form of a circular state in the limit of large principal quantum number $n = l \approx l + 1$ [16], with the Kepler frequency $\omega_{cl} = 1/n^3$ equal to the frequency of the circularly polarized field ω. For $\mathcal{E} = 0$ we also have $E_0 + \omega = -1/2l^2 - (l-1)\omega$, which is the energy of the circular state in the rotating frame.

This proves formally that Gaussian Trojan wave packets are adiabatically connected with circular states in the large angular momentum limit, which permits us to treat quantum numbers of the hydrogenic spectrum as continuous variables. The weak point of this argument is that in the limit $\mathcal{E} \to 0$, as the wave function (5) approches the circular state, it is not well confined in the coordinate ϕ and therefore the harmonic approximation, which was used in its derivation, is not well justified for small field values. Numerical results show however, that the prediction of an adiabatic connection is correct.

3. Identification of Trojans

We have solved the stationary Schrödinger equation with the Hamiltonian (1) in the aligned ($l = m$) states basis $\{\psi_{nl}\}$, which corresponds approximately to a two dimensional model of hydrogen [17]. The wave function expansion

$$\Psi_E(\mathbf{r}) = \sum_{nl} c_{nl}(E)\psi_{nl}(\mathbf{r}), \tag{6}$$

leads to the following matrix equation, which can be solved numerically

$$\mathcal{E}\sum_{n'l'} x_{nl}^{n'l'} c_{n'l'}^j = [E^j(\mathcal{E}) - E_n + l\omega)]c_{nl}^j. \tag{7}$$

The hydrogenic dipole matrix elements $x_{nl}^{n'l'}$ are known analytically [18] and $E_n \equiv -1/2n^2$. Figs. 1 and 2 show discrete eigenvalues E^j as functions of the scaled electric field $\mathcal{E}\omega^{-4/3}$ for $n = 10$ and $n = 20$, with $\omega = 1/n^3$. The eigenvalues obtained from the harmonic Hamiltonian (2) are obviously given by

$$E_{m_+,m_-,m_0}(\mathcal{E}) = E_0 + (m_+ + \frac{1}{2})\omega_+ - (m_- + \frac{1}{2})\omega_- + (m_0 + \frac{1}{2})\omega_0. \tag{8}$$

The harmonic structure (two doublets in Fig. 1) barely appears for $n = 10$ and can be clearly identified for $n = 20$ (two triplets in Fig. 2). The spacing between doublets and triplets can be associated with the larger frequency $\omega_+ \approx \omega$ and the

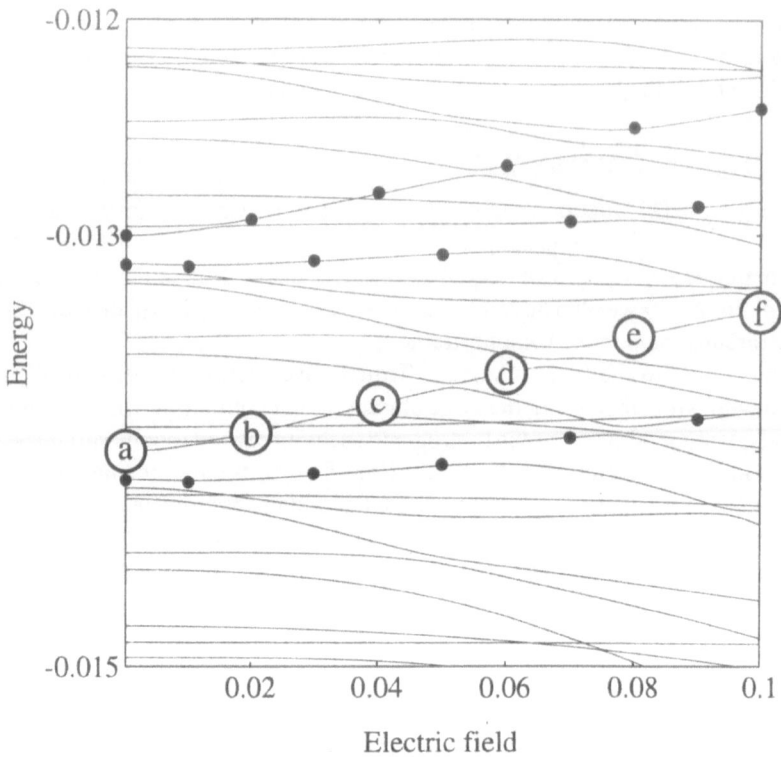

Figure 1. The eigenvalues of the Hamiltonian [1] found from the numerical solution of the equation [7] for $\omega = 1/10^3$ as functions of scaled electric field. Black points identify doublets of energies associated with the harmonic part of the spectrum. Large grey points follow Gaussian Trojan energy line (see Fig. 3).

spacing between lines creating those doublets and triplets can be associated with ω_- which is approximately proportional to $\mathcal{E}^{1/2}$. A more sophisticated nonlinear analytic approach [15] shows that our use of aligned states in expansion (6) gives only part of the harmonic spectrum associated with $m_0 = 0$ in the harmonic Hamiltonian. The Gaussian Trojan line, with the wave functions approximated by (4) (large grey points), can be identified because of the almost linear dependence on the external field $\mathcal{E}$, which originates from the classical contribution to the energy, E_0, which also increases almost linearly with the strength of the field. The plots shown in Figs. 3 and 4 of the two dimensional versions of the wave function (6) confirm that the eigenfunctions for the eigenvalues along this line do have Gaussian-like shapes and represent well confined probability droplets. The corresponding states for $\mathcal{E} = 0$ can be identified as circular states with Kepler frequency equal to ω.

Note that the search for Gaussian-like wave packets directly in the spectral data

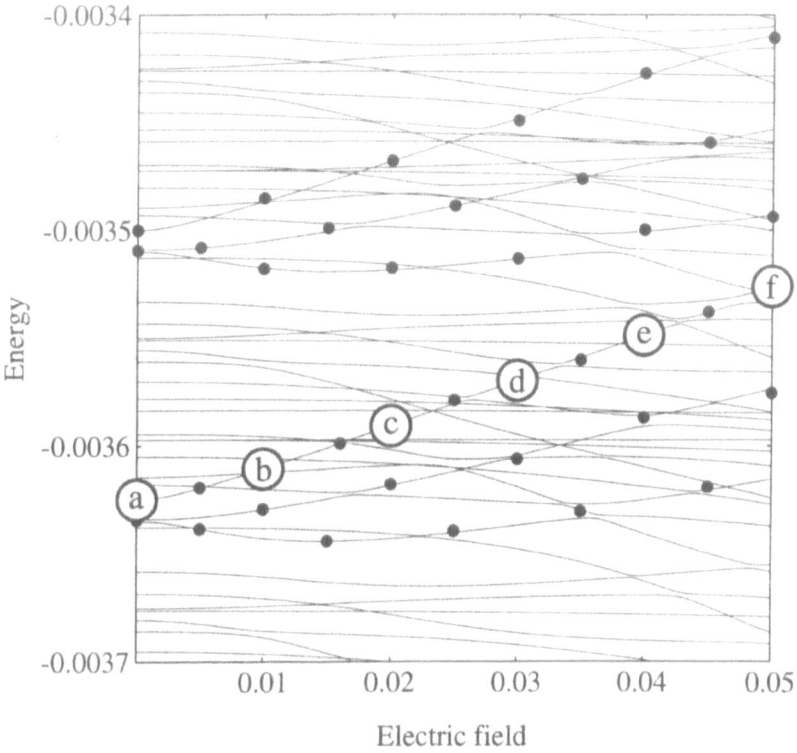

Figure 2. The eigenvalues of the Hamiltonian [1] found from the numerical solution of the equation [7] for $\omega = 1/20^3$ as functions of scaled electric field. Black points identify triplets of energies associated with the harmonic part of the spectrum. Large grey points follow Gaussian Trojan energy line (see Fig. 4).

obtained from the numerical solution of equation (7) would be almost impossible without the analytic theory, because of the complexity of the dressed spectrum.

4. Summary

The adiabatic connection between a Gaussian Trojan wave packet and a particular circular state implies an obvious method of preparation of the packet. The hydrogen atom should be prepared in the circular state with principal quantum number n and then a circularly polarized field with the frequency $\omega = 1/n^3$ should be switched on quasi-adiabatically. The Trojan energy line has a series of avoided crossings and the turn-on should also be fast enough to cause the packet state to remain on this line [19]. Results of a corresponding numerical experiment, which we have reported elsewhere [15], fully confirm this prediction.

We conclude that two successful analytic theories [2, 15], which explain the

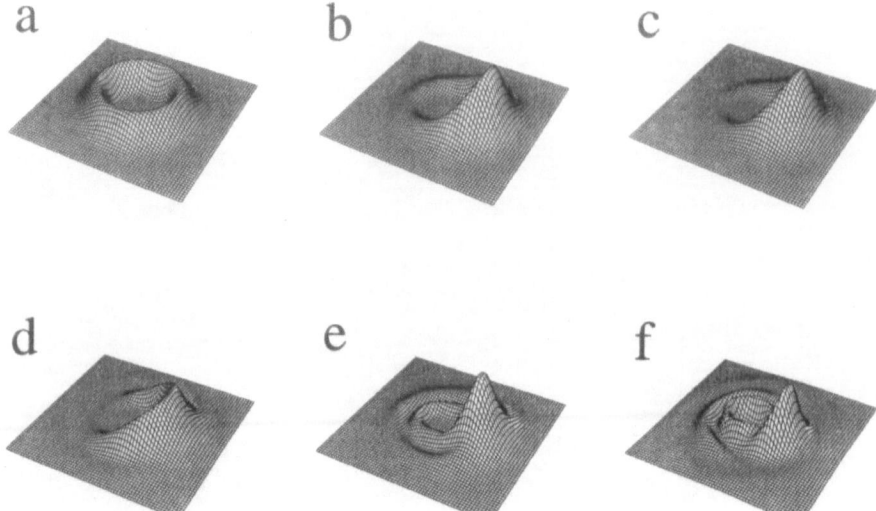

Figure 3. Selected eigenfunctions for $\omega = 1/10^3$ (probability density) from Gaussian Trojan line corresponding to grey points in Fig. 1.

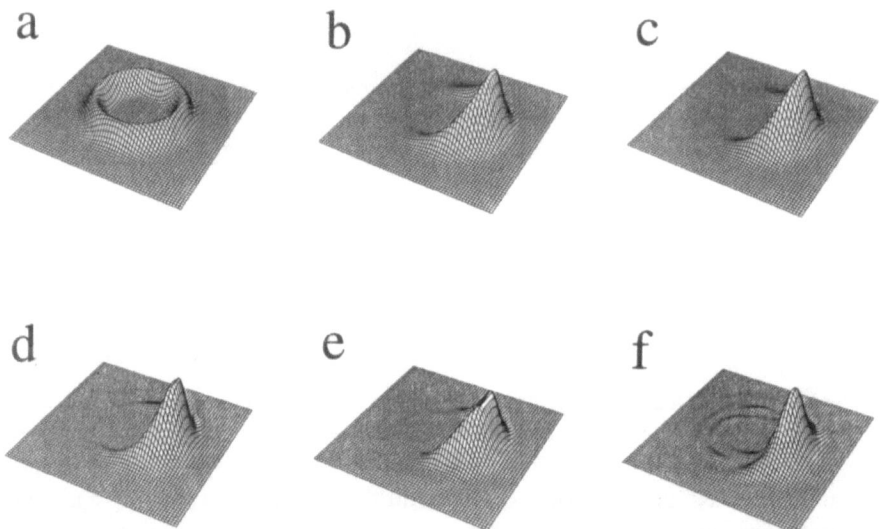

Figure 4. Selected eigenfunctions for $\omega = 1/20^3$ (probability density) from Gaussian Trojan line corresponding to grey points in Fig. 2.

properties of this part of the Floquet spectrum and the existence of Gaussian-like Trojan packets, suggest a new experimental method of generation of quantum wave packets which are both radially and angularly confined in a controllable way.

*Research supported by the U.S. National Science Foundation under grants PHY-9408733 and INT-9311766.

References

1. H. Klar, Z. Phys. D **11**, 45 (1989).
2. I. Bialynicki-Birula, M. Kalinski and J. H. Eberly, Phys. Rev. Lett. **73**, 1777 (1994).
3. See for example: F. R. Moulton, *An Introduction to Celestial Mechanics* (Macmillan, New York, 1914) (reprinted by Dover, New York, 1970).
4. I. Bialynicki-Birula and J. Mycielski, Ann. Phys. **100**, 62 (1976).
5. D. Farrelly, E. Lee and T. Uzer, Phys. Rev. Lett **74**, 972 (1995) (Comment).
6. I. Bialynicki-Birula, M. Kalinski and J. H. Eberly, Phys. Rev. Lett. **75**, 973 (1995) (Reply to ref. 5).
7. D. Farrelly and T. Uzer, Phys. Rev. Lett **74**, 1720 (1995).
8. M. Kalinski, J. H. Eberly and I. Bialynicki-Birula, Phys. Rev. A **52**, 2460 (1995).
9. R. G. Hulet and D. Kleppner, Phys. Rev. Lett. **51**, 1430 (1983).
10. W. A. Molander, C. R. Stroud and J. A. Yeazell, J. Phys. B **19**, L461, (1986).
11. J. Hare, M. Gross, and P. Goy, Phys. Rev. Lett. **61**, 1938 (1988).
12. P. Nussenzveig, F. Bernardot, M. Brune, J. Hare, J. M. Raimond, S. Haroche, and W. Gawlik, Phys. Rev. A **48**, 3991 (1993).
13. R. J. Brecha, G. Raithel, C. Wagner, and H. Walther, Opt. Comm. **102**, 257 (1993).
14. C. H. Cheng, C. Y. Lee, and T. F. Gallagher, Phys. Rev. Lett. **73**, 3078 (1994).
15. M. Kalinski and J. H. Eberly, Phys. Rev. A, (submitted for publication).
16. L. S. Brown, Amer. J. Phys. **41**, 525 (1973).
17. X. L. Yang, S. H. Guo, and F. T. Chan, Phys. Rev. A **43**, 1186 (1991).
18. H. A. Bethe and E. E. Salpeter, *Quantum mechanics of one and two-electron atoms*, (Springer-Verlag, 1957).
19. C. Zener, Proc. Roy. Soc. **A 137**, 696 (1932).

PHOTOIONIZATION AND STABILIZATION OF RYDBERG ATOMS WITH HIGH VALUES OF ORBITAL MOMENTUM

M.V. FEDOROV[†] and A.A. KILPIO
[†] *General Physics Institute, Russian Academy of Sciences,*
38 Vavilov St., 117942 Moscow, Russia
[‡] *Moscow Institute of Physics and Technology*
9 Institutskiy per., Dolgoprudny 141700 Moscow Region, Russia

The results of the experiment by Muller *et al.* [1] are interpreted in terms of the model of interference stabilization. A small difference between the ac Stark shift of Rydberg levels and the shift of the threshold of the continuum is taken into account. Under very reasonable conditions this difference is shown to provide the bound-continuum dipole matrix elements to be large enough for the main criterion of the model of interference stabilization to be fulfilled.

1. Introduction

1.1 EXPERIMENT

The present investigation is motivated by the first experiment [1] that has confirmed unambiguously the existence of stabilization of Rydberg atoms induced by a strong ionizing laser field. In this experiment , as shown in *Figure* 1, Ne atoms were excited initially to the circular 5g Rydberg state, i.e. to the state with maximal possible values of the electron orbital momentum ℓ and its projection m_ℓ ($\ell=m_\ell=n-1$,where n is the principal quantum number, n=5). Then, such atoms were ionized by a strong linearly polarized light field with the wavelength $\lambda=620$nm and two different values of the pulse duration: $\tau=1$ps and $\tau=0.1$ps. The electron yield measured in these two cases in the same range of the light fluence was shown to be lower in the case of a shorter pulse, i. e. of a stronger field. This is an undoubtful manifestation of the field-induced stabilization. However, interpretation of this experiment raises many questions, some of which are discussed below.

1.2 HIGH-FREQUENCY STABILIZATION.

There are two main rather well-known mechanisms of the strong-field stabilization. One of them is so-called high-frequency or adiabatic stabilization [2]. This model is based on the transformation to the frame that oscillates like a free

45

H. G. Muller and M. V. Fedorov (eds.), Super-Intense Laser-Atom Physics IV, 45–53.
© 1996 *Kluwer Academic Publishers.*

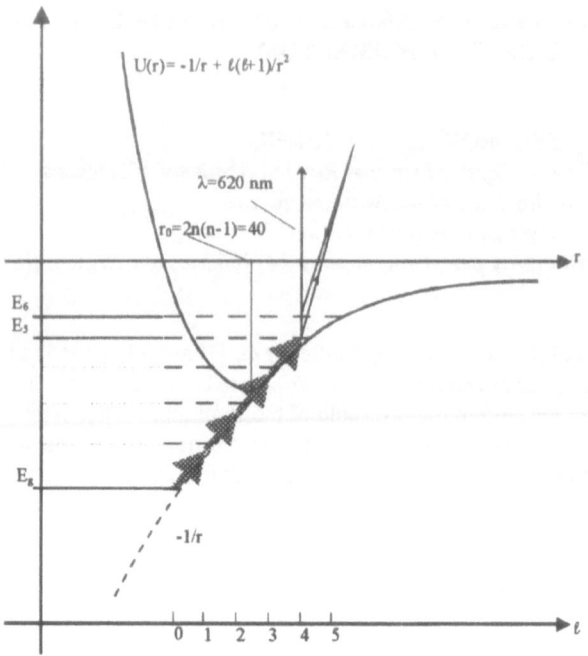

Figure 1. Potential energy U(r) and schemes of levels and transitions in a Rydberg atom, excited initially to the circular 5g-state (n=5, ℓ=m$_\ell$=4).

classical electron in a light field. In this frame, the entire interaction of the electron with the atomic potential and light field is characterized by the time-dependent potential energy

$$U(r,t) = U[\mathbf{r} - \alpha_0 \cos\omega t] \tag{1}$$

where U(r) is the field-free atomic potential, $\alpha_0 = \varepsilon_0/\omega^2$ is the classical free-electron quiver motion amplitude; ε_0 and ω are the field-strength amplitude and frequency of a strong light field correspondingly. Atomic system of units with $h = |e| = m = 1$ is used throughout the paper if other units are not indicated explicitly. In the high-frequency limit only the stationary part of the time-dependent potential U(r,t) is assumed to be most important. This part is known as the Kramers-Henneberger potential

$$U_{KH}(r) = \overline{U(\mathbf{r} - \alpha_0 \cos\omega t)} = \frac{\omega}{2\pi} \int_0^{2\pi} dt\, U(\mathbf{r} - \alpha_0 \cos\omega t) \tag{2}$$

Typical structures of the field-free ($U(r)$) and Kramers-Henneberger ($U_{KH}(r)$) potentials are shown in *Figure* 2. In contrast with $U(r)$, the Kramers-Henneberger potential has a double-well structure with wells separated by $2\alpha_0$ from each other.

Stationary states of the Kramers-Henneberger Hamiltonian describe adequately a stabilized part of atomic population, if only the oscillating part of the total time-dependent potential energy $U(r,t)$ (1) plays a really small role.

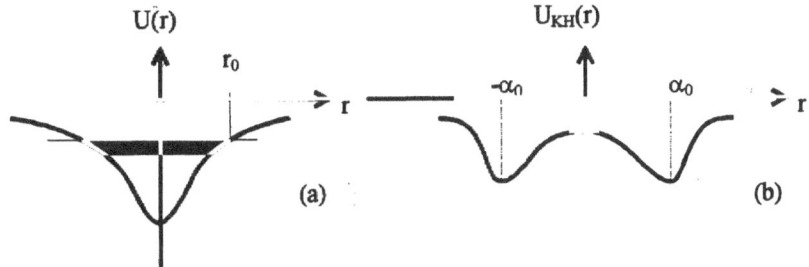

Figure 2.Field-free (a) and Kramers-Henneberger (b) atomic potentials.

For Rydberg states, there are two reasonable conditions of applicability of such an approximation: (i) the high-frequency condition

$$\omega > |E_n(\varepsilon_0)| \tag{3}$$

where $E_n(\varepsilon_0)$ is the energy of the n-th level of the Kramers-Henneberger Hamiltonian and (ii) the condition providing significant difference between the field-free and the Kramers-Henneberger potentials

$$\alpha_0 > r_0 \tag{4}$$

where $r_0 = 2n(n-1) = 40$ (for n=5) is the radius of the circular Rydberg state (see *Figure* 1).   Under the conditions of the experiment [1] ($I_{max} = 1.2 \times 10^{14}$ W/cm^2), $\alpha_0 \approx 19$, and the inequality (4) is not satisfied. This means that in this case the Kramers-Henneberger potential can hardly be considered as a good approximation for description of $U(r,t)$ (1). For this reason, interpretation of the experiment [1] in terms of the model of a high-frequency stabilization does not seem to be convincing.

1.3 INTERFERENCE STABILIZATION

The model of interference stabilization [3] is based on the idea about Raman-type transitions via continuum that provide coherent redistribution of atomic population between neighboring Rydberg levels. Subsequent transitions from these levels to the continuum can interfere with and partially cancel each other, resulting in suppression of ionization or atomic stabilization. The Raman-type transition can be efficient if the rate of ionization Γ_i calculated in the first-order perturbation theory (Fermie golden rule) becomes larger than the spacing between neighboring Rydberg levels (equal to $1/n^3$)

$$\Gamma_i > 1/n^3 \tag{5}$$

where

$$\Gamma_i = 2\pi \left| \left\langle E \left| \frac{d\varepsilon_0}{2} \right| n \right\rangle \right|^2 \tag{6}$$

and d is atomic dipole moment.

It is assumed usually that the condition (5) can hardly be fulfilled for circular Rydberg states because the corresponding matrix elements in Eq. (6) are too small. For this reason, the model of interference stabilization is not considered usually to be valid for interpretation of the experiment [1]. However, we assume that these conclusions are not necessarily correct. In the following section we will examine the dipole matrix element in Eq. (6) for circular Rydberg states. Then, in Section 3, we will formulate the conditions under which the experiment [1] can be interpreted in terms of the model of interference stabilization .

2. Dipole matrix elements for circular Rydberg states.

As it is shown in *Figure* 1, owing to the selection rules $\Delta\ell=1$ and $\Delta m_l=0$ in a linearly polarized field [4], the circular Rydberg state with $\ell=m_l=n-1$ is connected only with states of the continuous spectrum with $\ell=n$ and $m_l=n-1$. In the case of hydrogen atoms, the corresponding matrix elements are well known and they are given by the Gordon formulae [4]. Though rather complicated, in the specific case under consideration these formulae can be reduced to a much simpler form:

$$d_{En} \equiv \left\langle E \left| d_z \right| n \right\rangle = F(n) f_n(\xi), \tag{7}$$

where $\xi=kn$, d_z is the projection of the vector d on the polarization vector ($d_z \equiv d\varepsilon_0 / \varepsilon_0$), $k = \sqrt{2E}$ and $E=E_n+\omega$ are the electron momentum and energy in the continuum. The functions $f_n(\xi)$ and $F(n)$ can be shown to be given by

$$f(\xi) = \frac{1}{(1+\xi^2)^{n+2}} \exp\left\{ 2n\left(1 - \frac{\text{arctg}(\xi)}{\xi} \right) \right\} \tag{8}$$

and

$$F(n) = 4\sqrt{\pi}\,\frac{\sqrt{(2n+1)!}\;n^{n+2}}{\sqrt{2n}(n-1)!e^{2n}\,\Gamma(n+3/2)}.\qquad (9)$$

For large n, Eq. (9) can be further simplified with the help of the well-known Stirling formula:

$$F(n) \approx \frac{2\sqrt{2}}{\pi^{1/4}}\,n^{1.75}\,e^{-n(1-\ln 2)}\qquad (10)$$

The exact (Eq. (9)) and approximate (Eq. (10)) functions are shown in the *Figure* 3 by the solid and broken curves. These curves are rather close to each other. The curve F(n) has the maximum at $n \approx 6$, and it's maximal value is rather large ($F_{max} \approx 25$).

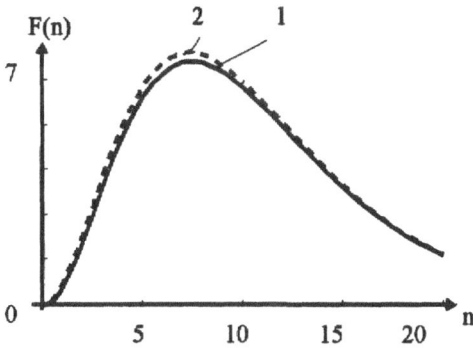

Figure 3. Exact (1) and aproximate (2) values of F(n).

The curves $f_n(\xi)$ are shown in *Figure* 4. These functions are normalized so that $f_n(0)=[f_n(\xi)]_{max}=1$. This means that exactly at the threshold ($\xi =E=0$), the dipole matrix elements $d_{nE}\big|_{E=0}$ are determined by the only function F(n) and, hence, they are rather large in a large interval of n. In dependence on ξ, the functions $f_n(\xi)$ are not small only if $\xi \le 1$, or $k \le n$, and $|E| \le 1/2n^2$. Hence, the range of energies E in the continuum, where the matrix elements d_{nE} are not small, is of the same order of magnitude as the atomic-electron binding energy $|E_n|$.

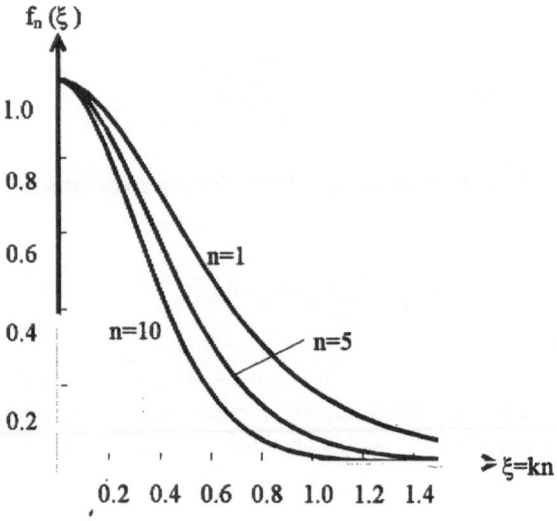

Figure 4. The functions $f_n(\xi)$ (Eq. (8)) for n=1, 5, 10.

3. Interpretation of the experiment [1] in terms of the model of interference stabilization.

By substitution λ=620nm and ω=0.073 into the energy conservation law

$$\frac{k^2}{2} = E_n + \omega \tag{11}$$

we find that for n=5 the product kn is equal to 1.6. For this value of ξ =kn, Eqs. (7)-(9) yield $d_{nE}\big|_{n=5} = 0.125$. At light intensity I=1.2×10^{14} W/cm^2 ($\varepsilon_0 \approx 10^{-1}$) this estimate of the dipole matrix element and Eq. (6) give $\Gamma_i \approx 10^{-4}$ and $\Gamma_i n^3 \approx 10^{-2}$ <<1. This estimate looks like a confirmation of the above-mentioned conclusion that interference stabilization can not explain the results of the experiment [1]. However, in our opinion, both this conclusion and its derivation can be not quite correct.

In fact, the energy conservation law (Eq. (11)) has to be corrected to take into account ac Stark shift of levels. In a light field the threshold of the continuous spectrum of photoelectrons is located at $E = \varepsilon_0^2 / 4\omega^2$, i. e. the threshold is shifted from $\dot{E}$=0 by the mean free-electron quiver energy. This term has to be added to the left-hand side of Eq. (11). On the other hand each atomic level E_n gets in a field the ac Stark shift equal to $-\frac{1}{4}\alpha_n(\omega)\varepsilon_0^2$, where $\alpha_n(\omega)$ is the dynamical polarizability. This shift to be added to the right-hand side of Eq. (11) to give

$$\frac{\varepsilon_0^2}{4\omega^2}+\frac{k^2}{2}=E_n+\omega-\frac{1}{4}\alpha_n(\omega)\varepsilon_0^2 \qquad (12)$$

For high Rydberg levels their dynamical polarizabilities $\alpha_n(\omega)$ are close to $-1/\omega^2$ and the corresponding shifts are close to the mean free-electron quiver energy $\varepsilon_0^2/4\omega^2$. However, the equality $\alpha_n(\omega)=-1/\omega^2$ is not exact. Let us assume that, rigorously,

$$\alpha_n(\omega)=-\frac{1}{\omega^2}+\delta\alpha_n \qquad (13)$$

where $\delta\alpha_n \ll 1/\omega^2$. Now the energy conservation law (12) is reduced to

$$\frac{k^2}{2}=E_n+\omega-\frac{1}{4}\delta\alpha_n(\omega)\varepsilon_0^2 \qquad (14)$$

As it shown qualitatively in *Figure* 5, the difference between the shift of the threshold ($\varepsilon_0^2/4\omega^2$) and the ac Stark shift of the level E_n can decrease significantly the photoelectron momentum k to provide smaller kn and the larger value of the dipole matrix element d_{nE}. For example, if this difference is sufficient to provide kn=0.9, this gives the $d_{nE} \approx 1$ and $\Gamma_i n^3 \approx 1$. This is just that condition which is necessary for realization of the model of interference stabilization. Let us use the condition kn=0.9 to find $\delta\alpha_n$ from Eq. (14):

$$\delta\alpha_n=\frac{4}{\varepsilon_0^2}\left(-\frac{k^2}{2}-\frac{1}{2n^2}+\omega\right)\approx 15 \qquad (15)$$

for $\varepsilon_0 \approx 10^{-1}$, ω =0.073 and n=5. As compared to the first term (ω^{-2}) on the right-hand side of Eq. (13) the found addition to the polarizability is not too large:

$$\omega^2\,\delta\alpha=\frac{\varepsilon_0^2/4\omega^2+\alpha_n\varepsilon_0^2/4}{\varepsilon_0^2/4\omega^2}\approx 8\% \qquad (16)$$

Such a small difference between the mean free-electron quiver energy and the actual ac Stark shift of the fifth Rydberg level can occur in reality. To confirm or reject this assumption one has to calculate numerically the polarizability $\alpha_n(\omega)$ for Ne atoms with n=5, $\ell=m_\ell=4$ and for ω=0.073 (λ=620 nm). Such specific calculations can hardly be

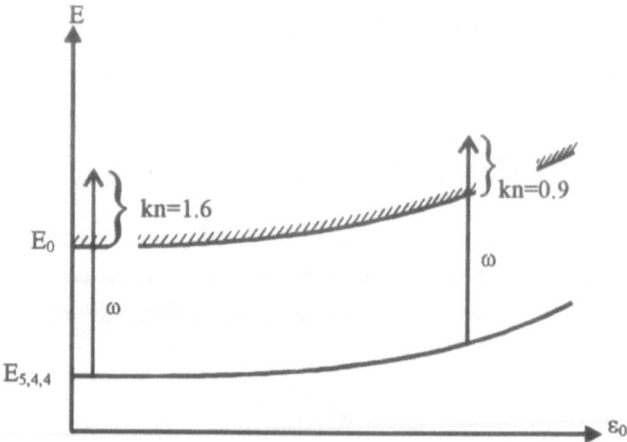

Figure 5. Ac Stark shifts of the threshold and Rydberg level E_5 with $l=m_l=4$.

found in any existing publications. We hope, however, that they can be done and will be done soon. Here, it's worth metioning that the required value of $\delta\alpha$ [Eq. (16)] agrees perfectly with qualitative estimate of the expected correction to the polarizability. By assuming that $\alpha(\omega)$ can be presented as a series in powers of ω^{-2}, we find a rather natural estimate

$$\omega^2 \, \delta\alpha = \left(\frac{E_n}{\omega}\right)^2 \approx 7.5\% \tag{17}$$

If exact calculations of $\alpha_n(\omega)$ will confirm the estimates of Eqs. (15) - (17), this will be a clear indication that the experiment [1] can be explained in terms of the model of interference stabilization.

4. Conclusion

As a resume, let us repeat briefly our two main statements concerning the interpretation of the experiment [1]. First, we doubt that this experiment can be interpreted correctly in terms of the model of high-frequency or adiabatic stabilization. The key-point for our objections is a rather small value of the free-electron quiver motion amplitude α_0 as compared to the radius of the circular Kepler orbit r_0. For this reason, the Kramers-Henneberger potential $U_{KH}(r)$ is hardly different from the field-free atomic potential $U(r)$ in the most important region of electron-nucleus distances r ($r \sim r_0$).

Second, we assume that an adequate interpretation of the experiment [1] can be given in terms of the model of interference stabilization. There, the key-point consists of the hypothesis that the ac Stark shift of the 5th Rydberg level with $\ell = m_\ell = 4$ is not equal exactly to the mean free-electron quiver energy. The difference is assumed to be as large as 8% (Eq. (16)). This difference provides a sufficiently small energy of photoelectrons in the continuum and, consequently, a sufficiently large value of the dipole matrix element d_{nE} for the main criterion of the model of interference stabilization ($\Gamma_i\, n^3 \geq 1$) to be fulfilled.

5. Acknowledgment

The authors acknowlege the support of the Russian Fund of Basic Research (grant # 93-02-15001) and International Science Foundation (grant # M6I000).

6. References

1. De Boer, M., Hoogenraad, J., Vrijen, R., Noordam, L. and Muller, H. (1993) Indications of high-intensity adiabatic stabilization in neon, *Phys. Rev. Lett.* v.71, p. 3263
2. Vos, R. and Gavrila, M. (1992) Effective stabilization of Rydberg states at current laser perfomances, *Phys. Rev. Lett.* v. 68, p. 170
3. Fedorov, M. V. and Movsesian, A. M. (1988) Field-induced effects of narrowing of photoelectron spectra and stabilization of Rydberg atoms, *Journ. Phys.* B, 21, L155-L158; Fedorov, M. V. (1993) Interference stabilization of Rydberg atoms in a strong ionizing field, *Las. Phys.*, 3, pp. 219-240
4. Landau, L. D., Lifshits, E. M. (1977) Quantum Mechanics, Oxford, Pergamon.

STABILITY WINDOWS IN IONIZATION VIA RYDBERG STATES

A. Wójcik and R. Parzyński
Quantum Electronics Laboratory, Institute of Physics, A. Mickiewicz University, Grunwaldzka 6, 60-780 Poznań, Poland

We study analytically the effect of stabilization in two-photon ionization from a well - isolated s state through the subthreshold series of p Rydberg states within a model which assumes the degenerate Raman transfer of population via continuum to both f and higher-angular-momentum ($L=5,7, ...$) Rydberg states.

1. Introduction

Stabilization against ionization is an unusual effect in non-perturbative dynamics of an atom exposed to intense-laser field [1,2]. Rigorous approach to the problem of stabilization needs solving, by numerical means, the time-dependent Schrödinger equation for a real atom interacting with radiation [3,4]. Quite different in methodology are approximate but fully analytical approaches to this problem [5-7] which, however, deal with model atoms rather than with real ones. Though based on models, the analytical approaches are often able to recover the main tendencies observed in the ab initio numerical calculations and to explain the physics of these tendencies. It is our aim in this paper to apply the model-based approach to the problem of nominal two-photon ionization from a well-isolated s state via the family of p Rydberg states, including the possible effect of migration of population via continuum to f and higher-angular-momentum Rydberg states [8].

2. The model and starting equations

The model under study is schematically depicted in Fig. 1. Its essence is the inclusion of the whole chain of Rydberg quasicontinua of different angular momenta L. Any two consecutive quasicontinua from the chain are mutually linked by degenerate two-photon Raman transition through the continuum of an appropriate angular momentum. Only the first p quasicontinuum is directly coupled by one-photon transition to the lower-lying initial s state.

We label by "0" the initial state and ascribe to it the Schrödinger population amplitude b_o. By $j = 1$ (L=1), 2 (L=3), 3 (L=5),..., N (with $N \to \infty$, in general) we denote Rydberg quasicontinua, and by n the states within a given quasicontinuum, so b_{nj} stands for the Schrödinger amplitude of the nth state in the jth quasicontinuum. Aiming at an analytical solution for the above model we choose an idealized rectangular

55

H. G. Muller and M. V. Fedorov (eds.), Super-Intense Laser-Atom Physics IV, 55–64.
© *1996 Kluwer Academic Publishers.*

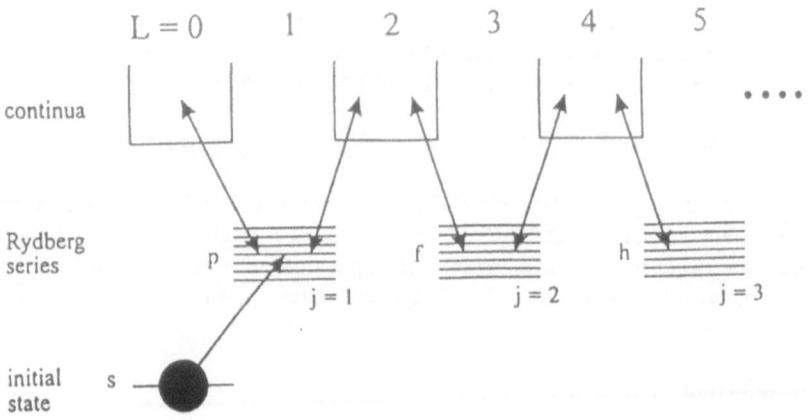

Figure 1. The model of ionization under study.

time envelope for the laser pulse. With such a choice, the standard procedure is the transformation of the Schrödinger equation to the Laplace domain whose variable we denote by s. Labelling by tilde the Laplace transforms of the Schrödinger amplitudes, we start with an infinite set of coupled algebraic equations for $\tilde{b}_o$ and $\tilde{b}_{nj}$. We write these equations in the rotating wave approximation and under the neglect of continuum-continuum transitions. Moreover, we assume that all Rydberg states from a quasicontinuum of a given angular momentum L are equally coupled to the neighbouring states in the model. This assumption is believed [5] to work well if the principal quantum number of intermediate Rydberg states is high enough ($n \gg 1$).

Under these assumptions the starting equation for the initial state is

$$s\tilde{b}_o = 1 - i\Omega K_1 , \qquad (1)$$

whereas the equation for the first (j=1) Rydberg quasicontinuuum looks like

$$(s - i\delta_{n1})\, \tilde{b}_{n1} = -i\Omega\, \tilde{b}_o - D_{11}K_1 - D_{12}K_2 , \qquad (2)$$

while for further quasicontinua ($j \geq 2$) we have

$$(s - i\delta_{nj})\, \tilde{b}_{nj\geq2} = -D_{jj}K_j - D_{j\,j-1}K_{j-1} - D_{j\,j+1}K_{j+1} , \qquad (3)$$

where

$$K_j = \sum_n \widetilde{b}_{nj} , \qquad (4)$$

Above, δ_{nj} is the laser frequency detuning from resonance with the nth state in the jth Rydberg quasicontinuum, Ω is the resonance Rabi frequency for the transition from the initial state to the first quasicontinuum, D_{jj} is the Raman coupling via the continuum between any two Rydberg states from the same quasicontinuum, and $D_{j\,j\pm1}$ is that between any Rydberg states from the neighbouring quasicontinua. The Raman couplings are defined as:

$$D_{j\,j\pm1} = (1 + iq_{j\,j\pm1}) \sqrt{\Gamma_j \Gamma_{j\pm1}}/2 \qquad (5)$$

with q_{ij} being the Fano-like parameter, and Γ_j the partial ionization rate from a state in the jth quasicontinuum to the continuum of a given L.

3. The continued - fraction solution

The essential point is that the set of Eqs (1)-(3) is structurally similar to the set obtained by Deng and Eberly [9] for their model of a different phenomenon, namely, above-threshold ionization with infinite sequence of continua. Formally, their equations can be obtained from ours by neglecting the diagonal Raman couplings ($D_{jj} = 0$), and replacing the remaining off-diagonal Raman couplings by the usual one-photon free-free couplings, precisely $D_{j\,j\pm1}$ by $iV_{j\,j\pm1}$. Due to the above stated structural similarity, we adopt the solution procedure of Deng and Eberly to the problem of our interest. Following their line, we divide Eq.(2) by $s-i\delta_{nj}$ and Eq.(3) by $s-i\delta_{nj}$, and then sum the results over all Rydberg states n, obtaining an alternative set for $\widetilde{b}_o$ and K_j ($j=1,2,\dots,N$). The set obtained can be solved exactly by the method of subsequent eliminations. The closed-form solution is conveniently expressed in terms of continued fractions G_j being generalization of the fractions introduced originally by Deng and Eberly [9] in their model of above-threshold ionization. For all j (also $j=1$), the continued fractions G_j relevant to our problem result from the recurrence formula

$$G_j = \cfrac{1}{1 + P_j D_{jj} \left(1 - \cfrac{D_{j\,j+1}^2}{D_{jj}} P_{j+1} G_{j+1} \right)} , \qquad (6)$$

fulfilling the boundary condition

$$G_N = \frac{1}{1 + P_N D_{NN}} , \qquad (7)$$

58

where

$$P_j = \sum_n \frac{1}{s - i\delta_{nj}} . \qquad (8)$$

Starting with this boundary condition, we create step by step G_{N-1} from G_N, G_{N-2} from G_{N-1}, ..., and finally G_1 from G_2.

In terms of the above G_j, the Laplace solutions for the Schrödinger amplitudes of the initial and Rydberg states are:

$$\tilde{b}_o = \frac{1}{s + \Omega^2 P_1 G_1} \qquad (9)$$

and

$$\tilde{b}_{nj \geq 1} = \frac{1}{s - i\delta_{nj}} \frac{K_j}{P_j} , \qquad (10)$$

where

$$K_{j \geq 1} = (-1)^j i\Omega \left(\prod_{k=1}^{j-1} D_{k\ k+1} \right) \left(\prod_{k=1}^{j} P_k G_k \right) \tilde{b}_o . \qquad (11)$$

For $j = 1$, the first product in Eq.(11) needs to be replaced by 1.

4. Time-dependent amplitudes

We transform analytically the obtained Laplace solution to the time domain, adopting equidistant Bixon-Joertner structure for Rydberg quasicontinua. Obviously, this structure roughly imitates what we meet in high Rydberg states only. For such states we put the level spacings independent of the quasicontinuum index ($\Delta_j = \Delta$ in, e.g., the hydrogen atom due to the actual orbital degeneracy). For mathematical simplicity, we also assume exact static one-photon resonance with a selected state in the first quasicontinuum, ensuring multi-photon Raman resonances as well. These assumptions allow us to apply the standard coth representation for P_j [9], namely, $P_j = P = (\pi/\Delta)(1+\mu)/(1-\mu)$, where $\mu = \exp(-Ts)$, with $T = 2\pi/\Delta$ having the sense of the classical Kepler period. With this P we follow the method of Stey and Gibberd [10], i.e., we expand both $\tilde{b}_o$ and K_j/P_j in a power series of μ and then transform to the time domain each term of this expansion. Assuming the laser pulse to be never longer than twice the Kepler period, we cut the expansion after the second term. The transformation of these two terms to the time domain $\tau = t/T$ results in the following analytical Schrödinger amplitudes:

$$b_o = e^{-2\pi\rho\tau} - 2\pi^2 u \left(2G_1(0) + \left(\frac{dG_1}{d\mu}\right)_{\mu=0} \right)$$

$$\times (\tau-1) \, e^{-2\pi\rho(\tau-1)} \, \theta(\tau-1) \, ,$$

$$(12)$$

and

$$b_{nj} = A_j \left\{ g_n(\tau) \, e^{i2\pi n\tau} \prod_{k=1}^{j} G_k(0) \right.$$

$$+ \left[\left(2\,(j-1) \prod_{k=1}^{j} G_k(0) + \left(\frac{d}{d\mu} \prod_{k=1}^{j} G_k \right)_{\mu=0} \right) g_n(\tau-1) \right.$$

$$\left. - 2\pi^2 u \left(2G_1(0) + \left(\frac{dG_1}{d\mu}\right)_{\mu=0} \right) p_n(\tau-1) \prod_{k=1}^{j} G_k(0) \right]$$

$$(13)$$

$$\left. \times e^{i2\pi n(\tau-1)} \theta(\tau-1) \right\} ,$$

where

$$A_j = i2\,(-\pi)^j \sqrt{u} \prod_{k=1}^{j-1} (D_{k\,k+1}/\Delta) \, , \qquad (14)$$

$$p_n(\tau) = (g_n(\tau) - \tau e^{z_n\tau})/z_n \, , \qquad (15)$$

$$g_n(\tau) = (1 - e^{-z_n\tau})/z_n \, , \qquad (16)$$

$$z_n = 2\pi(\rho + in), \quad \rho = \pi u G_1(0), \quad u = (\Omega/\Delta)^2, \quad (17)$$

$G_k(0)$ is the value of the fraction at $\mu = 0$, and $\Theta(x)$ is the Heaviside step function. We stress that these analytical Schrödinger amplitudes hold only for pulses of normalized duration $\tau \leq 2$, since we have retained only two terms in the Stey-Gibberd expansion. For $\tau < 1$ one should replace the Heaviside function by zero, while for $1 \leq \tau < 2$ by 1.

Having this analytical solution, in general we apply computer summation over state populations to calculate the total amount of population W_j transferred to the jth quasicontinuum by laser pulse. By subtracting from 1 the populations in all quasicontinua as well as the population left in the initial state, we arrive at the ionization probability. However, in the case of short pulse ($\tau < 1$) the actual discrete structure of the quasicontinua is not resolved effectively since the laser bandwidth exceeds the state separation. When calculating in this case the population transferred to a given quasicontinuum, we can integrate the squared modulus of b_{nj} over state energies instead of summing over state populations. It results in

$$W_j \overset{\tau<1}{=} |A_j \prod_{k=1}^{j} G_k(0)|^2 \, \frac{1 - e^{-4\pi\tau Re(\rho)}}{4\pi Re(\rho)} \, . \qquad (18)$$

5. Illustration of the population transfer and stability windows

Detailed numerical analysis of the above analytical solution proves that at the end of the pulse the redistribution of population over discrete states and ionization depend critically on the relation between different coupling strengths in the model (Ω, Γ_j), i.e., on the transition dipoles and laser intensity.

Not aiming at generality, we shall point to one potentially possible, specific effect. To show it, let us assume the laser pulse to be circularly polarized (the s continuum to be inactive) and to have its duration equal to half of the Kepler period ($\tau = 0.5$). It is assumed also that the excitation rate $R = 2\pi\Delta u$ of the first (p) Rydberg quasicontinuum from the initial s state is 4π times larger than the ionization rate Γ_1 from this quasicontinuum ($R = 4\pi\Gamma_1$). An additional assumption is that each subsequent partial ionization rate in the bound-free coupling chain from the first to the last quasicontinuua is 10 times smaller than the previous one. Then, under $q_{ij} = q_{ij'} = q = 1$ taken for all Fano parameters in the model, the results are as shown in Fig.2.

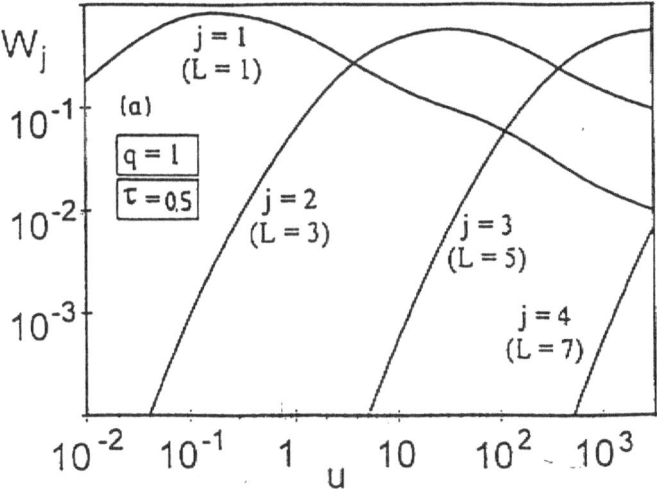

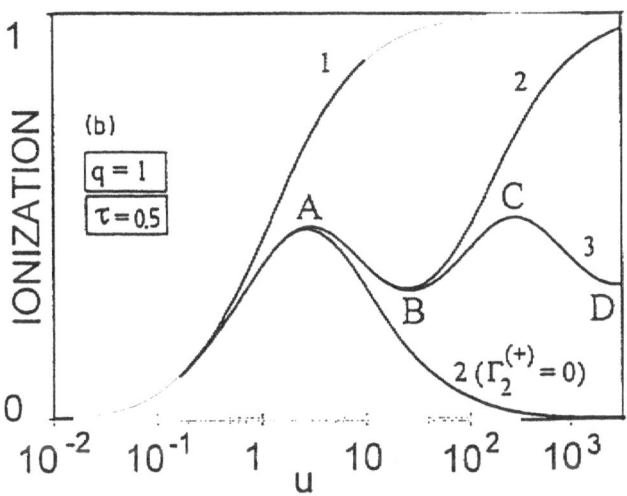

Figure 2. Total population W_j transferred to the jth Rydberg quasicontinuum by short pulse of $\tau=0.5$ (a), and ionization probability (b), versus laser intensity parameter $u=(\Omega/\Delta)^2$. For the coupling conditions, see text.

Fig.2a shows the dependence of the total population W_j trapped in the subsequent Rydberg quasicontinua with L=1,3,5 and 7, respectively, on the intensity parameter u at the end of the pulse. Under the above assumed coupling strengths, all other quasicontinua in our model got negligible population in the intensity range considered. As a matter of fact, this figure is the illustration of the transfer of population from the isolated initial state to higher-angular-momentum Rydberg states via the atomic continuum. With increasing intensity the maximum population is seen to be moved from the first p quasicontinuum to the second f quasicontinuum and then to higher-angular-momentum quasicontinua.

Fig.2b shows the effect of the above mentioned transfer of population to higher-angular-momentum Rydberg states on ionization. Curve 3 in this figure is the one obtained including all four Rydberg quasicontinua in the model (L=1,3,5,7).In practice, this curve differs by no means from the curve which would be obtained by neglecting the Rydberg quasicontinuum with L=7. There are two striking features of curve 3. Firstly, the maximum ionization probability is seen never to reach 1 and it is suppressed to the value of about 0.5. Secondly, there are two stability windows in the intensity intervals from A to B and from C to D, respectively, in which ionization decreases despite of increasing intensity. These windows give rise to the oscillatory behaviour of ionization probability versus intensity. Comparing Figs 2b and 2a, one can see that the ends of the stability windows (B and D) strictly correspond to the maximum transfer of population to the appropriate Rydberg quasicontinua, namely, to the L=3 and L=5 quasicontinua, respectively.

To show the effect of subsequent Rydberg quasicontinua on ionization more distinctly, we included in Fig.2b additional curves marked by 2, $2(\Gamma_2^{(+)} = 0)$ and 1, respectively. Curve 2 is the one obtained with the inclusion of only two Rydberg quasicontinua (L=1 and 3). Curve $2(\Gamma_2^{(+)}=0)$ results when the above two quasicontinua are included but at the same time the ionization channel from the L=3 quasicontinuum to the continuum with L=4 is neglected. Finally, curve 1 is obtained when taking into account only one Rydberg quasicontinuum (L=1), namely, the p quasicontinuum being the intermediary in the transition from the initial s state to the L=2 continuum. We would like to stress that curve 1, showing no features of suppression and stabilization, is representative for all situations when the p bound - d continuum coupling is much stronger than all other bound-free couplings in the chain from the first to the last Rydberg quasicontinua. Curves 1 and 3, when compared, show most distinctly what can happen under specific relation between the coupling strengths in the chain. Instead of saturation of ionization at the level of 1 one can expect substantial ionization suppresion with accompanying stability windows, due to transfer of population via continuum to higher-angular-momentum Rydberg states.

Under the same coupling conditions as previously, we show in Fig.3 the redistribution of population over discrete states produced by a long pulse of $\tau = 2$, i.e., the longest one allowed by our analytical solution. This figure shows that at relatively low intensities only the initial s state and the p Rydberg states directly coupled to it are important in the model. As seen, the population oscillates between the initial s state and the p states with increasing intensity, before the initial state gets completely depleted. Near the intensity at which this depletion takes place, the f series begins to be populated and then the population of f series increases at the expense of the p-series population. At the intensity nearly ensuring the maximum population of f series, the transfer to the h series gets turned on.

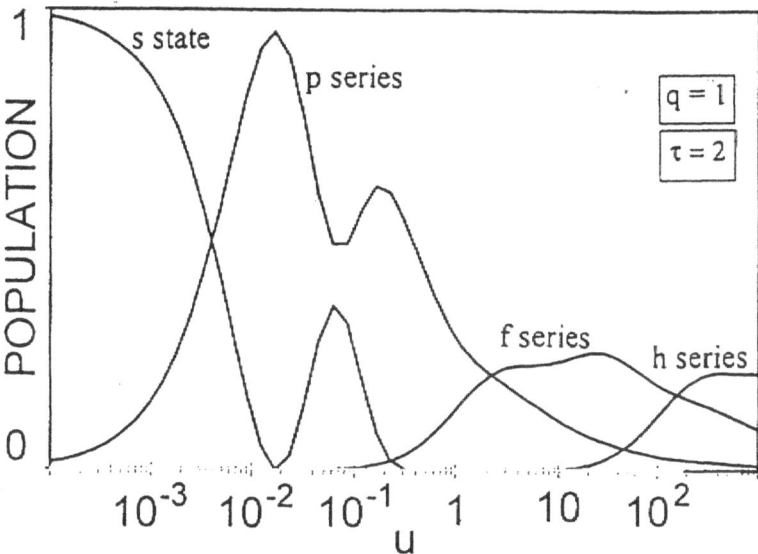

Figure 3. Redistribution of the population over discrete states after long pulse of $\tau=2$ versus intensity parameter u, under the same coupling conditions as in Fig.2.

6. Summary

In this article we have indicated a potentially possible effect which under specific coupling conditions can occur in intense-laser ionization of an atom via high Rydberg states ($n \gg 1$). It consists in formation of stability windows due to transfer of population via continuum to higher-angular-momentum Rydberg states. We would like to mention that the transfer of population to higher-angular-momentum states resulting from our analytical model is qualitatively consistent with what Huens and Piraux [8] observed in their numerical experiment on the hydrogen 2s-state two-photon ionization via the vicinity of the 8p Rydberg state.

References

1. Burnett,K., Reed,V.C., and Knight,P.L. (1993) Atoms in ultra-intense laser fields, *J.Phys.B: At.Mol. Opt.Phys.* **26**, 561-598.
2. de Boer,M.P., Hoogenraad,J.H., Vrijen,R.B., Constantinescu,R.C., Noordam,L.D., and Muller,H.G. (1994) Adiabatic stabilization against photoionization: An experimental study, *Phys.Rev.* **A50**, 4085-4098.
3. Faisal,F.H.M., and Dimou,L. (1994) Non-perturbative dynamics of atoms in strong laser fields: Adiabatic stability of hydrogen atom, *Acta Phys.Polon.* **A86**, 201-211.
4. Im,K., Grobe,R., and Eberly,J.H. (1994) Photoionization of the hydrogen 4s state by a strong laser pulse: Bare-state dynamics and extended-charge-cloud oscillations, *Phys.Rev.* **A49**, 2853-2860.
5. Fedorov,M.V. (1993) Interference stabilization of Rydberg atoms in a strong ionization field, *Laser Phys.* **3**, 219-240.
6. Ivanov,M.Yu. (1994) Suppression of resonant multiphoton ionization via Rydberg states, *Phys,Rev.* **A49**, 1165-1170.
7. Wójcik,A., and Parzyński,R. (1995) Suppression of two-photon ionization via Rydberg states, *Phys.Rev.* **A51**, 3154-3163; (1994) Rydberg-atom stabilization against photoionization: An analytically solvable model with resonance, *Phys.Rev.* **A50**, 2475-2489.
8. Huens,E., and Piraux,B. (1993) Dynamical stabilization of atoms in intense laser pulses accessible to experiment, *Phys.Rev.* **A47**, 1568-1571.
9. Deng,Z., and Eberly,J.H. (1985) Multiphoton absorption above ionization threshold by atoms in strong laser fields, *J.Opt.Soc.Am.* **B2**, 486-493.
10. Stey,G.C., and Gibberd,W. (1972) Decay of quantum states in some exactly soluble models, *Physica* **60**, 1-26.

DYNAMICS OF CLASSICAL SYSTEM WITH SHORT-RANGE POTENTIAL IN AN INTENSE ELECTROMAGNETIC WAVE FIELD

A.M. POPOV and O.V.TIKHONOVA
Institute of Nuclear Physics,
Moscow State University,
199899, Moscow, Russian Federation

1. Introduction

In the last few years it has been demonstrated in numerical quantum and classical experiments that atomic systems can become stable under very strong laser pulses [1-3]. This phenomenon called stabilization or suppression of ionization is characterized by a decreasing ionization rate with increasing laser intensity. The quantum stabilization mechanisms are widely discussed in many reports [1,2,4]. The atom-field interaction could be correctly described by a classical approach in the case of the high-lying Rydberg states of the atom. Then the correlation between quantum and classical results can be observed [5-7] . But there are a few bound states in the system with short-range potential only . Therefore the similarity of quantum and classical calculation results is not obvious in this case.

In this paper we will review the characteristic features of the interaction process between strong laser field and one dimensional classical model system with short-range potential . The model system was chosen to describe the negative hydrogen ion. The comparison with results of quantum calculations for the same system will be made [8].

2 . The one-dimensional model of the system

The negative hydrogen ion could be well considered as a single-electron system with one-dimensional short-range potential . The employed potential was chosen suitable both for classical and quantum calculations [1]:

$$V(X) = V_o \cdot \frac{\exp(-\sqrt{x^2/a^2+16})}{\sqrt{x^2/a^2+6.27}} \qquad (1)$$

Here $V_o = -676$ eV, a is the Bohr radius and the potential well depth is approximately -1.95 eV. As the ionization potential of H⁻ is 0.75 eV, the electron energy was chosen equal to 0.75 eV. The system could be described as an electron oscillating near X=0 with the amplitude of $X_{max}=1.48$ Å corresponding to electron energy value . Because of such electron energy, electron oscillations are anharmonic with the period of about T=1.68 fs.

H. G. Muller and M. V. Fedorov (eds.), Super-Intense Laser-Atom Physics IV, 65–71.
© 1996 Kluwer Academic Publishers.

Though the oscillations of the electron with the energy near the potential well bottom consider to be harmonic . The interaction between the system and laser pulse could be described by Newton's equation using dipole energy approach:

$$m \frac{d^2X}{dt^2} = -\frac{dV}{dX} + eE(t) \cos(\omega t + \phi) \qquad (2)$$

Here E(t) is the electric field amplitude, ω is the laser frequency, ϕ is the initial field phase and m is the mass of electron. The laser pulse had a Gaussian form:

$$E(t) = E_o \exp\left(-\tfrac{1}{2}\left(\frac{t - t_o}{\tau}\right)^2 \right) \qquad (3)$$

with a width of $2\tau = 10$ fs. The time moment to corresponded to the maximum field value was 15 fs . We integrated the Newton's equation using the forth-order Runge-Kutta routine during the time interval from 0 to 30 fs. The laser frequency corresponded to $h\omega = 5$ eV, so the relation $\omega\tau \gg 1$ was fulfilled and the adiabatic field turn on regime took place. It should be emphasized that laser frequency is much greater than the own oscillation frequency of the system in the potential (1).

The intensity of laser radiation was varied in a wide range from 10^{12} to 10^{16} W/cm^2. The laser field is of super-atomic value beginning from laser intensity of about $3\cdot10^{13}$ W/cm^2. The ionization occurs if the electron energy at the end of the laser pulse is positive. It was found that ionization process depends dramatically on the initial phase of electron oscillations corresponded to the initial electron coordinate X_o. Therefore the ionization probability must be averaged over the initial condition set. The probability for the electron to have a given initial coordinate should be taken into account. The probability density for every X_o set value was calculated by two different ways. At first , the probability density for initial X_o coordinate $\rho(X)$ was described in terms of classical

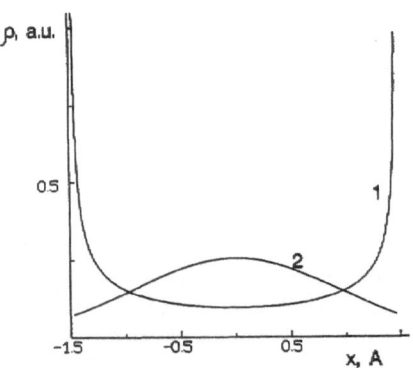

Figure 1. Classical and quantum probability densities

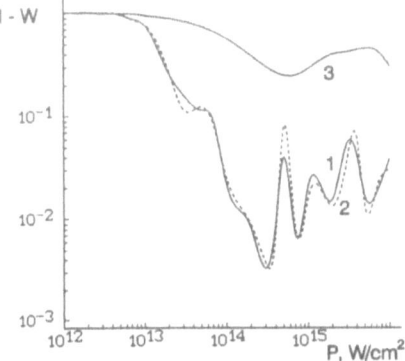

Figure 2. (1-W) versus peak laser intensity.

dynamic of the system. For the potential used the classical $\rho_1(X)$ could be presented by following equation,

$$\rho_1(X) = \frac{2}{T} \sqrt{\frac{m}{-2(V(X) + I)}} \cdot \tag{4}$$

Here T is the period of anharmonic oscillations within potential (1) and I = 0.75 eV is the ionization potential of the system. This distribution is demonstrated at Fig. 1 (curve 1). Another presentation for probability density was chosen correspondent to the quantum probability of the system with the same potential . As the quantum probability density $\rho_2(X)$ the squared wave function was chosen. This wave function is the solution of stationary Schrödinger equation for state energy value equal to -0.75 eV. The coefficient of about 1.13 was put to provide the normalized density in the classically allowed region:

$$\int_{-X_{max}}^{X_{max}} \rho_2(X) \, dx = 1 \, . \tag{5}$$

Fig. 1 (curve 2) presents the quantum probability density $\rho_2(X)$.

3. Results and discussion

As W is the ionization probability (1-W) means the probability for the electron to stay within the system potential . The dependence (1-W) on the laser pulse intensity for the classical calculations using different probability distributions $\rho_1(X)$ and $\rho_2(X)$ are presented at Fig. 2 (curves 1, 2). A good similarity could be observed because of the quite uniform distribution of the stable trajectories over the whole initial coordinate interval. It can be seen that the ionization probability increases sharply beginning from the laser intensity of about P = $3\cdot10^{13}$ W/cm^2 and reaches the maximum value at P = $3\cdot10^{14}$ W/cm^2. The resistance of the system to ionization process increases under laser

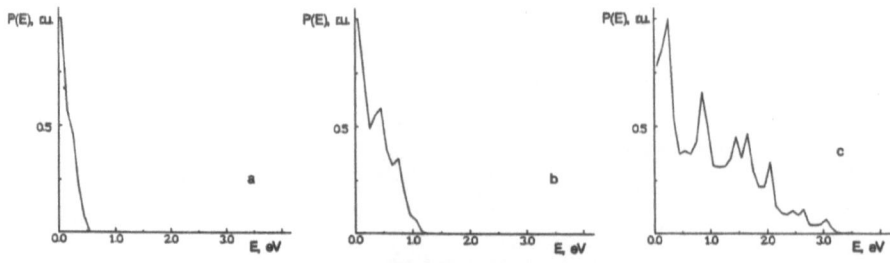

Figure 3. Electron energy spectra.

intensity values above P = $3 \cdot 10^{14}$ W/cm^2. This fact means suppression of ionization and P = $3 \cdot 10^{14}$ W/cm^2 seems to be stabilization threshold.

The dependence (1-W) on the laser intensity shows nonmonotonous behavior above the stabilization threshold. The results of quantum calculations for the system dynamic were obtained by direct numerical integration of nonstationary Schrödinger equation. The dependence (1 -W) on the laser intensity for quantum calculations is given on the Fig. 2 (curve 3). Both classical and quantum results are in qualitative agreement. There is the stabilization regime in both cases. But the quantum system is much more stable than the classical one. The nonmonotonous behavior of (1-W) above the stabilization threshold is a common feature for both curves and was also found by Geltman [9]. The electron energy distribution in the case of the system ionization was studied for different laser intensity values. Fig.3 a,b,c represent data obtained for P = $7 \cdot 10^{12}$ W/cm^2, $3 \cdot 10^{13}$ W/cm^2, 10^{15} W/cm^2, respectively.

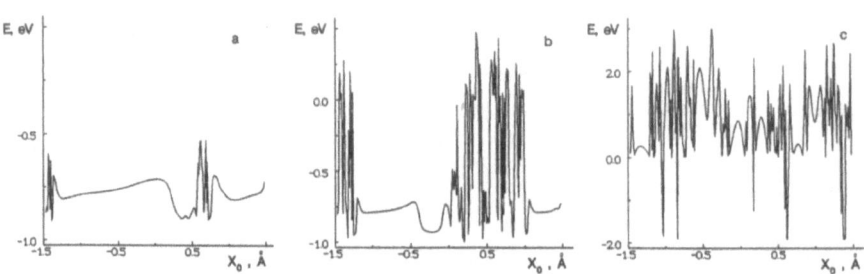

Figure 4. Final electron energy versus initial coordinate X_o.

It can be seen that energy spectrum width increases with the intensity increase. This fact means an increase of amount of high energy electron arising during ionization process. It should be emphasized that the maximum value of electron energy for a given laser intensity in classical calculations is less than first above threshold ionization peak energy for quantum calculations. Moreover a peak structure was found in classical electron energy distribution . The number of peaks rises with the increase of the laser intensity. To understand the reason of such peak structure the dependence of the final electron energy on the initial electron coordinate X_o was studied. Data obtained for different laser intensity values $P = 3 \cdot 10^{12}$ W/cm^2, $7 \cdot 10^{12}$ W/cm^2, $3 \cdot 10^{15}$ W/cm^2 are presented on Fig.4a, b, c, respectively. A set of regular and irregular areas for $E(X_o)$ dependence can be seen at these pictures . The comparison between these data and electron energy distribution

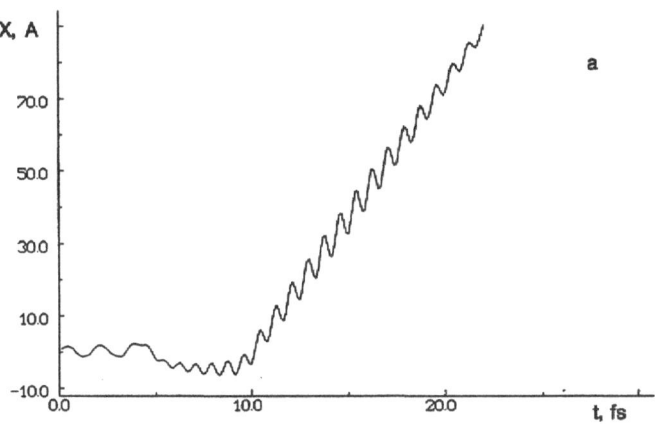

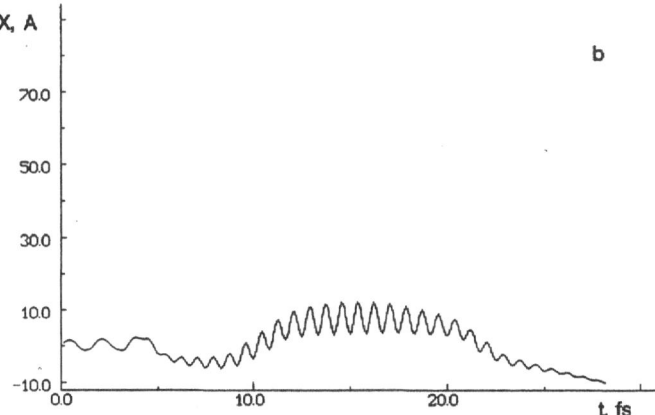

Figure 5. The divergence of initially close trajectories

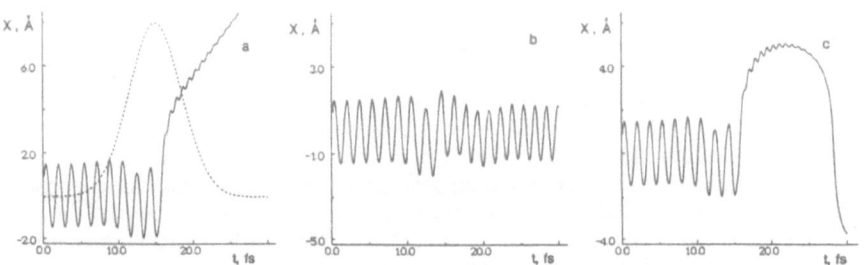

Figure 6. Electron trajectories for low laser intensity.

leads to the conclusion that there is the correspondence between regular area and one of the peaks with respective energy value. Irregular motion of electron leads to the exponential divergence between initially close trajectories [10] (Fig. 5).

Different types of the electron trajectories X(t) were investigated during the interaction between the system and the laser pulse. It was found that the ionization occurs near the maximum of laser pulse under low laser intensities (Fig. 6a, broken curve is the laser pulse envelope). In this case the oscillation amplitude caused by the laser field is quite small in comparison with that for free electron oscillation. The stable trajectory under the same intensity value is presented at the Fig. 6b. The electron motion seems to be finite during the whole laser pulse duration and represents the oscillations near X=0 free electron oscillation with the frequency approximately equal to that for free electron oscillation in the potential. The trajectory at the Fig. 6c was calculated for the same laser intensity and represents the intermediate realization between stable trajectory and ionization.

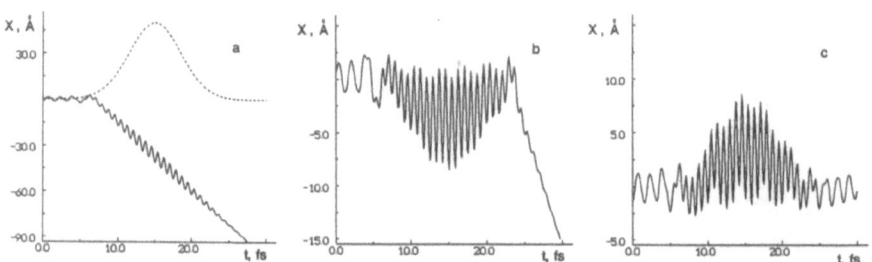

Figure 7. Electron trajectories for high intensity.

Another type of trajectories could be obtained under high laser intensities. Figures 7a and 7b show that under laser intensity $3 \cdot 10^{15}$ W/cm^2 ionization occurs during laser pulse turn-on or turn-off time. The amplitude of oscillations caused by laser field is great especially in the middle of the laser pulse, so the electron could be considered being free. This fact leads to existence of the stable trajectories X(t) under high laser intensity (Fig. 7c).

4. Conclusions

The interaction of the strong electromagnetic field with classical model system with short-range potential was studied. The phenomenon of ionization suppression was observed. The similarity between results of classical and quantum calculations was found. This fact allows to conclude that classical approach can qualitatively describe the system behavior . Some new system properties were found in comparison with quantum calculations . The system demonstrates stochastic behavior characterizing by divergence of initially close electron trajectories . So the system dynamic seems to be incidental . It is likely a classical system feature only.

5. References

1. Grobe, R. and Fedorov, M.V. (1993) Polychotomy, spreading, and relativistic drift in strong-field photodetachment, *Laser Phys.* **3**, 265.
2. Gaida, M. and Piraux, B. (1994) Ionization of an excited hydrogen atom by high frequency circularly polarized pulsed field, *Phys. Rev. A* **50**, 2528.
3. Grobe, R. and Law, C.K. (1991) Stabilization in superintense fields: a classical interpretation, *Phys.Rev.A* **44**, R4114.
4. Gavrila, M. and Kaminski, J. (1984) *Phys. Rev. Lett.* **52**, 613.
5. Nefedov, A.L. (1991) The stabilization of a classical atom in strong alternating field, *ZhETF* **100**, 803.
6. Rzążewski, K. and Piraux, B. (1993) Circular Rydberg orbits in circularly polarized microwave radiation, *Phys. Rev. A* **47**, R1612.
7. Gaida, M. , Grochmalicki, J., Lewenstein, M. and Rzążewski, K. (1992) Stabilization of atoms in ultrastrong laser fields: a classical approach, *Phys. Rev. A* **46**, 1638.
8. Volkova, E.A. , Popov, A.M. and Tikhonova, O.V. (1995) Three-dimensional model of negative hydrogen ion ionization in a superstrong electromagnetic wave field, *JETP* **108**, in press.
9. Geltman, S. (1994) Short pulse model atom studies of ionization in intense laser fields, *J. Phys. B* **27**, 1497.
10. Lichtenberg, A,J. and Liebeman, M.A. (1983) Regular and stochastic motion, Springer-Verlag, NY-Heidelberg Berlin.

INTENSE FIELD STABILIZATION OF ATOMIC HYDROGEN: CLASSICAL ENSEMBLE

C. C. SUNG and A. T. ROSENBERGER*
Department of Physics
University of Alabama in Huntsville
Huntsville, Alabama 35899

and

S. D. PETHEL and C. M. BOWDEN
Weapons Sciences Directorate, AMSMI-RD-WS-ST
Research, Development, and Engineering Center
U. S. Army Missile Command
Redstone Arsenal, Alabama 35898-5248, USA

1. Introduction

Based in part on earlier success in drawing analogy between quantum and classical approaches to intense-field stabilization (IFS) [1–4], and the supposition that in the correspondence limit of high Rydberg states and large field amplitude the classical and quantum approaches should exhibit distinct similarities, we here address two rather subtle aspects in detail. The first is the effect of the ramp-up time R on the probability of ionization P. Although it is generally agreed [4–11] that P depends on R, a quantitative description and interpretation based upon a systematic approach seems not to have appeared. The second aspect is the ionization lifetime, τ, in the field amplitude steady-state regime. Ionization lifetimes have been estimated from essentially two somewhat different approaches. In adiabatic ionization [12–17], τ is the lifetime of a state that is well defined by a specific set of quantum numbers, since it is a pure state of the effective Kramers-Henneberger (K-H) potential [18], evolving from a pure atomic state as the field amplitude is increased adiabatically. On the other hand, τ has been analyzed [18–22] with respect to a state that is a coherent superposition of many atomic Rydberg states. Despite the different assumptions made in the two approaches which would appear to preclude a quantitative comparison, they remarkably lead to the similar conclusion that the lifetime, τ, varies as a power of the field amplitude, ϵ: $\log \tau / \log \epsilon \approx 2\text{--}3$, with a rather complicated m-dependence, where m represents the z-component of the angular momentum. Here we demonstrate that the major features in the transient and steady-state regimes can be understood and well represented in terms of classical dynamics.

For this purpose we analyze classical ensemble dynamics to study, in detail, ionization probability as a function of field amplitude, frequency, and ramp-up time for laser-field excitation of atomic hydrogen. We demonstrate the existence of an optimal

---

* Permanent Address: Department of Physics, Oklahoma State University, Stillwater, Oklahoma 74078-0444, USA

H. G. Muller and M. V. Fedorov (eds.), Super-Intense Laser-Atom Physics IV, 73–82.
© 1996 *Kluwer Academic Publishers.*

ramp-up rate in the transient regime, and show that this can be adequately explained in classical-mechanical terms. This is the main result of this paper. The calculated ionization lifetime, after the amplitude of the optical field reaches steady state, is compared with results of several quantum-mechanical calculations and found to be in at least qualitative agreement.

The classical ensemble model and assumptions are presented in the next section. The IFS behavior in the transient regime of field turn-on is discussed in Section 3, and our results on ionization lifetime in the steady-state regime are presented in Section 4. A summary and conclusions are presented in Section 5.

2. The Model

In the presence of an external field $\varepsilon(t)$, at frequency ω, in the z-direction, the equations of motion from the classical Kepler model Hamiltonian in cylindrical coordinates (ρ,ω) are given by

$$\frac{d^2\rho}{dt^2} = - \frac{\rho}{\left(\rho^2 + z^2\right)^{3/2}} + \frac{m^2}{\rho^3}, \tag{1}$$

$$\frac{d^2z}{dt^2} = - \frac{z}{\left(\rho^2 + z^2\right)^{3/2}} + \varepsilon(t) \sin(\omega t+\theta), \tag{2}$$

where we use atomic units ($\hbar = e = m_e = 1$) and m is the angular momentum z-component. The field amplitude ramps up as $\varepsilon(t) = \varepsilon t/R$ for $t \leq R$ and then remains at the constant value $\varepsilon(t) = \varepsilon$ for $t > R$. Among the many parameters associated with Eqs. (1) and (2), some cause mainly statistical variations. Others, which are crucial for the final results, are to be studied here.

The ionization probability P depends on the phase θ in a manner similar to its dependence on the initial conditions. No attempt is made to average over θ since Eqs. (1) and (2) are solved for a random ensemble of initial conditions, and a detailed study is given for the role of the initial conditions in this paper. We therefore set $\theta = 0$ from the beginning.

The z-component of angular momentum, m, and the form of the potential are critical for discussion of IFS, as noted in the Introduction. A larger value of m generally enhances the suppression, as demonstrated earlier [4,5–17]. We use m = 0.75 unless otherwise stated.

The ω dependence is such that Eqs. (1) and (2) are invariant under the well-known scale transformation $t\omega = t_1$, $\varepsilon\omega^{-4/3} = \varepsilon_1$, $\rho\omega^{2/3} = \rho_1$, $z\omega^{2/3} = z_1$, $m^2 \omega^{2/3} = m_1^2$, where the subscript 1 denotes the value when $\omega = 1$, i.e., any change of ω yields identical results when all physical quantities are scaled accordingly. We note that the quantum mechanical ionization lifetime in the limit $\varepsilon \to \infty$ is a function of $\varepsilon/\omega^{4/3}$, which is in agreement with P being a universal function of $\varepsilon/\omega^{4/3}$ instead of ε/ω^2, as frequently implied in the literature. In view of the ω-scaling rule, we set $\omega = 1$ in the calculation. The classical approach here also is, of course, independent of the quantized selection rules which can be regarded as implicit in our calculation because of the following quantum scaling property (n is the principal quantum number): $n^2\rho \to \rho$, $n^2 z \to z$, $n^{-3}t \to t$, $n^4\varepsilon \to \varepsilon$, $n^{-3}\omega \to \omega$, and nm $\to$ m. For convenience, we choose initial energy in our calculation consistent with n = 1.

The initial positions are chosen such that $(z_0{}^2 + \rho_0{}^2) = 1$ with $2/3 \le \rho_0{}^2 \le 1$. Initial velocities $\dot{z}_0$ and $\dot{\rho}_0$ are chosen to give an initial energy $E_0 = -1/2$, corresponding to the hydrogen ground state (n = 1). We consider a particle ionized when ρ becomes greater than 2ε. The ensemble consists of a set of 200 particles with randomly chosen initial conditions, subject to the constraints noted above. The effects of different choices of initial conditions will be addressed later.

We note that our field $\varepsilon(t)$ does not have ramp-down, during which we would expect to find some additional ionization [4]. The role of ramp-down therefore is a separate matter, and will not be addressed in this study of the steady state in the limit t > R.

The region of applicability of classical theory must be addressed; our point of view is that by the correspondence principle, ε and the principal quantum number n must be large. We arbitrarily set $\varepsilon > 1$ to assume the validity of the classical approach. In our earlier work [1-3], ionization in classical physics was understood as a process whereby a particle takes advantage of its initial conditions to escape from the large time-dependent potential (K-H potential). We showed that some initial conditions induce the particles to reach points (ionization points) in the phase space at which an energetic kick from the time-dependent potential makes it kinematically impossible for them to remain in closed orbits. The term "gating" was used loosely to describe this mechanism [1]. How the ionization probability P depends on the various parameters in other than a one-dimensional model has not been fully quantified in a cohesive manner, and this we attempt to address in the next two sections.

3. Behavior in the Transient Regime

We find there are two regions of ionization in the transient regime t < R. (Time here is measured in units of the field period.) In the first region, for t < 1 when both ρ and z are small, the probability of ionization is completely determined by the initial conditions; the field and the Coulomb force can act coherently to drive the electron out of its atomic orbit. We have noticed that whether a member of the ensemble ionizes depends sensitively on the initial conditions; that is, the boundary in initial-condition phase space separating survivors from ionizers is not smooth and may well be fractal. Table 1 illustrates the sensitivity to initial conditions in this first region and Figure 1(a) shows a typical orbit for an ionized electron. In general, the second region of ionization takes place when $\varepsilon(t) \simeq 2$-3 (we call values in this range, which correspond to the maximum ionization rate or minimum ionization lifetime, ε_d); ρ may be small at first but becomes larger after several orbits when the centrifugal force overwhelms the Coulomb force to the extent that the electron cannot return. The term "ionization point" is used to describe this condition when $\varepsilon(t) \simeq \varepsilon_d$. When the electron survives the ε_d region, ionizations rarely appear again since the orbit radius becomes much larger and the centrifugal force cannot provide an efficient kick. This behavior is also demonstrated for a typical orbit in Fig.

1(b). The electron survives passage through the ε_d region, and in Fig. 1(c), the electron reaches an ionization point when $\varepsilon(t) \simeq 3$. In Figs. 1(b) and 1(c), the sinusoidal variation of the applied field is shown, to illustrate the result that the oscillation of the z-component of the electron's position is out of phase with that of the field. Figure 1 also illustrates that ρ is the position coordinate that is most illustrative of stability and/or ionization; as described in Ref. 25, the kick delivered by the nucleus to an electron oscillating in the z-direction will have its largest component perpendicular to the z axis, i.e., along ρ.

TABLE 1. Dependence of ionization on initial conditions. Percent ionized within the first field cycle for $\varepsilon = 8$ and $R = 2$ based on initial conditions. All cases use the same 100 random initial points with $m = 0.75$ and with the signs of $\dot{z}$, $\dot{\rho}$, and z varying as indicated.

Initial Conditions			% Ionized
$\dot{z}$	$\dot{\rho}$	z	
+	+	+	2
−	+	+	97
+	−	+	100
−	−	+	100
+	+	−	16
−	+	−	17
+	−	−	100
−	−	−	79

In the transient regime, the most important parameter determining the probability of ionization P is neither ε nor R but rather their ratio, the ramp-up rate ε/R, since, according to the previous discussion, the duration of the electron's stay around ε_d is the critical factor. Once the electron passes ε_d safely, the orbit radius gets larger and the electron becomes relatively immune to further increase in $\varepsilon(t)$. This is illustrated in Fig. 2, which shows that $P(\varepsilon/R)$ is almost independent of ε. In Fig. 2, P is determined by the fraction of the ensemble that does not survive ramp-up. It is obvious now why P is larger for $\varepsilon/R = 1$ than for $\varepsilon/R = 2$ for fixed ε: The slower ramp-up (larger R) keeps the electron in the vicinity of ε_d for a longer time as $\varepsilon(t)$ increases. Further, as ε/R becomes larger than about 2, P also increases, as discussed below for $\varepsilon/R = 4$. In this case, $\varepsilon(t)$ reaches 2, $\varepsilon = \varepsilon_d$, for $t = 1/2$ when the electron is still very close to the nucleus. The external field drives the electron strongly in the z-direction and the Coulomb attraction rapidly decreases; then ρ increases steadily to ionization. Between these two extremes, we find an optimum $\varepsilon/R \simeq 2$ which produces a minimum of ionization during ramp-up ($t \leq R$). Figure 3, showing that $P(\varepsilon)$ (probability of surviving ramp-up, as a function of the final ε) is essentially constant for a given value of ε/R, corroborates these results. The great variation of P over different ε and R and the minimum of P at ε/R = 2 seems to support this classical interpretation. In particular, these same features are qualitatively represented in the quantum-mechanical results of Refs. [6–11], where again small ε/R corresponds to too much time spent in the region of rapid ionization around ε_d, but where large ε/R can now be interpreted as producing a superposition of K-H states that includes unbound states, greatly increasing the probability of ionization.

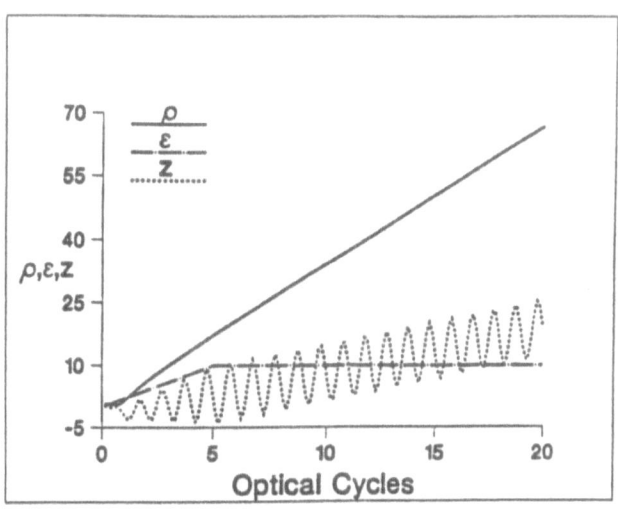

(a)

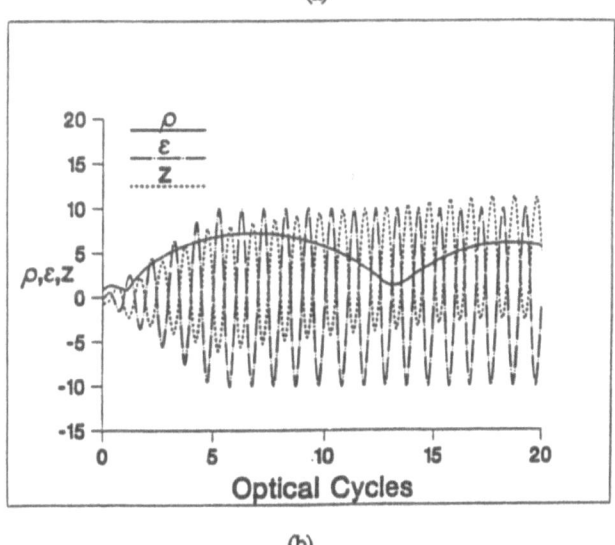

(b)

Figure 1. Here, ε/R = 2 (ε = 10, R = 5) and m = 0.75. (a) Field ε(t) and coordinates (ρ,z) as functions of time for a member of the ensemble that ionizes early in the transient regime. (b) Field ε(t) and coordinates (ρ,z) as functions of time for a member of the ensemble that survives ramp-up. (c) Field ε(t) and coordinates (ρ,z) as functions of time for a member of the ensemble that does not survive ramp-up.

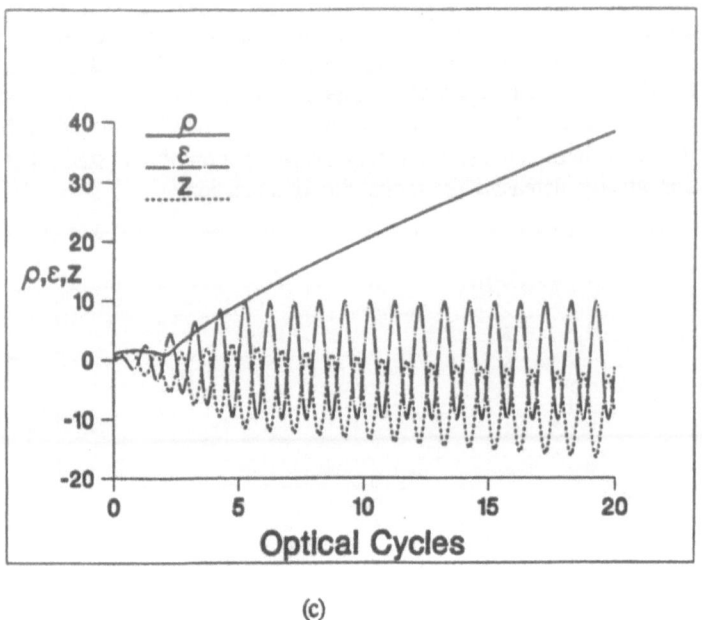

(c)

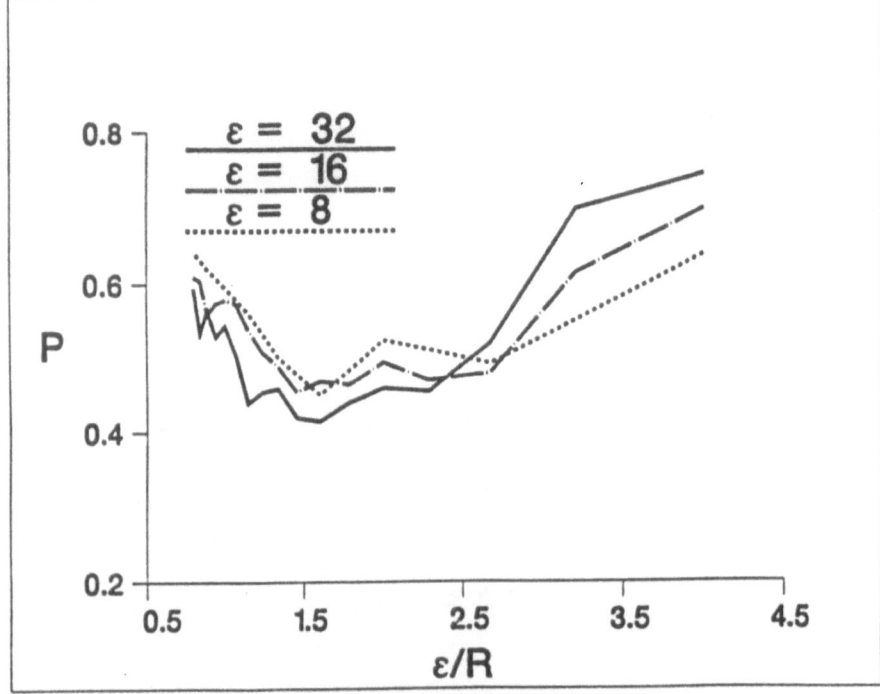

Figure 2. Probability of ionization during ramp-up as a function of the ramp-up rate, for m = 0.75 and for several values of the final field amplitude, ε.

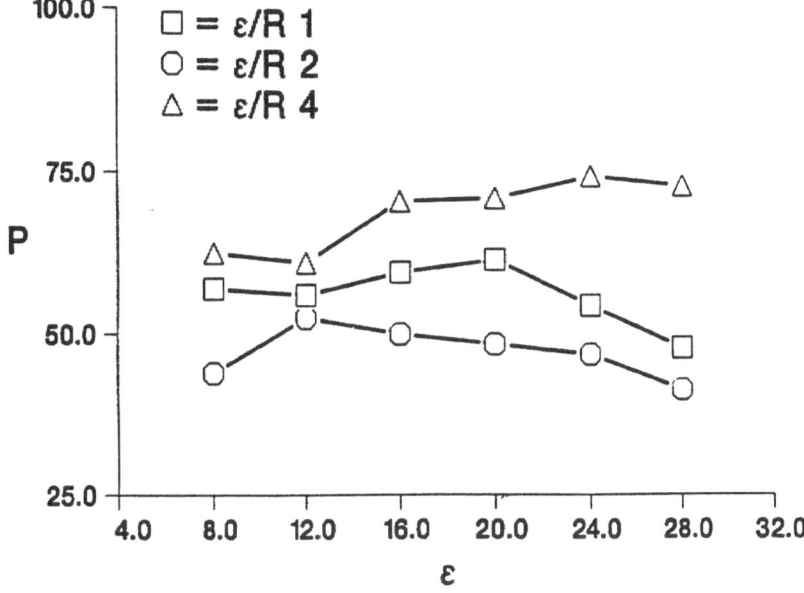

Figure 3. Probability of ionization as a function of the final field amplitude, for m = 0.75 and for several values of the ramp-up rate, ε/R.

4. Steady-State Ionization Lifetime

In the region (t ≥ R) where ε(t) = ε is a constant, the available ionization points are rather rare, especially when ε is large, and thus the rate of ionization is much smaller; in order for substantial additional ionization to be observed, t must get very large. We refer to this as the steady-state ionization regime. Here the orbits are determined by ε in large scale and by the initial conditions in minute scale. Our results show that for the electrons that survive ramp-up the ensemble average of ρ^2, $\langle \rho^2 \rangle$, is approximately proportional to ε^2 and almost independent of m. The qualitative behavior of the ionization lifetime, in a very crude approximation, can be obtained by using conservation of angular momentum:

$$T \cdot m = 2\pi \left\langle \rho^2 \right\rangle, \qquad (3)$$

Here, T is the period of the orbit. The ionization lifetime τ will be proportional to $N_0 T$, where N_0 is the number of orbits completed before the electron hits an ionization point. We emphasize that ionization points in the classical interpretation show under what conditions the orbits meet the initial conditions necessary to cause ionization. Using Eq. (3), if N_0 is independent of ε, we find $\tau \approx \varepsilon^2$, which is compared to the two distinct quantum approach predictions of $\tau \approx \varepsilon^2$ [18,19] and $\tau \approx \varepsilon^3$ [12]. However, our experience indicates that N_0 must have some ε– and m–dependence. How N_0 depends on ε or m is not easy to determine in view of the large number of parameters needed to define an ionization point. A crude estimate may be made by considering the minimum distance ρ_c. For $z = 0$, ρ_c is the distance of closest approach, where taking the left-hand side of Eq. (1) equal to zero gives $\rho_c \approx m^2$. The likelihood of hitting an ionization point increases as ρ_c decreases, so if $N_0 \approx \rho_c$, then $N_0 \approx m^2$, and $\tau \approx m\varepsilon^2$. This crude argument indicates that τ should increase with increasing m and ε. This is seen in our results for the steady-state ionization lifetime presented in Table 2, which shows a power-law dependence of τ on ε that may be summarized as $\log(\tau)/\log(\varepsilon) \simeq 2.9 \pm 1.2$ for m = 0.75 and $\log(\tau)/\log(\varepsilon) \simeq 0.5$ for m = 0.1, or $\log(\tau)/\log(\varepsilon) \simeq (4\pm1)$ m. These numerical results may be compared with quantum calculations that give $\log(\tau)/\log(\varepsilon) \simeq 8$ [15], $\simeq$ 2.3 [8], or $\simeq 2.2$ [16], for values of m (scaled form of m/n) closer to 0.75 than to 0.1. It should be emphasized that these comparisons of numerical results should be done with care, because the sampling in the classical and quantum approaches is different. Our results are based on the relatively small fraction of particles that survive the ramp-up without being ionized, and thus reflect a biased subset of the initial ensemble. Furthermore, the steady states used in the quantum calculations need considerable clarification. In the approaches in Refs. 18 – 22 and in Refs. 6 – 11, the particles are in coherent superpositions of Rydberg states and of K-H states, respectively. Such a superposition state cannot be described by a single set of quantum numbers. On the other hand, the adiabatic states [12–17] are K-H eigenstates and so each can be described by a single set of quantum numbers. Thus, comparison with the classical ensemble is not quantitatively meaningful. Our question is, to what extent can a quantum state be represented by a classical ensemble? Our steady states, being made up of the surviving subset of the ensemble, may not be a faithful representation of the initial distributions. It is not clear how we can emulate the quantum-mechanical states in the steady-state at this point. Resolution of this question will entail careful analyses of the initial phase-space distribution and the subset that survives ramp-up, and their correspondence to quantum states and superpositions of quantum states. In principle, although perhaps not easily accomplished in practice, it appears that as long as the quantum superpositions do not result in explicitly nonclassical distributions, the classical ensemble should permit calculations that reproduce the quantum results in the correspondence limit. Consequently, we believe the classical calculations presented here are strongly analogous to the quantum calculations of Refs. 18 – 22.

TABLE 2. Time, in optical cycles, required to ionize 30% of the particles surviving ramp-up, with ε/R = 2.

FIELD STRENGTH

		4	8	16
m	0.75	9	30	533
	0.1	3	4	6

5. Conclusions

To summarize, we have made use of classical ensemble dynamics in the correspondence limit to study IFS and ionization of atomic hydrogen. We have studied the process in detail in two regimes. The first regime consists of ramp-up, when the external field amplitude is growing linearly, and the second regime is steady state, when the external field amplitude remains constant.

In the transient regime of IFS, we have found that ionization is likely to take place either very early, when the electron is still close to the nucleus, or else when the field reaches a critical amplitude of $\varepsilon_d \simeq 2$ to 3, corresponding to a maximum ionization rate. These results show how it can be understood that in terms of classical mechanics there is an optimum choice of the ramp-up rate, $\varepsilon/R \simeq 2$, that yields the maximum probability of stabilization. This is the most important result of this paper.

In the steady-state regime, we have shown that IFS is a result of certain initial conditions in phase space that can avoid the ionization points at $\varepsilon(t) \simeq \varepsilon_d$, after which ionizations become rare events. We have shown that the ionization lifetime increases with increasing ε and m (z-component of angular momentum), and that the dependence is similar to that of several quantum versions of IFS.

Overall, our picture provides an interpretation of IFS in terms of classical mechanics. A detailed comparison of the classical and quantum pictures will require a study of the correspondence between the classical ensembles, both those prepared initially and the subset that survives ramp-up, with the corresponding quantum states. Our method seems to be naturally suited to a comparison with the previously published results of the quantum theory of Fedorov [18,19]; a better comparison with theories using K-H states [6–11, 12–17] may require another distribution of the initial conditions in phase space to sample the quantum probability distribution. At this time, however, we can say that both our theoretical and numerical classical results are in good qualitative agreement with quantum results to the extent that they can be compared directly. Comparison with the current results of Fedorov reported at this conference will require further assessment. Our conclusion is that classical mechanics is adequate and insightful for the study of certain of the major aspects of intense-field stabilization and ionization.

6. Acknowledgement

We thank Mr. M. Dombrowski, who pointed out the importance of the ramp-up during the course of the work reported in Ref. 4.

82

7. References

1. Ritchie, A. B., Bowden, C. M., Sung, C. C., and Li, Y. Q. (1990) Strong field ionization in classical and quantum dynamics, *Phys. Rev. A* **41**, 6114.
2. Bowden, C. M., Sung, C. C., Pethel, S. D., and Ritchie, A. B. (1992) Classical dynamics of strong-field ionization, *Phys. Rev. A* **46**, 592.
3. Bowden, C. M., Pethel, S. D., Sung, C. C., and Englund, J. C. (1992) Multicolor photoionization and classical dynamics, *Phys. Rev. A* **46**, 597.
4. Dombrowski, M., Rosenberger, A. T., and Sung, C. C. (1995) Intense-field stabilization and the range of the potential, *Phys. Lett. A* **199**, 204.
5. Su, Q., Eberly, J. H., and Javanainen, J. (1990) Dynamics of atomic ionization suppression and electron localization in an intense high-frequency radiation field, *Phys. Rev. Lett.* **64**, 862.
6. Reed. V. C. Knight, P. L., and Burnett, K. (1991) Suppression of ionization in superintense fields without dichotomy, *Phys. Rev. Lett.* **67**, 1415.
7. Kulander, K. C., Schafer, K. J., and Krause, J. L. (1991) Dynamic stabilization of hydrogen in an intense, high-frequency, pulsed laser field, *Phys. Rev. Lett.* **66**, 2601.
8. You, L., Mostowski, J., and Cooper, J. (1992) Suppression of ionization in one- and two-dimensional model calculations, *Phys. Rev. A.* **45**, 3203.
9. Grobe, R., and Fedorov, M. V. (1992) Packet spreading, stabilization, and localization in superstrong fields, *Phys. Rev. Lett.* **68**, 2592.
10. Volkova, E. A., Popov, A. M., and Smirnova, O. V. (1994), Stabilization of atoms in a strong field and the Kramers-Henneberger approximation, *Zh. Eksp. Teor. Fiz.* **106**, 1360 [Jetp **79**, 736].
11. Pont, M. and Gavrila, M. (1990) Stabilization of atomic hydrogen in superintense high-frequency laser fields of circular polarization, *Phys. Rev. Lett.* **65**, 2362.
12. Dörr, M., Potvliege, R. M., Proulx, D., and Shakeshaft, R. (1991) Multiphoton processes in an intense laser field V. The high-frequency regime, *Phys. Rev. A* **43**, 3729.
13. Dimou, L. and Faisal, F. H. M. (1992) Decay of metastable H atoms in intense excimer lasers, *Phys. Rev. A* **46**, 4442.
14. Vos, R. J. and Gavrila, M. (1992) Effective stabilization of Rydberg states at current laser performance, *Phys. Rev. Lett.* **68**, 170.
15. Potvliege, R. M. and Smith, P. H. G. (1993) Adiabatic stabilization of excited states of H in an intense linearly polarized laser field, *Phys. Rev. A* **48**, R46.
16. Dimou, L. and Faisal, F. H. M. (1994) Ab Initio results of adiabatic stability of the hydrogen atom, *Phys. Rev. A* **49**, 4564.
17. Fedorov, M. V. and Movsesian, A. M. (1989) Interference suppression of photoionization of Rydberg atoms in a strong electromagnetic field, *J. Opt. Soc. Am. B* **6**, 928.
18. Fedorov, M. V. (1992) Comments, *At. Mol. Phys.* **27**, 203.
19. Parker, J. and Stroud, C. R. (1990) Population trapping in short-pulse laser ionization, *Phys. Rev. A* **41**, 1602.
20. Yeazell, J. A., Mallalieu, M., and Stroud, C. R. (1990) Observation of the collapse and revival of a Rydberg electronic wave packet, *Phys. Rev. Lett.* **64**, 2007.
21. Burnett, K., Knight, P. L., Piraux, B. R. M., and Reed, V. C. (1991) Suppression of ionization in strong laser fields, *Phys. Rev. Lett.* **66**, 301.
22. Jensen, R. V. and Sundaram, B. (1993) Classical theory of intense-field stabilization, *Phys. Rev. A* **47**, 778.
23. Benvenuto, F., Casati, G., and Shepelyansky, D. L. (1992) Stability of Rydberg atoms in a strong laser field, *Phys. Rev. A* **45**, R7670.
24. Benvenuto, F., Casati, G., and Shepelyansky, D. L. (1993) Classical stabilization of the hydrogen atom in a monochromatic field, *Phys. Rev. A* **47**, R786.
25. Benvenuto, F., Casati, G., and Shepelyansky, D. L. (1994) Rydberg stabilization of atoms in strong fields: The "magic mountain" in the chaotic sea, *Z. Phys. B* **9**, 481.
26. Casati, G., Guaneri, I., and Mantica, G. (1994) Classical stabilization of periodically kicked hydrogen atoms, *Phys. Rev. A* **50**, 5018.

CLASSICAL MECHANICAL THEORIES OF INTENSE FIELD STABILIZATION

G. CASATI, I. GUARNERI and G. MANTICA
Center for the Study of Dynamical Systems
University of Como, via Lucini, 3
I-22100 Como, ITALY

Abstract
We review a few significant approaches to I.F.S. based on classical mechanics. We present some ideas for a new, complete theory of the classical dynamics of a Hydrogen atom in a strong, oscillating electric field.

1. Introduction

Classical Mechanics has been adopted by several authors [1-5] to describe the stabilization of Rydberg atoms in strong, oscillating electric fields, taking place as the field intensity surpasses a certain threshold [6]. The dynamical model describing this system is given by the Hamiltonian

$$H = \frac{p_\rho^2}{2} + \frac{p_\zeta^2}{2} + \frac{m^2}{2\rho^2} - \frac{1}{\sqrt{\rho^2 + \zeta^2}} + \varepsilon\zeta \cos \omega t, \tag{1}$$

where cylindrical coordinates ρ, ζ ($\rho^2 = x^2 + y^2$) and atomic units have been used, where m is the value of the angular momentum, and where ε and ω are the intensity and frequency of the external field. As the rule in classical mechanics, the motion of this dynamical system is highly complicated, with regions of chaos interwined to regular portions of the phase space. Reduction of this complexity to simple, quantitative prescriptions for I.F.S. has proven to be a challenging task, for the following difficulties.

The first is somehow ironical, for the Hydrogen atom in an external field is the battle-ground on which Quantum Mechanics defeated the classical theory: yet, a global analysis of this latter has been attempted only recently, particularly for what regards the large field limit, and much is still to be learned.

The second difficulty has more physical flavor: out of the rich cornucopia of dynamical phenomena, one needs to single out that (or those) more directly linked to ionization of the system. Care must be exerted in *defining* ionization at the classical level: bounded, yet extremely wide oscillations will likely be detected as ionization in a real experiment; changing the time scale will also likely change the ionization results in a manner which requires investigation.

H. G. Muller and M. V. Fedorov (eds.), Super-Intense Laser-Atom Physics IV, 83–95.

The third difficulty is even more linked to the experimental situation: the real atoms (originally in a state with well defined quantum numbers) experiences a switch-on process which brings the external field to its maximum value. Then, after a certain time-interval where the system is described by the ideal Hamiltonian (1) the field is switched off. How do the physical initial conditions map to and from the dynamical variables in (1)? Can ionization take place in the process?

Notwithstanding these difficulties, several approaches to a classical description of I.F.S. have been proposed. It is our purpose here to present some of them schematically, showing their pros and cons. We shall then propose a new approach to the global description of the system.

2. Importance of a Classical Description of I.F.S.

To analyze the effect of very large field intensities, the Hamiltonian (1) is conveniently expressed in a Kramers-Henneberger reference frame oscillating as a particle would do in the presence of the external field alone: $z := \zeta + \alpha \cos \omega t$, $\alpha := \varepsilon \omega^{-2}$, and

$$H := H_0 + R(\rho) + V(\rho, z, t) = \frac{p_\rho^2}{2} + \frac{p_z^2}{2} + \frac{m^2}{2\rho^2} - \frac{1}{\sqrt{\rho^2 + (z - \alpha \cos \omega t)^2}}. \quad (2)$$

In the above, we have denoted by R the repulsive centrifugal potential, and by V the attractive electrical potential. In the low field limit, the system (1,2) inherits K.A.M. tori from the $\varepsilon = 0$ case. Increasing ε these surfaces break, and for $\varepsilon > \frac{1}{50}\omega^{-\frac{1}{3}}$ its phase-space is largely chaotic: ionization is then produced by diffusion in energy (or the principal action variable n) [7].

This classical picture plays an important rôle also in the quantum mechanical behaviour of the system: on the one hand, the quantum analogue of classical resonance overlap has been derived, and the conditions by which it also leads to quantum ionization discussed [8]; on the other hand, the phenomenon of "quantum suppression of classical chaos" has been shown to lead to a quantum field threshold for ionization larger than the classical, and increasing with the frequency [7]. This theory has been confirmed by experiments [9]. Studying the classical dynamics of (2) is therefore more than an academical exercise–it could lead to the understanding of the real, quantum mechanical system. Let us pause no more and present some of the "classical theories" of I.F.S. we have mentioned.

3. The Floquet map of Jensen and Sundaram

To assess the stability of (2), Jensen and Sundaram (J.S.) [3] approximated the Floquet map, which effects the evolution of the system over one period of the external field. They considered the average of the time dependent potential appearing in (2):

$$V_0(\rho, z) := \frac{1}{T} \int_0^T \frac{1}{\sqrt{\rho^2 + (z - \alpha \cos \omega t)^2}} \, dt \quad (3)$$

ρ and z being held constant in the integration. $V_0(\rho, z)$ is the zeroth harmonic of the Fourier expansion of the time-dependent potential. Since this latter is a C^∞ function of t, high harmonics V_n must decay exponentially in n. Hoping that this decay sets in sufficiently late, J.S. assume that $V_n \simeq V_0$ for all n, thereby introducing a $\delta-$kick interaction, which is also extended to the centrifugal potential:

$$H_{JS} = \frac{p_\rho^2}{2} + \frac{p_z^2}{2} + T\,[V_0(\rho, z) + \frac{m^2}{2\rho^2}]\,\sum_k \delta(t - kT). \tag{4}$$

To further simplify the analysis J.S. disregard the oscillations in ρ (termed "stable") and fix the latter to $\hat{\rho}$, a minimum-value for the effective potential $V_{map} := V_0 + \frac{m^2}{2\rho^2}$. The equations of motion then become a map for z and its conjugate momentum p:

$$\begin{aligned} z_{n+1} &= z_n + Tp_{n+1} \\ p_{n+1} &= p_n - T\frac{\partial V_{map}}{\partial z}\big|_{z=z_n,\rho=\hat{\rho}} \,. \end{aligned} \tag{5}$$

Stability of this map is assessed studying the behavior around its fixed points. When $\alpha \ll 1$, there is only one such point, $z = 0, p = 0$, and $\hat{\rho} \simeq m^2$. This fixed point is stable (elliptic) for $\omega > \frac{3}{m^3}$. In the opposite case, $\alpha \gg 1$, the simple minimum of V_{map} splits in two z-symmetrical minima; stability analysis then requires

$$\varepsilon > 2\omega^{\frac{6}{7}}m^{-\frac{10}{7}}. \tag{6}$$

J.S. theory takes these equations as the parameter conditions for I.F.S.

Some remarks are in order. First, as shown in J.S., these estimates well describe motions trapped in the local minima of V_{map}. Yet, this can not be an exaustive explanation of the phenomenon of I.F.S.: in fact, to gauge the dynamical significance of these motions the volume of the corresponding portions of phase space needs to be estimated, as well as the probability of capture therein during the field switch-on. Secondly, even allowing the delta-function approximation, determining the stability of the system over a time-scale which is simply based on the period of field oscillations may be misleading. In fact, the crucial time scale in the problem is set—we shall see—by the period of the oscillations of the electron to and from the nuclear core, which may be larger by orders of magnitude. Finally, it turns out that stability of the system is determined mainly by oscillations in the ρ and not the z direction.

4. The Elementary Estimates of Benvenuto, Casati and Shepelyansky

Benvenuto, Casati, and Shepelyansky (B.C.S.) [4] described the phenomenon of I.F.S. by a series of elementary scaling considerations which have the virtue of capturing the physical effects in their large scale parameter dependence, beyond being transparent, and easily controllable. The main approximation on which B.C.S. theory rests is supposing that the time dependent potential $V(\rho, z, t)$ in eq. (2) produces an effective, time-independent potential as that of a charged, infinite thread:

$$V_{BCS}(\rho, z) = 2\sigma(z)\ln\left(\frac{\rho\omega^2}{\varepsilon}\right), \tag{7}$$

but with a charge density which varies with z as the time spent by the nucleus close to z: $\sigma(z) = [\omega^2/(\pi\varepsilon)][1 - (z\omega^2/\varepsilon)^2]^{-\frac{1}{2}}$. The reader can verify that this is an approximation to the true average potential (3) for $\rho\omega^2/\varepsilon < 1$, and $z \ll \alpha$.

Furthermore, B.C.S. assume that the main destabilizing effect on the motion derives from ρ oscillations, and therefore consider z as fixed, and $z \ll \alpha$. Then, the total potential– (7) plus centrifugal–has a minimum at $\bar{\rho} = \sqrt{\pi\varepsilon/2}(m/\omega)$, and the frequency Ω of the small ρ-oscillations at this minimum is $\Omega \sim \omega^2/(\varepsilon m)$. If this latter is much less than ω, according to B.C.S. the systems is adiabatically stable. This condition becomes, in the physical parameters

$$\varepsilon > \varepsilon_{\text{stab}} = \text{const.} \frac{\omega}{m} \tag{8}$$

This estimate is to be compared with J.S. eq. (6).

B.C.S. found good agreement between their result (8) and numerical integration of the exact equations of motion, for an ensemble of classical particles with the same values of action (quantum number n), orbital momentum l, and projection on the field direction m. A the same time, though, they noticed the existence of a second threshold, $\varepsilon_{\text{destab}}$, above which the I.F.S. disappears. They attributed this effect to the fact that in the initial distribution particles are characterized by $\rho < 2n_0^2$. When the potential minimum $\bar{\rho}$ is larger than this value, orbits cannot be trapped during the switch-on process which they included in the numerical simulations. They so obtained $\varepsilon_{\text{destab}} \approx \omega^2/(m^2 n_0^2)$.

B.C.S. theory shares some common virtues and downfalls of simple theories. Its main success is that it provides a sound explanation of the I.F.S. phenomenon over a large span in the parameters range. Its main weakness is that it over-simplifies the rich dynamics of the physical system it describes: we just mention here two points worth of further investigation. B.C.S. theory is severely limited by the validity range of eq. (7) to describe motions with $z \ll \alpha$ and therefore misses the interesting edge effects occurring at $z \sim \alpha$; at the same time, it is ineffective to describe the energy exchange occurring at the minimum of ρ when the motion is not confined to the potential minimum at $\rho = \bar{\rho}, z = 0$. We shall come back to these points in the following.

5. The Nucleus-in-a-Box model of Casati, Guarneri, and Mantica

Considerable complexity to the problem of I.F.S. is brought by the non-trivial properties of the average potential (3) when ρ tends to zero. Indeed, these properties are at the root of the appearance of a double-well potential, an effect which is commonly termed dichotomy, and is sometimes believed to be essential to stabilization. To escape this complexity, and at the same time to broaden the realm of the I.F.S. phenomenon, Casati, Guarneri, and Mantica (C.G.M.) [5] introduced a simplified model which does not show dichotomy, and yet it is characterized by I.F.S.

Looking back at the Hamiltonian (2), one can pictorially say that the nucleus oscillates harmonically on the ρ axis between $-\alpha$ and α. C.G.M. consider a different system, where this motion has constant absolute velocity, and is reflected elastically at the extrema. A nucleus in a box, that is. In this way, the edge effects are largely mitigated, and no dichotomy arises, for the average potential

$$V_{\text{CGM}}(\rho, z) = -\frac{1}{2\alpha} \ln \frac{\sqrt{\rho^2 + (z + \alpha)^2} + z + \alpha}{\sqrt{\rho^2 + (z - \alpha)^2} + z - \alpha} \qquad (9)$$

when added to the centrifugal term has a single minimum at $z = 0$, $\rho = \frac{m^2}{\sqrt{2}}[1 + (1 + \frac{4\alpha^2}{m^4})^{\frac{1}{2}}]^{\frac{1}{2}}$, for any value of the external field.

This model allows for a simple alternative interpretation which can also be taken as the basis for an efficient computational procedure. In fact, the nucleus-in-a-box motion described above can be obtained as a result of a K-H transformation of a non-monocromatic driving electric field. To obtain this field, C.G.M. replace the sinusoidal dependence by a train of Dirac delta-functions, of alternating signs, centered at the extrema of the sine function, and possessing the same integral over half a period. In formulae:

$$H_{\text{CGM}}(\vec{p}, \vec{r}, t) = \frac{p^2}{2} - \frac{1}{r} + \frac{2\varepsilon z}{\omega} \sum_{k=-\infty}^{+\infty} (-1)^k \delta\left(t - k\frac{T}{2}\right) \qquad (10)$$

where $\vec{r} = (x, y, z)$ and atomic units are used. Because of the delta-kicks, the motion induced by (10) over a period (the Floquet map) is the product of the action of four elementary maps: The first acts as the free Keplerian motion over a time $\frac{T}{2}$; the second suddenly changes p_z into $p_z + \frac{2\varepsilon}{\omega}$; next comes one more free Kepler motion, followed by one more kick changing p_z by $-\frac{2\varepsilon}{\omega}$.

To obtain realistic simulations, C.G.M. adopted microcanonically distributed initial conditions at $n = 1, l = 0.3, m = 0.25$: due to well-known scaling properties, cases with initial $n \neq 1$ can be reduced to this by using scaled variables $\omega n^3, \varepsilon n^4, l/n, m/n$. They defined ionization as the fraction of particles exiting a ball of radius $r = 500$ in a fixed time of 1000 Kepler periods at $n = 1$. First of all, C.G.M. investigated the rôle of the switch-on /off process: they found out that it is essential to I.F.S. In Fig. 1 the survival probability P_s (i.e. the relative number of non-ionized trajectories) is shown at fixed frequency versus the field intensity: the dramatical illustration of I.F.S. apparent for $\varepsilon \sim 50 \div 500$ completely disappears if the field is switched on suddenly.

The simplicity of this model, at the same time theoretical and numerical, allowed C.G.M. to perform an extensive investigation of the system dynamics. They sampled the ω, ε parameter space by 100 particles at each of 4800 locations: their results are presented in Fig. 2, which shows two wedge-shaped regions of high survival probability, the higher of which correspondig to I.F.S. These regions are separated by a "death valley" that becomes deeper when moving to higher frequencies. C.G.M. determined the boundaries of the I.F.S. region by a B.C.S. type analysis, which does

88

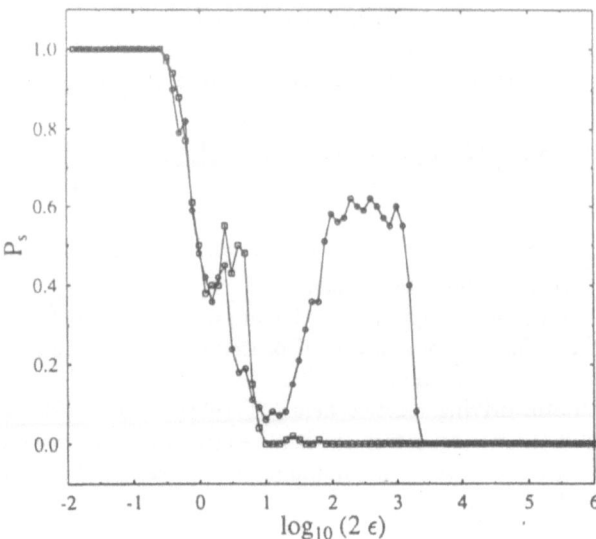

Figure 1: Survival probability P_s vs field intensity at $\omega = 2.5$. Switch on times are $t = 10$ (circles), $t = 1$ (squares)

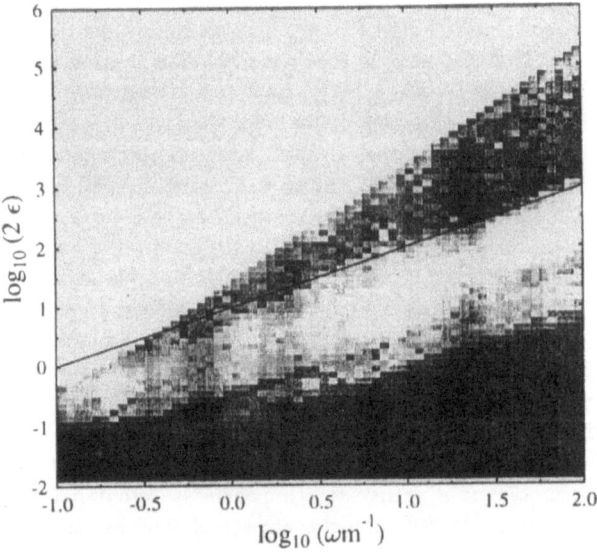

Figure 2: Survival probability in parameter space. Dark scale ranges from 0 (white) to 1 (black). The line has equation $\varepsilon = 5\omega/m$.

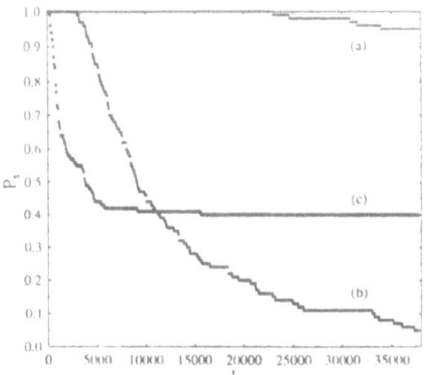

Figure 3: Stabilization probability P_s vs. time, for $\omega = 7.9$ and $\varepsilon = 0.2$ (a), $\varepsilon = 3.17$ (b), $\varepsilon = 500$ (c)

not suffer here from the inhomogeneity in the z direction: thanks to eq. (9), they obtain that small ρ-oscillations (hence, the system) are stable when B.C.S. relation (8) is verified. This gives the lower boundary of the I.F.S. region: To the contrary, the upper border follows a ω^2 law. Simple estimates suffice to explain this fact: since the potential well in the ρ direction is located around $m\varepsilon^{1/2}/\omega$, increasing ε it is shifted farer and farer from the z axis so that no orbit in the initial ensemble (which has a finite extent in ρ, independent of ε and ω) has a chance to be trapped in. So far, the nucleus-in-a-box model confirms the validity of the elementary estimates of B.C.S. Yet, it permits a closer investigation of the dynamics of the system, which sheds new light on the nature of the stabilization mechanicsm.

6. Dynamical details in the Nucleus-in-a-Box model

Of course, the ionization probability described in the previous section depends on the interaction time: making this longer, some of the surviving orbits contributing in the grey zones in Fig. 2 would most likely ionize, turning Fig.2 to overall paler tones of grey. Will its geography also change ? To answer this question C.G.M. studied the dependence on the interaction time of the survival probability, recorded at three different positions in the ε, ω plane of Fig.2: one deep in the lower stable region, another still in that region but periliously close to death-valley, and a third inside the IFS region. Fig. 3 shows that in the first two cases the survival probability decreases, almost negligibly in the first case, and more markedly in the second. Instead, in the third case this probability has a sharp initial descent which settles to a practically constant value. To explain these data, C.G.M. proposed the following theory. When increasing the field intensity from its zero value, perturbative (K.A.M. ?) stability

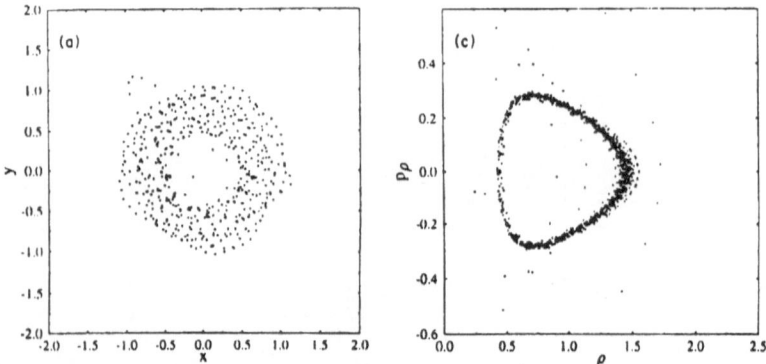

Figure 4: A stabilized orbit for $\varepsilon = 158$ and $\omega = 6.28$ projected on: the $x - y$ plane (a), the $\rho - p_\rho$ plane (c).

leaves the space to chaotic motion, and to diffusion in energy which eventually drives the system to ionization. Since the corresponding diffusion coefficient goes down when the field frequency ω increases, at fixed field intensity ε and interaction time the survival probability is found to increase with ω, as seen in Fig. 2. In summary, increasing the interaction time lowers the upper border of the stability region at low fields.

Contrary to this behaviour, the data of Fig. 3 suggest that the borders of the IFS region should not significantly change if one increases the interaction time. Fig.4 shows various representations of a typical stabilized orbit. Its projection on the initial plane fills an annulus-shaped region, which never comes very close to the nucleus. This is a sort of trapping which is evident in its (ρ, p_ρ) phase-space projection: after an initial transient the motion remains confined to a relatively narrow region, the shape of which is unambiguously tailored after the $z = 0$ orbits of the average Hamiltonian in the ρ direction. The width of this region is determined by the slow motion in the $z-$direction which encounters here no sharp inhomogeneity. C.G.M. computed numerically the maximal Lyapunov exponent for these orbits and found values of the order of the numerical error: motion in the IFS region is either stable, or very weakly chaotic at worst.

From this analysis it emerges that the centrifugal potential is essential in preventing energy exchange from the oscillating nucleus to the electron. Its role to stabilization is embodied in the border (8) via the angular momentum m. C.G.M. studied numerically the validity of this estimate for small m: Fig. 5 shows the survival probability P_s versus m at fixed ω/m: P_s is roughly stationary as predicted, but at small m a sharp fall-off is observed. This is a consequence of C.G.M. definition

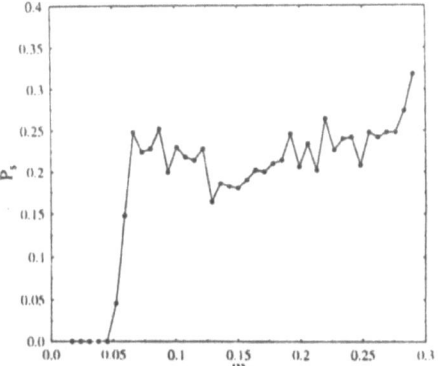

Figure 5: Survival probability vs. m at fixed $\omega/m = 1$, $\varepsilon = 0.629$.

of ionization: as m tends to zero at fixed ω/m, α becomes very large, and eventually the slow z–motion carries the electron beyond the maximal distance $r \sim 500$. Obviously the value of r is quite arbitrary: yet, it shows that proper "experimental" considerations must be taken into account when treating the I.F.S. phenomenon.

7. The Major League Theory[10]

Returning to the sinusoidal driving case, the difficulties exposed at the end of Sect. 4 must be faced. In this section we present some preliminary results which will hopefully pave the way to a more global understanding of the classical stabilization phenomenon.

Diffusion in energy which sets in after K.A.M. stability is broken is induced by close encounters between electrons and the oscillating nucleus: electrons move on almost perfect Keplerian ellipses in spite of the external field, except for when they come close to the nucleus, where they experience an energy change which ejects them on different ellipses. Eventually, some of these ellipses degenerate into hyperbolas, and the system ionizes: "death valley" is the place where this phenomenon is global. Therefore, stabilized orbits are not only those trapped in the local minima of the average electrical potential–whether in J.S. or B.C.S. picture–but also and foremost those living on unperturbed Kepler ellipses for most of the time, and bouncing off the centrifugal barrier without increasing sensibly their energy. For a quantitative description of I.F.S. it is therefore crucial to compute the energy transfer occurring at these instants.

This is readily accomplished writing Hamilton's equation $\frac{dH}{dt} = \frac{\partial V}{\partial t}$ (H and V as

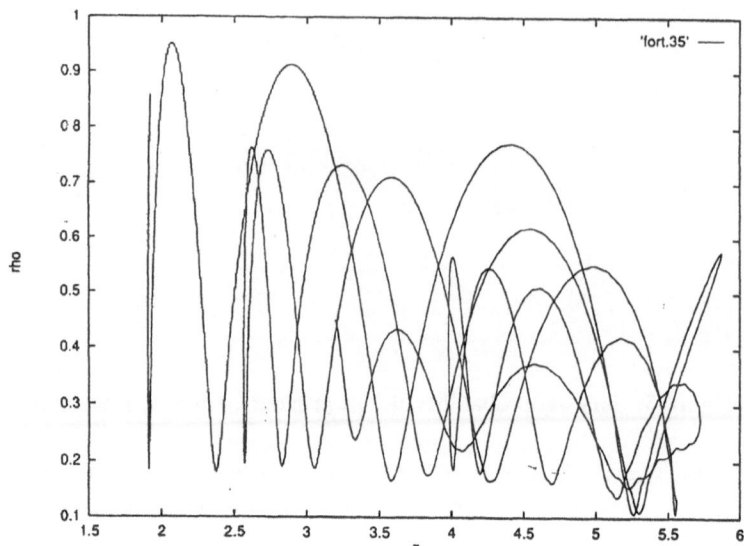

Figure 6: Motion in the z-ρ plane, initial point at 3.2, .45. Here $\omega = 30$, $\alpha = 5.55$, $m = .12$. Notice how the motion gets closer to the z axis at the strike-zone $z \sim \alpha$.

in eq. (2)). Therefore, the energy change over an approach to the nucleus, ΔH, is

$$\Delta H := \int_{-\tau}^{\tau} \frac{\partial V}{\partial t}\, dt = -\alpha\omega \int_{-\tau}^{\tau} k(\rho, z, t) \sin \omega t\, dt, \qquad (11)$$

where $[-\tau, \tau]$ is the bounce time interval, and where the kick function k is:

$$k(\rho, z, t) := \frac{\zeta}{\sqrt{(\rho^2 + \zeta^2)^3}}, \qquad \zeta := z - \alpha \cos \omega t. \qquad (12)$$

Before proceeding further, we remark that the B.C.S. [4] estimate of the integral (11,12) is $\omega^2 \varepsilon^{-2} \rho_0^{-2}$, ρ_0 being the minimum value of ρ as the z axis is approached. In a further work Shepelyansky [11] shows that ΔH is exponentially small in the parameter ω/Ω, that is $\varepsilon m \omega^{-1}$; Yet his analysis crucially depends on the inequality $\rho_0 > \alpha$, which is not at all typical in I.F.S.

We separate the analysis of (11,12) in various steps. First, remark that ρ and z in these equations are functions of time–indeed, they must be the *exact* solutions of the equations of motion. Clearly, this request is excessive for any sensible approximation scheme. Our first simplification is then to consider as negligible the z component of momentum at the impact (the electron heads at right angle towards the z axis) and to approximate the ρ dependence as due solely to the centrifugal wall:

$$\begin{aligned} z &= \alpha \cos \phi \\ \rho &= \sqrt{\rho_0^2 + q^2(t - t_0)^2} \end{aligned} \qquad (13)$$

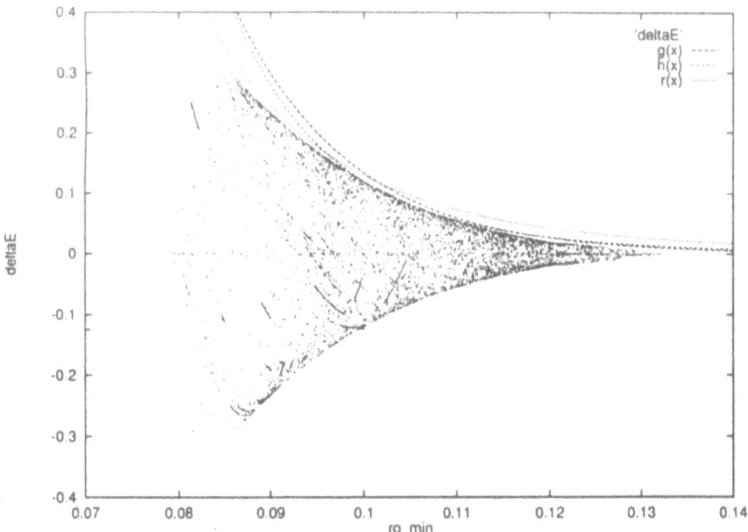

Figure 7: Energy change ΔH vs. ρ_0 for a series of rebounds (dots) with $\phi = 2.9$, $\omega = 30$, $\alpha = 5.55$, $m = .12$. The fitting line h has the ρ_0^{-8} dependence.

where t_0 is the impact time, ϕ measures the impact coordinate, and the momentum q is equal to $\frac{m}{\rho_0}$. We will moreover assume that $\omega \gg 1, \alpha \gg 1$.

Second, we notice that when the nucleus "passes close" to the incoming/outgoing electron, i.e. when $\omega t_j^{\pm} = \pm\phi + 2\pi j$, ζ has a zero, and in that neighbourhood the function $k(\rho, z, t)$ has two lobes, one positive and the other negative, which would exactly cancel each other if it were not for the time dependence of ρ given by eq. (13). If $\arccos\frac{\rho}{\alpha} \ll \min(\phi, \pi - \phi)$ (which is typically the case in the strong interaction region close to $\rho_0 \ll \alpha$) the two lobes of k die out much before the next two arise–the time-ordered sequence being $\ldots, t_j^+, t_{j+1}^-, t_{j+1}^+, \ldots$. In this case, we can evaluate the local integral at t_j^+, ΔH_j^+, by the linear approximation $\zeta \simeq \alpha\omega(t - t_j^+)\sin\phi$, and obtain $\Delta H_j^+ \simeq 2\frac{q}{\rho(t_j^+)\alpha\omega\sin\phi}$. Then, let us consider rebounds taking place close to $z = \pm\alpha$, for these are those really effective (the strike-zone in our model–Fig. 6): let us say $\phi \sim \pi$. Under these conditions, the two contributions ΔH_j^+ and ΔH_{j+1}^- come at very close times, have opposite signs, and again they would exactly cancel if it were not for the variation in ρ. Letting $t_j = \frac{1}{2}(t_j^+ + t_{j+1}^-)$ be the mean time of this one-two action, we have:

$$\Delta H_j^+ + \Delta H_{j+1}^- \simeq -4\frac{q^2}{\alpha\omega^2\rho(t_j)^2}\frac{\pi - \phi}{\sin(\pi - \phi)} \tag{14}$$

Finally, let us consider the cumulative effect of these contributions with their

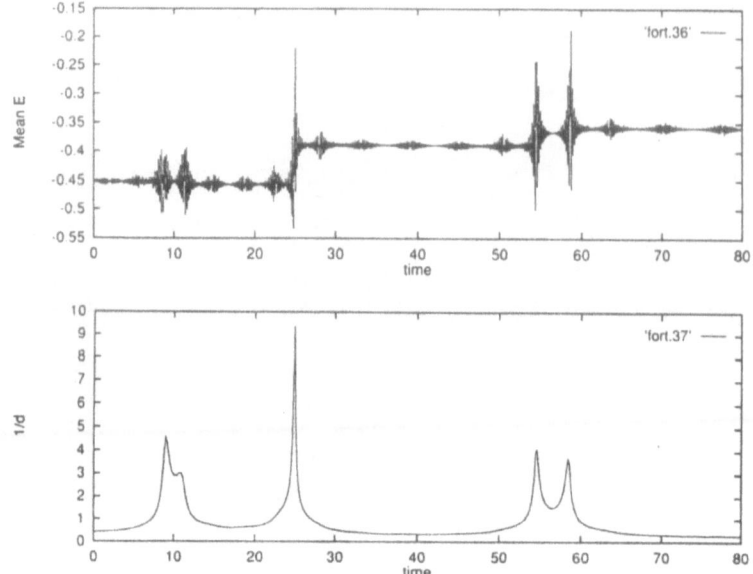

Figure 8: For the motion of Fig. 6, we plot: a) the average energy $H_0 + R + V_0$ vs time; b) inverse of the distance d from the electron and the closest singular point $z = \alpha, \rho = 0$. Coincidence between strike-zone rebounds and energy change is apparent.

signs:

$$\Delta H = 4 \frac{q^2}{\alpha \omega^2} \frac{\pi - \phi}{\sin(\pi - \phi)} \sum_j \frac{\text{sign}(j)}{\rho_0^2 + q^2(\tau + jT)^2} \tag{15}$$

where $\tau < T$ is the time lag between the hit time t_0 and the closest t_j before it. The summation in (15) is to be taken over a convenient set of indices–indeed, it automatically truncates because of the increase in the denominator when ρ gets large. The complete result is not difficult to find; yet a simple approximation obtained considering the last j term before the collision, and the first after, already contains the main physics. Letting $\psi := \omega \tau$ we have:

$$\Delta H \simeq 4 \frac{q^2}{\alpha \omega^2} \frac{\pi - \phi}{\sin(\pi - \phi)} \frac{q^2}{\rho_0^4} \frac{2\pi}{\omega} \left(\frac{2\pi}{\omega} - \frac{2\psi}{\omega} \right) = 16\pi(\pi - \psi) \frac{(\pi - \phi)}{\sin(\pi - \phi)} \frac{m^4}{\alpha \omega^4 \rho_0^8}. \tag{16}$$

The above relation is crucial: its foremost characteristic is the dependence on the high negative power of ρ which we verified numerically in Fig. 7: at fixed physical parameters ρ_0 determines whether the hit has the power to be a home-run, a triple, a double or merely a single. Timing of the swing is contained in the factor $(\pi - \psi)$, which can also make ΔH negative, while the rôle of the pitch angle ϕ in eq. (16) seems minor. Nothing more inaccurate, for indeed at fixed physical parameters and pitch energy E, ρ_0 is very much dependent on ϕ: from the average potential V_0 (3) we find that in the typical dichotomic case ρ_0 is much larger for ϕ close to 0, or π–the strike area: see Fig. 6, and 8.

These considerations can form the basis of a comprehensive theory of I.F.S. which we plan to develop. Yet, we can show here that they generalize the simple B.C.S. analysis. In fact, adopting a crude approximation of V_0 we find that $\rho_0 \sim \varepsilon^{\frac{1}{2}} m \omega^{-1}$, irrespective of ϕ, while the energy required for ionization from the bottom of the potential scales as $E \sim \alpha^{-1}$. The system is certainly unstable when this energy gap can be made out in a single swing (16) (either the ball-park is too small, or the bat is corked..): writing out this condition immediately leads to eq. (8).

8. References

1. Q.Su and J.H.Eberly, *Phys.Rev.* A **43** , 2474 (1991); R.Grobe and C.K.Law, *Phys. Rev.* A **44**, 4114 (1991).
2. J.Grochmalicki, M.Lewenstein and K.Rzazewski, *Phys. Rev. Lett.* **66** 1038 (1991).
3. R.V.Jensen and B.Sundaram, *Phys.Rev.* A **47**, 1415 (1993); *Phys.Rev.* A **47**, R778 (1993).
4. F.Benvenuto, G.Casati and D.L.Shepelyansky, *Phys. Rev.* A **47**, R786 (1993); *Zeits. Phys.* B **94** 481-486 (1994).
5. G. Casati, I. Guarneri and G. Mantica, *Phys. Rev.* A **50** 5018-5024 (1994).
6. S.Geltman and M.R.Teague, *J.Phys.* B **7**, L22 (1974); J.I.Gersten and M.H.Mittleman, *ibidem* **9** 2561 (1976); M.Gavrila and J.Z.Kaminsky, *Phys. Rev. Lett.* **52**, 614 (1984); M.V.Fedorov and A.M.Movsesian, *J.Opt.Soc.Am.* B **6**, 928 (1989); Q.Su,J.H.Eberly, and J.Javanainen, *Phys. Rev. Lett.* **64**, 862 (1990); M.Pont and M.Gavrila, *Phys. Rev. Lett.* **65**, 2362 (1990); K.C.Kulander, K.J.Schafer and J.L.Krause, *ibid* **66**, 2601 (1991); K.Burnett, P.L.Knight, B.R.M. Piraux, and V.C.Reed, *ibid* **66**, 301 (1991); V.C.Reed, P.L.Knight, and K.Burnett, *ibid* **67**, 1415 (1991);
7. G. Casati, I. Guarneri and D.L. Shepelyansky, *I.E.E.E. Jour. Quant. Electr.* **24**, 1420 (1988).
8. G. Mantica, *I.E.E.E. Jour. Quant. Electr.* **24**, 1453-1460 (1988).
9. J.Bayfield, G.Casati, I.Guarneri, and D.Sokol, *Phys. Rev. Lett.* **63** (1989) 364;
10. The bizzarre title of this section and its content are solely to be blamed on the third author.
10. D.L. Shepelyansky, *Phys. Rev.* A (1995).

STRONG FIELD ATOMIC DYNAMICS

L. F. DIMAURO[a], K. C. KULANDER[b], P. AGOSTINI[c], K. J. SCHAFER[d], B. WALKER[a], AND B. SHEEHY[a]
(a) Brookhaven National Laboratory, Upton, NY 11973
(b) Lawrence Livermore National Laboratory, Livermore, CA 94551
(c) CE Saclay, 91191 Gif Sur Yvette, France
(d) Louisiana State University, Baton Rouge, LA 70803

1. Introduction

The study of atoms exposed to strong laser fields has defined an important area in atomic, molecular, and optical physics. These investigations have led to the discovery of many new phenomena, i.e. above-threshold ionization and high harmonic generation, each of which has been extensively studied. However, since its inception over two decades ago, this discipline has strived for a unified view of all strong-field laser-atom interactions. The lack of progress was hampered in part by the limited experimentally accessible dynamic range which was unable to probe the regions necessary to discriminate between the predictions of various theoretical models. However in recent years, there has been considerable progress made in developing a universal model which describes the single electron dynamics in a strong laser field. This recent progress has been propelled by new advances in laser technology which extend the experimental sensitivity by a few orders of magnitude and powerful numerical time-dependent methods capable of accurately describing the experimental results. In this paper we will discuss the recent efforts of our group in developing and evaluating a quasi-classical view of all strong-field interactions.

2. The Classical View

Strong-field ionization results from the interaction with both the electromagnetic and Coulomb fields. However, once the electron is sufficiently displaced from the core, the electromagnetic field dominates. This forms the basis for a two-step view [1] of ionization in which the electron (amplitude) is (1) initially promoted to the continuum (bound-free) and (2) subsequently evolves under the dominance of the external field (free–free). In the first approximation, one can simplify the

97

H. G. Muller and M. V. Fedorov (eds.), Super-Intense Laser-Atom Physics IV, 97–108.
© 1996 *U.S. Government.*

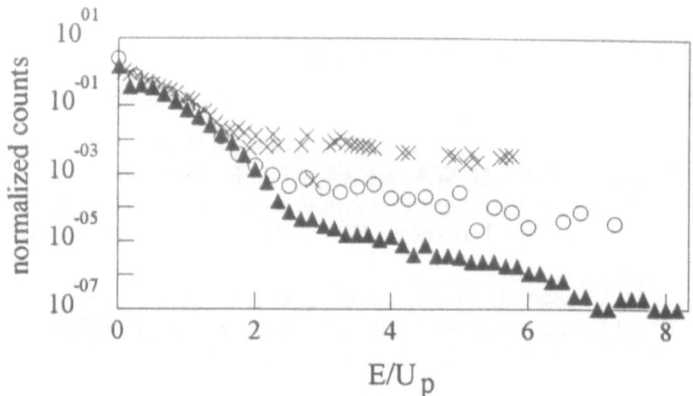

Figure 1. The photoelectron energy spectrum of helium in scaled energy units at different intensities for 100 fs, 0.78 μm excitation. The intensities are 0.2 (crosses), 0.4 (circles), and 1.0 (triangles) PW/cm^2.

problem by treating both the electron and field classically. The success of this amazingly simple idea [2-4], dubbed the Simpleman's model, has formed the intuitive basis for understanding strong-field phenomena. The model assumes that an electron is created in the field, $\varepsilon_o \sin \omega t$, at a time t_o with some initial velocity v_o and a corresponding kinetic energy $T_o = v_o^2 / 2$. It then acquires through its oscillatory motion in the field an extra kinetic energy term (quiver or ponderomotive energy) which is the classical counterpart of the ATI process.

For the purposes of this discussion the electron's initial velocity and therefore the energy is assumed to be zero in the quasi-static (long wavelength) limit [4] where tunnel ionization defines the initial conditions. In this limit, the electron acquires a maximum cycle averaged energy of $2U_p$. Figure 1 shows the photoelectron energy spectrum for helium at three different intensities. These spectra were recorded using 100 fs pulses from a titanium sapphire laser system operating at a kilohertz repetition rate. Qualitatively, the spectra agrees with the above model with ≥ 99% of the electron emission confined between zero and $2U_p$. However, it is the small fraction of electrons with energies $> 2U_p$, unaccountable by the Simpleman's model, which forms the basis of a more complete strong-field model.

The fundamental feature of this model is just a natural extension of the Simpleman model. Although the electron motion is dominated by its interaction with the field, it is oscillatory, and some classical trajectories return to the vicinity of the core. Thus, a physical picture [5,6] emerges in which electrons promoted into the continuum near the nucleus at arbitrary times during the optical cycle undergo two types of classical orbits: those that return to the region of the core and those that do not. Those electrons (approximately half) which return to the core after about a half cycle of free propagation, can undergo elastic or inelastic collisions. It is this fundamental scattering event which produces the relevant dynamics and energetics necessary to account for the existence of high harmonic radiation, high energy electrons present in Fig. 1, and two electron ionization.

The rescattering model offers a simple and clear explanation of the mechanism producing high harmonic radiation: harmonics are generated when the electron recombines with the core at the return time. Simple analysis shows that the maximum instantaneous energy of the electron upon return is $3.17U_p$, in remarkable agreement with the well known $3U_p + E_o$ cutoff rule [7]. High energy electrons are produced when the electron elastically scatters. Again, simple analysis shows that the maximum cycle averaged energy can be as large as $10U_p$ for large angle scattering, i.e. backscattering. Furthermore, two electron ionization [6] can occur if the returning electron's instantaneous energy is equal to or greater than the binding energy of a core electron, thus producing an e-2e inelastic event. Consequently, all the dynamics are contained in the half cycle of free propagation in the field prior to the relevant rescattering event.

Naturally all quantum aspects are lost in the classical models. However, it is not difficult to build some of these concepts into the visualization of the dynamics. A more complete view [8] should embrace the concept of a wave packet propagating in the field along a classical trajectory. As it propagates, the wave packet accumulates phase and spreads in all dimensions. The returning wave packet spreads for at least a half of an optical cycle before rescattering. This implies that the propagation spread (wavelength dependent) will diminish the effectiveness of rescattering. Analysis of time-dependent quantum mechanical calculations shows that the amount of transverse spread, α_t, is consistent with a freely propagating gaussian wave packet and is given in atomic units by $\alpha_t = (\alpha_0^2 + (2t/\alpha_0)^2)^{1/2}$, where α_0 is the width at t=0. Clearly these quantal aspects will strongly influence the rescattering event.

The results presented in this paper have their genesis in an effort for evaluating the validity of these simple classical ideas.

3. Electron Distributions and Scattering "Rings"

A significant fraction of our knowledge about strong-field ionization is derived from studies of the momentum characteristics of the photoelectron. For light linearly polarized along the z axis, the angular emission is expected to become more strongly peaked along this axis as the ATI order increases, due to the propensity rule that favors increasingly higher angular momentum states as additional photons are absorbed. This expectation is also compatible with the Simpleman prediction which views the acceleration (drift) of the tunneled electrons along the laser's electric field direction. Since the faster electrons are emitted before or after the extrema of the field, they are more likely to emerge closer to the z axis.

A series of *kilohertz* experiments [9,10] which examined the angular emission of the photoelectrons demonstrated that these simple ideas are incomplete and provided a window on the role of rescattering. Figure 2 shows photoelectron spectra of xenon recorded at two intensities using 50 ps, 1 μm light. The distributions show a propensity for producing higher energy ATI peaks with increasing intensity, as well as the presence of a change in slope above 25 eV for the highest intensity. The inserts are polar plots of the angular dependence for different ATI orders. The data (crosses) are fit be a sum of even Legendre polynomials represented by the solid lines. The angular distributions (AD) for the first several peaks are consistent with the above mentioned expectations. However, as the order increases further the ADs for several peaks exhibit a strong off-axis emission which has been dubbed *scattering rings* [9]. Beyond the "rings" the ADs return to the strongly aligned shape. Furthermore, the two plots show that the energy at which the "rings" appears is strongly intensity dependent, shifting to lower energy as the intensity decreases. The intensity dependence for the energy window for which the "ring" structures appear scales as 8-9U_p.

Another means of examining this effect is shown in Fig. 3 for a plot of the xenon electron distributions at different angles relative to the laser polarization direction. The experiment uses 100 fs, 0.78 μm excitation at an intensity of 70 TW/cm^2. The distribution shows that around 8U_p the count rate at 30° and 15° become comparable. This would manifest itself as a weak scattering ring in the polar representation. Other inert gas atoms at various wavelengths demonstrate similar behavior near 8U_p providing evidence supporting the universal nature of this effect. In fact the shape of the distributions suggest some form of quantum interference [11].

Numerical solution [9,12] of the time-dependent Schrodinger equation (TDSE) using a single active electron (SAE) approximation results in remarkable agreement with the distributions shown in Figs. 2 and 3. Furthermore, the off-axis structures including the degree of their strength are clearly reproduced, once again demonstrating the appropriateness of this method for quantitatively describing the strong field ionization of inert gas atoms. However, most importantly, the SAE calculation provides a crucial clue to the origin of the scattering "rings". Since this is a single electron approximation then the source of the "rings" must be contained in the single electron dynamics.

Examination of the classical rescattering model provides some additional insight into the physical origin of the scattering "rings". If one assumes that upon return to the core the electron elastically scatters into some angle, θ_s, then one obtains a simple relationship [13] between the detected electron's energy, T, and the scattering angle,

$$T = 2U_p[(0.5 + \cos^2 \omega t_o) + 2(1 - \cos\theta_s)\cos\omega t_r(\cos\omega t_r - \cos\omega t_o)] \quad (1)$$

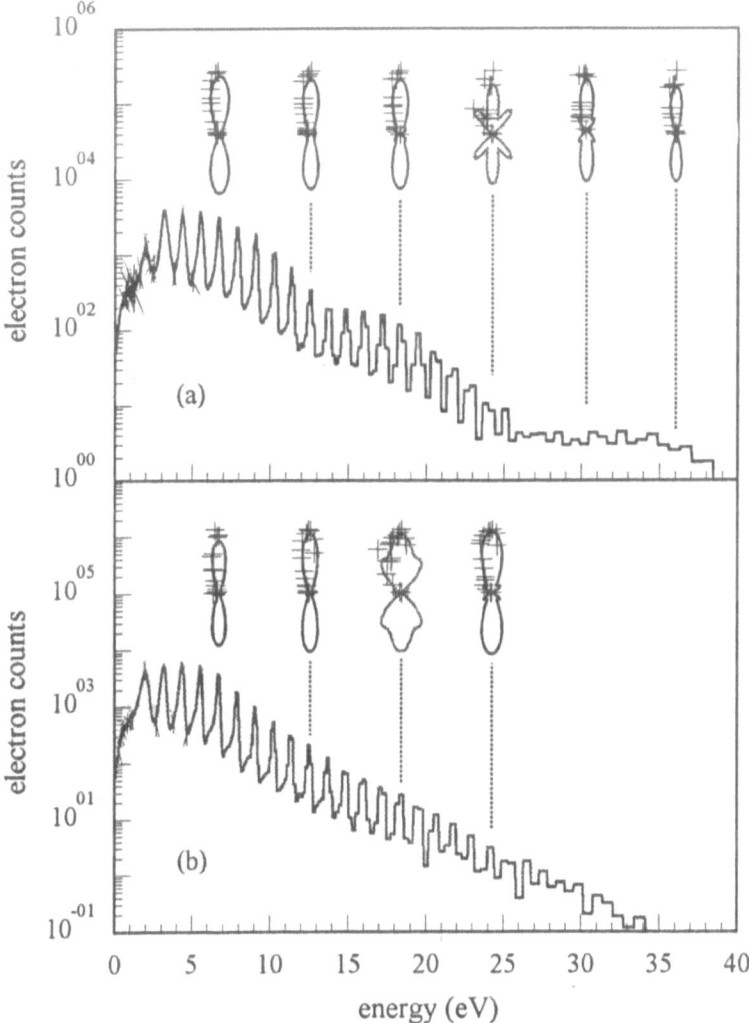

Figure 2. The photoelectron energy spectrum of xenon with 50 ps, 1.05 μm excitation at (a) 30 TW/cm² and (b) 19 TW/cm². The insets are polar plots of the ADs for selected ATI peaks. The solid line results from a Legendre fit to the raw data (crosses). The laser polarization is vertical.

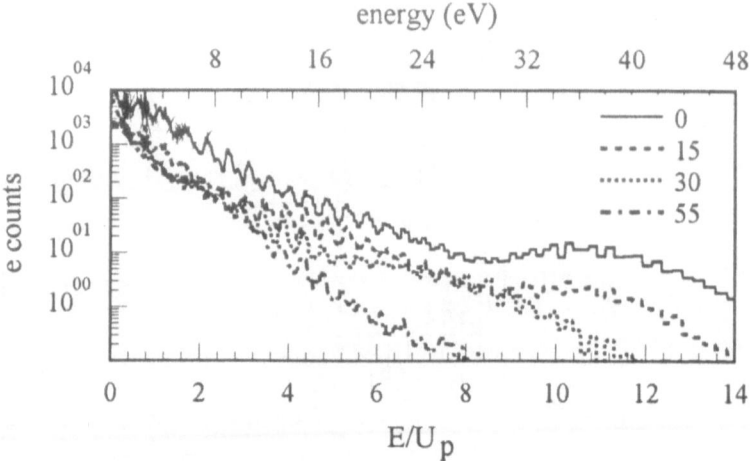

Figure 3. The angularly resolved photoelectron energy spectrum of xenon for 70 TW/cm^2, 100 fs, 0.78 µm excitation. The top and bottom scales are absolute and scaled energy units, respectively. The caption is the viewing angle (degrees) relative to the polarization direction.

where t_o and t_r are the initial and return times, respectively. If the electrons are backscattered, $\theta_s = \pi$, the energy can reach $10U_p$, as shown in Fig. 4. The insert is a polar plot of the electron emission with $8U_p$ energy, notice that the emission is off-axis at an angle similar to the xenon data. Thus, the model predicts that structures created at high energies, i.e. $8U_p$, result from large angle scattering. However, the rescattering model also predicts off-axis structure which is continuous in energy and has no on-axis emission, unlike the experimental evidence which is confined to a small energy region around $8U_p$. Such details could indicate the neglect of the quantum nature of the scattering event in the classical model and/or the invalid comparison with ionization in the multiphoton or mixed regime.

A quasi-classical calculation by Lewenstein et al. [14] using a model short range potential in a generalized strong field approximation confirms the importance of the rescattering dynamics by separately analyzing the classical trajectories in the continuum. Their results clearly show that the electrons with energies greater than $6U_p$ are solely due to backscattering. Additionally, the evolution of the time-dependent wave function reveals that after a half-cycle of free propagation in the field the returning wave packet interferes strongly near the core. Furthermore, some fraction of the packet continues to evolve against the field direction indicative of backscattering.

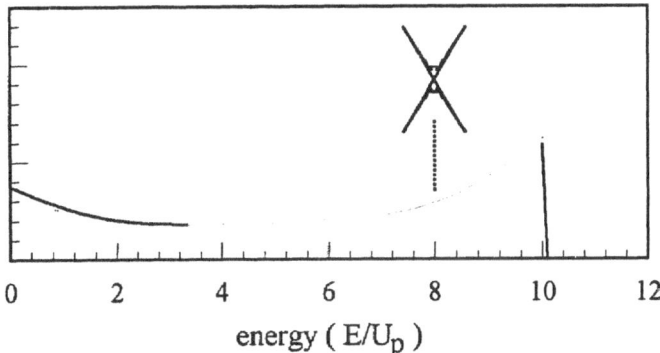

energy (E/U_p)

Figure 4. Cycle averaged electron distribution for backscattering ($\theta=\pi$). The inset is the AD resulting from electrons with $8U_p$ energy. The laser polarization is vertical.

The rescattering model's lack of predictive power in describing some of the essential experimental features can in part be attributed to the physically meaningful range of intensities accessed in the experiment. For the majority of inert gas atoms ionized with 100 fs, near visible pulses the ionization mechanism is predominantly multiphoton or at best in the mixed regime. Consequently, a well grounded comparison should be made with a purely tunneling atom. Such a scenario can be best represented by the ionization of helium with 100 fs, near visible pulses. Figure 1 shows the photoelectron spectrum (PES) at three different intensities ranging from 0.2 PW/cm^2 to saturation. The PES are normalized with respect to the ponderomotive energy associated with the peak intensity of the pulse. Analysis has shown that over this intensity range helium ionization evolves from multiphoton to pure tunneling at its saturation intensity of 0.8 PW/cm^2. Two features are clear, first the majority of electrons are emitted between $0-2U_p$ at all intensities consistent with the Simpleman model. Second, the fraction of electrons with energies greater than $2U_p$ decreases as $I^{-2.3}$. This reduction can be interpreted as arising from two sources, (1) the decrease in the elastic cross-section and/or (2) the decline in the fraction of multiphoton for effectively producing hot electrons.

Additional evidence can be seen in Fig. 5 which plots the angular resolved photoelectron spectra and representative ADs for helium ionized by 100 fs, 0.78 μm pulses. The bulk of electrons between $0-2U_p$ show strongly peaked angular distributions along the polarization direction, consistent with the Simpleman model. However, the distributions above $2U_p$ are strikingly different. Here the distributions are nearly flat in energy and have a broad angular distribution. In fact, these spectra appear to be a superposition of two independent distributions. Unlike the earlier studies in high-Z rare gas atoms, there is *no* restricted energy

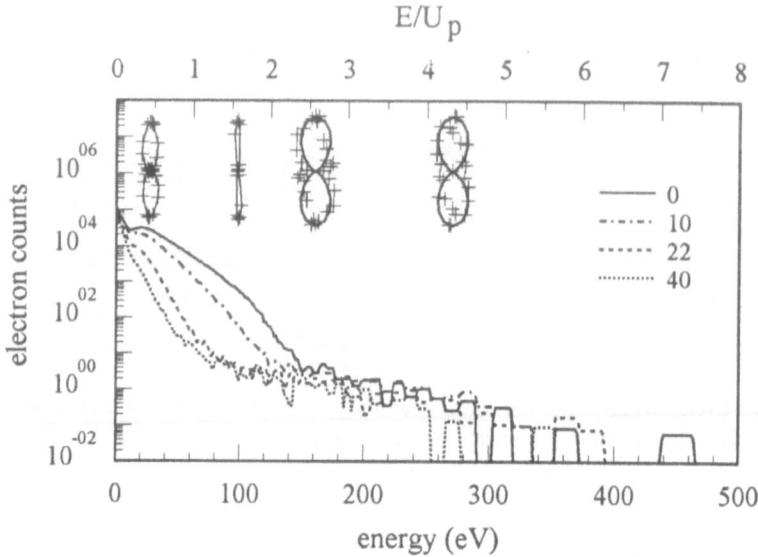

Figure 5. Angularly resolved photoelectron spectrum of helium at saturation for 0.78 μm light. The insets are the ADs at indicated scaled (top) or absolute (bottom) energies.

region where off-axis structure appears. In fact, some preliminary comparison with a calculation involving Coulomb scattering of a gaussian wave packet shows that the helium data is more consistent with a rescattering view. This result could denote the clear difference between the ionization regimes (initial conditions), with helium being the only case of pure tunneling. Consequently, the manifestation of a well defined energy window around $8U_p$ in the mixed regime can indicate that the necessary components for quantum interference are both multiphoton and tunneling in character.

4. Strong-Field Double Ionization

The first observation of multiphoton double ionization was reported in the midseventies by Suran and Zapesochnyi [15] in alkaline-earth atoms. This report was stimulated by the observation that the doubly charged production was several order of magnitude larger than predicted by assuming *sequential* nonresonant single electron dynamics within the context of lowest-order perturbation theory. Delone et al. [16] proposed a solution to this problem in which two electrons are first driven above the first ionization threshold through two-electron immediate states followed by rapid excitations above the two-electron limit via a dense manifold of quasi-continuum doubly excited states. This mechanism was dubbed *"direct"* excitation in sharp contrast to the single electron sequential dynamics. Many studies followed using tunable lasers and electron spectroscopy [17] which investigated the MPI of alkaline-earth atoms. The conclusion was that "direct" ionization was not responsible for the anomalous ratio but instead involved single electron ejection of excited electronic states of the ion, as well as resonant enhancements. This same higher-order sequential mechanism was later shown [18,19] to be responsible for the observed "direct" double MPI of xenon atoms [20] with short pulse, 0.5 µm excitation.

Helium atoms [21] exposed to an intense pulse of 100 fs, 0.8 µm light exhibit a behavior similar to the above cases, that is a premature production of doubly ionized helium which is strongly intensity dependent [8]. The major difference for helium is that it ionizes via tunneling while the other cases are multiphoton. This implies that the initial conditions determined at the time the electron is promoted to the continuum should result in very different ionization dynamics. In fact, in the intensity regime used in the experiment, all helium excited states can be field ionized by the peak electric field of the laser. Thus helium becomes an important test case for our understanding of multielectron ejection.

Figure 6 shows the measured ratio (circles) of $He^{2+}(NS)/He^{+}$ for 100 fs, 0.78 µm excitation over an intensity range where the total rate changes by 8 orders of magnitude. Note that over this same intensity range the ratio exhibits a gentle slope, scaling as $I^{-1.3}$, and a saturation value of 0.0020(3). The SAE rate was found to agree with the experimental He^{+} production over 12 orders of magnitude while the ac-tunneling result agreed near saturation but underestimated the rate at low intensity. This indicates that as the intensity is lowered, multiphoton ionization becomes increasingly important. The solid line in Fig. 6 is derived from the ratio of the calculated ac-tunneling rate to the SAE rate and shows striking agreement with the measured value. This results implies that strong-field two-electron ejection depends upon the tunneling dynamics of the first electron. Although this does not define the mechanism, it does provide an important insight into the underlying physics.

As discussed earlier, the rescattering model has intuitive explanation for two electron ejection via e-2e scattering. Furthermore, consistent with the above conclusion, this model is validated by tunnel ionization. The energetics require that the electron's kinetic energy upon return exceed the binding energy of the core electron. For helium this equates into $3.2U_p \geq 54.4$ eV, the ionization potential of helium ion, for two electron ejection. Consequently a threshold exists at $U_p = 17$ eV or an intensity of 0.3 PW/cm^2. No such threshold behavior is observed in the 0.78 µm experiment which extends down to 0.14 PW/cm^2. Additionally, by allowing for the possibilities that two-electron ionization may result from collisional excitation of the second electron to excited states (41 eV) which rapidly field ionizes and that the core potential is suppressed by the field at

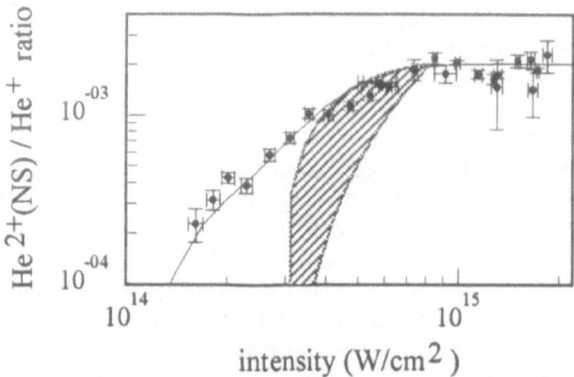

Figure 6. Intensity dependence of the He2+(NS)/He+ ratio for 0.78 μm excitation. Error bars indicated one standard deviation. Solid line is calculated; see text for details. The hashed area is calculated using the rescattering model; see text for details.

the rescattering time, this threshold is reduced to 0.23 PW/cm^2. This can be quantified by expressing the e-2e rescattering rate as $\propto \sigma(\varepsilon)F(\varepsilon,I)$, where σ is the field free e-2e inelastic cross-section and $F(\varepsilon,I)$ is the intensity dependent distribution of returning electrons with energy, ε. The result is shown in Fig. 6 by the hashed area which is bound on the left side by assuming that $F(\varepsilon,I)$ is a delta function at $3.2U_p$ and on the right by a flat distribution between 0 and $3.2U_p$. Furthermore, the calculated curves are fit to agree with the experimental ratio at saturation which implies an assumed transverse wave packet width of $3-4a_o$ *at the rescattering time*. However, analyses of the SAE wave packet shows that a realistic initial width is $3-4a_o$. Consequently the spreading of the wave packet due to a half cycle of free propagation in the field yields a rescattering width of $30-40a_o$. Using this width, an upper limit on the He^{2+}(NS)/He^+ ratio at saturation is 1.5×10^{-4}, more than an order of magnitude smaller than the experimental ratio (0.002). Clearly, the double ionization due to e-2e rescattering is much smaller than the measured yields over the entire dynamic range of the experiment.

One possible source of the discrepancy could be due to the use of field free cross sections in the calculation. Kulander et al. [22] using an approximate 1-D two electron model examined the influence of the field on the e-2e cross section. The result showed little effect on the inelastic cross section since it is the short range physics which dominates the process.

The physics of two electron ejection remains an unresolved issue since our best quantitative tests via the rescattering model fails. Walker et al. [8] have suggested that the physics of the process may be more like single photon double ionization involving a correlated threshold behavior or shakeoff process.

Consequently, a theoretical treatment using appropriate two-electron wave functions is warranted. Recently, a calculation [23] of helium double ionization using two electron Volkov states has shown that final state correlation are important in this process and represents an important first step for uncovering the underlying physics. Unfortunately this calculation which embraces all the above mentioned physics, is unable to distinguish between the different mechanisms.

5. Conclusion

Considerable progress has been achieved over the past three years in understanding the single electron dynamics in strong-field atomic ionization. TDSE methods have shown to be extremely powerful and accurate in quantitatively describing the experiments. The quasi-classical rescattering model has provided an useful foundation for viewing the course dynamics. However, the degree of quantum influence is yet unclear, especially as it pertains to our evaluation of control methods. Also, two or multiple electron dynamics remains an open question and poses some new and important future challenges.

This research was carried out in part at BNL under Contract No. DE-AC02-76CH00016 with the U.S. Department of Energy and supported by its Division of Chemical Sciences, Office of Basic Energy Sciences, and in part under the auspices of the DOE at LLNL under Contract No. W-7405-ENG-48. P.A. acknowledges travel support from NATO Contract No. SA.5-2-05(RG910678).

References

1. Becker, W., Schlicher, R.R., and Scully, M.O. (1986) *J. Phys. B* 19, L785.
2. van Linden van den Heuvell, H.B. and Muller H.G. (1988) in S.J. Smith and P.L. Knight (eds.), *Multiphoton Processes*, Cambridge University Press, London, pp. 25.
3. Gallagher, T.F. (1988) *Phys. Rev. Lett.* 61, 2304.
4. Corkum, P., Burnett, N., and Brunel F. (1989) *Phys. Rev. Lett.* 62, 1259.
5. Schafer, K.J., Yang, B., DiMauro, L.F., and Kulander, K.C. (1993) *Phys. Rev. Lett.* 70, 1599.
6. Corkum, P. (1993) *Phys. Rev. Lett.* 71, 1994.
7. Krause, J., Schafer, K.J., and Kulander, K.C. (1992) *Phys. Rev. Lett.* 68, 3535.
8. Walker, B., Sheehy, B., DiMauro, L.F., Agostini, P., Schafer, K.J., and Kulander, K.C. (1994) *Phys. Rev. Lett.* 73, 1227.
9. Yang, B., Schafer, K.J., Walker, B., Kulander, K.C., Agostini, P., and DiMauro, L.F. (1993) *Phys. Rev. Lett.* 71, 3770.
10. Paulus, G.G., Nicklich, W., and Walther, H. (1994) *Europhys. Lett.* 27, 267.
11. Paulus, G.G., Nicklich, W., Xu, Huale, Lambropoulos, P., and Walther, H. (1994) *Phys. Rev. Lett.* 72, 2851.

108

12. Kulander, K.C., Schafer, K.J., and Krause, J.L. (1993) in M. Gavrila (ed.), *Atoms in Intense Laser Fields*, Academic Press, New York, pp. 247.
13. Paulus, G., Becker, W., Nicklich, W., and Walther H. (1994) *J. Phys. B* **27**, L703.
14. Lewenstein, M., Kulander, K.C., Schafer, K.J., and Bucksbaum P.H. (1995) *Phys. Rev. A* **51**, 1495.
15. Suran, V.V. and Zapesochnyi, I.P. (1975) *Sov. Tech. Phys. Lett.* **1**, 420.
16. Delone, N.B., Suran, V.V., and Zon, B. (1983) in S.L. Chin and P. Lambropoulos (eds.), *Multiphoton Ionization of Atoms*, Academic Press, New York, pp. 235.
17. DiMauro, L.F. (1993) in T.N. Chang (ed.), *Many-Body Theory of Atomic Structure and Photoionization*, World Scientific Press, New Jersey, pp. 297.
18. Walker, B., Mevel, E., Yang, B., Breger, P., Agostini, P., and DiMauro, L.F. (1993) *Phys. Rev. A* **48**, R894.
19. Charalambidis, D., Lambropoulos, P., Schroder, H., Faucher, O., Xu, H., Wagner, M., and Fotakis, C. (1994) *Phys. Rev. A* **50**, R2822.
20. L'Huillier, A., Lompre, L.A., Mainfray, G., and Manus, C. (1983) *Phys. Rev. A* **27**, 2503.
21. Fittinghoff, D.N., Bolton, P.R., Chang, B., and Kulander, K.C. (1992) *Phys. Rev. Lett.* **69**, 2642.
    Kulander, K.C., Cooper, J., and Schafer, K.J. (1995) *Phys. Rev. A* **51**, 561.
22. Fasial, F.H.M. and Becker, A. (1995) in H. Kleinpoppen (ed.), *Invited Papers to Peter Farago Symposium*, Plenum Press, in press and this conference proceedings.

QUANTUM AND CLASSICAL MECHANICS OF RESCATTERING EFFECTS IN HIGH-ORDER ABOVE-THRESHOLD IONIZATION AND HIGH-HARMONIC GENERATION

W. BECKER[1,2], B. GOTTLIEB[1], M. KLEBER[1], A. LOHR[1],
G. G. PAULUS[3], and H. WALTHER[3,4]

[1] Physik-Department T30, Technische Universität München,
85747 Garching, Germany

[2] Center for Advanced Studies, Department of Physics and Astronomy,
University of New Mexico, Albuquerque, NM 87131, USA

[3] Max-Planck-Institut für Quantenoptik, 85748 Garching, Germany

[4] Sektion Physik, Ludwig-Maximilians-Universität München,
85747 Garching, Germany

1. Introduction

Providing simulations to laser-atom interactions at high intensity has been and still is a challenge to develop more and more accurate and efficient methods for the solution of the time-dependent Schrödinger equation. Much progress has been made in recent years, but to this date these methods are still very time-consuming and not very user friendly, particularly so in three dimensions. For a review of the state of the art, see [1, 2]. Moreover, even a successful description of some experimental data in this framework does not always and not necessarily satisfy the physicist's desire for an intuitively appealing picture of "what is really going on". In addition, reasonably simple models having some suggestive and, ideally, predictive power are vitally important. For in several problem areas, such as those involving two-color fields, the parameter space is just too large for a shotgun approach to be feasible to find the interesting regions.

In this paper we will present and discuss various such models. They extend from the fully quantum mechanical zero-range potential model atom all the way to the radical attempt to extract physical understanding from the consideration of very few selected classical orbits in the presence of just the laser field. It is fascinating to observe that the last mentioned approach has met with some success, not only for the very strong fields that are typical of the tunneling regime, but equally well for the much weaker fields that characterize the multiphoton regime. Particularly in the latter we are dealing with phenomena on the nanometer scale, and classical orbits should have lost any meaning. We will try to embed these classical orbits into the full quantum mechanics. We will not attempt to present a

109

H. G. Muller and M. V. Fedorov (eds.), Super-Intense Laser-Atom Physics IV, 109–121.
© 1996 Kluwer Academic Publishers.

completely consistent picture where every model has its well defined place. Rather, some of the models are in conflict with others, and their ultimate significance if any will have to be left open.

2. High-Harmonic Generation in a Zero-Range Potential

The zero-range binding potential to be written down compactly as the (regularized) delta-funtion potential

$$V(\mathbf{r}) = \frac{2\pi}{m\kappa}\delta(\mathbf{r})\frac{\partial}{\partial r}r \tag{1}$$

has exactly one bound ($l = 0$) state with energy $E_0 = -\kappa^2/2m$. This potential is a most radical approximation to a real atom for the situation where just one bound state and the continuum are expected to dominate what happens. The potential (1) provides a very convenient theoretical laboratory for laser-atom physics. For it allows, in the presence of an external time-dependent field with arbitrary polarization, for an almost analytical solution in the form of the quasi-energy wave function. The method of solution as well as applications, particularly to high-harmonic generation (HHG), have been extensively discussed in earlier work to which we refer for details [3].

We will calculate the S-matrix element for emission of exactly one photon with frequency Ω and polarization ϵ while the atom returns to its ground state after the field has been turned off again in the distant future. It is

$$S_\epsilon(\Omega) = \epsilon^* \cdot \int dt e^{i\Omega t} < \Psi^{(+)}(t)|e\mathbf{r}|\Psi^{(-)}(t) > \tag{2}$$

where the wave functions

$$< \mathbf{r}|\Psi^{(\mp)}(t) >= \int d^3r' dt' G^{(E)}_{\text{ret/adv}}(\mathbf{r}t, \mathbf{r}'t')V(\mathbf{r}')\psi_0(\mathbf{r}'t') \tag{3}$$

are explicitly related to the wave function $\psi_0(\mathbf{r}, t)$ of the free atom and $G^{(E)}_{\text{ret/adv}}$ denotes the retarded/advanced propagator of an otherwise free electron in the presence of the laser field in the length gauge. That is, the integration over t' is restricted to $t > t'$ and $t < t'$, respectively. In earlier work the expectation value of the dipole moment at time t in place of the S-matrix element (2) has been used. The former differs from the latter by the bra $< \Psi^{(+)}|$ being replaced by $< \Psi^{(-)}|$. The differences will be discussed in detail elsewhere [4]. The interpretation to be given below holds, in principle, only for the S-matrix element. However, quantitative differences between the two formulations are minute.

If the path integral representation of the propagators is used then the S-matrix element (2) can be evaluated almost by inspection and cast in the form [5]

$$S_\epsilon = \frac{2\pi i e \kappa}{m^2} \int_{-\infty}^{\infty} dt'' \int_{t''}^{\infty} dt' \epsilon^* \cdot \mathbf{r}(\Omega; 0t', 0t'') G_{\text{ret}}^{(E)}(0t', 0t'') e^{-i|E_0|(t'-t'')} \quad (4)$$

where

$$\mathbf{r}(\Omega; \mathbf{r}'t', \mathbf{r}''t'') = \int_{t''}^{t'} dt e^{i\Omega t} \mathbf{r}(t; \mathbf{r}'t', \mathbf{r}''t''), \quad (5)$$

$$G_{\text{ret}}^{(E)}(\mathbf{r}'t', \mathbf{r}''t'') = \left(\frac{im}{2\pi(t'-t'')}\right)^{\frac{3}{2}} \exp(iS[\mathbf{r}(t; \mathbf{r}'t', \mathbf{r}''t'')]) \theta(t' - t''), \quad (6)$$

$$S[\mathbf{r}(t; \mathbf{r}'t', \mathbf{r}''t'')] = \int_{t''}^{t'} dt \, L[\mathbf{r}(t; \mathbf{r}'t', \mathbf{r}''t'')]. \quad (7)$$

Here $\mathbf{r}(t; \mathbf{r}'t', \mathbf{r}''t'')$ denotes the classical orbit at time t of a free electron in the laser field that runs from position $\mathbf{r}''$ at time t'' to $\mathbf{r}'$ at a later time t', and S is the classical action of this orbit. In any field, for any given $\mathbf{r}'', t''$ and $\mathbf{r}', t'$, there is exactly one such orbit. The S-matrix element (4) is constructed from only those and all of those orbits that start from and return to the origin, i.e. the site of the potential. It sums over their Fourier transforms, and each orbit is weighted with the propagator $G^{(E)}(0t', 0t'')$ (which specifies the probability amplitude for propagation from $(0t'')$ to $(0t')$) and the exponential $\exp(-i|E_0|(t'-t''))$. A closely related form of the dipole matrix element has been given by Lewenstein et al. [6].

Looking at the S-matrix element (4) we realize that the semiclassical model of Kulander et al. [7] and Corkum [8] has just fallen in our laps: The electron is injected in the continuum at time t'' at the origin and starts on its classical orbit (5) which returns there at the later time t'. At that time the electron recombines with the ion and disappears from the scene.

The fact that the S-matrix element (4) can be constructed from the above mentioned classical orbits does not imply that HHG is a classical process. The contributing orbits are summed with complex weights. Hence, different orbits can and do interfere. For example, since generally the orbits start with a nonzero velocity, there are orbits that start from and return to the origin even for a circularly polarized laser field. It can be shown, however, that all of these orbits sum to zero as it should be [5], yielding no harmonic emission for a circularly polarized driving field. In what follows we will show that the immediate physical significance of the classical orbits is strongest near the cut-off of the plateau. To this end we investigate the variation of the phase of the integrand in Eq.(4).

For a monochromatic field with frequency ω, all of the terms of the Fourier transform (5) of the classical orbit are proportional to $\exp(i(\Omega \pm \omega)\tau)$ with either $\tau = t'$ or $\tau = t''$. The phase of the integrand of the S-matrix element (4) is

$$\Phi(t', t'') \equiv S[\mathbf{r}(t; 0t', 0t'')] - |E_0|(t' - t'') + (\Omega \pm \omega)\tau. \quad (8)$$

For the action (7), the derivatives with respect to t' and t'' yield the final and initial kinetic energies, respectively,

$$\frac{\partial S}{\partial t'} = -\frac{m}{2}\mathbf{v}'^2, \quad \frac{\partial S}{\partial t''} = \frac{m}{2}\mathbf{v}''^2, \tag{9}$$

where $\mathbf{v} = d/dt\ \mathbf{r}(t; 0t', 0t'')$ and $\mathbf{v}' = \mathbf{v}(t')$ and $\mathbf{v}'' = \mathbf{v}(t'')$. Taking $\tau = t'$ we have

$$\frac{\partial \Phi}{\partial t'} = -\frac{m}{2}\mathbf{v}'^2 - |E_0| + \Omega \pm \omega, \tag{10}$$

$$\frac{\partial \Phi}{\partial t''} = \frac{m}{2}\mathbf{v}''^2 + |E_0|. \tag{11}$$

A measure of the total variation of the phase at a point (t', t'') is furnished by

$$(\Delta\Phi)^2 = \left(\frac{\partial \Phi}{\partial t'}\right)^2 + \left(\frac{\partial \Phi}{\partial t''}\right)^2. \tag{12}$$

For real t' and t'', $\Delta\Phi$ can never be zero meaning there is no stationary phase point. However, we still expect the dominant contributions to the integral to come from regions of t', t'' where the variation $\Delta\Phi$ is slowest. Equations (10,11) show that the minimum value possible for $\Delta\Phi$ is $|E_0|$. This suggests that the plateau scales down in intensity with increasing $|E_0|$ as indeed it does. The minimum is assumed for harmonic frequencies Ω such that $|E_0| \leq \Omega \leq |E_0|+\max(m\mathbf{v}'^2/2)$ and for an initial velocity $\mathbf{v}'' = 0$. Recall that for a monochromatic linearly polarized laser field, the maximum of the kinetic energy of the returning classical electron is the well known $3.17U_p$ [7, 8] where U_p is the ponderomotive potential of the laser field. Hence the minimum can only be assumed when the harmonic frequency is within the plateau region, $|E_0| \leq \Omega \leq |E_0| + 3.17U_p$. Therefore, we notice that the S-matrix element (4) clearly distinguishes the plateau region from that below and beyond. For more general field configurations, Eqs.(10,11) still hold, but their analysis is more complicated. It can be shown [5] that a necessary condition for a maximal kinetic energy upon return is that the initial and the final velocities are perpendicular, $\mathbf{v}' \cdot \mathbf{v}'' = 0$. For linear polarization this enforces that $\mathbf{v}'' = 0$. In general, however, this is not so as it would prevent the electron from returning. If the ellipticity of the laser field increases from zero at constant intensity, then $|\mathbf{v}'|$ decreases while $|\mathbf{v}''|$ increases. Consequently, the minimum value of the variation $\Delta\Phi$ increases and the harmonic intensities decrease.

The above considerations are particularly relevant near the end of the plateau. This is because in that case the kinetic energy of the returning electron is near its maximum and this is afforded only by a very small region of the (t', t'') space. Hence, this is the case where interference effects in the coherent superposition of the various orbits are least important, and the classical orbits have their greatest significance. At lower harmonic frequencies, these interference effects are much more pronounced. They are virtually unpredictable and likely to be responsible for the very rugged structure of the calculated plateau.

3. Wave Mechanical Aspects of the Classical Orbits

The physical significance of the classical orbits discussed in the previous Section is restricted by the fact that HHG is determined by their coherent superposition, by means of the integration over the initial and final times t'' and t'. In this Section we will proceed differently: we will be concerned with the region near the end of the plateau and will try to identify only *one* particular classical orbit as especially relevant. However, we will associate with it a quantum mechanical wave packet that is subject to spreading about its center which is given by this classical orbit. This particular orbit is not necessarily such that it takes the electron back to its starting point. We first look at the case of an atom in a constant electric field and based on that will draw conclusions for the laser field.

3.1. IONIZATION IN A CONSTANT STATIC ELECTRIC FIELD [9]

The zero-range potential in a constant electric field F allows for a completely analytic solution in terms of Airy functions [10]. Owing to the quasi-stationarity of the problem we can just read off the spreading of the ionized wave packet transverse to the direction of the field. The tranverse width ρ where the current has dropped to $1/e$ of its value on axis is (for not too small z)

$$\rho^2 = \left(\frac{2}{m|E_0|}\right)^{\frac{1}{2}} z. \tag{13}$$

where z denotes the distance from the atom. Notice that ρ is independent of the applied field F. Qualitatively the same expression results from much more sophisticated descriptions [11]. It is in good agreement with currents measured in Scanning Tunneling Microscopy [12].

One may be tempted to model the situation by an electron that starts at the position of the atom with zero velocity and rolls down the inclined plane created by the electric field. We assume a Gaussian shape for the corresponding quantum mechanical wave packet, so its width at a later time t is

$$a(t)^2 = a(0)^2 + (t/ma(0))^2 \tag{14}$$

where $t^2 = 2z/eF$. If we envision $a(0)$ to be of the order of the dimension of the atom then in most situations the width $a(t)$ is much larger than ρ, even at macroscopic distances from the atom. Hence such a model is not applicable, the ionized electron never behaves like a free electron.

Or does it? Let us modify the model [9]. We now have the electron start at the end of the tunnel, at $z_0 = |E_0|/eF$, with zero velocity. At z_0, we read off its width from the exact solution, identifying $a(z_0) \equiv \rho(z_0)$. From there on we assume free spreading according to Eq.(14), that is

$$a(z)^2 = a(z_0)^2 + \frac{2}{eF(ma(z_0))^2}(z - z_0). \tag{15}$$

One may check immediately that this $a(z)$ agrees with the transverse width ρ of the exact solution written down above. Hence we have succeeded in deriving a two-step model for ionization in a constant electric field.

3.2. IONIZATION IN A LASER FIELD PLUS A TRANSVERSE STATIC ELECTRIC FIELD

In a time-dependent situation, we can no longer proceed in such a straightforward fashion. The interpretation of the quasi-energy wave function is made difficult by the fact that the electron after ionization may linger on near the atom for many periods. In order to shed light on this case we will perform a numerical experiment and then try to model its outcome. We apply a monochromatic linearly polarized laser field and superpose a constant electric field F tranverse to the former. Formally, for the odd harmonics (there are even harmonics now, too) the S-matrix element (4) is only changed by an additional term $e^2 F^2 (t' - t'')^3/24$ in the action (7). Details will be reported elsewhere [4]. Figure 1 displays a typical harmonic spectrum. For increasing field strength F, the brightness of the plateau decreases, while the harmonics below and beyond the plateau are affected to a much lesser degree.

We will try to model this situation with a Gaussian wave packet that starts at the origin with zero velocity and a certain width $a(0)$ to be determined later. Owing to the static field, this orbit never returns to the origin, in contrast to the orbits in the S-matrix element (4). Rather, when it has returned to $x = 0$ in the direction of the laser field (x, say) it is deflected by some amount $y = eFt_R^2/2$ in the direction of the static field (y, say) where t_R is the return time in the x-direction. We are interested in the end of the plateau, so we consider those orbits that return with maximal kinetic energy in the x-direction, and we have tacitly assumed that, for a not too strong static field, the motions in the two directions, under the condition of maximal kinetic energy in the x-direction, are essentially decoupled. Under these conditions, for a monochromatic laser field, t_R still has its standard value of $\omega t_R = 4.08$ [3]. We proceed similarly to Ref.[13]. We calculate the overlap $O(F)$ of a Gaussian at the origin with an unperturbed width of $1/\kappa$ which simulates the unperturbed ground state with another Gaussian wave packet describing the returning electron. Upon its return it has aquired a width of $a(t_R)$ given by Eq.(14) in terms of $a(0)$. For the harmonics near the end of the plateau, the square of this overlap should provide a measure of their rate of emission.

The question remains which value to take for the initial width $a(0)$. We have investigated various possibilities, and the following choice provides an optimal fit of the numerical experiment. We take what would be the width of the current density in a static electric field (whose magnitude F is identified with the peak amplitude F_L of the laser field) at the end of the tunnel. That is, we choose

$$a(0)^2 \equiv \rho(z = |E_0|/eF_L)^2 = \sqrt{2|E_0|/m(eF_L)^2} \equiv \rho_T^2, \qquad (16)$$

where $\rho(z)$ was defined in Eq.(13). We are unable to justify this choice in any rigorous way. For the case of Figure 1, a quantitative comparison is summarized in the inset. We are using a procedure similar to Ref.[13]. We compare the quantity $R = |O(F = 0)/O(F)|^2$, which is calculated as described above, with the corresponding intensity ratio of the strongest harmonics near the end of the plateau from the numerical experiment. Notice how sensitive the results of the model are to the precise value of ρ_T. The accuracy of the fit which we have checked in a variety of situations for values of the Keldysh parameter between 0.3 and 3 (which covers all situations of experimental interest thus far) suggests that this model has a fair overlap with reality. In the *tunneling regime*, the success is not surprising. One would have been tempted to set the electron free at the end of the tunnel rather than at the origin, but the difference between these two positions is small in this case. Taking the initial width $a(0)$ over from the static field case (13) is equally suggestive. The remarkable fact is that the model works so well in the *multiphoton regime* for field parameters where the classical orbit of the electron never even comes close to the end of the tunnel. It appears that the parameter (16) may have some significance beyond its derivation for the static electric field.

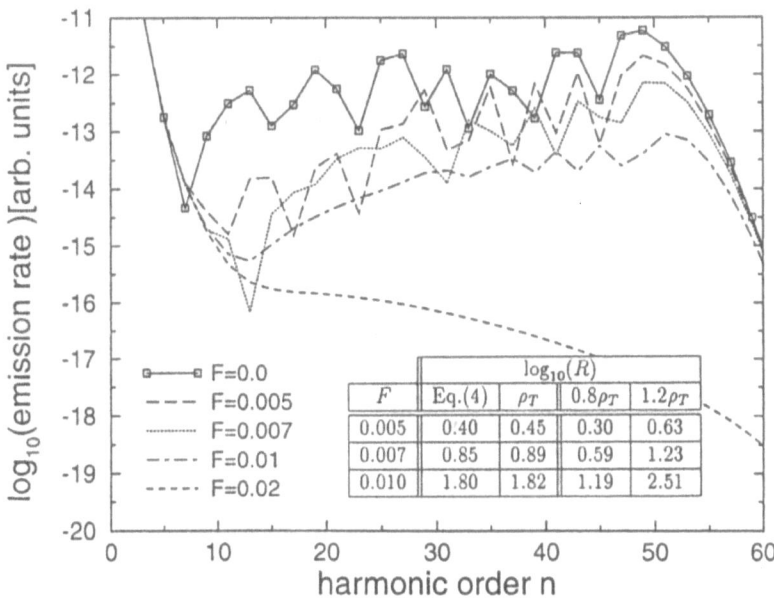

Figure 1. High-harmonic spectrum for the model atom with $|E_0| = 0.58$ a.u. in a laser field with $F_L = 0.058$ a.u. and $\omega = 0.043$ a.u. in the presence of an additional transverse static electric field for various field strengths F. Only the odd harmonics are given. The table inset lists the intensity ratio of the strongest harmonics near the end of the plateau as depicted and calculated from Eq.(4) (1st column) and compares to the ratio R calculated from the model described in the text (2nd column). Columns 3 and 4 give the results of the same model, but with ρ_T replaced by $0.8\rho_T$ and $1.2\rho_T$, resp. For the

parameters used, $\rho_T = 4.3$a.u. and the Keldysh parameter is $\gamma = 0.8$. The end of the tunnel is at 10a.u. and the orbit has a maximal excursion of 31a.u., both counted from the position of the atom.

Below, the same table is reprinted for different values of the parameters, $F_L = 0.01$a.u. and $\omega = 0.02$a.u., this time in the multiphoton regime with $\gamma = 2.1$, in order to demonstrate that the model still applies. Now, the maximal excursion of 29a.u. is still deeply in the tunnel which ends at 58a.u., and $\rho_T = 10.4$.

F	\multicolumn{4}{c}{$\log_{10}(R)$}
	q.m.	ρ_T	$0.8\rho_T$	$1.2\rho_T$
0.002	1.64	1.53	1.12	1.78
0.003	3.51	3.44	2.53	4.00
0.004	5.75	6.11	4.50	7.12

Models similar to this have been employed in the discussion of whether double ionization of helium can be understood in terms of a picture where the returning electron causes the second ionization [8, 13, 14, 15]. Indeed, an estimate like this has put the validity of this picture into serious doubt [14, 15]. In the tunneling regime, for the parameters of Ref. [15] ($F_L = 0.15$a.u.), the value of our parameter ρ_T roughly agrees with the one employed there. However, for smaller fields, ρ_T significantly exceeds atomic dimensions so that in that case the role of quantum mechanical wave packet spreading appears to be less than anticipated.

4. High-Order Above-Threshold Ionization

Soon after the discovery of high-order harmonic generation its relation to above-threshold ionization (ATI) became a topic of lively debate. Recent measurements of the energy spectra [16, 17] and angular distributions [16, 18] of ATI for very high electron energies (high-order above-threshold ionization; HATI) have clarified the connection: Different electrons are responsible for ordinary ATI on the one hand and for HATI and HHG on the other. It is the electrons that return to the ion that have the potential to generate both HHG and HATI. In what follows we will first discuss a simple fully quantum mechanical model that yields the essential features of HATI in good agreement with the data. We will then extract from it a completely classical model which will turn out still to contain some features of the observed spectra. In particular, we will apply the classical model to make predictions for HATI in the presence of a superposition of a linearly polarized driving field and its second harmonic.

4.1. A QUANTUM MECHANICAL MODEL

We will decompose ATI as a two-step process much in the same way as we were able to derive it for ionization in a static electric field in Section 3.1. We assume that some source distribution $\sigma(\mathbf{r}, t)$ injects electrons in the continuum into the lowest possible state corresponding to the absorption of the minimum number N of photons. The matrix element that describes the rescattering process responsible for HATI is [19]

$$M_{\mathbf{p}}^{(1)} = -i \int d^3r' dt' d^3r'' dt'' \psi_{\mathbf{p}}^{(V)}(\mathbf{r}'t')^* V(\mathbf{r}') G_{\text{ret}}^{(E)}(\mathbf{r}'t', \mathbf{r}''t'') \sigma(\mathbf{r}''t'') \qquad (17)$$

with

$$\psi_{\mathbf{p}}^{(V)}(\mathbf{r}t) = \exp\left(-\frac{i}{2m} \int^t d\tau (\mathbf{p} - e\mathbf{A})^2\right) \exp(i(\mathbf{p} - e\mathbf{A}(t)) \cdot \mathbf{r}) \qquad (18)$$

the Volkov solution. A choice of the source distribution that is appropriate for the multiphoton regime is

$$\sigma(\mathbf{r}t) = C I^{N/2} \delta(\mathbf{r}) e^{-i(N\omega - |E_0|)t}. \qquad (19)$$

The physical content of the matrix element (17) is obvious. The source σ puts an electron in the continuum at the origin with an energy overshoot of $N\omega - |E_0|$. The electron is then propagated by the propagator $G^{(E)}$ (Eq.(6)) back to within the range of the binding potential $V(\mathbf{r})$ for which we use again the zero-range potential (1). This is followed by rescattering whereupon the electron moves away from the potential but keeps interacting with the laser field. This last part is described by the Volkov solution (18). Since we assumed a zero-range source in addition to the zero-range potential the spatial integrations in the matrix element (17) are taken care of. Calculations [19] based on the matrix element (17) have led to good agreement with the experimental data [16, 17, 18].

In order to get an estimate of the effect of the remaining two integrations over time we again consider the total phase of the integrand of Eq.(17). It is

$$\Phi(t', t'') = S[\mathbf{r}(t; 0t', 0t'')] + \frac{1}{2m} \int^{t'} d\tau (\mathbf{p} - e\mathbf{A}(\tau))^2 - (N\omega - |E_0|)t'' \qquad (20)$$

with the derivatives

$$\frac{\partial \Phi}{\partial t'} = \frac{m}{2} \mathbf{v}'^2_{\text{out}} - \frac{m}{2} \mathbf{v}'^2, \qquad (21)$$

$$\frac{\partial \Phi}{\partial t''} = \frac{m}{2} \mathbf{v}''^2 - (N\omega - |E_0|). \qquad (22)$$

Here $m\mathbf{v}'_{\text{out}} = \mathbf{p} - e\mathbf{A}(t')$ denotes the velocity of the electron right after rescattering and $\mathbf{v}'$ and $\mathbf{v}''$ have the same meaning as above below Eq.(9). Unlike the corresponding phase (8) in HHG the phase (20) does have a stationary point. This is not because the two processes are different, but rather due to the fact that here we used the source to inject the electron into the continuum with a positive energy. In contrast, in the case of the HHG S-matrix element (4) the electron started from the ground state and ionization into the continuum was part of the S-matrix element. When we set Eqs.(21) and (22) equal to zero we get the classical limit of HATI, cp. the analogous situation in the case of potential scattering in a laser field [20, 21]. Equation (22) yields the condition that the initial energy of the electron correspond to the energy overshoot $N\omega - |E_0|$ while Eq.(21) demands that kinetic energy be conserved at the instant where the electron rescatters.

4.2. A CLASSICAL MODEL

Equation (21) does not determine the rescattering angle. We assume the laser field with vector potential $\mathbf{A} = \hat{x}A(t)$ to be in the x-direction and define the scattering angle θ_0 (at the instant of rescattering) with respect to the negative x-axis. For a zero-range potential, scattering is isotropic. Hence θ_0 is a random quantity with a uniform distribution. Then

$$v'_{\text{out},x} = |\mathbf{v}'|\cos\theta_0 = p\cos\theta - eA(t'), \tag{23}$$

$$v'_{\text{out},y} = |\mathbf{v}'|\sin\theta_0 = p\sin\theta. \tag{24}$$

Here θ denotes the *observed* scattering angle outside the field which is not equal to θ_0 (except in the direction of the laser field) since the electron after rescattering is further deflected by the laser field. Since the canonical momentum is conserved, $\mathbf{v}'$ and $\mathbf{v}''$ are related by $m\mathbf{v}' = m\mathbf{v}'' + e(\mathbf{A}(t'') - \mathbf{A}(t'))$. This allows for the calculation of the kinetic energy of the rescattered electron outside the pulse (assuming short-pulse boundary conditions). In principle, the magnitude of the initial velocity $\mathbf{v}''$ is determined by Eq.(22) and its direction is random. Since $(m/2)\mathbf{v}''^2 < \hbar\omega$ and since we are primarily interested in the high-energy part of the spectrum we may let $\mathbf{v}'' = 0$ and obtain [22, 23]

$$E_{\text{kin}} = \frac{\mathbf{p}^2}{2m} = \frac{e^2}{2m}\left\{(A(t''))^2 + 2A(t')(A(t') - A(t''))(1 \pm \cos\theta_0)\right\}, \tag{25}$$

where the upper (lower) sign holds for $A(t'') > A(t')$ $(A(t') > A(t''))$. Backscattering corresponds to $\theta_0 = 0$ $(\theta_0 = \pi)$. The observed scattering angle θ can be calculated in a similar fashion [22, 23].

It is clear from Eq.(25) that maximal kinetic energies occur for backscattering. They are

$$E_{\text{kin,max}} = \frac{e^2}{2m}(2A(t'') - A(t'))^2. \tag{26}$$

For a monochromatic linearly polarized field this maximal kinetic energy is $10.007U_p$, much higher than the maximal kinetic energy of $3.17U_p$ of the returning electron. The reason is obvious: if the returning electron backscatters at the right instant it may be accelerated by the field by up to another half period and acquire a significant additional energy. Backscattered HATI electrons come out to be emitted into a narrow cone with an opening angle of about 30°, in good agreement with the experimental data of [18].

4.3. CLASSICAL TWO-COLOR HIGH-ORDER ABOVE-THRESHOLD ION-IZATION

In the preceding Section we have left open the functional form of the vector potential of the laser field. In what follows we will apply the results to the two-color field

$$E(t) = \omega A(\cos \omega t + \cos(2\omega t + \phi)), \tag{27}$$

where the fundamental and the second harmonic have a well defined relative phase ϕ. The two-color environment introduces several new features. Generally, the total ionization rate depends on the electric field of the laser while the energy distribution of the emitted electrons depends on its vector potential. For one color, electric field and vector potential have the same functional form, and this is responsible for the fact that the shape of the electron energy distribution (on the scale of U_p) is largely independent of the field intensity. For a two-color field, depending on the value of the relative phase ϕ, the field and the vector potential may look very different. As a consequence, the electron energy distribution depends dramatically on the relative phase *and* on the intensity. For example, for initial times t'' that lead to the highest HATI energies the field $E(t'')$ may be weak compared to its peak value. The highest HATI energies (in terms of U_p) will then become more and more suppressed when the laser intensity is raised into the tunneling regime. This effect would provide a very clean signature of whether emission occurs in the multiphoton or in the tunneling regime.

Figure 2 displays an ATI spectrum as a function of the observed scattering angle θ and the electron energy scaled in units of U_p, for a relative phase of $\phi = 2\pi/3$. Several features are very noticeable in comparison with the corresponding one-color spectra: (i) the electron energies extend up to $21U_p$ for backscattering; (ii) there is a pronounced forward/backward asymmetry; (iii) there are several series of electron energies which cut off at different energies. They give rise to several side lobes at different angles in the fixed-energy angular distributions. A more detailed discussion of these effects can be found in Ref.citePBW. The angular cut-offs at fixed energy displayed in Fig.2 are very sharp. This has already been observed in the one-color case [22] and is in qualitative agreement with the measured angular distributions of Ref. [18]. To what extent this classical model may serve as a dependable guide to the real world must be left open at present.

The total emission rate (integrated over all angles) into an ATI peak allows for a stringent test of the importance of rescattering. It can be shown that, for the field (27), in the absence of rescattering this rate is invariant against $\phi \rightarrow -\phi$ while this invariance is destroyed by the rescattering contributions. This holds both classically and in quantum mechanics. Evidence of the relevance of rescattering in this context has been seen in a recent experiment [24].

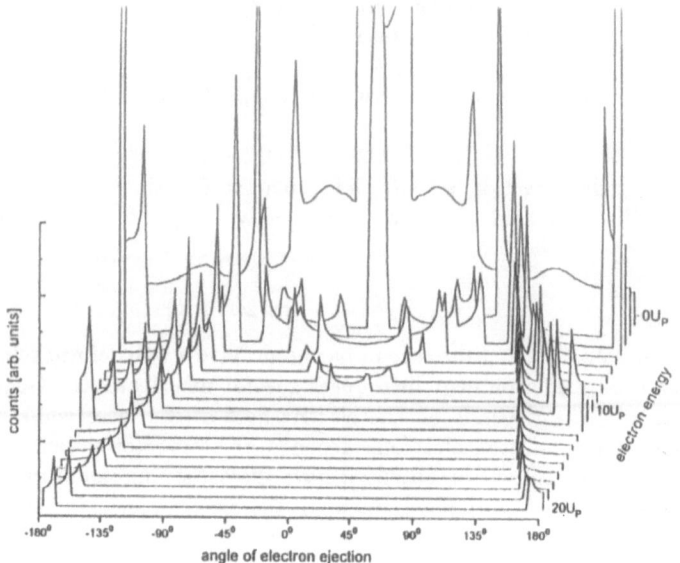

Figure 2. ATI angular distributions for fixed energy with respect to the observed angle θ for the field (27) with a relative phase $\phi = 2\pi/3$. The distribution ought to be symmetric about a scattering angle of $0°$. Deviations form this symmetry are due to the finite sampling size in the numerical procedure. The figure corresponds to the multiphoton regime where the electrons are injected uniformly in time. When the intensity increases into the tunneling regime, the part of the spectrum with scattering angles exceeding $130°$ gets more and more suppressed, first the high-energy series (which extends up to $21U_p$) and subsequently the lower-energy series (up to $12U_p$). The part of the spectrum for angles below about $90°$ is hardly affected.

References

[1] *Atoms in Intense Fields* ed. by M. Gavrila (1992), Advances in Atomic, Molecular, and Optical Physics, Supplement, Academic, London.

[2] *Super-Intense Laser-Atom Physics,* ed. by B. Piraux et al. (1993), NATO ASI Series B316, Plenum, New York.

[3] Becker, W., Long, S., and McIver, J.K. (1994), Phys. Rev. A50, 1540-60.

[4] Lohr, A., Becker, W., and Kleber, M., to be published.

[5] Becker, W., Lohr, A., and Kleber, M. (1995), Quantum Semiclass. Opt. 7, 423-48.

[6] Lewenstein, M., Balcou, Ph., Ivanov, M. Yu., L'Huillier, A., and Corkum, P. B. (1994), Phys. Rev. A49, 2117-32.

[7] Kulander, K. C., Schafer, K. J., and Krause, J. L. (1993), Ref.[2], pp.95-110.

[8] Corkum, P.,B. (1993), Phys. Rev. Lett. **71**, 1994-7.

[9] Gottlieb, B. (1995), Doctoral Dissertation, Technical University of Munich.

[10] Gottlieb, B., Kleber, M., and Krause, J. (1991), Z. Phys. A**339**, 201-9.

[11] Lang, N. D., Yacoby, A., and Imry, Y. (1989), Phys. Rev. Lett. **63**, 1499-1502.

[12] Fink, H.-W. (1988), Phys. Scr. **38**, 260-4.

[13] Dietrich, P., Burnett, N. H., Ivanov, M., and Corkum, P. B. (1994), Phys. Rev. A**50**, R3585-8.

[14] Walker, B., Sheehy, B., DiMauro, L. F., Agostini, P., Schafer, K. J., and Kulander, K. C. (1994), Phys. Rev. Lett. **73**, 1227-30.

[15] Kulander, K. C., Cooper, J., and Schafer, K. J. (1995), Phys. Rev. **51**, 561-8.

[16] Yang, B., Schafer K. J., Walker, B., Agostini, P., and DiMauro, L. F. (1993), Phys. Rev. Lett. **71**, 3770-3.

[17] Paulus, G. G., Nicklich, W., Xu H., Lambropoulos, P., and Walther, H. (1994), Phys. Rev. Lett. **72**, 2851-4.

[18] Paulus, G. G., Nicklich, W., and Walther, H., Europhys. Lett. **27**, 267-72.

[19] Becker, W., Lohr, A., and Kleber, M. (1994), J. Phys. B**27**, L325-32; **28**, 1931.

[20] Bunkin, F. V., and Fedorov, M. V. (1965), Zh. Eksp. Teor. Fiz. **49**, 1215-21 [Sov. Phys. JETP **22**, 844-7].

[21] Kroll, N. M., and Watson, K. M. (1973), Phys. Rev. A8, 804-9 .

[22] Paulus, G. G., Becker, W., Nicklich, W., and Walther, H. (1994), J. Phys. B**27**, L703-8.

[23] Paulus, G. G., Becker, W., and Walther, H., to be published.

[24] Weihe, F. A., Dutta, S. K., Korn, G., Du, D., Bucksbaum, P. H., and Shkolnikov, P. L. (1995), Phys. Rev. A**51**, R3433-6.

MPI OF ATOMS IN INTENSE LASER PULSES

Angular Distributions of Photoelectrons and Atoms Surviving in Excited States

J. GAUER and D.FELDMANN ,
Fakultät für Physik, Universität Bielefeld
Universitätsstrasse 25
D-33615 Bielefeld, Germany

1. Introduction

The energy spectra of photoelectrons from multiphoton ionization in intense short laser pulses show peaks which can be attributed to resonantly enhanced ionization processes [1,2]. These resonances are induced by AC-Stark shifts leading to enhanced ionization at certain intensities in the spatio-temporal volume of the focused laser pulse when the energy of an excited state is tuned into resonance with a multiple of the photon energy. The energy of these peaks $E_{kin}(e^-)$ within the lowest ATI-order (below the energy of one photon $h\nu$) is simply related to the intensity dependent binding energy [IP (I) - $E_{n,\ell}(I)$] of the resonantly participating state:

$$E_{kin}(e^-) = m \cdot h\nu - IP\ (I) - E_{n,\ell}(I)$$

where IP (I) denotes the intesity (I) dependent ionization energy and $E_{n,\ell}(I)$ the excitation energy of an state with quantum numbers n and ℓ. For the experimental conditions which will be considered below the number of photons m ,which connects the excited states with the lowest continuum, is one in most cases.

It is often possible to determine the principal quantum number of a resonant state by assuming that high lying Rydberg states exhibit an AC-Stark shift parallel to the shifted ionization limit.
Information about the angular momentum of such a state can to some extent be derived from the angular distribution of the energy selected photoelectrons.
This will be shown for several examples below.
A short remark will follow which adresses the question of atoms surviving intense short laser pulses in excited states.

123

H. G. Muller and M. V. Fedorov (eds.), Super-Intense Laser-Atom Physics IV, 123–1
© 1996 *Kluwer Academic Publishers.*

124

2. Angular Distributions of Photoelectrons

Angular distributions bare the signature of the partial waves by which the wavefunction of the outgoing electron can be described.

For linearly polarized light in dipol approximation and a single electron the well known selection rules : $\Delta\ell = \pm 1$ and $\Delta m = 0$ can be applied for each photon absorption step.

According to Fano's propensity rule ($\Delta\ell = +1$)-excitation transitions are usually stronger than those with ($\Delta\ell = -1$). And according to Bethe's rule this also holds for transitions in the continuum.

2.1 RESULTS OBTAINED WITH NS-LASER PULSES AT INTENSITIES BELOW 10^{13} W/CM^2

2.1.2 *Three-Photon Ionization Of Strontium*

One example is two-photon resonant ionization of Strontium atoms. Here one can obtain atomic structure information, when isolated resonances can be excited as intermediate states of an ionization process.. Figure 1 shows angular distributions of the photoelectrons from two-photon resonant three-photon ionization processes. By comparison of theoretical and experimental angular distributions, the configuration mixing of participating states could be determined [3]. Spin changing transitions in Strontium can easily be identified because the selection rule $\Delta m = 0$ is broken . An example is shown in figure 1 second graph.

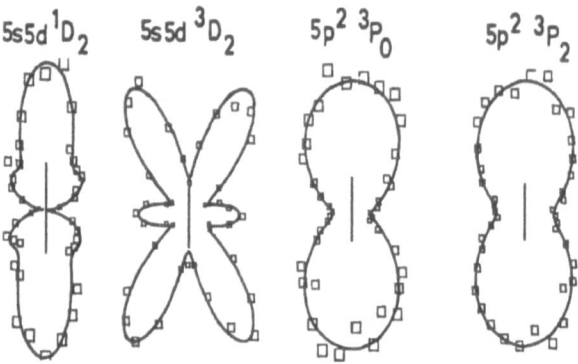

Figure 1. Angular distributions of photoelectrons from two-photon resonant three-photon ionization of strontium atoms.

2.1.2 *MPI of Atomic Hydrogen at 532 nm*

Other examples for which the results of perturbative calculations are realistic are the angular distributions of the ATI-peaks from MPI of atomic Hydrogen by long (ns)-laser pulses at wavelength of 532nm , which are shown in figure 2.

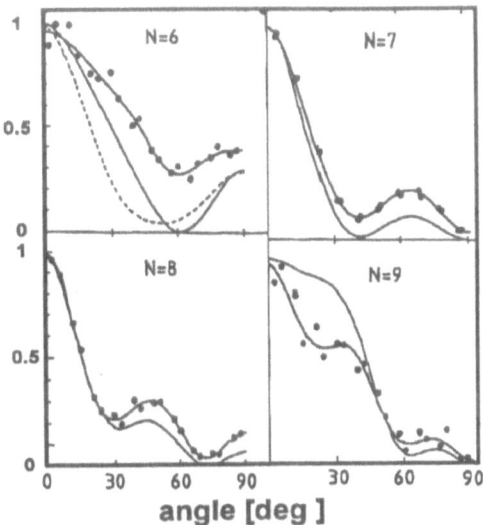

Figure 2.   Angular distributions of photoelectrons from MPI of atomic hydrogen at 532 nm for the four lowest ATI-orders after absorption of N=6, 7, 8 or 9 photons. The dots are experimental results and they are connected by a fit to a sum of Legendre polynomials. The other full line curves are the results of different perturbative calculations [4,5] which agree with each other.

The experimental distribution of the lowest ATI-peak (N = 6) is distorted by background signal and ponderomotive scattering, but never the less it exhibits significant d-wave character. The angular momentum quantum number of the outgoing electron-wave is only two after a six-photon absorption process.
This may be understood by the energy level diagram. The 2s-state at the 4-photon-level and the 3p-state at the 5-photon level seem to enhance the ionization process strongly, although they are not on exact resonance. However they are the only excited states of the proper parity which are available near these energies.--
The angular distributions of the four successive ATI-orders (figure 2) clearly show that Bethe`s rule of increasing angular momentum by photoexcitation in the continuum also holds for these multi-photon processes.

2.2. MPI IN INTENSE SHORT SUB-PS LASERPULSES

More complicated is the experimental situation for short, intense pulses, when resonances cannot be avoided [1,2] .

We shall here consider experimental results obtained at dye laser wavelengths around 600nm,with pulse intensities between about 10^{13} W/cm^2 up to some 10^{14} W/cm^2 and pulse durations of 0.4 ps.

Also under these conditions, angular distributions of photoelectrons can be used to get information about the resonantly participating states. However, these states are no longer bare atomic states. They can be strongly modified by the radiation field.

In many cases the binding energy of a resonantly participating excited state is less than the energy of one (or two) photon(s). Therefore, the corresponding lowest order ATI-peak from such a resonantly enhanced process should show an $(\ell +1)$ signature, when ℓ is the angular momentum of the resonant state.

Several examples show this behaviour for (nf)-states of Xenon and Hydrogen.[6,7]

2.2.1. *A change of the MPI-order identified by angular distributions*

By a change of the laser wavelength towards lower photon energies the order of the ionization process (the lowest number of photons neccessary for ionization) can increase. When Rydberg states serve as resonances which shift with intensity parallel to the ionization limit such an increase of the order of the ionization process cannot be seen as significant shifts of peaks in the spectrum. This can be seen in figure 3 which shows energy spectra of photoelectrons from Xenon at two different wavelengths.

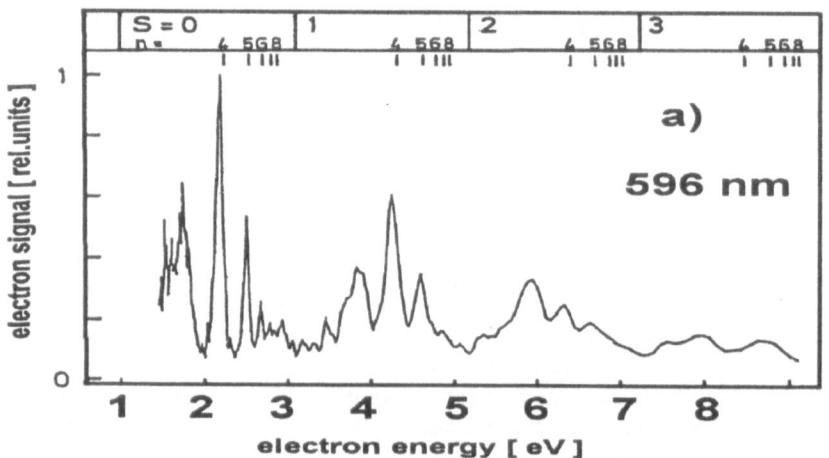

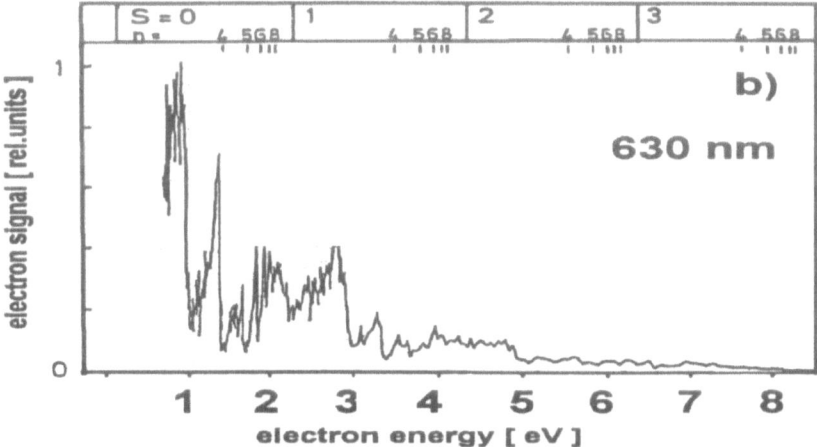

Figure 3. Energy spectra of photoelectrons from MPI of Xenon at two different wavelength. The expected peakpositions for resonant contributions by states with principal quantum numbers n are indicated on the upper scale, where also the ATI-orders are indicated by S-values.

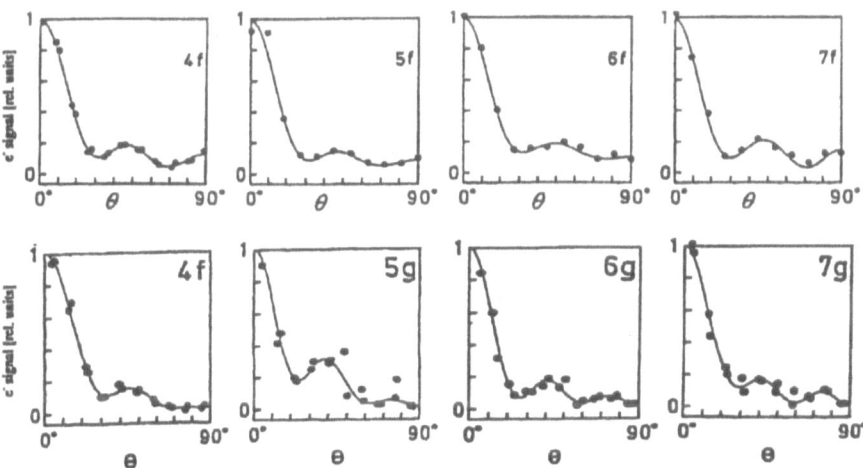

Figure 4. Angular distributions of photoelectrons within the first ATI-order (S = 0) corresponding to the peaks labeled by n in figure 3 for the same two wavelengths: upper part 596 nm, lower part 630 nm.

However, the angular distributions of the first ATI-order peaks shown in figure 4 change significantly with the wavelength.
At the shorter wavelength of 596nm the angular distributions of the outgoing electrons exhibit g-wave character which indicates that nf-states resonantly participate at an energy of one photon below the ionization limit. For the longer wavelength 630nm the typical small maxima at 90° have changed to minima, and the number of undulations indicates dominant h-wave contributions. This means that g-states have now taken over the role of the f-states at the longer wavelength.- Except for the 4f resonance whose angular distribution does not change.
From this change of angular distributions we can conclude an increase of the order of the resonant ionization process via states with principal quantum numbers n > 5 when the wavelength changes from nm to nm.

2.2.3. *Accumulation of angular momentum with increasing ATI-order*
Another fact, which we observe for many cases, is the preference of $(\Delta\ell = +1)$-transition in the continuum according to Bethe`s rule. Obviously this holds also for intensities up to 10^{14} W/cm^2 .Figure 5 shows an experimental result .The increase of angular momentum can be observed as an increase of the number of half-undulations in the angular distributions with ATI order of a series of peaks belonging to the same resonantly enhanced process. Because of limited angular resolution this trend can be followed in detail only for the first few ATI-orders.

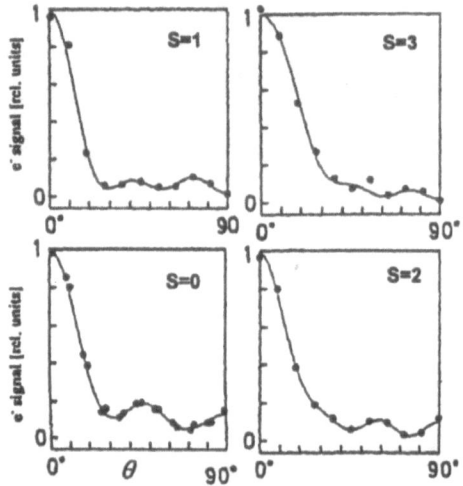

Figure 5. Angular distributions for successive ATI-orders of 4f-resonant ionization processes. S gives the number of excess photons absorbed in the continuum

2.2.4. *Peculiar behaviour of the " 4p "-level at higher intensities*

We have one case which looks complicated at a first view:

In the energy spectrum of photoelectrons from MPI of Hydrogen we have a small peak which can be attributed to a 4p-resonant ionization process by comparison with theoretical calculations. From this state the next photon leads into the continuum we therefore had expected to see a d-wave angular distribution. However the experimentally observed distribution shown in figure 6 has g-wave character like its much stronger neighbour, the 4f-resonant peak.

This observation was confirmed by a calculation [8].

This leads us to conclude, that this nominal 4p-state is no longer a p-state, when it is exposed to the radiation intensity at which it takes part in a resonantly enhanced ionization process leading to the peak in the energy spectrum. It seems to contain strong admixtures from f-states. The most probable mechanism seems to be 2-photon coupling with the neighbouring 4f- and 5f-states.

This is an application similar to the low intensity structur investigation of excited states of Strontium, however, here a change of the state-character is monitored which is induced by the strong radiation field.

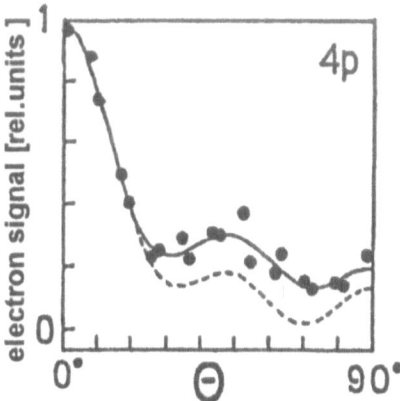

Figure 6. Angular distribution of photoelectrons from a "4p" resonant ionization prozess

2.3 ANGULAR DISTRIBUTIONS AS AN INDICATOR OF A NEW PROCESS

The extra lobes, showing up at high order ATI-processes are one example. The figure. 7 shows data for Xenon obtained at 1064nm with 10ns multimode YAG-laser

130

pulses. Much more detailed experiments have been performed by other groups [9]. Rescattering of an oscillating electron has been proposed to explain these lobes.

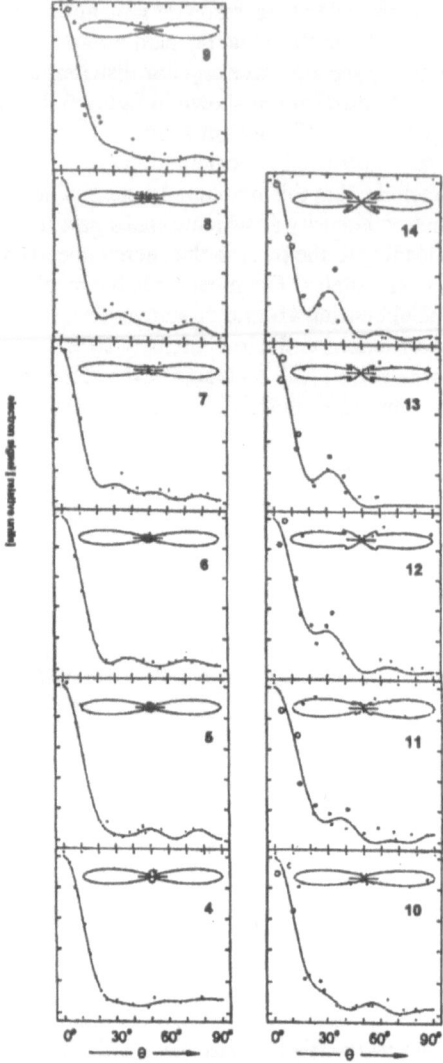

Figure 7.   Angular distributions of photoelectrons at higher ATI orders indicated by S from MPI of Xenon by long (10ns) pulses at a wavelength of 1064nm.

3. Atoms surviving a short, intense laser pulse in excited states:

In an experiment by de Boer et al. [10] it has been found that Xenon atoms can survive intense pulses of about 100fs duration in excited states. These excited atoms were detected by photo-ionization using a delayed longer laser pulse of a different colour.

A theoretical calculation [11] showed that this excited state population should be much lower when longer pulses of 500fs duration are applied,..-like those used in our experiments. Also Floquet type calculations [7] , give lifetimes below for lower nf-states of hydrogen at the intensity at which they resonantly participate in ionization at wavelengths around 600 nm.

We have tried to find a signal from atoms surviving in excited states using the same methode as the authors mentioned above [10] . The only differences were the pulse duration of our laser of 500 fs and the use of a simple TOF spectrometer instead of a magnetic bottle.

In a first attempt we did not find a reliable additional signal corresponding to the "post-ionized" excited atoms on the background of the spectrum originating from the strong short laser puls. The signal statistics would have allowed us to observe the expected extra peaks if 3% of the atoms had survived in excited states.

In a second attempt we improved the detectability of the photoelectrons from postionization by applying an electric extraction field during the short laser pulse and switched it off 20ns later and 10 ns before the ns-pulse for photoionization of excited atoms 30 ns was applied. Figure 8 shows the reults for Xenon and Hydrogen. This method does not allow a quantitative comparison between directly ionized atoms and those surviving in excited states. From our negative results of the first experiment we can only give an upper limit that less than 3% of the atoms survive in excited states.

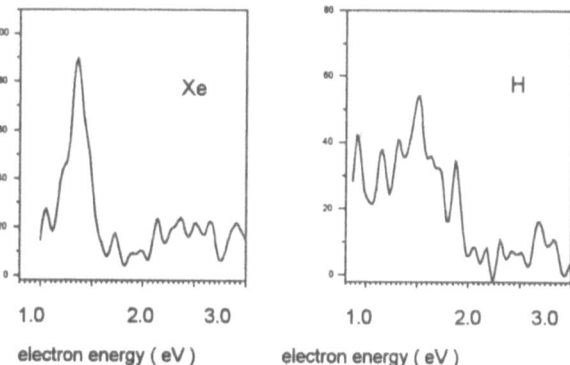

Figure 8. Energy spectra of photoelectrons from "postionization" by delayed ns-laser pulses for Xenon and Hydrogen. For details see text.

4. Conclusions

The examples described above show that angular distributions of energy selected photoelectrons give more insight into the atomic structure especially when the atom is exposed to a strong laser field. In several cases the states which are resonantly participating can be better characterized.

5. References:

1. Freeman, R. R., Bucksbaum, P. H., Milchberg, H., Darack, S., Schumacher, D. and Geusic, M.E. (1987) Above-threshold ionization with subpicosecond laser pulses, *Physical Review Letters* **59**, 1092.
2. Feldmann, D. (1990) Multiphoton ionization of atomic hydrogen in intense sub-ps laser pulses, *Comments on Atomic and Molecular Physics* **XXIV**, 311.
3. Agostini, P., Petite, G. and L'Hullier, A. (1988) Multiphoton spectroscopy of doubly excited, bound, and autoionizing states of strontium, *Physical Review A* **38**, 6165-79   and
   Lambropoulos, P., Tang, X., Feldmann, D. and Welge, K. H. unpublished
4. Gontier, Y., Rahman, N. K. and Trahin, M. (1988) Angular distribution of above threshold ionization of atomic hydrogen, *Europhysics Letters* **5**, 595.
5. Krack, G., Marxer, H., Broad, J. T. and Briggs, J. S. (1988) Multiphoton ionisation of atomic hydrogen and the angular distribution of photoelectrons, *Zeitschrift für Physik D* **8**, 103.
6. Rottke, H., Wolff, B., Tapernon, M., Welge, K. H. and Feldmann, D. (1990) Resonant multiphoton ionization of xenon in intense sub-ps-laser pulses, *Zeitschrift für Physik D* **15**, 133-139.
7. Rottke, H., Wolff-Rottke, B., Feldmann, D., Welge, K. H., Dörr, M., Potvliege, R. M. and Shakeshaft, R. (1994) Atomic hydrogen in a strong optical radiation field, *Physical Review A* **49**, 4837-51.
8. Gontier, Y. and Trahin, M. (1990) Angular distribution of photoelectrons from multi-resonant ionization of atomic hydrogen at 608 nm, *Journal of the Optical Society B* **7**, 463
9. Yang, B., Schafer, K. J., Walker, B., Kulander, K. C., Agostini, P. and DiMauro, L. F. (1993) Intensity dependent scattering rings in high order above-threshold ionization, *Physical Review Letters* **71**, 3770-3.
10. de Boer, M. P., and Muller, H. G. (1992) Observation of large Population in excited states after short-pulse multiphoton ionization, *Physical Review Letter* **18**, 2747-50.
11. Agostini, P. and DiMauro, L. (1993) Space localization and bound state population in short pulse resonant multiphoton ionization, *Physical Review A* **47**, R4573-6

STURMIAN-FLOQUET CALCULATIONS IN HYDROGEN

R. M. POTVLIEGE
Physics Department
University of Durham
Durham DH1 3LE, England

1. Introduction

The time-independent Floquet approach [1] is the natural framework for studying multiphoton processes that do not depend crucially on the temporal variation of the intensity of the incident laser field. Accurate *ab initio* Floquet calculations in complex atoms are now possible within this approach [2], although they are still difficult and very demanding in computer power owing to the necessity of taking electronic correlation into account. However, accurate calculations in atomic hydrogen have been possible for a number of years [3].

A large body of results have been collected for hydrogen, mostly through calculations employing complex Sturmian basis sets. The computation of level shifts and total rates of ionization on such bases are numerically stable for this atom, and in many cases can be performed on an ordinary workstation. Their main limitation, besides that inherent to any time-independent description, is the relatively small number of coupled partial waves (typically less than 20) that it is possible to retain in the expansion of the wave function in spherical harmonics without having to deal with excessively large matrices. The method is therefore not practical for tackling, e.g., ionization by ultra-intense fields in the visible or the infrared. Nonetheless, it permits a detailed, quantitative study of many processes of interest, in a large range of frequency and intensity. Progress has been made along this line of research since Sturmian-Floquet calculations were last surveyed in some detail [4], in particular in regards to adiabatic stabilization [5-6], the appearance of light-induced states [7], ionization in the tunneling regime [8], ionization by coherent superpositions of two harmonics of a same laser beam [9-12], and microwave ionization of Rydberg states [13]. This paper is a mini-review of some of the more significant recent results. It is mainly concerned with "high-frequency" phenomena, but results in the multiphoton regime are also described in order to illustrate the scope of the Sturmian-Floquet approach. The method is sketched in Section 2, omitting many relevant details; a more complete account is given in Ref. [4].

H. G. Muller and M. V. Fedorov (eds.), Super-Intense Laser-Atom Physics IV, 133–142.
© 1996 Kluwer Academic Publishers.

2. The Sturmian-Floquet approach

We consider an atom of hydrogen or a one-electron ion, with reduced mass μ, undergoing multiphoton ionization under the effect of a classical, single-mode monochromatic field, homogeneous over atomic dimensions, with polarization vector e, electric field vector $\mathbf{F}(t)$ and vector potential $\mathbf{A}(t)$:

$$\mathbf{F}(t) = F_0 \, \text{Re}[\mathbf{e} \exp(-i\omega t)] \qquad \mathbf{A}(t) = (c/\omega) \, F_0 \, \text{Im}[\mathbf{e} \exp(-i\omega t)]. \tag{1}$$

We can pass to the time-independent framework by seeking solutions to the time-dependent Schrödinger Equation in the form of a product of a secular phase factor and a function periodic in time with same period as the incident field,

$$\Psi(\mathbf{r}, t) = e^{-iEt/\hbar} \sum_{N=-\infty}^{\infty} e^{-iN\omega t} \psi_N(\mathbf{r}). \tag{2}$$

This reduces the Schrödinger Equation to the following system of coupled time-independent equations (H_0 denotes the field-free atomic Hamiltonian):

$$[E + N\hbar\omega - H_0]\psi_N = V_+\psi_{N-1} + V_-\psi_{N+1}. \tag{3}$$

The "velocity gauge", where the interaction of the atom with the field is

$$V(t) = (-e/\mu c)\mathbf{A}(t) \cdot \mathbf{p} \equiv V_+e^{-i\omega t} + V_-e^{i\omega t}, \tag{4}$$

is the appropriate gauge for carrying out the calculations described below [14]. No term in A_0^2 appears in the coupling, because this term can be removed exactly by a gauge transformation which does not affect the value of any observable physical quantity as long as the field can be considered as being spatially homogeneous. Ponderomotive acceleration of the photoelectrons, originating from the macroscopic variation of F_0, can be taken into account classically in a separate calculation.

The relevant physical solutions of the system (3) are regular at $r = 0$ and behave for $r \sim \infty$ as a superposition of outgoing waves in the open channels and damped ingoing waves in the closed channels:

$$\psi_N(\mathbf{r}) \sim \sum_M f_{MN}(\widehat{\mathbf{r}}) \, r^{iZ/k_M} \, e^{ik_M r}/r, \tag{5}$$

with $k_M = [(2\mu/\hbar^2)(E + M\hbar\omega)]^{1/2}$. Thus the quasienergy $E \equiv E_0 + \Delta - i\Gamma/2$ is a complex eigenvalue which depends on the parameters of the incident field. The real part of E is the field free energy of the initial state, E_0, shifted by the ac-Stark shift Δ. $\Gamma/\hbar$ is the total ionization rate, integrated over all directions of the emergent photoelectron and summed over all channels.

The system (3), truncated to a finite number of equations, can be solved by expanding the harmonic components $\psi_N(\mathbf{r})$ on a discrete basis set consisting of spherical harmonics and complex radial "Sturmian" functions $S_{nl}^\kappa(r)$:

$$S_{nl}^\kappa(r) = N_{nl}^\kappa \, r^{l+1} e^{i\kappa r} L_{n-l-1}^{2l+1}(-2i\kappa r) \tag{6}$$

The parameter κ may be complex and N_{nl}^κ is an unimportant normalization factor. (Expanding on complex Sturmians is equivalent to transforming the equations by complex scaling and expanding the wave functions on a real basis.) Projecting Eq. (3) onto that basis yields a large, sparse, complex symmetric matrix eigenvalue equation for the quasienergy. The boundary conditions (5) are implemented implicitly through the choice of $\arg(\kappa)$. Indeed, any well-behaved function of r which vanishes as r^{l+1} for $r \sim 0$ and behaves as $r^\nu \exp(ikr)$ for $r \sim \infty$ can be expanded in terms of the S_{nl}^κ with expansion coefficients that vanish for $n \sim \infty$ provided that $|\arg(\kappa) - \arg(k)| < \pi/2$. Thus we choose $\arg(\kappa)$ somewhere in the range where $|\arg(\kappa) - \arg(k_M)| < \pi/2$ for all channels. The quasienergies correspond to eigenvalues that are stable with respect to sensible changes in the basis set, and in particular to changes in $\arg(\kappa)$. In practice, good stability in the total ionization rate can often be achieved with no more than about 30 radial functions in each partial wave; an example relevant to Section 3 is given in Fig. 1. The Sturmian basis is equally suitable for dealing with screened Coulomb potentials of the form $r^m \exp(-r/a)$, $m \geq -1$, in one-electron models of complex atoms.

The amplitudes f_{MN} can be extracted from the harmonic components by integration over space, in analogy with the familiar integral representation of the scattering amplitude for scattering by a potential. This yields partial rates for ionization into specific ATI channels and angular distributions. Although it is facilitated by the convergence of the coefficients of the Sturmian expansion which was mentioned above, the calculation often requires much larger basis sets than for obtaining converged total ionization rates, and is therefore expensive in computer time.

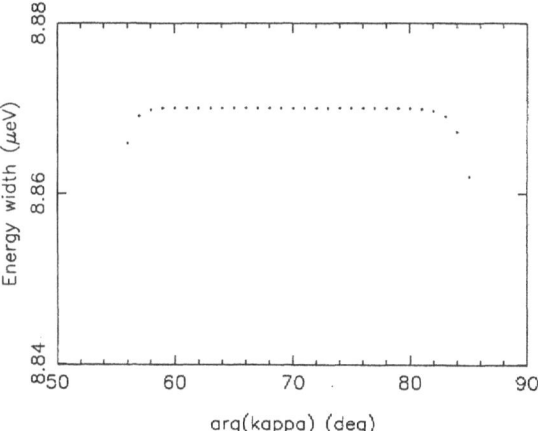

Figure 1. Energy width of the light-induced state of Fig. 5 as calculated with a set of 30 Sturmian functions in each partial waves, vs the the argument of the parameter κ. The wavelength is 19.5 μm, the intensity is 10^{10} W/cm^2, and $\mu = 1$ a.u. 23 harmonic components were retained in Eq. (2), each including 6 partial waves. $|\kappa^2|/2 = 0.005$ a.u.

A large series of differential rates have been computed in this way, for a few wavelengths between 596 nm and 630 nm and many different intensities, in order to test the predictions of the time-independent theory against an extensive set of measurements made in Bielefeld a few years ago. A detailed comparison of theory and experiment has been reported in Ref. [15]. Fig. 2 shows a representative photoelectron energy spectrum. All the main resonance peaks seen in the experiment can be assigned to particular dressed states with known zero-field limits. Although the Sturmian-Floquet results generally are in rather good agreement with the data, they fail to reproduce the observed branching ratios between the ATI peaks. (Compare for example the magnitude of the peaks at 1.2 eV and 3.2 eV, both corresponding to the 7-photon resonance $1s$-$4f$.) This discrepancy might be due to transfer of population between resonant dressed states, which was neglected in the calculations. The theory assumed that ionization proceeded from a single dressed state at all intensities, with the system following a diabatic curve in the quasienergy diagram, on account of the narrowness of the energy gaps between resonant states at avoided crossings. Clearly, a time-dependent calculation of the branching ratios for a pulse duration comparable to that in the experiment would be of great interest [16].

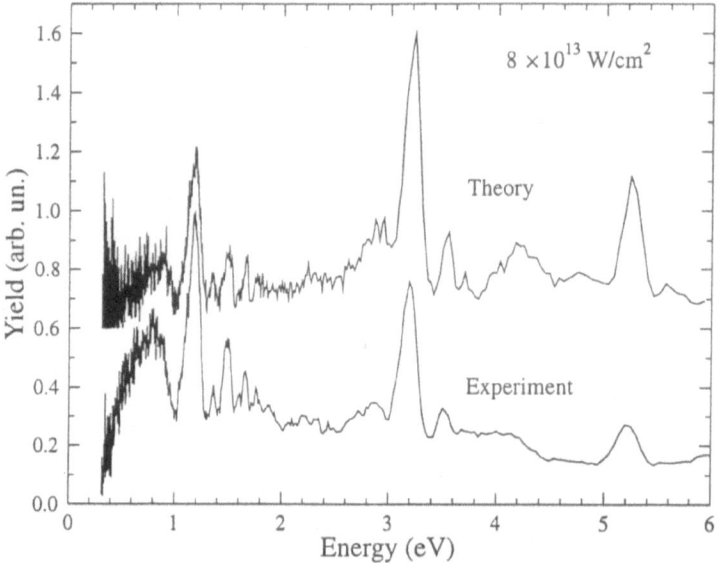

Figure 2. Energy spectrum of photoelectrons for ionization of H($1s$) by 0.5 ps pulses of wavelength 608 nm and peak intensity 8×10^{13} W/cm^2 [15]. The theoretical curve has been shifted upwards by 6 units. Spatio-temporal distribution of intensity, depletion of the target, ponderomotive acceleration of the photoelectrons, laser fluctuations, and detector binning and angular acceptance are taken into account in the theoretical results.

3. Two-colour multiphoton ionization

The Sturmian-Floquet theory can be extended immediately to the interesting situation where the atom is irradiated simultaneously by two harmonics of the same laser field. There is an important difference between this case and that of a superposition of two fields with incommensurable frequencies, or generally that of two fields acting incoherently. When the two frequencies are in a rational ratio, the initial state of the atom may be coupled to a given final continuum state in several ways which differ by the net number of low-frequency and high-frequency photons that are absorbed during the reaction. The different pathways interfere if the superposition is coherent, which makes the ionization rate dependent on the relative phase of the two fields. The phase dependence is insignificant, though, if the order of the harmonics are such that the interference requires the exchange of a large number of photons.

A sample of Sturmian-Floquet results is displayed in Fig. 3, for a coherent superposition of the first (fundamental) and third harmonics of a linearly polarized field. The total electric field is $\mathbf{F}(t) = [F_1 \cos(\omega t) + F_3 \cos(3\omega t + \phi)]\mathbf{e}$. In general, but not always, the total ionization rate is enhanced dramatically by even a small admixture of the harmonic field, the rate being much larger than the arithmetic sum of the rates of ionization by each field acting independently. As seen from the figure, the phase dependence can be particularly strong when the high frequency field is relatively weak, and can result in a significant *decrease* in the ionization rate. The decrease is in agreement, qualitative and even quantitative for weak harmonic intensity, with the prediction of the two-colour tunneling theory of Perelomov and Popov [17]. Phase-dependent interferences affect the multiphoton ionization a great deal more in the case of Fig. 3 than for other ratios of frequencies [12].

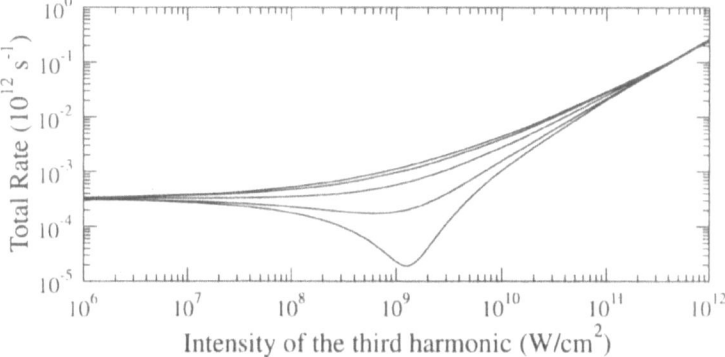

Figure 3. Rate of ionization of $H(1s)$ irradiated by a superposition of a field of wavelength 616 nm and intensity 1×10^{13} W/cm^2 and its third harmonic, vs the intensity of the latter [9,12]. From top to bottom, $\phi = 0$, 45 deg, 90 deg, 135 deg, and 180 deg.

4. Light-induced states

Because of the ac-Stark shift, the real part of the quasienergy varies with the intensity. It may happen that it becomes an integral multiple of the photon energy at some intensity I_{thr}: the system encounters a multiphoton threshold at I_{thr}. A better understanding of what happens at this point can be gained by allowing the wave function to take on an unphysical behaviour at large distance — much in the same way that the mathematical properties of functions of a real variable are illuminated by allowing the variable to take complex values — and to consider physical and unphysical solutions of the Floquet system on the same footing. A multiphoton channel either closes or opens as the intensity sweeps through I_{thr}. Thus at I_{thr} one of the outgoing waves in Eq. (5) should be swapped with an ingoing wave for the wave function to describe a physical state on both sides of the threshold. However, the wave function does not vary discontinuously with the intensity — its outgoing wave components remain outgoing across I_{thr}, and its ingoing wave components remain ingoing. Therefore the wave function has an unphysical character on (at least) one side of I_{thr}. (Where unphysical, the wave function is associated with a "shadow pole" of the S-matrix for laser-assisted scattering [7].) In most cases, when a physical state becomes unphysical an unphysical state becomes physical at about the same intensity. Dressed bound states shifting in the field can be followed as a succession of particular solutions of Eq. (3) replacing one another in this fashion at every multiphoton threshold.

However, some physical Floquet states, similar in every other respects to usual dressed states, do not correspond to any physically realizable discrete states below a certain threshold intensity. The appearance of such light-induced states is illustrated in Fig. 4, for a one-dimensional potential well supporting a single bound state in the absence of field [7]. Its binding energy $|E_0|$ is such that one-photon detachment is possible only if the wavelength is shorter than about 900 nm. In Fig. 4(a) the real part of the quasienergy follows the ground state energy of the "dressed potential" (the static, cycle-averaged part of the interaction in the oscillating Kramers-Henneberger frame) at 266 nm and 532 nm — the shorter the wavelength the closer the agreement, as should be expected. Provided the intensity is high enough, the quasienergy can be followed continuously while increasing the wavelength from 532 nm to above 900 nm; however, the Floquet state ceases to have a physical zero-field limit when the photon energy becomes smaller than $|E_0|$. In Fig. 4(a) a light-induced state associated with the ground state appears at a quiver amplitude $\alpha_0 \approx 2$ a.u. at 1064 nm and 3 a.u. at 2128 nm. (The relative closeness of the curves for these two wavelengths may surprise, since α_0 is not expected to be a relevant dynamical parameter outside the high-frequency regime.) As shown in Fig. 4(b), light-induced states associated with the excited states of the dressed potential are also found in Floquet calculations, even at long wavelength. Those in fact are associated with shifted resonance poles or antibound state poles of the system, not with shadow bound state poles.

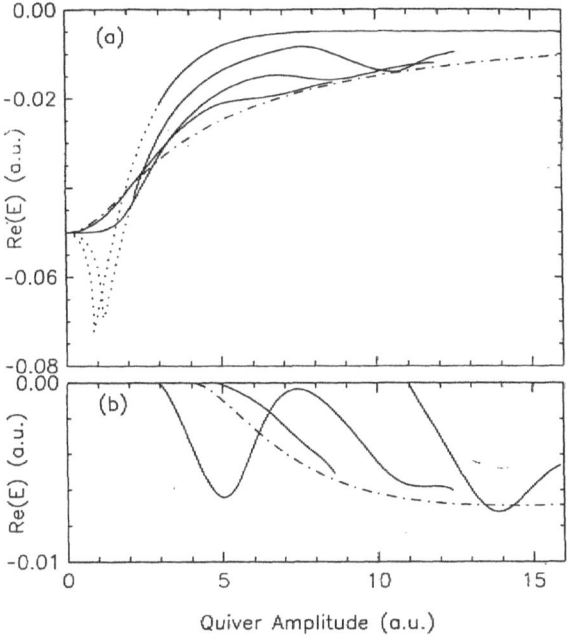

Figure 4. Appearance of light-induced states in photodetachment of an electron from a 1-d square potential well for various wavelengths λ of the incident field [7]. (a) The real part of the quasienergy, in atomic units, vs the quiver amplitude $\alpha_0 = eF_0/(\mu\omega^2)$. The curve is dotted where the associated wave function has unphysical asymptotic behaviour. From top to bottom at $\alpha_0 = 5$ atomic units, $\lambda = 2128$ nm, 1064 nm, 532 nm, and 266 nm; the chained curve represents the ground state energy of the dressed potential. (b) Same as (a) but for light-induced states corresponding to the first excited state supported by the dressed potential. From left to right at Re$(E) = 0$, $\lambda = 532$, 266, and 1064 nm.

Recent high-frequency calculations have raised the intriguing possibility of the appearance in H$^-$ of light-induced states similar to those of Fig. 4(b) [18]. It is clear that they have no counterpart in hydrogen as the Coulomb potential does not support resonance or antibound states. However, light-induced states akin to those of Fig. 4(a) are found in Sturmian-Floquet calculations in hydrogen, at moderate and high intensities, over a wide range of frequencies. Those associated to low-lying states have a very large decay width (typically several eV near their appearance intensity for those associated with the 1s state). This makes it unlikely they would be observable in experiments. In contrast, light induced states with a sufficiently large angular momentum have a very narrow width and are well separated in energy both from the threshold and from the adjacent states (Fig. 5).

140

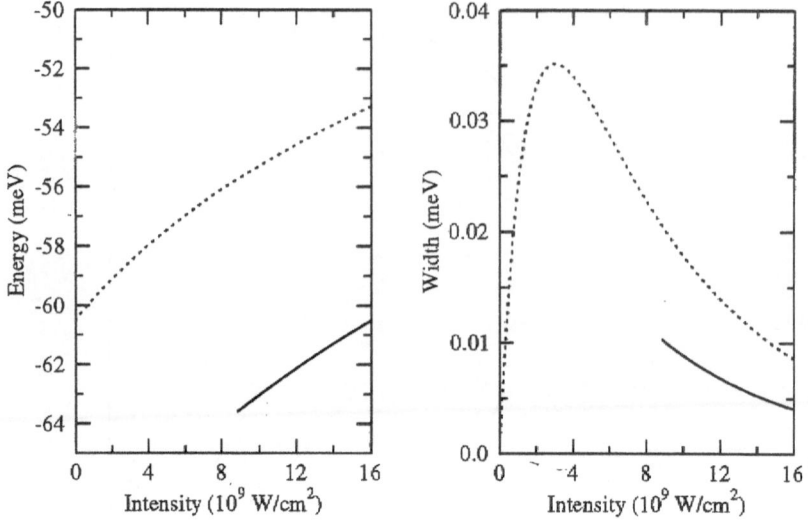

Figure 5. Appearance of a light-induced state in hydrogen irradiated by a linearly polarized field of 19.5 μm wavelength corresponding to the Rydberg state with n=14, l=m=13 (solid curves). Dotted curves: results for the state with n=15, l=m=13.

5. Adiabatic stabilization

Adiabatic stabilization of the "circular" Rydberg state with principal quantum number $n = 15$ is manifest in Fig. 5. The rate of ionization from that state first increases as the intensity increases, saturates at about 3×10^9 W/cm^2 and then decreases monotonically. The rate of ionization is the same at 1.6×10^{10} W/cm^2 than at an intensity 50 times smaller. In Floquet calculations for linear polarization, adiabatic stabilization is found for any state which, like this circular state, is not resonantly coupled to lower manifolds and can decay by one-photon ionization. The effect is only a theoretical curiosity for low lying states, owing to their extremely short lifetime in the field at the onset of stabilization, but it is quite real, and indeed observable [19], for highly excited states whose decay rate remains small at all intensities. A systematic study of adiabatic stabilization in states with large angular momentum [5] has yielded an "empirical" formula which, to a good approximation, gives the intensity I_{min} where the lifetime of the atom reaches a minimum: For linear polarization, $I_{min} \approx m(l + 1 - m)! \, \omega^3 \, I_0$, with ω expressed in atomic units and $I_0 = 3.51 \times 10^{16}$ W/cm^2. For example, this formula predicts that in Fig. 6, the half-life should be a minimum at 1.4×10^{13} W/cm^2, while the actual minimum occurs ar 1.25×10^{13} W/cm^2. Remarkably, only the angular and magnetic quantum numbers of the initial state appear in the formula, not the principal quantum number. The formula is not accurate, however, if the photon energy is too close to $|E_0|$. The ratio $\hbar\omega/|E_0|$ is 4.2 in the case of Fig. 6. The ratio is only 1.05 in the case of Fig. 5, for which the empirical formula predicts a value of $I_{\min}$ that is twice too large.

The way $I_{\min}$ scales with the frequency is easy to understand by a dimensional argument: in the high frequency regime, the characteristic energies in the problem are the ponderomotive energy and the photon energy, and their ratio is proportional to the intensity divided by ω^3. The dependence of $I_{\min}$ in l and m is not trivial. That $I_{\min}$ increases with m and $l-m$ supports Pont's and Shakeshaft's interpretation of adiabatic stabilization as a kinematical effect originating from the reduction of the angular phases space into which the photoelectron can be ejected readily [20].

Much less effort has been devoted so far to the case of circular polarization. At high frequency the ionization rates of H(1s) and H(2s) decrease in strong field in a way similar to linear polarization. However, the evidence is inconclusive for high-lying circular states. Existing Sturmian-Floquet calculations indicate that their ionization rate may stop decreasing at ultra-high intensity; however, these calculations may not be fully converged. In any case, the behaviour of these Rydberg states in circular polarization is quite different than in linear polarization: their ionization rate peaks at a much higher intensity and their ac-Stark shift is opposite in sign (Fig. 6).

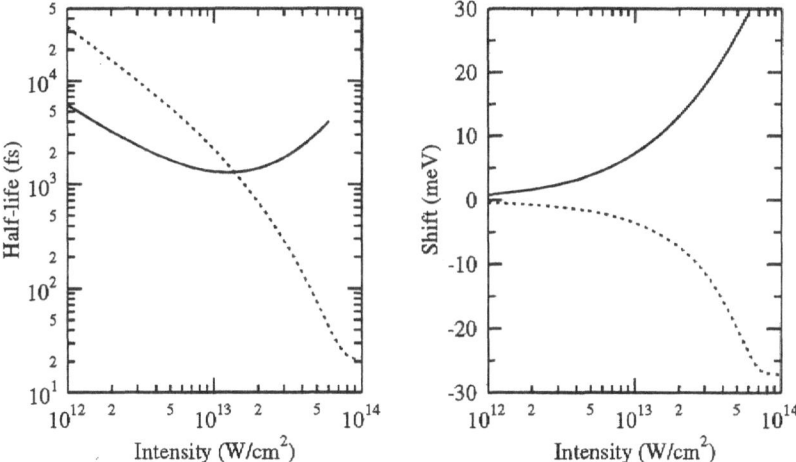

Figure 6. Half-life ($\hbar \log 2/\Gamma$) and ac Stark shift (measured from the continuum threshold) of H($n=7, l=m=5$) in a field of wavelength 1064 nm, either linearly polarized (solid curves) or circularly polarized (dotted curves), vs the intensity of the field.

142

6. References

1. Shirley, J.H. (1965) *Phys. Rev.* **138**, B979; Ritus, V.I. (1966) *Zh. Eksp. Teor. Fiz.* **51** 1544 [*Sov. Phys. — JETP* **24** 1041 (1967)]; Zel'dovich, Ya B. (1966) *Zh. Eksp. Teor. Fiz.* **51**, 1492 [*Sov. Phys. — JETP* **24**, 1006 (1967)].
2. See e.g. the paper by M. Dörr in this volume.
3. Chu, S.-I and Reinhardt, W.P. (1977) *Phys. Rev. Lett.* **39**, 1195; Maquet, A., Chu, S.-I and Reinhardt, W.P. (1983). *Phys. Rev. A* **27**, 2946.
4. Potvliege, R. and Shakeshaft, R. (1992) *Adv. At. Mol. Phys. Suppl.* **1**, 373.
5. Potvliege, R. and Smith, P.H.G. (1993) *Phys. Rev. A* **48**, R46.
6. Buchleitner, A. and Delande, D. (1993) *Phys. Rev. Lett.* **71**, 3633.
7. Fearnside, A.S., Potvliege, R.M. and Shakeshaft, R. (1995) *Phys. Rev. A* **51**, 1471.
8. Pont, M., Potvliege, R.M., Shakeshaft, R. and Teng, Z.-j. (1992) *Phys. Rev. A* **45**, 8235.
9. Potvliege, R.M. and Smith, P.H.G. (1991) *J. Phys. B* **25**, L641.
10. Potvliege, R.M. and Smith, P.H.G. (1992) *J. Phys. B* **25**, 2501.
11. Pont, M., Potvliege, R.M., Shakeshaft, R. and Smith, P.H.G. (1992) *Phys. Rev. A* **46**, 555.
12. Potvliege, R.M. and Smith, P.H.G. (1994) *Phys. Rev. A* **49**, 3110.
13. Buchleitner, A., Delande, D. and Gay, J.-C. (1995) *J. Opt. Soc. Am. B* **12**, 505.
14. Howland, J.S. (1980) in J.A. DeSanto, A.W. Sáenz and W.W. Zachary (eds.), *Mathematical Methods and Applications of Scattering Theory*, Springer-Verlag, Berlin, pp. 163-168.
15. Rottke, H., Wolff-Rottke, B., Feldmann, D., Welge, K.H., Dörr, M., Potvliege, R.M. and Shakeshaft, R. (1994) *Phys. Rev. A* **49**, 4837.
16. The temporal variation of the intensity in the Bielefeld experiment has been taken into account in calculations of the resonance structures within the first ATI peak [Gontier, Y. and Trahin, M. (1992) *Phys. Rev. A* **46**, 1488]. The first ATI peak is taller than the second peak, in agreement with the data, in energy spectra recently obtained by solving the time-dependent Schrödinger Equation at 608 nm [private communication with B. Piraux]; however, the pulse duration in the calculation (a few tens of optical cycles) was much shorter than in the experiment.
17. Perelomov, A.M. and Popov, V. (1967) *Zh. Eksp. Teor. Fiz.* **52** 514 [*Sov. Phys. — JETP* **25** 336)].
18. Muller, H.G. and Gavrila, M. (1993) *Phys. Rev. Lett.* **71**, 1693.
19. Deboer, M.P, Hoogenraad, J.H., Vrijen, R.B., Noordam, L.D. and Muller, H.G. (1993) *Phys. Rev. Lett.* **71**, 3263; Deboer, M.P, Hoogenraad, J.H., Vrijen, R.B., Constantinescu, R.C., Noordam, L.D. and Muller, H.G. (1994) *Phys. Rev. A* **50**, 4085.
20. Pont, M. and Shakeshaft, R. (1991) *Phys. Rev. A* **44**, R4110; Shakeshaft, R. (1992) *Comm. At. Mol. Phys.* **28**, 179 (1992).

ELECTRONS IN STRONG FIELD IONIZATION

H.R. REISS

The American University

Washington, DC 20016-8058, USA

1. Introduction

From all of the effort devoted in recent years to the nonperturbative calculation of strong field ionization of atoms, there have emerged three approaches which have had practical success in application to current laboratory environments. The first of these, and the simplest to employ, is the tunneling technique. Then there is the direct numerical solution of the Schrödinger equation with a high-speed computer. Finally, there is the amalgam of tunneling and multiphoton methods called the SFA (Strong Field Approximation). Other methods, such as those based on the Floquet property of periodic potentials, and the HFA (High Frequency Approximation), have important practical limitations. Each of these methods has strengths and limitations which are briefly discussed below. Then the foundations of the SFA are reviewed, and some of its recent successes in practical application are exhibited. In particular, it appears to be capable of reproducing experimental photoelectron spectra up to very high energies, and it has some ability to exhibit the effects of the revisiting of the atom by the photoelectron after ionization.

2. Overview of Theoretical Methods

2.1. TUNNELING

Tunneling methods were the first to be applied to strong field ionization[1, 2, 3], and they remain of great practical importance[4]. The tunneling description of ionization has the advantage of conceptual and structural simplicity, and the ability to describe ion yields in strong field experiments over a wide range of intensities employed in the laboratory. However, tun-

H. G. Muller and M. V. Fedorov (eds.), Super-Intense Laser-Atom Physics IV, 143–152.
© *1996 Kluwer Academic Publishers.*

neling descriptions are limited to low frequency fields, and are subject to both lower and upper limits on the intensities for which they are valid. The lower limit comes from the requirement that the laser field effects should dominate the atomic binding, which can be stated as $U_p \gg E_B$, where U_p is the ponderomotive energy of the detached electron in the laser field, and E_B is the electron binding energy in the initial atomic state. This limit is often stated as $\gamma \ll 1$, where γ is the Keldysh parameter, which can be written as $\gamma = (E_B/2U_p)^{1/2}$. Less well known is the upper limit on intensity, coming from the fact that ionization must not be "over the top" of the barrier. This can be stated as $F \ll (E_B/2)^2$, where F is the amplitude of the laser electric field, or alternatively as $F_{laser} \ll F_{atomic}$. In practical terms, this last limit means that tunneling methods are limited to a few times $10^{15}\, W/cm^2$.

2.2. FLOQUET METHODS

Floquet methods applied to strong field ionization, as exemplified by the work of Shakeshaft and coworkers[5], has the feature that it is very computer intensive. This has limited its practical applicability mostly to high frequencies. Current laser experiments are essentially all at low frequencies, except for a limited class employing Rydberg states.

2.3. HIGH FREQUENCY APPROXIMATION

The HFA, associated mostly with Gavrila[6], makes use of a Kramers-Henneberger transformation to a frame oscillating with the atomic electron in the field, followed by the assumption of high frequency to permit averaging over the atomic potential. It is difficult to appraise the effect of this averaging procedure, other than to note the limitation to high frequency.

2.4. NUMERICAL INTEGRATION OF THE SCHRÖDINGER EQUATION

This approach, developed most notably by Kulander and his associates[7], would appear to provide a definitive solution to the difficulties associated with strong field theories, and it has, in fact, proven to be very powerful. Nevertheless, fundamental computer limitations curtail its application to low frequencies at very high intensity. For example, direct computer solutions cannot deal with the high energy photoelectron spectra to be discussed below. Relativistic applications have not yet been attempted, since strong fields make "large-component, small-component" relativistic approximations impossible[8].

2.5. STRONG FIELD APPROXIMATION

Though the SFA was originally designed for short range potentials[9], it was realized that the method could be extended to atomic ionization as long as the laser field dominated atomic binding, or $U_p \gg E_B$. The method is based on an initially exact transition amplitude in which the approximation is made that the final detached state of the electron has its behavior dominated by the laser field. For circular polarization, strong-field final state corrections have been found[10], so that the method apparently gives completely reliable results for this case. For linear polarization, accuracy assessments are not yet clear, but excellent results will be shown below.

In a dichotomy reminiscent of the wave-particle duality in quantum mechanics, the SFA shows tunneling properties in some applications, and multiphoton properties in others. There appear to be no frequency limitations, and agreement with high frequency calculations of Gavrila is excellent[11]. Low frequency results have long been known to be in good agreement with experiment[12]. The only requirement is for strong fields.

3. Brief SFA Fundamentals

Comments in the literature and in referee reports make it clear that misperceptions about the SFA are still widespread after fifteen years. The most common of these will be remarked upon.

3.1. OVERLAP ONTO A NON-INTERACTING STATE

All present strong-field laser experiments make use of pulsed lasers. This means that measurements of the products of the interaction are made in a spatial and temporal environment in which the laser field is not present. Thus, it is appropriate to impose boundary conditions on the transition amplitude in which overlaps are made onto reference states free of the laser interaction. An exact transition amplitude can then be written in either of the two equivalent forms

$$
\begin{aligned}
(S-1)_{fi} &= -i \int_{-\infty}^{\infty} dt\, (\Phi_f, H_I \Psi_i) \\
&= -i \int_{-\infty}^{\infty} dt\, (\Psi_f, H_I \Phi_i),
\end{aligned}
\tag{1}
$$

where Φ is a "reference" state satisfying an equation of motion containing the atomic potential, but lacking the laser interaction terms H_I, and Ψ satisfies the complete Schrödinger equation including all atomic and laser contributions. The second form of the transition amplitude follows from time reversal invariance of the Schrödinger equation. Specifically, for the

146

boundary conditions which obtain in pulsed laser experiments, it is not correct to require both initial and final states in the above transition amplitudes to be fully interacting states.

3.2. INITIAL STATE IN THE SFA

The SFA is based on the second of the two equivalent expressions above, and employs the approach where $H_I = -\mathbf{A} \cdot \mathbf{p} + A^2/2$. It thus requires that the initial state be free of the laser field. This is a great strength of the SFA, in that it permits an essentially exact treatment of the initial bound atomic state, regardless of the field intensity. A large number of commentators have inappropriately regarded the lack of interaction in the initial state as a shortcoming rather than an advantage of the SFA formalism.

3.3. MISCELLANEOUS PROPERTIES

The SFA, although it 'is related to the tunneling method of Keldysh[1], has no tunneling restriction inherent in the formalism. It is therefore free of the

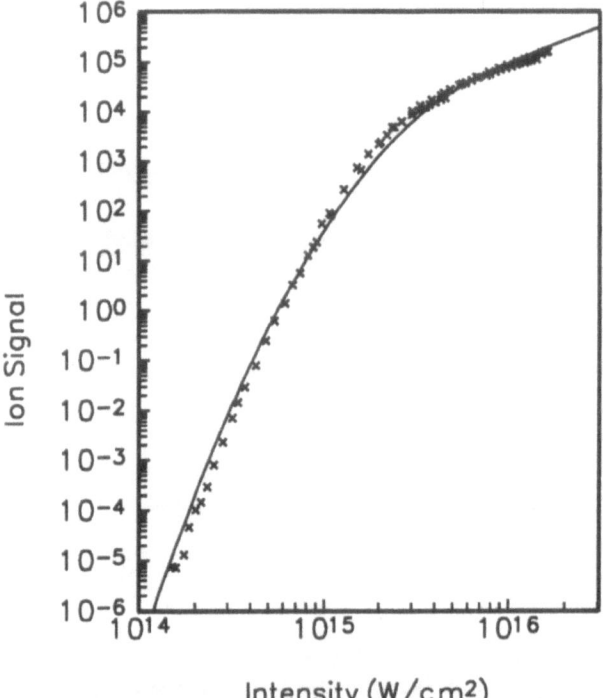

Figure 1. Application of the SFA to single ionization of helium.
upper limit on intensities that tunneling implies, and can be extended into

the stabilization regime without difficulty[13].

We also remark that the S-matrix genesis of the SFA means that all transition influences are included that occur between preparation of the experimental state and the final observation of the products of the interaction. This includes oscillations of the photoelectron in the laser field after ionization, subject to the SFA approximation that Coulomb effects are neglected in the final state. Thus the SFA includes "revisiting" effects in the final state, but not rescattering. The revisiting is structurally inherent in the SFA because the final state is a Volkov state. An example will be shown below.

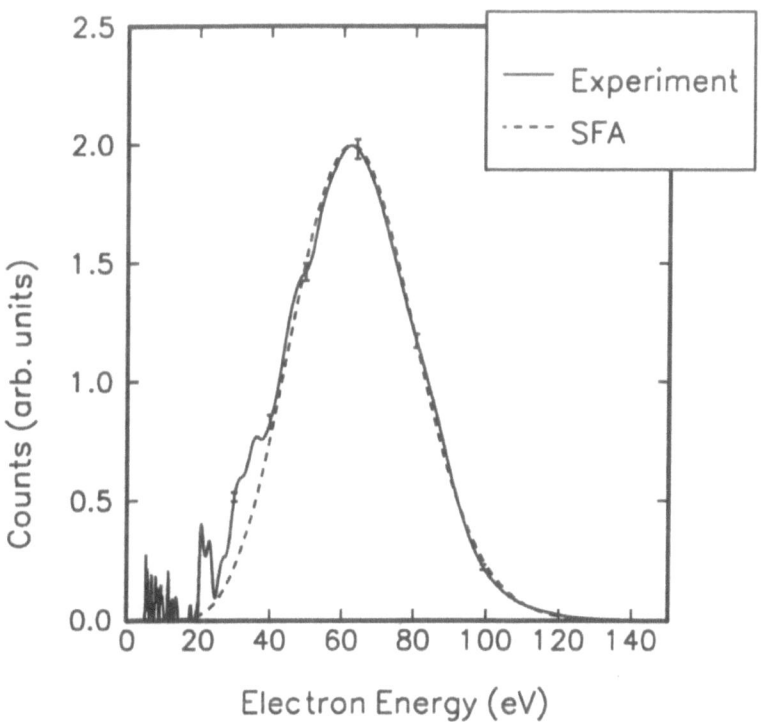

Figure 2. Circular polarization electron spectrum from helium.

4. Samples of SFA Applications

4.1. TOTAL ION YIELD

We refer to Fig.1, which shows the application of the SFA to the experiments by Walker et al.[14] on the single ionization of helium by a linearly polarized laser. The experiment has the feature of extending over a remarkable range of ion yields. The only parameter taken to be adjustable in the

application of the SFA is the peak laser intensity. Figure 1 follows from
the hypothesis that the peak intensity is 1.9 times larger than stated in
the experimental paper. This seems generally to be the case with the SFA
in application to linear polarization. It must regarded as a shortcoming, in
that the theoretical fit following from numerical solution of the Schrödinger
equation requires only an increase by a factor of 1.15 in the peak intensity[14]. Nevertheless, the excellent apparent fit of the SFA over the entire
range of ion yields is a feature present also in other applications of the SFA
to strong-field ion yields.

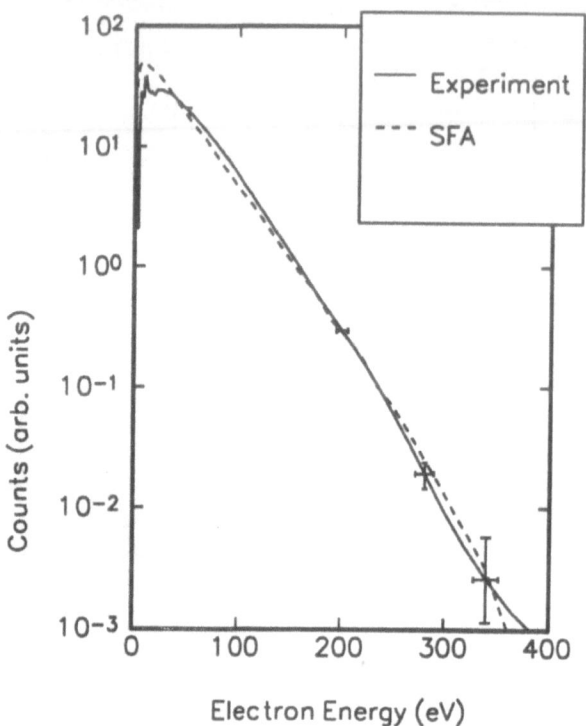

Figure 3. Linear polarization electron spectrum from helium.

4.2. PHOTOELECTRON ENERGY SPECTRA

The application of the SFA to the energetic photoelectron spectra measured
by Mohideen et al.[15], is especially interesting in that the authors were
unable to obtain a satisfactory explanation by any theory for the energetic
electrons measured. The problem is beyond the reach of numerical solution
of the Schrödinger equation, and the authors attempted to use the SFA,
but with only limited success. The excellent agreement shown in Figs.2 and
3 for circular and linear polarization, respectively, follows from the demand

that the SFA be applied with full account taken of the spatial and temporal distribution of laser intensities actually present in the focused laser pulse. It is often also important to use a truly accurate initial state wave function in the SFA. The results shown in all the figures contained in this paper are done with analytical Hartree-Fock atomic wave functions.

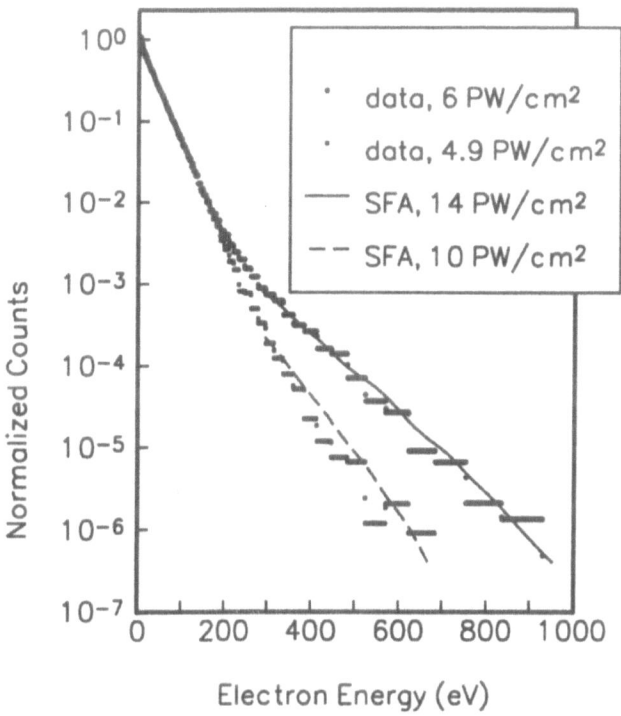

Figure 4. Energetic photoelectron spectra from He$^+$.

Figure 2 for the case of circular polarization is especially important in that it was done with the SFA including a final-state Coulomb correction now available[10] for strong, circularly polarized fields. The agreement found is truly excellent in that it would be virtually impossible with an inaccurate theory to give simultaneously the correct position of the peak and of the width of the familiar bell-shaped circular polarization spectrum, as well as the correct description of the high-energy tail. The total success of the Coulomb-corrected SFA appears to verify that it is a complete theory for ionization by circularly polarized strong fields.

The excellent agreement of the SFA with the spectrum for ionization with linear polarization shown in Fig.3 is reinforced by Fig.4, showing the capability of the SFA to predict spectra out to about $1\,keV$. The data in Fig.4 are for double ionization of helium[14]. The SFA is applied only

150

to the high energy part of the spectra (beyond the "knee") which can be ascribed[16] to sequential double ionization. Application of the SFA to direct double ionization is still in process of development[17].

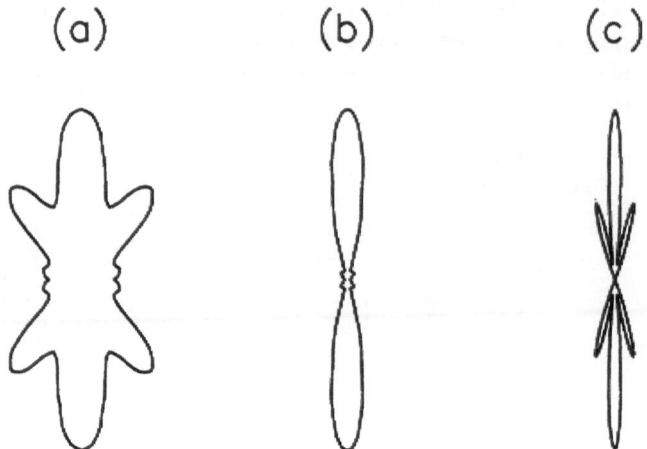

Figure 5. Angular distribution at $s = 20$ for (a) experiment, (b) Cray calculation, (c) SFA.

4.3. RINGS IN ANGULAR DISTRIBUTIONS

The experiments showing "rings" in the angular distribution of photoelectrons at high ATI orders[18] are especially interesting in that these rings

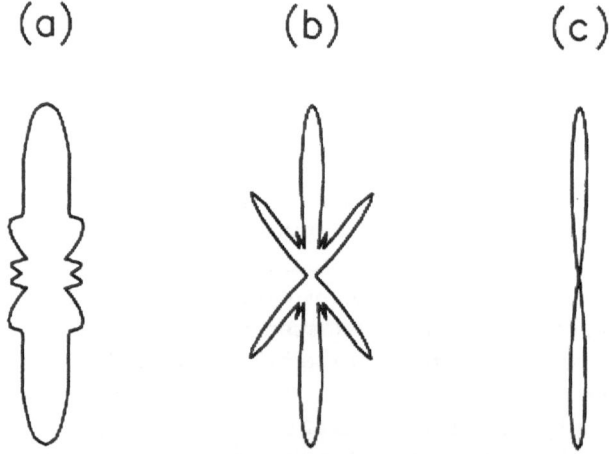

Figure 6. Angular distribution at $s = 25$ for (a) experiment, (b) Cray calculation, (c) SFA.

can be ascribed to final-state oscillations of the electrons in the laser field. The experiments were done at an intensity which would normally be regarded as too low to permit application of the SFA. However, the high ATI order explored makes it at least possible that the kinetic energy of these electrons might be sufficiently in excess of the binding energy that the SFA might have some utility. Figures 5 and 6 show angular distributions at ATI orders of 20 and 25. These orders are selected for examination because the experiments show the most fully developed ring structure at ATI order $s = 20$, whereas the Cray calculations show the rings at $s = 25$. The SFA is interesting in that it agrees with the experiment about the correct ATI order for the rings, although the angle for the ring is too small.

The most important point is that the SFA is capable of predicting a ring structure, indicating the presence of final-state electron oscillations.

5. Summary

All theoretical methods intended for the description of atoms in strong fields have limitations. The methods with the broadest applicability are the tunneling approach[4], the numerical solution of the Schrödinger equation[7], and the SFA[9]. For very high intensity, tunneling methods cease to be applicable as the ionization process goes over the top of the barrier. Numerical methods fail because they exceed the capacities of the fastest machines for very strong fields at low frequencies and for certain details such as energetic photoelectron spectra. The SFA gains strength and simplicity as the fields get stronger. For strong field ionization with circular polarization, the SFA with Coulomb correction[10] appears to be a complete theory. For ionization with strong linearly polarized fields, the SFA gives remarkably good agreement with experiments for energetic photoelectron spectra as well as for total ion yield, with the *caveat* that the SFA seems to require somewhat higher peak laser intensities than stated by the experimentalists. Coulomb corrections to the SFA with linear polarization are being sought in collaboration with Prof. V.P. Krainov, and A. Shabaev. Revisiting of the atom by the ionized electron is inherent in the SFA method.

References

1. Keldysh, L.V. (1964) Ionization in the field of a strong electromagnetic wave, *Zh. Eksp. Teor. Fiz.* **47**, 1945-1957 [*Sov. Phys JETP* **20**, 1307-1314].
2. Nikishov, A.I. and Ritus, V.I. (1966) Ionization of systems bound by short-range forces by the field of an electromagnetic wave, *Zh. Eksp. Teor. Fiz.* **50**, 255-270 [*Sov. Phys JETP* **23**, 168-177].
3. Perelomov, A.M., Popov, V.S., and Terent'ev, M.V. (1966) Ionization of atoms in an alternating electric field, *Zh. Eksp. Teor. Fiz.* **50**, 1393-1409 [*Sov. Phys JETP* **23**, 924-934].

152

4. Ammosov, M.V., Delone, N.B., and Krainov, V.P. (1986) Tunnel ionization of complex atoms and of atomic ions in an alternating electromagnetic field, *Zh. Eksp. Teor. Fiz.* **91**, 2008-2013 [*Sov. Phys JETP* **64**, 1191-1194].
5. Potvliege, R.M. and Shakeshaft, R. (1992) Nonperturbative treatment of multiphoton ionization within the Floquet framework, in M. Gavrila (ed.), *Atoms in Intense Laser Fields*, Academic Press, Boston, pp. 373-433.
6. Gavrila, M. (1992) Atomic structure and decay in high frequency fields, in M. Gavrila (ed.), *Atoms in Intense Laser Fields*, Academic Press, Boston, pp. 435-510.
7. Kulander, K.C., Schafer, K.J., and Krause, J.L. (1992) Time-dependent studies of multiphoton processes, in M. Gavrila (ed.), *Atoms in Intense Laser Fields*, Academic Press, Boston, pp. 247-300.
8. Reiss, H.R. (1990) Relativistic strong-field ionization, *J. Opt. Soc. Am. B* **7**, 574-586.
9. Reiss, H.R. (1980) Effects of an intense electromagnetic wave on a weakly bound system, *Phys. Rev. A* **22**, 1786-1813.
10. Reiss, H.R. and Krainov, V.P. (1994) Approximation for a Coulomb-Volkov solution in strong fields, *Phys. Rev. A* **50**, R910-R912.
11. Reiss, H.R. (in press) High-frequency, high-intensity photoionization, *J. Opt. Soc. Am. B.*
12. Corkum, P.B., Burnett, N.H., and Brunel, F. (1989) Above-threshold ionization in the long-wavelength limit, *Phys. Rev. Lett.* **62**, 1259-1262.
13. Reiss, H.R. (1992) Frequency and polarization effects in stabilization, *Phys. Rev. A* **46**, 391-394.
14. Walker, B., Sheehy, B., DiMauro, L.F., Schafer, and Kulander, K.C. (1994) Precision measurement of strong field double ionization of helium, *Phys. Rev. Lett.* **73**, 1227-1230.
15. Mohideen, U., Sher, M.H., Tom, H.W.K., Aumiller, G.D., Wood, O.R., Freeman, R.R., Bokor, J., and Bucksbaum, P.H. (1993) High intensity above-threshold ionization of He, *Phys. Rev. Lett.* **71**, 509-512.
16. DiMauro, L.F., *private communication.*
17. Bauer, J., Ivanov, M., Rzazewski, K., and Reiss, H.R. (in press) SFA applied to the one-dimensional two-electron model atom, *J. Phys. B.*
18. Yang, B., Schafer, K.J., Walker, B., Kulander, K.C., Agostini, P., and DiMauro, L.F. (1993) Intensity-dependent scattering rings in high-order above-threshold ionization, *Phys. Rev. Lett.* **71**, 3770-3773.

CIRCULAR DICHROISM AND RELATED EFFECTS IN MULTIPHOTON TRANSITIONS

N. L. MANAKOV
Department of Theoretical Physics, Voronezh State University
Voronezh, 394693, Russia

1. Introduction

In the early experiments on interaction of laser radiation with atoms and molecules, the linear-polarized radiation was used as a rule, and polarization dependence of cross sections was not investigated. But it is well known, that a polarization state of laser beam is often significant (and may be crucial) in multiphoton phenomena. As a rule the polarization effects result in the depedence of cross sections in absolute value of light field ellipticity degree only (see, e. g. [1]). However, in special cases the interesting effect of "circular dichroism" may be observable, i. e., difference in N-photon transitions cross sections $d\sigma$ at simultaneous change of signes of circular polarization degrees ξ_i for all (incoming and emitted) photons, $i = 1, \ldots, N$. The consequence of this effect is, in particular, the symmetry reduction of the multiphoton ionization angular distributions in elliptical polarized field, which was observed firstly in [2] and discussed later in number of works. The simplest example of dichroism is the splitting of atomic levels in elliptically-polarized field in the sign of magnetic quantum number M, which is proportional to ξ (see, e. g. [3]). So, the circular dichroism is a specific polarization effect and its study is worthwhile for investigation of new features in light field-atom interaction processes.

Below some of general conditions which cause dichroism in bound-bound, bound-free and free-free transitions are discussed and a number of examples are considered. Since the circular dichroism is not a strong field effect, we consider the low intensity light fields and use the perturbative approach for the atom-field interaction. On the other hand, the nonperturbative account of atomic potential action on bound and unbound electrons is necessary for the accurate studies of light polarization effects. For simplicity we ignore spin effects and suppose that the target atoms are randomly oriented.

H. G. Muller and M. V. Fedorov (eds.), Super-Intense Laser-Atom Physics IV, 153–162.
© *1996 Kluwer Academic Publishers.*

2. Generalities

The necessary conditions for nonzero dichroism may be stated on the grounds of the general arguments of space and time symmetry. The dichroism is caused by the presence in problem of axial time-odd vector

$$\mathbf{D} \equiv \xi \frac{\mathbf{k}}{|\mathbf{k}|} = i[\mathbf{e}^* \times \mathbf{e}], \tag{1}$$

for light wave with nonzero circular polarization degree $-1 \leqslant \xi \leqslant 1$. Here $\mathbf{e}$ and $\mathbf{k}$ are the complex polarization vector and wave vector of monochromatic field with electric vector

$$\mathbf{F}(\mathbf{r}, t) = F \operatorname{Re}\{\mathbf{e} \exp[i(\mathbf{kr} - \omega t)]\}, \qquad (\mathbf{e}\,\mathbf{e}*) = 1.$$

The parameter ξ may arise in the cross section only as a result of interference between different (partial) parts of the total transition amplitude and the dichroism in all cases have the interference origine.

Let all photons in the problem have different polarization vectors $\mathbf{e}_i$. In this case two following conditions are necessary for nonzero dichroism:

1. The axial vector $\mathbf{D}_i$ may be involved in $d\sigma$ only in product with another axial vector. This new vector should be constructed from other vectors of the problem.

2. The T-odd vector $\mathbf{D}_i$ may be involved in $d\sigma$ only with another "T-odd" quantity Γ. In general case the imaginary part of above-mentioned partial transition amplitude stands for Γ parameter.

In resonant processes the width of excited resonant level may be such parameter. For continuum states the skew-Hermitian parts of amplitude are propotional to $\sin \delta_l$, where δ_l are the scattering phases of electron on atom (for laser-assisted scattering) or atomic core (for ionization). For above-threshold N-photon processes the total amplitude involves the amplitudes of $(n < N)$-photon processes in skew-Hermitian part etc. As it follows from the general unitarity relation for amplitude [4], in all cases the skew-Hermitian part is connected with the amplitudes of other physical processes than the present one. We call all these processes as "dissipative" processes, because they lead to dissipation of light energy in the propagation of light through atomic medium [5,6]. In this terms the dichroism is caused by the irreversible (T-odd!) nature of the dissipation processes.

If two or more identical photons with polarization vector $\mathbf{e}$ are involved in process, the parameter ξ arises in cross section not only from vector (1), but also from the combination $(\mathbf{ea})^{2k}$ (in more general case the combination is $\{(\mathbf{ea})(\mathbf{eb})\}^n$) which has the imaginary part proportional to ξ. Here $\mathbf{a}$ and $\mathbf{b}$ are a vectors of problem. We used the following general representation of complex polarization vector $\mathbf{e}$:

$$\mathbf{e} = \frac{\mathbf{e}_x + i\gamma\mathbf{e}_y}{\sqrt{1+\gamma^2}}, \quad -1 \leqslant \gamma \leqslant 1, \quad \mathbf{ee}^* = 1. \tag{2}$$

Then

$$(\mathbf{ee}) = (\mathbf{e}^*\mathbf{e}^*) = l = \frac{1-\gamma^2}{1+\gamma^2}, \quad \xi = \frac{2\gamma}{1+\gamma^2} \tag{3}$$

are the degrees of linear and circular polarization respectively [4]. It is evident that $l^2 + \xi^2 = 1$ for full polarized light. Let φ be the angle between vector $\mathbf{a}$ and plane of polarization, θ, the angle between the main axis (Ox) of polarization ellipse and the projection of $\mathbf{a}$ to the plane of polarization. Then we have

$$(\mathbf{ep})^2 = \mathrm{Re}(\mathbf{ep})^2 + i\,\mathrm{Im}(\mathbf{ep})^2,$$

$$\mathrm{Re}(\mathbf{ep})^2 = \frac{1}{2}\sin^2\varphi(l + \cos 2\theta), \quad \mathrm{Im}(\mathbf{ep})^2 = \frac{1}{2}\xi\sin^2\varphi\sin 2\theta \tag{4}$$

$$|\mathbf{ep}|^2 = \frac{1}{2}\sin^2\varphi(1 + l\cos 2\theta).$$

Note, that the above-discussed statement 2. is valid here also, because by T-inversion we have $\mathbf{e} \to \mathbf{e}^*$ and so $\mathrm{Im}(\mathbf{ep})^2$ changes the sign. Since the combination $(\mathbf{ep})^2$ is involved in cross section with factor $l = (\mathbf{e}^*\mathbf{e}^*)$ only, it is clear also that in the case of identical photons the dichroism is zero for pure circular polarization $(\xi = \pm 1)$ and exists for elliptical light with $0 < l < 1$. In the case of two different vectors $\mathbf{a}$, $\mathbf{b}$ (e. g., in the electron scattering they are momenta $\mathbf{p}$, $\mathbf{p}'$ of incoming and scattered electrons) this statement is not true.

So, in any concrete multiphoton transition the general structure of circular dichroism terms may be determined on the grounds of space-time invariance arguments. We consider below the dichroism effects in typical multiphoton processes.

3. Bound-bound transitions

3.1. TWO-PHOTON PROCESSES

It is well known that in the dipole approximation for interaction of atoms with field the cross sections involve only the scalar products of polarization vectors $|\mathbf{e}_1\mathbf{e}_2|^2$ and $|\mathbf{e}_1\mathbf{e}_2^*|^2$ and dichroism disappears, in accordance with statement 1. of Sec. 2. So, the electric quadrupole or (and) magnetic dipole effects which are proportional to wave vectors $\mathbf{k}_1$ and $\mathbf{k}_2$, should be taken into account. The quantities ξ_1 and ξ_2 are pseudoscalars and only two following combinations of vectors may be involved in the cross sections in products with T-odd "dissipative" parameters $\Gamma_{1,2}$:

$$I_1 = \xi_1\,\mathrm{Re}\{(\mathbf{e}_2[\mathbf{n}_1 \times \mathbf{n}_2])(\mathbf{e}_2^*\mathbf{n}_1)\}, \quad I_2 = \xi_2\,\mathrm{Re}\{(\mathbf{e}_1[\mathbf{n}_2 \times \mathbf{n}_1])(\mathbf{e}_1^*\mathbf{n}_2)\}. \tag{5}$$

The expressions (5) determine the dichroism effects in arbitrary two-photon transitions between bound states of a randomly oriented systems. I_1, I_2 may be expressed also in terms of imaginary parts of following combinations of vectors, which arise at regular method of cross section calculations with the use of irreducible tensor operator techniques [6]

$$E_1 = (e_1 e_2^*)(e_1^* n_2)(e_2 n_1),$$
$$E_2 = (e_1 e_2)(e_1^* n_2)(e_2^* n_1),$$
$$I_{1,2} = \operatorname{Im}(E_2 \pm E_1). \tag{6}$$

The dichroism in light scattering by gases is studied in detail in [6]. In particular, the consequence of dichroism is the emergence of circular polarization of the scattered radiation in the scattering of photons with linear polarization. The phenomenon is similar to the emergence of ellipticity in the light reflected from an absorbing-medium surface (a metal) and is caused by the skew-Hermitian part of the scattering amplitude.

Here we present the results for dichroism in two-photon excitation of atom in two noncollinear laser beams ($k_i = \frac{\omega_i}{c} n_i$, ω_i, e_i are wave vectors, frequencies and polarization vectors; $i = 1, 2$) [7]. Let ω_1 (or ω_2) be resonant to the intermediate atomic level with the same parity as for initial ($|i\rangle$) and final ($|f\rangle$) levels (dipole-forbidden resonance). A transition amplitude in this case is the sum

$$A_{fi} = A_{fi}^{(d)} + A_{fi}^{(r)}$$

of nonresonant dipole-dipole amplitude $A_{fi}^{(d)}$ and resonant dipole-forbidden part $A_{fi}^{(r)}$. In general case amplitude $A_{fi}^{(r)}$ includes magnetic-dipole and electric-quadrupole interactions and depends on vectors n_1 and n_2. The transition rate R is proportional to $|A_{fi}|^2$ and contains the following interference terms (between amplitudes $A_{fi}^{(d)}$ and $A_{fi}^{(r)}$):

$$R_{1,2}^{as} \approx \frac{\Gamma_r}{\Delta^2 + \Gamma_r^2/4} I_{1,2}.$$

Here $I_{1,2}$, the above-discussed "dichroism" combinations of vectors, Γ_r, the width of resonant level, $\xi_{1,2}$, the circular polarization degrees of photons 1 and 2. Thus, for linearly polarized photon 1 the transition rate R is different at $\xi_2 = \pm 1$ and vice versa. As it was mentioned above, the effect is caused by width Γ_r, which is a time-odd parameter and compensates the change of signs $I_{1,2}$ at time inversion (e_i, $n_i \to e_i^*$, $-n_i$).

For example, we give below the accurate expression of $|A_{fi}|^2$ for S-S transition with intermediate quadrupole resonance at D-level (for simplicity, the fine-structure of levels is neglected):

$$|A_{fi}|^2 = F_1^2 \alpha_1^2 |e_1 e_2|^2 + \frac{2F_1 F_2 \alpha_1 \alpha_2}{\Delta^2 + \Gamma_r^2/4} \left[\left(|e_1 e_2|^2 (n_1 n_2) + \operatorname{Re} E \right) \Delta - \frac{\Gamma_r}{4}(I_1 + I_2) \right] +$$
$$\frac{F_2^2 \alpha_2^2}{\Delta^2 + \Gamma_r^2/4} \left[|e_1 n_2|^2 + |e_2 n_1|^2 + (n_1 n_2)(|e_1 e_2|^2 (n_1 n_2) + 2 \operatorname{Re} E) \right], \tag{7}$$

where $\alpha_{1,2}$ refer to radial dipole and quadrupole matrix elements respectively, $\Delta = E_i + \omega_1 - E_r$ is the resonance detuning, $E = (\mathbf{e}_1\mathbf{e}_2)(\mathbf{e}_1^*\mathbf{n}_2)(\mathbf{e}_2^*\mathbf{n}_1)$, $I_1 + I_2 = 2\,\mathrm{Im}\,E$. At resonance all terms in (7) have considerable magnitudes and the dichroism

$$\triangle R = R(\xi_1, \xi_2) - R(-\xi_1, -\xi_2)$$

may be measured in experiments similar to [8]. Such measurements may be very effective for determination of level width or the magnitudes and relative phases of atomic matrix elements.

3.2. MULTIPHOTON BOUND-BOUND TRANSITIONS

In this case the multipole corrections are not necessary and the dichroism terms contain only polarization vectors $\mathbf{e}_i$ of photons.

The general analysis similar to the two-photon scattering [6] is possible for three photon transitions with all different photons $\mathbf{e}_1$, $\mathbf{e}_2$, $\mathbf{e}_3$. In dipole approximation such transitions are possible between the initial $|i\rangle \equiv |n_i l_i m_i\rangle$ and final $|f\rangle \equiv |n_f l_f m_f\rangle$ states with opposite parities ($l_f = l_i \pm 1$, $l_i \pm 3$). For l_i, $l_f > 0$ the four different dichroism terms arise, which are proportional to the imaginary parts of the following four combinations Φ_α, $\alpha = 0 \div 3$, of vectors $\mathbf{e}_1$, $\mathbf{e}_2$, $\mathbf{e}_3$ (see eq. (6)):

$$\Phi_0 = (\mathbf{e}_1\mathbf{e}_2^*)(\mathbf{e}_2\mathbf{e}_3^*)(\mathbf{e}_1^*\mathbf{e}_3)$$
$$\Phi_i = \Phi_0(\mathbf{e}_i \to \mathbf{e}_i^*), \; i = 1, \, 2, \, 3. \tag{8}$$

In particular, if two photons are linearly polarized ($\mathbf{e}_1 = \mathbf{e}_1^*$ and $\mathbf{e}_2 = \mathbf{e}_2^*$) we have

$$\Phi_2 = \Phi_1 = \Phi_0, \quad \Phi_3 = \Phi_0^*$$

and only one dichroism term remains with vector combination of the type

$$\mathrm{Im}\,\Phi_0 = -\frac{1}{2}(\mathbf{D}_3\,[\mathbf{e}_1 \times \mathbf{e}_2])(\mathbf{e}_1\mathbf{e}_2). \tag{9}$$

For four and more different photons the general results are very cumbersome, and only some special cases may be simply analyzed.

So, at high harmonics generation by atoms the polarization vector of atom at the frequency $N\omega$ in two lowest orders of the pump intensity I has the form [9,10]

$$\mathbf{P} = I^{\frac{N-1}{2}}\big(\chi_0\mathbf{e} + I(\chi_1\mathbf{e} + I\chi_2\mathbf{e}^*)\big)I^N. \tag{10}$$

Here χ_i are the atomic susceptibilities which may be complex in the vicinities of resonances. In this case the intensity I_N of N-th harmonic with polarization vector $\mathbf{e}'$ contains the dichroism term which is proportional to

$$I^N\,\mathrm{Im}\big[(\chi_0 + I\chi_1)\chi_2^*\big]I^{N+1}\Phi_{\mathrm{dichr.}},$$

where

$$\Phi_{\text{dichr.}} = 2\,\text{Im}\{(\mathbf{ee'})(\mathbf{ee'^*})\}$$
$$= \xi l' \sin 2\theta. \tag{11}$$

Here l' is the linear polarization degree of harmonic, θ, the angle between the main axes of pump and harmonic polarization ellipses. The planes of polarization for the pump and harmonic coincide, as it is necessary for the generation process. In this case we have the identical photons and, in contrast to (9), the dichroism vanishes at $\xi = \pm 1$ (see Sec. 2).

4. Bound-free transitions (multiphoton ionization)

As it is evident from the rotational invariance arguments, in the case of N identical photons the angular distribution of photoelectrons in $\mathbf{p}$ direction ($|\mathbf{p}| = 1$) contains only vector combinations of types $|\mathbf{ep}|^2$ and $(\mathbf{ep})^2$. In this case the maximum in $\mathbf{p}$ term is $|\mathbf{ep}|^{2N}$ and minimum is ~ 1 for even N and $\sim |\mathbf{ep}|^2$ for odd N. The dichroism terms are proportional to $\text{Im}(\mathbf{ep})^2$ (see (4)).

In particular case of 2-photon ionization from initial state $|nL_iM_i\rangle$ with arbitrary L_i the cross section after the averaging over M_i-projections has the form

$$\frac{d\sigma}{d\Omega_{\mathbf{p'}}} = aI^2 \left[A + \text{Im}\,B\,l\,\text{Im}(\mathbf{ep})^2 + \text{Re}\,B\big(2P_2(|\mathbf{ep}|) + \xi^2 P_2(\cos\varphi)\big) \right.$$

$$\left. -B_2\left(P_2(|\mathbf{ep}|) - \xi^2 P_2(\cos\varphi)\right) + B_4\left(P_4(|\mathbf{ep}|) - \frac{5}{12}\xi^2\left(P_2(\cos\varphi) - \frac{7}{10}\right)\right)\right].$$

Here a, numerical dimensional coefficient, $P_L(x)$, the Legendre polynomial,

$$A = \frac{1}{3}l^2(R_{0L_i})^2 + \frac{1}{5}\left(1 - \frac{1}{3}l^2\right)\sum_{L=L_i,\ L_i\pm2}|R_{2L}|^2,$$

$$B = (-1)^{L_i+1}\frac{1}{3}\sqrt{\frac{2}{5}}\sum_{L=L_i,\ L_i\pm2}(-1)^{\frac{L_i-L}{2}}\sqrt{2L+1}\,C^{20}_{L0L_i0}R_{0L_i}R_{2L}e^{-i(\delta_{L_i}-\delta_L)},$$

$$B_g = (-1)^{L_i}\frac{g}{g+1}\sqrt{\frac{2}{7}}\left(\frac{5}{4}\right)^{\frac{g-2}{4}}\sum_{L,\ L'=L_i,\ L_i\pm2}(-1)^{\frac{L-L'}{2}}\sqrt{(2L+1)(2L'+1)}\ \times$$

$$\times\ C^{g0}_{L0L'0}\begin{Bmatrix}2 & L & L_i\\ L' & 2 & g\end{Bmatrix}R_{2L}R_{2L'}\cos(\delta_{L'}-\delta_L),$$

R_{pL} is the reduced two-photon matrix element:

$$R_{pL} = \sqrt{\frac{2p+1}{2L+1}}\sum_{l'=L_i\pm1}\begin{Bmatrix}1 & 1 & p\\ L & L_i & l'\end{Bmatrix}\langle L_i\|\,\mathbf{r}\,g_{l'}\,\mathbf{r}\,\|L\rangle,$$

$g_{l'}$ is the radial atomic Green function with orbital momentum l', $(\mathbf{ep})^2$ and $|\mathbf{ep}|^2$ see in eq. (4).

For $L_i = 0$ our result coincides with the result of [11]. The detailed analysis of N-photon case and numerical results will be published elsewhere.

5. Free-free transitions (laser-assisted electron-atom scattering)

In free-free transitions of electron-atom systems (for example, the stimulated bremsstrahlung and absorption) the polarization effects have not been analyzed in detail [12]. However, the circular dichroism in these processes has a large magnitude and should be observed experimentally.

For elastic scattering ($|\mathbf{p}| = |\mathbf{p}'|$) the dichroism was discovered firstly in analytical calculations of elastic Coulomb scattering in a low-intensity radiation field [13]. The dichroism term in cross section has the form

$$\frac{d\sigma(\xi)}{d\Omega_{\mathbf{p}'}} - \frac{d\sigma(-\xi)}{d\Omega_{\mathbf{p}'}} = \xi\left(\frac{\mathbf{k}}{|\mathbf{k}|}[\mathbf{n} \times \mathbf{n}']\right)\varphi(p,\theta). \tag{12}$$

Here $\varphi(p,\theta)$ is a scalar function of momentum p and scattering angle θ.

The inelastic scattering was examined in [14]. Below we discuss briefly the results for one-photon emission and absorption. As it is shown in [14] the one-photon amplitude $A(\mathbf{p},\mathbf{p}')$ has the form

$$A(\mathbf{p},\mathbf{p}') = (\mathbf{e}^*\mathbf{n})f(p,p',\cos\theta) - (\mathbf{e}^*\mathbf{n}')f'(p,p',\cos\theta) \tag{13}$$

with

$$f'(p,p',\cos\theta) = -f(p',p,\cos\theta). \tag{14}$$

Here f, f' are invariant (scalar) amplitudes which contain the sum of products of radial matrix elements, Legendre polynomials $P_L^k(\cos\theta)$ with $k = 0$, 1 and factors $\exp\left[i\left(\delta_L(p) - \delta_{L'}(p')\right)\right]$, $\delta_{L,L'}$ are the partial phases of elastic electron-atom scattering. Polarization and angular dependence of cross section is

$$\frac{d\sigma}{d\Omega_{\mathbf{p}'}} = \frac{p'}{p}|A(\mathbf{p},\mathbf{p}')|^2 = \frac{p'}{p}\left\{|f|^2|\mathbf{en}|^2 + |f'|^2|\mathbf{en}'|^2 - \right.$$
$$\left. - 2\operatorname{Re}(ff'^*)\operatorname{Re}\left[(\mathbf{e}^*\mathbf{n})(\mathbf{en}')\right] + \operatorname{Im}(ff'^*)\xi\left(\frac{\mathbf{k}}{|\mathbf{k}|}[\mathbf{n}\times\mathbf{n}']\right)\right\}. \tag{15}$$

So, the dichroism in this case has the same form as (12) with

$$\varphi(p,p',\theta) = 2\frac{p'}{p}\operatorname{Im}\left(f^*(p,p',\cos\theta)f(p',p,\cos\theta)\right). \tag{16}$$

Corresponding results for scattering with one photon absorbtion is given by (15) with substitutions $\omega \to -\omega$ and $\xi \to -\xi$.

The equation (15) shows that the term with ξ is of interference origine and does not appear in the first Born approximation, because the amplitude in this case depends only upon the momentum transfer $\Delta\mathbf{p} = \mathbf{p} - \mathbf{p}'$ (formally this is the consequence of the hermicity of the amplitude in the first Born approximation). The dichroism disappears also when $\theta \to 0$ (together with $[\mathbf{p} \times \mathbf{p}']$) and also when vectors $\mathbf{k}$, $\mathbf{p}$, $\mathbf{p}'$ are coplanar. As it is evident from (16), the dichroism vanishes in a low-frequency limit ($p \to p'$). For the case of slow electron this fact follows also from the accurate formulae for f [14]. Thus, polarization anomalies did not appear not only in the Fedorov and Bunkin's results [15] on multiphoton SBSA, obtained in the Born approximation, but also in Kroll and Watson's results [16], without Born approximation, but valid only in the low-frequency limit. Except for extraordinary cases all the terms of $\dfrac{d\sigma}{d\Omega_{\mathbf{p}'}}$ in eq. (15) have comparable magnitudes.

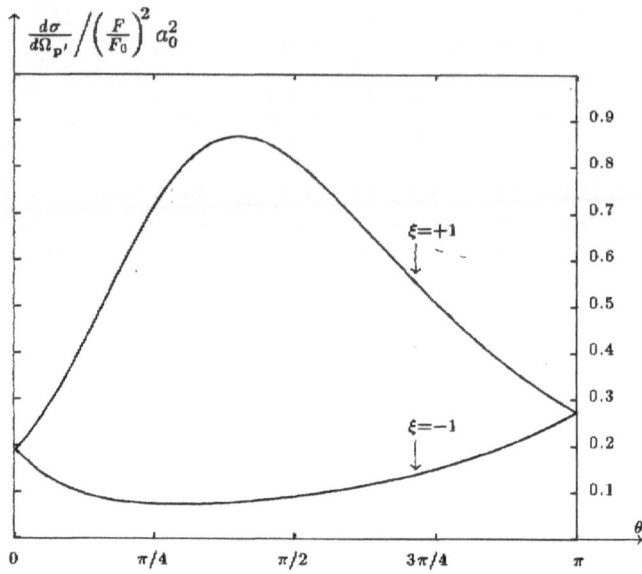

Figure 1. The angular dependence of one-photon Coulomb scattering cross section for right ($\xi = 1$) and left ($\xi = -1$) circular polarization of laser field.

Let us now consider the vanishing of dichroism in the first Born approximation as an immediate consequence of general arguments of Sec. 2. Indeed, in this case only one channel of scattering is open, that is, the direct transition $\mathbf{p} \to \mathbf{p}'$ with emission or absorption of photon. Then the amplitude $A^{(1)}$ has not the skew-Hermitian part. The simple analysis shows that in the second Born approximation the amplitude $A^{(2)}$ has the imaginary parts which are the products of first order amplitude $A^{(1)}$ and the amplitudes of elastic electron-atom scattering with momenta $\mathbf{p}$ and $\mathbf{p}'$ respectively. So, the additional channels (with re-scattering) are open in this case. Those are the "dissipation channels" and so the dichroism is nonzero.

In a similar way the dichroism at multiphoton ionization vanishes in the plane-wave approximation for the wave function of ejected electron. But already in the first Born approximation the re-scattering channels are open and the dichroism appears.

The analytical results for the one-photon Coulomb scattering are presented in [14]. Figure 1 elucidates the numerical value of dichroism and the angular dependence of $\frac{d\sigma}{d\Omega_{\mathbf{p}'}}$ at $\xi = \pm 1$, fixed $E = \frac{\mathbf{p}^2}{2m} = 1$ a. u. and $\hbar\omega/E = 0.9$. The direction of $\mathbf{k}$ is parallel to $[\mathbf{p} \times \mathbf{p}']$. It is clearly seen that the dichroism is large and should be observed in standard experiments. The change $\xi \rightarrow -\xi$ is evidently equivalent to the change $\mathbf{n}' \rightarrow -\mathbf{n}'$. So, for exprimental dichroism measurements the observation of difference in cross sections at fixed ξ and two opposite (orthogonal to $\mathbf{n}$) directions $\mathbf{n}'$ and $-\mathbf{n}'$ may be preferable.

As in the one-photon case, for N-photon scattering, described in low-intensity field by the N-order perturbation theory, the parameter φ in eq. (12) is determined also by interference between real and imaginary parts of invariant amplitudes. All kinematical properties of dichroism for the one-photon case and its smallness in Born- and low-frequency limits are also valid here.

6. Atomic orientation effects due to dissipative processes at multiphoton transitions

As we discussed above, the photon polarization anomalies in concrete multiphoton transition arise because this transition occurs against the background of an open dissipation channel, which leads to the nonhermicity of transition amplitude. Similarly, the dissipative effects caused also new atomic orientation effects in the interaction of light with atoms.

These effects are proportional to the product of Γ and atomic angular momentum $\mathbf{J}$, both of which are the T-odd quantities. It is very important that the light fields may be linearly-polarized (or nonpolarized) in this case.

The optical orientation (dc magnetization) of atomic gases in two linearly polarized nearly resonant optical fields was considered in [17]. The mean angular momentum $\bar{\mathbf{J}}$ (orientation vector) in this case has the form

$$\bar{\mathbf{J}} \approx \Gamma(\mathbf{e}_1 \mathbf{e}_2)[\mathbf{e}_1 \times \mathbf{e}_2].$$

In [18] the new effects in light scattering by oriented atoms and in orientation of unpolarized atomic target due to the light scattering when channels of dissipation for the light energy are open were analysed. Thus in scattering of linearly polarized or unpolarized light the degree of linear polarization and the angular distribution of scattered light depends on the initial atomic orientation. On the other hand the orientation of unpolarized atoms appears in scattering of unpolarized photon without registration of the scattered photon polarization. These effects are caused by terms $\sim \Gamma(\mathbf{J}[\mathbf{n}_1 \times \mathbf{n}_2])$ in cross section $d\sigma$. The orientation vector is perpendicular to the scattering plane, and the effect can be observed in a coincidence experiment.

The dissipation-induced orientation (polarization) effects must be significant in other processes such as hyper-Raman scattering and in processes of two-photon excitation and ionization also. The circular dichroism arises also in the double

162

ionization. For the one-photon double photoeffect the dichroism was analysed firstly in [19] and the results are presented in the form of series of bipolar harmonics of vectors **p** and **p'** of ejected electrons. In [14] this result was simplified and dichroism was presented in the form

$$(d\sigma)_{\text{dichr.}} \sim \xi(\frac{\mathbf{k}}{|\mathbf{k}|} [\mathbf{p} \times \mathbf{p'}]).$$

The experimental observation of dichroism in multiphoton double ionization may be useful for the separation of direct and stepwise mechanisms of multiphoton ionization.

Finally, we noted, that new pecularities in circular dichroism and orientation effects arise in the presence of weak dc field $\mathbf{F_0}$. This field reduces the central symmetry of medium (for linear in $\mathbf{F_0}$ approximation) or induces anisotropy (for quadratic approximation) and therefore results in new features of above-mentioned effects. The detailed analysis of this problem will be presented elsewhere.

References

1. Delone, N.B. and Krainov, V.P. (1994) *Multiphoton processes in atoms*, Springer-Verlag, Berlin.
2. Bashkansky, M., Bucksbaum, P.H., and Schumacher, D.W. (1988) *Phys. Rev. Lett.* **60**, 2458–2462.
3. Manakov, N.L., Ovsiannikov, V.D., and Rapoport, L.P. (1986) Atoms in a laser field, *Phys. Rep.* **141**, 319–433.
4. Berestetskii, V.B., Lifshitz, E.M., and Pitaevskii, L.P. (1982) *Quantum electrodynamics*, Pergamon Press, Oxford.
5. Manakov, N.L. and Fainshtein, A.G. (1984) *Zh. Eksp. Teor. Fiz.* **87**, 1552–1564.
6. Manakov, N.L. (1994) *Zh. Exp. Teor. Fiz.* **106**, 1286–1305.
7. Manakov, N.L. and Meremjanin, A.V. (1995) Circular dichroism in resonant two-photon, two-color excitation of atoms, in *Contributed papers of 5-th ECAMP, part II*, Edinburgh, p. 635.
8. Cook, L., Olsgaard, D., Havey, M., and Sieradzan, A. (1993) *Phys. Rev. A* **47**, 340–349.
9. Manakov, N.L. and Ovsiannikov, V.D. (1980) *Zh. Eksp. Teor. Fiz.* **79**, 1769–1778.
10. Fainshtein, A.G., Manakov, N.L., Ovsiannikov, V.D., and Rapoport, L.P. (1992) Nonlinear susceptibilities and light scattering on free atoms, *Phys. Rep.* **210**, 112–222.
11. Kassaee, A., Rustgi, M.L., and Long, S.A.T. (1988) *Phys. Rev. A* **37**, 999–1002.
12. Gavrila, M. (1989) Free-free transitions of electron-atom systems in intense radiation fields, in F.A. Gianturco (ed.), *Collision Theory for Atoms and Molecules*, Plenum Press, New York, pp. 139–189.
13. Fainshtein, A.G., Manakov, N.L., and Marmo, S.I. (1994) *Phys. Lett. A* **195**, 358–361.
14. Manakov, N.L., Marmo, S.I., and Volovich, V.V. (1995) Circular dichroism in laser-assisted electron-atom scattering, *Phys. Lett. A*, in press.
15. Bunkin, F.V. and Fedorov, M.V. (1965) *Zh. Eksp. Teor. Fiz.* **49**, 1215–1221.
16. Kroll, N.M. and Watson, K. (1973) *Phys. Rev. A* **8**, 804–809.
17. Manakov, N.L. and Fainshtein, A.G. (1986) Optical orientation of atoms in two linearly polarized nearly resonant optical fields, in *Contributed papers of All-Union Seminar on optical orientation of atoms and molecules*, Leningrad, pp. 103–105.
18. Agre, M.Ya. and Manakov, N.L. (1995) Atomic orientation effects in light scattering due to dissipative processes, *J. Phys. B.*, in press.
19. Berakdar, J. and Klar, H. (1992) Circular dichroism in double photoionization, *Phys. Rev. Lett.* **69**, 1175–1177.

DYNAMICS OF MULTIPHOTON MOLECULAR IONIZATION AND DISSOCIATION

K. C. KULANDER(a), F. H. MIES(b), AND K. J. SCHAFER(c),
(a) Lawrence Livermore National Laboratory, Livermore, CA 94551
(b) NIST, Gaithersburg, MD 20899
(c) Louisiana State University, Baton Rouge, LA 70803

1. Introduction

In a strong laser field a molecule can eject electrons, producing a molecular ion, or dissociate to neutral or ionized fragments. The disposition of the energy transferred to the molecule from an intense pulse varies with the laser wavelength, intensity and pulse shape. It also depends strongly on the initial state of the molecule. In order to explore the molecular excitation dynamics fully, it is very useful to be able to follow the time evolution of the total wave function throughout the pulse. Unfortunately the high dimensionality of even the simplest molecular system makes it difficult to completely describe and display the complex interplay between the different internal modes. For this reason we have developed a simple model for the hydrogen molecular ion which includes only the two most important degrees of freedom, the internuclear separation and the distance from the center-of-mass of the nuclei to the electron. Thus the nuclear and electronic degrees of freedom are treated on an equal footing. We can solve numerically the Schrödinger equation for this model in a pulsed laser field to a high degree of accuracy. This allows us to plot the 2d time dependent density distribution of the wave function which greatly facilitates our understanding of this problem.

2. The Model

It is often very useful to explore the behavior of simplified models in order to understand real but intractably complicated systems. One frequently employed simplification is to reduce the dimensionality of the system to the most active or most important coordinates. Because a strong, linearly polarized laser field tends to transfer its energy to electrons along the direction of polarization, many useful and informative calculations for atomic systems have been carried out by confining the electronic motion to a single dimension [1]. The first to use this approximation for a multi-electron system were Pindzola, Griffin and Bottcher (PGB) [2] who

H. G. Muller and M. V. Fedorov (eds.), Super-Intense Laser-Atom Physics IV, 163–169.
© *1996 Kluwer Academic Publishers.*

investigated the dynamics of a model for helium in which the electrons were constrained to move only along the polarization axis. This reduced the problem to a manageable two spatial dimensions so numerically exact calculations could be carried out. They compared the full calculations to those using Hartree and SAE approximations [3]. Additionally they were able to visualize the excitation and ionization dynamics because the 2d wave function evolution could be followed completely using contour or three dimensional plots. In this work [2] PGB demonstrated and explained the failure of the Hartree approximation to accurately represent the time-dependent dynamics in the system.

To study molecules we have adapted this idea to reduce the spatial representation of the wave function for H_2^+ to two coordinates: R, the distance between the protons and z, the distance from the center-of-mass of the nuclei to the electron. This leads to a Hamiltonian given in atomic units by

$$H_0 = -\frac{1}{2\mu}\frac{\partial^2}{\partial R^2} - \frac{1}{2}\frac{\partial^2}{\partial z^2} + \frac{1}{\sqrt{R^2 + q_n}} - \frac{1}{\sqrt{(z - R/2)^2 + \tilde{q}_e}} - \frac{1}{\sqrt{(z + R/2)^2 + q_e}} \quad (1)$$

where μ is the reduced mass of the two nuclei and q_n (q_e) is a screening parameter which softens the Coulomb singularity when the nuclei (electron and nucleus) coincide. This model also takes into account the observed fact that molecules in high intensity laser fields are found to quickly align along the polarization direction [4].

We solve the time-dependent Schrödinger equation for this collinear model of H_2^+ in a pulsed field on a two-dimensional finite difference grid. The flux reaching the boundaries of the grid is absorbed using mask functions [5] to obtain either the dissociation probability at the large R-boundary, or ionization if the flux leaves the integration volume at large $|z|$. Varying the wavelength, pulse shape and peak intensity we can follow the differences in the evolution of the system by monitoring the structure of the wave function density distribution for a variety of conditions. We can also investigate the accuracy of the Born-Oppenheimer (BO) approximation by fixing R, calculating electronic eigenfunctions and treating the nuclear motion on coupled, adiabatic electronic states [6,7].

In Fig. 1 we show the two lowest BO potentials for the Hamiltonian in Eq. (2) for the screening parameter choices, $q_n = 0.03$ and $q_e = 1.0$. We obtain these by fixing R in Eq. (1) and solving the 1d Schrödinger equation for the electronic eigenfunctions. We find the expected bound σ_g ground state and the repulsive σ_u state. The equilibrium separation is slightly larger than in the real system, but for this choice of screening parameters the binding energy of the ground state and the vibrational constants are quite accurate.

BO molecular states are constructed by combining the adiabatic electronic states with the vibrational eigenfunctions of these potentials. We also obtain exact (fully coupled) ground and excited states by propagating the time-dependent Schrödinger equation in imaginary time with the full two-dimensional potential.

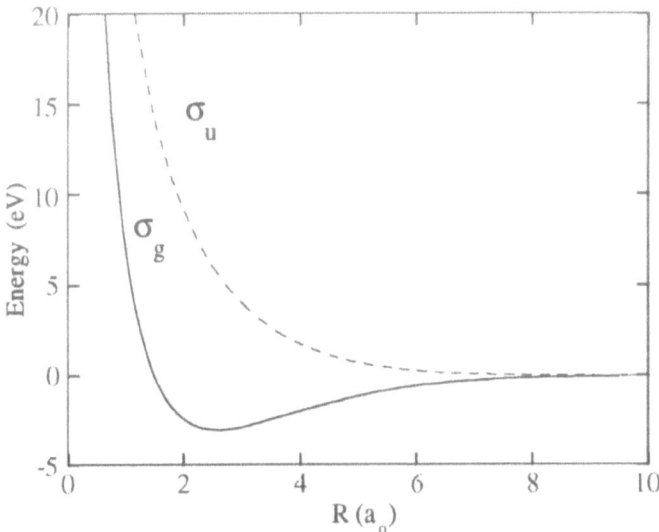

Figure 1. The lowest two Born-Oppenheimer potential energy curves for collinear H_2^+.

We then calculate to overlap of the approximate and exact wave functions. In this way we find that the BO states for the lowest several vibrational levels are very accurate when compared with the fully coupled states.

Fig. 2 shows a snap-shot of the time dependent wave function at the end of 15 optical cycles of a 770 nm laser whose intensity rose over 12.5 cycles with a $\sin^2$

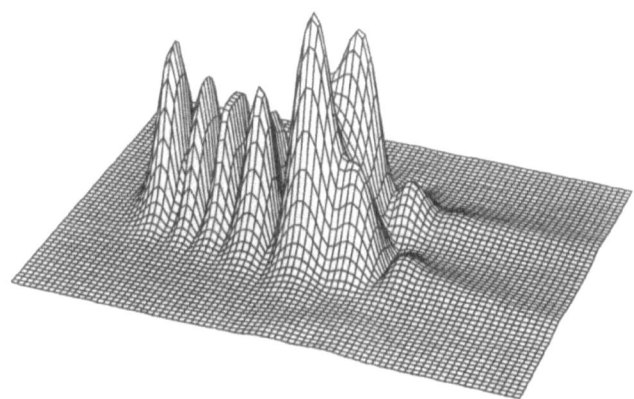

Figure 2. Wave function density distribution for collinear H_2^+ in an 8×10^{13} W/cm^2 770 nm laser field. The horizontal axis is R, the internuclear separation, and the vertical axis, z, the electronic coordinate.

ramp then was held constant at a value of 8×10^{13} W/cm^2. The structure of the initial $\nu = 4$ state is clearly evident, along with a large amplitude centered near 5 a_0 where the avoided crossing between the BO dressed (adiabatic) states occurs. This shows the presence of a laser-induced bound state [6]. Dissociation is represented by the weak density trails moving off toward the right side of the grid. The electron density follows the two separating nuclei. Little evidence for ionization, which would appear as flux moving toward the front or rear portions of the grid, is visible. However, substantial ionization is occurring. The velocity of the emitted electron is high enough that its density is difficult to see on the scale of this figure.

Observing the time evolution of this wave function we see the laser-induced state clearly inhibits dissociation. The small-R part of the wave function is actually quite different from the initial state both because of the interaction with the ungerade state which has modified the shape of the lower (mostly) gerade BO well and because other vibrational states have become populated. We find, by projecting the wave function onto the two lowest BO states, that there is also a small fraction of the total probability distributed among the excited electronic states.

The time dependent radial density distribution can be obtained from the wave function by averaging the total probability density over the electronic coordinate:

$$P(r,t) = \int |\Psi(R,z,t)|^2 dz \qquad (2)$$

$P(r,t)$ for the case discussed above is shown in Fig. 3. Initially we see the multi-peaked distribution of the $\nu = 4$ initial state. This structure is distorted as the intensity rises toward its maximum at 12.5 cycles and dissociation becomes evident. The evolution of the radial probability shows that there is a resonance between the vibrational state located within the σ_g well and a trapped state centered near the avoided crossing. Just after the ramp a substantial amount of probability begins to dissociate, but some of it is caught within the adiabatic dressed well. After several cycles, this leaks back into the "diabatic" state which resembles the $\nu = 4$ initial state with some fraction of that previously trapped being lost to dissociation. This probability shift between the adiabatic and diabatic states is repeated several times during the interval shown with a period of approximately 12 - 13 cycles. During the trapping periods there is clearly a marked inhibition of dissociation which ends as the probability flows back into the diabatic state. An estimate of the velocity of the dissociating fragments can be obtained from the slope of the dissociating flux in this figure. From this we conclude that there is a dominant component which has a fragmentation energy of approximately 1 eV. This is consistent with the absorption of two photons by the initial state.

We can estimate the distribution of internuclear separations at which ionization occurs by binning the flux crossing the large $|z|$ boundaries. Taking the time derivative of the probability flux reaching the z-gobbler within 1 a_0-wide R-bins, we can obtain the time and position dependent rates of the ionization. There is some

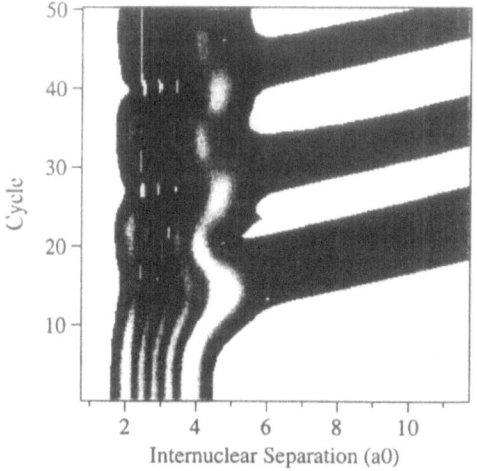

Figure 3. Time dependent radial d density distribution, $P(R,T)$, for collinear H_2^+ in a 770 nm laser field which ramps over 12.5 cycles to a then constant intensity of 8×10^{13} W/cm^2.

finite amount of time during which the nuclei begin to separate (Coulomb explode) while the electron propagates to the boundary. Therefore there is a slight increase in the radial distance at which we "detect" this flux compared to where it was initially freed. Because of the large velocity of the emitted photoelectron at this intensity, this displacement is less than 1 a_0.

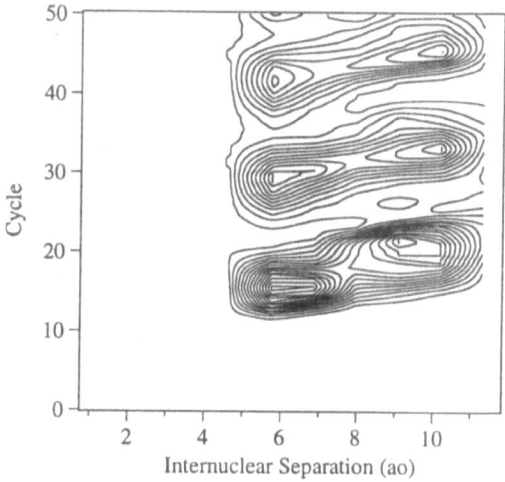

Figure 4. Time dependent ionization rate as a function of the internuclear separation for collinear H_2^+ in a 770 nm laser field which ramps over 12.5 cycles to a then constant intensity of 8×10^{13} W/cm^2.

The results of this flux analysis for this case are shown in Fig. 4 where the contours indicate the R-dependent ionization rate distribution. These partial rates have been scaled by the time-dependent norm. The R-dependent rate exhibits a broad maximum for large values of the internuclear separation resulting in ionization being probable only during dissociation. The peak in the R-dependent rate is due to a combination of the ionization rate increasing and the probability density decreasing as the internuclear separation becomes larger. The ionization rate increases because the electronic binding energy is a decreasing function of R. As has been shown by Posthumus et al. [8] and Seideman et al. [9] there is a very pronounced increase in the ionization rate for internuclear separations within the observed range shown here due to enhanced tunneling ionization. The bare nucleus on one side of the dissociating molecule works in concert with the electric field of the laser to lower the effective potential barrier, allowing the electron localized near the other nucleus to escape. Therefore the rate can significantly exceed that for the isolated atom. These fully coupled calculations show how significant this effect is. Even though only a small amount is dissociating at any time, all the ionization is produced in this region. Recent experiments on multi-electron diatomic systems [10] have shown that the Coulomb explosion to produce charged fragments occurs at bond lengths which are much larger (up to a factor of two) than the equilibrium bond length. Our calculated ionization distributions are consistent with these observations. The dissociative flux which survives this region of enhanced ionization would most likely escape the focal volume to be measured as an ion and a neutral atom - i. e., multiphoton dissociation.

3. Conclusion

This simple model of a molecular system has allowed us to probe in detail the dynamics of multiphoton excitation, ionization and dissociation. We were able to treat the nuclear and electronic degrees of freedom on an equal footing to investigate the transfer of absorbed energy from the electron to the nuclei and demonstrate the effects of nuclear motion on the absorption process. The model showed the recently observed large R dominance in Coulomb explosion experiments. Because the time dependent wave function is restricted to two dimensions we were also able to follow the time evolution of the wave function completely using surface and contour plots.

Acknowledgment

This research was carried out in part at the Lawrence Livermore National Laboratory under Contract No. W-7405-ENG-48.

References

1. Su, Q. and Eberly, J. H. (1991) Model atom for multiphoton physics, *Phys. Rev. A* **44**, 5997-6008.
2. Pindzola, M. S., Griffin, D. C. and Bottcher, C. (1991) Validity of time-dependent Hartree-Fock theory for the multiphoton ionization of atoms, *Phys. Rev. Lett.* **66**, 2305-2307.
3. Kulander, K. C. (1987) Time dependent Hartree-Fock theory of multiphoton ionization: Helium, *Phys. Rev. A* **36**, 2726-2738.
4. Zavriyev, A., Bucksbaum, P. H., Muller, H. G. and Schumacher, D. W. (1990) Ionization and dissociation of H_2 in intense laser fields at 1.064-μm 532-nm and 355-nm, *Phys. Rev. A* **42**, 5500-5513.
5. Kulander, K. C., Schafer, K. J., and Krause, J. L. (1993) Time dependent studies of multiphoton processes, in M. Gavrila (ed.), *Atoms in Intense Laser Fields*, Academic Press, New York, pp. 247-300.
6. Bandrauk, A. D. and Sink, M. L. (1978) *Chem. Phys. Lett.* **57**, 569.
7. Guisti-Suzor, A., He, X., Atabek, O. and Mies, F. H., Above-threshold dissociation of H_2^+ in intense laser fields (1990) *Phys. Rev. Lett.* **64**, 515-518.
8. Posthumus, J. H., Frasinski, L. J., Giles, A. J. and Codling, K. (1995) Dissociative ionization of molecules in intense laser fields - a method of predicting ion kinetic energies and appearance intensities, *J. Phys. B* **28**, L349-L353.
9. Seideman, T., Ivanov, M. Yu. and Corkum, P. B. (1995) *Phys. Rev. A* (in press).
10. Codling, K., Frasinski, L. J., and Hatherly, P. A. (1991) The dynamics of molecules in intense laser fields probed by covariance mapping, in *Multiphoton Processes* G. Mainfray and P. Agostini, (eds.) CEA, Saclay, pp. 205-216 and Schmidt, M., Normand, D. and Cornaggia, C. (1994) Laser-induced trapping of chlorine molecules with pico- and femtosecond pulses, *Phys. Rev. A.* **50**, 5037-5045.

References

LASER-INDUCED COULOMB EXPLOSION OF SMALL LINEAR POLYATOMIC IONS

C. CORNAGGIA
C.E.A.
DRECAM/Service des Photons, Atomes et Molécules
Centre d'Etudes de Saclay, Bât. 522
F-91191 GIF-SUR-YVETTE
FRANCE

1. Introduction

Multiple ionization of molecules and subsequent multifragmentation are studied using a variety of experimental techniques such as inner shell excitation with synchrotron radiation, ion and electron impact collisions, beam-foil interaction and laser excitation [1]. In particular in beam-foil experiments, molecular ions are accelerated to a velocity of a few percents of the speed of light and pass through a thin solid film (typically 50 Å thick). The valence electrons are stripped from the molecule in less than one femtosecond (1 fs = 10^{-15} s) and the resulting system explodes due to the strong mutual Coulomb repulsion between the multicharged atomic ions. Recently, the laser-induced Coulomb explosion of molecules has become possible with the advent of picosecond and femtosecond lasers, which can deliver peak intensities in the 10^{15}-10^{16} W/cm² laser intensity range, where the laser electric field is comparable to the molecular field experienced by the valence electrons [2, 3]. However the time scales of the laser-molecule coupling are longer than in the case of the beam-foil experiments and are respectively given by the laser period τ = 2.7 fs at λ = 800 nm and the laser pulse duration T > 50 fs for the available commercial Ti:Sa laser systems. These time scales are no longer short compared to the molecular vibrations (> 10 fs) and in consequence, the electrons are not scattered off suddenly from a frozen molecular ion core. The laser pulse duration is long enough for the developments of electron-nuclei couplings which can affect the geometry of the molecule.

In this report, the response of polyatomic species to a strong femtosecond laser field is determined through the overall fragmentation pattern and the geometrical configuration of the exploding molecules. The multifragmentation can be a direct instantaneous explosion yielding multicharged atomic ions:

$$[ABC^{M+}] \rightarrow A^{z^*+} + B^{z'+} + C^{z+} + \text{Kinetic Energy Release} \qquad (1)$$

171

H. G. Muller and M. V. Fedorov (eds.), Super-Intense Laser-Atom Physics IV, 171–180.
© 1996 *Kluwer Academic Publishers.*

or a sequence of unimolecular or laser-induced fragmentations of daughter molecular ions:

$$[ABC^{M+}] \rightarrow [AB^{N+}] + C^{Z+} + \text{Kinetic Energy Release 1} \qquad (2\text{-a})$$
$$[AB^{N+}] \rightarrow A^{Z'+} + B^{Z+} + \text{Kinetic Energy Release 2} \qquad (2\text{-b})$$

The experimental results show that the dominant mode of multifragmentation is a direct process with no intermediate daughter molecular ions production. In the case of diatomic molecules, much experimental data have been published in the litterature, since the pioneering work of Frasinski et al [2]. For molecules built with light atomic elements H, C, N and O, the measured total kinetic energy releases (E_{Exp}) are found to be 50% weaker than the kinetic energy releases expected from a pure Coulomb repulsion starting at the unperturbed molecular ion equilibrium internuclear distance (E_{Coul}) [3, 4]. In recent experiments performed with heavier molecules such as Cl_2 and I_2, the ratio E_{Exp}/E_{Coul} is measured to be 70 % [5, 6, 7]. The simplest interpretation of these results is to assume that the molecular ion is strongly perturbed by the laser field and explodes at a critical internuclear distance larger than its unperturbed equilibrium distance. This hypothesis has been confirmed by recent theoretical models based on field-ionization calculations [8-10] and non-pertubative integration of the Schrödinger equation [11-13], which predict a dramatic enhancement of the multiple ionization at internuclear distances larger than the equilibrium internuclear distances. The extension of the experiments to linear polyatomic systems will constitute conclusive tests for these theoretical works.

The second important goal of this work is to determine the geometry of the exploding molecule. Since the multiple ionization is predicted to occur at larger internuclear distances, the transition from the unperturbed nuclear configuration to the exploding one is expected to produce large amplitude nuclear motions. Moreover, the strong electronic polarizability induced by the intense laser field induces a torque on the molecular frame which aligns the molecule along the laser polarization direction [14-17]. In addition to this overall torque, the laser-induced anisotropic electronic distribution along the laser electric field will produce additional non uniform forces on the nuclei. In consequence, the electronic-nuclei couplings will produce deformations of the nuclear structure, which will be observed in the explosion of the molecule.

2. Experiment

The CEA/DRECAM laser is a commercial 130 fs pulse duration Ti:Sa laser system, which is operated at $\lambda = 790$ nm and focused intensities up to 2×10^{16} W/cm^2. The molecular multifragmentation into multicharged atomic ions is analyzed using time-of-flight mass spectrometry associated to statistical treatments of time-of-flight events. The ion detection set-up is a short drift tube Wiley and Mc Laren ion spectrometer [18] housed in a high vacuum chamber. The small dimensions of the spectrometer ensure no angular discrimination of the ejected ions using moderate collection electric fields in order to have a good time-of-flight resolution.

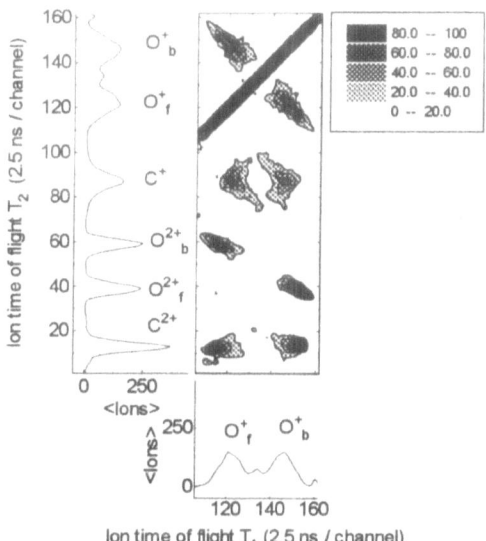

Figure 1. Double-correlation map of CO_2 recorded at $\lambda = 790$ nm and $I = 10^{15}$ W/cm² with the laser polarization direction parallel to the detection axis. The correlation coefficient $R^{(2)}(T_1, T_2)$ is multiplied by 1000 and represented using a five-level grey scale as a function of the horizontal T_1 and vertical T_2 time-of-flight axes.

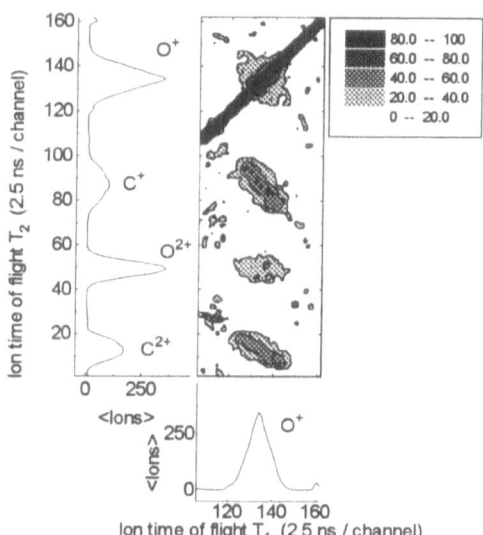

Figure 2. Double-correlation map of CO_2 recorded at $\lambda = 790$ nm and $I = 10^{15}$ W/cm² with the laser polarization direction perpendicular to the detection axis.

174

The spectrometer is operated so that the time of flight of an ion with initial momentum **P** is given by:

$$T(\mathbf{P}) - T(\mathbf{P} = 0) = P \cos(\theta_d) / (ZeF) \tag{3}$$

where P is the modulus of **P**, θ_d is the angle between the momentum **P** and the spectrometer cylindical axis, Z is the ion charge state number, e is the elementary charge and F is the first chamber collection electric field. Notice that only the component $P_d = P \cos(\theta_d)$ of **P** along the spectrometer axis is measured. However since the molecular frame gets aligned along the laser polarization direction, the experiments are performed at several angles between the laser polarization and detection axes in order to get the different components of the ions momenta. For instance, if $P_d = P \cos(\theta_d)$ is the detected component of **P** when the laser polarization is parallel to the detection axis, then the same event will give $P'_d = P \sin(\theta_d)$ in the perpendicular configuration.

In Fig. 1 and 2, the conventional time-averaged one-dimensional time-of-flight spectra are placed along the horizontal T_1 time-of-flight axis (bottom curve) and vertical T_2 time-of-flight axis (left curve). In the parallel configuration illustrated in Fig. 1, the CO_2^+ ions are aligned along the detection axis. In consequence, the oxygen ions are ejected toward (forward $O^{Z'+}_f$ ions) and backward the detector (backward O^{Z+}_b ions), while the middle carbon ions C^{Z+} ions exibit a single peak shape. In the perpendicular case, the time-of-flight spectra exhibit oxygen and carbon peaks with comparable widths, which give a first evidence for bending motions, since a stiff linear exploding molecule would produce narrower carbon peaks.

Often in particular for polyatomic species, one-dimensional time-of-flight spectra are insufficient to identify the fragmentation channels. Coincidences techniques are widely used for continuous or high repetition rate excitation sources. For relative low repetion rate lasers at 10 or 20 Hz, Frasinski et al have introduced an alternative and powerful technique called covariance mapping which is based on the correlated fluctuations of the ion signals coming from the same dissociation pathway [19, 20]. Statistical coefficients such as second-order $R^{(2)}(T_1, T_2)$ and third-order $R^{(3)}(T_1, T_2, T_3)$ correlation coefficients can be used to correlate respectively two ions with times of flight T_1, T_2 and three ions of times of flight T_1, T_2 and T_3. Since the times of flight are linear functions of the initial momenta of the ions (Eq. 3), this method produces two- and three-dimensional momenta correlation spectra. In Fig. 1 and 2, the central maps exhibit a wide variety of correlation peak shapes. For instance in the parallel configuration (Fig.1), the C^+/O^+ correlation peaks look like corners of a parallelogram and correspond to correlations of the C^+ ion with the forward and backward components of the O^+ ions. In the perpendicular configuration (Fig. 2), only a single correlation peak appears in the middle of the above-mentionned parallelogram. This comparison between the measurements of the same physical event in both configuration, shows the complementarity of the experimental data and the fact that the degeneracy of the ion momenta measurements can be overcome. Finally with circular polarization, the C^+/O^+ correlation peak appears as the whole parallelogram, since all the components of the momenta along the spectrometer axis are detected [21].

3. Direct instantaneous explosion

For linear carbon chains such as C_2H_2 or C_3H_4, the molecular contibutions to the time-of-flight spectra come from the single charge cations and doubly charged dications. No daughter molecular ions such as CH^+, C_2H^+ are detected. This is a clear signature of the direct explosion of C_2H_2 via $H^+ + C^{Z''+} + C^{Z'+} + H^+$ fragmentation channels and of the C_3 carbon chain of C_3H_4 via $C^{Z''+} + C^{Z'+} + C^{Z+}$ channels [22]. For CO_2, the situation is a little more complicated because of the presence of the CO^+ daughter molecular ion in the time-of-flight spectrum in addition of CO_2^+ and CO_2^{2+} ions. The CO^+ ions belong to three channels:

$$[CO_2^+] \rightarrow CO^+ + O + \quad (E < 1\ eV) \tag{4-a}$$
$$[CO_2^{2+}] \rightarrow CO^+ + O^+ + 5\ eV \tag{4-b}$$
$$[CO_2^{3+}] \rightarrow CO^+ + O^{2+} + 9\ eV \tag{4-c}$$

From double- and triple-correlation experiments, we detect $O^{Z''+} + C^{Z'+} + O^{Z+}$ channels such as for instance:

$$[CO_2^{6+}] \rightarrow O^{2+} + C^{2+} + O^{2+} + 60\ eV \tag{5-a}$$
$$[CO_2^{9+}] \rightarrow O^{3+} + C^{3+} + O^{3+} + 130\ eV \tag{5-b}$$

In the case of a sequential multifragmentation involving the CO^+ daughter ion, low energy single- or multicharged oxygen ions coming from channels (4-a-b-c) are expected to appear in the $O^{Z''+} + C^{Z'+} + O^{Z+}$ channels. However, only high energy oxygen ions are detected, and thus come from a direct instantaneous explosion.

For a particular channel $[ABC^{M+}] \rightarrow A^{Z+} + B^{Z+} + C^{Z+}$, the kinetic energy release spectrum exhibits only one maximum at E_{Exp}, which is called briefly the kinetic energy release of the channel. For linear molecular ion species composed of light atomic elements H, C, N and O, we observe basically two propensity rules:

$$\Delta Z = -1, 0, 1 \tag{6-a}$$
$$E_{Exp} = 0.5 \times E_{Coul} \tag{6-b}$$

The ΔZ values represent the differences between the charge states of the atomic multicharged fragments, which belong to a same fragmentation channel. This observed charge-symmetric fragmentation propensity rule is due to the fact that molecular multiple ionization occurs at the lowest ionization potentials. Let us take an example with the next channel observed from the $[C_3^{5+}]$ transient ion, which decays into the $C^{2+} + C^{2+} + C^+$ channel. The energy required to ionize C^+ is 24.4 eV and is significantly lower than the 47.9 eV ionization potential of the C^{2+} ion. In consequence, the removal of one electron from $[C_3^{5+}]$ which gives the $[C_3^{6+}]$ transient ion, is expected to produce the charge-symmetric output channel $C^{2+} + C^{2+} + C^{2+}$ rather than the asymmetric channels $C^{3+} + C^{2+} + C^+$ or $C^{2+} + C^{3+} + C^+$. Besides this energetic arguments, an other reason for the charge-symmetric fragmentation is the

176

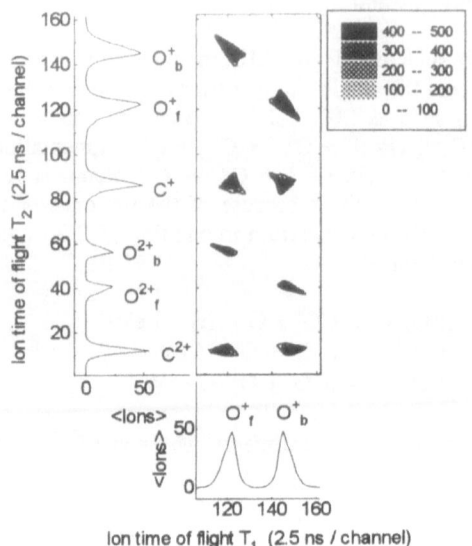

Figure 3. Calculated double-correlation map with CO_2^+ ions oriented along the detection axis and associated with the experimental data of Fig. 1.

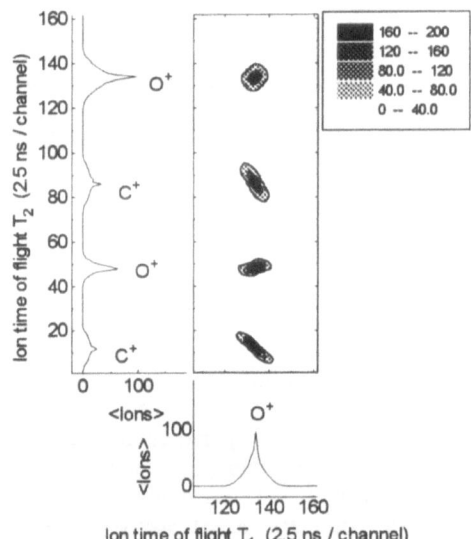

Figure 4. Calculated double-correlation map with CO_2^+ ions oriented perpendicular to the detection axis and associated with the experimental data of Fig. 2.

redistribution of the electronic charge density along the nuclear structure due to the oscillating high laser field so that the average charge per atom remains very close.

In Eq. 6-b, E_{Coul} is the kinetic energy release expected from a simple Coulomb repulsion starting at the ABC^+ molecular ion internuclear distances Re^+. The measured lower kinetic energy releases and constant ratios E_{Exp}/E_{Coul} is a generalization of the observations with diatomic species [2, 3]. The simplest explanation is to consider that Coulomb explosion occurs at internuclear distances R_c^+ larger than the unperturbed ion equilibrium internuclear distance R_e^+ due to the $1/R$ dependence of the Coulomb potential. The ratio R_c^+/R_e^+ is given by $R_c^+/R_e^+ = E_{Coul}/E_{Exp} = 2$ in our case. Since these ratios remain constant whatever the charge state of the exploding channel, the successive ionization steps of the initial molecular ion do not modify the average geometry from one ionization event to the next one. It could be an indication for a collective multiple ionization. These observations are in good agreement with recent models based on field ionization [9, 10], and on quantum time-dependent calculations [11-13]. These different approaches predict that Coulomb explosion is dramatically enhanced at critical internuclear distances larger than the equilibrium internuclear distances. In addition these distances are found to be independent of the fragmentation channel charge state as it is observed in this work for polyatomic species.

4. Molecular geometry: evidence for large amplitude nuclear motions

The geometry of the exploding molecular ion is analyzed using the measured atomic ions momenta in comparison with the calculated momenta from a repulsive Coulomb potential in the molecular frame:

$$V[\{R_i\}] = \sum_{i<j} Z_j Z_i / |R_j - R_i| \qquad (7)$$

where $\{R_i\}$ are the positions vectors of the repelling atomic ions. Assuming a negligible molecular ion kinetic energy before the explosion, the final momenta $\{P_i\}$ are only functions of the initial atomic positions. For instance for CO_2^+ ions, the momenta depend only on the initial $R = R(C-O)$ internuclear distance and $\gamma = \gamma(OCO)$ angle and can be represented by their modulus $P(R, \gamma)$ and angle $\beta(R, \gamma)$ with the Z axis of the molecular frame.

The laser-molecule interaction is easier to follow in the frame attached to the laser field, which can be represented by the unit vector $e_z = e_p$ parallel to the laser field. The alignement of the molecular frame for initial linear ions does not change the initial isotropic azimuthal angle distribution $D_\phi(\phi)$ in the laser frame, but produces a polar angle distribution $D_\theta(\theta)$ strongly peaked along the e_p vector. Finally, the measured momenta in the laboratory frame are obtained after two frame transformations:

$$P_d = P(R, \gamma) \{ \cos(\theta_{pd}) \cos[\theta + \beta(R, \gamma)] - \sin(\theta_{pd}) \sin[\theta + \beta(R, \gamma)] \} \qquad (8)$$

178

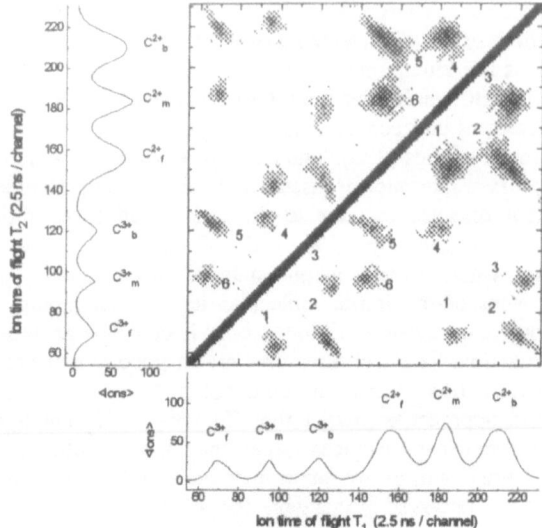

Figure 5. Double-correlation map recorded for allene at $\lambda = 790$ nm and $I = 5 \times 10^{15}$ W/cm^2 with the laser polarization direction parallel to the detection axis.

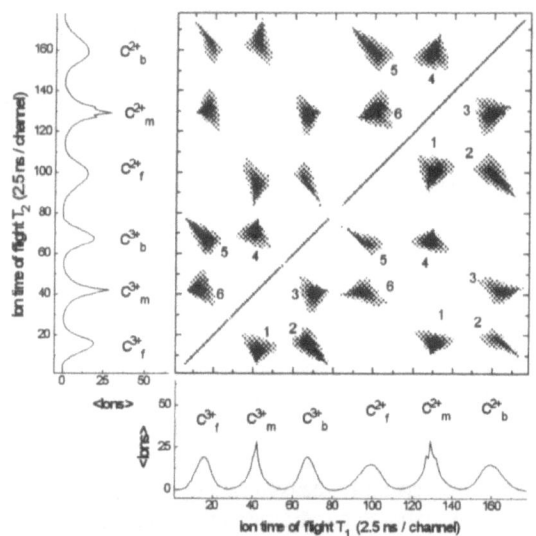

Figure 6. Calculated double-correlation map of allene associated to the experimental data of Fig. 5.

where θ_{pd} is the angle between the spectrometer and laser polarization axes. The initial positions of the atomic ions are given by a set of parameters (R, γ, θ) in the case of a triatomic molecule. The momenta correlations intensities are related to the (R, γ, θ) parameters distributions $D_R(R)$, $D_\gamma(\gamma)$ and $D_\theta(\theta)$, which are taken as independent triangular distributions centered respectively on R_c, γ_c, and θ_c with Half-Width-Half-Maximum ΔR, $\Delta\gamma$ and $\Delta\theta$.

For CO_2, a good agreement is obtained for $R_c = 2 R_e^+(C-O)$, $\gamma_c = 0$ and $\theta_c = 0$ associated to broad atomic positions distributions $D_R(R)$, $D_\gamma(\gamma)$, where $R_e^+(C-O)$ is the C-O internuclear distance of the unperturbed CO_2^+ ion. Fig. 3 and 4 represent the calculated double-correlation map associated to the experimental data of Fig. 1 and 2. In consequence, the molecule explodes for an average elongated linear configuration along the laser electric field and the large introduced HWHM ΔR and $\Delta\gamma$ give clear evidence for large amplitude stretching motions between $R_{min} = R_e^+(C-O)$ and $R_{max} = 3 R_e^+(C-O)$ and bending motions up to $\Delta\gamma_{max} = 2 \Delta\gamma = 40°$. The intense laser field induces a strong electronic polarizability, which involves bound and freed electrons from the molecular ion core. The Coulomb interaction between the displaced electrons along the laser electric field and the nuclei produces stretching and bending forces which are responsible for the observed large amplitude nuclear motions [21].

In order to show that these small-scale experiments could become a tool for stereostructure studies of molecules, studies were performed with the allene molecule which has has a linear C_3 chain in its ground state while the $C_3H_3^+$ and $C_3H_4^+$ ions are are known to exhibit linear, bent or cyclic carbon geometries in their ground states or a few eV above. In these experiments the $C_3H_3^+$ and $C_3H_4^+$ cations are produced at the beginning of the laser pulse and undergo Coulomb explosion for higher instantaneous laser intensities. Fig. 5 and 6 represent respectively the experimental and calculated double-correlation maps. An overall good agreement is obtained for an average linear elongated structure of the C_3 carbon chain [22].

5. Conclusion

In conclusion, the multifragmentation of small linear polyatomic ions is shown to be a direct process with no intermediate daughter molecular ion formation. The Coulomb explosion occurs on an average for elongated linear structure in good agreement with recent theoretical models developped in the case of diatomic molecules. Moreover large amplitude stretching and bending motions are observed using momenta correlations calculations in comparison with the experimental data. Further experiments are planned to analyze in detail the role of the dynamic electronic polarizability induced by the intense laser field, and the resulting electronic-nuclei couplings which are responsible for the observed large deformations of the nuclear structure.

180

6. Acknowledgements

The author is pleased to acknowledge P. D'Oliveira and P. Meynadier for operating the CEA/DRECAM laser system and M. Bougeard and E. Caprin for their skilled technical assistance.

7. References

1. Naaman R. and Vager Z. (1987) *The Structure of Small Molecules and Ions*, Plenum Press, New York and London, and references therein.
2. Frasinski L.J., Codling K., Hatherly P.A., Barr J., Ross I.N. and Toner W. (1987) Femtosecond dynamics of multielectron dissociative ionization by use of picosecond laser, *Phys. Rev. Lett.* **58**, 2424-2427.
3. Cornaggia C., Lavancier J., Normand D., Morellec J., Agostini P., Chambaret J.P. and Antonetti A. (1991) Multielectron dissociative ionization of diatomic molecules in an intense femtosecond laser field, *Phys. Rev* A **44**, 4499-4505.
4. Cornaggia C., Normand D. and Morellec J. (1992) Role of the molecular electronic configuration in the Coulomb fragmentation of N_2, C_2H_2 and C_2H_4, *J. Phys.* B **25**, L415-L422.
5. Dietrich P., Strickland D.T., Laberge M. and Corkum P.B. (1993) Molecular ions in intense laser fields, *Molecules in Laser Fields* edited by Bandauk A. (Marcel Dekker, New York), 181-216.
6. Hatherly P.A., Stankiewicz M., Codling K., Frasinski L.J. and Cross G.M. (1994) The multielectron dissociative ionization of molecular iodine in intense laser fields, *J. Phys.* B **27**, 2993-3003.
7. Schmidt M., Normand D. and Cornaggia C. (1994) Laser-induced trapping of chlorine molecules with pico- and femtosecond pulses, *Phys. Rev.* A **50**, 5037-5045.
8. Brewczjk M and Frasinski L.J. (1991) Thomas-Fermi-Dirac model of nitrogen ionized by strong laser fields, *J. Phys.* B **24**, L307-L312.
9. Posthumus J.H., Frasinski L.J., Giles A.J. and Codling K. (1995) Dissociative ionization of molecules in intense laser fields: a method of predicting ion kinetic energies and appearance intensities, *J. Phys.* B **28**, L349-L353.
10. Posthumus J.H., Giles A.J., Thompson M.R., Frasinski L.J. and Codling K. (1995) Small molecules in intense laser fields, *Super Intense Laser Atom Physics* IV in this issue.
11. Seideman T., Yu. Yvanov M. and Corkum P.B. (1995) Role of electron localization in intense laser field molecular ionization, submitted to *Phys. Rev. Lett.*
12. Kulander K.C., Mies F.H. and Schafer K.J. (1995) Dynamics of multiphoton ionization and dissociation, *Super Intense Laser Atom Physics* IV in this issue.
13. Zuo T. and Bandrauk A.D (1995) Charge-resonance-enhanced ionization of diatomic molecular ions by intense lasers, *Phys. Rev.* A **52**, 1-4.
14. Hatherly P.A., Frasinski L.J. Codling K., Langley A.J. and Shaikh W. (1990) The angular distribution of atomic ions following the multielectron ionisation of carbon monoxyde, *J. Phys.* B **23**, L291-L296.
15. Strickland D.T., Beaudoin Y., Dietrich Y. and Corkum P.B. (1992) Optical studies of inertially confined molecular iodine ions, *Phys. Rev. Lett.* **68**, 2755-2758.
16. Normand D., Lompré L.A. and Cornaggia C. (1992) Laser-induced molecular alignment probed by a double-pulse experiment, *J. Phys.* B. **25**, L497-L503.
17. Dietrich P., Strickland D.T., Laberge M. and Corkum P.B. (1993) Molecular reorientation during dissociative multiphoton ionization, *Phys. Rev.* A **47**, 2305-2311.
18. Wiley W.C. and Mc Laren I.H. (1955) Time-of-flight mass spectrometer with improved resolution, *Rev. Sci. Instrum.* **26**, 1150-1157
19. Frasinski L.J., Codling K. and Hatherly P.A. (1989) Multiphoton multiple ionisation of N_2 probed by covariance mapping, *Physics Letters* A **142**, 499-503.
20. Codling K., Frasinski L.J., Hatherly P.A. and Stankiewicz (1990) New triple coincidence techniques applied to multiple ionisation of molecules, *Physica Scripta* **41**, 433-439.
21. Cornaggia C. (1995) Experimental evidence for large amplitude nuclear motions in the laser-induced Coulomb explosion of CO_2^+ ions, submitted to *Phys. Rev. Lett.*
22. Cornaggia C. (1995) Carbon geometry of $C_3H_3^+$ and $C_3H_4^+$ molecular ions probed by laser-induced Coulomb explosion, submitted to *Phys. Rev.* A.

SMALL MOLECULES IN INTENSE LASER FIELDS;
IS THERE A PLACE FOR STABILISATION?

J H POSTHUMUS, A J GILES, M R THOMPSON,
L J FRASINSKI AND K CODLING

J J Thomson Physical Laboratory,
The University of Reading, Whiteknights, PO Box 220,
Reading RG6 6AF, UNITED KINGDOM

1. Introduction

At focused intensities in excess of 10^{14} W/cm^2 the ionization of *atoms* by infrared radiation can be understood in terms of tunnelling, with ionization rates given by a classical field ionization model [1,2]. It was pointed out by Frasinski *et al* [3] that by studying the multiple ionization of *diatomic molecules*, the dynamics of the multielectron dissociative ionization (MEDI) process can be investigated on a timescale of femtoseconds (the dissociation timescale), even if a laser pulse of picosecond duration is used. It is reasonable to assume that after ionization beyond the first or second charge state, the molecule starts to dissociate and subsequent ionization steps will therefore occur at increasingly large internuclear distances. Hence by measuring the dissociation energies of the various fragment ion pairs it should be possible to deduce the ionization and dissociation pathways of the molecule.

Two specific consequences of sequential ionization during dissociation at the leading edge of the laser pulse are: i) Short pulses should generally yield higher dissociation energies than long pulses, the highest possible energy being the Coulomb explosion energy, $E_e = Q_1 Q_2 / R_e$, where Q_1 and Q_2 are the charges of the fragments and R_e is the equilibrium internuclear distance in the neutral molecule. ii) Pulses which are much longer than the dissociation time (loosely defined as the time to reach an internuclear distance of $2R_e$) are expected to show post dissociative ionization (PDI), i.e. ionization at a large internuclear distance. This PDI process results in dissociation energy for the ion pairs with an energy typical for pairs with lower charges.

The first experiments on MEDI [3, 4] supported the field-ionization Coulomb explosion model sketched above, i.e. the dissociation energies increased with increasing charge states and they were considerably (~50 %) below the Coulomb explosion energy, indicating that ionization occurred whilst the molecule was dissociating. The introduction of covariance mapping [5], a triple coincidence technique for pulsed sources, was a big step forward with respect to the assignment of the dissociation partners [6].

Doubts arose about the validity of the field ionization model when experiments

181

H. G. Muller and M. V. Fedorov (eds.), Super-Intense Laser-Atom Physics IV, 181–191.
© *1996 Kluwer Academic Publishers.*

performed with different pulse lengths turned out to yield surprisingly similar ion spectra [7, 8], culminating in the experiment by Schmidt *et al* [9]. They showed that Cl_2 subject to pulses of either 130 fs at 790 nm or 2 ps at 610 nm, with the same peak intensity of 10^{15} W/cm^2, yield chlorine ion time-of-flight (TOF) spectra which are virtually identical. Moreover, for all ion pairs, the dissociation energy is always a specific fraction (e.g. ~70% for Cl_2, ~70% for I_2, ~45% for N_2) of the Coulomb explosion energy Q_1Q_2/R_c [9, 10]. It thus appears that the molecule relaxes to a critical distance, R_c, where it stabilizes and subsequently ionizes with little or no change in internuclear distance. We recently developed a model [11] that accurately predicts R_c for any ionization stage of a diatomic (or linear triatomic) molecule.

Schmidt *et al* [12] have recently performed a double-pulse experiment on Cl_2 using two 130 fs pulses, 1.5 ps apart. The second pulse had the same intensity (10^{15} W/cm^2) as the peak of a previously used 2 ps pulse. They observed PDI peaks in this experiment, but they did not observe any of these peaks in the 2 ps experiment. This, they suggest, is a further indication of stabilisation. We report here on our latest experiment on I_2 using short pulses with interferometrically controlled risetime [13]. But first we will give some more details on the field-ionization Coulomb explosion model [2,14] and the stabilisation model [9,11].

2. Theory

2.1 INERTIAL CONFINEMENT

The field-ionization Coulomb explosion model was introduced quite some time ago [2], but has never been fully exploited. It treats the laser E field and the molecule classically and the E field is applied along the internuclear axis of a diatomic molecule. We focus our attention on the outer electron and combine the remaining electrons and nuclei into two point-like atomic ions. Along the axis the double-well potential is given by (we use atomic units throughout this paper, unless explicitly stated otherwise)

$$U = -\frac{Q_1}{|x+R/2|} - \frac{Q_2}{|x-R/2|} - Ex \tag{1}$$

where E is the laser electric field amplitude. As in [14] we approximate the electron energy level by the expression

$$E_L = \frac{(-E_1-Q_2/R)+(-E_2-Q_1/R)}{2} \tag{2}$$

where E_1 and E_2 are the known ionization potentials of the atomic ions; these are lowered by the Coulomb potential of a neighbouring ion, Q/R. When the electric field is increased, the electron can go over the barrier and the molecule ionizes; this defines the threshold for ionisation. Figure 1(a) shows the ionisation sequence. We now allow the molecule to dissociate along Coulombic potential curves, i.e. the internuclear distance to increase. From (2) we note that the energy level rises for increasing internuclear distance, see figure 1(b), but in addition the laser field now acts over a longer distance. These two effects lower the threshold intensity.

At a certain internuclear distance the inner barrier becomes higher than the energy level and localizes the electron. Figure 1(b) illustrates what can happen now. As the field turns on, the Stark shift, which is given approximately by $\pm ER/2$, raises the level in the left hand well and lowers the level in the right hand well. Since we want to find the threshold intensity we keep increasing the field. At a certain field strength the level is shifted upwards sufficiently to overcome the inner barrier. Now this level is again delocalized and stops shifting. A further increment of the field is needed to lower the outer barrier below the electron energy level and produce ionisation. For even larger internuclear distances a third situation occurs, illustrated in Figure 1(c). This is similar to the second situation except that the inner barrier is much higher and when the

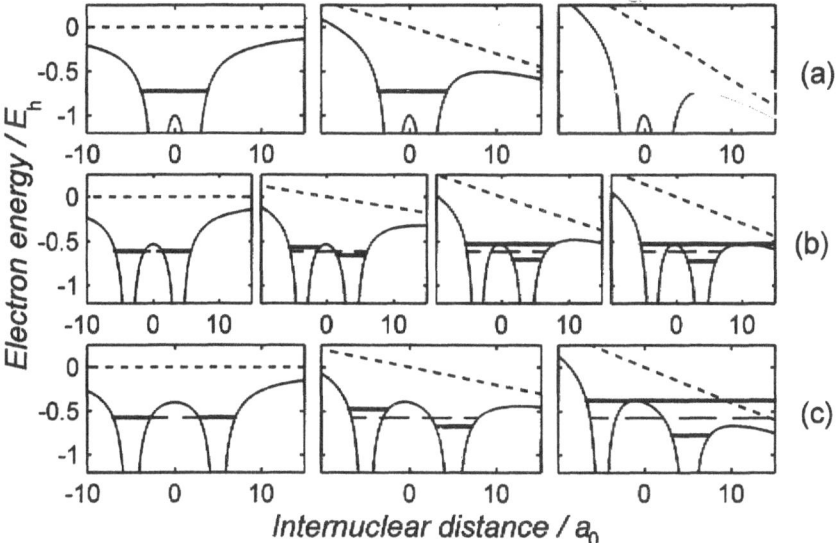

Figure 1. Field ionisation of diatomic molecules. (a) $R_e < R < R_c$ (b) $R_c < R < R_{min}$ (c) $R > R_{min}$

electron can finally go over it, the outer barrier is already lower than the inner one and the molecule ionizes immediately.

The resultant threshold intensities are drawn as solid curves in Figure 2 for several charge states of Cl_2 as a function of internuclear distance. The ranges of R over which the above three situations apply are defined by the two kinks in each of the curves. Codling et al [2] pointed out that there is an optimum ion separation which maximises the ionization process, R_{min}; we see these points as minima of the solid curves. However, it was realized only recently [11], that the optimum internuclear distances are almost the same for all charge states. This may well explain why all ion pairs appear to originate from the same critical internuclear distance, R_c [9, 10, 14].

To check this we have recently modelled a 100 fs sech^2 pulse of 10^{15} W/cm^2 focal intensity and calculated classical trajectories. Depending on the exact position in the focal volume, different tracks are followed on the Intensity versus distance plots; this is illustrated by the three short dashed curves in Figure 2. The curves, all starting at the

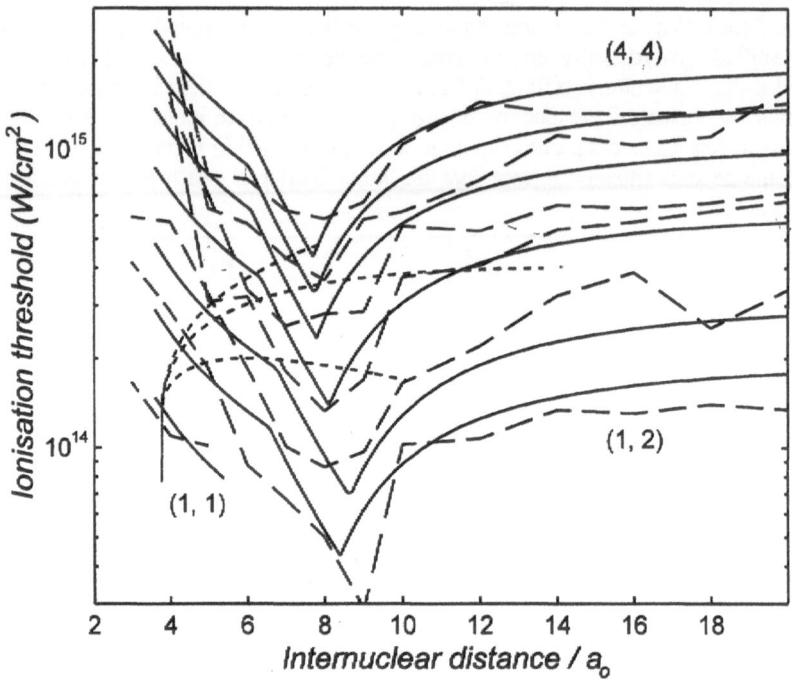

Figure 2. Solid curves: over-the barrier field-ionization. Short dashed: 1E15, 4E14 and 2E14 W/cm^2 pathways. Long dashed: Intensities for which Γ=5E13 s^{-1}.

equilibrium distance R_e= 3.8 a$_o$, show trajectories for positions in the focal volume where the maximum intensity reaches values of (top to bottom) 10^{15} (the exact centre of focus), 4 10^{14} and 2 10^{14} W/cm^2. Weighting each trajectory according to the spatial intensity distribution (Gaussian beam profile) we can construct an energy spectrum.

In Table 1, we compare the calculated dissociation energies (column 3) with the experimental values for pulses of 130 fs length [12]. The agreement is remarkably good. Thus far, the model fails to predict the correct branching ratios, e.g. (1, 1) is totally absent. This is even worse for 2 ps pulses, where the model only finds (2, 2) and (2, 3). However, we expect a substantial improvement if tunnelling is introduced into the model. Tunnelling is likely to fill in the empty spaces in Table 1, correctly for the low charge states, but the higher charge states for 2 ps will show PDI, i.e. if there are any (3, 3), (3, 4) and (4, 4) pairs, they will probably have a dissociation energy close to 26.4 eV, in contrast to experiment [12] which shows an ion energy spectrum for 2 ps virtually identical to the one for 130 fs.

TABLE 1. Dissociation energies in eV for Cl_2. Exp = experiment [9]; FICE = field-ionization Coulomb explosion model, based on threshold intensities; MC = Monte Carlo classical trajectory calculations, based on ionization rates; Stab = stabilisation model [11].

Channel	E_{Exp} [12] 130 fs	E_{FICE} 100 fs	E_{FICE} 2 ps	E_{MC} 100 fs	E_{MC} 2 ps	E_{Stab}
(1, 1)	5.0 ± 0.5			7.1	7.1	4.4
(1, 2)	10.7 ± 0.8	11.0		11.4 ± 0.9	10.3 ± 0.7	9.0
(2, 2)	20.1 ± 1.0	19.3	19.3	19.4 ± 1.7	15.1 ± 2.9	16.1
(2, 3)	30.2 ± 1.5	28.0	26.4	28.6 ± 2.2	20.1 ± 4.3	26.7
(2, 4)	41.0 ± 5.0					37.6
(3, 3)	44.3 ± 3.0	42.2		40.4 ± 3.2	14.1 ± 4.4	40.2
(3, 4)	60.5 ± 5.0	55.5		52.0 ± 3.9	11.6	55.3
(4, 4)	78.2 ± 8.0	71.0		64.9 ± 4.2		72.3

Seideman et al [14] have introduced a method to estimate ionization rates for molecules as a function of charge state and internuclear distance. This is done by solving the time-dependent Schrödinger equation for one electron in a softcore potential on a one-dimensional grid

$$U(x,t) = -\frac{Q}{\sqrt{(x+R/2)^2+a^2}} - \frac{Q}{\sqrt{(x-R/2)^2+a^2}} - xEf(t)\sin\omega t \qquad (3)$$

for fixed internuclear distance R and fixed laser E field. $f(t)$ is a function that ensures a smooth turn-on. For chlorine a value of $a = 2$ is appropriate. By terminating the grid with absorbing potentials and monitoring the decay as a function of time, they obtained ionization rates, Γ.

We have adopted this method and calculated a large data-matrix of ionization rates $\Gamma(Q, R, I)$ for all relevant charges Q, at many different values of R and laser intensity I. For asymmetric pairs, such as (2, 3), we have used $Q = (Q_1+Q_2)/2$. The long-dashed curves in Figure 2 show at what intensities an ionisation rate of $\Gamma = 5\ 10^{13}$ s^{-1} (lifetime 20 fs) is reached. The close agreement with the solid curves shows that the above "inertial confinement" model is essentially the same as the classical field ionization model [2].

Using this large matrix of ionization rates it is possible to simulate an experiment by performing classical trajectory Monte Carlo calculations. For each trajectory a point in the focal volume is selected by random numbers and the laser intensity $I(t)$ for this trajectory as a function of time is determined. Time is started when the laser intensity is still very low and is incremented in appropriately small steps Δt. For each step, Δt is multiplied by the relevant Γ, found by interpolation of the closest points in the data-matrix, and compared with a new random number between 0 and 1. If the random number is smaller than this value, the molecular charge increases by 1. We start with $(0, 1)$ and assume symmetric ionization, i.e. the next pair is $(1, 1)$, not $(0, 2)$, then come $(1, 2)$, $(2, 2)$, etc. From $(1, 1)$ on, we assume that the molecule dissociates along Coulombic repulsive curves and the internuclear distance $R(t)$ is calculated as a function of time. After sufficient trajectories a spectrum is constructed.

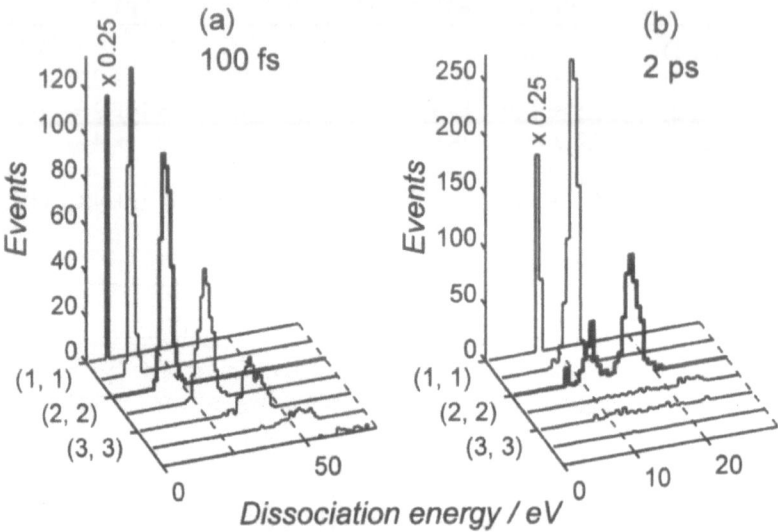

Figure 3. Spectra for pulses of (a) 100 fs and (b) 2 ps pulses of 10^{15} W/cm² focal intensity, simulated by Monte Carlo classical trajectory calculations.

Columns 5 and 6 of Table 1 show the dissociation energies, E_{MC} for 100 fs and 2 ps pulses of 10^{15} W/cm² focal intensity. The energies for 100 fs agree quite well with experiment, and furthermore the branching ratios are reasonable. However the 2 ps pulse confirm the limitations of the inertial confinement model: the really high charge states are not reached and an appreciable amount of PDI is observed. This is illustrated in Figure 3(a) and (b), for pulses of 100 fs and 2 ps respectively. We see that at 100 fs the energies of the various ion pairs increase as predicted by the field ionisation Monte Carlo model, but at 2 ps the results disagree with experiment, (the experimentally determined energies at 2 ps are very similar to those at 100 fs, see Table 1). One can see that for the $(2, 2)$ channel there are, apart from the expected peak at approximately 17 eV, extra peaks at 7 and 11 eV, indicating PDI from $(1, 1)$ and $(1, 2)$ respectively.

2.2 LASER-INDUCED STABILISATION

In the stabilisation model [9,11] it is assumed that after the first ionization, the singly charged molecule starts to relax due to bond-softening [15], but instead of dissociating, it enters the bond-hardening [16, 17] regime, at about R_c. In the same way that bond-hardening occurs in H_2^+, it should also occur in heavier molecules under the appropriate conditions of laser intensity and rise time. The single electron approach seems unable to provide effective binding for highly ionized molecules. However, in the present high intensity regime it is possible that several electrons oscillating between the wells can provide additional bonding. Schmidt et al [9] suggest that transitions between charge resonance states may be important for stabilisation.

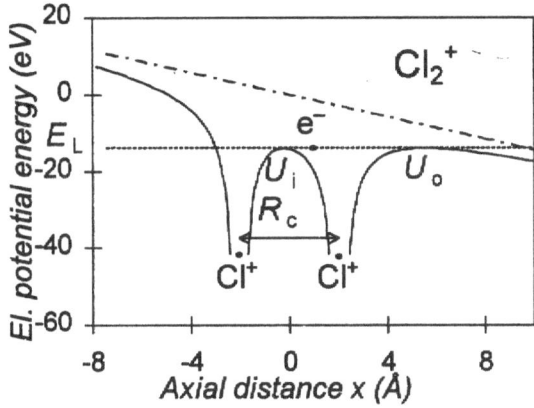

Figure 4. Stabilisation model. The critical internuclear and the threshold intensity are such that the inner as well as the outer potential barrier touch the electron energy level.

Figure 4 illustrates how to determine R_c. We use formulas (1) and (2) for the double-well potential and the electron energy level. At the equilibrium distance the inner barrier, U_i, (see Figure 4) is well below the energy level and therefore the electron can move freely between the two wells. But as the ions move apart due to bond-softening, the inner barrier rises and at a certain, critical internuclear distance, it starts to impede the motion of the electron. The electron starts to lag behind the field until it is in total antiphase, i.e. a transition to bond-hardening has been made. Clearly, the critical distance must be close to the point where the inner barrier touches the energy level. At R_c, the electric field is allowed to increase to a value, E_c, such that the electron can go over the outer barrier, U_o, (see Figure 4) and the molecular ion is field-ionized. After ionization, the next electron has a lower energy level (higher binding energy), but since the molecule is now more positively charged the inner potential barrier is also lower. It turns out that R_c is virtually the same for every charge state. The last column of

Table 1 shows the dissociation energies calculated from $E_{Stab} = Q_1 Q_2 / R_c$.

In summary, there are at present two models that attempt to explain the fact that the ion dissociation energies are given by the formula $Q_1 Q_2 / R_c$ for all values of Q_1 and Q_2. The inertial confinement model envisages uninhibited separation of the fragment ion pairs as they sequentially ionize. The reason that the ion pairs have such energies is that the sequence of ionization processes is extremely rapid at or near R_c. The ionization rates peak at around R_c [14] and fall off as the ions move further apart. This model agrees with the experimental data on Cl_2 at 130 fs because the laser rise time is similar to the dissociation time. The model cannot explain the 2 ps data; computer modelling shows very few ions of charge state higher than 2+ and substantial PDI for 2+ ions, but experimentally the spectrum is virtually identical to the one obtained at 130 fs.

The stabilisation model, on the other hand, assumes that the atoms, or ions of low charge stage, relax to a distance R_c at which point they are stabilised by the laser field i.e. bond-hardening occurs. However, it seems unlikely but perhaps not impossible that highly-ionised molecular ions (up to Cl_2^{8+}) could be stabilised by an oscillating E field; this would require many electrons to move together in antiphase with the laser field.

Since both models invoke multiple ionization close to R_c, they cannot easily be distinguished by measuring the kinetic energies of the fragment ions. Nevertheless, what happens between the equilibrium distance R_e and R_c is crucial. If the stabilisation model is correct, it means that some time is required to form the bond-hardened state, because the singly-charged molecule has to relax via bond-softening in order to reach R_c. If the pulse rise time is fast, further ionization can occur before R_c is reached and thus the field-ionization inertial confinement model should apply. We have recently suggested that an experiment involving the interferometric control of the laser pulse rise time might shed light on the situation. The rise time can be varied in a well-controlled manner, whilst leaving the other characteristics of the laser unaffected [13].

3. Experiment

The experiment consists of a femtosecond laser, Michelson interferometer and ion time-of-flight (TOF) spectrometer. The laser system consists of a Spectra Physics argon-ion-pumped Ti:sapphire laser amplified by a 3-stage dye amplifier pumped by a mode-locked, Q-switched, Nd:YAG laser [18]. The laser wavelength is 750 nm and the pulse width (FWHM) about 50 fs. The interferometer can vary the rise time in a precise way by using constructive and destructive interference. The ion TOF spectrometer incorporates two drift tubes, one to obtain branching ratios using covariance mapping [5], the other, somewhat longer, to give good energy resolution. The laser E-field lies along the drift tube axes and, because of their anisotropic angular distribution, the fragments ions are emitted either towards (forward ions) or away from (backward ions) the detector. This leads to the double-peaking seen in figure 5.

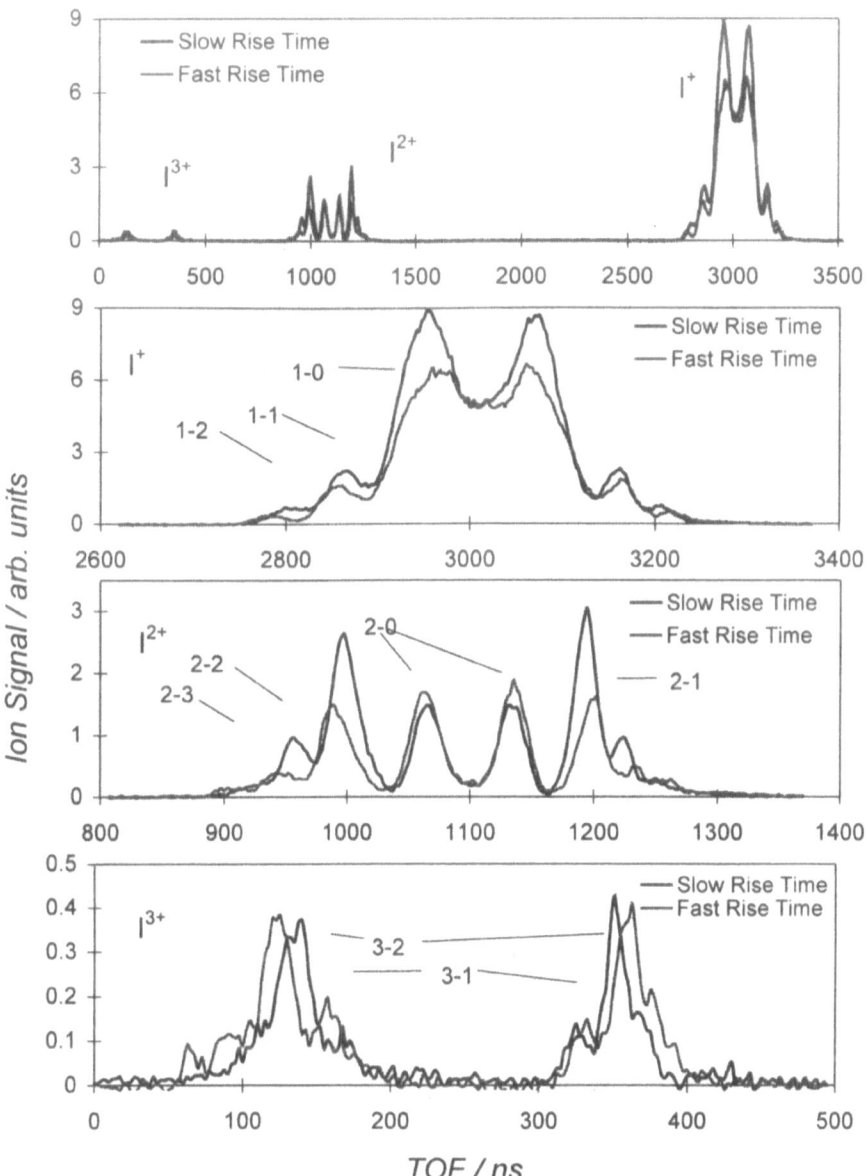

Figure 5. TOF spectra of Iodine ions.

190

4. Results and Discussion

The effect of change in rise time by interferometric means [19] is illustrated in figure 5 for a focused laser intensity of about 10^{15} W/cm^2 and a pulse length of 50 fs. The top TOF spectrum is expanded in the lower figures to give greater detail. The various groups of ions show the forward-backward doubling mentioned above. We note that in general the energies of the various ion pairs increase with decreasing pulse width (a detailed analysis can be found in [19]). The shifts appear to be consistent with the field-ionization inertial confinement models. Another interesting feature of the spectra is the enhancement of the symmetric channels (2, 1) and (2, 2) for the slow rise time. However, this enhancement is not at the expense of the asymmetric channels. For example, the (2, 1) channel does not replace the (3, 0) channel (In fact the (3, 0) channel is not observed in either spectrum). One should bear in mind that angular distributions are more peaked along the laser E-field for longer pulses [10] and therefore the ions are more effectively collected.

5. Conclusions

Two approaches have been used to attempt to explain the MEDI process, inertial confinement and laser-induced stabilisation; both utilise R_c as an important parameter. We have shown that R_c, the critical internuclear separation, is independent of the stage of ionization and reproduces experimental dissociation energies for the various fragment ion pairs [11]. In the first model R_c is the point at which ionisation rates are greatly enhanced, i.e. the molecule is simply inertially confined, whereas in the second it is the position at which stabilisation followed by ionisation occurs.

We have performed careful computer modelling of the MEDI process, taking into account the intensity distribution throughout the focal volume and assuming rapid sequential ionisation at R_c [14]. These model calculations agree with experiment in the case of Cl$_2$ with a pulse 130 fs. However, at 2 ps there is disagreement in that high charge states are totally absent and substantial PDI is present in the model, whereas the experiment yields a spectrum similar to the one measured with 130 fs pulses.

Since in both of these models the dynamics between R_e and R_c is critical, we have performed experiments to probe this region using interferometric control of laser pulse rise time. What we observed can be explained in terms of simple field ionization combined with Coulomb explosion and inertial confinement. However, that is not to say that under the appropriate conditions of laser rise time and dissociation time stabilisation could not occur.

In answer to our question "Is there a place for stabilisation?" , the jury is still out. A model based on inertial confinement, although conceptually more satisfying, seems unlikely to be able to yield identical ion spectra for 130 fs *and* 2 ps for Cl$_2$, even if substantial modifications are introduced. On the other hand, the laser-induced stabilisation model, which can explain the experimental results obtained to date, requires a rather exotic multielectron stabilisation mechanism.

We are indebted to W Shaikh and A J Langley from the Rutherford Appleton Laboratory for the successful operation of the laser and their assistance with the experiment. This work was made possible due to financial support from the Engineering and Physical Science Research Council

6. References

1. Augst, S., Strickland, D., Meyerhofer, D. D., Chin, S. L., and Eberly, J. H. (1989) *Phys.Rev. Lett.* **63**, 2212-5
2. Codling, K. , Frasinski, L. J., and Hatherly, P. A. (1989) *J. Phys. B.* **22**, L321-7
3. Frasinski, L. J., Codling, K., Hatherly, P. A., Barr, J., Ross, I. N., and Toner, W. T. (1987) *Phys. Rev. Lett.* **58**, 2424-7
4. Normand, D., Cornaggia, C., Lavancier, J., Morellec, J., and Liu, H. X. (1991) *Phys. Rev. A* **44**, 475-82
5. Frasinski, L. J., Codling, K., and Hatherly, P. A. (1989) *Science* **246**, 1029-31
6. Frasinski, L. J., Codling, K., and Hatherly, P. A. (1989) *Phys. Lett.* **142A** , 499
7. Cornaggia, C., Lavancier, J., Normand, D., Morellec, J., Agostini, P., Chambaret, J. P., Antonetti, A. (1991) *Phys. Rev. A* **44**, 4499-4505
8. Codling K., Frasinski, L. J., and Hatherly, P. A. (1991) *Multiphoton Processes* ed G. Mainfray and P. Agostini (Saclay: CEA)
9. Schmidt, M., Normand, D., Cornaggia, C. (1994) *Phys. Rev. A* **50** 5037-45
10. Hatherly, P. A, Stankiewicz, M., Codling, K., Frasinski, L. J., and Cross, G. M. (1994) *J. Phys. B.* **27**, 2993-3003
11. Posthumus, J. H., Frasinski, L. J., Giles, A. J., and Codling, K. (1995) *J. Phys. B.* **28**, L349-53
12. Schmidt, M., Normand, D., Lewenstein, M., and D'Oliveira, P. (1995) *Light Induced Stabilization of Molecules in Intense Laser Fields,* to be published
13. Giles, A. J., Posthumus, J. H., Thompson, M. R., Frasinski, L. J., Codling, K., Langley, A. J., Shaikh, W., and Taday, P. F. (1995) *Interferometric shaping of femtosecond laser pulses,* appearing in Optics Comm.
14. Seideman, T., Ivanov, M. Yu., and Corkum, P. B. (1995), *The Role of Electron Localization in Intense-Field Molecular Ionization,* to be published
15. Bucksbaum, P. H., Zavriyev, A., Muller, H. G., and Schumacher, D. W. (1990) *Phys. Rev. Lett.* **64**, 1883
16. Giusti-Suzor, A., and Mies, F. H. (1992) *Phys.Rev. Lett.* **68**, 3869
17. Zavriyev, A., Bucksbaum, P. H., Squier, J., and Saline, F. (1993) *Phys.Rev. Lett.* **70**, 1077
18. Langley, A. J., Noad, W. J., Ross, I. N., and Shaikh, W. (1994) *Appl. Opt.* **33** 3875
19. Posthumus, J. H., Giles, A. J., Thompson, M. R., Frasinski, L. J., Codling, K., Langley, A. J., and Shaikh, W. (1995) *Interferometric control of the disssociation dynamics of Iodine,* to be published

SINGLE-ATOM AND PLASMA PROCESSES AT INTENSITIES UP TO 10^{18} W/cm²

H. ROTTKE, J. LUDWIG, M. DÖRR,
P. V. NICKLES, M. SCHNÜRER, M. P. KALACHNIKOV,
T. SCHLEGEL, N. DEMCHENKO*, AND W. SANDNER
Max-Born-Institut
Rudower Chaussee 6, P.O. Box 1107, D-12474 Berlin, Germany
Physical Lebedev Institute, Moscow, Russia

1. Introduction

The interaction of high intensity laser radiation with single atoms and molecules and dense media has attracted much interest during the last years. Experiments and theory on single particle-light interaction aim at the basic understanding of the processes in a regime where perturbation theory fails to account for the interaction. Experimental investigations in this field have mainly been done in an intensity range up to about 10^{16} W/cm² with pulsed radiation in the spectral range from the IR up into the UV. The pulse widths used range down to about 100 fsec. In dense media the effort aims at the understanding and control of plasma generation and laser-plasma interaction at extreme light intensities of 10^{18} W/cm² and beyond, with applications for example in new schemes for x-ray lasers, intense incoherent x-ray sources, or charged particle acceleration in the large plasma wake-fields generated. At these high intensities relativistic effects in the electron motion in the oscillating light field are beginning to be noticeable. Relativistic effects are governed by the Lorentz invariant ratio U_p/mc^2, where U_p is the ponderomotive energy of the electron and mc^2 the electron rest energy. When this ratio becomes of order unity relativistic effects set in.

In atomic systems important discoveries made in the high light intensity range are above threshold ionisation (ATI) [1], and the importance of resonances, induced by large relative AC-Stark shifts of bound atomic states, in 'non-resonant' multiphoton ionization (MPI) [2]. The role resonances play could only be revealed after intense light pulses in the sub-psec time domain had become available. Only under these circumstances the ponderomotive acceleration of low energy photoelectrons is small enough to enable the detection of the kinetic energy they are born with in the photoionization process. But just this is essential to detect the resonant ionization processes as resonance enhancements in the photoelectron yield at specific electron kinetic energies. Experimental work on MPI of atomic hydrogen was able to confirm the observations made on ATI and on the role resonances play and made possible the

193

H. G. Muller and M. V. Fedorov (eds.), Super-Intense Laser-Atom Physics IV, 193–207.
© *1996 Kluwer Academic Publishers.*

first detailed comparison between experiment and theory [3,4]. Recent investigations on atoms try to reveal the role of electron-electron correlation in strong field ionization with experiments done on helium [5,6]. In the long wavelength optical tunneling regime the present interest aims at rescattering mechanisms of the photoelectron on the ion core after it has been accelerated in the strong radiation field. This process seems to be responsible for high order harmonic generation in this regime and certain features in the photoelectron kinetic energy spectra and angular distributions [7,8,9].

Molecules in a high intensity radiation field give rise to new dynamical phenomena through the additional nuclear degrees of freedom. Nuclear motion is influenced in an indirect way through the modification of the electonic charge cloud in the strong light field. Two recent reviews by Giusti-Suzor et al. [10] and Codling and Frasinski [11] give a detailed overview over this field. Of special experimental and theoretical interest is the hydrogen molecule H_2 and its singly charged ion H_2^+. High light intensity phenomena specific for molecules are the process of Coulomb explosion of at least doubly charged species after localisation of at least one elementary charge on each of the final fragments. Details of the interaction of specific electronic states with intense laser fields are studied specifically on the H_2^+ $1s\sigma_g$ ground and $2p\sigma_u$ dissociative first excited state. The electric dipole coupling between these states gives rise to phenomena like bond softening, above threshold dissociation (ATD), and laser induced bound states for the nuclear motion (see [10] and references cited there).

Investigation of the interaction of highly intense laser pulses with solid matter is attracting the interest, because it is now possible to create short-lived plasmas of high temperature and density. Due to the high laser intensity more or less all the processes during the creation of the plasma, like energy coupling, hydrodynamics and radiation processes, are strongly influenced by the strong laser field and the ponderomotive pressure. The first question arising here concerns the absorption of laser energy by the small scale hot plasma with steep density gradient. In plasmas generated by longer pulses with moderate flux densities ($I_L \sim 10^{14}$ W/cm^2), collisional absorption mechanism is dominant. With higher laser intensities, collisionless absorption processes become more important. Strong resonance absorption was deduced in hot plasmas (see e.g. [12]). Furthermore it was found that laser energy resonantly absorbed in the plasma critical density region gives rise to the generation of high temperature electrons [13]. A similar situation can be expected when a solid target is irradiated with short intense light pulses. Additionally the picture complicates due to strong plasma turbulences, which may be caused by the competition between the large ponderomotive and thermal forces in the dense plasma. There are now several theoretical models describing the scenario of the energy coupling on the base of PIC- and hydro-codes [14-18]. Furthermore, experimental investigations of the real plasma absorption in the intensity range up to ~10^{17} W/cm^2 have been done with laser pulse parameters available at the different laser installations [19]. For still higher intensities up to 10^{18} W/cm^2 such measurements were carried out recently [20]. It followed that particularly pulse duration, polarization and contrast ratio remarkably influence the overall absorption. As predicted by the theory [15,16], resonance processes in the plasma enhanced by ponderomotive steepening of the field distribution can effectively accelerate free electrons to extremely high energies.

The yield and the energy scaling of these hot electrons as a function of laser intensity is of great importance. This interest is mainly caused by the fact, that corresponding to a new proposal of a fast fusion igniter [21] the short intense laser pulse generated hot electrons should be able to heat the fuel to the necessary burning temperature resulting in a enhancement of the gain factor by more than one order of magnitude and reducing the demand on symmetry of the irradiation. On the other hand the hot electrons can penetrate into the solid and create K-line emission in the X-ray range as well as Bremsstrahlung emission in the hard X-ray region [20, 22]. Both types of emission will become important as short X-ray pulses suited well for X-flash radiography [23]. Therefore investigations are to be done to study the dependence of the hard X-ray emission on laser parameters as well as hot electrons in more detail.

The experiments we want to discuss in two parts are first, investigations into the ionization/dissociation dynamics of H_2/D_2 at 527 nm and about 0.5 psec pulse width in an intensity range up to ~5 x 10^{14} W/cm^2. They extend experiments with long laser pulses (>50 psec) at this wavelength reported by Zavriyev et al. [24] and Yang et al. [25] into the short pulse regime, where for example resonances in MPI of H_2 can be revealed and the saturation intensity of all processes is shifted to much higher values. Second, laser plasma interaction experiments will be presented, with the plasma generated through coupling of 1053 nm radiation into Al- and Ta-solid targets. Laser intensities reached on the target are 10^{18} W/cm^2 in pulses with 1.5 psec and 0.7 psec pulse width.

2. Experimental

For the experiments on the hydrogen molecule we use a Kerr-lens mode locked Ti:Sapphire laser system pumped by a CW Argon ion laser as primary light source. The pulses from this laser are amplified in a Ti:Sapphire regenerative amplifier using the chirped pulse amplification technique (CPA). Pumping of the amplifier is done by the second harmonic of a Q-switched Nd:YAG laser. The output of the amplifier consists of pulses at 1053 nm center wavelength with a repetition rate of 10 Hz, about 1.2 mJ pulse energy, and ~0.6 psec pulse width after final compression. This laser is the front end of a large Ti:Sapphire-Glass laser system used for the laser plasma interaction experiments.

Radiation at 527 nm is generated by frequency doubling the output of the regenerative amplifier in a Lithium-Iodate crystal. The energy of these pulses reached ~0.5 mJ at a pulse width similar to the IR pulses. Focusing of this beam into our vacuum chamber allowed us to reach a peak intensity of about 5 x 10^{14} W/cm^2 in a nearly Gaussian focal spot.

Analysis of the kinetic energy of photoelectrons and H^+ /D^+ ions generated is done in a field free time of flight spectrometer with the time measurement performed by a multiple stop time to digital converter. The spectrometer has an energy resolution of ~20 meV near 1 eV kinetic energy. It is installed in an UHV vacuum vessel with a base pressure of 2×10^{-10} mbar. Molecular hydrogen is introduced into the system through a precision leak valve.

The experiments on plasma absorption and hard X-ray emission were carried out at our CPA glass laser system [26,27]. The main parameters of the laser system are the following: pulse duration 1.5 - 2 ps, energy on target 1.5 J and a contrast ratio 500 ps before the pulse peak better than 10^9 or 10^{10} with an additional nonlinear absorbing dye cell. Moreover with the dye cell in place also the temporal pulse shape strongly changes and becomes steeper as well (Fig. 1). The intensity on target with an f/2 aspheric lens in the vacuum target chamber was determined to 10^{18} W/cm². Recently we have also studied the intensity dependence of hard X-ray emission using shorter laser pulses of a new CPA Ti:Sapphire-glass laser system. The results are given in ref. [20]. This hybrid laser uses at the front end the ~1050 nm Ti:Sapphire oscillator (TSUNAMI type) and the Ti:Sapphire regenerative amplifier described above. This system delivers 0.7 ps pulses with 2.5 J pulse energy and a focused intensity up to 3×10^{18} W/cm².

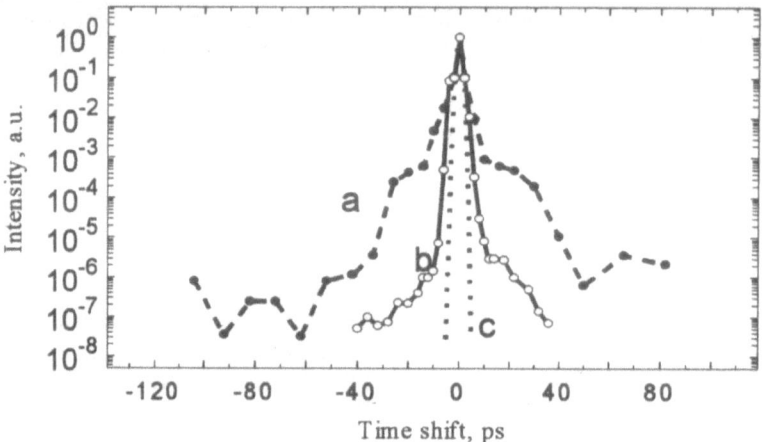

Figure 1. CPA glass laser pulse shape (third order correlation traces with 2 ps FWHM). a-dashed line: pulse after grating compressor. b-solid line: pulse with steepened front due to nonlinear absorption in an additional dye cell. c-dotted line: 2 ps Gaussian shape

In the plasma absorption experiment an Al plasma was created by irradiating solid Al targets. To determine the dependence of absorption on laser intensity, polarization, and angle of incidence on the target as well as on contrast ratio of the laser pulse the energy scattered by the plasma was measured by a calibrated Ulbricht sphere. This sphere is supplied with diode detectors. A calorimeter was used to record the backward reflected laser light. The plasma dynamics was studied in dependence on laser parameters by measuring the spectral structure of the laser light reflected back at the plasma critical surface with a grating spectrograph with CCD camera read out [18]. Analysis of the hard x-ray emission depending on laser intensity/energy and angle of incidence were done for flat tantalum and aluminum targets by measuring the hard

X-rays using a scintillator, an ionization chamber, and different thermo luminescence detectors (TLD). The TLD's were fabricated and calibrated by the Physikalisch-Technische-Bundesanstalt Braunschweig (PTB). The energy window of the TLD was selected by different filters in front of them. Values for the hot electron energy and yield dependence on laser energy were evaluated within the framework of a Bremsstrahlung model from the measured hard x-ray dose.

3. Results and discussion

3.1. MPI OF MOLECULAR HYDROGEN

Starting with the total yield of ions Fig. 2 shows the dependence of the dissociation ratio, i. e. the ratio of H^+ yield to the sum of the H^+ and H_2^+ yields $H^+/(H^+ + H_2^+)$, on the intensity of the 527 nm laser radiation up to a maximum intensity of 4×10^{14} W/cm². The pulse width in this experiment was 0.77 psec. Up to an intensity of ~1.3×10^{14} W/cm² the ratio shows a fast rise from about zero at 1×10^{13} W/cm² up to 0.6, then the ratio saturates at about 0.8.

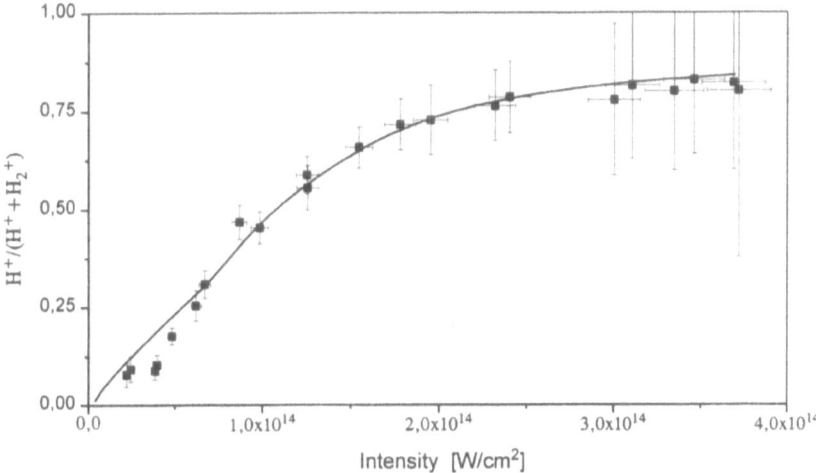

Figure 2. H₂ dissociation ratio after irradiation with 0.77 psec pulses at 527 nm center wavelength. The full
line is the result of a simulation based on a simple rate model for the processes.

We tried to simulate this behaviour using a simple rate approach to gain information on the saturation behaviour of the individual processes. To do this we divided the ionization/dissociation processes into the steps:

$$H_2 \rightarrow H_2^+ + e^-$$
$$H_2^+ \rightarrow H^+ + H(1s)$$
$$H(1s) \rightarrow H^+ + e^-$$

That is, in analogy to previous investigations [10], we assume atomic ion generation only via the intermediate step of H_2^+ generation. Because of lack of information on the intensity dependence of H_2 ionization and H_2^+ dissociation we assumed pure power laws for these rates. For H(1s) ionization we used an averaged rate based on Floquet calculations. The result of this simulation is shown as the full curve in Fig. 2.

The asymptotic value of the ratio for large intensity, ~0.8 in this case, is a measure for the relative sizes of the spatial volume where only H^+ remains at the end of the pulse, to the volume where an appreciable amount of H_2^+ is left. This ratio in turn is determined by the relative magnitudes of the saturation intensities for H_2^+ generation and H_2^+ dissociation and the effective orders of non-linearity of these processes, because both processes are saturated above ~1.3 x 10^{14} W/cm^2. The simulation gives a saturation intensity for H_2^+ dissociation which is nearly the same as that for H_2 MPI. This means that 'preparation' of the main amount of H_2^+ ions through MPI of H_2 is at high intensity (~1 x 10^{14} W/cm^2) under our short pulse experimental conditions. We do not have the situation that this preparation is finished at low intensity where the ion is not yet strongly perturbed by the radiation field. Preparation is preferentially at high intensity into probably heavily perturbed ionic states. Ionization of the dissociation product H(1s) is definitly saturated under our experimental conditions for intensities above 7 x 10^{13} W/cm^2. Despite this we do not see any characteristic (i. e. resonance structure) indication of this process in our photoelectron spectra above this intensity.

Fig. 3 shows two photoelectron kinetic energy spectra taken at the intensities 6.5 x 10^{13} W/cm^2 and 2 x 10^{14} W/cm^2, respectively. The laser wavelength was 527 nm and the laser pulse width 0.7 psec. The spectrum taken at the lower intensity shows electrons clearly grouped into several ATI orders ranging from 1 to 5, i. e. up to 4 photons are absorbed above the H_2 ionization threshold. The shape of all ATI peaks is nearly the same. This indicates that after absorption of photons above the H_2 ionization threshold ionization is into the same vibrational states of the ion as for the lowest order ATI peak. The whole excess energy is carried away by the photoelectron.

Within the lowest ATI peak a clear resonance substructure, consisting of 3 resolved resonances, is visible. These resonances also appear in the higher order peaks but are not so clearly resolved. This structure is unambiguously generated by excited states of the H_2 molecule shifting into resonance through the AC-Stark effect at definite intensities within the laser pulse. Approximating the AC-Stark shift of the excited states by the ponderomotive shift of a free electron and neglegting a shift of the H_2 $X^1\Sigma_g^+$ and H_2^+ $1s\sigma_g$ ground electronic states the resolved resonant excited states may tentatively be identified as states of a single excited electron with principal quantum numbers n = 4, 5, 6. Atomic hydrogen resonances may be excluded through the following arguments. First, the resonance structure becomes more pronounced with decreasing peak intensity of the laser pulses where the dissociation ratio is yet small. Second, a calculation showed that the pronounced atomic hydrogen n = 4 resonance, in the H(1s) MPI process (see above) should appear at a kinetic energy distinct from all observed resonance positions in Fig. 3.

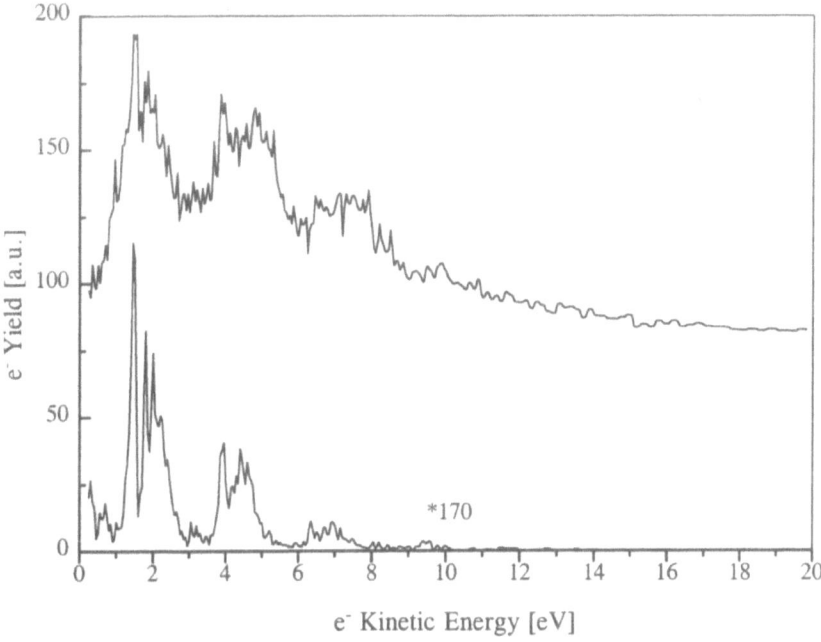

Figure 3. H$_2$ photoelectron kinetic energy spectra at 527 nm excitation wavelength and 0.7 psec pulse width. The upper spectrum was taken at 2 x 10^{14} W/cm^2 pulse peak intensity and the lower one at 6.5 x 10^{13} W/cm^2.

The resonant states are ungerade H$_2$ Rydberg states shifted into 7-photon resonance. Ionization is then accomplished through absorption of one further photon. This mechanism is corroborated by long pulse low intensity photoelectron spectra measured by Yang et al. [25]. The final H$_2^+$ vibrational distribution visible in their photoelectron spectra at pulse peak intensities above 10^{13} W/cm^2 indicates the presence of resonances with H$_2$ Rydberg states with low vibrational quantum number v. The presence of the Rydberg series in our lower intensity photoelectron spectrum indicates that ionization of the excited bound states proceeds predominantly via $\Delta v = 0$ transitions into low v ionic vibrational states. Otherwise we would have seen either a different resonance structure or no resonances at all.

At the higher intensity (upper curve in Fig. 3) again different ATI peaks are visible in the energy range from 0 eV up to 10 eV. The peak ponderomotive shift of about 5.2 eV at this intensity amounts to about 2.2 photon energies (hv = 2.355 eV). In the lowest two ATI orders the n = 4 resonance is still visible, the other smaller resonances have vanished. Other processes seem to dominate the H$_2$ ionization now. The structure of the second ATI peak clearly differs from the first one, it is broader and seems to appear on top of a broad background. This may indicate that in this ATI order the H$_2^+$ ion core is left behind in a different vibrational distribution than in the lowest order one. The H$^+$/D$^+$

kinetic energy spectra may give rise to an alternative way to interpret the change in the photoelectron spectra with increasing laser intensity.

Fig. 4 shows the kinetic energy distribution of fragment D^+ ions after irradiation of D_2 with 527 nm light at peak intensities 1.0×10^{14} W/cm^2, 1.5×10^{14} W/cm^2, and 3.0×10^{14} W/cm^2, respectively, and 0.75 psec pulse width. The lowest intensity is chosen below the saturation point of D_2 MPI and D_2^+ dissociation and the highest one above all saturation levels. As can be seen in the figure the shape of the distribution changes dramatically with increasing intensity. At the lowest intensity the spectrum is dominated by a quite sharp peak located at an energy of ~0.3 eV. At ~1.5 eV a broad peak just begins to rise. It grows with increasing intensity, shifts to a slightly higher kinetic energy, and broadens to higher energies. At 3.0×10^{14} W/cm^2 the high energy peak dominates the spectrum. In between these two peaks a third one is found at ~0.7 eV which is best developed at 1.5×10^{14} W/cm^2. A similar evolution with laser intensity of the H^+ kinetic energy distribution was observed by Giusti-Suzor et al. at 780 nm excitation wavelength and 150 fsec pulse duration [10].

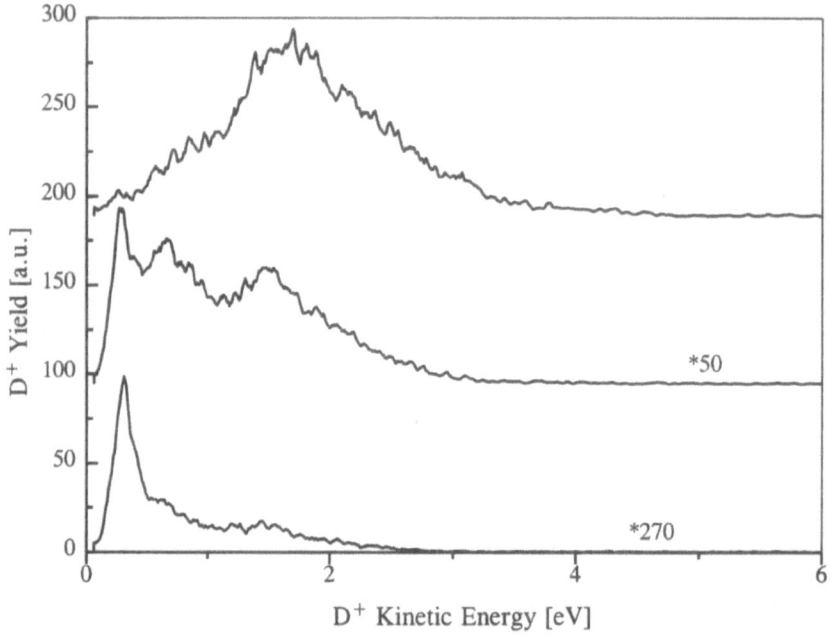

Figure 4. D^+ ion kinetic energy spectra at 527 nm excitation wavelength and 0.75 psec pulse duration. The spectra were taken at 1.0×10^{14} W/cm^2 (lower one), 1.5×10^{14} W/cm^2 (middle), and 3.0×10^{14} W/cm^2 (upper) pulse peak intensity.

At 1.0×10^{14} W/cm^2 our D$^+$ kinetic energy spectrum resembles that from a 'long' pulse experiment of Zavriyev et al. at the same intensity at a slightly different wavelength (532 nm, pulse width ~70 psec) [24]. Following their interpretation of the spectrum the dominant low energy peak is attributed to dissociation by the bond softening mechanism. The just appearing yield maximum near 1.5 eV may be generated by above threshold dissociation (ATD) where first three photons are absorbed by the D$_2^+$ ion to excite it to the 2pσ_u electronic state at a small internuclear distance. Then during dissociation stimulated one photon emission at a larger internuclear distance brings the ion back to the electronic ground state. In this way it finally dissociates with the effective absorption of two photons. This mechanism alone is certainly not sufficient to explain the high energy ion yield maximum near 1.5 eV at higher laser intensities where it dominates the spectrum because this would mean that the D$_2^+$ ion is prepared up to an energy corresponding to very high unperturbed ionic vibrational states through MPI of D$_2$.

With increasing laser intensity dissociation through bond softening becomes unimportant. Other mechanisms resulting in higher energy D$^+$ ions now dominate. One creates a peak near 0.7 eV (at 1.5×10^{14} W/cm^2) which survives as a low energy shoulder of the high kinetic energy yield maximum at 1.5 eV (at 3×10^{14} W/cm^2). This peak does not fit to the dissociation mechanisms mentioned above.

To the high energy peak (at 1.5 eV) in the spectrum the dominant contribution may come from Coulomb explosion after ionization of D$_2^+$ from light induced bound vibrational states [28]. The detailed mechanism in our case is different from that discussed by Zavriyev et al. [28]. Possible candidates which can evolve into these states are the unperturbed D$_2^+$ v = 0,1 vibrational states. These two states are lying just below the threshold for 1-photon dissociation if no rotation is excited. Therefore they cannot dissociate through bond softening. If also the 3-photon dissociation rate of these states is small they can remain bound up to high light intensities. When the light intensity reaches about 1×10^{14} W/cm^2 the lower adiabatic electronic state formed from the diabatic 1sσ_g state and the 2pσ_u state dressed with one photon develops a second shallow minimum near an internuclear distance R = 5 au (see fig. 3 in ref. [10]). Alltogether a double minimum potential is created with the inner minimum nearly at the same position where it appears for the unperturbed 1sσ_g electronic ground state. The unperturbed v = 0,1 vibrational states evolve into bound states of this double minimum potential. Because the perturbed potential is rather wide more than two bound states may exist with increasing laser intensity. Also with intensity increasing above 1×10^{14} W/cm^2 the potential barrier between the minima decreases more and more meaning that all vibrational wave functions gain an appreciable amplitude over all classically allowed internuclear distances up to about 8 au.

Preparation of the light induced states may either proceed in a direct way at high laser intensity or through excitation into nearly unperturbed v = 0,1 vibrational states at low intensity which then evolve into the strongly perturbed states. The first process probably plays a role because saturation of H$_2$/D$_2$ ionization first sets in at about 10^{14} W/cm^2. Because the light induced bound states survive up to high intensities efficient ionization from these states becomes possible. D$_2^+$ ionization leads to Coulomb explosion of the bare deuterons. The kinetic energy of the deuterons depends on the internuclear distance where ionization happens. Because the ionization threshold decreases with increasing internuclear distance R it is most probable at large R. If one assumes ionization in the range from R ~ 4 au up to R ~ 8 au (range of the outer potential minimum) the kinetic energy of the ions from Coulomb explosion falls into the range from ~1.7 eV up to ~3.5 eV. In our experiment the high energy peak in the D^+ kinetic energy spectrum partly falls into this energy range. But the maximum appears at a slightly lower energy (~1.5 eV). Thus it seams that these laser induced bound ionic states cannot account for the whole high energy peak.

3.2. SUMMARY

Concluding this part, our short pulse H_2/D_2 ionization/dissociation experiment at an excitation wavelength where up to now only 'long' pulse experiments had been reported reveals new features in the excitation processes. 7-photon resonances with n = 4, 5, 6 Rydberg states of the molecule dominate the photoelectron spectrum up to 1.5 x 10^{14} W/cm^2. The D^+ kinetic energy distributions show a strong dependence on laser intensity above 10^{14} W/cm^2. At low intensity bond softening is a pronounced feature giving rise to slow dissociation products. With increasing intensity this mechanism becomes unimportant. The spectrum shifts to faster D^+ ions which may partly be produced either through ATD or Coulomb explosion after ionization of laser induced bound ionic states. A complete analysis of the processes responsible for the spectra was not yet possible.

3.3. LASER PLASMA INTERACTION

3.3.1. Absorption

We have investigated the laser light absorption in solid targets for the first time at high intensities up to 10^{18} W/cm^2. The absorption was studied by measuring the integral reflectance as a function of angle of incidence and temporal shape of ultra-bright laser pulses in the intensity range 5 x 10^{16} - 10^{18} W/cm^2. A theoretical model was developed to simulate the experiments numerically. The model contains one dimensional hydrodynamics in a plane geometry together with Maxwell's equations for the laser field. According to the initial data of the experiments, relativistic electron motion as well as oblique incidence are supposed. The simulated propagation of electromagnetic fields in dense small-scale plasmas shows the origin of large resonance fields in electron plasma waves at the critical surface [18]. The high ponderomotive pressure taken into account in the numerical model strongly affects the hydrodynamic motion of the plasma. Acceleration of free electrons in the electrostatic wave fields due to nonlinear Landau damping is probably the dominant energy dissipation process. This process is also considered calculating resonant absorption.

Absorption efficiencies for some angles of incidence for s- and p-polarized laser light are shown in Fig. 5. In order to prevent backreflection of laser light into the laser system only a limited range of incidence angles between 20-60 degrees was probed. The curve for p-polarization has a well pronounced maximum around 40-50 degree of incidence typical for resonance absorption. The maximum value of absorption is about 50%. We would like to emphasize the good agreement of both experimental and theoretical findings concerning the dependence of absorption on the angule of incidence.

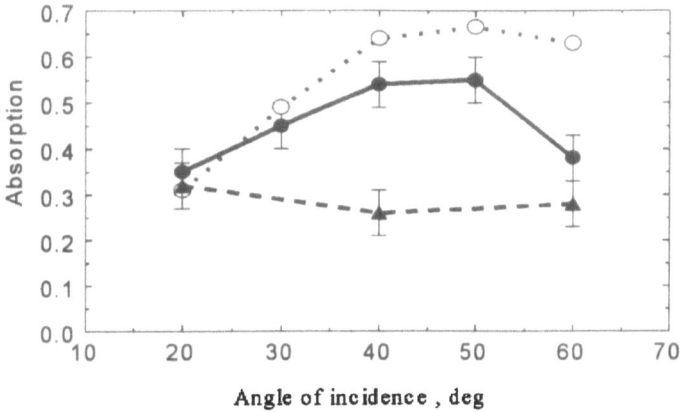

Figure 5. Dependence of absorption values on the angle of incidence of the laser pulse ($\lambda_L = 1053$ nm, $\tau_L = 1.5$ ps, $I_L = 7 \times 10^{17}$ W/cm^2, Al-target). solid line: p-polarized laser light (exp.). dotted line: p-pol. (theo.). dashed line: s-pol. (exp.)

Another interesting point is the high absorption values of 30% for s-polarized light. At a first glance this seems surprising in a plasma with high temperatures of $T_e = 1$-3 keV because inverse Bremsstrahlung should be very small. But in fact, we always have a superposition of s- and p- polarization due to the finite aperture of the focusing optics and hence a field component parallel to the density gradient. Furthermore besides resonance absorption additional collective processes may occur [14]. Investigating the dependence of absorption on the level of laser intensity, we could not find significant changes in the absorption behaviour varying the laser flux density. In Fig. 6 one can see that for p-polarized light the absorption value taken at the optimum angle of incidence of 45 degree is about 40% - 50% in the whole intensity range from 5 x 10^{16} W/cm^2 to 10^{18} W/cm^2. Therefore a remarkable part of laser energy can also be coupled into the plasma in case of short ultra-bright laser pulse irradiation. It is worth mentioning that the 'clean' pulse generated with an additional dye cell after the pulse compressor is absorbed more efficiently than the laser pulse with non-steepened front. This follows also from our theoretical model. The origin for this difference is probably a strong laser light scattering process for the non-steepened pulse. This result is favourable for the short pulse driven plasma as a source of X-radiation as well as for the fast igniter proposal. Absorption investigations at still higher intensities than 10^{18} W/cm^2 are to be carried out in the near future to learn more about resonance and scattering processes

204

under conditions where extremely high ponderomotive pressure and relativistic electron motion have to be taken into account.

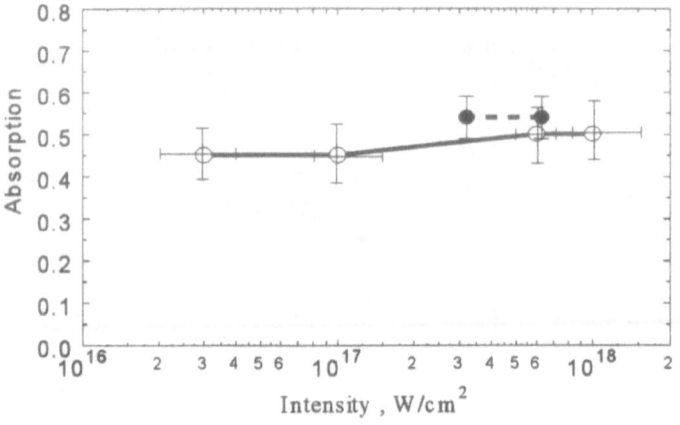

Figure 6. Intensity dependence of the absorption for p-polarized laser pulse ($\lambda_L = 1053$ nm, $\tau_L = 1.5$ ps, angle of incidence 40 degree, Al-target). solid line: pulse without steepening. dotted line: steepened "clean" pulse

3.3.2. Hard X-ray emission and hot electrons

Angular dependence. As mentioned in the previous paragraph hot electron generation is assumed to be an effective process dissipating the resonantly absorbed laser energy. It can be studied via hard X-ray spectroscopy because these energetic electrons emit hard Bremsstrahlung when decelerated in the solid target. In Fig. 7 results concerning the angular dependence of the hard X-ray emission detected at photon energies above 50 keV (Al-filter, transmission at ~10 mm) are shown for p-polarized incident laser light. The hard X-rays also peak at angles between 40 to 50 degrees. This maximum however is much more peaked as compared to the resonance absorption maximum. The X-ray intensity is changing by a factor of more than 10 within the investigated angular range. Moreover, our observed dependence of X-ray yield as a function of the irradiation angle shows a more pronounced maximum in the hard X-ray emission than previously observed in the softer (~1 keV) spectral range [29]. The resonance field at the critical surface of the plasma in the case of maximum resonance generates suprathermal electrons which can emit in free-free transitions extremely hard X-ray photons. The softer X-ray photons are caused by fast electrons with lower energies produced near the field resonance or at weaker fields. Summarizing these results, we can state that absorption values and X-ray data together with the numerical simulation show the tight link in the transformation of the absorbed laser energy into hard X-rays via hot electron generation.

Absolute dose measurements. The absolute dose of X-ray emission is a parameter very important for a better understanding of the transformation processes from laser photons into hard X-ray photons. It is also essential for using the short energetic X-ray pulses as an extremely fast and bright flashlight for investigations of ultra fast dynamical changes in matter. For maximum hard X-ray emission and a comparision to other results [21, 22] tantalum targets have been irradiated at 45 degrees of incidence. Behind a 1.7 mm thick Al-filter, which has an $E_{cut\ off} \sim 25$ keV in the X-ray range and blocks highly energetic

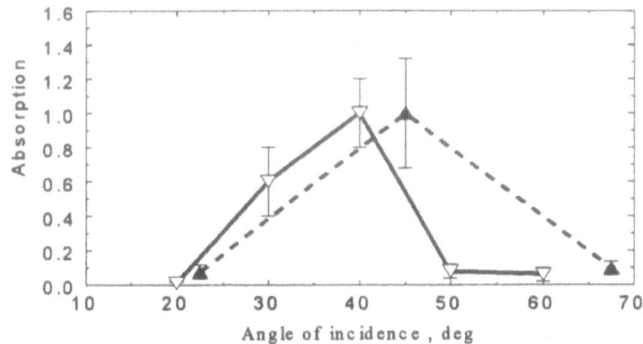

Figure 7. Angular dependence of the hard X-ray emission (photon energy >50 keV, $\lambda_L = 1053$ nm, $\tau_L = 1.5$ ps, $I_L = 5 \times 10^{17}$ W/cm², p-polarized light, Al-target). Detection with ionization chamber (solid line), with scintillator (dashed line)

electrons, the total integral X-ray dose was measured with TLD detectors from the Physikalisch Technische Bundesanstalt which are sensitive up to MeV photons. The TDL's were placed 8 cm away from the plasma source. A dose of 0.05 mSv (1 mSv = 0.1 rem) was measured per shot with an average laser pulse energy of $E_L = 670$ mJ which corresponds to a laser flux density of $I_L = 7 \times 10^{17}$ W/cm² at irradiation conditions described above.

Using the absolute TLD calibration data from the PTB one can estimate the total X-ray photon flux (fluence). With different filters in front of two detector channels we have obtained different X-ray signals. This signal ratio can be interpreted with an X-ray spectrum produced by Bremsstrahlung emission from hot electrons with a temperature of $T_h = 40$ keV assuming a Maxwellian kinetic energy distribution. In one channel behind a 1.7 mm Al-filter this maximum of the X-ray spectrum is at 95 keV which is used as an averaged value for the photon energies in the following rough estimation. It is worth mentioning that there are some experimental results using thicker Al-filters and Al- targets which could indicate that an overall description of the radiation field based on a single Maxwellian hot electron distribution is only a first approximation, which does not include all details of the electron energy distribution in a hot dense plasma produced by an ultra-bright laser pulse. Nevertheless, we will continue to apply this model for the following approach.

From the given PTB-calibration at 95 keV photon energy one obtains a dose/ fluence of 0.35 pSvcm². Corresponding to our irradiation geometry and assuming isotropic emission one obtains a hard x-ray emission pulse energy of

$$E_x = (0.05 \text{ mSv} / 0.35 \text{ pSvcm}^2) * 4\pi * (8 \text{ cm})^2 * 95 \text{ keV} = 1.75 \text{ mJ}$$

which corresponds to 0.26% of the incident laser energy E_L. An estimate of the conversion efficiency ε from the hot electron kinetic energy E_h to Bremsstrahlung energy E_x, assuming a Maxwellian electron distribution, following [30] gives:

$$\varepsilon = E_x/E_h = 1.5 * 1.1 * 10^{-9} * Z * V$$

($Z = 73$ (Ta atomic number), $V = 40$ keV (electron energy)). For Ta one gets an $\varepsilon = 0.48\%$. From both conversion efficiencies we estimate an $E_h/E_L = 54\%$. This value indicates that as an upper boundary about ~ 50% of laser energy at our irradiation conditions is converted into hot electrons during the primary absorption process. This means nearly the whole absorbed laser energy (see Figs. 5, 6) is transferred into hot electrons.

3.3.3 Summary
We have investigated some aspects of the connection between plasma absorption, the production of suprathermal electrons, and the generation of hard X-ray emission in case of short pulse laser interaction with solid targets at extremely high laser intensities up to 10^{18} W/cm^2. The overall absorption is considerably high (about 50%) and resonant absorption plays a dominant role for p-polarized laser light which can be concluded from angular characteristics of the absorption and hard X-ray emission. The absolute dose measurements for hard X-rays have shown that a remarkable large fraction (up to 50%) of the laser energy is transformed into hot electrons as it was predicted in our [20] and other [15,16] simulations.

Part of this work was supported by the Bundesministerium für Forschung und Technologie BMFT.

4. References

1. Agostini, P., Fabre, F., Mainfray, G., Petite, G., and Rahman, N. K. (1979), *Phys. Rev. Lett.* **42**, 1127
2. Freeman, R. R., Bucksbaum, P. H., Milchberg, H., Darack, S., Schumacher, D., and Geusic, M. E., (1987) Above-threshold ionization with subpicosecond laser pulses, *Phys. Rev. Lett.* **59**, 1092
3. Rottke, H., Wolff, B., Brickwedde, M., Feldmann, D., and Welge, K. H. (1990) Multiphoton ionization of atomic hydrogen in intense subpicosecond laser pulses, *Phys. Rev. Lett.* **64**, 404
4. Rottke, H., Wolff-Rottke, B., Feldmann, D., Welge, K. H., Dörr, M., Potvliege, R. M., and Shakeshaft, R.(1994) Atomic hydrogen in a strong optical radiation field, *Phys. Rev. A* **49**, 4837
5. Fittinghoff, D. N., Bolton, P. R., Chang, B., and Kulander, K. C. (1992) Observation of nonsequential double ionization of helium with optical tunneling, *Phys. Rev. Lett.* **69**, 2642
6. Walker, B., Sheehy, B., DiMauro, L. F., Agostini, P., Schafer, K. J., and Kulander K. C. (1994) Precision measurement of strong field double ionization of helium, *Phys. Rev. Lett.* **73**, 1227
7. Lewenstein, M., Balcou, Ph., Ivanov, M. Yu., L'Huillier, Anne, and Corkum, P. B., Theory of high-harmonic generation by low-frequency laser fields, *Phys. Rev. A* **49**, 2117
8. Baorui Yang, Schafer, K. J., Walker, B., Kulander, K. C., Agostini, P., and DiMauro, L. F. (1993) Intensity-dependent scattering rings in high order above-threshold ionization, *Phys. Rev. Lett.* **71**, 3770

9. Paulus, G. G., Nicklich, W., Huale Xu, Lambropoulos, P., and Walther, H. (1994) Plateau in above threshold ionization spectra, *Phys. Rev. Lett.* **72**, 2851
10. Giusti-Suzor, A., Mies, F. H., DiMauro, L. F., Charron, E., and Yang, B. (1995) Dynamics of H_2^+ in intense laser fields, *Jour. Phys. B* **28**, 309
11. Codling, K., and Frasinski, L. J. (1993) Dissoziative ionization of small molecules in intense laser fields, *Jour. Phys. B* **26**, 783
12. Kruer, W. L. (1988), in *The physics of laser plasma interactions*, Addison Wesley; Hora, H. (1981), in *Physics of laser driven plasmas*, John Wiley & Sons, New York
13. Forslund, D. W., Kindel, J. M., and Lee, K. (1977) Theory of hot electron spectra at high laser intensity, *Phys. Rev. Lett.* **39**, 284; Priedhorsky, W., Lier, D., Day, R., and Gerke, D. (1981) Hard X-ray measurements of 10.6 μm laser-irradiated targets, *Phys. Rev. Lett.* **47**, 1661
14. Brunel, F. (1987) Not-so-resonant, resonant absorption, *Phys. Rev. Lett.* **59**, 52
15. Wilks, S. C., Kruer, W. L., Tabak, M., and Langdon, A. B. (1992) Absorption of ultra-intense laser pulses, *Phys. Rev. Lett.* **69**, 1383
16. Gibbon, P., and Bell, A. R. (1992) Collisionless absorption in sharp-edged plasmas, *Phys. Rev. Lett.* **68**, 1535
17. Demchenko, N. N., Rozanov, V. B., and Stenchikov, G. L. (1980), *Sov. Phys. JETP* **51**, 703
18. Kalachnikov, M. P., Nickles, P. V., Schlegel, T., Schnürer, M., Billhardt, F., Will, I., and Sandner, W. (1993) Dynamics of laser-plasma interaction at 10^{18} W/cm², *Phys. Rev. Lett.* **73**, 260
19. Fedosejevs, R., Ottmann, R., Sigel, R., Kühnle, G., Szatmari, S., and Schäfer, F. P. (1990) Absorption of subpicosecond UV laser pulses in high density plasma, *Appl. Phys. B* **50**, 79; Meyerhofer, D., Chen, H., Delletrez, J. A., Soom, B., Uchida, S., and Yaakobi, B. (1993) Resonance absorption in high-intensity contrast, picosecond laser-plasma interactions, *Phys. of Fluids B* **5**, 2584
20. Schnürer, M., Kalachnikov, M. P., Nickles, P. V., Schlegel, T., Sandner, W., Demchenko, N. N., Nolte, R., and Ambrosi, P. (1995) Hard X-ray emission from short laser plasma, *Phys. Rev. Lett.* (in press)
21. Kmetec, J. D., Gordon III, C. L., Macklin, J. J., Lemoff, B. E., Brown, G. S., and Harris, S. E. (1992) MeV X-ray generation with a femtosecond laser, *Phys. Rev. Lett.* **68**, 1527
22. Tabak, M., Hammer, J., Glinsky, M. E., Kruer, W. L., Wilks, S. C., Woodworth, J., Campell, E. M., and Perry, M. D. (1994) Ignition and high gain with ultrapowerful lasers, *Phys. Rev. Lett.* **1**, 1626
23. Herrlin, K., Svahn, G., Olsson, C., Pettersson, H., Tillman, C., Perso, A., Wahlström, C. G., and Svanberg, S. (1993) Generation of X-rays for medical imaging by high-power lasers, *Radiology* **189**, 65
24. Zavriyev, A., Bucksbaum, P. H., Muller, H. G., and Schumacher, D. W. (1990) Ionization and dissociation of H_2 in intense laser fields at 1.064 μm, 532 nm, and 355 nm, *Phys. Rev. A* **42**, 5500
25. Yang, B., Saeed, M., DiMauro, L. F., Zavriyev, A., and Bucksbaum, P. H. (1991) High-resolution multiphoton ionization and dissociation of H_2 and D_2 molecules in intense laser fields, *Phys. Rev. A* **44**, R1458
26. Billhardt, F., Kalachnikov, M. P., Will, I., and Nickles, P. V. (1993) A high contrast picosecond-terawatt Nd:glass laser system with fiberless chirped pulse amplification, *Opt. Commun.* **98**, 99
27. Kalachnikov, M. P., Nickles, P. V., Will, I., Billhardt, F., and Schnürer, M. (1994) A high contrast all-glass picosecond terawatt CPA laser system, *Laser and Particle Beams* **12**, 463
28. Zavriyev, A., Bucksbaum, P. H., Squier, J., and Saline, F. (1993) Light-induced vibrational structure in H_2^+ and D_2^+ in intense laser fields, *Phys. Rev. Lett.* **70**, 1077
29. Teubner, U., Bergmann, J., van Wonterghem, B., and Schäfer, F. P. (1993) The dependence of X-ray emission on absorbed laser intensity in a laser produced plasma generated by a high intensity ultrashort pulse, *Phys. Rev. Lett.* **70**, 794
30. McCall, G. H. (1982) Calculation of X-ray Bremsstrahlung and characteristic line emission produced by a Maxwellian electron distribution, *J. Phys. D* **15**, 823

A TIME OF FLIGHT PHOTOELECTRON SPECTROMETER FOR THE ANALYSIS OF FULLERENS

A. Yu. ELIZAROV

A. F. Ioffe Physico-Technical Institute, 194021, St Petersburg, Russia

ABSTRACT: A time-of-flight photoelectron spectrometer for analysis of gases is present. A sample of fullerens and atom Ba with nonresonant multiphoton ionization using these ionization methods is analyzed. Results of energy measurements are presented.

1. Introduction

The technique combining laser ionization and photoelectron spectroscopy to study fullerens has been very convenient for different application. Here we shortly describe the apparatus and present the firs results multiphoton ionization of fullerens.

2. Experimental Apparatus

The photoelectron spectrometer was designed and build as shown in Fig. 1.[1], where: (1)—electrodes were constructed from high purity copper covered by gold. Each grids (2) was perfumed from platinum mesh. Drift tube (3) was coated with aquadag. The ions produced by photoionization were detected by two microchannel plates. The Earth magnetic field were shielded by three μ-metal tubes (4). The magnetic field inside the spectrometer were not exceed 2 mG. The $E(1) = 0V$, $E(2) = 0.5$ V/cm.

Pulsed Q-switched Nd.YAG laser delivered pulses of radiation with the wavelength 1.06 μm, pulse duration of about 20 ns, pulse repetition rate 12.5 Hz, and output energy 0.2 J. KDP crystal was used for frequency doubling of the laser radiation. The energy conversion efficiency to the second harmonic was about 20%. A small portion of this radiation was reflected to a photodetector to provide synchronizing pulses for an automatic data registration system.

The radiation from laser was focused by lens with a focal length 12 cm. at the beam of fullerens moving between the grids of the photoelectron spectrometer.

An fullerens beam was formed by an aperture 0.5 mm. in diameter in a side wall of cylindrical molybdenum crucible which was heated by a tungsten helix to a temperature of 600 C. The distance from the aperture to the focused laser beam was 25 mm. The fullerens beam was further formed by a aperture 3 mm. in diameter. An electric field was applied to the grid electrodes.

H. G. Muller and M. V. Fedorov (eds.), Super-Intense Laser-Atom Physics IV, 209–211.
© *1996 Kluwer Academic Publishers.*

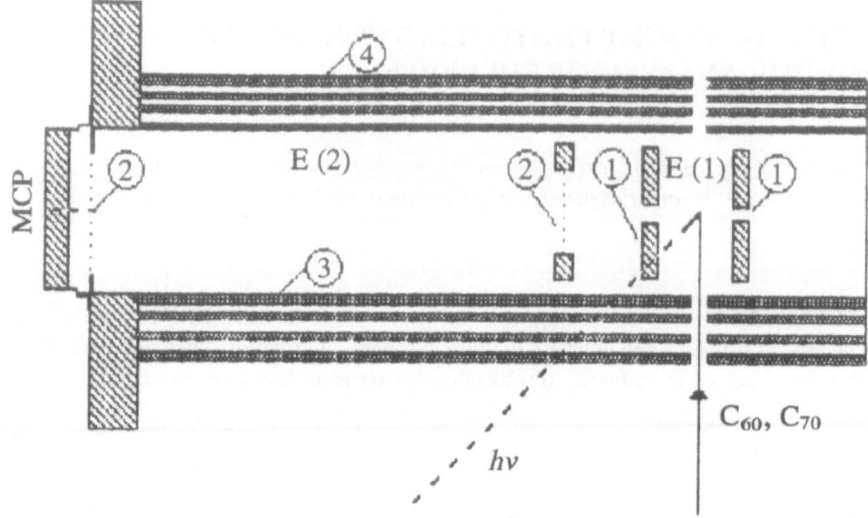

Fig. 1. Schematic view of the photoelectron spectrometer. (1) electrodes, (2) grids, (3) drift tube, (4) three mu metal tubes.

3. Results

For making energy calibration of electron spectrometer the three photon ionization of atom Ba have been made. The two photon excitation via the $6p^2(^1S_0)$ intermediate state was perform [2]:

$$Ba\left[6s^2\left(^1S_0\right)\right] \xrightarrow{2\omega(581.9\text{ nm})} Ba^*\left[6p^2\left(^1S_0\right)\right]$$

$$Ba^*\left[6s^2\left(^1S_0\right)\right] \xrightarrow{\omega(581.9\text{ nm})} Ba^+\left[6s\left(^2S_{1/2}\right)\right],5d(^2D_{3/2,5/2})$$
$$+e^-\left[\epsilon p_{1/2},\epsilon f_{3/2},\epsilon f_{5/2}\right] \qquad (1)$$

The photoelectron spectra of process (1) is shown in Fig. (2a). The energy of exciting state and energy of ground state Ba^+ are known [3] and as result the calibration of electron spectrometer was fulfilled. In the first demonstration of our device, we investigated a sample containing approximately 50% of fullerene C_{60} and 50% of C_{70} by using multiphoton ionization. For ionization of sample the second (530 nm.) harmonic of the Nd YAG laser have been used [4].

The ionization potentials of C_{60} is 7.61 eV [5], so four-photon ionization of fullerens was expected. Fig. (2b) shows a photoelectron spectrum of the multiphoton ionization of fullerens. The laser light power density used for the photoelectron spectrum in Fig.(2b) is approximately 20 MW/cm^2. From this spectrum have been extracted the difference between of ionization potential of C_{60} and C_{70}. The ionization potential of C_{70} is 8.1±0.1 eV.

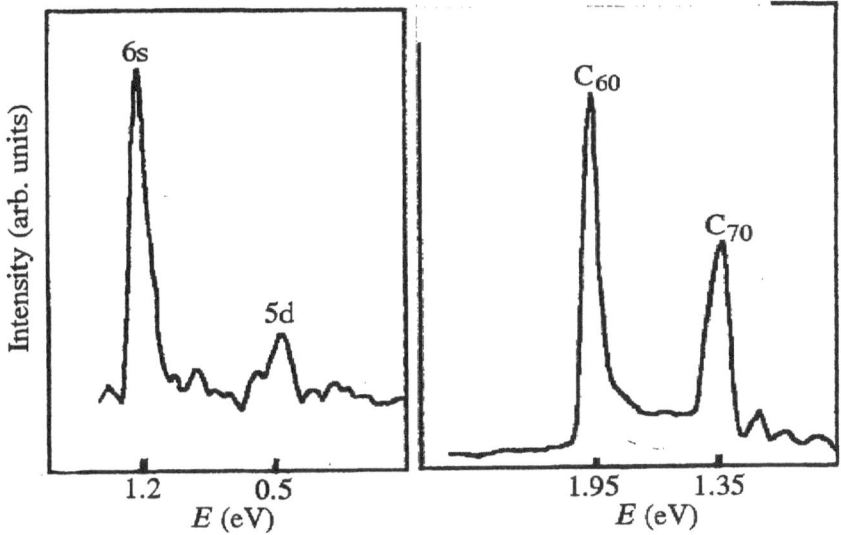

Fig. 2. (a) Photoelectron spectrum obtained from ionization of Ba. (b) Photoelectron spectrum obtained from ionization of C_{60}, C_{70}.

4. Acknowledgment

The authors want to thank Prof. G. Djuzhev for fullerens. We are grateful for financial support from Foundation for Intellectual Collaboration and RFFI (No. 95-03-09326).

5. References

1. Elizarov, A.Yu. (1995) Angular distribution of photoelectrons ejected from Ba by-polarized radiation, Pis'ma v Zh. Eksp. Teor. Fiz., 62, 23–26 (Russian).
2. Cherepkov, N.A. and Elizarov, A.Yu.(1991) Two-photon excitaton of Ba atoms and absolute measurements of $\sigma^{(2)}$, J. Phys.B, 24, 4169–4179.
3. Lichtenberger, D.L., Jatcko, M.E., Nebesny, K.W., Ray, C.D., Huffman, D.R., Lamb L.D. (1991) In Mater. research society :Boston, Vol 206, 673–678.
4. Moore, C.E. (1958), Atomic energy levels. NBS, Washington.
5. Bobashev, S.V., Dubensky, B.M., Elizarov, A.Yu., Korshunov, V.V. (1994) A time-of-flight mass-spectrometer for analysis of fullerenes, Mol. Mat.14, 155–158.

CORRELATION AND ITS MEASURE

Kazimierz Rząƶewski
Center for Theoretical Physics and *College of Science,*
Al. Lotników 32/46, 02-668 Warsaw, Poland

1. Introduction

Most of the theoretical work on strong field ionization to date has been devoted to single electron processes while   most experiments are performed with atoms other than hydrogen. So, inevitably, the attention is turning to the processes in which more than one electron is taking part. This shift of interest is well represented in the present volume.

The most important property of the multiparticle quantum state is its correlation. Typically correlation is associated with the extent to which predictions of a specific approximation scheme disagree with exact results. In atomic physics the correlation is usually associated with the concept of correlation energy. This often refers to the amount by which a simple single Hatree-Fock multiparticle wave-function overestimates the exact energy of the multielectron quantum state. Such notion is certainly useful in the discussion of the stationary states [1]. It is not adequate if we want to talk about the correlation in the time dependent process such as ionization.

On the other hand, in the discussion of the fundamental concepts of quantum mechanics one stresses the existence of peculiar quantum correlation between the subsystems often referred to as the *entanglement* . The most famous entangled state is the one that appears in the Einstein, Podolsky, Rosen paradox [2]. It is the singlet state of two spin 1/2 particles:

$$|\Psi> = \frac{1}{\sqrt{2}}\left[|\uparrow>|\downarrow> -|\downarrow>|\uparrow>\right].\tag{1.1}$$

Much more complicated two particle states are possible if each particle has more states to occupy, like in the case of translational degrees of freedom for a particle moving in space. The question then arises: How to measure a degree of correlation in a way which characterizes the given quantum state and not its representation or some specific observable, such as the energy?

H. G. Muller and M. V. Fedorov (eds.), Super-Intense Laser-Atom Physics IV, 213–220.
© 1996 *Kluwer Academic Publishers.*

In a recent letter [3] we have introduced such a measure. It is based on the intuitive notion of correlation of statistical variables as the deviation of the corresponding probability distributions from the simple product. In this article we discuss the notion of degree of correlation, give several examples and illustrate the notion with several applications.

2. Measure of Correlation of Multielectronic States

The convenient starting point is the decomposition of the exact two-particle wave-function, which in the case of Bose-Einstein statistics is known as the Schmidt decomposition [4]:

$$\Psi(x_1, x_2) = \sum_j D_j \varphi_j(x_1) \varphi_j(x_2), \quad < \varphi_j | \varphi_k > = \delta_{jk} . \tag{2.1}$$

Note the single sum in (2.1). The basis φ_j is not universal. On the contrary, it is determined by the two particle wave-function Ψ. The functions φ_j are called natural orbitals. The squares of expansion coefficients D_j are probabilities of finding j-th natural orbital in the state Ψ. In fact they add-up to unity if the state Ψ is normalized. The *average* probability $|D_j|^2$ is then given by $\sum_j |D_j|^4$. Its inverse gives the effective number of natural orbitals in (2.1). We define the degree of correlation K as:

$$K = \left[\sum_j |D_j|^4 \right]^{-1} \tag{2.2}$$

Its obvious minimal value is 1 if the sum in (2.2) has only one term. Of course in this case the state is not correlated. It is easy to check that if there are N equally probable orbitals in (2.1) then $K = N$.

The coefficient K may be easily computed from the reduced, single particle density matrix $\rho(x_1, x_2) = \int dx \, \Psi(x_1, x) \Psi^*(x_2, x)$. Indeed, using the decomposition (2.1) one easily gets:

$$K = \left[Tr[\rho^2] \right]^{-1}. \tag{2.3}$$

This is the most important formula. Several niec properties of K follow from (2.3). It also suggests some generalizations.
1. It shows how to compute K without knowing the individual D_j.

2. It offers an alternative, informational interpretation of K. Since the trace of the square of the density matrix determines the degree of statistical mixing of the state, and our starting point was the two-particle pure state, the degree of correlation is a measure of the loss of information due to integration over the coordinates of one particle. Obviously we loose a lot of information this way if the state was strongly correlated. Hence, the produced state is strongly mixed.

3. Our K is invariant under unitary transformations. It does not depend on the representation of the wave-function. It is the same computed in the position, momentum or any other representation of the wave-function.

4. Among unitary transformations there are also gauge transformations. K is then gauge invariant. It unambiguously characterizes the quantum state although the specific calculations (such as the numerical solution of the Schrödinger equation for strong field ionization problems) are typically carried out in some particular gauge. This way the parameter K has a good meaning also while the atom is under the influence of the laser pulse.

5. The formula (2.3) suggests some obvious generalizations. In the case of more then two particles we may still construct the reduced single particle density matrix. Then the formula (2.3) may be used without any modification for more than two particles and also for fermions. In the last case the minimum value of K for n fermions is n . This last fact is easy to understand since the simplest fermionic n-particle wave-function requires n orthonormal orbitals combined into a single Slater determinant.

6. For more then two particles complex entanglements may occur at higher then two-particle level. In fact also in the stationary case one often distinguishes the genuine three particle correlation. They may be found by constructing separately two and single particle density matrices:

$$\rho_2 = \int \Psi(x,y,z)\, \Psi^*(x',y',z)\, dz$$

(2.4)

$$\rho_1 = \iint \Psi(x,y,z)\, \Psi^*(x',y,z)\, dy\, dz$$

Then one can define two correlation coefficients: $K_{1,2} = \left[Tr(\rho_{1,2}^2)\right]^{-1}$. They satisfy the following inequality: $1 \le K_2 \le K_1$. Now the state has genuine three particle correlation if $K_2 \gg 1$ and $K_1 \approx K_2$. If also $K_1 \gg K_2$, the state has true two particle correlation as well. The above construction has obvious generalization to bigger number of particles.

3. Simple Examples

In this section we present several simple examples of calculation of coefficient K.

1. Suppose the wave-function of two bosons is a product of a center-of-mass function $\Phi(\vec{R})$ which extends over large volume V and a relative coordinate function $\varphi(\vec{r})$, which extends over small volume v. Of course this is a highly correlated state of two particles. If we know the position of one particle, we also know where to look for the other one. It is clear that the reduced density matrix extends over the volume V. In fact the degree of correlation of such a state is of the order of $V/_v$.

2. A general two particle wave-function constructed with the help of two (not necessarily orthogonal) orbitals

$$\Psi(x_1, x_2) = N[\psi_1(x_1)\psi_2(x_2) + \psi_1(x_2)\psi_2(x_1)] \tag{3.1}$$

has the degree of correlation:

$$1 \leq K = \frac{2(1+u^2)^2}{u^4 + 6u^2 + 1} \leq 2; \quad u = |(\psi_1|\psi_2)|, \tag{3.2}$$

which is equal to 1 for parallel orbitals and equal to 2 for orthogonal orbitals.

3. In the multiconfiguration Hartree-Fock method the two particle wave-function of the singlet spin state may be written as

$$\Psi(x_1, x_2) = \sum_{i,j} A_{ij}\, \varphi_i(x_1)\varphi_j(x_2), \tag{3.3}$$

where the single particle orbitals φ_i are the states in a selfconsistent Hartree-Fock potential: $\varphi_i : i = 1, ..., N$, $(\varphi_i|\varphi_j) = \delta_{ij}$. The degree of correlation does not depend on the form of each orbital and is fully determined by the matrix $\hat{A}$. In fact this matrix itself is the wave function in the φ_i basis. Hence the $N \times N$ reduced density matrix $\hat{\rho} = \hat{A}^+\hat{A}$ gives $K = \left[Tr[\hat{\rho}^2]\right]^{-1}$.

4. Not so Simple Examples

The first question is: are there any highly correlated states in isolated atoms. We have investigated this question in a recently published paper [5] . We found that all bound states of helium atom have degree of correlation either a bit higher than 1 (the ground state) or very close to 2 (all other states). We are always close to the two limiting cases of the formula (3.2). At most two natural orbitals are needed to adequately describe the bound states of helium.

The situation is different for heavier atoms with two electrons in the outer shell. The doubly excited states and the single excited electron Rydberg series coexist in such atoms. We have investigated in some detail the $4snd^1D_2$ configuration in Ca. The degree of correlation for the Rydberg states with principal quantum number $n=5$ to 25 is shown in Fig. 1. As we see it is larger than 2 for many states and reaches its maximum (K close to 3 in the vicinity of the embedded doubly excited state. The multichannel Hartree-Fock method based on the Dirac equation was used here. The two particle wave-function was expanded on the single particle orbitals of Hartee-Fock potential according to the formula (3.3).

In Ref. 5 we presented a hypothesis that the anomalously correlated two electron states carry the traces of chaos, which may be found in the dynamics of the two electron classical Coulomb problem.

From the above mentioned example it may be conjectured that also the autoionization resonances have high degree of correlation.

Introducing the measure of correlation we had in mind mostly the time dependent processes. Due to the numerical difficulties the two electron processes are usually

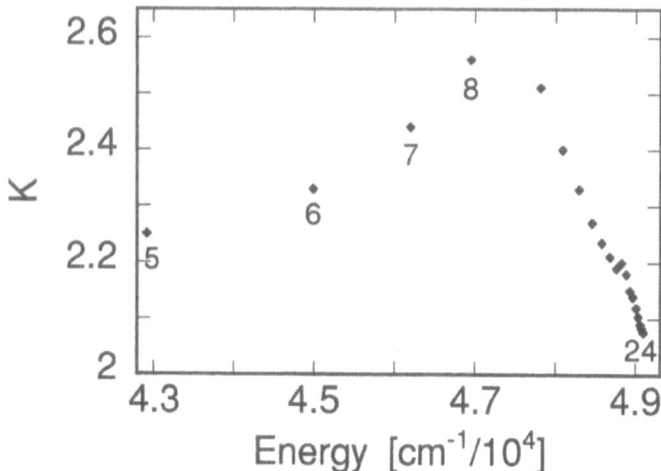

Figure 1. The degree of correlation K versus the energy of the Rydberg state of $4snd^1D_2$ series of Ca, with n changing from 5 to 25 [5].

studied with the help of one dimensional models. Of particular importance is the two electron extension of the smoothed Coulomb potential studied extensively in Rochester [6,7].

The two particle, one dimensional Hamiltonian modeling the H⁻ ion is

$$H = \frac{p_1^2}{2} + \frac{p_2^2}{2} - \frac{1}{\sqrt{x_1^2+1}} - \frac{1}{\sqrt{x_2^2+1}} + \frac{1}{\sqrt{(x_1-x_2)^2+1}}. \tag{4.1}$$

Its ground state is a symmetric function of coordinates and has $K=1.1$. We studied the symmetric time dependent functions of x_1 and x_2 only. They would be multiplied by the antisymmetric function of spin variables.

In the time dependent calculations we see the important difference between a more conventional basis expansion of the wave-function and the expansion (2.1) on the natural orbitals. In the first case the basis is typically time independent and only the expansion parameters depend on time. In the latter case not only the expansion coefficients but also the natural orbitals change with time.

In the first time - dependent calculation we have studied an inelastic scattering of electron on the hydrogen atom. Various excited final states of the hydrogen atom are correlated in this case with the final state of the scattered electron. This case is described in some detail in the paper of R. Grobe in this volume [8].

In the most important calculation we have traced the time dependence of K for the ionization of the "negative ion" (4.1) by a 20 optical cycles laser pulse. We have monitored the time development of the wave-function after the laser turn-off for additional 10 optical cycles.

The results for the three distinctly different regimes of parameters (frequencies and amplitudes) are shown in Fig. 2. Note a smooth time dependence in all three cases. No oscillations with the optical period are present.

The curve (a) describes a process of one photon photodetachment by relatively weak field. The field is strong enough to detach completely the outer electron in the 20 cycles but weak enough not to change the state of the inner electron. The degree of correlation changes from the initial value of 1.1 to almost 2. We have the situation described by the formula (3.2). The only correlation built during the evolution in this case comes from the indistiguishibility of particles. We do not know which of the two electrons is ionized. So it would be possible to construct an essential states model of this ionization process using merely two orbitals.

The next regime is still characterized by the single electron detachment but the other electron is active. It undergoes the Rabi oscillations because the laser light is resonant with the bound-bound transition of the core electron. It introduces additional correlation between the electrons. The effect is known as the coherence transfer [8]. Note that the degree of correlation oscillates in tune with the Rabi oscillations of the inner electron.

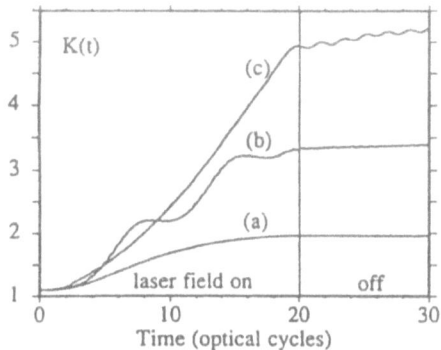

Figure 2. The correlation K as a function of time for three different regimes in laser intensity and frequency ω. The laser pulse has a two-cycle linear turn-on and turn-off and a constant field strength $\mathcal{E}$ between the second and the eighteenth cycle. (a) The inner electron is passive and the outer electron is 98.4% photodetached under the absorption of a single photon ($\mathcal{E} = 0.005$ au, $\omega = 0.08$ au). (b) The inner electron is mainly bound but very active and the outer electron is 85.5% photodetached ($\mathcal{E} = 0.05$ au, $\omega = 0.395$ au). (c) Both electrons are active and ionize rapidly with a detachment probability 85.3% and a double-ionization probability of 31.3% ($\mathcal{E} = 0.5$ au, $\omega = 1.0$ au) [3].

In the last example, (high photon frequency and stronger field) we enter the regime of rapid double ionization. Many different mechanisms of double ionization contribute to the correlation in this case. Note that K may reach values as high as 5. Since the ionized electrons continue to interact, changing their wave-function, the degree of correlation continues to grow even after the end of the laser pulse.

5. Conclusions

We have presented a definition of the degree of correlation. We argued that our definition has many attractive features. But there is no unique definition of the degree of correlation. We have shown that our parameter K may successfully help to characterize the two particle states. The degree of correlation of the bound states is larger than the minimal one if the two particles interact with each other. It may be time dependent for the scattering processes described by the collisions of the localized wave-packets. If the system is under the influence of the time dependent forces, even acting separately at each particle, like the force of the electric field in the light pulse, the degree of correlation also changes in time. In all cases studied the degree of correlation increases in time. It is still a challenge to find a case of external influence which would reduce the amount of correlation in the interacting two particle system.

6. References

1. For a review see: U. Fano, *Rep. Prog. Phys.* **46**, 97-165 (1983)
2. A. Einstein, B. Podolsky, and N. Rosen, *Phys. Rev.* **47**, 777 (1935)
3. R. Grobe, K. Rzążewski, and J.H. Eberly, *J.Phys.* B, **27**, L503, (1994)
4. E. Schmidt, *Math. Ann.* **63**, 433, (1906)
5. M.Yu. Ivanov, D. Bitouk, K. Rzążewski, and S. Kotochigova, *Phys. Rev.* **52**, 149 (1995)
6. For a review of single electron model see: J.H. Eberly, J. Javanainen, and K. Rzążewski, *Phys. Rep.* **204**, 333 (1991)
7. R. Grobe, J.H. Eberly, *Phys. Rev. Lett.* **68**, 2905 (1992); *Phys. Rev.* **47**, RC1605, (1993); *Phys. Rev.* A, **48**, 623 (1993)
8. R. Grobe, article in the present volume.

MODEL TWO-ELECTRON ATOMS AND IONS IN INTENSE SHORT LASER PULSES

R. GROBE
Department of Physics
Illinois State University
Normal, IL 61790   USA

J.H. EBERLY
Department of Physics and Astronomy
University of Rochester
Rochester, NY 14627   USA

We review some of our recent work on the response of two-electron atoms to strong laser fields. Reduced dimensional model systems have the advantage that exact fully correlated two-electron wave functions can be obtained numerically. This approach allows for an unambiguous testing of approximate theories and a microscopic insight into ionization mechanisms. We will discuss the relevance of electron-electron correlations for the generation of high harmonics and for the applicability of the hyperspherical-coordinate and Hartree-Fock approaches and show how correlations can grow in inelastic electron-atom scattering.

1. Introduction

There has been a lot of progress in the last ten years in understanding the dynamics of atoms in strong laser fields.[1] Experimental progress has been made partly due to the availability of new high power lasers with very high repetition rates allowing, e.g., for highly resolved photoelectron spectra. Some theoretical breakthroughs have occured due to extensive software development accompanied with faster and more powerful computers. This book provides many examples of this fascinating progress.

221

H. G. Muller and M. V. Fedorov (eds.), Super-Intense Laser-Atom Physics IV, 221–231.
© 1996 *Kluwer Academic Publishers.*

Most of the theoretical work has been restricted to investigations of atoms with only a single electron. In some cases this restriction is unimportant because the dynamical response of more complicated multi-electron systems depends on the optical response of only one of the electrons. However, in the last few years lasers have approached such high intensities that the participation of several electrons is either readily observed or must be seriously considered.

The Coulombic electron-electron repulsion, or correlation interaction, adds a critical degree of complexity to any theoretical treatment. Even though structural studies of the energy levels and transition rates of multielectron atoms have been actively pursued for many years, research which investigates the non-perturbative dynamical response of these atoms to laser fields is just in an early stage.

In 1991 we began to investigate strong laser interactions of reduced dimensional two-electron systems. These systems have the advantage that the full time-dependent Schrödinger equation can be solved exactly on a computer. Therefore, exact fully correlated two-electron wave functions are numerically accessible. Of course, when three dimensional aspects of the dynamics are important and this approach cannot be relied on, the reduced dimensional dynamics can be thought of, usually without serious error, as 3D dynamics for which only two angular momentum channels are permitted. This approach has become a remarkably flexible tool to explore e-e interaction effects without the need for any further simplyfying assumptions or approximations.

At the SILAP III meeting in Han-sur-Lesse we have reviewed predictions for negative ions, detachment rates and ionization rates including aspects of stabilization.[2] Since then our work on two-electron physics has further progressed. In this volume we want to give four examples of the advantages of direct numerical simulations for two-electron atoms. As the wave functions are exact they can serve as fruitful testing grounds for approximate theories. We have checked the applicabilty of the Hartree-Fock and the hyperspherical coordinate approach for several two-electron atoms. Two-electron wave functions are available with an almost arbitrary spatial and temporal resolution and this allows for a microscopic view of the ionization mechanism. We have exploited this aspect and have investigated the influence of e-e collisions on the generation of higher harmonics in scattered light spectra. As the wave functions are unambiguous they can serve as testing grounds for new concepts. We have proposed a direct and operator-independent measure to determine the degree of correlation embodied in a wave function. Numerical simulations were used to establish whether this criterion is indeed appropriate to distinguish between several regimes of two-electron response.

Analytical approximate approaches are typically based on some a priori knowledge of the phenomenon under discussion. In most cases one knows in advance what one wants to calculate and what type of approximations are appropriate. Direct simulations, however, do not depend on any a priori knowledge and are therefore suited to discover unexpected phenomena. The coherence transfer effect is probably the best example of such a discovery. Several experiments have been motivated by this discovery and confirmed our theoretical predictions.

2. Model two-electron atoms

Two-electron systems are characterized by a nucleus of positive charge Z fixed at the origin at x=0 and two electrons whose spatial coordinates are x_1 and x_2. Depending on the choice for the nuclear charge Z, this model can be thought of as a one-dimensional analog of a negative ion, a neutral atom or a positive ion,

$$H_0 = P_1^2/2 + P_2^2/2 + ZV(x_1) + ZV(x_2) - V(x_1-x_2) \qquad (2.1)$$

where the same quasi-coulombic soft core potential

$$V(x) = -1 / \sqrt{(1 + x^2)} \qquad (2.2)$$

describes the attractive electron-proton interaction as well as the mutual electron-electron repulsion. The properties of this potential and its consequences in one-electron systems have been widely studied. [3] Here and elsewhere we will generally use atomic units, for which $e=m=\hbar=1$.

The time-dependent interaction of the two-electron system with the laser field is described by the Schrödinger equation

$$i\partial\Psi(x_1,x_2,t)/\partial t = [H_0 + (x_1+x_2) \mathcal{E}(t) \sin \omega t] \Psi(x_1,x_2,t) \qquad (2.3)$$

where $\mathcal{E}(t)$ is the laser field envelope and ω the frequency. The amplitude has been turned on and off smoothly, in some cases the envelope was $\sin^2(\pi t/T)$ and in some cases we chose a trapezoidal pulse shape with a two-cycle ramp. The time-dependent Schrödinger equation (2.3) has been solved numerically on a 1024×1024 spatial lattice grid to obtain the two-electron wave function $\Psi(x_1,x_2,t)$. The wave functions are available numerically with arbitrary precision. Our numerical methods have been described in Ref. 4.

3. Hartree-Fock approach and the hyperspherical-coordinate approximation

Many of the theoretical efforts to study the structural properties of multi-electron atoms rely on the Hartree-Fock approximation[5] or on the hyperspherical-coordinate approach[6]. We have investigated how well these two approximation schemes work to describe reduced dimensional atoms. We will begin our discussion with the Hartree-Fock approach and present both methods in a notation appropriate to our 1D system.

In the Hartree-Fock approximation the wave function is described by the product of two single-electron orbitals:

$$\Psi_{HF}(x_1, x_2) = \psi_n(x_1)\, \psi_m(x_2) \pm \psi_m(x_1)\, \psi_n(x_2) \qquad (3.1)$$

Using the calculus of variation one can numerically determine the single electron orbitals $\psi_n(x)$ which extremalize the energy. These orbitals have to satisfy an effective nonlinear Schrödinger equation. We have solved this equation using an iterative procedure to find the Hartree-Fock energies for various two-electron atoms.[7] We show in Table 1 the ground state energies for the isoelectronic series of one-dimensional two-electron atoms beginning with the negative ion.

TABLE 1. Ground-state energies (a.u.) of the isoelectronic series of 1D two-electron atoms starting with H^-.

Nuclear charge	exact	Hartree-Fock	hyperspherical
$Z = 1$	- 0.731	- 0.692	- 0.726
$Z = 2$	- 2.238	- 2.224	- 2.185
$Z = 3$	- 3.896	- 3.888	- 3.797
$Z = 4$	- 5.615	- 5.610	- 5.472
$Z = 5$	- 7.371	- 7.367	- 7.187

Another frequently used multi-electron approximation scheme is based on hyper-spherical coordinates. For one-dimensional systems these coordinates reduce to the usual polar coordinates: $R = \sqrt{(x_1^2 + x_2^2)}$ and $\tan \alpha = x_1/x_2$. The applicability of this approach in three-spatial dimensions relies on the quasiseparability of the Hamiltonian in these collective coordinates. This separability corresresponds to a slow motion along the hyper-radial coordinate and a relatively faster motion along the α-direction. We have systematically tested this approach and found that it is relative reliable in one spatial dimension.[8] In the fourth column of Table 1 we show the ground state energies computed from the hyperspherical method.

Both Hartree-Fock and hyperspherical methods seem to give a good agreement with the exact energies. The maximum error is below 5% in all cases. It is interesting to note that the ground state energy of the negative ion ($Z=1$), which is the most strongly correlated among all two-electron systems, is better predicted by the hyperspherical method than by the Hartree-Fock approach. To understand this we have to examine the corresponding wave functions. We have plotted in Figure 1 contour plots of the spatial probability of the ground state wave function for the negative ion. The exact distribution shows indentations along the diagonal $x_1=x_2$ line, when the electrons are on the same side of the nucleus. The dents in the wave functions can be viewed as a quantum manifestation of the electrons' preference to avoid each other. Due to the symmetry imposed on the Hartree-Fock wave function [Eq. (3.1)], the ground state cannot show this asymmetry (Fig. 1c). The hyperspherical coordinates, however, allow for such an asymmetry as is presented in the corresponding contour plot [Fig. 1b]. One can expect that the hyperspherical coordinate approach in 1D could work better the more correlated a two-electron system is. This conjecture is in agreement with the computed energies in Table 1.

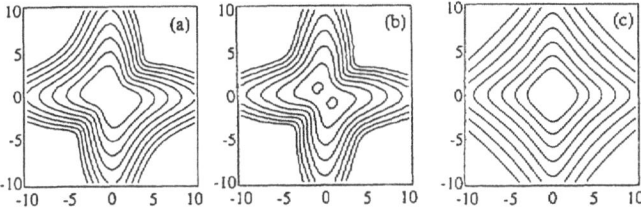

Figure 1. Contour plot of the spatial distribution of the ground state $|\Psi(x_1,x_2)|^2$ of the negative ion (a). It is compared to that calculated within the Hartree-Fock approximation (c) and the hyperspherical-coordinate approach (b). The eight contours correspond to $|\Psi(x_1,x_2)|^2 = 10^{-c}$, for c=2, 2.5, 3, ...,5.5.

The hyperspherical coordinate appproach has led to classification schemes for highly correlated two-electron states by providing novel sets of quantum numbers. Furthermore this approach introduces the concept of potential curves which are helpful to investigate two-electron transitions. In Figure 2 we show the relevant lowest-lying potential curves for the He-atom. The energy separation between potential curves of same symmetry is generally an indication of the quality of the hyperspherical approach. We found that the lower lying curves were sufficiently spaced from each other to justify the method.

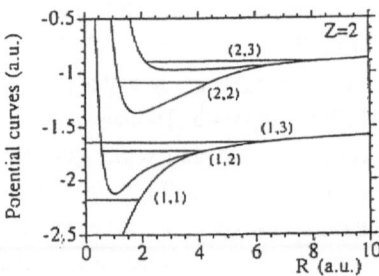

Figure 2. The relevant lowest-lying potential curves for the He-atom for two-electron wave functions which are even with respect to electron permutation. The energies of the ground state as well as the first few single and doubly ecited states are presented. We have used independent-electron labels.

4. Importance of e-e collisions in higher harmonic generation?

In this chapter we investigate the validity of the single-active electron approximation used frequently to compute higher harmonic spectra for noble gases. To isolate the effect of electron-electron collisions, we have simulated the dynamics of the negative ion under a somewhat artificial condition[9]: In the Schrödinger equation (2.3) only one of the two electrons was coupled to the external laser field by ignoring the coupling $x_2\, \mathcal{E}(t) \sin \omega t$. In this case the two electrons are distinguishable and any dynamical activity of the uncoupled electron (the one with coordinate x_2) has to come directly from collisions with the other electron. The laser frequency was chosen to be $\omega=0.08$ a.u. to correspond to 1-photon detachment for the outer electron but 9-photon ionization for the inner electron. The regime we have investigated ($\mathcal{E}=0.02$ a.u.) is characterized to a relatively rapid single ionization, but an almost negligible double ionization during the 12-cycle pulse.

As a diagnostic tool we have computed the individual electron's acceleration d^2/dt^2 $<x_i>$ as shown in Figure 3. The continuous line is the response of the optically coupled electron. The dashed curve shows the acceleration of the uncoupled electron. It is remarkable that the oscillations of the electrons are of similar magnitude. This probably indicates that electron-electron collisions play an important role and should not be neglected in this regime. Indeed, the light spectrum which is just the Fourier transformation of the total acceleration contains more harmonics for the case in which both electrons are directly coupled to the field compared to the case where only one electron is coupled (single-active electron approximation). Our results suggest that the single-active electron approximations have to be applied with great care even in regimes of laser intensity and frequency, for which single ionization is the dominant decay mechanism.

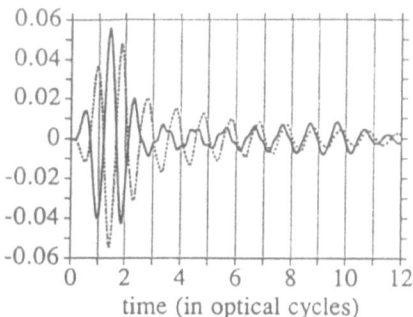

time (in optical cycles)

Figure 3. The acceleration of each electron for a somewhat artificial dynamics in which only the electron with coordinate x_1 is directly coupled to the laser field. The continuous line shows the acceleration of the coupled electron and the dashed line corresponds to the one for the uncoupled electron.

5. Which of two multi-particle states is more correlated?

In this chapter we will review how our wave functions can be used to establish a recently proposed criterion to measure the degree of correlation. This degree can be calculated directly from a given wave function without the need to compute expectation values of energies or analyzing other arbitrarily chosen quantities.

Everybody will agree that the least correlated wave function for a bosonic two-particle system would be the simple product of two identical single-particle wave functions. Obviously, the more of these single-particle orbitals that are necessary to synthesize the wave function, the more correlated is such a state. The Schmidt-Everett decomposition[10,11] gives precisely this decomposition in the form of a sum of several product states:

$$\Psi(x_1, x_2, t) = \sum_\alpha D_\alpha(t)\, \Phi_\alpha(x_1, t)\, \Phi_\alpha(x_2, t) \tag{5.1}$$

Note that in contrast to an expansion of the state in an arbitrary basis, the above summation includes only a single index. These special states Φ_α allowing for a such a unique expansion are called natural orbitals or canonical states. The canonical states depend uniquely on the wave function. If the wave function evolves in time the canonical states will change.

We have proposed to use the effective number of canonical states as a quantitative measure for the degree of correlation.[12] More precisely, the parameter K defined as

$$K(t) \equiv [\Sigma_\alpha |D_\alpha(t)|^4]^{-1} \tag{5.2}$$

is a direct measure of the number of effectively nonzero probabilities $|D_\alpha(t)|^2$. Note that K can be computed quite conveniently directly from the wave function without the need of explicitly constructing any canonical basis states. For a two-particle wave function it can be shown rigorously that K can be obtained from the trace of the square of the effective single-particle density matrix. This density matrix is obtained from the total multi-particle density operator by tracing out all other particles but one. The parameter K can be easily computed for all fermionic and bosonic multi-particle states. It has been recently proposed as a criterion for quantum chaos for a single quantum state.[13]

We have computed K for a variety of different physical situations. The correlation analysis for the above mentioned ground states of the isoelectronic series indicates, that in agreement with our expectations, the negative ion ground state has the largest K (=1.10) whereas the correlation decreases with increasing nuclear charge, as both electrons are screened more and more from each other due to the stronger nuclear attraction.

Naturally K need not to remain constant in time, and we have also calculated the time-dependence of the correlation parameter K for ionization processes for wide variety of regimes of predominantly single and double ionization. As a general rule we found that K varies sufficiently to distinguish between different dynamical regimes. For more details see the article by K. Rzazewski in this volume.

As an example of how electron-electron correlations can grow as a function of time, we have computed K for time-resolved inelastic electron-atom scattering. We present this example here also to demonstrate the flexibility of the direct numerical approach to these model two-electron atoms. An incoming (Gaussian) electron wave packet is scattered of a neutral atom which is initially in its ground state. When the incoming wave packet overlaps spatially with the bound electron of the target atom several exchange processes[14] take place. Part of the kinetic energy of the incoming electron can be used to excite the target electron into a higher lying bound state. The inset in Fig 4. shows how the various bound state probabilities of the target atom change during the scattering event. The degree of correlation grows from K=2 to 6.4 corresponding to an increase in the effective number of outgoing channels. Obviously the scattered electron and the target electron are highly correlated due to their energy exchange.

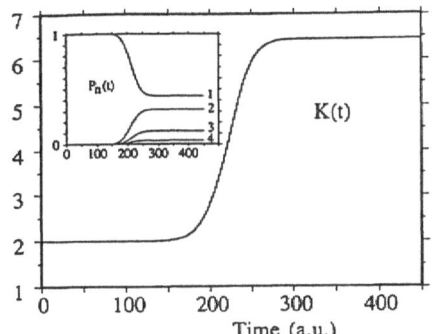

Figure 4. Growth of correlation during inelastic electron-atom scattering. The inset shows the corresponding time dependence of excited target level populations.

6. Coherence transfer by two-electron correlation

In this last section we review theoretical progress for the socalled coherence transfer effect[15] leading to an Autler-Townes continuum-continuum splitting. In 1992 our numerical simulations have shown, that the kinetic energy spectrum of an ionized outer electron can be strongly influenced if the laser frequency is resonant with a bound-bound transition of a second electron in the core. We have predicted that when a two-electron system is excited at the core-resonance frequency, the photoelectron spectrum will become doubly peaked with a peak spacing directly given by the Rabi frequency of the core transition. This prediction has been confirmed in experiments by the DiMauro group at Brookhaven[16] as well as the Muller group in Amsterdam[17].

Our first work on coherence transfer[15] dealt with a negative-ion example. It introduced the effect and established the minimum set of states required for an essential states theory that could be implemented completely analytically. The approximations leading to essential states theories often require careful testing and justification and we have used our ab initio wave functions to do this, finding good agreement. Subsequently, we have treated resonant excitation of the ionic core of a neutral atom. Depending on the specific atomic species, other states might become important and need to be included to allow for more precise predictions.[18]

In Figures 5 we sketch the more complicated situation for the reduced-dimensional helium atom, for which some of these additional states need to be included. The relevant part of its energy structure is shown in the Figure. For a complete discussion of the general theory, we refer the reader to an article by S.L Haan et al.[19]. To illustrate how the core coherence can modify the photoelectron spectrum for this complicated atom we compare in Fig. 5a the spectrum with the (artificial) case, for which the core-transtion has been neglected (dashed curve). It clear that due to the core-resonance the spectrum becomes appreciably more complicated. Fig.5b shows the spectra obtained from the full-scale numerical simulation. The agreement between the exact spectrum (b) and the analytical prediction (a) shows that the essential states approach introduced in Ref. 15 is also reliable when applied to more complicated situations.

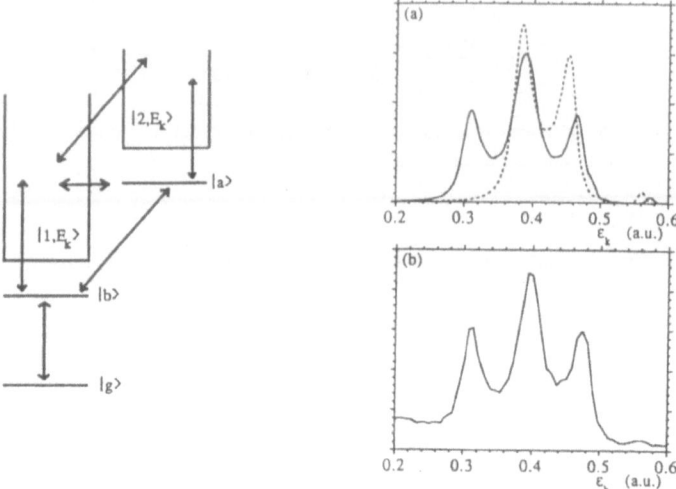

Figure 5. Relevant energy levels for the 1D-helium atom. We show only the nearby resonant bound states as well as one relevant autoionizing state. (a) Photoelectron energy spectra as predicted by the essential states theory when applied to 1D-helium. To illustrate the importance of the coherence transfer effect, the dashed line correspond to the case in which the resonant core coupling has been (artificially) omitted. To verify the theoretical predictions of (a) we show in (b) the exact photoelectron energy spectrum obtained from the direct solution of the full time-dependent Schrödinger equation (2.3).

7. Summary and Conclusions

In summary we have given four examples of recent theoretical advances obtained by using exact wave functions for a variety of reduced dimensional two-electron systems. In some cases the wave functions have led to new sets of experiments, in other cases they have helped in better understanding approximate theories in order to gain microscopic insight into the complicated dynamical processes. As fully three-dimensional ab initio calculations for two-electron systems in the intense field regime seem to be out of range of present computers, one can expect that for the near future these reduced dimensional model systems will remain helpful theoretical tools in exploring atom-laser dynamics.

We would like to acknowledge collaborations with A. Artemyev, S.L. Haan, D.G. Lappas and K. Rzazewski. This research was supported by the National Science Foundation through PHY89-20108, PHY92-00542 and PHY94-08733. We also acknowledge assistance with computing resources from the Pittsburgh Supercomputing Center.

References

1. For review, see articles in *"Multiphoton Processes"* edited by G. Mainfray and P. Agostini (CEA Press, Paris, 1991); in *"Atoms in Intense Laser Fields"*, edited by M. Gavrila (Academic, Orlando, 1992); in *"Super-Intense Laser Physics"*, ed. by B. Piraux, A. L'Huillier and K. Rzazewski, NATO ASI Series Vol. 316 (Plenum Press, Amsterdam, 1993); and in *"Multiphoton Processes"*, ed. by D.K. Evans and S.L. Chin, Series in Optics and Photonics Vol. 6 (World Scientific, 1994).

2. J.H. Eberly and R. Grobe, in *"Super-Intense Laser Physics"*, Ref. 1, p. 445.

3. Q. Su and J.H. Eberly, Phys. Rev. A. 43, 2474 (1991); J.H. Eberly, R. Grobe, C.K. Law and Q.Su in *"Atoms in Intense Laser Fields"*, Ref. 1, p. 301.

4. R. Grobe and J.H. Eberly, Phys. Rev. A 48, 4664 (1993).

5. Hartree-Fock techniques are discussed in, e.g., C. Froese Fischer, *"The Hartree-Fock method for Atoms"*, (Wiley, New York, 1977) and for time-dependent HF methods see, e.g., K.C. Kulander, K.J. Schafer and J.L. Krause in *"Atoms in Intense Laser Fields"*, Ref. 1, p. 247.

6. For a review see, e.g., C.D. Lin, Adv. At. Mol. Phys. 46, 97 (1983).

7. S.L. Haan, R. Grobe and J.H. Eberly, Phys. Rev. A 49, 378 (1994).

8. A. Artemyev, R. Grobe and J.H. Eberly, Phys. Rev. A 51, 155 (1995).

9. D.L. Lappas, R. Grobe and J.H. Eberly, Intl. J. Nonlin. Opt. (in press).

10. E. Schmidt, Math. Ann. 63, 433 (1906).

11. H. Everett III, Rev. Mod. Phys. 29, 454 (1957).

12. R. Grobe, K. Rzazewski and J.H. Eberly, J. Phys. B27, L503 (1994).

13. M.Yu. Ivanov, D. Bitouk, K. Rzazewski and S. Kotochigova, Phys. Rev. A, 52, 149 (1995).

14. D.L. Lappas, R. Grobe and J.H. Eberly, Phys. Rev. A (in preparation).

15. R. Grobe and J.H. Eberly, Bull. Am. Phys. Soc. 37, No.4, 1191 (1992) and *Phys. Rev. A* 48, 623 (1993).

16. B. Walker, M. Kaluza, B. Sheehy, P. Agostini and L.F. DiMauro, Phys. Rev. Lett. 75, 633 (1995); see also the article in this book by B. Walker, B. Sheehy, M. Kaluza, L.F. DiMauro, M. Trahin, and P. Agostini.

17. N.J. van Druten and H.G. Muller, J. Phys. B (submitted).

18. R. Grobe and S.L. Haan, J. Phys. B27, L735 (1994); L.G. Hanson, J. Zhang and P. Lambropoulos, Europhys. Lett. 30, 81 (1995) and comment by S.L. Haan and R. Grobe, Europhys. Lett. (subm.).

19. S.L. Haan, M. Bolt, H. Nymeyer and R. Grobe, Phys. Rev. A 51, 4640 (1995).

MULTIPLY CHARGED NEGATIVE IONS OF HYDROGEN INDUCED BY A SUPERINTENSE HIGH-FREQUENCY LASER FIELD

ERNST VAN DUIJN, M. GAVRILA AND H.G. MULLER

FOM-Institute for Atomic and Molecular Physics, Kruislaan 407, 1098 SJ Amsterdam, The Netherlands

1. Introduction

It has been shown theoretically that nature does not allow stable atomic multiply charged negative ions in vacuo [1]. Moreover, there is no experimental evidence for the existence of atomic multiply charged negative ions (AMCNI) [2]. However, in the last decade it has been shown, both theoretically and experimentally, that the character of atomic systems changes drastically in a radiation field. For example, whereas distribution of the ground-state wave function of an electron in a Coulombic potential is spherically symmetric around the nucleus, in a superintense high-frequency laser field this wave function is extremely distorted. In case of linear polarization, the wave function splits up into two non-overlapping parts, an effect also referred to as dichotomy of the electronic wave function [3]. For circular polarization, the distortion of the wave function leads to toroidal shaping [4].

Another striking feature, resulting from the interaction between the atom and the radiation field, is the existence of states that do not exist outside this radiation field. These so-called light-induced states were shown to exist for example in H^-, in the presence of a superintense high-frequency laser field. In vacuo, H^- exhibits only one loosely bound state in the symmetry manifold ^1S_g ($L = 0, S = 0$, even parity), for which the detachment energy is small, $D = 0.751$ eV. In a superintense high-frequency laser field, however, H^- does have excited states as well [6]. Appearently it is possible, with the help of a radiation field, to create stable atomic states that do not exist outside this field. If a radiation field is the appropriate environ-

233

H. G. Muller and M. V. Fedorov (eds.), Super-Intense Laser-Atom Physics IV, 233–244.
© 1996 Kluwer Academic Publishers.

ment in which the character of an AMCNI of hydrogen changes such that it does exhibit bound states is the question we will answer positively in this chapter.

The classical picture of how an AMCNI might exist, consists of a set of N ($N > 2$) electrons oscillating in the radiation field with the frequency of this radiation field. The nucleus is assumed to be infinitely heavy, and therefore does not follow this so-called quiver motion. If the quiver motion, which is the same for all N electrons, has a large amplitude $\alpha_o = I^{1/2}/\omega^2$ (we use atomic units throughout this paper), it is possible for each of the N electrons to be near the nucleus for part of the quiver period and to be far away from the other electrons during whole the period. This sharing of the attractive interaction with the nucleus might create the possibility for N electrons to be bound to one single proton.

By increasing the amplitude of the quiver motion, an extra electron might share the attraction to the proton, since the total repulsive energy decreases more rapidly than the attractive energy with inceasing amplitude (see below). If the frequency of the radiation field is large compared to the spacing of the electronic levels, the wave functions of all electronic levels, bound as well as continuum, are forced to oscillate in the field. Therefore, it is appropriate to use the rest frame of a freely oscillating electron as reference frame. This frame, also referred to as the Kramers-Henneberger frame (K-H-frame) is connected to the lab-frame , for arbitrary polarization labelled by the ellipticity parameter δ, via a space translation over $\vec{\alpha}(t) = \alpha_o \left[\hat{e}_1 \cos(\omega t) + \tan \delta \hat{e}_2 \sin(\omega t) \right]$, where $\hat{e}_1$ and $\hat{e}_2$ denote perpendicular unit vectors in the plane perpendicular to the propagation direction of the radiation. In the K-H-frame, the proton oscillates with the frequency of the radiation field and with amplitude α_o. If the frequency of the radiation is very high compared to all electronic spacings, the electrons cannot follow the motion of the proton. In this high- frequency limit, the electrons effectively feel a potential generated by the charge on the orbit traced out by the motion of the proton, averaged over one period of the laser. Since it is this limit in which it has been shown that there exist light induced excited states of H^-, we are interested in high frequencies when we are looking for a radiation field in which an AMCNI might exhibit bound states.

In the remaining part of this chapter we will first discuss the theory that is used as a basis for our calculations. In order to emphasize the importance of the shape of the positive charge distribution in the K-H-frame for creating an AMCNI, two special types of polarization, circular and linear, will be discussed in more detail. The use of polychromatic radiation in order to reduce the required value of α_o to create an AMCNI is illustrated for circular polarization. Finally, a numerical calculation on H^{2-} in a linearly polarized laser field is presented.

2. Theory

We use high-frequency Floquet theory (HFFT) for describing the interaction between the ion and the laser field. One of the major advantages of using HFFT is that the Schrödinger equation becomes time-independent. However, the price we pay for removing the time-dependence is that we have to handle now with an infinite set of coupled time-independent Schrödinger equations for all Floquet components . Since this has been described extensively elsewhere (see for an overvieuw [5]), we will only briefly comment on this. We make the dipole approximation, neglecting retardation effects $(\vec{A}(\vec{r}, t) = \vec{A}(t))$ and magnetic terms $(\vec{B} = \nabla \times \vec{A})$. In the K-H-frame the Schrödinger equation to be solved for the N-electron wave function is the following

$$\left[\sum_{i=0}^{N} \left\{ -\frac{1}{2}\Delta_i + V_o(\alpha_o; \vec{r}_i) + \frac{1}{2}\sum_{j\neq i}\frac{1}{|\vec{r}_i - \vec{r}_j|} \right\} - (E + n\hbar\omega) \right] \phi_n =$$

$$-\sum_{i=0}^{N}\sum_{m\neq n} V_{n-m}(\alpha_o; \vec{r}_i)\phi_m, \qquad (1)$$

where the ϕ_n are the Floquet components of the N-electron wave function and V_n is the n-th Fourier component of the space-translated Coulomb potential,

$$V_n(\alpha_o; \vec{r}) = \frac{1}{2\pi}\int_0^{2\pi} e^{in\omega t}\frac{1}{|\vec{r} + \vec{\alpha}(t)|}dt. \qquad (2)$$

Note that the zeroth Fourier component of the potential on the left-hand side of eq. 1 is the space-translated potential time averaged over one period of the laser. Eq. 1 can be solved by iteration [5]. For high frequencies this iteration procedure converges rapidly, the contribution of the k-th iteration is an order $1/\omega$ smaller than that of the $(k-1)$-th iteration. In the limit $\omega \to \infty$, we are left with the zeroth order solution in $1/\omega$, giving rise to the following Schrödinger equation,

$$\left[\sum_{i=0}^{N} \left\{ -\frac{1}{2}\Delta_i + V_o(\vec{r}_i, \alpha_o) + \frac{1}{2}\sum_{j\neq i}\frac{1}{|\vec{r}_i - \vec{r}_j|} \right\} \right] \Phi = W(\alpha_o)\Phi, \qquad (3)$$

where Φ is now the zeroth order Floquet component of the N-electron wave function after zero iterations. Eq. (3) obviously has real eigenvalues W, showing the stability with respect to photoionization of the ion in the high-frequency limit. Moreover, it should be mentioned that the eigenvalues only depend on the intensity and frequency of the laser via the parameter α_o.

The solutions of Eq. 3 depend, through V_o, very strongly on the motion of the proton in the K-H frame. It will be shown that for creating an N-fold multiply charged negative ion at as low value of α_o as possible, it is important that the motion of the proton is such that it has zero velocity at N points in space. This can be achieved by using a bichromatic laser field, consisting of two circularly polarized laser beams with different amplitude. Since the motion of the proton is determined by the polarization of the laser field, we will treat two special cases for the polarization. First, we will concentrate on the circularly polarized case, for which we will compare the results with the bichromatic radiation. Second, we treat the case of a linearly polarized laser field, for which we use a self-consistent-field procedure to calculate the energy and wave function of the ground state of the simplest type of an AMCNI, H^{2-}.

3. Circular Polarization

As mentioned above, the symmetry of V_o depends on the polarization of the laser. For a circularly polarized laser field, V_o is the potential generated by a circular charge , in the plane perpendicular to the propagation axis, of radius α_o with a homogeneous charge density $1/(2\pi\alpha_o)$. Due to the symmetry of V_o, each symmetry manifold is labelled with the quantum number Λ (the absolute value of the total angular momentum with respect to the polarization axis), parity P (g or u) and total spin S. (For more details, see [4]).

If an atomic multiply charged negative ion of hydrogen were to exist in a circularly polarized laser field, the N (N>2) electrons would arrange themselves into a configuration that minimizes the total repulsive energy, while attached to the positive charge circle due to the logarithmic singularity the attractive potential has on the circle. In this particular electron configuration the centers of the N one-electron wave functions are located each on a corner of an N-sided polygon, resulting in a total repulsive energy

$$E_{rep}(\alpha_o) = \frac{N}{4\alpha_o} \sum_{k=1}^{N-1} \left(\sin \frac{k\pi}{N} \right)^{-1} + O(\alpha_o^{-3/2}). \qquad (4)$$

The corrective term of order $O(\alpha_o^{-3/2})$ results from the extension of the N one-electron wave functions *along* the circle (corrections to the repulsive energy due to motion *transverse* to the charge circle scale as $O(\alpha_o^{-5/2})$. The spread along the charge circle scales as $\alpha_o^{3/4}$, so the extension of the one-electron wave functions relatively decreases as $\alpha_o^{-1/4}$.

With an expression for the binding energy of an electron to a charge circle, we can find the value of α_o needed to bind N electrons to one single proton by comparing the total energy of the N- and $(N-1)$- electron configuration. This binding energy for large α_o is given by [4]

$$E_{bind}(\alpha_o) = -\frac{1}{2\pi\alpha_o}\left[\ln\frac{64\alpha_o}{\pi} - 0.35987\right] + O((\ln\alpha_o/\alpha_o)^2). \qquad (5)$$

It should be mentioned that this scaling law is derived for one electron bound to the proton ("toroidal shaping"), whereas we use it here for N electrons that are not toroidally shaped but have a limited extent along the circle. In the latter case, as mentioned above, the correction to the energy asymptotically scales as $\alpha_o^{-3/2}$, so it can be neglected compared to the binding energy to the circle. The constant on the right-hand side is determined by the symmetry manifold under consideration, in this case the σ_g manifold.

In the right column of table 1 the minimum value of α_o to bind N electrons to one proton in a monchromatic laser field, $\alpha_o(N)$, is shown. From these values we see that the correction term on the binding energy is negligible. $\alpha_o(N)$ grows rapidly to values for which the theory we use is not valid, since relativistic effects start to play a role. The main reason for this rapid growth is the logarithmic scaling behaviour of the binding energy, leading to an exponential behaviour of $\alpha_o(N)$. Moreover, the relaxation energy due to the redistribution of the $(N-1)$ electrons after one of the N electrons is detached is large, increasing $\alpha_o(N)$.

By using a bichromatic radiation source the rapid growth of $\alpha_o(N)$ can be suppressed. If a left- resp. right-handed circularly polarized fundamental is accompanied by its (N-1)th harmonic, right- resp. left-handed polarized, with an $(N-1)$ times smaller amplitude, the proton has zero velocity at N equally spaced points on a circle with radius α_o [9]. Since the positive charge distribution becomes inhomogeneous, the electrons can not move along the positive charge distribution as freely as for an homogeneous charge distribution, reducing the relaxation energy. In figure 1 the electron configuration is depicted for $N = 3$. For large α_o, the potential around these points is defined by the "end-point" potential [5]. Since the binding energy to an end-point potential scales according a power law, $E_{bind} \sim \alpha_o^{-2/3}$, $\alpha_o(N)$ increases as a power instead of exponentially as in the monchromatic case, reducing $\alpha_o(N)$ tremendously, as shown in the right column of table 1 [10].

4. Linear Polarization

A linearly polarized monochromatic laser field induces a V_o that is generated by a line charge with an inhomogeneous charge density. If the polarization axis is defined as z-axis, the charge of the proton is smeared out

TABLE 1. The minimum value of α_o, required to bind N electrons, calculated in the high-frequency approximation, for monochromatic (left column) and polychromatic (right column) circular polarization. For the monochromatic (polychromatic) radiation, the scaling laws for the wave function and energy, in the limit for large α_o, for circular (linear) polarization are used.

N	$\alpha_o(N)$ monochromatic	$\alpha_o(N)$ polychromatic
3	$1.6 \ 10^2$	$6.4 \ 10^1$
4	$3.7 \ 10^4$	$4.3 \ 10^2$
5	$1.5 \ 10^7$	$2.0 \ 10^3$
6	$9.7 \ 10^9$	$7.2 \ 10^3$

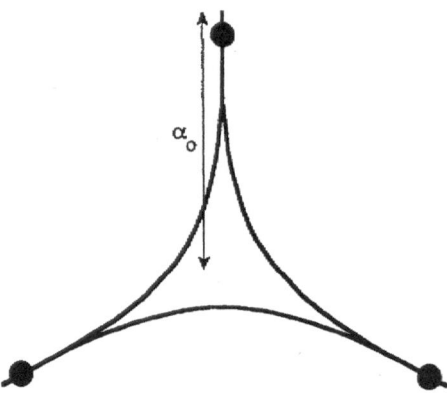

Figure 1. Schematic picture of the positive charge distribution (line) in the bichromatic (fundamental and second harmonic) case for $N = 3$ (the electrons are depicted as dots). The charge density is inhomogeneous and singular at the endpoints.

along the line segment $z = [-\alpha_o, +\alpha_o]$, with a non-uniform charge density,

$$\sigma = \frac{1}{\pi\sqrt{\alpha_o^2 - z^2}}. \tag{6}$$

The charge density has an integrable singularity at $z = \pm\alpha_o$, since the proton has zero velocity at these points. Moreover, the charge density on the line segment $z = [-\alpha_o, +\alpha_o]$ is at least twice as large as that for circularly polarized light, since the proton passes this interval twice per period of the laser (For a more detailed description of V_o in case of linear polarization, see [3]). As for circular polarization, parity and the total angular momentum projected on the z-axis are good quantum numbers. However, the quantization axis is now perpendicular to the propagation direction, whereas they are parralel when the polarization is circular. The positive charge is concentrated at two spatially well-separated $(2\alpha_o)$ points for monochromatic radiation, whereas one needs bichromatic radiation to achieve this when the polarization is circular.

Let us consider now the simplest example of an AMCNI of hydrogen, H^{2-}. If H^{2-} were to exhibit bound states in a linearly polarized monochromatic laser field, we expect the one-electron wave functions of two of the three electrons to be concentrated around the endpoints, at $z = \pm\alpha_o$ (we consider the ground state only). From symmetry considerations, the one-electron wave function of the third electron is confined to be symmetric in the $z = 0$-plane. A favorable position for the third electron would be near the charge line at $z = 0$, since there the attractive potential with the line charge has a singularity. Therefore, a linearly polarized laser source is a good candidate as an external radiation source to change the character of the ion such that three electrons can be bound to one proton $(N = 3)$, resulting in a light stabilized AMNCI, H^{2-}.

We performed a Hartree calculation for determining the existence of H^{2-} as described above. We do not expect the one-electron orbitals to have any overlap, since simple point-charge calculations show that the detachment energy (defined as the energy required to detach one of the three electrons and leave H^- in the ground state) becomes positive only for values of α_o far in the regime where H^- is dichotomized ($\alpha_o > 100$, whereas H^- dichotomizes for $\alpha_o > 20$). Therefore, this Hartree calculation is similar to an *unrestricted* Hartree-Fock calculation. A *restricted* Hartree-Fock calculation would give a very poor result, since, for minimizing the repulsive energy, the electrons should be described by different orbitals.

The results of the calculation are shown in figure 2, which shows the negative of the detachment energy of the ground state of H^{2-} in the Σ_g symmetry manifold as a function of α_o. From these results one can see that a doubly charged negative ion of hydrogen does exist in a superintense high-frequency laser field for $\alpha_o > 155$. For $\alpha_o < 155$, the character of the state of H^{2-} is a shape resonance. The repulsive interaction due to the electrons around $z = \pm\alpha_o$ and the attractive interaction with the positive line charge together create a potential barrier for the electron around $z = o$ in the

240

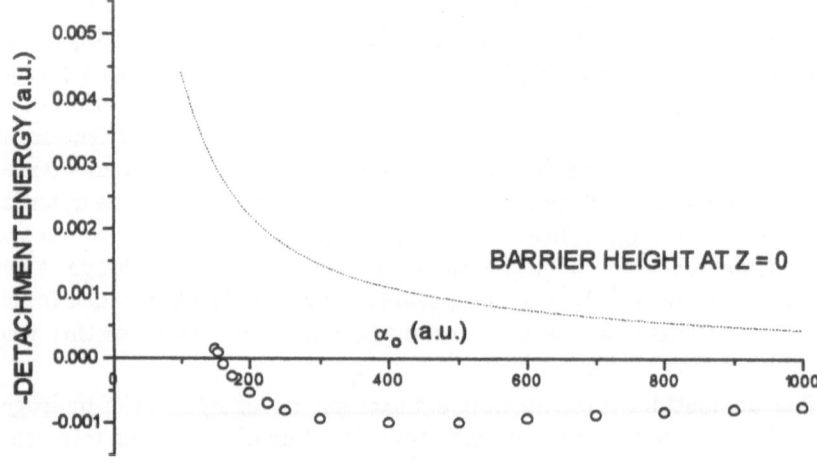

Figure 2. Negative of the detachment energy for the ground state of H^{2-} (open dots) in the Σ_g-manifold as a function of α_o. For values of $\alpha_o > 155$ (a.u.), H^{2-} exhibits a bound state. For $\alpha_o \simeq 400$, the maximum detachment energy is reached, $D_{max} = 0.0272$ eV. For α_o slightly smaller than 155, states with negative detachment energy lie far below the top of the potential barrier (dashed line), giving rise to shape resonances.

direction perpendicular to the polarization axis. Due to the logarithmic singularity of the potential on the line charge, a potential well is generated along the line charge. The width of this well increases with α_o. While α_o increases, a metastable state is created for the electron around $z = 0$. If α_o increases even more, this metastable state will turn into a bound state as α_o becomes larger than $\alpha_o(N = 3)$. For $\alpha_o > \alpha_o(N = 3)$, the one-electron orbitals of the electrons around $z \pm \alpha_o$ behave according to the scaling laws for the "end-point" potential to within a few percent [5]. This polarization of the orbitals towards $z = 0$, which increases the total repulsive energy, shifts $\alpha_o(N)$ towards higher values than one would obtain in the lowest order approximation ($O(\alpha_o^{-1})$; next higher order term is $O(\alpha_o^{-5/3})$), in which the two of the three electrons are treated as point charges fixed at $z = \pm\alpha_o$. For those values of α_o for which the detachment energy is positive, the three one-electron orbitals have zero overlap, as expected. In figure 3 we show the three-electron wave function of H^{2-} at $\alpha_o = 155$. As α_o increases, the one-electron orbitals become more and more spatially separated. Note

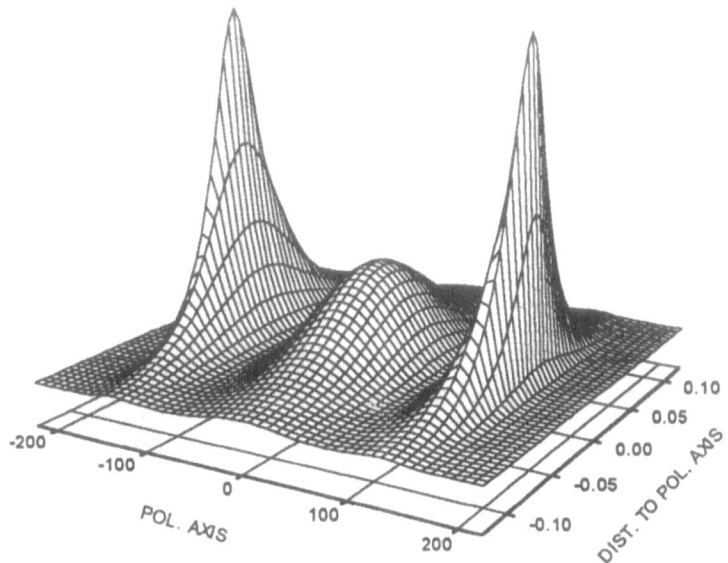

Figure 3. The three-electron wave function of H^{2-} in a linearly polarized (along z-axis) laser field at $\alpha_o = 155$. Note there is no spatial overlap between the one-electron orbitals. The orbital of the electron centered around $z = 0$ is more diffuse in the z-direction than the orbitals at $z \pm \alpha_o$, since the potential along the polarization axis near $z = 0$ is harmonic with a small force constant. Despite the repulsive interaction with the electron around $z = 0$, the wave functions of the electrons at $z \pm \alpha_o$ are slightly polarized towards $|z| < \alpha_o$ due to the attraction those electrons feel with the positive line charge.

that the value of α_o required to create a doubly charged negative ion of hydrogen is of the same order of magnitude for both linearly and circularly polarized light.

4.1. DICHOTOMY

Let us take a closer look at the potential around $z = 0$, generated by the positive line charge and the electrons at the end-points (for a moment we approximate the expectation value of z for the electrons at the end points to be $< z > = \pm \alpha_o$. Close to the line charge, this potential is given by

$$V(z, \rho) = \frac{-2}{\pi \sqrt{\alpha_o^2 - z^2}} \left[\ln \frac{4(\alpha_o^2 - z^2)}{\alpha_o} - \ln \rho \right] + \frac{2\alpha_o}{\alpha_o^2 - z^2} + O((\rho/\alpha_o)^2) \quad (7)$$

Due to the logarithmic singularity in the ρ-direction, the excitation energy in the ρ-direction is much larger than that in the z-direction. Therefore we can adiabatically separate the potential in ρ and z. Using the scaling behaviour of the eigenvalue in the logarithmic potential, we get the following potential for the motion in the z-direction,

$$V(z) = \frac{-2}{\pi\sqrt{\alpha_o^2 - z^2}} \left[\ln \frac{4(\alpha_o^2 - z^2)}{\alpha_o} + \frac{1}{2} \ln \frac{2}{\pi\sqrt{\alpha_o^2 - z^2}} - E_i \right] + \frac{\alpha_o \pi}{\alpha_o^2 - z^2},$$

(8)

where E_i is the energy of the i-th level in the $\ln\rho$ potential. Since the electron we are interested in is located around $z = 0$, we can expand eq.8 around $z = 0$. This leads to

$$V(z) = \frac{2}{\pi\alpha_o} \left[E_i + \pi - \ln\sqrt{\frac{32\alpha_o}{\pi}} \right] + \frac{1}{2}\omega^2 z^2 + O((\frac{z}{\alpha_o})^4),$$

(9)

where

$$\omega^2 = \frac{2E_i + 4\pi + \frac{3}{2} - \ln\frac{32\alpha_o}{\pi}}{\pi\alpha_o^3}.$$

(10)

From eq.(10) we see that the electron is approximately harmonically bound in the z-direction. Moreover, ω^2 becomes negative if α_o increases. The value of α_o for which this happens, depends via E_i on which state we consider in the $\ln\rho$ potential, ground state or excited state. For the ground state, $E_i = 0.17999$; the corresponding value of α_o for which ω becomes negative for the ground state is therefore 180.000. For values of $\alpha_o > 180.000$, the potential around $z = 0$ therefore consists of two minima instead of one. As α_o increases, these minima become deeper, and for a certain value of α_o they will both be deep enough to have a bound state. For that value of α_o the one-electron wave function of the electron around $z = 0$ dichotomizes. We found that this happens for extremely large values of α_o, n.l. $\alpha_o > 2.000.000$. Note that relativistic effects not necessarily play a role here ($\beta = \alpha_o\omega/137$), since as the intensity increases the frequency required for the high-frequency approximation to be valid decreases rapidly.

4.2. STABILIZATION

There are several reasons why the electron around $z = 0$ is more stable against photoionization than the two electrons around the end-points of the line charge. Since the binding energy of the middle electron is much smaller than that of the outer two electrons, it might happen that for a certain fixed frequency the high-frequency approximation holds for the electron around $z = 0$, whereas it does not for the outer two electrons. In that case, the quasi-energy of the outer two electrons has a much larger imaginary part than the electron in the middle.

243

Moreover, since the proton passes the origin $z = 0$ twice per period of the laser, the electron at $z = 0$ effectively "sees" a frequency that is twice the frequency "seen" by the outer two electrons. This is another reason why the middle electron can be in the high-frequency regime whereas the outer two are not.

To lowest order in $1/\omega$, the expression for the total decay rate is given by

$$\Gamma = 2\pi \sum_{m \neq 0} \sum_k |\langle \phi_o | V_m | k \rangle|^2 \delta(W_m - E_k). \tag{11}$$

Since the V_m have parity m, all $V_m(z = 0) = 0$ for m odd. So the first contribution to the decay rate in the sum in Eq.(11) for the middle electron comes from the $m = 2$-term. This also reflects the fact that the electron around $z = 0$ sees frequency twice the frequency seen by the outer two electrons.

For these reasons we can conclude that the lifetime of H^{2-} is mainly determined by the lifetime of the electrons near the end-points, i.e. the underlying H^- ion. It has been shown that the lifetime of H^- equals half the lifetime of hydrogen in the same field [12]. Since atomic hydrogen is fully dichotomized at those values of α_o for which H^{2-} exists, we can apply this result to find an expression for the lifetime of H^{2-}. So the n-photon ionization rate of H^{2-} is slightly larger than twice the ionization rate of hydrogen in the same field:

$$\Gamma_n^{(H^{2-})} \simeq 2\Gamma_n^{(H)}, \tag{12}$$

Since we know that atomic hydrogen adiabatically stabilizes against ionization, we see from Eq.(12) that the AMCNI H^{2-} also adiabatically stabilizes.

5. Acknowledgements

This work is part of the research program of the "Stichting voor Fundamenteel Onderzoek der Materie (FOM)" and was made possible by financial support of the "Nederlandse Organisatie voor Wetenschappelijk Onderzoek (NWO)"

References

1. E. Lieb, Phys. Rev. Lett.52, 315 (1984).
2. D.R. Bates, Adv. At. Mol. Opt. Phys. 27, 1 (1990), Sec. I D.
3. M. Pont, N. Walet, M. Gavrila and C.W. McCurdy, Phys.Rev.Lett. 61, 939 (1988); M. Pont, N. Walet and M. Gavrila, Phys.Rev. A 41, 477 (1990).
4. M. Pont, Phys. Rev. A40, 5659 (1989).
5. M. Gavrila in "Atoms in Super Intense Laser Fields", edited by M. Gavrila (Academic, New York, 1992) p. 435.

6. H.G. Mulller and M. Gavrila, Phys. Rev. Lett. **71**, 1693 (1993).
7. N.J. van Druten et.al., to be published.
8. M.P. de Boer et.al., Phys. Rev. Lett. **71**, 3263 (1993).
9. The creation of a Lissajou-like positive charge distribution in the K-H- frame by two laser sources with commensurate frequencies was proposed by M. Lewenstein et.al. in Proceedings of a NATO Advanced Research Workshop on "*Super Intense Laser-Atom Physics*", edited by B.Piraux, A. L'Huilllier and K. Rząžewski (Plenum Press, New York, 1993), p. 425.
10. The change in total binding resp. repulsion energy, between the (N-1) - and N-electron configuration for polychromatic radiation, scales as $[(N-1)\alpha_o]^{-2/3}$ resp. $\sum_{k=1}^{N-1}\left[2\sin\frac{k\pi}{N}\left(\alpha_o - 2.3\left\{(N-1)\alpha_o\right\}^{1/3}\right)\right]^{-1}$.
11. W. Kolos and C.C.J. Roothaan, Rev. of Mod. Phys. **32**, 205 (1960); M. Abramowitz and I.A. Stegun, "*Handbook of Mathematical Functions*" (National Bureau of Standards, 1970), sec 21.
12. M. Gavrila and J. Shertzer, submitted to Phys. Rev. A.

FLOQUET THEORY OF MULTIPHOTON PROCESSES
IN MULTIELECTRON ATOMS

M. DÖRR
Max-Born-Institut
D-12474 Berlin

Abstract.
Calculations in several atomic systems (H, He, H⁻, Ne, Ar) using the R-matrix-Floquet approach have shown that the method gives accurate *ab initio* nonperturbative results for multiphoton processes in strong laser fields. A general introduction and overview to this theory is given and recent progress using these methods is reported, including results on electron–proton scattering in a laser field, multiphoton detachment of H⁻ and laser-induced degeneracies.

1. Introduction

A high-accuracy theory for atoms in strong laser fields must include an accurate description of the atom in the absence of the field and take into account the strong modification due to the field. Such a theory has been developed over the last years within an international collaboration, leading to the R-matrix-Floquet method.

The Floquet or quasi-stationary picture is central to this approach : although the evolution of an atom in a strong laser pulse is inherently a time-dependent process, a description in terms of quasi-stationary states of an atom in a laser field has several advantages. First, the accurate computation of the quasi-stationary solution is much faster than a fully time-dependent calculation. Second, the Floquet description allows an interpretation of the evolution of an atom in a laser pulse of variable intensity in terms of energy level shifts and crossings and of effective ionization rates. Partial rates into the various ATI channels and the corresponding angular distributions can also be calculated. The quasi-stationary theory is applicable even for very

H. G. Muller and M. V. Fedorov (eds.), Super-Intense Laser-Atom Physics IV, 245–255.

short pulses [1, 2] and also in the presence of crossings. In the latter case, however, several states of the atom will in general be populated through the crossings and Floquet theory together with an appropriate time-dependence can be used to describe such processes [3]. Thus, the Floquet approach is a nonperturbative theory in the field intensity which applies up to very high field intensities, and it also incorporates correctly the low intensity limit, where perturbation theory is applicable.

An accurate ab-initio description of atomic processes involving multi-electron systems is given by the R-matrix theory [4], which has been very successful in studies of electron-atom scattering and photoionization (that is, single photon ionization, within a perturbative approach). In the present context, the word atom is meant to include positively or negatively charged ions. The R-matrix theory is also widely applied to molecules and an extension of the present R-matrix-Floquet theory to molecules is planned. At present, our calculations are restricted to single ionization; multiple ionization can be calculated as a sequential process. It appears that in almost all circumstances (unless there are particular intermediate resonances [5]) single ionization is much more probable than double ionization. Work is in progress on a double-continuum R-matrix approach [6], which will allow the study of simultaneous double photoionization.

In the next section, the theory is briefly presented, with emphasis on the discussion of the physical ideas rather than completeness, since the theory has been presented in detail before [7, 8, 9]. In section 3, a few illustrative results are presented. Again, this section cannot be comprehensive, since many calculations are currently in progress, at various places in Europe. These are mentioned in section 4.

2. Theory

2.1. FLOQUET ANSATZ

Consider a laser field with vector potential

$$\mathbf{A}(t) = \hat{\epsilon} A_0 \sin \omega t, \tag{1}$$

where ω is the angular frequency and $\hat{\epsilon}$ is the polarization vector. The electric field is $\mathcal{E}(t) = \hat{\epsilon} \mathcal{E}_0 \cos \omega t$, with $A_0 = -c\mathcal{E}_0/\omega$.

This classical field is coupled to an atomic system composed of a nucleus of atomic number Z and $N + 1$ electrons to yield the time-dependent Schrödinger equation, in atomic units,

$$i\frac{\partial}{\partial t}\Psi(\mathbf{X}_{N+1}, t) = \left[H_{N+1} + \frac{1}{c}\mathbf{A}(t)\cdot\mathbf{P} + \frac{N+1}{2c^2}\mathbf{A}^2(t) \right]\Psi(\mathbf{X}_{N+1}, t). \tag{2}$$

Here H_{N+1} is the field-free atomic Hamiltonian

$$H_{N+1} = \sum_{i=1}^{N+1} \left(-\frac{1}{2}\nabla_i^2 - \frac{Z}{r_i} \right) + \sum_{i>j=1}^{N+1} \frac{1}{r_{ij}} \tag{3}$$

and $\mathbf{P} = \sum_{i=1}^{N+1} \mathbf{p}_i$ is the total momentum operator. The symbol $\mathbf{X}_{N+1}$ denotes the totality of the $N+1$ electrons spatial $\mathbf{r}_i$ and spin σ_i coordinates, $\{\mathbf{x}_1, \ldots \mathbf{x}_{N+1}\}$, where $\mathbf{x}_i = \{r_i, \hat{\mathbf{r}}_i, \sigma_i\}$.

In order to solve this time-dependent equation we use the periodicity of the driving field and introduce the Floquet ansatz

$$\Psi(\mathbf{X}_{N+1}, t) = e^{-iEt} \sum_{n=-\infty}^{+\infty} e^{-in\omega t} \psi_n(\mathbf{X}_{N+1}) \tag{4}$$

which leaves us with a system of time independent coupled equations

$$(H_{N+1} - E - n\omega)\psi_n + D_{N+1}(\psi_{n-1} + \psi_{n+1}) = 0, \tag{5}$$

where D_{N+1} is the dipole operator.

This Floquet approach allows to calculate quasi-stationary atomic states, which are the "analogues" of the stationary (bound or resonance) states for the field-free problem. In order to obtain the solution to the full time-dependent situation, where the laser pulse is turned on and off, one must perform a Floquet calculation for each intensity and subsequently integrate over the spatio-temporal pulse shape.

2.2. COMPLEX ENERGY SOLUTIONS

The energy E defined above is complex in our case since the system decays under the influence of the laser field [10]. Thus

$$E = E_0 + \Delta - i\Gamma/2, \tag{6}$$

where E_0 is the field-free energy, Δ an energy shift, and Γ the width or decay rate of the resonance. When considering the product $\Psi(\mathbf{r}, t)^*\Psi(\mathbf{r}, t)$ we see that for fixed r this is proportional to $\exp(-\Gamma t)$. Such a decay always appears when coupling a bound state to a continuum (or to several continua). Here the laser field can ionize the atom and this can be viewed as a system of coupled channels, corresponding to absorption or emission of photons, which include open channels in which the atom ionizes. In principle the field couples infinitely many channels (also in angular momentum) but in practice usually convergence can be achieved with a finite number (just as for field free scattering there is convergence with the number of angular momenta included). The channel energies in the Floquet approach are thus related by $E_n = E_0 + n\omega$ and consequently if the energy becomes complex so do the channel momenta [11]. Strictly speaking, we are calculating poles of the resolvent operator $(z - H)^{-1}$, which have an observable effect only if close enough to the physical (real) energy axis which implies that Γ must not be too large (in particular $\Gamma < \omega$). These eigenvalues and associated eigenfunctions are smoothly connected to the field-free eigenvalues which are on the real energy axis – or to autoionizing states, which have a finite width even in the absence of a laser field, see subsection 3.3. Since in the open channels the momenta have a negative imaginary part, the wavefunction is not normalizable in the usual sense. It is however normalizable by considering the analytic continuation to the real energy axis or equivalently by considering the dual space wavefunction [12], which leads to a normalization integral without complex conjugation. An easy example consists of a complex symmetric matrix, which has complex eigenvalues and eigenvectors, which are orthogonal under a scalar product defined without complex conjugation.

2.3. R-MATRIX APPROACH

The key idea of the R-matrix theory [4] is to subdivide configuration space into an internal and an external region by a sphere of radius a, the physical picture being that only one electron is leaving the atom, while all N other (core) electrons remain bound to the nucleus. This allows a considerable simplification of the problem : within the finite internal region we must solve the full multielectron problem, while in the outer region exchange with the bound electrons can be neglected and we have only a single electron interacting with the laser field and the residual core via a multipole potential.

For the solution of the multielectron problem in the inner region, an R-matrix basis expansion is chosen, of the form

$$\psi_{kn}(\mathbf{X}_{N+1}) = A \sum_{\Lambda i \ell} \bar{\phi}_i^\Lambda(\mathbf{x}_1, ..., \mathbf{x}_N, \hat{\mathbf{r}}_{N+1}, \sigma_{N+1}) r_{N+1}^{-1} u_\ell^\Lambda(r_{N+1}) a_{i\ell kn}^\Lambda, \quad (7)$$

where A is the antisymmetrisation operator, the u_ℓ^Λ – called "continuum orbitals" – are radial basis functions which are non-vanishing on the boundary of the internal region ($r_{N+1} = a$), and $a_{i\ell kn}^\Lambda$ are coefficients.

The channel functions $\bar{\phi}_i^\Lambda(\mathbf{x}_1, ..., \mathbf{x}_N, \hat{\mathbf{r}}_{N+1}, \sigma_{N+1})$ are formed by coupling the configuration interaction core wavefunctions $\phi_i(\mathbf{x}_1, ..., \mathbf{x}_N)$ with the spin-angle functions of the scattered or ejected electron (which depend on $\hat{\mathbf{r}}_{N+1}$ and σ_{N+1}) to give a state with quantum numbers $\Lambda \equiv \lambda L S M_L M_S \pi$. Here L is the total orbital angular momentum quantum number, S the total spin quantum number, M_L and M_S are the corresponding magnetic quantum numbers and π is the parity of the $(N+1)$-electron system. The quantity λ serves to specify the remaining quantum numbers required to completely define the channel.

As a simple non-trivial example, consider the H^- ion : here we have $Z = 1$ and $N = 1$, $i.e.$ we have a single electron in the residual "core" system (atomic hydrogen), left over after the detachment process. We have used two different models for the core, each defined by a set of three orbitals or basis functions in expansion (7) forming the core wavefunctions ϕ_i. In general, for a multi-electron core, the ϕ_i consist of a configuration-interaction superposition of products of the basis orbitals.

For studying detachment at low frequencies (photon energies below or slightly above the detachment threshold), an approximation which models correlation effects accurately and also includes the full polarizability of the hydrogen atom ground state is required. This (pseudostate) model comprises the three orbitals called $1s$, $\overline{2s}$ and $\overline{2p}$ whose radial functions are $\phi_{1s}(r) = 2re^{-r}$, the physical hydrogenic 1s orbital, while $\phi_{\overline{2s}}(r) = (2\sqrt{3}\,r - 4\sqrt{3}\,r/3)e^{-r}$, and $\phi_{\overline{2p}} = \sqrt{32/129}(r^2 + r^3/2)e^{-r}$ are so-called pseudo-orbitals. This model yields an attachment energy of -0.02538 a.u. As has been remarked by Liu, Gao and Starace [13], it is important in order to obtain accurate cross sections to semi-empirically shift the attachment energy to the exact value. This corrects not only the threshold positions, but more importantly the multiphoton detachment rates, which are very sensitive to the exact attachment energy. Thus, although our method is very different from that of [13], we use a very similar semi-empirical shift, which is a small correction, and obtain results in close agreement with their low frequency perturbative results.

Pseudostates are not appropriate to study autoionizing resonances below excited core thresholds. In this energy region we must use the physical (hydrogenic) $n^* = 2$ core states 2s and 2p, together with the 1s. This basis results in poorer accuracy at low energies (the polarizability is then given as 2.96 a.u.), but this is not important at higher energies in the vicinity of or above the $n^* = 2$ region. This model yields an attachment energy of 0.02214 a.u. We do not adjust the attachment energy to the exact energy in this case since the shift is negligible compared with the photon energy.

The pseudostate model can be completed in a consistent way by introducing higher excited pseudo-orbitals [14]. The physical orbital model can also be augmented by adding higher excited hydrogenic orbitals. It must however be noted in this case that the expansion is formally not complete since the continuum of the core is not included.

In practice, additional quadratically integrable terms are added to expansion 7 : $\sum_{\Lambda i} \chi_i^{\Lambda}(\mathbf{X}_{N+1}) b_{ikn}^{\Lambda}$. The χ_i^{Λ} are configurations built out of the basis orbitals, which vanish at the boundary of the internal region.

In the external region, $r_{N+1} > a$, a single-particle multichannel problem must be solved. There are standard methods available, such as R-matrix propagators and asymptotic expansions, which however require modification when a laser field is present [9].

2.4. R-MATRIX-FLOQUET THEORY

The inner region problem in the presence of the laser field is solved using the R-matrix basis expansion above and the Floquet Hamiltonian is written in the length gauge, since the convergence of the R-matrix expansion is much faster than in the velocity gauge.

It should be noted that our method goes beyond usual (even beyond symmetry-dependent) model potentials, because in our case the core is not static. Exchange is fully taken into account and the core can be distorted and excited by the field and by the interaction with the outgoing electron and thus participates actively. This is what we mean by "fully correlated".

The external region solution proceeds as described in detail in [9]. Here we use the velocity gauge for the outer electron for propagating the R-matrix from the inner region boundary to the asymptotic radius. The R-matrix is transformed to the velocity gauge at the radius of the internal region; this transformation is local, and is given by a unitary matrix. The outgoing electron asymptotic wavefunction must be defined in the acceleration frame, since in order to define partial rates the channels must be asymptotically decoupled, which is not the case in the length or velocity gauges. The asymptotic channels are defined by (1) the kinetic energy of the outgoing electron (2) the field-distorted eigenstate of the residual core in the field and (3) the asymptotic angular momentum of the outgoing electron, which is coupled to the remaining core. We assume that the ionization occurs within the field : we are not considering here the effects of varying laser intensity during the ionization process. An electron emerging with an average kinetic energy of 2 eV travels a distance of 30 a.u. within 2 fsec, which is far enough to be considered as effectively ionized. Care must be taken for low energy electrons in ultrashort pulses and also for processes with large excursion amplitude in low frequency fields (the latter are at the limits of validity of Floquet theory and better described in a stationary field picture, possibly using semiclassical methods – see other contributions in this volume).

As the field increases, more and more Floquet blocks and angular momenta are required to ensure convergence. Thus, for the strongest field couplings considered, a total of 10 Floquet blocks and 6 angular momenta were used. This results in the diagonalization of a Hamiltonian of dimension 1700 and 110 channels in the external region. Moreover, as the field increases in intensity, the asymptotic expansion (see [9]) must be made at increasingly larger radial distances to ensure convergence, the outer radius at which the asymptotic expansion is computed being 40 or 50 a.u. Hence the propagation of the R-matrix must be done over a larger radial range.

3. Results

3.1. ELECTRON–PROTON SCATTERING IN A LASER FIELD

We have performed many calculations of differential and total scattering cross sections both for net absorption of zero photons and for absorption or emission of a number of photons from the field [15]. The field induces resonances due to the temporary capture of the projectile electron into atomic hydrogen bound states [16]. Such resonances are presented in figure 1. At the frequency and energies considered, the $n = 3$ resonances are through one photon while the $n = 2$ resonances are through two photons. Several features are evident in the figure : even at the rather low intensity, the cross-section is dominated by resonance structures, which become more prominent in the backwards scattering direction. The appearance of the resonances is strongly geometry-dependent. Depending on the incoming and outgoing angles, some resonances can be suppressed. All bound sublevels can appear in the resonance structure, but some may be hidden since they are overlapping. At zero field all states of an n–manifold in hydrogen are degenerate but the laser field breaks the symmetries and removes the degeneracies.

3.2. MULTIPHOTON DETACHMENT OF H⁻

In figure 2, the perturbative polarizability coefficient α of the shift of the bound state is shown, the shift being

$$\Delta(\omega, \mathcal{E}_0) = \mathcal{E}_0^2 \alpha(\omega)/4. \qquad (8)$$

The dashed line labelled E_P gives the ponderomotive energy $E_P = \mathcal{E}_0^2/4\omega^2$. Experimentally relevant is the binding energy of the state, that is the energy difference between the (quasi-) bound state and the lowest (1s) threshold, which also shifts with respect to the double ionization threshold, due to the laser field. The shift of the core thresholds turns out to be quite small at high and at low frequency. Thus at low frequency the detachment energy is dominated by the ponderomotive shift term (which tends to increase the detachment energy). In the high frequency limit we have obtained values for the detachment energy that indicate ranges of validity of the high-frequency theory.

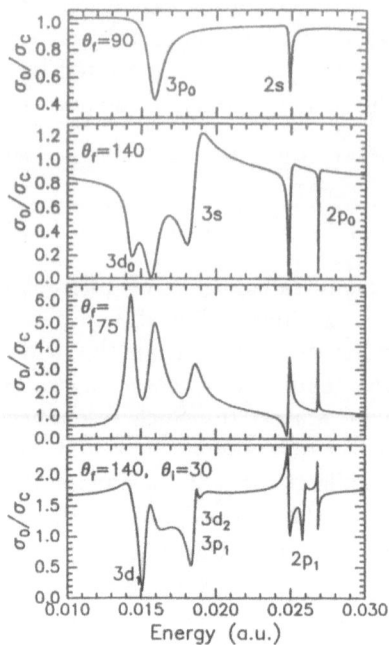

Figure 1. Multiphoton resonances in electron-proton scattering in a laser field of angular frequency of $\omega = 0.074$ a.u. and intensity 10^{12} W/cm^2. The ratio of the differential cross section σ_0 with net absorption of zero photons to the field-free Coulomb cross section σ_C is shown for incoming direction along the laser polarization axis (except bottom figure, where the incoming direction is $\theta_i = 30$ degrees from the axis), and outgoing direction θ_f with respect to the polarization axis, indicated on the figures, in degrees.

We have performed calculations exhibiting nonperturbative effects, such as channel closing [17] and ATD. We obtain very good agreement with previous work at low frequency for perturbative field intensities [14]. We compare with the modeling of the experiment [18] reported by Wang, Chu and Laughlin [19]. In figure 3, we show the rates in the threshold regions for the highest intensities considered in [19] and it can be seen that there is excellent quantitative agreement between the two theoretical approaches.

3.3. LASER-INDUCED DEGENERACIES

Since our method can treat autoionizing resonances we have investigated what happens when a bound state and an autoionizing state are brought into near resonance by one or more photons. Such processes have been investigated within effective atomic models [21, 22]. We have studied a one-photon resonance in Ar and a two-photon resonance in H$^-$ [23]. The

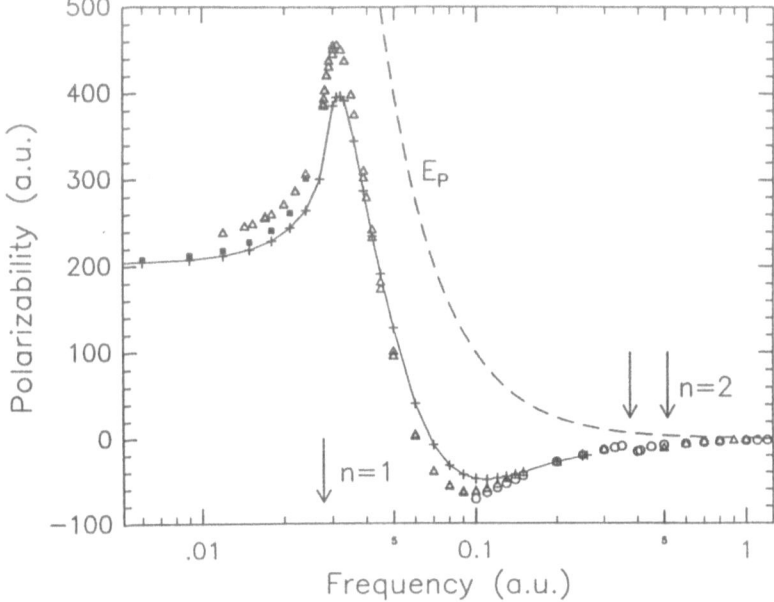

Figure 2. Polarizability of the H⁻ ion. Circles and triangles : present R-matrix-Floquet results; crosses connected by a solid line : Nicolaides et al [20]; squares : exact results.

results for the one-photon resonance in Ar are presented in figure 4. There are two slightly asymmetric structures, within each of which there is a degeneracy of two complex Floquet quasienergies (corresponding to the ground and the autoionizing states). It can be observed that population trapping, which is usually given as total by effective models, is only very weak for the left degeneracy point, since the intensity is higher and other multiphoton channels become important which destroy perfect trapping. It must be noted that the autoionizing resonance is also "stabilized" at certain frequencies. Adiabatic passages around the degeneracies can be imagined, which may exhibit the associated geometric phase.

4. Outlook and Acknowledgments

Development of the methods and software and computation of the results for the R-matrix-Floquet method has been a joint effort by several people, working in different places in Europe. It is my pleasure to mention my collaborators in the work reported here : C J Joachain, N Kylstra, O Latinne, at the Université Libre de Bruxelles, P G Burke and J Purvis, at The Queen's University in Belfast, C J Noble, at the CCLRC Daresbury Laboratory, and M Terao-Dunseath, at the Université de Rennes I.

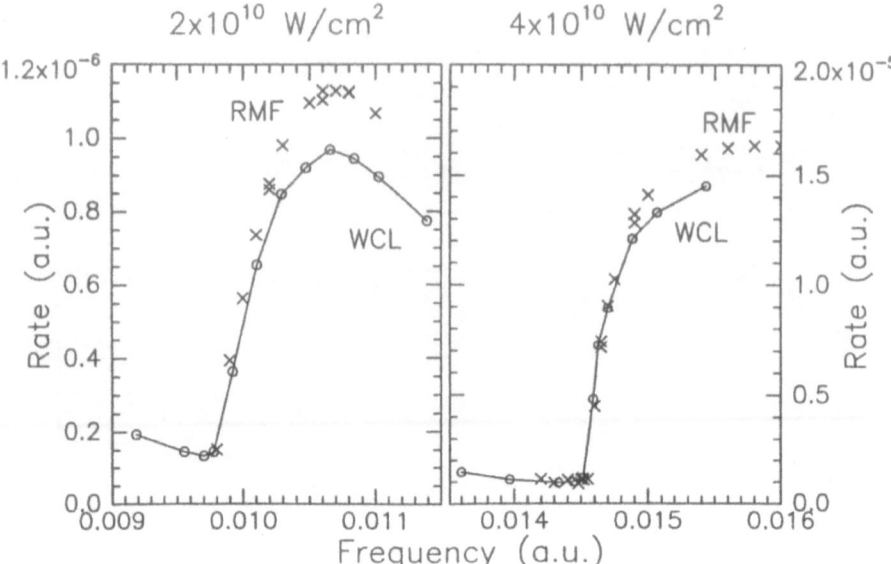

Figure 3. Rates, in a.u., for multiphoton detachment of H⁻ near the shifted thresholds for 2-3 photon detachment (right) and for 3-4 photon detachment (left), for the intensities indicated on top of the figures. Crosses (RMF) : present R-matrix-Floquet results; Circles (WCL) : Wang et al [19].

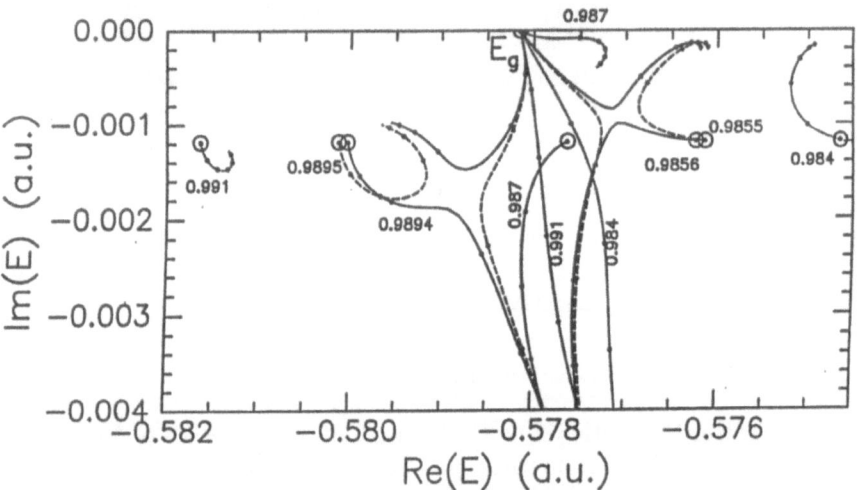

Figure 4. Complex energy trajectories versus intensity, for fixed frequencies indicated next to the curves, near the one-photon resonance between the Ar ground state and the lowest $^1P^o$ autoionizing state.

This project is supported through the EC HCM contracts Nos. ER-BCHRXCT 920013, 930346, 390459 and 940470. The author is presently supported by a Habilitations-Stipendium of the Deutsche Forschungsgemeinschaft.

Others curretly involved in the R-matrix-Floquet collaboration are : M Bensaid (Brussels), A Cyr (Rennes), H van der Hart and P Glass (Belfast), G Woeste (Daresbury) and A Fearnside (Durham University). Current developments include laser-assisted electron-atom scattering for multielectronic targets, harmonic generation, new methods and algorithms in the R-matrix approach, and porting the software to parallel computer architectures.

References

1. M Dörr O Latinne and C J Joachain, *Phys Rev* A, to be published
2. M Dörr, J Purvis, M Terao-Dunseath, P G Burke, C J Joachain and C J Noble, preprint (1995)
3. M Gatzke, R B Watkins and T F Gallagher, *Phys Rev* A **51**, 4835 (1995)
4. K J Berrington and P G Burke, *R-matrix theory* (1994)
5. N J van Druten, R Trainham and H G Muller, *Phys Rev* A **50**, 1593 (1994)
6. K M Dunseath, M Terao-Dunseath, M LeDourneuf and J M Launay, *Proceedings of the ECAMP IV conference* (1995)
7. Proceedings of "Super-Intense Laser-Atom Physics III", B Piraux ed. (Plenum, New York, 1993)
8. P G Burke, P Francken and C J Joachain, *J Phys* B **24**, 761 (1991)
9. M Dörr, M Terao-Dunseath, J Purvis, C J Noble, P G Burke and C J Joachain, *J Phys* B **25**, 2809 (1992)
10. C J Noble, M Dörr and P G Burke, *J Phys* B **26** 2983 (1993)
11. A J F Siegert, *Phys Rev* A **56** 750 (1939)
12. A Bohm, *Quantum Mechanics* (Springer, New York, 1993)
13. C-R Liu, B Gao and A F Starace, *Phys Rev* A **46**, 5985 (1992)
14. H van der Hart, *Phys Rev* A **50** 2508 (1994)
15. M Dörr, M Terao-Dunseath, J Purvis, P G Burke, C J Joachain and C J Noble, *J Phys* B, to be published
16. M Dörr, P G Burke, C J Joachain, C J Noble, J Purvis and M Terao-Dunseath, *J Phys* B **26**, L275 (1993)
17. J Purvis, M Dörr, M Terao-Dunseath, C J Joachain, P G Burke and C J Noble, *Phys. Rev. Lett* **71**, 3943 (1993)
18. C Y Tang, H C Bryant, P G Harris, A H Mohagheghi, R A Reeder, H Sharifian, H Toutounchi, C R Quick, J B Donahue, S Cohen and W W Smith, *Phys. Rev. Lett* **66**, 3124 (1991)
19. J Wang, S-I Chu and C Laughlin, *Phys Rev* A **50**, 3208 (1994)
20. C A Nicolaides, T Mercouris and N A Piangos, *J Phys* B **23**, L669 (1990)
21. P Lambropoulos and P Zoller, *Phys Rev* A **24**, 379 (1981)
22. K Rzazewski and J H Eberly, *Phys Rev Lett* **47**, 408 (1981)
23. O Latinne, N J Kylstra, M Dörr, J Purvis, M Terao-Dunseath, C J Joachain, P G Burke and C J Noble, *Phys. Rev. Lett* **74**, 46 (1995)

THE CORRELATION EFFECTS OF MULTIELECTRON ATOM IN INTENSE ELECTROMAGNETIC FIELD

L.P.RAPOPORT, A.F.KLINSKIKH*, V.V.MORDVINOV
Voronezh State University
394693 Universitetskaya pl., 1, Voronezh, Russia,
     tel: (0732) 55-65-52
Voronezh State Agrarian University
394087 ul.Michurina 1, Voronezh, Russia tel: (0732) 56-48-43

1. Introduction

It is a wellknown fact, that correlation effects without field beyond the framework of Hartree-Fock approximation for the multielectron atom

$$E_{\text{exact}} = E_{\text{H-F}} + \Delta E_c,$$

where E_{exact} is an exact energy, $E_{\text{H-F}}$ is the energy, calculated in the Hartree-Fork approximation, ΔE_c is the correlation energy.

The particular feature of the correlation effect problem for multielectron atom in the intense electromagnetic field is that at the some strength F_0 of external field the interaction of valent electron with the field V_{e_v-f} can grow up to the value of valent and core electron interaction $V_{e_v-e_{cor}}$ [1]:

$$V_{e_v-f} \approx V_{e_v-e_{cor}}.$$

The occuring effects are usually described in the context of the Hartree-Fock approximation for the atom and the effective dipole approximation for atom interaction with an electromagnetic field. The atom quasienergy therewith can be expressed in the form:

$$\varepsilon_{\text{exact}} = \varepsilon_{\text{H-F}} + \Delta \varepsilon_c + \Delta \varepsilon_{ef}.$$

The term $\Delta \varepsilon_{ef}$ corresponds the correlation effects, induced by the field and disappeares without the field.

The aims of the paper are:

1) discussion of the theoretical scheme for taking into account the valent and core electron interaction with the intense field in the uniform approach;

2) to obtain the closed equation system for the eigenfunctions and eigenvalues of the multielectron atom in the circulary polarized electromagnetic field;

H. G. Muller and M. V. Fedorov (eds.), Super-Intense Laser-Atom Physics IV, 257–266.
© *1996 Kluwer Academic Publishers.*

3) discussion of the possible methods for solving and analizing the system and estimation of the correlation effects;

4) discussion of the parameters, determining field induced correlation effects in the multielectron atom.

Now we would like to discuss different approaches that have been suggested for the problem.

Amusia with collaborators has conducted considerable and purposeful seach of the collective effects at single-electron ionization of the complex atoms in the framework of the perturbation theory by interaction with external field[2]. They have showed it theoretically and later it was disclosed experimentally that the correlation effects are essential for electrons of internal unfilled shells and for the inert gas atoms.

The valent electron correlation effects for the problems of multiphoton ionization, exemplified by Xe atoms, have been studied in [3]. The interaction with a field was considered in the framework of the perturbation theory. RPA (random phase approximation) was used for the description of correlation.

The complexity of taking into account the correlation effects for multiphoton processes in a such way is not only in replacing the interaction with the effective one. It is necessary to allow for Feynman diagram, the amplitude of which is similar by the structure to the corresponding diagrams of single-electron approximation and coincide with it, when the core polarization is neglected. Shaded triangle depicts effective dipole operator in RPA(fig.1). In the same order by the electromagnetic field there appear graphics of the other type (fig. 2), taking into account the contribution to the transition amplitude of photon absorbtion by core with the subsequent excitation transition. Evidently the calculation of a such an amplitude is extremely complicated and, as we know, it has not been conducted yet[4].

Figure 1   *Figure 2*

For the uniform electron gas without field correlation effects have been taken into account, as known, by Gell-Mann and Brakner[5]. Parameter, determining effect value, is the ratio of the sphere radius, volume of which is equal to the volume corresponding to the one electron at definite density, to

Bohr radius a_B :

$$r_s = \frac{\left(\frac{3}{4\pi n}\right)^{1/3}}{a_B} = \frac{r_0}{a_B}.$$

It is natural to suggest, that for the atom in the intense field there appear a new length parameter, which is similar to this one: $\alpha_0 = \frac{F}{\omega^2}$ is the amplitude of the classic oscillation in the field. Therefore the relative role of the correlation effects, induced by the field, has to be characterized qualitatively by the ratio of the parameters

$$r_s = \frac{r_0'}{a_B} \quad \text{и} \quad r_F = \frac{\alpha_0}{a_B},$$

i.e. by the quantity

$$\frac{r_F}{r_s} = \frac{\alpha_0}{r_0'}$$

(without field this parameter disappears), here r_0' is the characteristic size (for free electrons it is the Fermi sphere radius).

2. The Hamiltonian of multielectron atom in circularly polarized laser field

Let atomic Hamiltonian takes the form (atomic units):

$$\widehat{H}_a = \sum_{k=1}^{N} \left(\frac{\widehat{\mathbf{p}}_k^2}{2} - \frac{Z}{r_k} \right) + \frac{1}{2} \sum_{k \neq q=1}^{N} \frac{1}{|\mathbf{r}_k - \mathbf{r}_q|} \tag{1}$$

where $\widehat{\mathbf{p}}_k = -i\frac{\partial}{\partial \mathbf{r}_k}$ is the momentum of k-th electron, $\mathbf{r}_k$ is the radius-vector of k-th electron, N is the number of electrons in the atom, Z is the core charge. Interaction of atomic electrons with the electromagnetic circularly polarized wave, which is of the strength

$$\mathbf{F}(t) = Re(\vec{\lambda}e^{i\omega t})F_0, \tag{2}$$

(where $\vec{\lambda} = -\frac{1}{\sqrt{2}}(\mathbf{e}_x \pm i\mathbf{e}_y)$ is a unit polarization vector ω is a frequency, F_0 is an amplitude) is determined in the dipole approximation by the expression:

$$\widehat{H}_{\text{int}} = \sum_k \mathbf{r}_k \mathbf{F}(t). \tag{3}$$

In this case the time dependent Schrödinger equation

$$i\frac{\partial \Psi_0}{\partial t} = \left(\widehat{H}_a + \widehat{H}_{\text{int}} \right) \Psi_0 \tag{4}$$

can be reduced to the time-independent form. In the beginning we use the unitary transformation [6]

$$\Psi_0 = \prod_{k=1}^{N} U(\mathbf{r}, t)\widetilde{\Psi}, \tag{5}$$

where

$$U(\mathbf{r}, t) = \exp\left[-i\mathbf{A}(t)\mathbf{r}_k - \frac{i}{2}\int_{-\infty}^{t} \mathbf{A}^2(t')dt'\right], \tag{6}$$

and

$$\mathbf{A}(t) = \int_{-\infty}^{t} \mathbf{F}(t')dt'.$$

We use the variable replacement

$$\vec{\xi} = \mathbf{r}_k + \vec{a}(t) = \mathbf{r}_k + \int_{-\infty}^{t} \mathbf{A}(t')dt', \tag{7}$$

to bring equation (4) to the form:

$$\left[i\frac{\partial}{\partial t} - \sum_{k=1}^{N}\left\{-\frac{\Delta_{\xi_k}}{2} - \frac{Z}{|\vec{\xi}_k - \vec{a}(t)|}\right\} + \frac{1}{2}\sum_{k \neq q=1}^{N}\frac{1}{|\vec{\xi}_k - \vec{\xi}_q|}\right]\widetilde{\Psi} = 0. \tag{8}$$

Subsequent unitary rotation transformation

$$\widehat{D}(\alpha, \beta, \gamma)\widetilde{\Psi} = \Psi e^{-iEt}, \tag{9}$$

where

$$\widehat{D}(\alpha, \beta, \gamma) = e^{-i\alpha\widehat{J}_z} e^{-i\beta\widehat{J}_y} e^{-i\gamma\widehat{J}_z}, \tag{10}$$

with $\alpha = \beta = 0$ and $\gamma = \pm\omega t$ let us put (8) in the time-independent form. At this unitary transformation

$$\frac{1}{|\vec{\xi}_k - \vec{a}(t)|} \rightarrow \frac{1}{|\vec{\xi}_k - \vec{a}_0|}, \tag{11}$$

where $\alpha_0 = F/\omega^2$ is the amplitude of classic oscillations in the wave, vector $\vec{a}_0 = \alpha_0\mathbf{e}_y$.

Multielectron time-independent Hamiltonian for atom+field system takes the form

$$\widehat{H}_{a-f} = \sum_{k=1}^{N} \left\{ -\frac{\Delta_{\xi_k}}{2} - \frac{Z}{|\vec{\xi}_k - \vec{\alpha}_0|} + \omega(\widehat{L}_z)_k \right\} + \frac{1}{2} \sum_{k \neq q = 1}^{N} \frac{1}{|\vec{\xi}_k - \vec{\xi}_q|} \qquad (12)$$

Therewith E is the exact atom quasienergy in the electromagnetic field:

$$\widehat{H}_{a-f}\Psi = E\Psi \qquad (13)$$

It is important to note, that the Hamiltonian potential (12) has the Coulomb asymptotic at $r \rightarrow \infty$. It is rather essential for the nonperturbative Hartree-Fock calculation scheme for "atom+field" system. Similarly to the Hartree-Fock procedure for atoms we can formulate the problem of the best electron functions for the states of the system.

Let us consider multielectron time-independent Hamiltonian (12). We express the length Hamiltonian values in the units of r_0

$$\xi_k = \xi' r_0'$$

$$\widehat{H}_{a-f} = \sum_{k=1}^{N} \left\{ -\frac{1}{r_0'^2} \frac{\Delta_{\xi_k}}{2} - \frac{Z}{r_0' \left| \vec{\xi}_k - \frac{\vec{\alpha}_0}{r_0'} \right|} + \omega(\widehat{L}_z)_k \right\} + \frac{1}{2r_0'} \sum_{k \neq q = 1}^{N} \frac{1}{|\vec{\xi}_k' - \vec{\xi}_q'|} \qquad (14)$$

It is clear, that the kinetic energy is proportional to $r_0'^{-2}$, the exchange energy is proportional to $r_0'^{-1}$ and, hence, according to Gell-Mann and Brakner the correlation energy is proportional to $\ln(r_0')$. The induced by field effects therewith depend on the parameter

$$\frac{\alpha_0}{r_0'} = \frac{F}{\omega^2 r_0'}.$$

Consequently, in the intense electromagnetic field with $\alpha_0 \leq a_B$ such sort of correlation should be rather essential, in any case till the system can be regarded as "atom+field", but not as "plasma+field".

3. The self-consistent field approximation

The Hartree-Fock equation system can be found from (13) with Hamiltonian (12) by wellknown variational principle with additional requirement that the single-electron functions have to be normal.

It leads to the system of the integro-differential Hartree-Fock equations

$$\left\{ -\frac{\Delta_{\xi_k}}{2} - \frac{Z}{|\vec{\xi_k} - \vec{a_0}|} + \omega(\widehat{L}_z)_k \right\} \varphi_i(\vec{\xi}) + \qquad (15)$$

$$\sum_{k=1}^{N} \int \frac{d\vec{\xi'}}{|\vec{\xi} - \vec{\xi'}|} \varphi_k^*(\vec{\xi'}) \left[\varphi_k(\vec{\xi'})\varphi_i(\vec{\xi}) - \delta_{s_i s_k} \varphi_i(\vec{\xi'})\varphi_k(\vec{\xi}) \right] = \mathcal{E}_i \varphi_i(\vec{\xi})$$

The equations (15) determine in the self-consistent field approximation the best single-electron functions and eigenvalues $\mathcal{E}_i$ for the "atom+field" system. The interaction with field in the framework of the dipole approximation is allowed for exactly.

For the integrable-in-square wave functions φ_i

$$\int |\varphi_i|^2 d\vec{\xi} < \infty$$

the eigenvalues $\mathcal{E}_i$ determine the energy levels of "atom+field" system. If we formulate the problem with the outgoing wave boundary conditions, the eigenvalues $\mathcal{E}_i$ will be complex. The interpretation of complex values $\mathcal{E}_i$ is clear. The real part $\text{Re}\mathcal{E}_i$ determines the place of quasistationary level with the width $\Gamma_i/2$, which is equal to the imaginary part $\text{Im}\mathcal{E}_i$:

$$\Gamma_i/2 = \text{Im}\mathcal{E}_i.$$

The complex energy values can also be found by using the complex coordinate rotation and calculating the corresponding α-trajectories.

The quantity Γ_i is determined by the multiphoton ionization processes in the single-electron approximation, i.e. without taking into account the correlation effects.

In the foregoing sense the correlation effects will be taken into account.

For solving (15) we express φ_i in the form:

$$\varphi_{nlm}(\vec{\xi}) = \frac{1}{\xi} f_{nlm}(\xi) Y_{lm}(\Omega) \chi_\sigma \qquad (16)$$

Substituting (16) to (15), we derive the equation system:

$$\left\{ -\frac{1}{2}\frac{d^2}{d\xi^2} + W_{lm}(\xi) + \frac{Y_{nlm}(\xi)}{\xi} + \frac{l(l+1)}{2\xi^2} + \omega m - \mathcal{E}_{nlm} \right\} f_{nlm}(\xi) = X_{nlm}(\xi)$$

where

$$W_{lm}(\xi) = -Z \int \frac{|Y_{lm}|^2}{|\vec{\xi} - \vec{\alpha}_0|} d\Omega =$$

$$-Z(-1)^m \sqrt{4\pi}(2l+1) \sum_k \frac{Y_{l0}(\vec{\alpha}_0/\alpha_0)}{\sqrt{2k+1}} \frac{r_<^k}{r_>^{k+1}} \begin{pmatrix} l_1 & k & l_1 \\ 0 & 0 & 0 \end{pmatrix} \begin{pmatrix} l_1 & k & l_1 \\ -m_1 & 0 & m_1 \end{pmatrix}$$

corresponds to the "laser dressed" Coulomb potential ($r_< = \min(\xi, \alpha_0)$, $r_> = \max(\xi, \alpha_0)$),

$$Y_{nlm}(\xi) = \xi \sum_{n'l'm' \neq nlm} \int d\Omega \int d\vec{\xi}' \frac{\varphi_{nlm}^2(\vec{\xi})\varphi_{n'l'm'}^2(\vec{\xi}')}{|\vec{\xi} - \vec{\xi}'|}$$

corresponds to the self-consistent field of the other electrons,

$$X_{nlm}(\xi) = \sum_{n'l'm' \neq nlm} \int d\Omega \int d\vec{\xi}' \frac{\varphi_{nlm}(\vec{\xi})\varphi_{n'l'm'}(\vec{\xi})\varphi_{nlm}(\vec{\xi}')\varphi_{n'l'm'}(\vec{\xi}')}{|\vec{\xi} - \vec{\xi}'|}$$

corresponds to the exchange potential.

Here the part of the potential $Z/|\vec{\xi}_k - \vec{\alpha}_0|$ in the equation (15) which is not of centrally symmetrical character has not been taken into account. The preliminary results show that the contribution of this part is not more than 5% for a frequency $\omega \geq 1$ (a.u.). Therefore the numerical calculation results are presented only for this range of frequency, whereas the quantity α_0 is restricted by the limitation $\alpha_0 < a_B$.

4. Numerical results and discussion

Now we will consider a number of the results obtained for Xe atom in the intense circularly polarized laser field.

First, let us draw attension to the radial dependence of the potentials $W_{lm}(r)$ at different values of l,m (Fig.3). The potentials, determining the atom electron dynamics, form "zone".

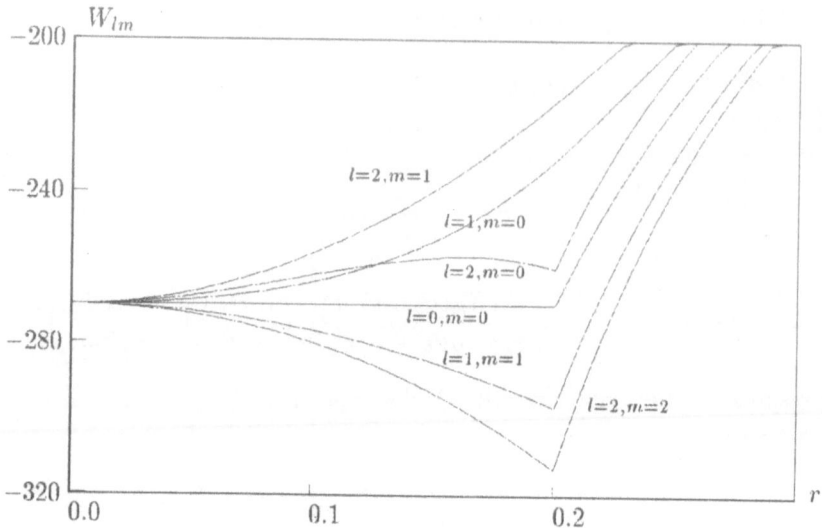

Figure 3. Atom Xe "laser dressed" potentials for $\alpha_0 = 0.2$.

The atomic levels shift takes place in relation to the parameter α_0 for different quantum numbers l,m of the levels. If $\mathcal{E}_0$ is nonperturbated energy value and $\mathcal{E}(\alpha_0)$ is the value of atomic level in field, than the function $f(\alpha_0) = \mathcal{E}/\mathcal{E}_0$ at different l,m are as shown in Fig. 4,5.

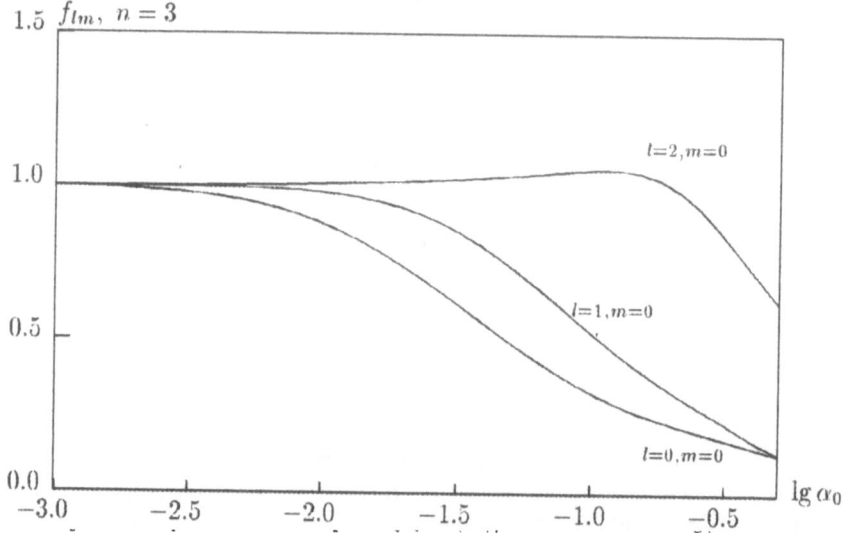

The relationship of the quantity f_{lm} is shown in Fig.4 as a function of $\lg(\alpha_0)$ for $\binom{m=0}{n=3}$ and different l=0,1,2. The corresponding values of α_0/a_B range from 10^{-4} up to 0.562.

Figure 5 shows the dependence of f_{lm} for $\left(\begin{smallmatrix} l=0 \\ m=0 \end{smallmatrix}\right)$ and n=1,2.

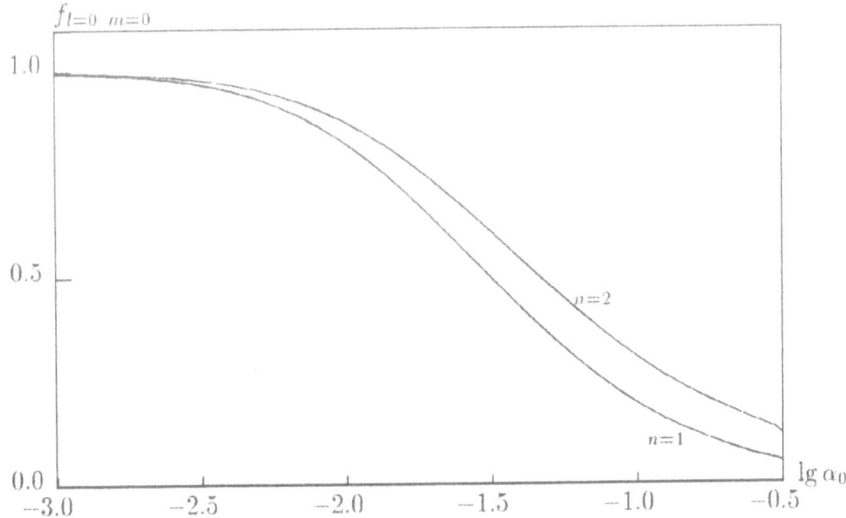

Figure 5. Dependence of the quantity $f=\mathcal{E}(\alpha_0)/\mathcal{E}_0$ as a function of $\lg(\alpha_0)$

The quatity $f_{l=0\ m=0}$ is present in Fig. 6 as a function of the principal quantum number n at $\lg \alpha_0 = -0.25$ $(\alpha_0/a_B = 0.562)$.

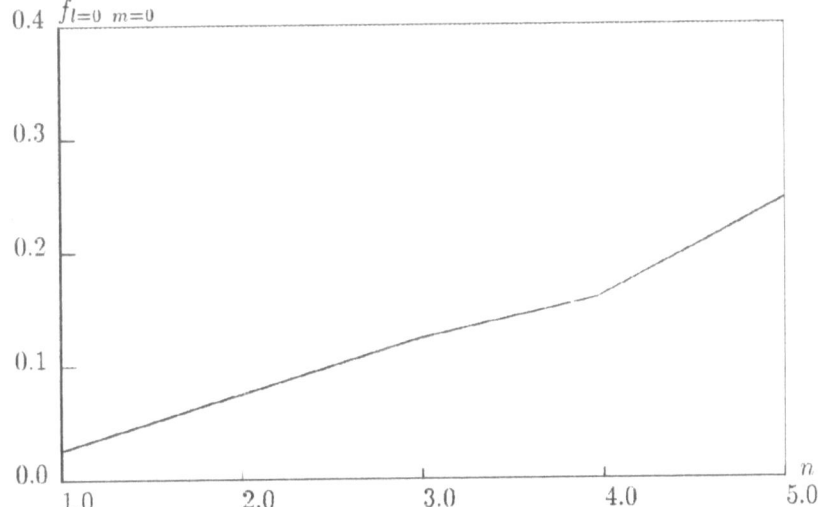

Figure 6. Dependence of the quantity $f=\mathcal{E}(\alpha_0)/\mathcal{E}_0$ as a function of n at $\lg \alpha_0 = -0.25$

It should be noted, that the $\lg \alpha_0 \to -0.25$ limit of f_{lm} tends for $m = 0$ and $l = 0, 1$ to the same value. Corresponding energy level shift is about 87%

266

in relation to nonperturbated level. Thus the atomic spectrum rearranges itself at $\alpha_0/a_B = 0.562$.

So the most stable electrons are the electrons with a large n. For them the spectrum reconstruction is about 6% at $\lg \alpha_0 \approx -0.2$, whereas it is equal to $97 - 98\%$ for $n = 1$.

Probably it can be explaned by the fact that the electrons with small n are more sensible to potential shape changing at small r, which takes place at $\lg \alpha_0 = -0.25$.

In the conclusion it should be noted, that the approach, developed in this work, is of certain importance for estimation of the correlation effects in multielectron atoms in intense electromagnetic field. The simplest method is to use calculated eigenfunctions in RPA-like diagrams:

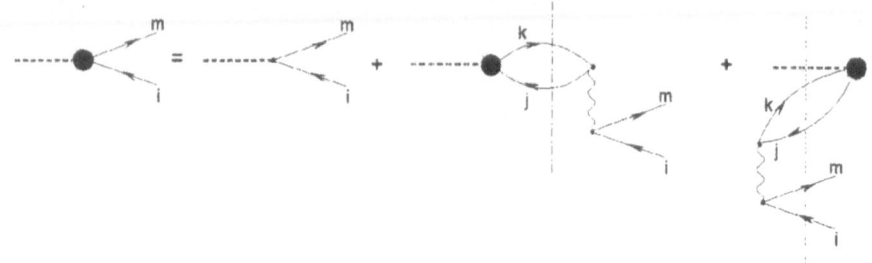

for the calculation of effective interaction matrix elements $(m|M(\omega)|i)$. It is the aim of the further investigations of the problem.

Acknowledgment

This work has been supported by the RFFI, Grant 94-02-03469

References

1. Delone,N.B. and Krainov,V.P. (1994) *Multiphoton processes in atoms*, Springer-Verlage.
2. Amusia,M.Ya. (1987) *Atomnii photoeffekt*, Nauka, Moscow.
3. Huillier,A.L. and Wendin,G. (1987) J.Phys. **B20**, p.37.
4. Rapoport,L.P. and Agre,M.Ya. (1987) Multiphoton processes, in Smith and Knight(eds), Proc. 4th Int. Conf. on Multiphoton Processes. Cambridge Stud. in Mod. Opt. 8, p.106.
5. March,N.H, Joung,W.H., Sampanthar,S. (1969) *The many-body problem in quantum mechanics*, Mir, Moscow.
6. Rapoport,L.P. (1994) Zh.Eksp.Teor.Fiz., **105**, p.534.

DENSITY-FUNCTIONAL APPROACH TO ATOMS IN STRONG LASER PULSES

C. A. ULLRICH, S. ERHARD and E. K. U. GROSS
Institut für Theoretische Physik
Universität Würzburg
Am Hubland
D-97074 Würzburg
Germany

1. Introduction

The basic idea of density functional theory (DFT) is to describe an interacting many-particle system exclusively and completely in terms of its density. The formalism rests on two basic theorems:

I. Every observable quantity can be calculated, at least in principle, from the density alone, i. e. each quantum mechanical observable can be written as a functional of the density.

II. The density of the interacting system of interest can be obtained as density of an auxiliary system of non-interacting particles moving in an effective *local* single-particle potential (the so-called Kohn-Sham potential).

In the original work of Hohenberg, Kohn and Sham [1, 2] these theorems were proven for the ground-state density of *static* many-body systems. On the basis of these theorems, DFT has provided an extremely successful description of ground-state properties of atoms, molecules and solids [3, 4]. The accuracy of approximations for the Kohn-Sham potential has steadily improved over the years and the currently best functionals yield ground-state properties in very close agreement with configuration interaction results [5].

DFT of time-dependent systems (TDDFT) is a more recent development [6]–[8]. The important theorems I and II stated above have been shown to hold true for the time-dependent density as well [6]. So far, TDDFT has been applied almost exclusively in the regime of linear response (for recent reviews see, e. g., Refs. [7]–[9]).

In this paper, TDDFT is used to study the interaction of atoms with very strong laser pulses having intensities that require a non-perturbative treatment of the laser field. Many authors have attacked this problem with a variety of techniques [10, 11]. Often the electronic motion has been treated classically or semiclassically. Most of the quantum mechanical work was either done on the hydrogen atom or

H. G. Muller and M. V. Fedorov (eds.), Super-Intense Laser-Atom Physics IV, 267–284.

on rare gas atoms within a single-active-electron model. The principal advantage of TDDFT is that it provides a fully quantum mechanical approach that incorporates correlation effects due to the electron-electron interaction in a systematic fashion. The interacting many-body problem is mapped on an auxiliary system of non-interacting particles moving in an effective time-dependent potential. Since this potential is *local* in configuration space the resulting numerical scheme is less involved than the time-dependent Hartree-Fock (TDHF) method. In section 2, we give an overview of TDDFT. In sections 3 and 4 numerical results on multiphoton ionization and harmonic generation of helium and neon will be presented.

2. Time-dependent density functional formalism

We study the time evolution of a system of N electrons governed by the time-dependent Schrödinger equation

$$i\frac{\partial}{\partial t}\Phi(t) = \hat{H}(t)\Phi(t) \tag{1}$$

(atomic [Hartree] units are used throughout). The total Hamiltonian, written in second quantized notation, is given by

$$\hat{H}(t) = \hat{T} + \hat{W} + \hat{V}(t) \quad, \tag{2}$$

where $\hat{T}$ is the kinetic energy of the electrons,

$$\hat{T} = \sum_{\sigma=\uparrow\downarrow} \int d^3r\, \hat{\psi}_\sigma^\dagger(\mathbf{r}) \left(-\frac{\nabla^2}{2}\right) \hat{\psi}_\sigma(\mathbf{r}) \quad, \tag{3}$$

and $\hat{W}$ is the mutual Coulomb interaction,

$$\hat{W} = \frac{1}{2}\sum_{\sigma,\sigma'} \int d^3r \int d^3r'\, \hat{\psi}_\sigma^\dagger(\mathbf{r})\hat{\psi}_{\sigma'}^\dagger(\mathbf{r}')\frac{1}{|\mathbf{r}-\mathbf{r}'|}\hat{\psi}_{\sigma'}(\mathbf{r}')\hat{\psi}_\sigma(\mathbf{r}) \quad. \tag{4}$$

The electrons move in an explicitly time-dependent external potential

$$\hat{V}(t) = \int d^3r\, v_{\text{ext}}(\mathbf{r}t)\hat{n}(\mathbf{r}) \quad, \tag{5}$$

where $\hat{n}(\mathbf{r})$ is the density operator

$$\hat{n}(\mathbf{r}) = \sum_{\sigma=\uparrow\downarrow} \hat{\psi}_\sigma^\dagger(\mathbf{r})\hat{\psi}_\sigma(\mathbf{r}) \quad. \tag{6}$$

Generally the external potential $v_{\text{ext}}(\mathbf{r}t)$ consists of a static contribution (e. g., the nuclear Coulomb potential) and a time-dependent part (e. g., the laser field).

Time-dependent density functional theory is based on the existence of an exact one-to-one mapping between time-dependent densities and external potentials. We investigate the densities $n(\mathbf{r}t)$ of electronic systems evolving from a *fixed* initial (many-particle) state $\Phi(t_0) = \Phi_0$ under the influence of different external potentials $v_{ext}(\mathbf{r}t)$. Each external potential leads, via solution of the time-dependent Schrödinger equation (1), to a time-dependent many-body wave function $\Phi(t)$. For a fixed initial state Φ_0, this defines a map

$$\mathcal{A} : v_{ext}(\mathbf{r}t) \longrightarrow \Phi(t) \tag{7}$$

between the external potentials and the corresponding time-dependent many-particle wave functions. By virtue of the density operator (6), a second map

$$\mathcal{B} : \Phi(t) \longrightarrow n(\mathbf{r}t) = \langle \Phi(t)|\hat{n}(\mathbf{r})|\Phi(t)\rangle \tag{8}$$

is established between the many-particle wave functions and the time-dependent densities. The heart of TDDFT is the proof of invertibility of the combined map $\mathcal{G} \equiv \mathcal{B} \circ \mathcal{A}$:

$$\mathcal{G} : v_{ext}(\mathbf{r}t) \longrightarrow n(\mathbf{r}t) \quad . \tag{9}$$

The invertibility of this map was first proven by Runge and Gross [6]. These authors demonstrated that two densities $n(\mathbf{r}t)$ and $n'(\mathbf{r}t)$ evolving from a common initial state Φ_0 under the influence of two potentials $v_{ext}(\mathbf{r}t)$ and $v'_{ext}(\mathbf{r}t)$ always become different infinitesimally later than t_0, provided that the potentials differ by more than a purely time-dependent function $c(t)$. The set of potentials for which invertibility can be shown comprises all potentials expandable in a Taylor series with respect to the time coordinate around the initial time t_0. Having established the existence of the inverse map

$$\mathcal{G}^{-1} : n(\mathbf{r}t) \longrightarrow v_{ext}(\mathbf{r}t) + c(t) \quad , \tag{10}$$

subsequent application of the map $\mathcal{A}$ tells us that the full many-particle wave function is a functional of the time-dependent density, unique up to within a purely time-dependent phase $\alpha(t)$:

$$\Phi(t) = e^{-i\alpha(t)}\Psi[n](t) \quad . \tag{11}$$

As a consequence, the expectation value of any quantum mechanical operator $\hat{Q}(t)$ is a *unique* functional of the density:

$$Q[n](t) = \langle \Psi[n](t)|\hat{Q}(t)|\Psi[n](t)\rangle \quad . \tag{12}$$

The ambiguity in the phase cancels out (provided that $\hat{Q}(t)$ contains no time derivatives). This proves theorem I stated in the introduction. Some quantities (such as harmonic spectra) are easily calculated from the time-dependent density, while other quantities (such as ATI spectra) are difficult to extract from the density. But, as a matter of principle, *all* physical observables are determined by the time-dependent density alone, once the initial many-body state Φ_0 is specified[1].

The 1–1 correspondence between time-dependent densities and time-dependent potentials can be established for any *given* interaction $\hat{W}$, in particular also for $\hat{W} \equiv 0$, i. e. for non-interacting particles. Therefore, if $n(\mathbf{r}t)$ is a *given* density,

---

[1] If the initial state Φ_0 is a non-degenerate ground state then the traditional Hohenberg-Kohn theorem [1] ensures that Φ_0 is a functional of the initial density n_0. Hence, in this case, Φ_0 need not be given explicitly, i. e., knowledge of n_0 is sufficient.

the potential $v(\mathbf{r}t)$ of *non-interacting* particles that reproduces the given density $n(\mathbf{r}t)$ is *uniquely* determined, $v(\mathbf{r}t) = v[n](\mathbf{r}t)$, i. e. the given density $n(\mathbf{r}t)$ can be calculated from

$$n(\mathbf{r}t) = \sum_{j=1}^{N} |\phi_j(\mathbf{r}t)|^2 \tag{13}$$

with the single-particle orbitals $\phi_j(\mathbf{r}t)$ satisfying

$$i\frac{\partial}{\partial t}\phi_j(\mathbf{r}t) = \left(-\frac{\nabla^2}{2} + v[n](\mathbf{r}t)\right)\phi_j(\mathbf{r}t) \quad . \tag{14}$$

Whether or not $v(\mathbf{r}t)$ actually *exists* for an *arbitrary* given density $n(\mathbf{r}t)$ is an open question in the time-dependent case[2]. But if it exists it is *unique*. If one chooses for $n(\mathbf{r}t)$ the density of the *interacting* system of interest (i. e. the density of Coulomb-interacting particles moving in the external potential $v_{\text{ext}}(\mathbf{r}t)$) then the potential $v[n]$ is termed the time-dependent Kohn-Sham (TDKS) potential. The latter is usually decomposed into the external potential, a time-dependent Hartree part and the so-called exchange-correlation (xc) potential:

$$v[n](\mathbf{r}t) = v_{\text{ext}}(\mathbf{r}t) + \int d^3r' \frac{n(\mathbf{r}'t)}{|\mathbf{r} - \mathbf{r}'|} + v_{\text{xc}}[n](\mathbf{r}t) \quad . \tag{15}$$

The xc potential is a *universal* functional of the density, i. e. it has the *same* functional dependence on n for *all* Coulomb systems, independent of the particular external potential v_{ext} of the system at hand. As in the static case, the great advantage of the TDKS scheme lies in its computational simplicity compared to other methods such as TDHF or time-dependent configuration interaction. The crucial feature of v_{xc} is that it is a *local* potential in configuration space in contrast, e. g., to the non-local TDHF potential.

The basic formalism is easily extended to spin-polarized systems [12]. In that case the xc potential depends on the spin densities

$$n_\sigma(\mathbf{r}t) = \sum_{j=1}^{N_\sigma} |\phi_{j\sigma}(\mathbf{r}t)|^2 \quad , \quad \sigma =\uparrow\downarrow \tag{16}$$

with $N = \sum_\sigma N_\sigma$, and the spin orbitals $\phi_{j\sigma}(\mathbf{r}t)$ satisfy the single-particle equations

$$i\frac{\partial}{\partial t}\phi_{j\sigma}(\mathbf{r}t) = \left(-\frac{\nabla^2}{2} + v_\sigma[n_\uparrow, n_\downarrow](\mathbf{r}t)\right)\phi_{j\sigma}(\mathbf{r}t) \tag{17}$$

with

$$v_\sigma[n_\uparrow, n_\downarrow](\mathbf{r}t) = v_{\text{ext}\sigma}(\mathbf{r}t) + \int d^3r' \frac{n_\uparrow(\mathbf{r}'t) + n_\downarrow(\mathbf{r}'t)}{|\mathbf{r} - \mathbf{r}'|} + v_{\text{xc}\sigma}[n_\uparrow, n_\downarrow](\mathbf{r}t) \quad . \tag{18}$$

In practice, the xc potential $v_{\text{xc}\sigma}[n_\uparrow, n_\downarrow](\mathbf{r}t)$ has to be approximated. The simplest

---

[2]This question is termed the *v-representability problem*. In the static case, the question could be answered in a satisfactory way. For a review of the static *v*-representability problem see, e. g., chapter 4.2 of Ref. [3].

possible form is the so-called adiabatic local density approximation (ALDA):

$$v_{\mathrm{xc}\sigma}^{\mathrm{ALDA}}[n_\uparrow, n_\downarrow](\mathbf{r}t) = \left.\frac{de_{\mathrm{xc}}^{hom}(n_\uparrow, n_\downarrow)}{dn_\sigma}\right|_{n_\sigma = n_\sigma(\mathbf{r}t)} , \qquad (19)$$

where $e_{\mathrm{xc}}^{hom}(n_\uparrow, n_\downarrow)$ is the xc energy per volume of the homogeneous spin-polarized electron gas. This approximation can be expected to be good only if the time dependence of the $n_\uparrow$ and $n_\downarrow$ is sufficiently slow. In practice, however, it gives quite good results even for cases of rather rapid time dependence. In the exchange-only case (to which we shall restrict ourselves in the following), one explicitly obtains

$$v_{\mathrm{x}\sigma}^{\mathrm{ALDA}}(\mathbf{r}t) = -(6n_\sigma(\mathbf{r}t)/\pi)^{\frac{1}{3}} . \qquad (20)$$

The ALDA is *local* both in space and time, i. e. $v_{\mathrm{xc}}(\mathbf{r}t)$ only depends on the density values at the very same time t and the very same location $\mathbf{r}$. Recently, a different time-dependent xc potential has been proposed which is tailored for the description of memory effects [13]. In this approximation $v_{\mathrm{xc}}(\mathbf{r}t)$ depends on the density values $n(\mathbf{r}'t')$ at other locations $\mathbf{r}'$ and earlier times $t' \leq t$. Both approximations have in common that they are based on results derived from the homogeneous electron gas.

We have recently developed a new method of constructing approximations of $v_{\mathrm{xc}\sigma}$ [8, 14, 15] which also in principle takes memory effects into account but does not make use of the theory of the homogeneous electron gas. The approach can be viewed as a time-dependent extension of the so-called optimized potential method (OPM). As before, the description of the time evolution of an N-electron system with a given initial state is made in terms of a set of time-dependent spin orbitals $\{\phi_{j\sigma}(\mathbf{r}t)\}$ obeying a single-particle Schrödinger equation analogous to Eq. (17). The difference compared to conventional density-functional schemes is that the time-dependent xc potential appearing in Eq. (18) is now given as a functional of the *orbitals* $\{\phi_{j\sigma}(\mathbf{r}t)\}$ rather than the spin densities. It is constructed by requiring the spin orbitals in Eq. (17) to make a given total quantum mechanical action functional $A[\{\phi_{j\sigma}\}]$ stationary. This condition leads to the following integral equation for the optimized xc potentials [14]:

$$i\sum_j^{N_\sigma} \int_{-\infty}^{t_1} dt' \int d^3r' \left[v_{\mathrm{xc}\sigma}^{\mathrm{OPM}}(\mathbf{r}'t')\phi_{j\sigma}^*(\mathbf{r}'t') - \frac{\delta A_{\mathrm{xc}}[\{\phi_{j\sigma}\}]}{\delta\phi_{j\sigma}(\mathbf{r}'t')} \right] \phi_{j\sigma}(\mathbf{r}t)K_\sigma(\mathbf{r}t, \mathbf{r}'t')$$

$$+ \quad c.c. \ = \ 0 \qquad (21)$$

with the kernel $K_\sigma(\mathbf{r}t, \mathbf{r}'t') = \sum_{k=1}^\infty \phi_{k\sigma}^*(\mathbf{r}t)\phi_{k\sigma}(\mathbf{r}'t')\,\theta(t - t')$. The functional $A_{\mathrm{xc}}[\{\phi_{j\sigma}\}]$ in Eq. (21) is the xc part of the total action functional and has to be approximated in practice. If all time-dependent correlation effects are neglected (x-only case) then A_{xc} is given by the usual TDHF expression

$$A_{\mathrm{x}} = -\frac{1}{2}\sum_\sigma \sum_{i,j}^{N_\sigma} \int_{-\infty}^{t_1} dt \int d^3r \int d^3r' \frac{\phi_{i\sigma}^*(\mathbf{r}'t)\phi_{j\sigma}(\mathbf{r}'t)\phi_{i\sigma}(\mathbf{r}t)\phi_{j\sigma}^*(\mathbf{r}t)}{|\mathbf{r} - \mathbf{r}'|} . \qquad (22)$$

The numerical implementation of the full time-dependent OPM scheme is an extremely demanding task: at each time step one has to solve the integral equation (21) for $v_{\text{xc}\sigma}^{\text{OPM}}$. For this reason, we have developed a simplified scheme similar to the one proposed by Krieger, Li and Iafrate (KLI) [16] for the static case which yields approximations to $v_{\text{xc}\sigma}^{\text{OPM}}$ as *explicit* functionals of the orbitals. These approximate xc potentials are given by [8, 14, 15]

$$v_{\text{xc}\sigma}^{\text{KLI}}(\mathbf{r}t) = w_{\text{xc}\sigma}(\mathbf{r}t) + \frac{1}{n_\sigma(\mathbf{r}t)} \sum_{j,k}^{N_\sigma} n_{j\sigma}(\mathbf{r}t) \left(\Pi_\sigma^{-1}(t)\right)_{jk} \int d^3r'\, n_{k\sigma}(\mathbf{r}'t) w_{\text{xc}\sigma}(\mathbf{r}'t) \quad (23)$$

with

$$w_{\text{xc}\sigma}(\mathbf{r}t) = \frac{1}{n_\sigma(\mathbf{r}t)} \sum_j^{N_\sigma} n_{j\sigma}(\mathbf{r}t) \frac{1}{2} \left(\frac{1}{\phi_{j\sigma}^*(\mathbf{r}t)} \frac{\delta A_{\text{xc}}[\{\phi_{j\sigma}\}]}{\delta \phi_{j\sigma}(\mathbf{r}t)} + c.c. \right)$$

$$- \frac{1}{n_\sigma(\mathbf{r}t)} \sum_j^{N_\sigma} n_{j\sigma}(\mathbf{r}t) \frac{1}{2} \left(\int d^3r\, \phi_{j\sigma}(\mathbf{r}t) \frac{\delta A_{\text{xc}}[\{\phi_{j\sigma}\}]}{\delta \phi_{j\sigma}(\mathbf{r}t)} + c.c. \right)$$

$$+ \frac{i}{4n_\sigma(\mathbf{r}t)} \sum_j^{N_\sigma}{}' \nabla^2 n_{j\sigma}(\mathbf{r}t) \int_{-\infty}^t dt' \left(\int d^3r\, \phi_{j\sigma}(\mathbf{r}t') \frac{\delta A_{\text{xc}}[\{\phi_{j\sigma}\}]}{\delta \phi_{j\sigma}(\mathbf{r}t')} - c.c. \right) \quad (24)$$

and $n_{j\sigma}(\mathbf{r}t) = |\phi_{j\sigma}(\mathbf{r}t)|^2$. The $N_\sigma \times N_\sigma$ matrix $\Pi_\sigma(t)$ in Eq. (23) is defined as

$$\Pi_{kj\sigma}(t) = \delta_{kj} - \int d^3r\, \frac{n_{k\sigma}(\mathbf{r}t) n_{j\sigma}(\mathbf{r}t)}{n_\sigma(\mathbf{r}t)} \quad . \quad (25)$$

The essential ingredient of the quantity $w_{\text{xc}\sigma}$ is the functional derivative $\delta A_{\text{xc}}/\delta \phi_{j\sigma}$ which can be calculated analytically once the approximation of A_{xc} is specified. For example, in the x-only case $w_{\text{xc}\sigma}$ becomes

$$w_{\text{x}\sigma}(\mathbf{r}t) = -\frac{1}{n_\sigma(\mathbf{r}t)} \sum_{j,k}^{N_\sigma} \left[\phi_{j\sigma}(\mathbf{r}t)\phi_{k\sigma}^*(\mathbf{r}t) \int d^3r'\, \frac{\phi_{k\sigma}(\mathbf{r}'t)\phi_{j\sigma}^*(\mathbf{r}'t)}{|\mathbf{r}-\mathbf{r}'|} \right.$$

$$\left. - n_{j\sigma}(\mathbf{r}t) \int d^3r \int d^3r'\, \frac{\phi_{j\sigma}(\mathbf{r}t)\phi_{k\sigma}^*(\mathbf{r}t)\phi_{k\sigma}(\mathbf{r}'t)\phi_{j\sigma}^*(\mathbf{r}'t)}{|\mathbf{r}-\mathbf{r}'|} \right] \quad . \quad (26)$$

The full x-only OPM potential constitutes the exact x-only limit of TDDFT. It is distinguished from TDHF by the fact that the OPM exchange potential is local and therefore numerically favourable. We emphasize that the x-only TDOPM should not be considered as a local approximation to TDHF. Apart from its numerical simplicity the x-only OPM is also *physically superior* to HF. This is most easily appreciated in the static limit: The static OPM orbitals (both occupied and unoccupied ones) are self-interaction free. By contrast, in HF only the occupied orbitals are self-interaction free while the unoccupied ones have a serious self-interaction error which causes them to be much too weakly bound. Since time-dependent external fields will cause transitions to the virtual orbitals (which are poorly represented in HF) we expect the x-only OPM to be more accurate than TDHF (even if the full OPM exchange potential is approximated by the KLI potential (23)). The KLI potential (23) is significantly more accurate but also numerically more involved than the ALDA potential (20). In section 4, TDKLI results will be compared with ALDA results for neon in strong laser pulses. For these calculations, the difference in CPU time between TDKLI and ALDA is about a factor of 3. Before that, in section 3, we illustrate our numerical procedure by considering various aspects of harmonic generation in helium within a time-dependent Hartree approach.

3. Harmonic generation and two-colour mixing in helium

The initial KS ground state of He is the doubly occupied 1s orbital. In this case, there are no exchange contributions and, for the time being, the correlation part of v_{xc} will be neglected. The TDKLI equation (17) then reduces to the time-dependent Hartree equation

$$i\frac{\partial}{\partial t}\phi_{1s}(\mathbf{r}t) = \left(-\frac{\nabla^2}{2} + \int d^3r' \frac{|\phi_{1s}(\mathbf{r}'t)|^2}{|\mathbf{r}-\mathbf{r}'|} + E_0 f(t) z \sin(\omega_0 t) - \frac{2}{r}\right)\phi_{1s}(\mathbf{r}t) \quad . \tag{27}$$

The index "1s" indicates that the time-dependent orbital $\phi_{1s}(\mathbf{r}t)$ initially was in the 1s state.

The laser field, assumed to be linearly polarized along the z direction, has been written in dipole approximation in the usual length form, with peak field strength E_0 and frequency ω_0. The envelope function, $f(t)$, is such that the laser is linearly ramped to its maximum amplitude over the first 10 cycles which is then held constant for another 10 cycles.

We solve this equation in cylindrical coordinates with a finite-difference scheme very similar to Kulander [17, 18], using a finite non-uniform grid as introduced by Pindzola et al. [19]. The spatial extent of the grid is about 20 a.u. × 60 a.u., and the initial helium ground state has an energy eigenvalue of $\epsilon_{1s} = -0.955$ Hartrees which is 3.9% off the exact Hartree-Fock value of -0.918 Hartrees. This error is due to the relatively coarse grid spacings in the vicinity of the nucleus, which is inevitable to keep the numerical effort tractable.

We simulate ionization by an absorbing grid boundary [17, 18] so that the norm of the wave function

$$N_{1s}(t) = \int_{\substack{\text{finite}\\\text{volume}}} d^3r \, |\phi_{1s}(\mathbf{r}t)|^2 \quad , \tag{28}$$

taken over the finite volume of the grid, decreases with time. The time-dependent norm $N_{1s}(t)$ refers to a singly occupied spin orbital. The probabilities for neutral, singly and doubly charged helium atoms can therefore be expressed as

$$P^0(t) = N_{1s}(t)^2 \tag{29}$$
$$P^{+1}(t) = 2N_{1s}(t)(1 - N_{1s}(t)) \tag{30}$$
$$P^{+2}(t) = (1 - N_{1s}(t))^2 \quad . \tag{31}$$

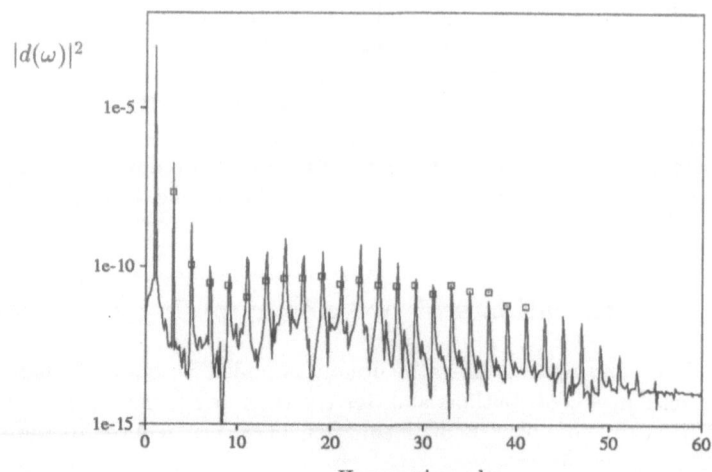

Figure 1: *Harmonic spectrum for He at $\lambda = 616$nm and $I = 3.5 \times 10^{14}$ W/cm^2. The squares represent experimental data taken from Ref. [20] normalized to the value of the 33rd harmonic of the calculated spectrum. The experiment was performed with a peak intensity of 1.4×10^{14}W/cm^2.*

Since $N_{1s}(t) = \int d^3r\, n(\mathbf{rt})/2$, the probabilities P^0, P^{+1} and P^{+2} as given by Eqs. (29)–(31) are explicit functionals of the density.

In order to investigate harmonic generation for the helium atom, we take the laser wave length 616 nm, which has been used for an experimental study by Miyazaki and Sakai [20]. They employed a dye laser with a pulse duration of 800 femtoseconds and a peak intensity of 1.4×10^{14} W/cm^2. The highest detected harmonic was the 41st, corresponding to a wavelength of 15 nm.

To obtain the harmonic spectrum, we calculate the induced dipole moment $d(t) = \int d^3r\, z\, n(\mathbf{rt})$ which is then Fourier transformed over the last 5 cycles of the constant-intensity interval. The square of the resulting Fourier transform, $|d(\omega)|^2$, has been shown [21] to be proportional to the experimentally observed harmonic distribution to within a very good approximation.

We performed calculations with different peak intensities and achieved the best agreement with experiment for $I = 3.5 \times 10^{14}$ W/cm^2, see Fig. 1. The discrepancy between this intensity and the experimental intensity of 1.4×10^{14} W/cm^2 might be due to the uncertainty of the experimentally determined peak intensity which can be as high as a factor of two.

To explain experimental harmonic generation data, Lambropoulos and coworkers [22] have performed numerical simulations for helium based on a single-active-electron model. Their aim was to clarify the role of He^{+1} in the harmonic generation process at different laser wavelengths. For this purpose, they calculated the harmonic spectra separately for neutral helium and for He^{+1} at the respective saturation intensities (i. e. those intensities for which about 5% of the populations of

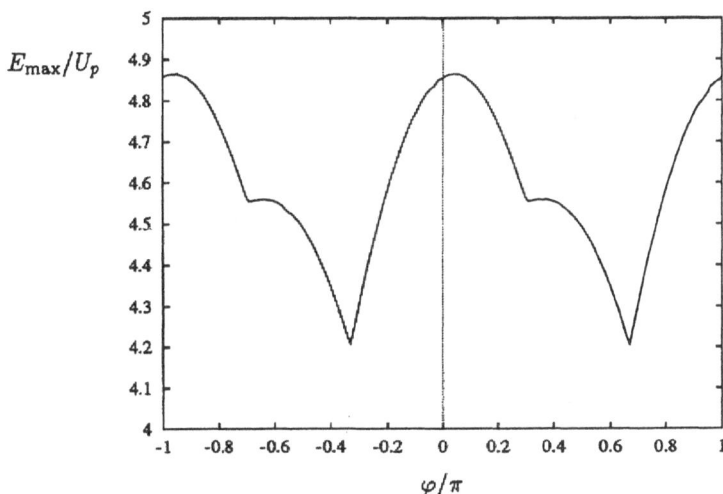

Figure 2: *Maximum return energy of a classical electron released from the nucleus in a two-colour laser field with a relative phase difference φ.*

the He atom or the ion are ionized during the laser pulses). The spectra were then compared with experiment in order to find out which part of the detected harmonic spectrum was caused by neutral He and which part by singly charged He. This method explains quite well an experiment performed at the wavelength 248 nm [23], but does not give a clear picture for the experimental data of Miyazaki and Sakai [20] at 616 nm: for the latter case, the intensities ($5 \times 10^{14}\,\mathrm{W/cm^2}$ for neutral He and $5 \times 10^{15}\,\mathrm{W/cm^2}$ for He^{+1}) leading to good agreement with experiment were out of the range of the experimental peak intensity of $1.4 \times 10^{14}\,\mathrm{W/cm^2}$.

If we calculate the probabilities for the charge states of He at 616 nm and $3.5 \times 10^{14}\,\mathrm{W/cm^2}$ after 20 cycles by using Eqs. (29)–(31), we obtain 99.93% probability for neutral He and only 0.07% for He^{+1}. We thus conclude that the spectrum shown in Fig. 1 is exclusively due to the neutral atom.

We also studied the harmonic generation for a helium atom in a strong two-colour laser field. The two lasers with frequencies ω_0 and $2\omega_0$, respectively, are operated with the same peak intensity and a constant relative phase difference φ. This results in a time-varying laser potential of the form

$$v_{\mathrm{laser}}(\mathbf{r}t) = E_0 f(t) z[\sin(\omega_0 t) + \sin(2\omega_0 t + \varphi)] \quad , \tag{32}$$

where both fields are linearly polarized along the z-axis. One expects to obtain harmonic distributions with a plateau region extending up to a cutoff which depends on the phase difference φ. In the one-colour case, this cutoff is approximately determined by the well-known $I_0 + 3.2U_p$ rule [24], where I_0 denotes the atomic ionization potential and U_p the ponderomotive shift.

To obtain an analogous rule for the two-colour case, we have calculated the maximum return energies E_{max} for a classical electron released at the nucleus in a

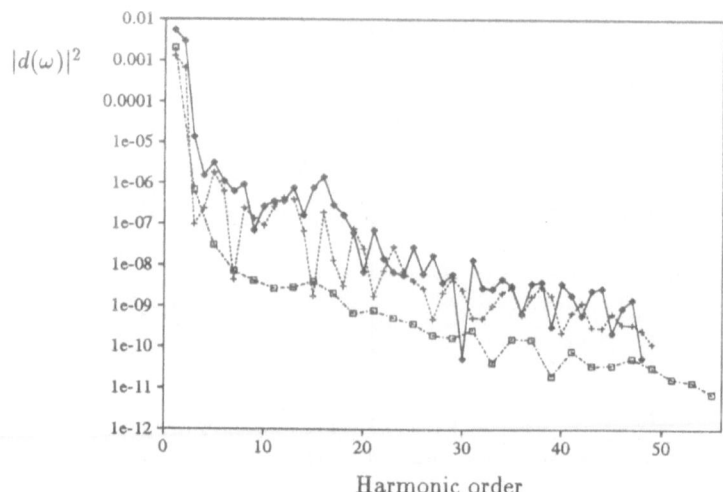

Figure 3: *Harmonic distribution for He in a two-colour laser field. The two wavelengths are 616 nm and 308 nm, and the intensity is 3.5×10^{14} W/cm² for both of them. Crosses are the results for $\varphi = 0$ and diamonds denote the values obtained with phase shift $\varphi = 0.7\pi$. For comparison, the squares indicate the harmonic distribution for He in a one-colour field with $\lambda = 616$ nm and $I = 7 \times 10^{14}$ W/cm².*

two-colour laser field. The cutoff is then given by $I_0 + E_{\max}$. In Fig. 2, $E_{\max}/U_p$ is plotted as a function of the phase shift, where U_p is now given by the sum of the ponderomotive potentials of the two individual laser fields. We have made several calculations for laser intensities in the range of 10^{13} to 10^{15} W/cm² and wavelengths between 200 and 1200nm. The curve plotted in Fig. 2 turned out to be insensitive to these variations of the laser parameters. The two-colour cutoff rule can thus be written as $I_0 + c(\varphi)U_p$, where $c(\varphi) = E_{\max}/U_p$ has its maximum value of 4.86 for $\varphi = 0.05\pi$ (mod π), whereas it has a minimum of 4.20 for $\varphi = 0.67\pi$ (mod π).

Calculated harmonic distributions induced by a two-colour field with different relative phases are shown in Fig. 3. The fundamental wave length is 616 nm and the intensity is 3.5×10^{14} W/cm² for both frequency components. We also show the one-colour spectrum for $\lambda = 616$ nm calculated with the *same* total intensity as the two-colour field, i. e. $I = 7 \times 10^{14}$ W/cm². In the two-colour spectrum, harmonics at all higher multiples (including even multiples) of the fundamental frequency ω_0 occur due to nonlinear mixing processes of the two fields [25]. We chose the phase differences $\varphi = 0$ and $\varphi = 0.7\pi$ which according to our semiclassical model (Fig. 2) lead to cutoff energies at the 50th harmonic and the 46th harmonic, respectively. These classical estimates are found to agree quite well with the full quantum mechanical calculations shown in Fig. 3.

Most of the harmonics produced by the two-colour field in the plateau region are one to two orders of magnitude more intense than those obtained in the one-colour calculation (see Fig. 3). Similar results have recently been found for hydrogen in a two-colour field [26]. One possible reason for this remarkable enhancement is that in a two-colour field one specific high-order harmonic can be generated by a large number of different mixing processes [25]. Several of the two-colour harmonics, on the other hand, are found to be strongly suppressed: the 7th and 15th harmonics for the case $\varphi = 0$, e. g., are even below their counterparts calculated in the one-colour field. For $\varphi = 0.7\pi$, however, this suppression does not occur. Other harmonics such as the 30th are suppressed for $\varphi = 0.7\pi$.

4. Neon: beyond the single-active-electron approximation

For atoms heavier than He, exchange terms are present in the TDKS equations. In this section we present a full TDDFT calculation for the neon atom. We have solved the TDKS equations with the TDKLI and ALDA potentials for the Ne valence electrons in a laser pulse with $\lambda = 248$ nm for two different intensities, $I = 3 \times 10^{15}$ W/cm^2 and 5×10^{15} W/cm^2. The 1s electrons have been frozen, i. e. we propagate only the 2s and 2p electrons by solving the TDKS equations, whereas the time evolution of the 1s electrons is given by

$$\phi_{1s}(rt) = \phi_{1s}(rt_0)\, e^{-i\epsilon_{1s}(t-t_0)} \quad . \tag{33}$$

We emphasize that the only approximation made in this frozen-core prescription is to write the frozen orbitals in the form (33). The exchange between the frozen orbitals and the other orbitals is fully included in the TDKLI or ALDA potentials. In this respect, our scheme differs from other frozen-core prescriptions such as, e. g., in Ref. [24]. In view of the high binding energy of the 1s electrons compared

	HF$^{(\text{exact})}$	KLI$^{(\text{exact})}$	LDA$^{(\text{exact})}$	KLI$^{(\text{grid})}$	LDA$^{(\text{grid})}$
$-\epsilon_{1s}$	32.77	30.80	30.24	35.13	34.47
$-\epsilon_{2s}$	1.930	1.707	1.266	1.951	1.522
$-\epsilon_{2p_0}$	0.8504	0.8494	0.4431	0.8098	0.4159
$-\epsilon_{2p_1}$	0.8504	0.8494	0.4431	0.8065	0.4126

Table 1: *Ne orbital energies (in Hartrees).*

to the other electrons (see Table 1), freezing only the 1s electrons is expected to be a very good approximation for the neon atom. Later we shall discuss the effect of additionally freezing electrons of the valence shell and only propagating the most loosely bound, i. e. the 2p$_0$ orbital.

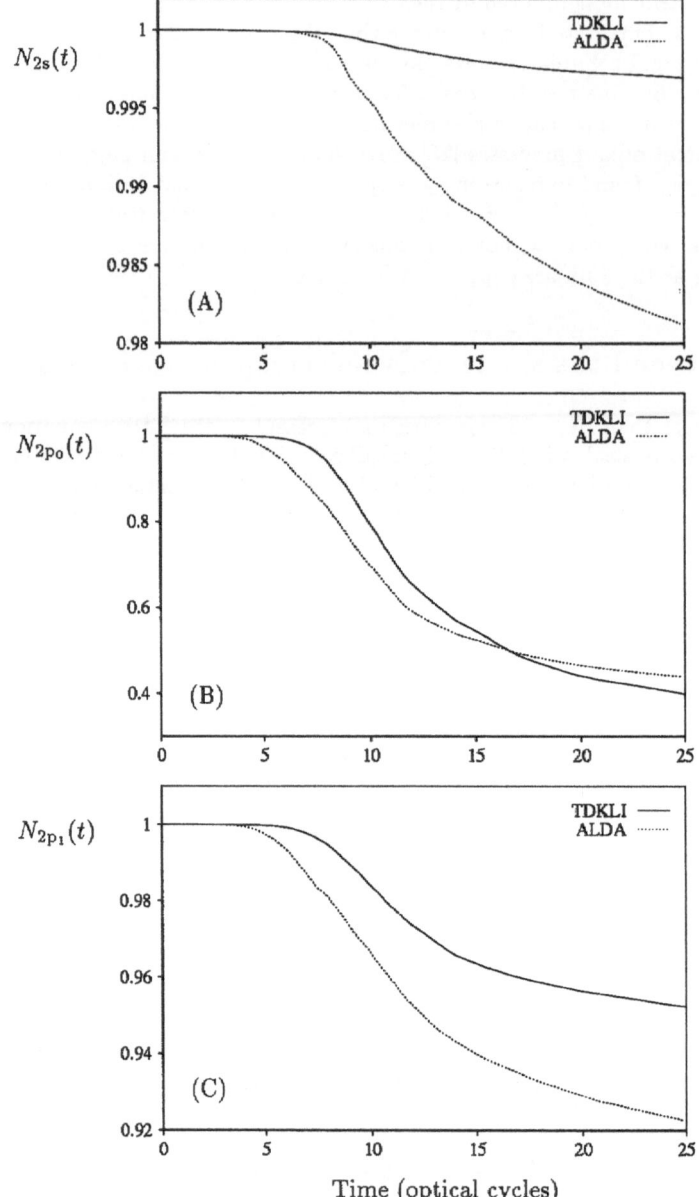

Figure 4: *Time evolution of the norm of the Ne 2s orbital (A), the Ne 2p$_0$ orbital (B) and the Ne 2p$_1$ orbital (C), calculated in the x-only TDKLI and ALDA schemes. Laser parameters: $\lambda = 248$ nm, $I = 3 \times 10^{15}$ W/cm^2, linear ramp over the first 10 cycles. One optical cycle corresponds to 0.82 femtoseconds.*

In order to assess the accuracy of the numerical procedure we first compare the energy eigenvalues resulting from the diagonalization of the stationary KS equation on our two-dimensional grid with the exact values from the literature [16, 27, 28], see Table 1. We find that there is a slight difference (about 3 mHartrees) between the eigenvalues of the $2p_0$ and $2p_1$ orbitals due to the different orientations of these orbitals in our cylindrical grid: the $2p_0$ orbital is oriented along the z-axis, the $2p_1$ orbitals perpendicular to it. In the limit of infinitesimally small grid spacings, this difference goes to zero. The deviation of the average of the 2p orbital energies from the exact value is 4.9% (41 mHartrees) for KLI and 6.5% (29 mHartrees) for LDA. The KLI results for the orbital binding energies (on the exact level as well as calculated on the rectangular grid) are found to be much closer to the HF results than the LDA energy eigenvalues. The first ionization potential of Ne in LDA is too small by almost 50%, whereas the exact KLI result reproduces the HF ionization potential within 1 mHartree (0.1%). We mention that the experimental ionization potential is 0.792 Hartrees.

Fig. 4 shows the norm of the Ne 2s, $2p_0$ and $2p_1$ orbitals for the laser intensity $I = 3 \times 10^{15}$ W/cm^2 (a very similar behaviour is found for $I = 5 \times 10^{15}$ W/cm^2). Once again the indices "2s", "$2p_0$" and "$2p_1$" denote the *initial* state of the otherwise fully propagated orbitals. The pulse has been linearly ramped over the first ten cycles and is then kept constant for another 15 cycles. As expected, the 2s orbital is the least ionized of the three orbitals (only 0.3% ionization for TDKLI and 1.9% for ALDA at the end of our calculation). A little surprising at first sight, the $2p_0$ and $2p_1$ orbitals differ by about an order of magnitude in their degree of ionization (60% for the $2p_0$ orbital compared to only 4.75% for the $2p_1$ orbital within TDKLI, and 56% for the $2p_0$ compared to 7.7% for the $2p_1$ orbital within the ALDA). This difference has been observed before by Kulander [29, 30] for the case of xenon (in a single-active-electron calculation). It is due to the fact that the $2p_0$ orbital is oriented along the polarization direction of the laser field, which makes it easier for the electrons to escape the nuclear attraction than for the case of the $2p_1$ orbital, which is oriented perpendicularly to the field polarization.

To explain the difference between the results obtained within the TDKLI and ALDA schemes shown in Fig. 4, we observe from Table 1 that it takes 5 photons to ionize the 2p orbitals in TDKLI compared to only 3 photons in ALDA. Similarly, it takes 11 photons to ionize the 2s orbital in TDKLI and only 9 in ALDA. The difference between the curves in Fig. 4A and C is thus hardly surprising. On the other hand, it seems quite unexpected that the ALDA and TDKLI curves cross in Fig. 4B so that the ALDA curve comes to lie *above* the TDKLI curve. This behaviour can be attributed to the fact that the other orbitals are ionized much more strongly in ALDA than in TDKLI, so that their electron density near the nucleus (and therefore their screening of the nuclear charge) is decreased. This makes it more difficult for the $2p_0$ electrons to escape within the ALDA scheme.

We have calculated the harmonic spectra for both sets of laser parameters. The distributions are displayed in Figs. 5A and B. We see that for the lower intensity, $I = 3 \times 10^{15}$ W/cm^2, the plateau extends up to the 23rd harmonic, whereas for $I = 5 \times 10^{15}$ W/cm^2 it goes up to the 33rd harmonic. We also observe that the

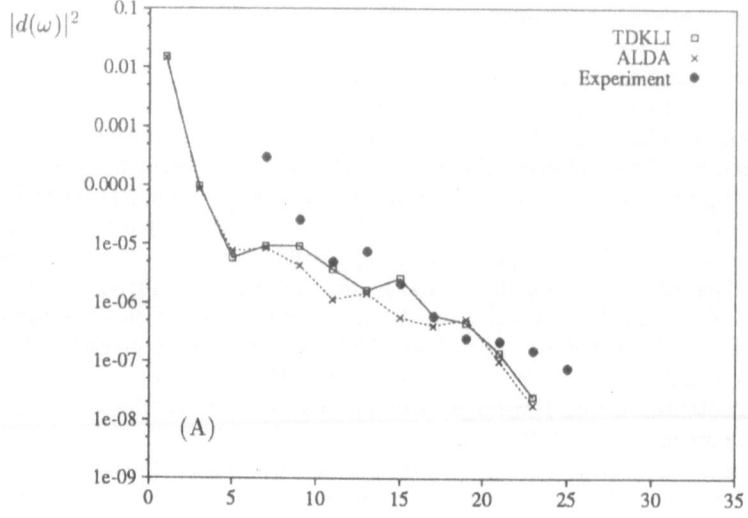

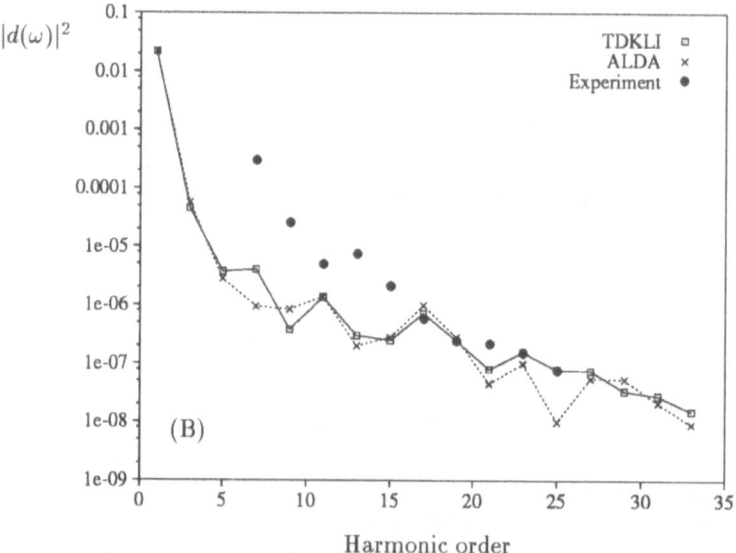

Figure 5: *Harmonic distributions for Ne ($\lambda = 248$ nm), calculated within the TDKLI and ALDA schemes at $I = 3 \times 10^{15}$ W/cm^2 (A) and $I = 5 \times 10^{15}$ W/cm^2 (B). The experimental data were taken at $I = 4 \times 10^{17}$ W/cm^2 [23].*

difference between the TDKLI and ALDA spectra is not very pronounced. We thus conclude that the strongly differing ionization energies of the individual orbitals (a factor of two for the 2p electrons) play only a minor role for the shape of the harmonic spectrum.

We compare our calculated harmonic distributions with experimental data by Sarukura *et al.* [23]. The experiment was performed with a KrF laser ($\lambda =$ 248 nm) at a pulse duration of 280 femtoseconds and a peak intensity of 4×10^{17} W/cm^2. The experimental peak intensity is two orders of magnitude higher than the intensities used in our calculations. It is to be expected, however, that the atoms have become completely ionized by the time the pulse reaches its maximum. The detected harmonic radiation must therefore have been induced during the rise time of the pulse, probably in an intensity range close to the intensities used in our calculations.

The experimental data points shown in Figs. 5A and B have both been normalized to the value of the 17th harmonic (within TDKLI) in Fig. 5A. At 3×10^{15} W/cm^2, we see that the calculated spectra can explain the measured harmonics 15 to 21, whereas the harmonics 17 to 25 (with the exception of the strongly suppressed 25th harmonic for ALDA and the a little less strongly suppressed 21st harmonic for both schemes) are explained by the spectrum at 5×10^{15} W/cm^2. Hence, our calculations show that the generation of the harmonics 15 to 25 is dominated by the intensity range covered in our calculations. We can match this part of the experimental harmonic distribution pretty well by a superposition of the two spectra with equal weights. This corresponds to the experimental situation where the harmonic photons generated on different positions in the laser focus (and, therefore, coming from regimes with different laser intensities) are superimposed in the detector.

In order to explain the same experimental data, Kondo *et al.* [31] have performed numerical simulations based on a simple atomic model (a single electron in a short-range model potential). They calculated the harmonic spectrum for neutral Ne at 4.5×10^{14} W/cm^2, for Ne^{+1} at 1.8×10^{15} W/cm^2 and for Ne^{+2} at 4.8×10^{15} W/cm^2. By a suitable superposition of these single-electron spectra, they reproduced the qualitative features of the experimental harmonic distribution. The authors attributed the harmonics above the 11th to Ne^{+1} and the harmonics above the 21st to doubly charged Ne.

We come to a similar conclusion by calculating the populations of the differently charged states as we did for the helium atom, see Eqs. (29)–(31). For the intensity 3×10^{15} W/cm^2, we find a slightly higher probability for Ne^{+1} than for Ne^{+2}. Thus, the harmonics up to the 21st are most probably caused by Ne^{+1}. At the higher intensity, we find that the doubly charged Ne ions are prevailing. We can therefore attribute the harmonics above the 21st to Ne^{+2}, in accordance with Ref. [31].

We also found that the dipole moment of the 2p$_0$ orbital alone leads to spectra looking very similar to the full harmonic spectra displayed in Figs. 5A and B. This brings us to the following question: To what extent is the harmonic motion of the 2p$_0$ electrons influenced by the motion of the other electrons? In order to study this question, we have performed an additional TDKLI calculation with the same

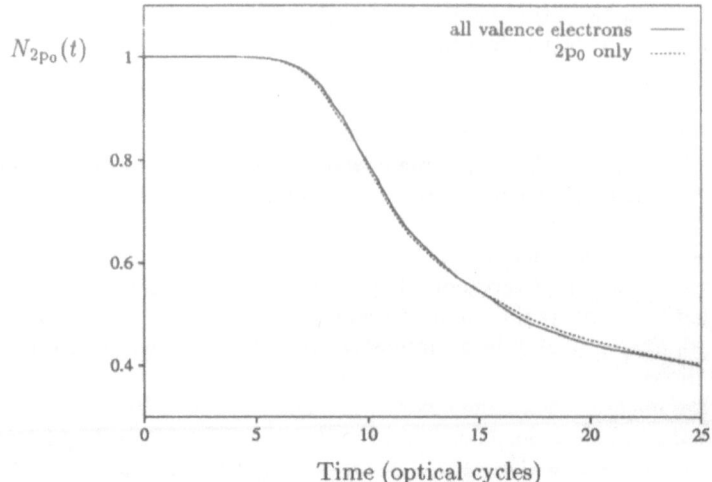

$N_{2p_0}(t)$

Figure 6: *Time-dependent norm of the $2p_0$ orbital, calculated with TDKLI in two different schemes (propagation of all valence electrons and of the $2p_0$ orbital only, respectively). The laser parameters are $\lambda = 248$ nm and $I = 3 \times 10^{15}$ W/cm^2.*

laser parameters as in Fig. 5A, this time with all but the $2p_0$ electrons frozen in their initial states. Fig. 6 shows a comparison of the time-dependent norm of the $2p_0$ orbital calculated in the original scheme (i. e. all electrons are propagated under the influence of the laser except the 1s electrons) and in the new frozen-core scheme (i. e. propagation of the $2p_0$ orbital only). The difference between the two curves is very slight, implying that the *ionization* of the $2p_0$ orbital can reliably be calculated with the new frozen-core prescription.

However, if we calculate the *harmonic spectrum* in the new frozen-core scheme, we find a strong deviation from the spectrum calculated in the original scheme. From the comparison of the two spectra in Fig. 7, we come to the conclusion that the effect of freezing the 2s and $2p_1$ electrons is twofold: First of all, the whole spectrum is slightly shifted towards lower values of $|d(\omega)|^2$. The second, more drastic effect is the appearance of a pronounced Lorentz-profile resonance peak just below the 7th harmonic. This resonance dominates the background of the spectrum, as becomes clearly visible in the region beyond the plateau, i. e. beyond the 21st harmonic, which can be fitted very well with a Lorentz curve. A very similar resonance phenomenon has been observed by Kulander and Shore [30] for the case of Xenon, where a single $5p_0$ electron was propagated only.

In the former computational scheme, where all valence electrons of the neon atom were fully propagated, the resonance had been suppressed due to the influence of the 2s and $2p_1$ electrons on the motion of the $2p_0$ electrons. In other words, the resonance is an *artefact* of the more restricted frozen-core approximation, where only the $2p_0$ orbital was propagated. This leads us to the conclusion that a reliable calculation of harmonic spectra requires simultaneously treating

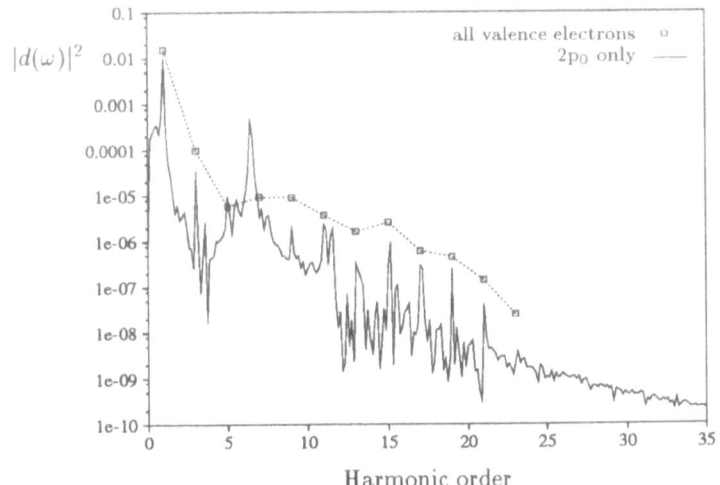

Figure 7: *Harmonic spectrum of Ne calculated by propagating the $2p_0$ orbital only. Squares: harmonic distribution calculated by propagating all valence electrons (both calculations were done in TDKLI). Laser parameters as in Fig. 6.*

the time evolution of *all* electrons belonging to the outermost atomic shell. Neglecting the mutual influence of the electrons on their harmonic motion, as done in the single-active-electron approximation, can lead to spurious resonance effects.

Acknowledgments

This work was supported in part by the Deutsche Forschungsgemeinschaft. We thank the Rechenzentrum der Universität Würzburg and the Landesrechenzentrum München for making their facilities available. S. E. and C. A. U. acknowlede with thanks a fellowship of the Studienstiftung des deutschen Volkes.

References

[1] P. Hohenberg and W. Kohn, Phys. Rev. **136**, B864 (1964)

[2] W. Kohn and L. J. Sham, Phys. Rev. **140**, A1133 (1965)

[3] R. M. Dreizler and E. K. U. Gross, *Density Functional Theory: An Approach to the Quantum Many-Body Problem* (Springer-Verlag, Berlin, 1990)

[4] *Density Functional Theory*, ed. by E. K. U. Gross and R. M. Dreizler, NATO ASI series B337 (Plenum Press, New York, 1994)

[5] T. Grabo and E. K. U. Gross, Chem. Phys. Lett. **240**, 141 (1995)

[6] E. Runge and E. K. U. Gross, Phys. Rev. Lett. **52**, 997 (1984)

[7] E. K. U. Gross and W. Kohn, Adv. Quant. Chem. **21**, 255 (1990)

[8] E. K. U. Gross, C. A. Ullrich, and U. J. Gossmann, in Ref. [4], p. 149

[9] G. D. Mahan and K. R. Subbaswamy, *Local Density Theory of Polarizability* (Plenum Press, New York, 1990)

[10] *Atoms in Intense Laser Fields*, ed. by M. Gavrila (Academic Press, Boston, 1992)

[11] *Super-Intense Laser-Atom Physics*, ed. by B. Piraux, A. L'Huillier and K. Rzążewski, NATO ASI Series B316 (Plenum Press, New York, 1993)

[12] K. L. Liu and S. H. Vosko, Can. J. Phys. **67**, 1015 (1989)

[13] M. Bünner, J. F. Dobson, and E. K. U. Gross, to be published

[14] C. A. Ullrich, U. J. Gossmann, and E. K. U. Gross, Phys. Rev. Lett. **74**, 872 (1995)

[15] C. A. Ullrich, U. J. Gossmann, and E. K. U. Gross, Ber. Bunsenges. Phys. Chem. **99**, 488 (1995)

[16] J. B. Krieger, Y. Li, and G. J. Iafrate, Phys. Rev. A **45**, 101 (1992)

[17] K. C. Kulander, Phys. Rev. A **35**, 445 (1987)

[18] K. C. Kulander, Phys. Rev. A **36**, 2726 (1987)

[19] M. S. Pindzola, T. W. Gorczyca, and C. Bottcher, Phys. Rev. A **47**, 4982 (1993)

[20] K. Miyazaki and H. Sakai, J. Phys. B: At. Mol. Opt. Phys. **25**, L83 (1992)

[21] A. L'Huillier, L. A. Lompré, G. Mainfray, and C. Manus, in Ref. [10], p. 139

[22] H. Xu, X. Tang, and P. Lambropoulos, Phys. Rev. A **46**, R2225 (1992)

[23] N. Sarukura, K. Hata, T. Adachi, R. Nodomi, M. Watanabe, and S. Watanabe, Phys. Rev. A **43**, 1669 (1991)

[24] K. C. Kulander, K. J. Schafer, and J. L. Krause, in Ref. [11], p. 95

[25] H. Eichmann, A. Egbert, S. Nolte, C. Momma, and B. Wellegehausen, W. Becker, S. Long, and J. K. McIver, Phys. Rev. A **51**, R3414 (1995)

[26] M. Protopapas, A. Sanpera, P. L. Knight, and K. Burnett, Phys. Rev. A **52** (1995) (in press)

[27] E. Clementi and C. Roetti, At. Data Nucl. Data Tables **14**, 177 (1974)

[28] T. Grabo, private communication

[29] K. C. Kulander, Phys. Rev. A **38**, 778 (1988)

[30] K. C. Kulander and B. W. Shore, J. Opt. Soc. Am. B **7**, 502 (1990)

[31] K. Kondo, T. Tamida, Y. Nabekawa, and S. Watanabe, Phys. Rev. A **49**, 3881 (1994)

HYPERSPHERICAL COORDINATES APPROACH TO ONE-DIMENSIONAL MODELS OF TWO-ELECTRON QUANTUM SYSTEMS

A.I. ARTEMYEV
General Physics Institute, Russian Academy of Sciences,
Vavilov St. 38, Moscow, 117942, Russia

R. GROBE
Department of Physics, Illinois State University,
Normal, IL 61790, USA

J.H. EBERLY
Department of Physics and Astronomy, University of Rochester,
Rochester, NY 14627-0171, USA

The hyperspherical coordinates method is applied to one-dimensional models of two-electron atoms and ions We have calculated excitation spectra and ground state wave functions. The results for the first several ground states energies as well as for the continuum thresholds are compared with their exact values, and with the values obtained within different modification of the Hartree-Fock approach. We analyze the spatial distribution of the ground state wave functions obtained within different approaches. The hyperspherical method stresses the electron correlation while the other methods underestimate them. The applicability and accuracy of the hyperspherical approach for the study of one-dimensional systems is discussed.

1. Introduction

The theoretical treatment of multi-electron systems is a problem of central interest in atomic physics. Evidence of the primary role of electron correlation in helium was first discovered in 1963 from the observation of the doubly excited $^1P_0$ levels by Madden and Codling [1]. These results were first understood in terms of linear combinations of 2s-np and 2p-ns excited states by Cooper, Fano and Prats [2] and in more detail by other authors later [3-5]. Significant progress have been done in 1968 when Macek suggested a new way to study the correlated behavior of the two helium electrons by introducing the hyperspherical basis [6]. He explained the main features of the helium spectrum, such as large differences of brightness and spectral width of various Rydberg series, and unified the results of previous investigations within his approach. Recent

285

H. G. Muller and M. V. Fedorov (eds.), Super-Intense Laser-Atom Physics IV, 285–294.
© *1996 Kluwer Academic Publishers.*

work applying the hyperspherical approach to study two-electron systems has been devoted to behavior of the wave functions at large distances, the estimation of errors due to non-adiabaticity, the study of the analytic properties of the wave-functions in the vicinity of adiabatic level-crossing [7-10], and the modification of adiabatic levels due to interaction with laser radiation [11].

Reduced-dimensional two-electron systems have been used in a wide variety of contexts to explore basic properties of the combined electron-electron and electron-nucleus interaction. Reduced dimensional systems resemble in many of their features the corresponding 3D systems, and have the advantage that the exact fully correlated wave functions can be computed numerically [12]. This allow for test of the accuracy of approximation schemes [13], to study the effects of single and double ionization [14,15]. Some studies led to discovery of new phenomena [16,17] based on strong electron-electron correlation and to new measures for degree of electron correlation [18].

In this paper we study one-dimensional two-electron ions and atoms by the hyperspherical approach and check the accuracy of this method for strongly correlated states. We describe the static properties of 1D two-electron systems and compare the energy level positions found within the hyperspherical approach with their exact values and with values found from different kinds of Hartree-Fock approximations.

2. The model and the hyperspherical coordinate approximation

Let us begin with a brief description of our two-electron systems. It is characterized by a nucleus of positive charge Z fixed at the origin at $x=0$ and the two electrons whose spatial coordinates are x_1 and x_2. Its bare Hamiltonian is (atomic units, a.u.)

$$\mathbf{H}(x_1, x_2) = \frac{p_1^2}{2} + \frac{p_2^2}{2} + Z\,V(x_1) + Z\,V(x_2) - Z\,V(x_1 - x_2) \qquad (1)$$

where the soft-core Coulomb potential has been chosen as

$$V(y) = -\frac{1}{\sqrt{y^2 + 1}} \qquad (2)$$

This potential has been extensively studied in the one-electron context, mostly for strong laser interactions [19], and its behavior is proven to be remarkably useful. In the present case it permits the two electrons to move past the nucleus and past each other, so that entire x-axes is available for both electrons. In Figure 1 we show the total two-electron potential as a function of coordinates x_1 and x_2. This potential, and thus the bare Hamilto-

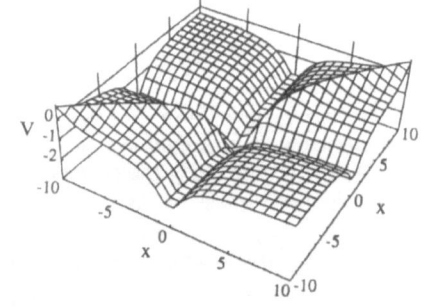

Figure 1. The two-electron potential Eq. (1)

nian, is invariant under the electron permutation and parity operators. Thus the time-independent Schrodinger equation for the eigenenergies and eigenfunctions can be solved separately in four different subspaces:

$$\Psi(x_1, x_2) = \pm\Psi(-x_1, -x_2) \quad \text{and} \quad \Psi(x_1, x_2) = \pm\Psi(x_2, x_1) \tag{3}$$

This property is important for applicability of the hyperspherical approximation discussed below.

We introduce the hyperspherical coordinates R and α :

$$R \equiv (x_1^2 + x_2^2)^{1/2} \quad \text{and} \quad \tan(\alpha) \equiv x_1/x_2 \tag{4}$$

The hyperspherical coordinates are collective coordinates for the electron pair and thus they are convenient for the description of electron-electron correlation. These coordinates have the property that the Hamiltonian in hyperspherical coordinates is quasi-separable. This corresponds to slow motion along the R coordinate and a relatively faster motion along the α coordinate. Thus the hyperspherical coordinates method resembles in many details the Born-Oppenheimer approximation for diatomic molecules. The Schrodinger equation $\mathbf{H}\Psi(x_1, x_2) = E\Psi(x_1, x_2)$ with the Hamiltonian (1) then reads

$$\mathbf{H}\Phi(R,\alpha) \equiv \left(-\frac{1}{2}\frac{\partial^2}{\partial R^2} + \mathbf{H}_\alpha\right)\Phi(R,\alpha) = E\ \Phi(R,\alpha) \tag{5}$$

where the angular Hamiltonian $\mathbf{H}_\alpha$ is defined as:

$$\mathbf{H}_\alpha = \frac{1}{R^2}\left(-\frac{\partial^2}{\partial\alpha^2} - \frac{1}{8} + R^2 V(R,\alpha)\right) \tag{6}$$

and the wave function $\Phi(R,\alpha)$ is defined as:

$$\Phi(R,\alpha) = R^{-1/2}\Psi(R\sin\alpha, R\cos\alpha) \tag{7}$$

As an approximation we will treat the dynamics in R-direction as frozen and neglect the derivatives in R in Eq. (5). It suggests that the approximate eigenfunction $\Phi(R,\alpha)$ of the Eq. (5) can be set by the ansatz:

$$\Phi(R,\alpha) = f(R)\varphi(R,\alpha) \tag{8}$$

where $\varphi(R,\alpha)$ is an eigenfunction of the angular Hamiltonian $\mathbf{H}_\alpha$:

$$\mathbf{H}_\alpha\ \varphi(R,\alpha) = u(R)\varphi(R,\alpha) \tag{9}$$

where u(R) denote the eigenvalue. $\mathbf{H}_\alpha$ depends on the "slow" variable R only parametrically and its spectrum is descrete because of periodic boundaries $\varphi(R,\alpha) = \varphi(R,\alpha+2\pi)$. We assume that all the mutually orthogonal states $\varphi_v(R,\alpha)$ are normalized with respect to α for all values of R. The assumption of "frozen" dynamics along the R-coordinate means that the derivatives of $\varphi_v(R,\alpha)$ along the angular coordinate are much bigger than those along the radial coordinate. The Schrodinger equation for the pure radial part f(R) of the wave function $\Phi(R,\alpha)$ Eq. (8) then reads:

$$\left(-\frac{1}{2}\frac{\partial^2}{\partial R^2} + u_v(R)\right)f_{vn} = E_{vn}\ f_{vn} \tag{10}$$

where E_{vn} and f_{vn} are the eigenenergies and eigenfunctions. As it is in the usual Born-Oppenheimer approach, the eigenvalue $u_v(R)$ of the Hamiltonian $\mathbf{H}_\alpha$ plays the role of an effective potential for Schrodinger equation in the radial direction. The energies E_{vn}

of the one-particle Hamiltonian (10) are approximate values of the energies of the original two-electron system (1).

We have computed the potential curves $u_v(R)$ by numerically diagonalizing $\mathbf{H}_\alpha$ on an angular grid with periodic boundaries for 400 different values of R. The potential curves for two-electron systems are quite similar for nuclear charges $Z=1,2,...5$. We present in Figure 2 the first few lowest lying potential curves for 1D negative helium atom ($Z=1$). Let us discuss the structure of these potential curves. For small values of R the lowest potential curve approaches minus infinity while the other potential curves tend to plus infinity in pairs. For large values of R the curves bunch into groups of 4 which approach the energies of the corresponding core system (the He^+-ion in the case $Z=2$). This behavior can be understood by analyzing the symmetry and asymptotic properties of the Hamiltonian $\mathbf{H}_\alpha$.

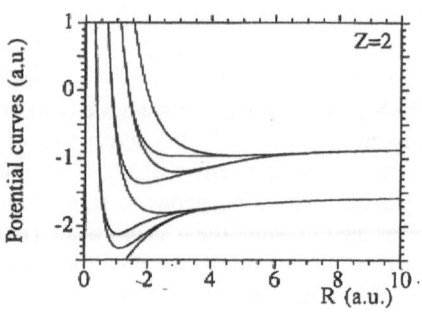

Figure 2. The potential curves $u_v(R)$ for the helium atom ($Z=2$)

For small values of R the angular Hamiltonian $\mathbf{H}_\alpha$ tends to the sum of α-kinetic energy operator plus a constant negative potential:

$$R^2 \mathbf{H}_\alpha \rightarrow -\frac{1}{2}\frac{\partial^2}{\partial \alpha^2} - \frac{1}{8} \qquad (R \rightarrow \infty) \qquad (11)$$

Its spectrum is purely descrete due to periodicity of the eigenfunctions. The lowest eigenvalue is $e_1 = -1/8$ and all others are pairwise degenerate: $e_{2v} = e_{2v+1} = v^2/2 - 1/8$. Correspondingly the eigenenergies of $\mathbf{H}_\alpha$ tend to plus or minus infinity as $R \rightarrow \infty$:

$$u_1(R) \rightarrow -\frac{1}{8R^2} \qquad u_{2v}(R) \sim u_{2v+1}(R) \rightarrow -\frac{v^2/2 - 1/8}{R^2} \quad \text{(for } v=1,2,3... \text{ and } R \rightarrow \infty) \qquad (12)$$

For large R the spectrum of $\mathbf{H}_\alpha$ is four-fold degenerate. This is due to the fact that the effective potential in $R^2 \mathbf{H}_\alpha$ has 4 minims, each of them corresponding to one of the two electrons close to the nucleus. The depth of these valleys tends to infinity with increase of R, corresponding to localization of one of the electrons near the nucleus while the other one is far away. The angular wave function can be localized in any of these valleys and, as they become deeper with increasing of R, the four-fold degeneracy becomes apparent. In the limit $R \rightarrow \infty$ the functions $u_v(R)$ converge to the ionization thresholds. This can be seen by rewriting $\mathbf{H}_\alpha$ in terms of a new variable $y = \alpha R$ and assuming a large value of R while leaving y finite. In this limit $\mathbf{H}_\alpha$ becomes $p^2/2 + V(y)$, which describes the core-electron system in the absence of the other electron. Its bound state energies are exactly the ionization thresholds of the corresponding two-electron system.

As we have mentioned above, the eigenstates of the two-electron Hamiltonian (1) can be classified as having different spatial parity and symmetry with respect to electron permutation. A corresponding classification for the potential curves $u_\nu(R)$ is based on the fact that the parity operator $(x_1, x_2) \to (-x_1, -x_2)$ acts in (R,α) coordinates as $(R,\alpha) \to (R,\alpha+\pi)$ while the electron permutation operator $(x_1, x_2) \to (x_2, x_1)$ is equivalent to $(R,\alpha) \to (R, \pi/2-\alpha)$. The symmetries of the eigenfunctions $\varphi_\nu(R,\alpha)$ alternate with increase of the state number ν, as follows:

electron permutation	spatial parity	
$\varphi_1(R,\alpha) = \varphi_1(R, \pi/2-\alpha)$	$\varphi_1(R,\alpha) = \varphi_1(R,\alpha+\pi)$	(13a)
$\varphi_2(R,\alpha) = -\varphi_2(R, \pi/2-\alpha)$	$\varphi_2(R,\alpha) = -\varphi_2(R,\alpha+\pi)$	(13b)
$\varphi_3(R,\alpha) = \varphi_3(R, \pi/2-\alpha)$	$\varphi_3(R,\alpha) = -\varphi_3(R,\alpha+\pi)$	(13c)
$\varphi_4(R,\alpha) = -\varphi_4(R, \pi/2-\alpha)$	$\varphi_4(R,\alpha) = \varphi_4(R,\alpha+\pi)$	(13d)

The symmetry of the higher lying potential curves are expected to change periodically in a similar way.

The symmetry properties of the eigenfunctions $\varphi_\nu(R,\alpha)$ are quite relevant to decide whether the hyperspherical approximation is appropriate to 1D two-electron systems. Indeed, the hyperspherical method described is applicable if the energy separation between the potential curves is large enough compared to the rate of change of the eigenfunctions $\varphi_\nu(R,\alpha)$ with respect to R:

$$\left\langle \varphi_\eta(R,\alpha) \left| \frac{\partial^2}{\partial R^2} \right| \varphi_\mu(R,\alpha) \right\rangle \ll \left| u_\eta(R) - u_\mu(R) \right|$$
$$\left\langle \varphi_\eta(R,\alpha) \left| \frac{\partial}{\partial R} \right| \varphi_\mu(R,\alpha) \right\rangle \ll \left| u_\eta(R) - u_\mu(R) \right|$$

(14)

The consideration presented above as well as Figure 2 indicate that both at small and at large values of R some potential curves become infinitely close to each other. However this does not lead to inapplicability of the hyperspherical approach because these curves belong to different symmetry classes and can not interfere with each other. The relevant potential curves that should be taken into account are those whose indices differ by 4.

Those curves are indeed separated and, in contrast to the 3D case we did not find any avoided crossings for the lower lying potential curves.

From now on we will restrict our analysis to spin singlet states corresponding to wave functions that are even under electron permutation. We discard from our discussion every second potential curve and use the labels $\nu=1,2,3,...$ to numerate the relevant potential curves $u_\nu(R)$ from below. When substituted into the radial equation (10) each potential curve $u_\nu(R)$ provides a set of eigenstates for the total system. The first few lower potential curves and eigenenergies of the radial Hamiltonian (10) for 1D helium atom

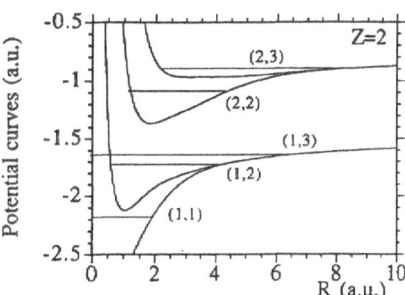

Figure 3. He atom relevant (even) potential curves and the total energy levels.

are presented in Figure 3. Each eigenstate is characterized by the number of the potential curve and by the state number within this potential curve. It is important to note that one can identify some of the states (v,n) in the hyperspherical picture with the states enumerated by two principal quantum numbers (n_1,n_2) in the independent electron excitation approach. We summarize these results in Table 1.

TABLE 1. Energy levels of the 1D helium atom obtained within the hyperspherical approach compared to the exact energy levels.

Exact energy (a.u.)	Hyperspherical energy (a.u.)	Hyperspherical state numbers (v,n)	Electron excitation state numbers (n_1,n_2)
-2.238	-2.185	(1,1)	(1,1)
-1.704	-1.725	(2,1)	(1,2)
-1.626	-1.629	(1,2)	(1,3)
-1.567	-1.570	(2,2)	(1,4)
-1.545	-1.542	(1,3)	(1,5)
-1.483	-1.483	$(1,\infty)$ and $(2,\infty)$	$(1,\infty)$
-1.045[*]	-1.085	(3,1)	(2,2)
?	-0.897	(4,1)	(2,3)
-0.857[*]	-0.864	(3,2)	(2,4)
-0.772	-0.772	$(3,\infty)$ and $(4,\infty)$	$(2,\infty)$

As it was mentioned above, the hyperspherical method yields the exact values of the ionization thresholds. The exact energies presented in Table 1 were obtained in [20] by use of the partitioned Hilbert space diagonalization approach. According to [20] the positions of energies of the autoionizing states, that are marked with *, depend on the details of the approximation scheme used to couple bound states with the continuum.

3. Comparison of the hyperspherical method with other approaches

In order to compare the hyperspherical method with other approaches we have computed the ground states energies and wave functions according to different approximation schemes and compared them to each other. Table 2 summarizes our results for the ground state energies of the first 5 ions of the isoelectronic series beginning with H^-. The second column shows the energy obtained from the hyperspherical coordinate approach described above. The third column reproduces the energies obtained from the Hartree-Fock (HF) approach in which the functional form of the wave function is restricted to a product of two identical single electron wave functions. $\Psi(x_1, x_2) = g(x_1)\, g(x_2)$. The fourth column is obtained from an approximation method which is similar to the traditional HF method, but the product wave functions are different an presented in hyperspherical coordinates $\Psi(x_1, x_2) = R^{-1/2}\, F(R)G(\alpha)$. We tried it in order to figure out the relation between the hyperspherical and Hartree-Fock approaches as well as to check if the hyperspherical version of the HF method can lead to better results then traditional HF calculations.

TABLE 2. Ground state energies of the first 5 ions of the isoelectronic series beginning with H⁻ obtained by different methods. The underlying marks the best approximation to the exact values.

Nuclear charge Z	Energy levels			
	Exact	Hyperspherical	HF-(x_1,x_2)[*]	HF-(R,α)
1	-0.731	<u>-0.726</u>	-0.692	-0.656
2	-2.238	-2.185	<u>-2.224</u>	-2.132
3	-3.896	-3.797	<u>-3.888</u>	-3.752
4	-5.615	-5.472	<u>-5.610</u>	-5.437
5	-7.371	-7.187	<u>-7.367</u>	-7.162

[*] - values from [12] and [20]

The ground states and their energies for both versions of the HF method have been found by imaginary time integration of the HF equations. The equations of motion for the HF wave functions $F(R)$ and $G(\alpha)$ were found by minimizing the time-dependent Raleigh-Ritz functional [21]. The general procedure for finding these equations is as follows. Typically, the two-electron Hamiltonian has the form $\mathbf{H}_x + \mathbf{H}_y + V_{xy}$ and the HF approach requires us to minimize the following functional:

$$\left\langle f(x)g(y) \left| -i\partial_t + \mathbf{H}_x + \mathbf{H}_y + V_{xy} \right| f(x)g(y) \right\rangle \tag{15}$$

We used two modifications of the HF method with x and y being either the usual electron coordinates x_1 and x_2 or the hyperspherical coordinates R and α. In the second case the resulting equations of motion are:

$$i\partial_t F(R) = -\frac{1}{2}\frac{\partial^2}{\partial R^2} + \left\langle G(\alpha) \left| V(R,\alpha) \right| G(\alpha) \right\rangle F(R)$$

$$i\partial_t G(\alpha) = -\frac{1}{2}\left\langle F(R) \left| R^{-2} \right| F(R) \right\rangle \frac{\partial^2}{\partial \alpha^2} G(\alpha) + \left\langle F(R) \left| V(R,\alpha) \right| F(R) \right\rangle G(\alpha) \tag{16}$$

These equations have been integrated numerically on (R,α) grid with 1024×256 points with hyperradius R changing from 0 to 30 a.u.

The hyperspherical method is accurate within 0.6% for the negative hydrogen ion and the accuracy decreases to 2.3% for the helium atom. Note that in the hydrogen ion the electron correlation is more important then it is in helium. When the nuclear charge becomes larger, the accuracy of the hyperspherical ground state energy is about 2.5% that is comparable to the HF method accuracy in (R,α). We conclude that the hyperspherical method provides more accuracy with increasing electron correlation (decreasing Z).

So far we have restricted our discussion to the bare energies. In the following we will also analyze the eigenfunctions. We focus on the hydrogen ion and compare the spatial electron density distribution of its ground state obtained in four different ways: through the approximate methods mentioned above, and via the exact one. The logarithmic contour plots are presented at Figure 4. Figure 4b shows the state as computed by the hyperspherical approximation. The agreement between both states especially impressive for larger values of R. For small R the hyperspherical state develops into two minims corresponding to both electrons localized on the opposite sides of the nucleus. This illustrates that the hyperspherical method exaggerates the electron correlation. The HF wave function calculated in (x_1, x_2) coordinates is presented at Figure 4c. It gives better approximation to the exact wave function than the HF ground state wave function calculated in (R, α) coordinates shown at Figure 4d.

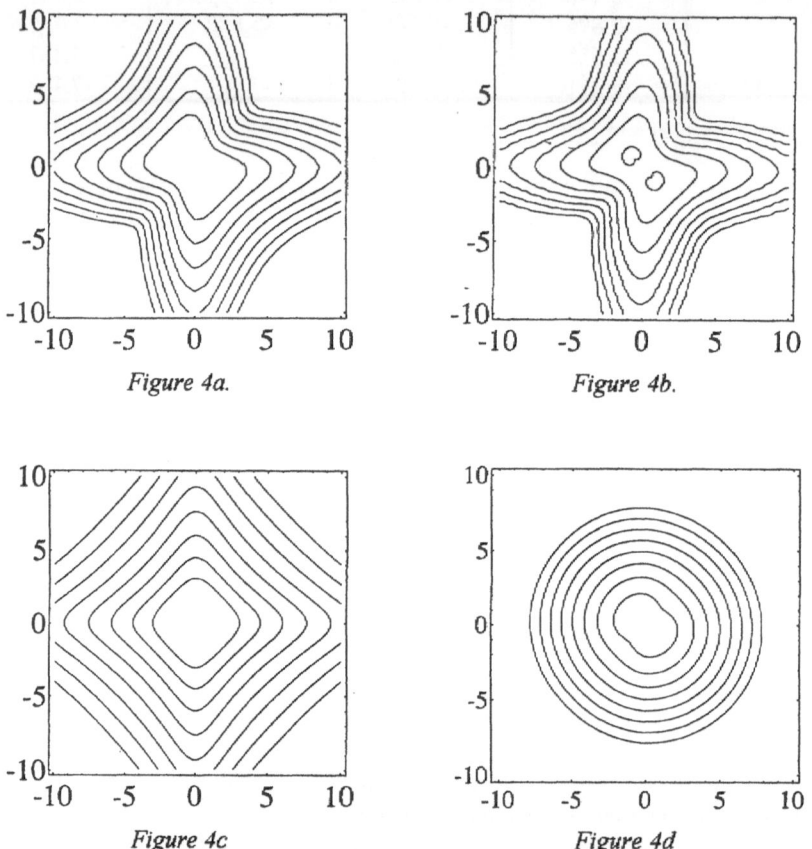

Figure 4a.   *Figure 4b.*

Figure 4c   *Figure 4d*

Figure 4. Contour plot of the exact spatial distribution of the ground state of the negative hydrogen ion $(Z=1)$ compared to that calculated within 3 different approximation schemes. Contours correspond to the $|\Psi(x^1, x^2)|^2 = 10^{-c}$, for $c = 2, 2.5, 3, \ldots 5.5$

a) exact electron density distribution   c) Hartree-Fock approximation

b) hyperspherical approximation   d) hyperspherical version of HF method

This is attributed to the fact that the α-dependence of $V(R,\alpha)$ is very different at various values of R. The hyperspherical method is based on the adiabatic separation of R and α, and this is different from the HF assumption of complete independence in R and α. Let us also mention that the hyperspherical wave function gives the best agreement with the exact one.

4. Conclusions.

We have applied the hyperspherical approach to study the bare properties of 1D two-electron systems. The results for the ground state energies indicate interesting features of the hyperspherical method that to the best of our knowledge have not been recognized before: the accuracy of the hyperspherical method increases with the increase of electron-electron correlation, or equivalently, with the decrease of the nuclear charge Z. This feature is opposite to the behavior of HF-like results. We also compared the spatial distribution of the ground state wave functions obtained within different approaches. It was found that the hyperspherical method exaggerates the profile of the shaping of the ground state wave function associated with the electron correlation. To the contrary, the Hartree-Fock-like methods provide the wave functions with a rounder shape than it should have.

The applicability of the hyperspherical approach to one-dimensional systems is suggested by the symmetry properties with respect to electron permutation and parity. It follows from the symmetry that the potential curves that might contribute to the non-adiabatic coupling are well separated by at least three other potential curves and have different asymptotic behavior at both small and large values of hyperradius R.

5. Acknowledgment

We would like to thank S.L. Haan and M.V. Fedorov for many useful discussions and M.K. Kalinski for his constant help with *Mathematica*. AA acknowledges the support of National Science foundation grant PHY94-08733, the partial support of International Science Foundation grant M1I300 and a fellowship of INTAS grant 93-2492 provided by the research program of International Center of Fundamental Physics in Moscow. RG and JHE acknowledge support by Division of Chemical Sciences, Office of Energy Research of the US Department of Energy.

6. References

1. Madden R.P. and Codling K. (1963) *Physical Review Letters* 10, 516 and (1965) *Astrophysical journal* 141, 364.
2. Cooper J.W., Fano U. and Prats F. (1963) *Physical Review Letters* 10, 518.

3. O'Malley T.F. and Geltman S. (1965) *Physical Review A* **137,** 1344.
4. Atick P.L. and Moore E.N. (1965) *Physical Review Letters* **15,** 100.
5. Lipski L. and Russek A. (1966) *Physical Review* **142,** 59.
6. Macek J. (1968) *Journal of physics B* **1,** 831.
7. Hatton G.H. (1976) *Physical Review A* **14,** 901.
8. Hornos J.E., MacDowel S.W. and Caldwell C.D. (1986) *Physical Review A* **33,** 2212.
9. Gusev V.V., Puzynin V.I., Kostrikin V.V., Kvitinsky A.A., Merkuriev S.P. and Ponomarev L.P. (1990) *Few Body Systems* **9,** 137.
10. Kosricin V.V. and Kvitinsky A.A. (1994) *J. Mathematical Physics* **35,** 47.
11. Klar H., Zoller P. and Fedorov M.V. (1984) *Physical Review A* **30,** 658.
12. Grobe R. and Eberly J.H. (1993) *Physical Review A* **48,** 4664.
13. Pindzola M.S., Griffin D.C. and Bottcher C. (1991) *Physical Review Letters* **66,** 2305.
14. Grobe R. and Eberly J.H. (1992) *Physical Review Letters* **68,** 2905.
15. Grobe R. and Eberly J.H. (1993) *Physical Review A* **47,** RC1605.
16. Grobe R. and Eberly J.H. (1993) *Physical Review A* **48,** 623;
17. Grobe R., Eberly J.H. and Haan S.H. (1993) *J. Physics B,* submitted.
18. Grobe R., Rzhazhevski K. and Eberly J.H. (1994) *J. Physics B,* in press.
19. Eberly J.H., Grobe R., C.K. Law and Q. Su. (1992) in Gavrila M. (ed.), *Atoms in strong laser fields,* Academic Press, Orlando, p. 323.
20. Haan S.L., Grobe R. and Eberly J.H. (1994) *Physical Review A* **50,** 378.
21. Kerman A.K. and Koonin S.E. (1976) *Annals of Physics* **100,** 332.

CONTINUUM-CONTINUUM AUTLER-TOWNES SPLITTING IN CALCIUM

B. WALKER, B. SHEEHY, M. KALUŽA AND L. F. DIMAURO
Chemistry Department
Brookhaven National Laboratory, Upton, USA

AND

M. TRAHIN, P. AGOSTINI
Service des Photons, Atomes et Molécules
CE Saclay, 91191 Gif Sur Yvette, France

1. Introduction

Strong-field ionization of two-electron atoms can result in scenarios in which the electron-electron correlation plays an important role. It was recently suggested[1] that a splitting similar to the Autler-Townes effect[2] would occur when two ionization continua are resonantly coupled in two-electron atoms. This is obviously at variance with the case of one electron atoms where coupling between continua does not induce oscillations but instead leads to exponential decay of one continuum into another. The special case considered by Grobe and Eberly is that of a strong radiation field resonantly coupling two ionic states (i.e. a core transition). Formally, the states which are coupled are continuum states (two-electron states in which one electron is in a continuum state), but nevertheless the corresponding photoelectron peak is split. Physically, the reason for this is that the electron-electron interaction transfers the energy shift of the core electron to the outgoing electron and has been dubbed "coherence transfer" by Ref. [1].

One simple way to see this effect is to think that the final ionic state is split by the resonant (core)-interaction, thus the outgoing electron sees two asymptotic energy limits separated by the Rabi frequency $\Omega = \mu_{\pm} \mathcal{E}/\hbar$, where $\mu_{\pm}$ is the ionic dipole and $\mathcal{E}$ the electric field. To emphasize the fact that it is actually two continua that are coupled, one can talk about continuum-continuum Autler-Townes splitting. Dynamically, the outer elec-

H. G. Muller and M. V. Fedorov (eds.), Super-Intense Laser-Atom Physics IV, 295–304.
© 1996 *Kluwer Academic Publishers.*

tron is being ionized and, at the same time, the core-electron is driven by a Rabi oscillation. Note that the splitting would reflect directly on Rydberg states as well[3]. Actually, the time-evolution of a Rydberg wave-packet under strong coupling of the core-electron gives rise to very intersting effects as discussed by Hanson and Lambropoulos[4]. For the phenomenology of strong-field optical resonance for two-level systems, the reader is referred to the literature[5]. We just summarize the general behavior of the photo-electron energy spectrum "on" resonance. At low intensity, the spectrum would consist of a single energy peak, as the intensity increases the peak will be symmetrically split by Ω and proportional to the square root of the intensity. One should also recall that "on" resonance, the states are actually a linear superpositions of bare states, thus any labelling of the split components by bare state quantum numbers is arbitrary.

Experimentally, two-photon ionization of calcium around the core resonance 4s-4p (393.5 nm) offers, in principle, an ideal realization of this situation. The strong ionic dipole moment (approximately 1.5 atomic units) yields, for an intensity of 300 GW/cm^2, easily observable Rabi splitting of about 120 meV. Furthermore, the wavelengths needed conveniently corresponds to the second harmonic of a titanium sapphire laser. The first observation of such a splitting has been reported in a recent Letter[6]. Although a number of the observed features are in agreement with the simple prediction discussed above, even a superficial inspection of the data reveals a number of significant differences. The most conspicuous are an *asymmetric* splitting and the presence of extra peaks in the energy spectra. The former is easily understood if one takes into account more precisely the atomic structure of Ca. For instance, Hanson et al.[7] have shown that the *neutral atomic* resonance $4s^2 - 4s4p$ modifies significantly the spectrum through the 4s4p-(4s,ϵ) coupling. A similar effect can be assigned to the interaction between the 4p and 5s continua. The presence of additional peaks may be traced to the influence of the fine structure for the 4p ionic-state. We show that its role is large at low intensity and diminishes around saturation.

2. Experiment

A frequency-doubled, regeneratively amplified titanium sapphire laser produces tunable (380-405 nm), 180 fs pulses. The pulse bandwidth ($\sim$ 15 meV) is less than twice the transform limit and the intensity fluctuations are $\leq$ 6%. Spectral measurements were made on the fundamental light with a monochromator and an optical multichannel analyzer calibrated with a krypton arc lamp. The spectral resolution was 0.5 nm. The calcium was produced in an 775 K atomic beam and background contamination was less than 0.01%. Various lenses with f-numbers ranging from 7 to 25 focused

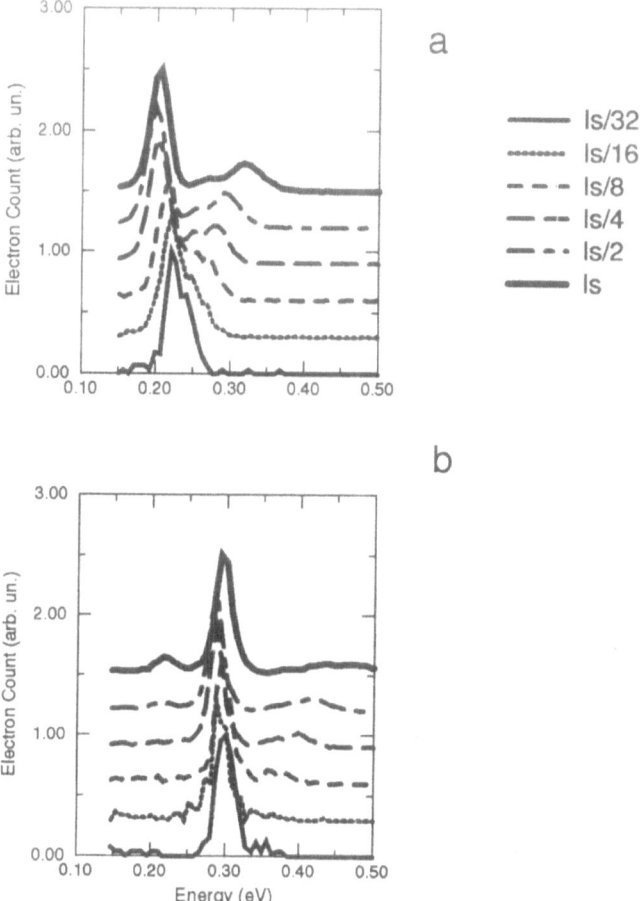

Figure 1. Experimental spectra (a) "on resonance" ($\lambda = 393.5nm$) and (b) "off resonance" ($\lambda = 388.1nm$) for different laser intensities labelled in fraction of the saturation intensity $I_s = 3 \times 10^{11}$ W/cm^2.

the light into the atomic beam. The laser's confocal length exceeded the atomic beam's cross-sectional length, ensuring a flat intensity distribution in the interaction volume. Electron energy analysis was performed with a time-of-flight spectrometer with 2π solid angle collection.

Figure 1 shows the intensity dependence of the photoelectron energy spectra at constant wavelength. In Fig. 1a the laser is tuned "on resonance" with the ionic $4s_{1/2}$-$4p_{3/2}$ transition (393.5 nm) for intensities ranging from about 10^{10} to 3×10^{11} W/cm^2. At the lowest intensity only one peak emerges at the expected energy for the two-photon ionization, with a small shoulder evident on the high energy side. As the intensity increases, the main feature is red-shifted while the shoulder develops into new structures on the high energy side becoming progressively blue-shifted. In fact, the blue shifted structure resolves into a clear doublet, whose relative amplitude switches depending upon the intensity.

The laser is tuned "off resonance" in Fig. 1b. Besides the trivial shift due to the change in photon energy, the intensity dependence of the spectra is somewhat different: the main peak is basically unshifted, a weak component is increasingly blue-shifted and at the highest intensity, a new feature appears on the red side of the main peak. The qualitative behavior of the spectra as a function of the wavelength and intensity are certainly reminiscent of the predictions of the Autler-Townes model. However, other couplings could give rise to analogous splittings. For instance, the observed splitting could be related to a Rabi coupling of atomic *bound* states rather than continua[8]. This is ruled out immediately by inspection of the electron peak leaving the ion in the 3d states: this peak is *not* split at any intensity or wavelength. Furthermore, only a quantitative comparison may distinguish between the resonant splitting and "ordinary" Stark shifts. A simple plot of the energy difference between the two main peaks versus intensity shows a square root dependence which signifies a resonant shift. However, a more detailed theory is necessary in order to account for the obvious differences between the observed spectra and the predictions of the simple Rabi model.

3. "Essential states" calculation

The theory, beyond the simple model[1], has been worked out by several groups[7, 9, 10]. All calculations rely on the "essential states" approximation. We follow here an equivalent appoach: by using projection operators, the eigenstate space is partitioned into the essential states subspace and the complementary space. The choice of the "essential states" is guided by the neutral and ionic calcium energy level scheme and the photon energy. We have retained the following states: $4s^2\ {}^1S_0$, $4s4p\ {}^1P_1$ bound states and $4s^2 S_{1/2}$, $4p^2 P_{3/2}$, $4p^2 P_{1/2}$ and $5s^2 S_{1/2}$ continua in CaI. Projecting the time-dependent Schrödinger equation on the essential states yields a set of coupled integro-differential equations which read:

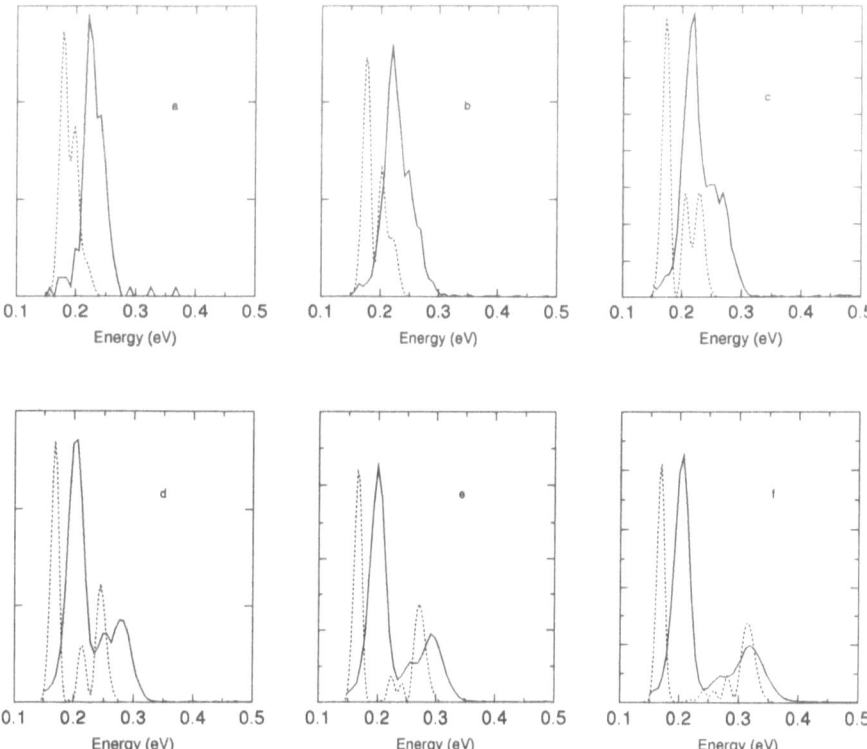

Figure 2. Calculated (dashed) and observed (solid) electron energy spectra at 393.5 nm for different intensities: (a) 10^{10} W/cm^2, (b) 2×10^{10} W/cm^2, (c) 4×10^{10} W/cm^2, (d) 8×10^{10} W/cm^2, (e) 1.5×10^{11} W/cm^2 and (f) 3×10^{11} W/cm^2.

$$i\dot{a}_0(t) = E_0\ a_0(t)\ +\ \Omega_{01}(t)a_1(t)\ +\ \int d\epsilon'\ g_{02}(\epsilon',t)\ a_2(\epsilon',t) \qquad (1)$$

$$i\dot{a}_1(t) = (E_1 - \omega)\ a_1(t)\ +\ \Omega_{01}(t)a_0(t)\ +\ \int d\epsilon\ g_{12}(\epsilon',t)\ a_2(\epsilon',t) \qquad (2)$$

$$i\dot{a}_2(\epsilon,t) = (E_2\ +\ \epsilon\ -\ 2\omega)\ a_2(\epsilon,t)\ +\ \Omega_{02}(t)\ a_0(t)\ + \qquad (3)$$
$$\Omega_{12}(t)\ a_1(t)\ +\ \Omega_{23}(t)\ a_3(\epsilon,t)\ +\ \Omega_{24}(t)\ a_4(\epsilon,t)\ +\ \Omega_{25}(t)\ a_5(\epsilon,t)$$

$$i\dot{a}_3(\epsilon,t) = (E_3\ +\ \epsilon\ -\ 3\omega)\ a_3(t)\ +\ \Omega_{03}(t)\ a_0(\epsilon,t)\ + \qquad (4)$$
$$\Omega_{23}(t)\ a_2(\epsilon,t)\ +\ \Omega_{35}(t)\ a_5(\epsilon,t)$$

$$i\dot{a}_4(\epsilon,t) = (E_4\ +\ \epsilon\ -\ 3\omega)\ a_4(\epsilon,t)\ +\ \Omega_{04}(t)\ a_0(t)\ + \qquad (5)$$
$$\Omega_{24}(t)\ a_2(t)\ +\ \Omega_{45}(t)\ a_5(\epsilon,t)$$

$$i\dot{a}_5(\epsilon,t) = (E_5\ +\ \epsilon\ -\ 4\omega)\ a_5(\epsilon,t)\ + \qquad (6)$$
$$\Omega_{25}(t)\ a_2(\epsilon,t)\ +\ \Omega_{35}(t)\ a_3(\epsilon,t)\ +\ \Omega_{45}(t)\ a_4(\epsilon,t)$$

Where the labels "0", "1",...,"5" refer to the 6 essential states in the above order, ϵ is the photoelectron energy, and t the time.

The continuum-continuum couplings Ω_{ij} are approximated by the corresponding single-electron ionic dipoles using the independent-electron coupling ansatz[1]:

$$< i, \epsilon |H'(r_1, r_2, t)|j, \epsilon' >=\ \Omega_{ij}(t)\delta(\epsilon\ -\ \epsilon') \qquad (7)$$

and the bound-free couplings $g_{ij}(\epsilon, t)$ are defined as:

$$g_{i,j}(\epsilon, t) = < a, a|\ H(r_1, r_2, t)|a, \epsilon > . \qquad (8)$$

The dependence of the $\Omega's$ on ϵ has been dropped by making the usual assumption of a "flat" continua. Eqns. (1) and (2) are reduced to differential equations by standard methods. The CaI bound-bound dipoles are taken from the tables[11] and the bound-free matrix elements are those used by Hanson et al.[7, 12]. The CaII dipoles were re-calculated using a non-relativistic approximation of the core polarization[13]. The resulting differential equations are numerically integrated using a fourth-order Runge-Kutta method for a gaussian pulse with FWHM of 180 fs.

A comparison between the calculated and experimental spectra is shown in Fig. 2 and Fig. 3 for the "on" and "off" resonant cases, respectively. The

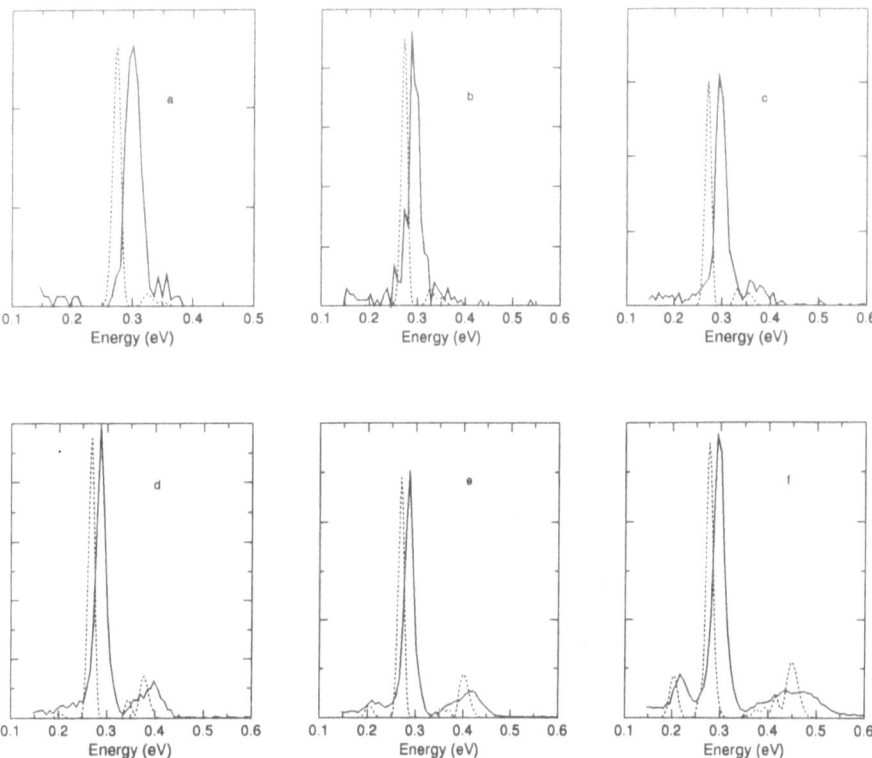

Figure 3. Same as Fig. 2 except at 388.3 nm.

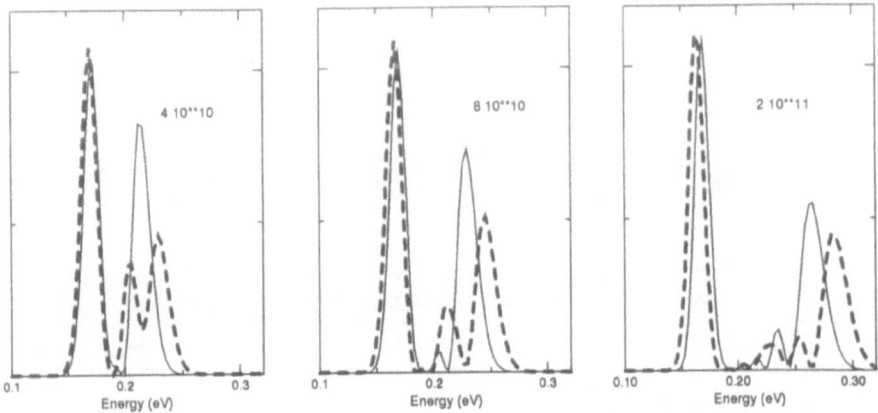

Figure 4. Calculated electron energy spectra for three peak intensities, with (dashed lines) and without (solid) the $p_{1/2}$ continuum (fine structure). The intensity is indicated in each plot in W/cm^2.

overall agreement is excellent (the slight systematic shift of the experimental spectra with respect to the calculated ones is probably due to a contact potential inside the spectrometer). In order to check in more details the role of the fine structure, the $(4p_{1/2}, \epsilon)$ continuum has been removed from the above equations. The result is shown in Fig. 4 for three intensities. For intensities below 10^{11} W/cm^2, the fine structure is responsible for the splitting of the high energy peak into two components. Note also the differents shifts. Above 10^{11} W/cm^2, the fine structure manifests itself only in the splitting. This confirms our previous calculation[6] and is further illustrated in Fig. 5 which shows the peak energies versus intensity for the "on resonance" case.

The $(5s_{1/2}, \epsilon)$ influences the system through the generalized Rabi frequencies $\Omega_{i5}, (i = 2, 3, 4)$. The two-photon Rabi frequency Ω_{25} is taken as the non-resonant part of the two-photon coupling (ie excluding the 4p intermediate states). This coupling is weak and at the two-photon resonance wavelength ($\lambda = 383.4$nm) produces basically no effect. At 393.5 nm (the

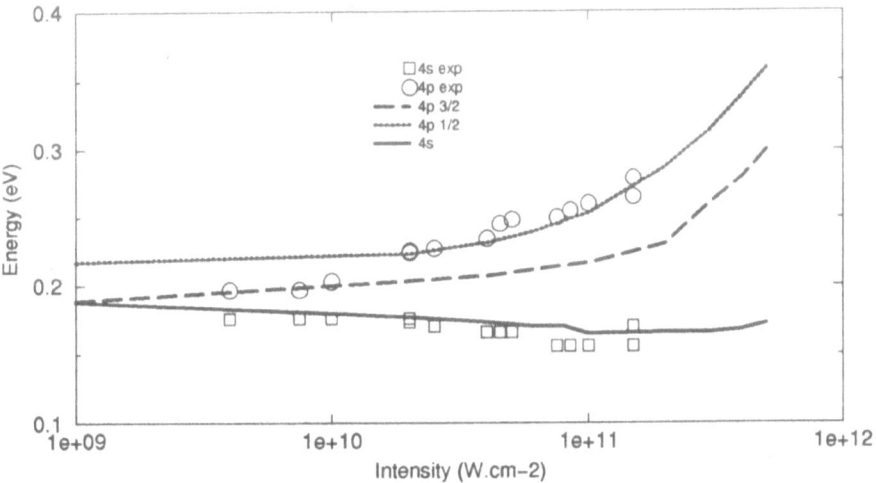

Figure 5. Calculated and experimental peak positions versus intensity at $\lambda = 393.5$ nm. Note that the experiment for the highest energy peak first agrees better with the calculated $4p_{3/2}$ then with the $4p_{1/2}$ positions.

4s-4p resonance), on the other hand, the $5s_{1/2}$ strongly repels the $4p_{3/2}$ through (Ω_{35}, therefore the corresponding electron peak has *more* energy. This effect combined with the influence of the neutral resonance produces the observed splitting. The role of the neutral resonance was first stressed by Hanson et al.[7].

4. Conclusions

In conclusion, the continuum-continuum Autler-Townes splitting[1] occuring in two-electron atom when the radiation frequency is close to a core-resonance has been experimentally observed in the two-photon ionization of calcium. However, the experimental observations cannot be described by a model involving only one discrete state coupled to two coupled continua. A reasonable description is obtained by including the neutral resonance and four continua. Finally, let us remark that the effect of the core resonance on the outgoing electron energy relies entirely on the electron-electron correlation[1]. As pointed out by Hanson et al.[7]: "if the two electrons were non-interacting particles, nothing unusual would be expected to happen". Interestingly enough, the most important coupling in this problem, namely Ω_{23}, is obtained through an independent-electron approximation.

Acknowledgments

This work was carried out at Brookhaven National Laboratory under contract No. DE-AC02-76CH00016 with the U.S. Department of Energy and supported by its Division of Chemical Sciences, Office of Basic Energy Sciences and at the Service des Photons,Atomes et Molécules, CEA Saclay, France, under partial support from NATO Collaborative Research Grant No. SA.5-2-05(RG910678).

References

1. R. Grobe and J. H. Eberly, Phys. Rev. A **48**, 623 (1993).
2. S. L. Autler and C. H. Townes, Phys. Rev. **100**, 703 (1955).
3. F. Robicheaux, Phys. Rev. A **47**, 1391 (1993).
4. L. Hanson et. al., Phys. Rev. Lett. **74**, 5009 (1995).
5. L. Allen and J. H. Eberly, *Optical Resonance and Two-level Atoms* John Wiley & Sons, New York (1975).
6. Barry Walker, M. Kaluža, B. Sheehy, P. Agostini and L. F. DiMauro, Phys. Rev. Lett. **75**, 633 (1995).
7. L. Hanson, Jian Zhang and P. Lambropoulos, Europhys. Lett. **30**, 81 (1995).
8. P. L. Knight, J. Phys. B **11** L511 (1978).
9. R. Grobe and S. L. Haan, J. Phys. B **27**, L735 (1994).
10. S. L. Haan, M. Bolt, H. Nymeyer and R. Grobe, Phys. Rev. A **51**, 4640 (1995).
11. Atomic Transition Probabilities **2**, 22 Nat. Bur. Stand. (US) (1969).
12. L. Hanson, private communication.
13. M. Poirier Z.Phys.D **25**, 117 (1993).

WAVE PACKET DYNAMICS IN TWO-ELECTRON ATOMS: INFLUENCE OF A STRONGLY DRIVEN CORE RESONANCE

LARS G. HANSON[1,2] AND P. LAMBROPOULOS[1,3]

1. *Max-Planck-Institut für Quantenoptik*
Hans-Kopfermann-Str. 1, D-85748 Garching, Germany

2. *Niels Bohr Institute, Ørsted Laboratory*
Universitetsparken 5, DK-2100 København Ø, Denmark

AND

3. *Foundation for Research and Technology Hellas, I. E. S. L.*
P.O. Box 1527, Heraklion 71110, Crete, Greece

When a short laser pulse coherently excites a superposition of Rydberg states, a radial wave packet is formed. On a short time scale the wave packet obeys the laws of classical mechanics, as the center of the probability distribution oscillates between the inner and outer classical turning radius of the Kepler orbit. In the limit of high effective radial quantum number ν, the period of this motion agrees with the semiclassical orbit time $T_{cl}(\text{a.u.}) = 2\pi\nu^3$ (e.g. 19 ps for $\nu = 50$) calculated from the binding energy $W_\nu(\text{a.u.}) = -1/(2\nu^2)$. As time evolves the wave packet spreads on a time scale determined by the spread in radial linear momentum [14, 2].

The dynamics of wave packets of electrons bound to atoms [2] or more general potentials [11, 20, 3] is an important current topic. We consider a relatively simple problem: A wave packet orbits the structured core of a two-electron atom being manipulated by an external field. Despite the simplicity, the setup represents a quantum zoo in itself where many interesting and rare species can be studied at close-up distance: These include the correspondence principle, interference and entanglements of the classical position of one electron with the quantum state of another. From a more pragmatic point of view, the results are very interesting as they represent a realistic method of controlling wave packet dynamics.

H. G. Muller and M. V. Fedorov (eds.), Super-Intense Laser-Atom Physics IV, 305–316.

306

1. Strongly Driven Isolated Core Excitations

Suppose an electron in a two-electron atom[1] is highly excited by a laser pulse of frequency ω_a and leaves the core region as a wave packet (fig. 1). The core left behind is only slightly perturbed by the far-away electron between scattering events. During this period, the core can be prepared in a certain state using a laser pulse of frequency ω resonant with a transition of the ion-like core. If not too strong, the field will not directly affect the wave packet, which is far away from other objects, and this process is therefore referred to as *isolated core excitation* [4].

A wave packet returning to an unexcited core must scatter elastically assuming the energy of the electron to be inappropriate for exciting other states. Upon the return of the wave packet to an excited core, elastic scattering is again a possibility. However, a very probable result is that the wave packet gains additional energy from the excited core. Since the electron was loosely bound, the additional energy will make it free leaving an unexcited ion behind. This *autoionization* process involves at least two electrons and is therefore confined to the core region.

Rabi oscillations of the population between the ground and excited states of the core will occur with a frequency proportional to the field amplitude. Depending on the instantaneous phase of the Rabi oscillation, the core might be excited or unexcited upon the return of the wave packet. Autoionization can only happen in the first case causing a loss of atomic population. This system is therefore expected to exhibit interesting beating phenomena involving the classical orbiting and the Rabi oscillation.

In general, the core can be in a superposition of the excited and unexcited states, which again can be entangled with the state of the highly excited electron. As we will see, this adds to the rich spectrum of interesting effects.

We have described the formalism elsewhere [6]. It basically consists in setting up a set of amplitude equations for the states shown in fig. 2. These are solved numerically for various pulse shapes.

In order to make the Rabi and orbit frequencies equal, $\Omega = \omega_{cl}$, a rather moderate intensity is required. For example, an intensity of $I = 0.5$ MW/cm^2 is needed for the $4s$-$4p$ transition of calcium to match a wave packet centered around the $\nu = 50$ Rydberg state.

The configuration interaction is responsible for the autoionization and important aspects of our results, and we will discuss it in some detail. The

[1]For clarity, we take calcium as a model atom, even though we will *not* give quantitative predictions pertaining to calcium. The formalism is more clear, when cast in terms of a real atom, which furthermore seems to be a good candidate for experiments. The coupling constants cited here were readily available from our work in connection to core-resonant ionization [8, 7].

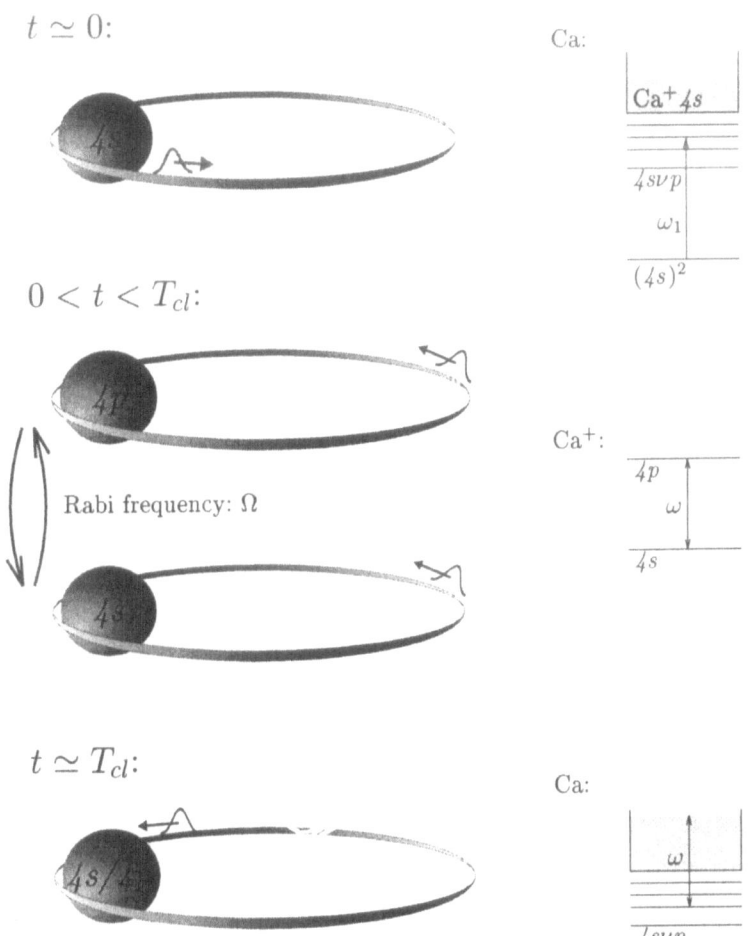

Figure 1. The situation we consider, with calcium chosen as a model atom. A short pulse of frequency ω_1 excites a wave packet of $4s\nu p$ states from the $(4s)^2$ groundstate. While the wave packet is away, a second pulse of frequency ω resonant with the ionic $4s$-$4p$ transition is applied. Returning to the nucleus, the wave packet will scatter from an excited or unexcited core depending on the phase of the Rabi oscillation. Autoionization causing loss of atomic population, might only occur in the first case, when the energy is above the ionization threshold.

autoionization widths $\gamma(\bar{\nu})$ scale as $\bar{\nu}^{-3}$ as expected from classical arguments: Autoionization is a two-electron process confined to the core region, which an electron passes once per orbit time. Thus, the autoionization per orbit, $\Gamma \equiv \gamma(\bar{\nu})T_{cl}(\bar{\nu}) = \gamma(\bar{\nu})2\pi\bar{\nu}^3$, is independent of $\bar{\nu}$ for highly excited Rydberg states. A wave packet with small angular momentum will normally

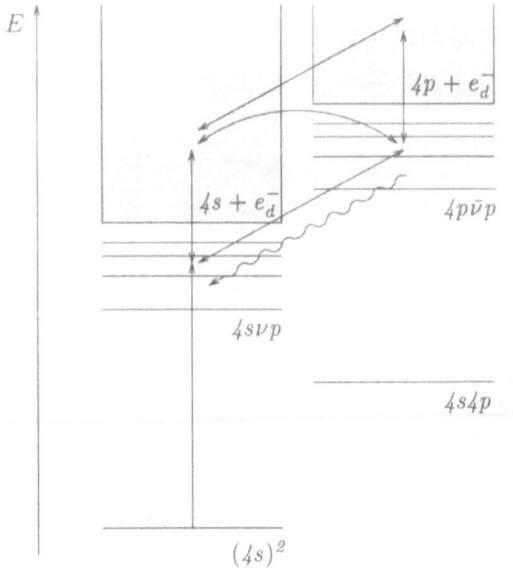

Figure 2. Level scheme. Straight arrows signify field couplings while the curved is configuration interaction. The wavy line is incoherent population transfer to the lower multiplet due to spontaneous emission in the core.

scatter only a few times from an excited core before being lost. Thus Γ is of the order of one.

1.1. RELATED WORK

Our work is a continuation of the already extensive work on Rydberg atoms in core-resonant laser fields, which we shortly review here.

The isolated core excitation scheme [4] has been used with success to obtain atomic parameters in a number of cases, e.g. [9]. Story, Duncan and Gallagher [16] related this to wave packet dynamics by measuring a time dependence of the probability for core transitions after exciting a wave packet. Wang and Cooke have done extensive work describing long range effects of the core transition, [23, 22], and by calculating the decay of an autoionizing wave packet [21]. Especially the latter work is interesting to us, as they demonstrate that the decay only happens whenever the wave packet passes the core.

All this work was related to weak core-resonant fields. Lambropoulos and Zoller [10] described the effect of a strong field driving a transition between a bound and an autoionizing state. The work of Robicheaux [15] is closely related to ours. His idea was to do spectroscopy on an atom in a strong core-resonant field, and he did indeed predict interesting effects when the orbit time corresponding to the Rydberg state was comparable to the Rabi period. One of his conclusions was, however, that it would be exceedingly difficult to observe experimentally.

In the course of our work, van Druten and Muller [19, 18] and also Zobay and Alber [25] worked on related subjects. We will return to their work.

2. A Cavity Analogy

For the discussion to come it is useful to introduce an analogy based on conventional laser theory. The wave packet orbiting the unexcited core resembles a pulse traveling in an optical cavity (fig. 3). The mirrors of the cavity limit the longitudinal extent of the field modes, while the atomic limitation is set by the centrifugal and Coulomb barrier. In the Rydberg atom, several energy eigenfunctions must be excited to form a localized wave function just as several modes are excited when a laser is operated in a pulsed way.

Atom	Cavity analogy
Rydberg levels	Field modes
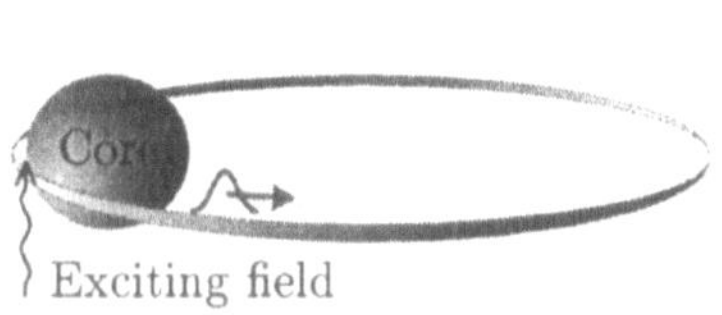	
Core in $4s$	Shutter open
$4p$	closed
Loss process: Autoionization	Loss process: Absorption
Rabi oscillation $\Rightarrow$ Mode locking, non-spread	Periodic opening of shutter $\Rightarrow$ Mode locking, non-spread

Figure 3. The cavity analogy.

In order for the analogy to be appropriate, we must imagine the cavity filled with a dispersive medium. This introduces a difference in the energy spacing of the cavity modes, causing a spread of a traveling pulse with time similar to the spread of the wave packet in a Rydberg atom. This analogy is not new [12]. However, it becomes a useful tool in our case, where the wave packet orbits a structured core. The natural analog to the autoionization loss mechanism is a shutter inserted in one end of the cavity. When the shutter is open (core unexcited), the wave packet travels unattenuated, while a closed shutter (core excited) causes significant loss.

In our case, the shutter is oscillating between open and closed with the Rabi frequency. In general this does not allow a pulse to travel for long, since it will experience great loss. The exception occurs when the shutter is operated with a frequency, which is a multiple of the round trip frequency of the cavity, since a pulse with proper timing can move in the cavity without ever meeting a closed shutter. The tails of the pulse will, however, experience losses. From laser theory this is known to cause *mode locking by loss modulation* [24] manifested in a constant phase relation between the excited modes, as opposed to the free dispersive evolution.

This is exactly the mechanism being responsible for the effects which we encounter in the present case with one very important difference. A shutter in a cavity is a classical object, being either open, closed or semi-transparent. This is fundamentally different from the core being either un-excited, excited (corresponding to semi-transparent, since an electron might still scatter elastically), or in a *coherent superposition* these states. This introduces an important difference to the cavity case, adding to the spectrum of interesting effects.

3. A Non-Dispersing Decaying Wave Packet

In order to understand the general case, it is fruitful first to consider the case being as close to the cavity analogy as possible. We will refer to a pulse changing the phase of the Rabi oscillation by the angle π as a π-*pulse*. A train of π-pulses can be made experimentally by splitting and further sub-dividing an intense pulse of duration much less than T_{cl}. Since the resulting pulses are all alike, the desired Rabi phase shift can be obtained by varying the intensity of the initial pulse before splitting.

When incident on an Rydberg atom, such a train of equidistant π pulses will periodically change the state of the core abruptly between excited and ground state. It thus *discretizes* the Rabi oscillation. It takes, for instance, an optical path length of approximately 3 mm between successive pulses to make the state of the core flip twice per orbit time for a wave packet centered around the $\nu = 50$ Rydberg state.

The situation is depicted in fig. 4 and is in complete analogy to the cavity case with the shutter state being changed twice per round-trip time. The result is as expected from this: The wave packet does not spread, since the tails are cut off at each scattering from the core. It is slowly lost at a rate given by the normal spreading time of the wave packet rather than the short autoionizing time of a Rydberg state.

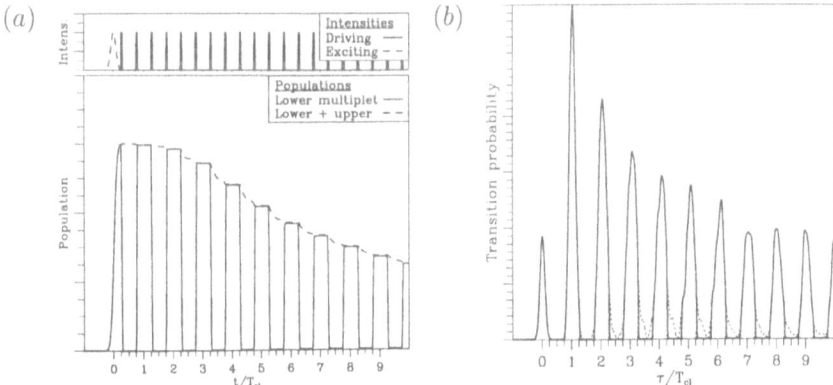

Figure 4. A wave packet is excited by a short pulse of duration $0.2\,T_{cl}$ (FWHM intensity) centered in energy around the $\nu = 50$ Rydberg state ($\Gamma = 1.0$, equal quantum defects). It is followed by a train of equidistant π-pulses driving the core transition. (*a*) The time evolution of the intensities and populations of the lower ($4s\nu p$) and upper ($4p\bar{\nu}p$) Rydberg series in arbitrary units. (*b*) The dependence of the corresponding transition probability (probe signal) as a function of the delay between the exciting pulse and a second short pulse mediating transitions from the $4s\nu p$- (solid curve) or $4p\bar{\nu}p$-series (dashed) to a tightly bound state. The wave packet is seen to stay localized due to the configuration interaction.

By varying the relative timing of the π-pulses, the wave packet can be kept more or less narrow if the natural autoionizing decay per orbit time is significant (it usually is). Unless the "pump" exciting the wave packet is kept running, it will however be lost at a rate which depends on the spreading time i.e. energy width of the wave packet. The probing scheme chosen here is for conceptual simplicity the one suggested in [1]. In an experiment the much more efficient Ramsey-like detection [13] is preferable.

4. Superpositions of Non-Dispersing Wave Packets

The proposed scheme is in itself interesting but there is additional insight to be gained by considering the situation where the Rabi oscillation happens smoothly. This leads to slowly decaying superpositions of non-dispersing wave packets in different phases of the orbiting entangled with the state of the core.

We consider the case where the core-resonant field is turned on after excitation of a spatially rather broad wave packet. Figure 5 shows the evolution of populations for different values of the autoionization coupling and delay between pulses. The field was chosen to make the Rabi- and orbit frequencies equal. Graph (a) and (b) differ only by the time at which the core-resonant field is turned on. As expected, the loss of population during the first orbit depends crucially on the relative phase of the orbiting and the Rabi oscillation. Decay occurs only when the wave packet has overlap with an excited core. In the cavity analogy, the two graphs correspond to the extreme cases of the pulse meeting a shutter which is (a) open and (b) closed. Even when the shutter is closed (core is excited), it is not perfect (since $\Gamma = 0.9$) causing only part of the wave packet to decay upon scattering.

In the graphs (c) and (d) the timing is chosen as before, but the autoionization coupling strength is increased so as to make the shutter nearly perfect, $\Gamma = 4.0$. The increased decay is obvious on a short time scale, but it is remarkable that the long-time effective decay rate actually decreases with increasing configuration coupling. Note also that the time scale for the decay is different than for the discretized Rabi oscillation mentioned above – the possibility of the core being in a superposition of states introduces an additional stabilization process.

A calculation of the radial probability density during the first orbit [5] reveals that in graph (c) the wave packet actually returns to the core after the first orbit being somewhat more narrow than when it was created –the tails are cut of. Similarly, only the tails of the wave packet survive the scattering in graph (d) forming two distinct wave packets. This would also be true with the same timing in the cavity case where the shutter is a classical object. However, the fundamental difference becomes apparent, as we realize that here $both$ tails orbit in optimal phase with the Rabi oscillation for surviving more orbits. That is, the position of the Rydberg electron is entangled with the state of the core, so that both wave packets are certain to scatter from an unexcited core. This can be illustrated by a special case occurring at specific times in graph (c) and (d): $|\psi\rangle = |inner; unexcited\rangle + |outer; excited\rangle$, i.e. a wave packet far from an excited core superposed with a wave packet scattering from an unexcited core. In general any superposition of localized wave packets each orbiting in optimal phase with the Rabi oscillation decayes slowly. The overall superposition may change in time, but for any core state a localized wave packet will always be found.

So far we have only considered the case of equal Rabi and orbit frequencies and zero detuning from ionic resonance. We must expect slow effective decay in the long-time limit whenever the Rabi-frequency is a multiple of the orbit frequency, since this allows wave packets to travel unattenuated. Figure 6(a) shows the effective decay rate of a wave packet as a function of the Rabi frequency in units of the orbit frequency. There are indeed

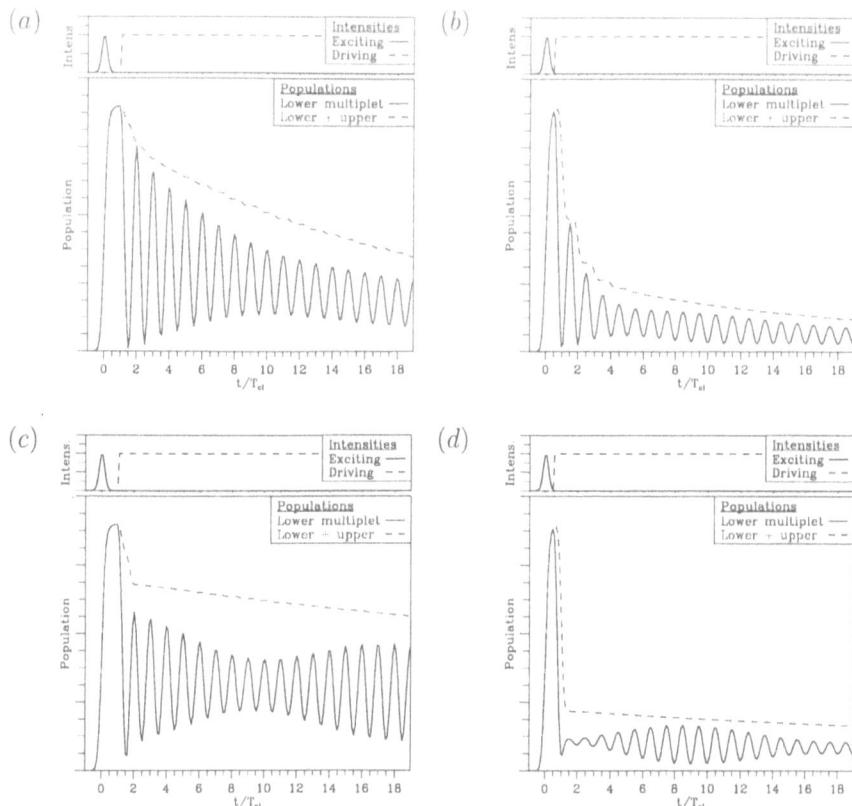

Figure 5. A wave packet is created by a pulse of duration $0.4\,T_{cl}$ centered in energy around the $\nu = 50$ Rydberg state. The quantum defects are both taken zero. The first two graphs (a,b) are calculated for an autoionization rate of $\Gamma = 0.9$ being increased to $\Gamma = 4.0$ in the last (c,d). The core resonant field is instantly turned on after a delay of (a,c) $1.0\,T_{cl}$ and (b,d) $0.5\,T_{cl}$. (a) Little decay occurs first during the orbit, since the center of the wave packet scatters from an unexcited core. On a longer time scale, the decay is signicicant but much smaller than the natural autoionizing decay. (b) Wave packet scatters from excited core causing significant but incomplete decay. The decay continues for more orbits but a slowly decaying part is left. (c,d) The increased configuration coupling causes more rapid decay during the first orbits, but increased stability on a longer timescale. The long time modulation of the oscillating population shows that not every phase of the problem is "locked".

prominent minima in the decay rate at $\Omega/\omega_{cl} = 0,1,2$. As the coupling gets stronger, the shutter is operated faster, and the structure becomes irregular with a decay which is in general decreasing. The details of the graph depend on the initial conditions.

314

Introducing a detuning from the ionic resonance has two effects: The Rabi frequency changes causing faster decay. However, as the detuning increases, less population is transferred to the autoionizing Rydberg series. Hence the decay decreases for large detunings. This is illustrated in fig. 6(b). The effect of a non-zero difference in quantum defect is from the two graphs seen to be most prominent off-resonance.

(a)   (b)

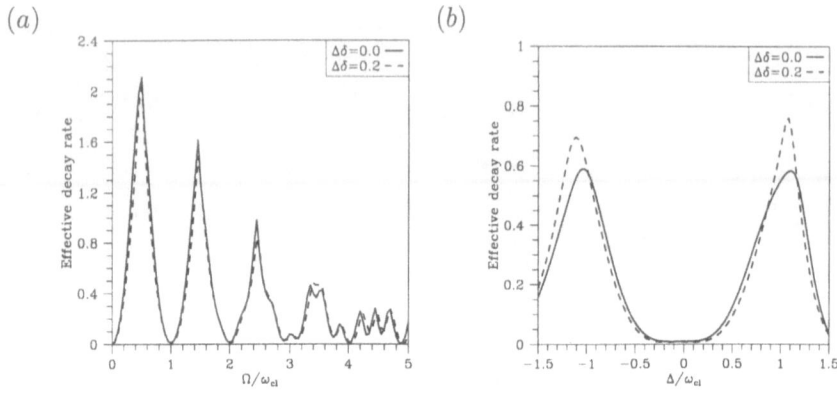

Figure 6. The effective decay rate per orbit as a function of (a) the Rabi frequency at zero detuning from ionic resonance and (b) the detuning for the zero detuning Rabi frequency equal to the clasical orbit frequency. The total bound population $P(t)$ is found at $t = 5\,T_{cl}$ and again at $t = 10\,T_{cl}$ after the fastest transients have decayed. We define the effective decay rate as the logarithm of the ratio of the two populations divided by the number of elapsed orbit times, i.e. $\Gamma_{eff} \equiv -\log(P(10\,T_{cl}))/P(5\,T_{cl}))/5$. For a purely exponential decay, this rate reduces to the normal decay rate, but this is not the case here. The graphs are plottet for two values of the difference in quantum defect. The other parameters are taken as in fig. 5 except that the driving field is turned on before excitation of the wave packet.

5. Probing

Observing wave packet phenomena requires a temporally narrow probe in comparison to the classical orbit time. A very nice feature of the more extended scheme we consider here, is that we do not necessarily need to add a narrow external probe: We already have one in the autoionization coupling being localized in the core region. The mere absence of decay is a signature of wave packet dynamics. Even measuring just the slow decay is, however, not a trivial task, since realistic pulse shapes tend to make the population decay: During the finite rise time the Rabi and orbit frequencies do not match. We have discussed this aspect and various solutions in detail elsewhere [5, 6].

Probing the non-dispersing character of the wave packets directly is an interesting task. The probing scheme mentioned above is *not* sufficient, but it can nevertheless be done. A *quantum non-demolition measurement* of the core state would tell us the state of the shutter and would leave us with a single wave packet (even from a classical point of view, we can not infer the location of the localized wave packet, if we do not know the state of the shutter). Another method is simpler by far: If the core-resonant field is turned off, part of the population will decay. The remaining population always survive in the form of a localized wave packet, which can be probed by normal techniques. This tells us that the the wave function really is a superposition of localized wave packets, entangled with appropriate states of the core.

6. Conclusion

We have demonstrated how wave packet dynamics can be controlled by means of an external field causing loss modulation. With a pulsed field discretizing the Rabi oscillation, a single wave packet can be kept from dispersing. We suggested a simple way of implementing this experiment. The picture got more complicated, as a constant core-resonant field was chosen. We found slowly decaying superpositions of localized wave packets each orbiting cores in different phases of the Rabi oscillation. The experiments are accordingly difficult, but within reach of present day technology as the required field strengths and time scales are quite moderate [5].

Van Druten and Muller developed the same formalism as us [19, 17]. In particular, they found a very similar set of differential equations. However, like the earlier work of Robicheaux [15], the focus was on spectroscopy rather than wave packet dynamics, and our work differs accordingly. The results are, nevertheless, of great interest to us, since actual experiments were carried out in magnesium [18]. Zobay and Alber [25] are treating the same problem as we have addressed here, but in a different formalism suitable for the case of a square core-resonant field. The emphasis is put on shake-up couplings and weak configuration interaction in combination with unmatched Rabi and classical orbit frequencies.

We are happy to acknowledge useful discussions with Morten Elk and Paul Maragakis. Thanks also go to the authors of reference [19] and [25] for making their articles available to us prior to publication and for lengthy discussions.

316

References

1. Alber, G., Ritsch, H., and Zoller, P. (1986) Generation and detection of Rydberg wave packets by short laser pulses, *Phys. Rev. A* **34**, 1058.
2. Alber, G. and Zoller, P. (1991) Laser excitation of electronic wave packets in Rydberg atoms, *Phys. Rep.* **199**, 231.
3. Broers, B., Christian, J. F., Hoogenraad, J. H., van der Zande, W. J., van Linden van den Heuvell, H. B., and Noordam, L. D. (1993) *Phys. Rev. Lett.* **71**, 344.
4. Cooke, W. E., Gallagher, T. F., Edelstein, S. A., and Hill, R. M. (1992) *Phys. Rev. A* **46**, 4347.
5. Hanson, L. G. (1995) Masters thesis, University of Copenhagen, Denmark.
6. Hanson, L. G. and Lambropoulos, P. (1995) Non-dispersing wave packets in two-electron atoms; atomic mode-locking by loss modulation, *Phys. Rev. Lett.* **74**, 5009.
7. Hanson, L. G., Zhang, J., and Lambropoulos, P. (1995) Theoretical interpretation of continuum-continuum Autler-Townes splitting (unpublished).
8. Hanson, L. G., Zhang, J., and Lambropoulos, P. (1995) Theory of core-resonant ionization, *Europhys. Lett.* **30**(2), 81.
9. Jaffe, S. M., Kachru, R., van Linden van den Heuvell, H. B., and Gallagher, T. F. (1985) Ba $6png$, $J = 3$ and $J = 5$ autoionizing states, *Phys. Rev. A* **32**, 1480.
10. Lambropoulos, P. and Zoller, P. (1981) *Phys. Rev. A* **24**, 379.
11. Marmet, L., Held, H., Raithel, G., Yeazell, J. A., and Walther, H. (1994). *Phys. Rev. Lett.* **72**, 3779.
12. Noordam, L. D. Ph.D. thesis, University of Amsterdam, The Netherlands.
13. Noordam, L. D., Duncan, D. I., and Gallagher, T. F. (1992) Ramsey fringes in atomic Rydberg wave packets, *Phys. Rev. A* **45**, 4734.
14. Parker, J. and Stroud Jr., C. R. (1986) Coherence and decay of Rydberg wave packets, *Phys. Rev. Lett.* **56**, 716.
15. Robicheaux, F. (1993) Atomic dynamics with photon dressed core states, *Phys. Rev. A* **47**, 1391.
16. Story, J. G., Duncan, D. I., and Gallagher, T. F. (1993) Spatially resolved transitions to autoionizing states, *Phys. Rev. Lett.* **71**, 3431.
17. van Druten, N. J. (1995) Ph.D. thesis, FOM-Institute for Atomic and Molecular Physics, The Netherlands.
18. van Druten, N. J. and Muller, H. G. (1995) Observation of Rydberg transitions induced by optical core dressing, *J. Phys. B (in press)*.
19. van Druten, N. J. and Muller, H. G. (1995) Rydberg transitions induced by optical core dressing, *Phys. Rev. A (in press)*.
20. Wals, J., Fielding, H. H., Christian, J. F., Snoek, L. C., van der Zande, W. J., and van Linden van den Heuvell, H. B. (1994) Observation of Rydberg wave packet dynamics in a coulombic and magnetic field, *Phys. Rev. Lett.* **72**, 3783.
21. Wang, X. and Cooke, W. E. (1991) Wave-front autoionization: Classical decay of two-electron atoms, *Phys. Rev. Lett.* **67**, 976.
22. Wang, X. and Cooke, W. E. (1992) Amplitude modulation of atomic wave functions, *Phys. Rev. A* **46**, R2201.
23. Wang, X. and Cooke, W. E. (1992) Wave-function shock waves, *Phys. Rev. A* **46**, 4347.
24. Yariv, A. (1988) *Quantum Electronics*, John Wiley and sons, New York.
25. Zobay, O. and Alber, G. (1995) Dynamics of electronic Rydberg wave packets in isolated-core excited atoms, *Phys. Rev. A* **52**, 541.

MECHANISM OF DIRECT DOUBLE IONIZATION OF HELIUM IN INTENSE LASER FIELDS

F.H.M. FAISAL and A. BECKER
Fakultät für Physik
Universität Bielefeld
D-33615 Bielefeld, Germany

1. Introduction

Surprisingly large signals for direct double ionization of He by intense laser light have been observed in a series of recent experiments [1-6]. The most dramatic observation of this effect is in the latest experiment by Walker et.at. [6], who measured the double ionization signal over a remarkable range of twelve orders of magnitude covering an intensity domain between 10^{14} W/cm^2 and 10^{16} W/cm^2 at a wavelength of 780 nm. Perhaps the most unexpected aspect of this result is that the observed double ionization signal *below* the experimental 'saturation intensity' (i.e. the intensity at which the initially present neutral target atoms in the interaction volume is depleted) is several orders of magnitude *greater* than that expected for the single ionization of the already once ionized atoms! The most interesting question is: what is the mechanism behind this anomalously large signal?

The first thing to note is that the process of interest is an extremely non-linear one: at least some 50 photons at an wavelength of 780 nm must be absorbed, by a single He atom, to overcome the unperturbed double ionization threshold of He, at 79.02 eV. In the research area of intense field atomic dynamics there are other examples which appeared to be equally surprising at the time when they were initially predicted theoretically or observed experimentally. To name only the most prominent among them: the above-threshold-ionization (ATI) and the adiabatic stabilization (AS) phenomena. The existence of ATI was discovered by Agostini et. al. [7] and soon confirmed by Muller et al [8]; it was predicted within the so-called KFR-theory [9-11], initially at a time long before the advent of strong enough laser fields necessary for its observation. AS was initially predicted using the so-called high frequency model [12,13] and subsequently by *exact numerical* solution of the Schrödinger equation for the H atom (3D) in intense laser fields[14-17]; a first experimental indication of its existence has been recently found by de Boer et.al. [18]. In the former case (ATI) a whole sequence of peaks is observed in the energy spectrum of the ejected electron, that lie above the Einsteinian photoeffect peak nearest to the ionization threshold; in the latter case (AS) the ionization probability per unit time can *decrease* with *increasing* intensities. These phenomena are now understood to be due to the highly non-linear nature of the atom-field interaction and at the same time they can be both analyzed in terms of the independent electron hypothesis for this interaction; generally this is done within an *active single electron* assumption for non-hydrogenic systems or without it only for the hydrogenic systems. Recent observations of the laser induced direct double ionization of He bring in a *new* element in intense field physics, namely that of a possible complete break-down of the active single electron hypothesis due to a potentially dominant role of electron-electron correlation for this fundamentally two-electron process. The theoretical challenge is

317

H. G. Muller and M. V. Fedorov (eds.), Super-Intense Laser-Atom Physics IV, 317–330.
© 1996 *Kluwer Academic Publishers.*

318

to simultaneously account for the *combined* influence of Coulomb correlation and field non-linearity during this process. The observation of laser induced direct double ionization in He atom is of considerable theoretical interest not only because it involves the simplest of the real two-electron atomic systems available but also because a detailed understanding of the process in this system is probably essential for a proper understanding of the response of many-electron systems (ranging from complex atoms to solids) to intense laser fields under similar conditions.

2. Two Heuristic Pictures

Before proceeding further let us briefly consider two recently proposed qualitative pictures for the double ionization of He in intense laser fields: (I) the so-called 'shake-off' picture, and (ii) the 'quasi-static tunneling and rescattering' picture. According to the first picture, proposed by Fittinghoff et.al.[1], at first one of the two electrons absorbs the laser photons and reaches the continuum very rapidly while the other electron, being unable to adjust to the resulting rapid change in the effective charge of the core, will be 'shaken off ' the atom. According to the second picture, suggested first by Corkum [19], at first one of the electrons will tunnel out of the atom with zero initial energy, due to a supposed quasi-static influence of the strong electric field of the laser, then it would behave classically as a free electron in the laser field and absorb the maximum possible classical energy, $E_{max.} = 3.2$ U_p, from the (say, linearly polarized) field during the first half-cycle and propagate away from the core. But in the next half-cycle, as the direction of the field will change, the electron will return near to the core and 'kick off ' the second electron by a 'e-2e'-like scattering process from the ground state of He^+ ion.

Note the following characteristic differences between these two pictures or models. The first model does not make any explicit distinction between the linear and circular polarizations of the laser field. The second model, in contrast, gives specific importance to the linear polarization for the assumed effective reencounter with the core electron, and thus predicts significant double ionization for the linear polarization but *not* for the circular polarization case. A second important difference is that mechanism (i) does not imply any specific 'threshold intensity' for the double ionization of Helium, but mechanism (ii) predicts a threshold (or cut off) intensity for double ionization corresponding to $E_{max} = 3.2$ Up = E_B = 54.4 eV, where $U_p = I/4\omega^2$ is the socalled 'quiver energy' and E_B is the binding energy of the He^+ core. A third difference concerns the frequency and time dependence; mechanism (i) does not restrict itself to a specific range of the frequency. It requires, however, a short enough ejection time of the first electron (of the inverse order of the binding frequency of the He^+ ionic core). Mechanism (ii), on the other hand, appears to be restricted to the lower range of available laser frequencies, in view of the requirement of quasi-static tunneling of the first electron, but the over all time scale involved could be larger in this model, e.g. of the order of the inverse photon frequency, than in the 'shake off' picture.

The experiments done with linear- and circular polarizations [1-6] showed strongly reduced signals for the circular polarization case. It also seemed that a heuristic combination of the so-called ADK tunneling rate [20] and the known 'e-2e' cross sections [21] could be used to fit [19] the initially measured rates of double ionization of He [13,15-17]. These facts generated much interest in favor of the second mechanism. But with the latest precision measurements of double ionization of He by Walker et.al. [18], that extended dramatically the observed range of signals by as much as 6-orders of magnitude (made possible by the development of kHz repetition rates of the 780 nm laser) and covered an intensity range of more than an order of magnitude, revealed *no* sign of any cut-off of the double ionization signal for intensities decisively below the 'cut-off intensity', required by the second model. Thus at present neither model (i) nor model (ii) can satisfactorily explain the observed signals of the double ionization of He in intense laser fields.

An exact numerical simulation of double ionization of a two-electron atom in intense laser fields poses formidable practical difficulties in view of the fact that the fundamental Schrödinger equation of the system consists of a (six + one)-dimensional partial differential equation; the large size of the space-time grid over which the wave function needs to be propagated in the configurational space (and time), in order to obtain experimentally relevant information, remains as yet practically untractable. More over, numerical solutions, despite their other virtues, are often rather poor in obtaining physical insights into the mechanism behind a new phenomenon. In view of these, in this paper we present a two-electron diagrammatic perturbation theory, appropriate for intense fields, and apply the leading terms of the associated S-matrix series to analyze the double ionization process under experimentally relevant conditions. Use of Feynman diagrammatic technique will prove to be particularly convenient in view of the complexity of the two-electron S-matrix series. The results of numerical calculations will be compared with the recent experimental data, that will allow us to identify the dominant mechanism behind the laser induced direct double ionization process.

3. Intense-Field Two-Electron Diagrammatic Perturbation Theory: Diagrams for Single- and Double Ionization of a Two-Electron Atom

In this section we develop and present an 'intense-field two-electron diagrammatic perturbation theory (IFTEPT)'. It will permit us to take *both* the highly non-linear electron-photon interaction as well as the electron-electron correlation into account. The usual perturbation theory for multiphoton ionization of atoms is based on the expansion of the wavefunction in terms of the amplitude of the light field (or equivalently, of the vector potential) and it is known to breakdown for field strengths above 10^{13} W/cm^2 for usual laser frequencies (e.g., p.51[22]). We shall, therefore, rewrite the perturbation theory in such a way that two-electron evolution of the He atom in the laser field can be analyzed systematically through the use of a six-dimensional intense field 'Volkov-cum-ionic-propagator', $G^0(t,t')$, as shown below.

The Schrödinger equation of the system "He atom + laser field" is (below we use, unless stated explicitly otherwise, Hartree a.u. $|e|=m=\hbar=1$):

$$id/dt \ \Psi(r_1,r_2;t) = H(t) \ \Psi(r_1,r_2;t) \tag{1}$$

where,

$$H(t) = H_i^0 + V_i(t) \tag{2}$$

is the total Hamiltonian of the system, and

$$H_i^0 = 1/2 \ p_1^2 + 1/2 \ p_2^2 - z/r_1 - z/r_2 + 1/r_{12} \tag{3}$$

is the Hamiltonian of the unperturbed He atom. Thus, the associated 'initial state interaction' is:

$$V_i(t) = v_1(t) + v_2(t)$$

where $v_j(t) = -1/c p_j.A(t) + 1/2c^2 A(t)^2,$ $\quad$ (j=1,2) $\qquad$ (4)

The total Green's function, $G(t,t')$, of the system is defined by

$$[id/dt - H(t)] \, G(r_1,r_1';r_2,r_2';t,t') = \delta(t-t') \, \delta \, (r_1-r_1') \, \delta(r_2-r_2') \qquad (5)$$

A formal solution of Eq.(1), evolving from an arbitrary initial state $\Phi^0_i (r_1,r_2) e^{-iE_i t}$, where

$$H_i^0 \, \Phi^0_i (r_1, r_2) = E_i \, \Phi^0_i (r_1, r_2) \qquad (6)$$

can be written down with the help of $G(t,t')$ as:

$$|\Psi_i (t)> = |\Phi^0_i (t)> + \, S \, dt' \, G(t,t')V_i (t') |\Phi^0_i (t)> \qquad (7)$$

(The validity of Eq.(7) can be verified at once by operating with [id/dt - H(t)] on the left and using the definition (5) of $G(t,t')$.)

The form of the intense field two-electron perturbation (IFTEPT) series we are seeking here is obtained by choosing the zeroth order part, $G^0(t,t')$, of $G(t,t')$ as that associated with the propagation of one electron in the Volkov-state and the other electron in the eigenstates of the ion. (For a one-electron system, clearly the propagation would be only in the Volkov state of the single electron, as in the well-known KFR-theory [1-3]). Thus, we split $G(t,t')$, as:

$$G(t,t') = G^0_f(t,t') + \, S dt_1 \, G^0_f(t,t_1)V_f(t_1)G(t_1,t'). \qquad (8)$$

where , we partition $H= H^0_f + V_f$, so that substituting in (7) we get

$$|\Psi_i (t)> = |\Phi^0_i (t)> + S dt' \, G^0_f (t,t')V_i (t') |\Phi^0_i (t)>$$

$$+ S dt' \, S dt_1 \, G^0_f (t,t_1)V_f(t_1)G(t_1,t')V_i(t')|\Phi^0_i (t')> \qquad (9)$$

And also in terms of an arbitrary interaction U^0, defined below, we can expand

$$G(t,t') = G^0(t,t') + \, S dt_1 \, G^0(t,t_1)U^0(t_1)G^0(t_1,t')$$

$$+ S dt_1 \, S dt_2 \, G^0(t,t_1)U^0(t_1)G^0(t_1,t_2)U^0(t_2)G^0(t_2,t') + ... \qquad (10)$$

where, $G^0(t,t')$ is defined by:

$$\{id/dt - H^0(t)\}G^0(r_1,r_2,t; r_1',r_2',t') = \delta(t-t') \, \delta \, (r_1-r_1') \, \delta(r_2-r_2') \qquad (11)$$

with, $\quad H^0 (t) = H_{ion}(r_2) + [1/2(p_1 - 1/cA(t))^2]$ $\qquad$ (12)

and the interaction

$$U^0(t) = H(t) - H^0(t) = 1/r_{12} + [-z/r_1 - 1/c \, p_2.A(t) + (1/2c^2)A(t)^2] \qquad (13)$$

where, $H_{ion}(r_2)$ is the unperturbed Hamiltonian of the He$^+$ ion and $[1/2(p_1 -1/cA(t))^2]$, that of the free electron interacting with the laser field. Thus, substituting (10) in (9) we get the expansion of the total wavefunction in the form:

$|\Psi_i(t)>=|\Phi^0_i(t)>$

$\qquad + S\,dt'\,G^0_f(t,t')V_i(t')|\Phi^0_i(t')>$

$\qquad + Sdt_1\,Sdt'\,G^0_f(t,t_1)V_f(t_1)G^0(t_1,t')V_i(t')|\Phi^0_i(t')>$

$\qquad + S\,dt_2\,Sdt_1\,Sdt'G^0_f(t,t_2)V_i(t_2)G^0(t_2,t_1)U^0(t_1)G^0(t_1,t')|\Phi^0_i(t')> +\ldots\ldots (14)$

Projecting on to the final eigenstate corresponding to the final Hamiltonian H^0_f, we get the S-matrix expansion:

$S(t,-00) = <\Phi^0_f(t)|\Psi_i(t)>$

$\qquad = S_0 + S_1 + S_2 + S_3 \ldots\ldots\ldots\ldots\ldots$

with

$\qquad S_0 = <\Phi^0_f(t)|\Phi^0_i(t)>$,

$\qquad S_1 = -iSdt' <\Phi^0_f(t')|V_i(t')|\Phi^0_i(t')>$,

$\qquad S_2 = -iSdt_1\,Sdt'<\Phi^0_f(t_1)|V_f(t_1)G^0(t_1,t')V_i(t')|\Phi^0_i(t')>$,

$\qquad S_3 = -iSdt_2\,Sdt_1\,Sdt'<\Phi^0_f(t_2)|V_i(t_2)G^0(t_2,t_1)U^0(t_1)G^0(t_1,t')V_i(t')|\Phi^0_i(t')>$,

$\qquad \ldots\ldots\ldots\ldots\ldots\ldots$ (15)

From the S-matrix series, Eq. (15), we observe that the first term, S_0, is merely a non-singular overlap between the initial and the final reference states and is negligible asymptotically, compared to the second and the successive terms (which all contain the singular delta function associated with the over all energy conservation, asymptotically at large times, at the end of the transition process). Hence the leading contributing terms in this expansion are the second, S_1, and the third, S_2, terms. (To the formally interested reader we may point out that an 'ordering' of the series arises, in terms of the potential U^0, only from the fourth term onward; hence the second and the third terms *together* form the non-trivial 'leading term' of the series, in the strict sense).

This is a new form of the expansion of the S-matrix series and is much more flexible than the well-known conventional expansion of the S-matrix series either in the 'prior' (initial state interaction) or in the 'post' (final state interaction) forms. The flexibility of the present expansion arises from the fact that the reference states in the initial and the final states do *not* have to belong to the same reference Hamiltonian and yet *any* desired propagator, G^0, can be introduced in the *intermediate* states of the expansion. Thus, in the present intense-field two-electron theory we shall choose the initial state to be that of the He-atom and the final state to be given by any asymptotic wavefunction (Coulombic included) describing the state of two ejected electrons with momenta k_a and k_b and the intermediate propagator, G^0, to represent the propagation of one electron in the Volkov state and the other electron in the complete set of virtual states of the residual ion. The propagator G^0 is obtained by solving Eq.(11) using standard techniques as:

$$G^0(r_1,r_2,t; r_1',r_2',t') = -i\theta(t-t') S_j \ (2\pi)^{-3} \ S_k \ \phi_j^{\ c} \ (r_2,t)\phi^v(k,r_1,t)\phi_j^{\ c*}(r_2',t')\phi^{v*}(k,r_1',t') \qquad (16)$$

where $\phi_j^{\ c} \ (r_2,t)$ and $\phi^{(v)} \ (k,r_1,t)$ are the jth eigenstate of the ion and the Volkov state of the free electron of momentum k, respectively. For the usual monochromatic radiation field (linear polarization), Eq. (16) has the explicit Floquet representation (c.f. [22-24]):

$$G^0(r_1,r_2,t; r_1',r_2',t') = -i \ \theta(t-t') S_N \ S_n \ \exp\{-iN\omega t - i(k^2/2+E_j+U_p-n\omega)(t-t')\}$$

$$.S_j \ (2\pi)^{-3} \ S_k \ [\phi_j^{\ c} \ (r_2)\phi^0(k, r1)J_{n-N}(\alpha_0.k;U_p/2\omega). \ J_n(\alpha_0.k;U_p/2\omega)\phi_j^{\ c} \ *(r_2')\phi^{0*}(k, r_1')$$

$$\qquad (17)$$

where $J(a;b)$'s are the generalized Bessel functions of two arguments (e.g. [22,chap.1;23]); $\phi^0(k,r_1)$ is the plane wave state (momentum k), and $E_B = -E_i$ is the total binding energy of the He atom.

In view of the complexity of the two-electron theory, we now analyse diagrammatically the leading contributions arising from the present intense-field S-matrix expansion (15):

i) The term S_1 gives rise to two diagrams giving contributions only in the case of allowed *single* photon transitions via the interaction v_1 or v_2, Eq. (4). This term therefore exactly reproduces the one-photon transition amplitude according to the well-known first order perturbation theory (and the associated Fermi golden rule), for example, for the double ionization of He by weak synchrotron radiation. In the case of two- or more photon transition, this term vanishes identically. The simple Feynman diagram associated with this term can be represented as:

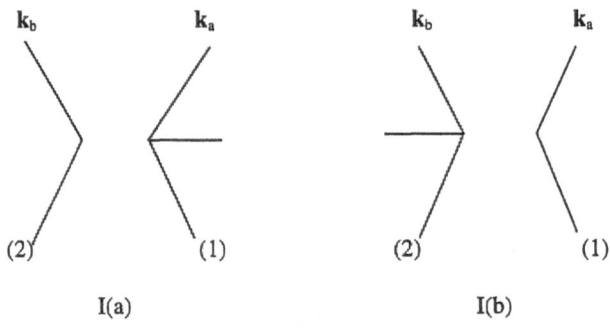

Fig.1 The two Feynman diagrams I(a), and I(b), arising from the first term S_1 of the IFTEPT for the one-photon double-ionization of a two-electron atom.

ii) The term S_2 includes the leading diagrams for the *non-linear* processes at any intensity. This term, in fact, brings in the electron-electron correlation for the first time in the expansion (15). There are in fact nine diagrams in this case which are shown and discussed below. Note that the first four diagrams, II(a), II(b), II(c) and II(d) are disjoint diagrams and, therefore, can not contribute significantly for simultaneous double ionization process which require a direct interaction joinig the electrons. The fifth and the sixth diagrams II(e) and II(f) are of the latter kind. Consider first the diagram II(e) which shows that the first electron interacts with the photon field and goes into an intermediate Volkov state of momentum **k**, while at the same time, the second electron propagates in the eigenstates, {j}, of the He$^+$ion. (The joint propagation in the intermediate states is governed by, as indicated, the Green's function G^0). This is the major conjoint diagram. During the propagation, the first electron shares energy with the second electron through the Coulomb correlation, $1/r_{12}$, until both of them have enough energy to escape together (with the respective momenta $\mathbf{k_a}$ and $\mathbf{k_b}$). We note that the diagram II(f) corresponds to the unlikely process in which the second electron interacts with the photon field causing the first electron to go into an intermediate Volkov state and thus constitutes a minor conjoint diagram which we shall neglect below.

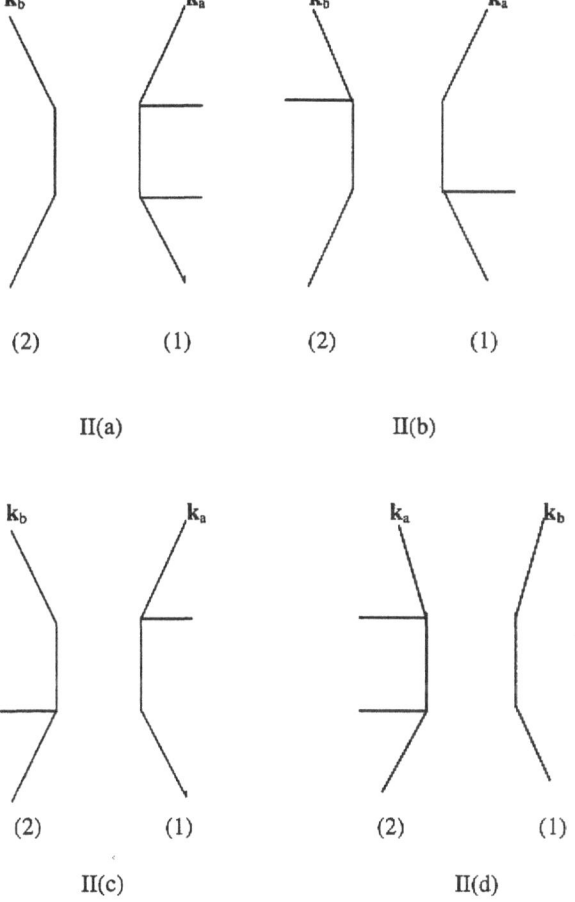

Fig. 2. The four disjoint Feynman diagrams, II(a), II(b), II(c) and II(d) for double ionization of He, which arise in the second leading second term S$_2$ of the IFTEPT. Note that the diagrams II(c) and II(d) arise simply from the diagrams II(a) and II(b) by exchanging the first interaction with electron (1) by that with electron (2). The time is assumed implicitly to run from below upward.

324

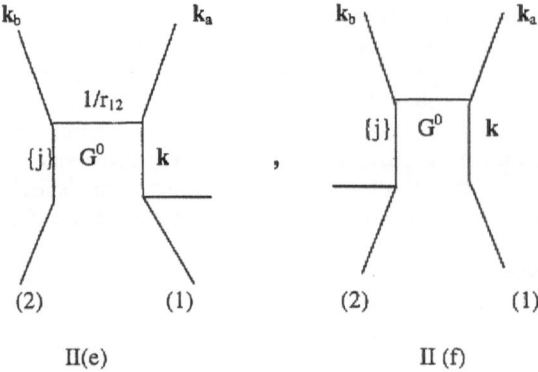

II(e)   II (f)

Fig.2 (contd.) The major, II(e), and the minor, II(f), conjoint Feynman diagrams for double ionization of He in an intense laser field. Note that while II(e) describes the likely amplitude for electron (1) to get into the intermediate Volkov state (momentum k) after iteracting with the field, the diagram II(f) correspons to the much less likely process of electron (1) to enter the Volkov state when electron (2) interacts with the field. The time is assumed implicitly to run from below upward.

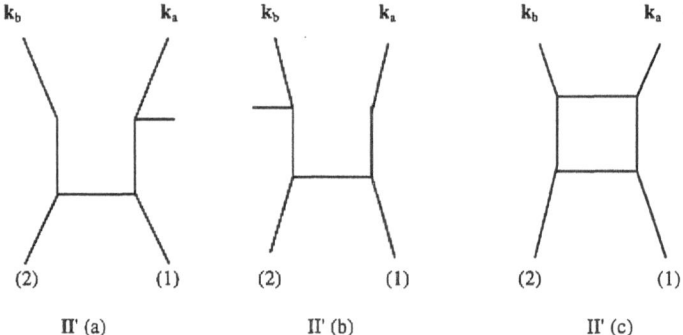

II' (a)   II' (b)   II' (c)

Fig.2 (contd.)   The three additional diagrams in S_2 involving the initial state correlation, (when the initial state is given by the product of independent one-electron orbitals); these diagrams do not apppear if the initial wavefunction is chosen to be a correlated wavefunction. The time is implicitly assumed to run from below upward.

The diagrams II'(a,b,c) appear explicitly in S_2 if the initial sate wavefunction is considered to be a product of two independent particle orbitals. Thus they do *not* arise if the initial state correlation is included in the (intial) wavefunction of the He atom, as may bee assumed in general.

It is immediately clear from the diagrams II(a)-II(d) that they are disjoint i.e there are no line joinig the two electron beyond the initial state and hence as a result the contribution from these diagrams requiring simultaneous escape of the two electrons is expected to be negligible compared to that from any diagram which involve interactions joining the electron lines beyond the initial state. Such conjoint diagrams in the leading order are given by II(e) and II(f) only. Note further that from these two conjoint diagrams, the diagram II(e) will dominate over the diagram II(f), as can be seen most clearly from the likely and unlikely sequence of interactions appearing respectively in these diagrams. Thus in II(e) one observes that the electron (1) interacts with the field which expectedly causes it to go over into the intermediate Volkov state (momentum **k**) . This circumstance is intuitively very likely. The diagram II(f), on the other hand, corresponds to the intuitively unlikely circumstance in which electron (2) interacts with the field but electron (1) to go over into the Volkov state. This is possible quantummechanically as a virtual case but is clearly may not dominate over the natural sequence of II(e). We are thus able to identify qualitatively, from diagrammatic considerations alone, that the diagram II(e) is the dominant diagram in the leading term of the new S-matrix series for the simultaneous double-escape process for a two-electron atom e.g. the He atom subjected to intense laser fields. We shall concentrate therefore in the rest of this work on diagram II(e) and bring it next into a form suitable for actual calculations. To this end we use the Floquet representation for the Green's function $G^0(t,t')$, Eq.(13), and carry out the double integrations over t_1 and t' (by suitably employing partial integrations and noting the vanishing surface terms) we can write it down in the form:

$$\text{Diag. II(e)} = -2\pi i \ \Sigma_N (E_a + E_b + E_B - N\omega) \ T^{(N)}, \tag{18}$$

$$\text{with, } T^{(N)} = \Sigma_n \ (2\pi)^{-3} S_k S_j \ \{ \ <\phi^c(\mathbf{k}_a, \mathbf{r}_1) \ \phi^c(\mathbf{k}_b, \mathbf{r}_2) | 1/r_{12} | \ \phi_j^c(\mathbf{r}_2) \ \phi^0(\mathbf{k}, \mathbf{r}_1)>$$

$$.(U_p - n\omega)(k^2/2 + E_j + E_B + U_p - n\omega + i0)^{-1}$$

$$.J_{n-N}(\alpha_0 \cdot \mathbf{k}); U_p/2\omega). \ J_n(\alpha_0 \cdot \mathbf{k}; U_p/2\omega) \ <\phi_j^c(\mathbf{r}_2)\phi^0(\mathbf{k}, \mathbf{r}_1) | \Phi_i^0(\mathbf{r}_1, \mathbf{r}_2)> \} \tag{19}$$

where, the angular brackets denote integrations with respect to the coordinates.

All the physically interesting quantities, such as the angular- and energy distributions of the ejected electrons, as well as the integrated total probability of double ionization, can now be readily expressed in terms of the basic expressions of the N-photon T-matrix elements above. Thus the five-fold angular correlation distribution, corresponding to one electron emering with the momentum $\mathbf{k}_a$ and at the same time the other electron is emerging with the momentum $\mathbf{k}_b$ is given by:

$$dW / dE_b \ d\Omega_a \ d\Omega_b = (2\pi)^{-5} \ \Sigma_N \ |T^{(N)}|^2 \ k_a k_b \tag{20}$$

with $T^{(N)}$ defined by Eq.(24); $k_a = (2(N\omega - E_B - k_b^2/2))^{1/2}$ is a real momentum determined by the energy conservation; $d\Omega_a$ and $d\Omega_b$ are the elements of the solid angles of ejection of the two electrons.

The total probability per unit time of double ionization per atom is obtained by integrations over the energy dE_b and the solid angles $d\Omega_a$ and $d\Omega_b$.:

$$\Gamma = \Sigma_N \ S \ S \ S dE_b \ d\Omega_a \ d\Omega_b \ (2\pi)^{-5} \ |T^{(N)}|^2 \ k_a k_b \tag{21}$$

4. Numerical Results and Comparison with Experiments

The integration of the double ionization amplitude, Eqs.(18,19), is non-trivial. Thus, on carrying out the six-fold integrations over the coordinates of the two electrons in the expression for the T-matrix element Eq.(19), we are left with three integrations over the intermediate momentum, **k,** of the Volkov state, an intermediate summation (integration) over the virtual states, {j}, of the He$^+$ ion, and a double infinite sum over the virtual photon numbers, n. Furthermore, after squaring this amplitude in the final expression, Eq.(21), for the total rate Γ, we must carry out the five-fold integrations over the solid angles of the two outgoing electrons and the energy of the second electron, E_b. Finally, an infinite sum over the actually absorbed final photon numbers, N, starting from the minimum number, N_0 (determined by the threshold), has to be carried out. These many-fold integrations (8-fold) and infinite summations (3-fold) present a formidable task and hence give in the present report give our preliminary estimates of the same using a combination of analytical and numerical methods along with a number of simplifications. Thus here we have specifically evaluated the on-shell intermediate amplitudes numerically exactly and have neglected the off-shell contribution. We found that the contribution from the intermediate ground state of the ion dominates greatly over that from the excited intermediate states. The latter contribution has been estimated by the closure technique (e.g.[22,§4.5]), and is found to be negligible. The initial state of the He atom is assumed to be the well-known simple Hylleraas wavefunction and plane waves of momentum k_a and k_b, instead of Coulomb waves, have been used to keep the calculations managable and the possibility of taking more sophisticated initial and/or final states with effective charges (e.g. [25]), is left open for the future investigations.

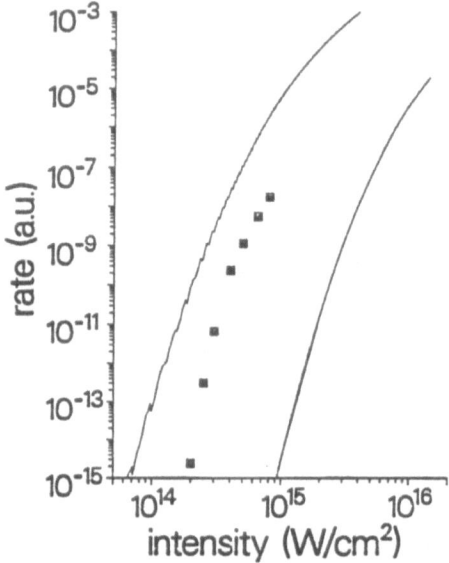

Fig.3 Theoretical rates of single ionization of He atom (left hand curve) and He$^+$ ion (right hand curve) obtained from one- and two-electron single ionization KFR-formula, *and* double ionization of He atom (solid squares) according to the present *ab initio* theory. The laser wavelength is 780 nm.

In Fig.3 we present the calculated theoretical rates of laser induced ionization for three different processes, namely (i) *single* ionization of (neutral) He using a direct generalization of the KFR-formula for single ionization for a two-electron atom, (ii) *double* ionization of (neutral) He using the present *ab initio* theory and (iii) *single* ionization of He$^+$ ion using the well-known KFR-formula. All of these calculations correspond to the laser wavelength of 780 nm, which has been used for the recent precision measurements of single- and double ionization of He by Walker et.al.[18]. The peak intensities of the field in our calculations are varied over an order of magnitude from about 10^{14} W/cm^2 upward. In Fig.3 the calculated rates of single ionization of He are shown by the full curve on the left hand side. The calculated rates for the double ionization He, obtained from the (conjoint) diagram II(e), are shown by the solid squares. The single ionization rates of He$^+$ ion are shown by the full curve on the right hand side. A comparison results of the present *ab initio* theory predicts, for the first time, a many orders of magnitude greater rates for double ioniztion in the intensity range between about 10^{14} W/cm^2 to about 10^{15} W/cm^2 than the probability of *single* ionization of He$^+$ ion (right hand curve). This is a highly counter-intuitive result, in view of the fact that the threshold of single ionization of He$^+$ ion is only 54.4 eV, while that of the double ionization of He atom is about 79 eV. This result clearly suggests that it is many orders of magnitude more probable to transfer over 79 eV of energy from the field to doubly ionize the neutral He atom than to transfer over only 54.4 eV to singly ionize the already once-ionized He atom! This anomalously high probability of double ionization of He atom is, in fact, what has been observed in a series of recent experiments [13-18] with intense laser fields, and most recently in the precision measurements by Walker et.al.[18] at 780 nm laser wavelength and for intensities starting with about 10^{14} W/cm^2 upwards. The same wavelength and the intensity region are chosen in the present calculations for the sake of comparison. These measurements are particularly significant as they cover an extraordinary range of some *twelve* orders of magnitude of signals, including those at intensities much below the saturation region.

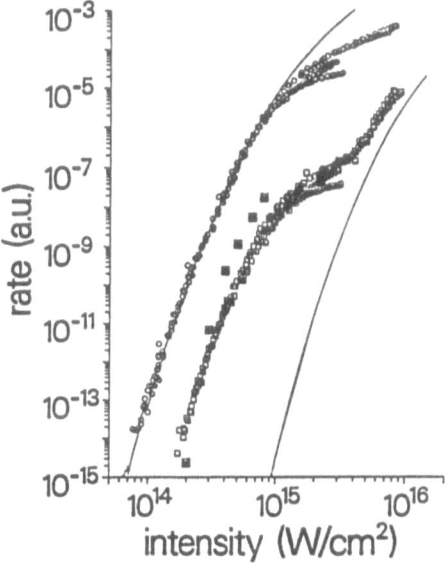

Fig.4. Comparison of the single ionization of He (left hand curve; KFR-model)), single ionization of He$^+$ (right hand curve; KFR-model) and double ionization (solid squares) rates of He atom obtained by approximate calculations based on the present *ab initio* theory, and the experimental single ionization yield (open circles) and the double ionization yield (open squares), as measured by Walker et.al.[18], in He gas. The laser wavelength corresponds to 780 nm. Note that the double ionization result like the experimental data show *no* sig n of an 'intensity threshold' of the process.

In Fig.4 we compare the results of the present approximate calculations with the said measurements [18]. In view of the arbitrary scale of the experimental signals we have simply rescaled the latter by setting the single ionization data-point at one intensity (chosen here to be I= 0.77×10^{15} W/cm^2) equal to our calculated value at that intensity. It should be noted, therefore, that this one-point match with the theory simultaneously rescales the entire set of experimental data involving *both* the single- and double ionization yields. We also point out that *no* additional adjustment of the data with respect to the intensity axis has been made above (as is some times done in view of the usual uncertainty in the intensity measurements in the experiments). This is the first time that the results of a theoretical prediction of the double ionization rates in the sub-saturation laser intensities (i.e. below ca. 0.8×10^{15} W/cm^2) is of the same orders of magnitude for the total rates of double ionization as observed experimentally. We note also that the calculated and the measured signals for the single ionizations are in general in good agreement with each other over a wide range of intensities. The result of the double ionization rates at the sub-saturation intensities is of particular interest since it extends to intensities lying far *below* the so-called 'threshold intensity' (at about 3×10^{14} W/cm^2 for the present wavelength of 780 nm) predicted by the 'quasi-static tunneling and rescattering' model discussed earlier. In fact, as can be seen from the comparison, the present theory, like the experimental data, does *not* support the existence of such a threshold intensity for the double ionization of He in intense laser fields.

5. Mechanism of Direct Double Ionization of He in Intense Laser Fields

This first orders of magnitude agreement of the predictions of the present theory with the experimental data for double ionization of (neutral) He at sub-saturation intensities allows us to unambiguously identify the main mechanism of the anomalous double ionization of He atoms in intense laser fields. It also provides us with a vivid intuitive picture of the mechanism of the process responsible for the predicted and observed anomalously large rates of the process. As discussed above the main contribution for the double ionization of He arises from the Feynman diagram II(e). A simple reading of this diagram itself, therefore, tells us what the dominant mechanism for the process is. This diagram distinguishes itself as being the dominant diagram in the leading term for the double-escape process of the present *ab initio* diagrammatic perturbation theory, which involves electron-electron correlation in the final state. Reading from below upward (i.e. in the assumed direction of the flow of time) we see that at first one of the two electrons interacts with the light field and absorbs *virtually* a large number of photons (and propagates in the Volkov state) while the other electron propagates *virtually* into the ionic states (mainly in the virtual ground state). At this intermediate stage of the process the crucial electron-electron correlation enters in action and mediates the sharing of the photon energy by one electron (which virtually absorbed a large number of photons by directly interacting with the field) with the other electron (and vice versa) until both of them get enough energy to escape together from the binding force of the nucleus. In this mechanism the Coulomb correlation plays a crucial dynamical role. Since otherwise, the electron that absorbs the large amount of photon energy from the field first, would escape alone *and* thus abort the double escape process. We add parenthetically that an explanation of the process in terms of tunneling alone would also fail due to the absence of electron-electron correlation in that mechanism. And, by construction, all types of 'single active electron' models, which have been very useful for the single ionization problems, remain out side the scope of a two-electron process. The present mechanism of the direct double ionization in intense laser fields is reminiscent of the model of 'e-2e'-like rescattering of a classical free electron in a laser field discussed in sec.2 above. It is worth observing, however, that the mechanism of double ionization presented here is fundamentally a *quantum* mechanical one which permits absorption of, in principle, an unlimited number of photons with varying probabilities at all intensities. This is also why there is no 'threshold intensity' for the process; a classical mechanism like the above mentioned model would in general break down due to a restriction on the maximum possible energy that an ejected electron can gain from the field and hence an 'intensity threshold', contrary to both experiment and the present quantum mechanism. The present theory shows clearly that the double ionization process in the sub-threhold intensities is, in fact, an example of a quantum collective phenomenon at its most rudimentary form: two-particle joint escape by energy sharing through the Coulomb correlation. Finally, note that the associated double ionization probability, at sub-saturation intensities, is *many* orders of magnitude than the single escape probability from the He$^+$ ion, and thus it constitutes one of the *largest* effects of electron-electron correlation observed in atomic physics.

6. Conclusions

To conclude this paper, we have developed an *ab initio* intense field two-electron perturbation theory and applied it to analyze the recently observed anomalously large probability of double ionization of He atom in intense laser fields. The double ionization processes in the leading term of this new intense field S-matrix expansion are discussed graphically and the dominant term is identified. Results of the priliminary evaluation of this diagram are compared with the experimental data. For the first time an orders of magnitude agreement is found between our priliminary calculations based on the new S-matrix theory and the experimental results for double ionization of He atom. The dominant mechanism for the double-escape is shown to be *quantum energy sharing through Coulomb correlation* in the final state. In this mechanism the energy due to the *virtual* absorption of a large number of photons by one electron, is transferred to the other electron (and vice versa) through the electron-electron *correlation*, until both electrons get enough energy to escape jointly. Thus, according to the present analysis the observed anomalously large signal of double ionization of He in, which is many orders of magnitude larger than the single ioization signals from the already once ionized He atom, turns out to be one of the *largest* effects of Coulomb correlation known in atomic physics.

7. Acknowledgments

We are thankful to L.F.DiMauro and P.Agostini for kindly making available to us their published data, on the single- and double ionization of He, in the numerical form, and for the stimulating correspondence on the results. We also take this opportunity to thank K.Kondo for kindly making available their data. This work was partially supported by the Deutsche Forschungsgemeinschaft, Bonn.

330

8. References

1. D.N.Fittinghoff, P.R.Bolton, B.Chang, and K.C.Kulander, Phys.Rev.Lett.69, 2642 (1992)
2. J.Peatross, B.Buerke and D.D.Meyerhofer, Phys.Rev. A47, 1517 (1993)
3. B.Walker, E.Mevel, B. Yang, P.Breger, J.P.Chambaret, A.Antonetti, L.F.DiMauro and P.Agostini, Phys.Rev. A48, R895 (1993)
4. K.Kondo, A.Sagisaka, T.Tomida, Y.Nabekawa and S.Watanabe, Phys.Rev.A48, R2531 (1993)

5. D.N.Fittinghoff, P.R.Bolton, B.Chang and K.C.Kulander, Phys.Rev. A49, 2174 (1994)
6. B.Waker, B.Sheehy, L.F.DiMauro, P.Agostini, K.J.Schafer and K.C.Kulander, Phys.Rev. Lett. 73, 1227 (1994)
7. P.Agostini, F.Fabre, G.Mainfray, G.Petite and N.K.Rahman, Phys.Rev.Lett. 42, 1127 (1979)
8. P.Kruit, J.Kimman and M.J. van der Wiel, J.Phys. B14, L597 (1981)
9. L.V.Keldysh, Zh.Ekps.Theor.Phys. 47, 1945 (1964) [Sov.Phys. JETP 20, 1307 (1965)]
10. F.H.M.Faisal, J.Phys.B5, L89 (1973)
11. H.R.Reiss, Phys.Rev. A22,1786 (1980)
12. J.I.Gersten and M.H.Mittleman, J. Phys. B9, 2561 (1976)
13. M.Gavrila and J.M.Kaminski, Phys.Rev. 52, 612 (1984); M.Pont and M.Gavrila, Phys.Rev.Lett. 65, 2362 (1990)
14. K.Kulander, K.J.Schafer and J.L.Krause, Phys.Rev.Lett. 66, 2601 (1991)
15. M.Dörr, R.M.Potvliege, D.Proulx and R.Shakeshaft, Phys.Rev. A43, 3729 (1991)
16. L.Dimou and F.H.M.Faisal, Phys.Lett A171, 211 (1992); Phys.Rev. A46, 4442 (1992); Laser Phys.3, 440 (1993); Acta.Phys.Pol. A86, 201 (1994); Phys.Rev. A49, 4564 (1994)
17. R.M.Potvliege and H.G.Smith, Phys.Rev. A48, R46 (1993)
18. M.P.deBoer, J.H.Hoogenraad, R.B.Vrijen, L.D.Noordam and G.H.Muller, Phys.Rev.Lett. 71, 3263 (1993)
19. P.B.Corkum, Phys.Rev.Lett. 71,1994 (1993)
20. M.V.Ammosov, N.B.Delone and V.P.Krainov, Zh.Eksp.Teor. Fiz. 91, 2008 (1986) [Sov. Phys. JETP 64, 1191 (1986]
21. H.Tawara and T.Koto, At. Data and Mol. Data Tables 36, 167 (1987)
22. F.H.M.Faisal, *Theory of Multiphoton Processes*, Plenum Press, N.Y. (1987), Chap.1
23. F.H.M.Faisal, *Floquet Green's Function Method for Radiative Electron Scattering and Multiphoton Ionization in a Strong Laser Field*, Computational Physics Reports, Vol.9, No.2 , North Holland, Amsterdam (1989)
24. A.Becker and F.H.M.Faisal, Phys.Rev. A50, 3256 (1994)
25. S. Jetzke and F.H.M.Faisal, J.Phys.B25, 1543 (1992)
26. M.R.Cervenan and N.R.Isenor, Opt. Commun. 13, 175 (1975)

NON-SEQUENTIAL TRIPLE IONIZATION OF ARGON ATOMS IN AN INTENSE 1053 nm LASER FIELD

A. TALEBPOUR, S. AUGST, Y. LIANG, S.L. CHIN, [+]Y. BEAUDOIN, [+]M. CHAKER
Center d' Optique, Photonique, et Laser (COPL), Dept. de Physique, Pav. Vachon, Universite Laval, Quebec, QC G1K 7P4, Canada
[+]*INRS-Energie et Materiaux, Montee St. Julie, Varennes, Quebec, Canada, J3X 1S2*

ABSTRACT - We investigate direct, non-sequential ionization of argon atoms using linearly polarized light at 1053 nm with 600 fs pulse length. We report for the first time the production of triply charged argon ions at intensities below the saturation intensity of the first charge state.

1. Background

Multi-photon ionization is a subject whose advancement follows closely that of laser technology. Its development represents a classic story in physical research that defies our imagination, challenges our mental capacity, and continuously creates new controversies after three decades of intensive activities in the field. The quantum mechanical theory of two photon absorption in the pre-laser years (Goeppert-Mayer 1931) was almost a pure exercise of quantum theory until the invention of the laser, when nonlinear optical phenomena became a reality in the laboratory. Following the invention of the Q-switched laser, multiphoton ionization of atomic gases was observed (Delone and Voronov 1965, 1966) which led to a large quantity of theoretical and experimental work in subsequent years. A review of the early development of the field can be found in Chin and Lambropoulos (1984). With the laser pulse becoming shorter and more intense, phenomena such as above threshold ionization, multiple ionization, tunnel ionization, high order harmonic generation, stabilization etc. were observed. Some of the most recent developments of these subjects can be found in Evans and Chin (1994). Hidden among these phenomena is the question of whether two or more electrons could be ejected simultaneously (or in a non-sequential way) as opposed to ejecting one electron after another during multiple ionization of atoms in a strong laser field. This question arose soon after the first observation of double ionization of alkaline earth atoms (Aleksakhin et. al. , 1979). L'Huillier et. al. (1983) were the firsts to have observed

331

H. G. Muller and M. V. Fedorov (eds.), Super-Intense Laser-Atom Physics IV, 331–336.

experimentally the signature of non-sequential (NS) double ionization as opposed to the mechanism of sequential ejection of two electrons from one atom. This experimental signature is represented by the observation of doubly charged ions in the intensity region below the saturation intensity of the first charged state. Subsequently, most other experimental results did not show this behavior because with hindsight , we now know that they did not go to low enough signal level to measure this phenomenon. This signature showed up in recent multiple ionization experiments of Augst *et. al.* (1991) and Auguste *et. al.* (1992). However no discussion was made then.

The renewed interest came from the unambiguous experimental observation of the signature of NS double ionization of helium by Fittinghoff *et. al.* (1992). This is followed by a number of new experimental and theoretical work published along with the proposal of various theoretical models to explain the behavior (Corkum, 1993 ; Walker *et. al.* , 1993 ; Kondo *et. al.* , 1993; Fittinghoff *et. al.* , 1994 ; Walker *et. al.*, 1994; Charalambidis *et. al.* , 1994 ; Dietrich *et. al.* , 1994 ; Becker and Faisal, 1994 ; Kulander *et. al.*, 1995). Two widely discussed models are the "shake-off "model (Fittinghoff *et. al.,* 1992) and electron re-scattering model (Curkum, 1993).

The "shake-off "model describes the NS process as a mechanism where one electron is ionized by the laser field and the departure of this electron is so rapid that the remaining electrons are "shaken up" enough to free an additional electron. This model conceptually is similar to β^- decay in nuclear physics, where during the process the nuclear charge increases by one unit. Since the decay time is very short (about 10^{-17} sec) , the electrons do not have enough time to adjust themselves to the new energy states, and therefore there is a certain probability that after the process some of the electrons are excited to states with higher energy or even ionized. Such an excitation can be justified by the uncertainty principle. In the case of ionization with high intensity field, when one electron is removed from the atom, its screening effect on the remaining electrons disapears and the net result is the same as increasing the nuclear charge by a certain amount. In principle, one might be able to evaluate the probability of double or triple ionization if the time duration of the ionization to first charge state was known. In our knowledge nobody has calculated this time interval. From this discussion, it follows that the shake-off model is wavelength dependent because the interaction or transition time is at least one cycle of the laser field oscillation. The model is not polarization dependent in principle, however, Walker *et. al.* (1994) have mentioned that the angular momentum that is absorbed by the electron with circularly polarized light could inhibit the transfer of energy to additional electron in the skake-off process. This would make the ionization enhancement less noticeable for cicularly polarized light than for linearly polarized light.

The re-scattering model proposed by Corkum (1993) is a model that has been used not only for NS ionization but it has also had a large degree of success in describing high harmonic generation and above threshold ionization (ATI). This model describes the NS ionization as a process whereby an electron is tunnel ionized. The electron then interacts with the laser field where it is accelerated away from the nuclear core. If the electron has been ionized at an appropriate phase of the field, it will return and pass by the position of the remaining ion half a cycle later where it can free additional electrons by electron impact. Only half of the electrons are released with the appropriate phase and the other half will never return to the nuclear core.

The electron impact ionization cross section is highly dependent on electron kinetic energy and drops dramatically for electron energies below the ionization potential of the bound electrons. The maximum and most probable kinetic energy that the returning electrons can have is 3.17 times the pondermotive potential (U_p) of the laser. This places a cutoff limit on the minimum intensity (intensity is proportional to U_p) where ionization due to re-scattering can occur. The production of ions at intensities lower than this cutoff intensity must be due to some process other than electron re-scattering. This lack of a cutoff intensity was indeed shown experimentally in the double ionization of helium (Walker *et. al.* 1994).

High order harmonic generation is so far ignored in the analysis of multiphoton/tunnel ionization. Yet, it always exisits in the intensity range below the saturation intensity of the creation of the first charge state (Li *et. al.* 1989). The contribution of high-harmonics to NS ionization could be in the form of absorption of a high harmonic photon in the presence of a strong laser field. Maquet *et. al.* (1995) has shown theoretically that the above threshold ionization (ATI) cross section is increased by several orders of magnitude when a high harmonic photon is absorbed by an atom in the presence of a strong laser field. The experimental results of Agostini *et. al.* (1995) indeed confirmed such an inhancement in the ATI cross section. It is thus reasonable to extend the above findings to strong field enhanced absorption of harmonic photons by an atom, leading to NS double or triple ionization. Because there are many high harmonics generated by an atom in the strong laser field, and each high harmonic could have many possible ways to combine with the strong laser field to provoke a NS ionization, it is possible that the superposition of these various channels could give rise to a significant (measureable) signal as observed by many experiments, including ours.

2. Present work

We have studied the NS ionization of argon using an Nd:Glass/Ti:Sapphire laser operating at 1053 nm with a pulse length of approximately 600 fs (Augst *et. al.* 1995). The results presented in figure 1 show evidence of NS triple ionization as well. A significant triply charged ion signal is present at laser intensities lower than the saturation intensity of both the second and the first charge states. This indicates the appearance of triply charged ions at intensities where there are still neutral atoms remaining in the laser focus. There are two possible channels through which the triply charged ions can be created: NS triple ionization of neutral atoms, or NS double ionization of singly charged ions. We conclude that both NS triple and NS double ionization is occurring. We have also seen the appearance of the fourth charge state at an intensity lower than the saturation intensity of the second charge state. This position is marked by an arrow in figure 1. Again, the 4+ ions probably result from both NS double and triple ionization.

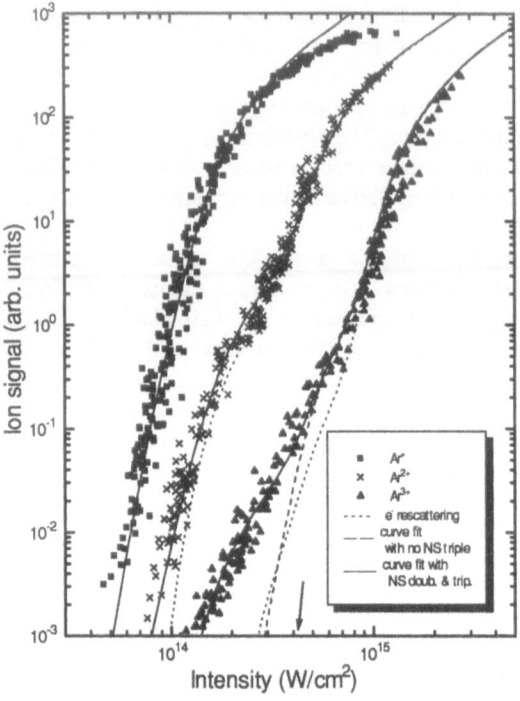

Figure 1. Argon ion signal as a function of incident laser intensity for a 600 fs 1053 nm linearly polarized laser palse. Each point represent a single laser shot. The calculated curves are shown as dotted line (ADK plus electron rescattering model), solid lines (ADK plus curve fit for NS double and triple ionization), and dashed lines (ADK plus fit for NS double ionization only). The experimentally measured intensities are multiplied by 1.2 for comparison to the calculatios.

Theoretical curves have been calculated for NS ionization models by including a channel for the production of doubly charged ions directly from neutrals and for triply charged ions from both the first charge state and from the neutral atoms. The populations of each charge state are described by four coupled rate equations. These equations are solved using the corrected ADK tunneling formula (Ammosov *et. al.* 1986) for sequential terms and remaining NS rates are estimated either by means of the re-scattering model or by numerically fitting.

The rescattering model calculations are shwon in figure 1 as dotted lines. The model produces significantly better agreement with data than a purely sequential model would, however, it clearly fails at the low intensity end of the third charge state and for the second charge state it can not completely account for the ion production in view of the fact that the calculation is an overestimate. This suggests that if the electron re-scattering process is contributing to the NS ionization, it is not the dominant factor.

We have also solved the coupled rate equations by using the corrected ADK formula for sequential ionization and using the remaining NS rates as free parameters in a curve fit. The solid lines in figure 1 shows the fit if all possible NS channels are included. However the data for the third charge cannot be fit without the inclusion of the rate for NS triple ionization as shown by the dashed line in figure 1. These suggest that NS triple ionization is occurring.

Due to time limitations on laser time allocated to this experiment it was not possible to obtain quantitative data for circular polarization, however, we have observed with circular polarization that the appearance of the second charge state does not occur until the signal from the first charge state has saturated. Likewise, the third charge state does not appear until the second charge state has saturated. These observations are consistent with the electron re-scattering model, any model involving resonont enhancement, and the shake-off model as discussed by Walker *et. al.* (1994).

We have studied the possibility of NS ionization of atoms by absorbing the harmonics, using a Ti:sapphire laser operating at 800 nm with a pulse length of 250 fs. The laser was focused using $f/100$ optics into an ultrahigh-vacuum chamber having a background pressure of 2×10^{-9} Torr. The chamber was filled with a mixture of krypton and xenon gases each with partial pressure of 1×10^{-4} Torr, using two separate precision leak valves. The intensity of the input field was set below the saturation intensity of Xe^+ and slightly above the appearance intensity of Xe^{2+}. For this intensity the Kr^+ signal is appreciable. If Xe^{2+} comes from absorption of high harmonics, then a fraction of Kr^+ signal must be due to the absorption of high harmonics produced by xenon atoms. In fact reducing the partial pressure of xenon had no appreciable effect on the observed Kr^+ signal.

3. Summary

In conclusion, our observation of triply charged argon atoms at laser intensities lower than the saturation intensity for both the first and second charge states are qualitatively consistent with a shake-off model, but no quantitative comparison can be made presently with this model. Ionization rates predicted by a re-scattering model are too low to account for the observed ion production. Finally, strong field enhanced absorption of high harmonics can not have significant contribution to the observed NS double or triple ionization.

4. Acknowledgement

We thank N.B. Delone, V.P. Krainov, M.V. Ammosov, S. Goreslavsky, F.A. Ilkov, A. Maquet, F. Faisal, and R. Shakeshaft for helpful discussions. This work is supported in part by NSERC, le Fonds-FCAR and NATO.

5. References

1. Aleksakhin, I.S., Delone, N.B., Zapesonchnyi, I.P., and Suran, V.V. (1979), *Sov. Phys. JETP*, English translation, **49**, 477.
2. Ammosov, M.V., Delone, N.B., and Krainov, V.P., (1986), *Sov Phys. JETP* **64**, 1191. corrections to some of the equations can be found in Ilkov, F.A., Decker, J.E., and Chin, S. L., (1992),), *J. Phys. B: At. Mol. Opt. Phys.* **25**, 4005.
3. Augst,S., Talebpour, A., Chin, S.L., Beaudoin, Y., and Chaker, M., (1995), *Phys. Rev. A*, August.
4. Augst, S., Meyerhofer, D.D., Strikland, D., and Chin, S.L., (1991), *J. Opt. Soc. Am. B*, **8**, 858.
5. Auguste, T., Monot, P., Lompre, L.A., Mainfray, G., and Manus, C.,(1992), *J. Phys. B: At. Mol. Opt. Phys.* **25**, 4181.
6. Agostini, P., *et. al.* (1995), "XUV-infrared multiphoton processes" invited paper, 15th International Conference on Coherence and Nonlinear Optics (ICONO), St. St. Petersborg, Russia, 27 June - 1 July, 1995.
7. Becker, A., and Faisal, F.H.M., (1994), *Phys. Rev. A*, **50**, R3256.
8. Charalambidis, D., Lambropoulos, P., Schroder, H., Faucher, O., Xu, H., Wagner, M., and Fotakis, C.,(1994), *Phys. Rev. A*, **50**, r2822.
9. Chin, S.L. and Lambropoulos, P. , eds. (1984), "Multiphoton ionization of atoms" , Academic Press, Toronto.
10. Corkum, P.B., (1993), *Phys. Rev. Lett.*, **71**, 1994.
11. Delone, N.B. and Voronov, G.S. (1965), *JETP Lett.* (English translation) **1** , 42.
12. Dietrich, P., Burnett, N.H., Ivanov, M., and Corkum, P. B., (1994), *Phys. Rev. A*, **50**, R3585.
13. Evans, D.K. and Chin, S.L., eds. (1994), "Multiphoton Processes", World Scientiphic, Singapore.
14. Fittinghoff, D.N., Bolton, P.R., Chng, B. and Kulander, K.C., (1992), *Phys. Rev. Lett.*, **69**, 2642.
15. Fittinghoff, D.N., Bolton, P.R., Chang, B., and Kulander, K.C., (1994), *Phys. Rev. A. Brief Reports* **49**, 2174.
16. Goeppert-Mayer, M. (1931), *Ann. Phys.* (leipzig) **9** , 273.
17. Kondo, K., Sagisaka, A., Tamida, T., Nabekava, Y., and Watanabe, S., (1993), *Phys. Rev. A*, **48**, R2531.
18. Kulander, K.C., Cooper, J., and Schafer, K.J., (1995), *Phys. Rev. A*, **51**, 561.
19. L'Huillier, A., Lompre, L.A., Mainfray, G., and Manus, C., (1983), *J. Phys. B : At. Mol. Phys.* **16**, 1363.
20. Li, X.F., L'Huillier, A., Ferray, M., Lompre, L., and Mainfray, G., (1989), *Phys. Rev. A*, **39**, 5751.
21. Maquet, A., Taieb, R., and Veniard, V., (1995), " Two-color multiphoton ionization of atoms with high order harmonics", invited paper, 15th International Conference on Coherence and Nonlinear Optics (ICONO), St. St. Petersborg, Russia, 27 June - 1 July, 1995.
22. Walker, B., Mevel, E., Yang, B., Breger, P., Chambaret, J.P., Antonetti, A., Dimauro, L.F., and Agostini, P., (1993), *Phys. Rev. A*, **48**, R 894.
23. Walker, B., Sheehy, B., Dimauro, L.F., Agostini, P., Schafer, K.J., and Kulander, K.C., (1994), *Phys. Rev. Lett.*, **73**, 1227.

A NEW MODEL FOR NON-SEQUENTIAL IONIZATION OF XENON ATOMS IN AN ULTRAFAST INTENSE Ti:Sapphire LASER FIELD

S.L.CHIN, A.TALEBPOUR, Y. LIANG, S. AUGST[+], AND C-Y CHIEN[*]

Centre d'Optique Photoniqe et laser (COPL), departement de Physique, Universite Laval, Quebec, Canada, G1K 7P4

[+]*Department of physics, Caltech, mail code103-33, Pasadena, California, 91125,USA*

[*]*Center for Ultrafast Optical Science, The University of Michigan, Ann Arbor, Michigan, USA*

Abstract

We observed structures in the Xe^+ ion vs. peak intensity plot using very stable 250 fs Ti-sapphire laser pulses and probing only the central section of the large focal volume of a f/100 focusing optics. These structures are related to the appearance of the Xe^{2+} and Xe^{3+} ions. A new scheme of non-sequential ionization related to stabilization is proposed.

Main text

The background of sequential versus non-sequential (NS) ejection of two or more electrons from an atom using intense laser pulses is already given in a separate paper in the present proceedings (Talebpour et al., 1995). The main conclusion in that paper is the

337

H. G. Muller and M. V. Fedorov (eds.), Super-Intense Laser-Atom Physics IV, 337–342.
© 1996 *Kluwer Academic Publishers.*

same as that of Augst *et. al.* (1995) and Walker *et. al.* (1994); i.e. NS ejection of two and three electrons from an atom cannot be described by all presently known models. These include the electron re-scattering model (Krause *et.al.* 1992; Corkum 1993), the shake-off model (Fittinghoff et al, 1992, Talebpour et al, 1995) and the model of enhanced absorption of high harmonics (Talebpour et al, 1995). The present paper was presented at the hot topics session at the SILAP IV meeting. It reported new and precise results on NS double and triple ionization of Xe atoms using transform limited ultrashort Ti-sapphire laser pulses. The experimental conditions are the following: λ= 800nm, $\Delta\lambda$= 12nm (e^{-2} width); repetition rate, 10 Hertz; pulse duration, 250 fs (e^{-2} width); focal length of the focusing lens, 100cm; beam diameter, 1cm; background pressure, 1×10^{-9} Torr; operating gas pressure ranging from 10^{-4} to 10^{-7} Torr.

Fig.1 shows the ion versus peak laser intensity on a log-log scale over 7 orders of magnitude. It represents the superposition of two independent sets of data taken at two separate times spaced by two weeks. Each point is a two-shots average. Single shot data shows the same smoothness in the plot. To the best of our knowledge, this represents the best plot of its kind since multiphoton ionization experiment was performed thirty years ago. To obtain such smooth data, we stabilize the operation of the laser system by controlling the temperature of the whole laboratory to ± 0.5 °C. This stabilizes the alignment precision of the laser beam such that once the laser is aligned no further alignment touch-up is necessary for many weeks to come. The spectrum of the laser as well as the pointing precision are always the same.

The first interesting observation of the smooth plots is that the first charged state Xe^{+} reveals some subtle structures. Should the plot be obtained under "normal" situations, the scattering of the points would be thicker than the structure. Any averaging technique would not reveal it. Such "fine" structures come about due to two experimentally chosen conditions. One is the above mentioned stability of the laser. The other is that we used a very long focal length lens (f=100 cm; f/number =100). In fact, the Rayleigh range of the beam is about 6 cm and our time-of-flight mass spectrometer collects only ions from the central section 1 cm in length (isointensity regions almost cylindrical). Normal experiments would collect all the ions from the whole Rayleigh range such that as the intensity of the

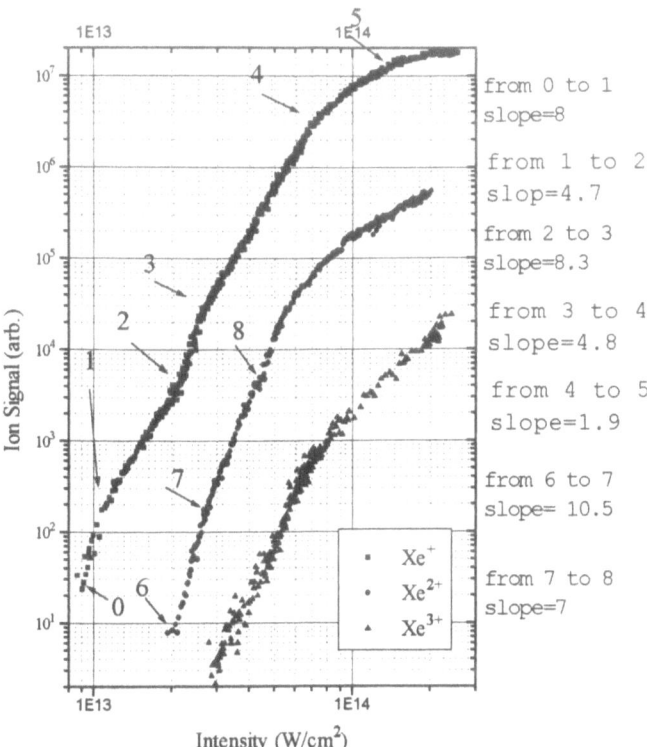

Fig.1 Triple ionization of Xe using 250 fs, 800 nm laser pulses. Note that the vertical ion number scale spans over seven orders of magnitude while the horizontal peak laser intensity scale spans over only one order of magnitude. The plot represents the superposition of two sets of data taken independently at two weeks interval. Each point is the average from two laser shots. The precision of the plot is extra-ordinary. This permits us to measure the various slopes shown in the figure without any ambiguity.

laser pulse increases, the fast expansion of the focal volume (Ma *et. al.* 1983) would result in masking any changes in the central part of the volume. Selecting the central zone that does not expand very fast minimizes such a masking effect and reveals the structure that might have been concealed. Previously in the MPI data published by Perry et. al. (1988), some similar small structure also occurs on the plot of the singly charged state of rare gas atoms using 586 nm dye laser pulses. This peculiarity was not emphasized probably because of the scatter of the data that could not justify a reasonable discussion.

The Xe^+ plot starts as a straight line (Fig.1, from 0 to 1) of slope 8 in the low intensity and thus low probability region. This corresponds exactly to theoretically expected eight photon ionization of Xe. From 1 to 2, the slope reduces to 4.7, indicating a reduction in the rate of ionization as the peak intensity increases. This is not saturation (or depletion of neutral atoms) in the center of the focal volume because the probability of ionization is still very low. The slope increases to 8.3 between 2 and 3 and reduces again to 4.8 between 3 and 4. After 4, saturation (depletion of neutral atoms) sets in. Clearly NS ionization (2 and 3 electron ejections) occurs below the saturation intensity at point 4.

The second observation is that Xe^{2+} starts to occur around the intensity at the turning point 2 of the Xe^+ plot while Xe^{3+} starts to occur beyond the intensity at the turning point 7 of Xe^{2+} plot. We note that the turning point 7 is a real one at which the slope of the plot reduces from 10.5 to 7 as the intensity increases. This latter change of slope is revealed because of the above mentioned stability of the laser and the selection of only the central section of the long focal volume.

We note that all previous and present experimental observations of NS ionization occur at a lower intensity range as compared to sequential ionization. Direct ionization of two or three ground state electrons into the continuum requires much higher intensity than sequential ionization. As such, any model that explains the observed NS ionization has to have some enhancement mechanism built into it. With this in mind, we propose the following new model. Multiphoton ionization occurs between points 0 and 1. As the intensity increases, the ionization edge is shifted into resonance with eight photons at point 1. Further increase in intensity allows many Rydberg states to go into resonance with eight photons. We note that the spectral width of the laser is wide ($\Delta\lambda = 12$ nm). Such resonances would lead to a dynamic interference stabilization of the atom (Fedorov and Movsesian, 1988; Jones and Bucksbaum, 1991; de Boer and Muller, 1992.) resulting in a reduction of the rate of ionization. During such a dynamic stabilization when the electron has not yet settled into an equilibrium large Rydberg orbital, a second electron from the core could be coupled to the first electron (Coulomb correlation) and the two of them could be ejected simultaneously at an appropriate intensity (point 2). The energy it takes to eject them is only slightly higher than the ionization potential of Xe^+. This would make it much easier as compared to ejecting two electrons from the ground state. The core electron could also be ejected alone leaving behind a Xe^+ Rydberg ion. This later process is more probable than the double ionization resulting in the enhancement of the Xe^+ ion signal between 2 and 3. At point 3, the ionization edge is shifted into 9 photon resonance at the peak laser intensity theoretically expected to be at 2.9×10^{13} W/cm^2. This agrees well with the experimental value at point 3. Further increase in intensity would bring the Rydberg states of some atoms into 9 photon resonance leading again to stabilization and the NS formations of Xe^{2+} as explained above. Further repetition will not be efficient because sequential ionization now becomes dominant.

Meanwhile, as the laser intensity increases the ejection of the second electron with the first still stabilized in the Rydberg state will also find its ionization edge passing through multiphoton resonance leading to a similar stabilization of this second electron. This is indicated by the turning point 7 on the Xe^{2+} plot beyond which the slope is reduced (indicating a reduction of ionization rate). Beyond point 7, Xe^{3+} starts to appear because we could now have two Rydberg electrons, both coupled to a third (core) electron. The simultaneous ejection of these three correlated electrons needs to overcome only an energy slightly higher than the ionization potential of Xe^{2+} and is thus relatively easy. Correlation of this core electron with only one Rydberg electron leading to their simultaneous ejection (formation of Xe^{2+} in the Rydberg state) is also possible. So is the ejection of only this core electron leading to the formation of Xe^+ ion with two Rydberg electrons.

In summary, a new model of NS double and triple ionization of Xe that involves stabilization of electrons in the Rydberg states and the correlation of such electrons to a core electron seems to be able to explain qualitatively all the experimental Features.

We thank Y. Gontier, H. Muller, D.D. Meyerhofer , F. Faisal and M. Perry, for fruitful discussions. The technical support of S. Lagace and J-P Giasson is appreciated. This work is supported by le Fond-FCAR, NSERC and NATO.

References

Augst, S.J., Talebpour, A., Chin, S.L., Beaudoin, Y., Chaker, M., (1995), *Phys. Rev. A*, August.

Corkum, P.B., (1993), *Phys. Rev. Lett.*, **71**, 1994.

de Boer, M.P., and Muller, H.G., (1992), *Phys. Rev. Lett.*, **68**, 2747.

Fedorov, M.V., Movsesian, A.M., (1988), *J. Phys. B.*, **21**, L155.

Fittinghoff, D.N. *et. al.* (1992), *Phys. Rev. Lett.*, **69**, 2642.

Jones, R.R., and Bucksbuam, P.H., (1991), *Phys. Rev. Lett.*, **67**, 3215.

Krause, J.L., Schafer, K.J., and Kulander, K.C., (1992), *Phys. Rev. Lett.*, **68**, 3535.

Ma, X.X., Xu, G.Y., Galarneau, P., and Chin, S.L., (1983), *Appl. Opt.* **22**, 2007.

Perry, M.D., Szoke, A., Landen, O.L., Campbell, E.M., (1988), *Phys. Rev. A*, **37**, 747.

Walker, B., Sheehy, B., Dimauro, L.F., Agostini, P., Shafer, K.J., and Kulander, K.C., (1994), *Phys. Rev. Lett.*, **73**, 1227.

Talebpour, A., Augst, S., Liang, Y., Chin, S.L., Beaudoin, Y., Chaker, M., (1995), " Non-sequential triple ionization of argon atoms in an intense 1053 nm laser field", Proc. SILAP IV, NATO-AWS, on the Volga, Russia, 4-9 Aug. 1995.

MULTIPHOTON PHYSICS WITH X-RAYS

G. A. KYRALA
Los Alamos National Laboratory
Los Alamos, NM 87544

1. Abstract

With the generation of intense bursts of short wavelength radiation during the interaction of ultra-intense laser beams with solid targets , the possibility of extending the study of the non-linear interactions of radiation with matter to shorter and shorter wavelength radiation becomes possible. One would hope that an extension of the techniques to the x-ray region would lead to new tools that are similar to those used in the visible and that can be widely applied. However, a basic knowledge of the relevant efficiencies of the properties of the materials, under these unusual conditions, is necessary before one can build such sources and devices. In this work we discuss the design of the first experiments on non-linear interactions using x-rays in the kilovolt energy range, and present some of the some of the early findings.

2. Introduction

The interaction of radiation with matter has been of great interest and relevance to atomic physics since the beginning of atomic theory. Before the invention of lasers, studies concentrated on the linear interactions of the radiation with material.[1] Since the invention of lasers, intense visible-wavelength radiation has become available and has allowed the study of nonlinear interactions of radiation with matter. The interactions have allowed, among other possibilities, the generation of harmonics of the incident radiation at much shorter frequencies. as well as the generation of characteristic radiation from the plasmas that are created by the interactions with solids. In both cases the generated radiation is short, of the order of the laser irradiation time itself. Using non-linear interactions for the generation of novel sources of tunable radiation has caused an explosion in the knowledge of atomic and molecular physics spectroscopy and led to an immense number of novel devices and

343

H. G. Muller and M. V. Fedorov (eds.), Super-Intense Laser-Atom Physics IV, 343–354.
© 1996 U.S. Government.

techniques that are used routinely in biological, forensic and analytic applications. One would hope that an extension of the techniques to the x-ray region would lead to similar tools that can be widely applied.

In what follows we will try to describe how we reached our experimental decision by surveying the possible x-ray sources, and then using the most intense source to design an experiment to utilize that source.

The objective of this research is to exploit the intense x-ray generation capability during the interaction of a suitable high power laser pulse with a solid target, such as of the Los Alamos Bright Source (LABS) [2]. The effort will start by demonstrating one of the non-linear processes. We will concentrate on the possibility of measuring the simplest nonlinear process, two-photon inner-shell ionization of chlorine, using keV X-rays generated from an aluminum target. An initial estimate shows that a measurable signal should be obtained using the Aluminum L_α line from the LABS plasma. The uncertainty in the calculation is largely due to novelty of the process. Theoretically, the problem is interesting because the closest intermediate states, in contrast to hydrogenic calculations, are not vacant. Thus, a measurement would be one of the first tests of this new approach to two-photon ionization.

3. Intense Source Regime:

How we define an intense x-ray source depends on the use of the source. In the present work we will judge a source by whether it can cause a measurable two photon signal during the lifetime of the radiation pulse. If, for scaling purposes, we use the two-photon ionization cross section for a hydrogenic ion of nuclear charge Z:[4]

$$\sigma^{[2]} (h\nu, Z) = Z^{-8} \sigma^{[2]} (h\nu/Z^2, 1) \qquad (1)$$

where $\sigma^{[2]} (h\nu/Z^2,1)$ is the two photon ionization cross section for hydrogen and $\sigma^{[2]} (h\nu, Z)$ is that for Z-ionized hydrogenic ion at the photon frequency ν. Then for an x-ray photon with an energy of a kilovolt, and ion with a nuclear charge Z of 10 the scaled hydrogenic photon energy $h\nu/Z^2$ is 10 eV [123 nm]. The two-photon ionization cross section then scales as:

$$\sigma^{[2]} (1 \text{ keV}, 10) = 3.9 \text{ x } 10^{-10} \sigma^{[2]} (10, 1) \qquad (2)$$

Using the hydrogenic result for the nonresonant processes, Klarsfeld [3] gives a cross-section that linearly depends on I, the X-ray irradiance of a coherent linearly polarized field:

$$\sigma^{[2]} / I(W/cm^2) = \alpha = 1.56 \times 10^{-41} \quad (cm^4/W) \tag{3}$$

For the propagation of the x-rays we can use the generalized propagation equation: $dI/dz = -k\,I\,n - \sigma^{[2]}\,I\,n = -k\,I\,n - \alpha\,n\,I^2$, where k is the linear absorption coefficient and n is the particle density. The solution to this equation:

$$I = I_0 \left[e^{-knz} \right] \left[\frac{1}{1 + I_0 \dfrac{\alpha}{k}(1 - e^{-knz})} \right] \tag{4}$$

has two parts. The first part is the attenuation by the normal linear processes, the second part is due to the non-linear two-photon attenuation coefficient. When the linear attenuation is negligible, i.e. $k \sim 0$, the solution reduces to:

$$I = \frac{I_0}{\left[1 + \alpha\,I_0\,z\,n \right]} \tag{5}$$

The number of electron vacancies, holes, created in the target material by an irradiance I is given by $\Delta N = \Delta E/(h\nu)$, where ΔE is the energy absorbed in time ΔT by an illuminated area ΔA, $\Delta E = \Delta I\,\Delta T\,\Delta A$ with the creation of a hole with a bound energy $h\nu$. Combining all the formulas gives:

$$\Delta N = \alpha\,n\,\Delta z\,\Delta A\,\Delta T\,I^2/(h\nu) \tag{6}$$

Using a solid target of thickness of one mean photon depth, [one micron thickness, or an equivalent gas target of path length 5×10^{18} at/cm^2] an area equal to the x-ray source area [100 μ^2], a pulse length of 1 psec, and a hole energy of 2 keV, then:

$$\Delta N = 2.44 \times 10^{-25}\,I^2 \tag{7}$$

An x-ray irradiance exceeding 2×10^{12} W/cm^2 is required to create one hole, by two photon ionization, during one pulse. This irradiance is a very strong function of the x-ray wavelength and nuclear charge Z. Of course in an actual experiment the number of observable events, ΔO, is much smaller than the number of generated holes, ΔO, must be large to overcome other sources of radiation, the observation efficiency that can be quite low, and the statistical uncertainty of counting. The number of observable event ΔO is equal to $\Delta N \, \Omega \, \varepsilon$. Where Ω is the solid angle fraction of the detector, and ε is the efficiency of observation. Assuming a unit efficiency, a solid angle fraction of 10^{-4} and 100 observable events in that solid angle, then one requires an irradiance exceeding 10^{15} W/cm^2 to produce those signals in the kilovolt x-ray range.

If we want to repeat the analysis for lower Z material we will look at the Z dependence. The hydrogenic cross section varies with 6-th power of the scaled frequency. Hence the attenuation of the x-rays by two photon absorption at that scaled frequency [say hv/Z^2=10] depends on Z^{-8}. Since the binding energy depends on Z^2, the required irradiance to give the same number of holes behaves as Z^5. For a 100 eV x-ray photon one has to perform the experiment in a beryllium target with Z=4, and the required irradiance, 5×10^{12} W/cm^2., is Z^{-5} time smaller than that at 1 keV

4. Intense Sources of Short Wavelength Radiation.

As mentioned in the introduction, various sources of intense x-ray radiation has been built. The most common sources are built around Synchrotron accelerators where the large facilities generate large average power, but with a small peak power. The maximum spectral brightness of synchrotrons for example is [5] 10^{12} photons/sec/mrad/eV at 100 eV which exceed that of a regular x-ray tube by many orders of magnitude.

Harmonic generation from a laser is another possible source. Many groups measured an unoptimized conversion efficiency of a PPM per harmonic. Thus for a 33 mJ , 100 fsec Ti:Saphire laser, the measured x-ray yield at 100 eV [the 66-th harmonic of a Ti:Saphire laser] is roughly 10^5 photon/mrad [6]. If one guesses that the x-ray pulse length is << 100 fsec [roughly 100 fsec/n, where n~100 is the harmonic number], and that the bandwidth << 1 eV [roughly equal to the laser bandwidth ~2Å@8000 Å~5×10^{-4} eV], then the actual spectral brightness is much higher than 10^{19} photons/sec/mrad/eV, and much higher than the peak spectral brightness of a synchrotron or an x-ray tube. If the laser runs at 1 KHz, then the time average flux is only 10^8 photons/sec/mrad.

Black body radiation is another source of radiation.. The radiance of a black body of temperature T is I=10^5 T^4 where the temperature is measured in eV. A black body of temperature .32 keV emits of 10^{15} W/cm^2, and have a peak radiance at hv=3 kT= 960 eV of 10^{23} X-rays/sec/cm^2/eV. Such conditions are not unusual, they commonly occur in laser-generated plasma. The main disadvantage of this source is the broad-band non selective nature of the radiation.

X-ray laser sources are also available at many wavelengths[28 to 428 nm] in the soft x-ray domain [7]. A conversion efficiency of a PPM is typical of the best double-passed x-ray lasers[8]. However since they use kiloJoule style lasers to produce lasing, an output radiance of 10^{14} W/cm^2/sr is available at 23 nm with divergence of 10 mr x 10 mr. Thus their irradiances can exceed 10^{31} Soft-Xray/Sr/cm^2 at an energy of 54 eV.

Line emission from laser generated plasmas is another source of intense radiation. For long laser pulse lengths, exceeding a nanosecond, the conversion efficiency has been measured for the hydrogenic and helium-like lines of many elements [9,10] the data for the He-like resonance line can be summarized as:

$$\zeta = 6554 \, (h\nu)^{-4} \text{ Photon/sphere/Joule} \qquad (8)$$

where the photon energy is measured in keV.. For an aluminum plasma the 1.6 keV He-like resonance line output can exceed 10^{14} photon per joule of incident laser light, which corresponds to a conversion efficiency of the laser light to x-ray lines of around 2%. Since the few nanosecond x-ray pulse length of these plasmas is relatively long the output radiance from a plasma irradiated by a joule of laser light is around 10^{12} W/cm^2. The way around this is the use of a short pulse laser, say shorter than a picosecond, to generate the x-rays.[11,12] The short pulse work of [11] that generated an intense source of x-ray radiation at the Aluminum He-like line of 1.6 keV motivated us to study whether we can use such intense radiation.

Choice Of Material
H- and He-like Aluminum

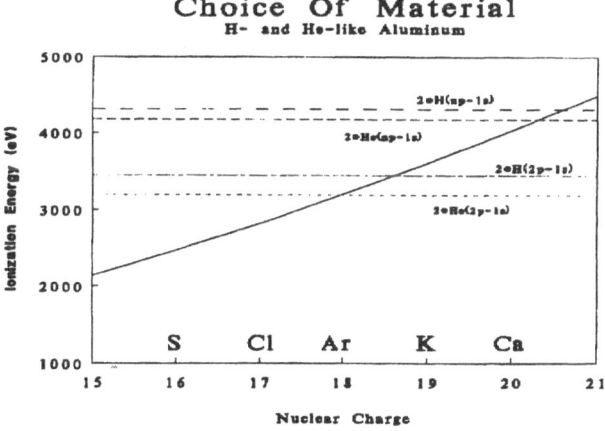

Figure 1.

The choice of material depends on the K-shell ionization energy and the X-ray line used.

5. Two-Photon Photoionization Physics Using Non-Coherent Sources:

Three atomic targets Cl, Ar, and K are possible candidates that allow the observation of two-photon K-shell ionization by the various lines from an aluminum plasma [Figure 1]. Because the He(2p-1s) resonance line is the most intense and has the smallest photon energy, the combination with a chlorine atom leads to the largest signal. The chlorine atom was the atom with the largest nuclear charge where two photon ionization was possible. Chlorine has a further advantage over smaller nuclear charge material where the cross sections were larger because of its lower sensitivity to photoionization by the rest of the plasma radiation.

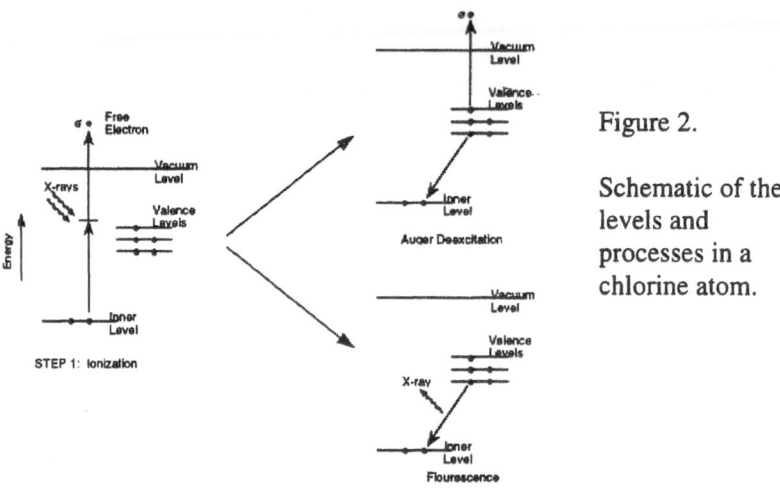

Figure 2.

Schematic of the levels and processes in a chlorine atom.

A simplified energy levels of chlorine is shown in[Figure 2] where the possible photoionization mechanisms are shown as well. A unique feature of photo ionization from the inner shell of a multi electron atom is the presence of intermediate energy levels states that are fully occupied and that are quite far off resonance. This is in contrast to two-photon ionization with visible light where electrons can be promoted to the intermediate energy-level states that are unoccupied, and where those intermediate states are not far off-resonance. The actual calculations of [13] showed that there are two paths to photoionization, the first is through exciting the intermediate levels then photoionizing that level, and the second through photoionizing an electron from the outer shell followed by exciting that level from the ground state of chlorine. Although each transition is far off resonance, the final state conserves energy, generates a hole in the inner K-shell core, and produces a free electron with an energy of 300 eV. This hole can be filled by emission of an Auger electron or a fluorescent x-ray at 2.62 keV The results of the calculation for the total two-photon ionization cross section, for a neutral chlorine atom with linearly polarized coherent radiation at hv=1.6 keV was

$$\sigma^{[2]} / I(W/cm^2) = \alpha = 1.032 \times 10^{-41} \quad (cm^4/W) \qquad (9)$$

in contrast the scaled hydrogenic result, at this particular energy, gives a coefficient that is about a factor of two lower,

$$\sigma^{[2]} / I(W/cm^2) = \alpha = 5 \times 10^{-42} \quad (cm^4/W) \qquad (10)$$

Since the light that we use is chaotic, non-coherent and occurs from all directions, one has to modify these estimates to account for the difference in behaviour from a laser induced two-photon ionization. The net result is to increase these estimates by a factor of two. This is still within the estimates used to judge the feasibility of the experiment. The next task is to create the necessary flux of radiation.

6. X-ray Source Design

Using a small KrF laser system Cobble et al[11]. measured a large conversion efficiency to the Aluminum x-ray radiation, nevertheless their x-ray-flux was short too low to perform this experiment. The cause of the large conversion efficiency, when compared to other laser based sources, was not well understood, but was suspected to be caused by the low contrast pulse that interacted with the target. Since, we have built a well diagnosed XeCl laser system where we could monitor, and to a certain extent control, the prepulse in the final laser output[14]. At the highest laser irradiance, we measured a conversion efficiency was 0.8 % of the laser energy into one aluminum line[15]. The reason for the large efficiency, compared to other measurements[16,17,18] was due to the presence of the prepulse that was controlled and that provided the appropriate conditions in front of the target for x-ray conversion.. The pulse length was also measured to be less than 5 psec [2], much longer than the interacting laser pulse, and of the order of the atomic ionization equilibrium times. While shorter x-ray pulses have been generated with high-contrast no prepulse lasers, their conversion efficiency were abysmal[16,17,18]. For these experiments, what matters is the x-ray flux, not the x-ray pulse length per se.[2, 17, 18]. Some other characteristics of the x-ray source are: a source diameter of less than 5 micrometers, thermal electron temperature of 300 eV, hot electron temperature of 12 keV, line width of few eV, and little hot electron generation. The x-ray radiance exceeded 5×10^{13} W/cm^2 in the Aluminum H-like resonance line at 1.6 keV. The flux, 3×10^{27} Xray/cm^2/Sr/eV, is large enough to be usable as a source for our experiment[Figure 3]. What remained was to design a target geometry that uses this source and removes the extraneous x-ray radiation that accompanies the required line.

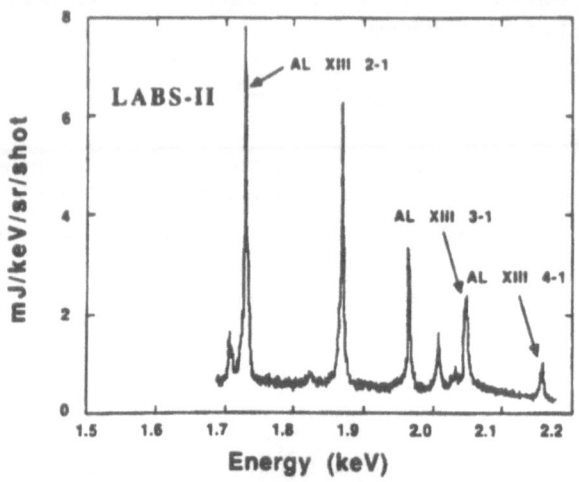

Figure 3.

Measured Spectrum from Aluminum at a XeCl laser irradiance 5×10^{18} W/cm^2

7. Target Hydrodynamic:

Positioning a chlorine target presented a challenge. The x-ray flux decreased rapidly with distance from the x-ray source, and the signal decreased as the fourth power of distance. Thus a gaseous target, though attractive, would not have produce a measurable signal because not enough density could have been placed close to the x-ray source without perturbing the laser interaction itself. A solid target NaCl provided a convenient solid target that could be deposited close to the source and had the added convenience of providing a null measurement through the presence of the sodium which requires 3 photons to give an x-ray signal at 3.384 keV. That signal would be used to measure a competing rate for the production of inner shell holes by hot electron collisions. The design of the target requires that we filter the extraneous x-rays, mostly that lie at low photon energies, thus a silicon substrate was needed to do two tasks. First, to filter the low energy x-rays completely, and attenuate the high energy x-rays, but transmit the aluminum x-rays. Second, and because of its crystal structure, to form a supersmooth optical surface where the aluminum was deposited [The depth of focus of the optical system was few tens of microns].

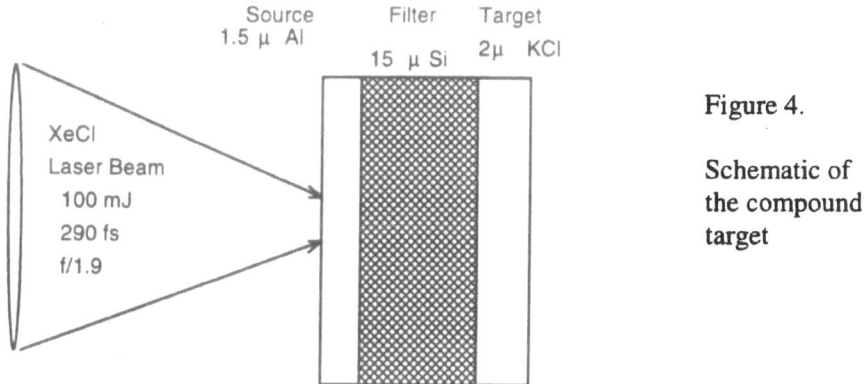

Figure 4.

Schematic of the compound target

Hydrodynamic calculation were carried out to find the appropriate thickness to be used in order to maximize the x-rays and to mitigate the effect of damage from the prepulse before the experiment could take place. The calculations used a radiation transport code that handled the laser radiation interactions including resonance absorption, above threshold ionization, multiphoton absorption, and that treated the hydrodynamics in 1-D [19]. Some of the results show that the shock from the prepulse, which occurs 3 ns before the main laser pulse, destroys the target well after the end of the x-rays from the main pulse are over. The final design is shown in Figure 4.

8. Measurement.

The resultant x-rays were measured with a filtered CCD camera that was used in the pulse height analysis mode. Each absorbed x-ray deposited its energy in a pixel or two, and 3.6 eV were required to create one free electron that was collected. The CCD camera was filtered by an 8 μm thin Nickel filter that absorbed the visible light and attenuated the intense Aluminum radiation by a factor of 8×10^{-11} .

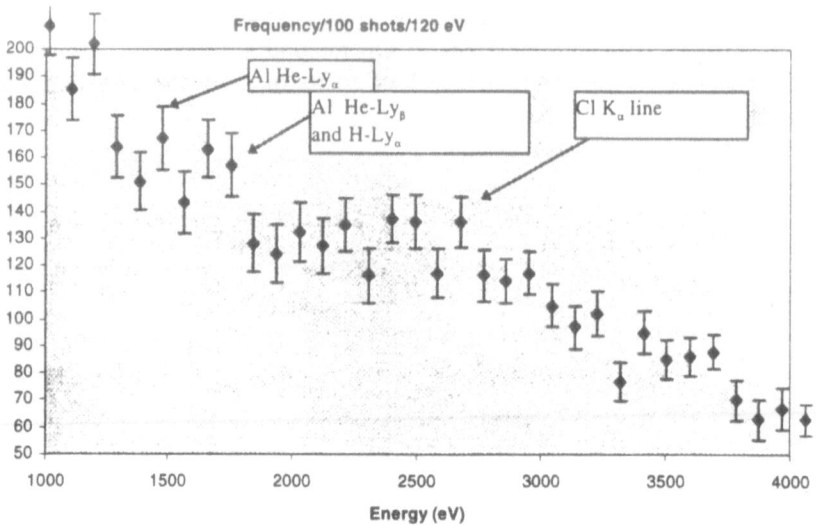

Figure 5. The measured x-ray spectrum from 100 laser shots, using 120 eV wide energy bins.

The result of 100 shots on the target were accumulated with the CCD. The pulse height distribution, excluding readout noise, and the background counts, is shown in Figure [5]. The error bars are the 1σ statistical error bars. The histogram has energy bins that are 120 eV wide about the resolution of the CCD detector, as measured with an electron beam source of x-rays. The histogram shows the Aluminum lines around the 1.6 keV bin as well as a weak signal where the Chlorine K_α line at 2.672 keV. With a bin that large, we do not observe the K_α from the Potassium target, indicating that the hot electron production rate of holes was less than that from the two-photon absorption rate. Other features of the spectrum is some energy at 2.2 keV which may come from the series limit in Al-XIII emission, and some emission at 2400 eV which may come from the late time emission of the Si XIV species. The Aluminum line at 1.48 keV can not be resolved.

The amplitude of the unresolved aluminum He-Ly$_\alpha$ line at 1.48 keV gives us a source strength of 10^{12} photon/shot roughly three times the estimate from flat crystal spectrometry. The amplitude of the Chlorine feature gives us a hole production rate of 1.1×10^6 vacancies per shot. The estimated number of vacancies is 3 times that predicted using crystal spectrometry and the theoretical cross sections. These measurements should be treated with care, since the x-ray pulse length was not measured accurately, the laser was generating multiple x-ray pulses, and the immense attenuation of the Nickel filter amplify any published attenuation cross setion measurements errors.

9. Conclusion

We have performed an experiment that should be viewed as proving the feasibility of the measurement of the two x-ray-photon absorption rate. Many difficulties must be overcome before a cross section measurement can be performed with any accuracy.

10. Acknowledgments:

I would like to acknowledge the many different contribution made by Gottfried Schappert to this work. D. Oro helped with the calibration of the CCD. J. Studebaker helped build many parts of the apparatus. G. L. Olson performed many of the calculation. A. J Taylor, J. Roberts and K. Hosak helped with the XeCl laser. Without these colleagues contributions this work would not have been feasible. The work was performed under the sponsorship of the U. S. A. Department of Energy, and the Los Alamos National Laboratory LDRD funding office.

11. References

1. Maria Goppert-Mayer, (1931) Ann. Phys. **9**, 273.
2. G. A. Kyrala, R. D. Fulton, J. A. Cobble, G. T. Schappert, and A. J. Taylor (1994), Generation, Amplification and Measurement of Ultrashort Laser Pulses, *Proceedings of the SPIE*, **2116**, 323-334.
3. W. Zernik, (1964) *Physical Review* **135**, A51.
4. S. Klarsfeld, (1970) *Lett. Nuovo Cimento* **3**, 395.
5. H. Winick and S. Doniach, (1980) *Synchrotron Radiation Research*, Plenum, N.Y.
6. K. Kondo et al. (1994), *Physical Review*, **A49**, 3881.
7. B. J. MacGowan, J. L. Bourgade, P. Combis, C. J. Keane, R. A. London, M. Louis-Jacquet, D. L. Mathews, S. Maxon, M. D. Rosen, G. Thiell, and D. A. Whelan, (1988), title, *in Short Wavelength Coherent Radiation: Generation and Applications*, Vol2 of OSA Proceedings series, R. W. Falcone and J. Kirz eds. (Optical Society of America, Washington, D. C.) p2.
8. Mathews et al, (1991), in *X-ray Lasers*, edited by G. Tallents ,IOP, Bristol, 205-210.

354

9. D. W. Phillion and C.J. Hailey (1986), Brightness and duration of x-ray line sources irradiated with intense 0.53 μm laser light at 60 ps and 120 ps pulse width. *Phys. Rev.* **A34**, 4886.

10. B. A. Hammell et al (1993), X-ray radiographic measurement of radiation driven shock and interface motion in a solid density material, *Physics of Fluids* **B5**,2259.

11. J. A. Cobble, G. T. Schappert, L. A. Jones, A. J. Taylor, G. A. Kyrala, and R. D. Fulton (1991), The interaction of a high irradiance, subpicosecond laser pulse with aluminum: The effect of the prepulse on X-ray production, *J. Applied Physics* ,**69**, 3369-3371.

12. M. Murnane, H. Kapteyn, and R. Falcone, (1989) Generation and application of ultrafast x-ray sources, *Science* **25**, 2417-2422.

13. J. Abdallah, L.A. Collins, G. Csanak, A. G. Petschek, and G. T. Schappert, (1995), Two-photon ionization of an inner shell electron of the chlorine atom, Accepted for publication in Zeit. Fur Physik.

14. A. J. Taylor, C. R. Tallman, J. P. Roberts, C. S. Lester, T. R. Gosnell, P.H. Y. Lee and G. A. Kyrala, (1990), High-intensity subpicosecond XeCl laser system, *Optics Letters* **15**, 39-41.

15. G. A. Kyrala,R. D. Fulton, E. K. Wahlin, L. A. Jones, G. T. Schappert, J. A. Cobble, and A. J. Taylor, (1992), X-ray generation by high irradiance subpicosecond lasers, *Applied Physics Letters* **60**, 2195-2199.

16. A. Ziegler, P. G. Burkhalter, D. J. Nagel, M. D. Rosen, K. Boyer, T. S. Luk, A. McPherson and C. K. Rhodes (1991), Opt. Lett 16, 1261-1264.

17. U. Teubner, G. Kuhnle, and F. P. Schafer (1991) Soft X-ray spectra produced by subpicosecond laser-double pulses, *Appl. Phys. Lett.* **59**, 2672-2678.

18. Z. Jiang, J. C. Kieffer et al (1995) X-ray spectroscopy of hot solid density plasmas produced by subpicosecond high contrast laser pulses at 10^{18}-10^{19} W/cm^2, *Physics of Plasmas* **2**, 1702-1711.

19. G. Olson, priv. comm.

Experimental Studies of X-Ray Emission Resulting From the Intense Irradiation of Atomic Clusters

T. DITMIRE, T. DONNELLY[*], R. W. FALCONE[*], and M. D. PERRY

Laser Program, Lawrence Livermore National Laboratory
P.O. Box 808, L-443
Livermore, CA 94550

[*]*Department of Physics*
University of California at Berkeley
Berkeley, CA 94720

1. Introduction

Much effort has gone into the understanding of the interaction of short ($\leq$ 1 ps), intense ($> 10^{15}$ W/cm^2) laser pulses with matter [1]. Many of these experiments have investigated the production of short wavelength radiation or the generation of energetic electrons. Significant progress has been made in the generation of bright, coherent soft x-rays in the 30 to 100 eV range through harmonic generation [2] or x-ray lasers [3, 4], and photons and particles with energies reaching the MeV range have been generated as well [5, 6]. These studies have included both underdense plasmas resulting from the irradiation of gas phase targets as well as high density plasmas resulting from solid targets [7].

High pressure gas jets produce a unique combination of both gas and solid phase components providing an interesting medium for the study of high intensity laser-matter interactions [8]. Solid density clusters form in a gas jet, resulting from the cooling associated with the adiabatic expansion of the gas into vacuum [9]. This cooling causes the gas to supersaturate and nucleate. Under appropriate conditions, when the gas jet stagnation pressure exceeds a few atmospheres, the clusters formed in the expanding jet can be quite large ($>10^4$ atoms per cluster) for gases such as Ar, Kr, N$_2$ and Xe [9]. In this paper, we report on experimental studies of the strong short wavelength emission from a gas that results when clusters are irradiated by an intense femtosecond laser. The collisional heating of large clusters in the laser field leads to the production of high charge states during the laser pulse and causes strong x - ray emission in the underdense plasma resulting after the heated clusters expand.

It has been shown previously that solid targets of porous Au-black, which are composed of ~ 100 Å gold clusters, are much more efficient at absorbing the incident laser energy than conventional flat gold targets due to the large surface-to-volume ratio of

H. G. Muller and M. V. Fedorov (eds.), Super-Intense Laser-Atom Physics IV, 355–370.
© *1996 Kluwer Academic Publishers.*

the clusters [10]. Intense irradiation of these Au-black targets gave rise to bright x-ray emission [10]. Strong x-ray emission from the irradiation of atomic clusters in a gas jet has been seen previously, as well [8], indicating that the absorption mechanism seen with the Au-black clusters may also work in the clusters formed in gas jets. Consequently, a plasma formed when an intense pulse traverses a gas jet target containing large atomic clusters might be expected to exhibit the efficient laser energy absorption characteristic of a solid target plasma but will have the radiative and kinetic properties of a low density plasma after the clusters expand. This enhanced absorption will result in plasma temperatures far beyond that expected from above threshold ionization (ATI) or inverse bremsstrahlung heating of a low density gas target alone.

Up to an irradiance of approximately 10^{17} W/cm^2, absorption of femtosecond pulses in a low density gas is primarily through strong field ionization. Radiation emitted by the plasma following the laser pulse is well described by conventional radiative and three-body recombination into charge states produced by strong field ionization of individual atoms. The plasma temperature will be determined by the single atom ATI energy distribution [11, 12]. Only when the laser intensity approaches 10^{18} W/cm^2 do other heating mechanisms, such as stimulated Raman scattering, become important [13].

The plasma produced by the irradiation of clusters will be quite different from that produced by a conventional gas target. Since the cluster diameter is smaller than a skin depth, the laser will uniformly heat the cluster by collisional inverse bremsstrahlung. We have calculated the heating of clusters with diameters in the 50-250 Å range irradiated by a 130 fs, 825 nm laser with peak intensity of between 10^{14} and 10^{17} W/cm^2 in a model that includes the effects of cluster expansion and electron free streaming during the laser pulse [14]. We find that cluster electron temperatures reach 1 to 3 keV for clusters of 100 Å irradiated by intensities between 10^{16} and 10^{17} W/cm^2. Our model also indicates that high charge states will be produced in the cluster before it expands through collisional ionization by hot electrons. The model predicts that charge states as high as Ar^{9+} and Kr^{14+} can be produced in the cluster with these intensities [14].

The hot cluster microplasmas will undergo rapid expansion into the surrounding underdense plasma during and after the laser pulse. We can estimate the time of this expansion if we assume the expansion is hydrodynamic and, therefore, described by the plasma sound speed. We estimate the expansion time by requiring the density to drop from the solid cluster density n_0 to the surrounding ambient density n_s. The resulting cluster expansion time is approximately

$$\tau_{ex} \approx r_0 \sqrt{\frac{m_i}{ZkT_e}} \left(\frac{n_0}{n_s}\right)^{1/3} \qquad (1)$$

where r_0 is the initial cluster radius, kT_e is the cluster electron temperature and m_i is the ion mass. For an argon cluster (which has an initial lattice spacing of approximately 3.8Å) with an initial radius of 50Å, an initial electron temperature of 1 keV, a $Z \approx 8$, and a surrounding bulk plasma with a density of 10^{18} atoms/cm^3, the expansion time is approximately 1 ps. After the cluster expands the plasma dynamics will be determined by the underdense uniform plasma.

2. Experimental Apparatus

With this picture of intense laser interactions with clusters in mind, we have performed a series of experiments in which a supersonic gas jet containing clusters is irradiated by femtosecond pulses from a Cr:LiSAF laser capable of producing 0.5 J pulses at 825 nm with 130 fs pulse duration. This laser has been described in detail previously [15, 16]. These pulses are focused into a vacuum chamber with an f/25 MgF_2 lens (f=100 cm). The focal spot of this lens contains roughly half of the energy in a 50 μm x 70 μm spot. The peak intensity is roughly 10^{17} W/cm^2 with 0.5 J of energy focused into the chamber.

The gas jet used in these experiments is a Mach 8 Laval nozzle that produces atom densities of 1 to 5 x 10^{18} atoms/cm^3 for backing pressures of 100 to 700 psi [17]. For all experiments the laser was focused ~ 1mm from the nozzle output which is 1.4 cm from the jet throat. We can estimate the extent of the atomic clustering for this jet from the empirical gas jet scaling parameter of Hagena [18] which is given by

$$\Gamma^* = k \frac{(d / \tan \alpha)^{0.85} p_0}{T_0^{2.29}} \qquad (2)$$

where d is the jet throat diameter in μm (150 μm for our jet), α is the jet expansion half angle (~5° for our jet), p_0 is the backing pressure in mbar, T_0 is the initial gas temperature (~298 K for our experiments) and k is a an empirical constant ($k \approx 2900$ for Kr, 1700 for Ar, 180 for Ne, and 4 for He [19]). Clustering begins when this parameter exceeds ~300 [9, 19] and large clusters (>10^4 atoms/cluster) predominate when $\Gamma^* > 5$ x 10^4 [9]. This parameter varied from ~ 1 x 10^4 to 1 x 10^5 for Ar and ~ 2 x 10^4 to 2 x 10^5 for Kr backing our jet with 100 to 600 psi respectively. It never exceeded 5000 for Ne and He in our experiments.

We diagnosed the x-ray emission from the gas jet through the use of two spectrometers monitoring the soft x-ray emission from 500 Å down to 25 Å on each shot. Both spectrometers utilize variable line spaced gold gratings with a grove density of 1200 lines/mm. One spectrometer monitored x-ray emission along the laser axis in the 500 - 170 Å spectral region. A 2000 Å Al filter blocked the direct laser light for this spectrometer. The second spectrometer monitored emission from 170 Å down to 25 Å in a direction perpendicular to the laser propagation. The spectral resolution of both spectrometers was roughly $\Delta\lambda/\lambda \sim 10^{-2}$. Each spectrometer could be mounted with either a CsI coated, 45 mm micro-channel plate detector, which yielded simultaneous information on the spectrum and the angular distribution of the radiation on each shot, or a Kentech x-ray streak camera, which permitted simultaneous measurement of the spectrum and the time history of the x-ray radiation. The temporal resolution of the streak camera was limited to approximately 10 ps.

358

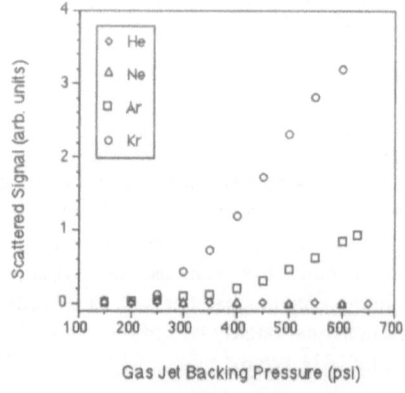

Figure 1: Measured Rayleigh scattered light signal as a function of backing pressure for pure He, Ne, Ar, and Kr.

3. Cluster Condensation Onset and Size Measurement By Rayleigh Scattering.

To experimentally confirm the presence of large clusters in our gas jet we used a technique of Rayleigh scattering [20]. To do this, the gas jet was probed with light from the frequency doubled LiSAF laser. The gas at the output of the jet nozzle was irradiated with approximately 1 μJ of 412 nm light. The centerline of the ~1 mm diameter flow was probed with a beam of approximately 400 μm in diameter. The 90° scattered light was collected with a lens and imaged onto the face of a photomultiplier tube.

Figure 1 shows the scattered light signal as a function of backing pressure for He, Ne, Ar, and Kr. No significant light scattering above the noise level is observed from either Ne or He over the range of backing pressures studied. The scattered light signal from the expansion of Ar and Kr, however, exhibits nonlinear growth with backing pressure, rising above the noise with as little as 150 psi backing the gas jet. This measurement qualitatively confirms our assertion that large clusters are formed in our gas jet when it is backed with Ar or Kr, and that neither He or Ne cluster to any significant degree in the jet.

The nonlinear dependence of the scattered signal is consistent with accepted scaling for the cluster size with backing pressure. Previous measurements of clustering in Ar have shown that the mean number of atoms per cluster, N_c, scales roughly like, $N_c \sim p^2$ [21]. Since the scattering cross section is given by the classical Rayleigh scattering cross section formula:

$$\frac{d\sigma}{d\Omega} = 2\pi \frac{r^6}{\lambda^4}\left(\frac{n^2-1}{n^2+2}\right) \tag{3}$$

then we can say that the observed scattered signal should scale as $S \sim n_{clust}N_c{}^2$ where n_{clust} is the density of clusters in the gas jet. In the regime of large cluster formation (when $\Gamma^* > 1000$) we can assume that 100% of the atoms have condensed into clusters [22]. This implies that if n_0 is the average gas density, which is presumably linear with backing pressure, then $n_{clust} = n_0 / N_c$. Thus, the scattered signal at the highest backing pressures should scale as $S \sim p^3$. This is in good agreement with the rise in the Ar data which scales as $S_{Ar} \sim p^{3.3}$, as well as the Kr data which grows as $S_{Kr} \sim p^{3.5}$.

We can make a quantitative estimate for the cluster size based on the scattered light levels. Using eq. 3 and an estimate for the refractive index of the cluster, the signal levels at the highest backing pressure (600 psi) suggest that the mean cluster size is $1 - 5 \times 10^4$ atoms for Ar (a diameter of 80 to 140 Å) and $0.5 - 3 \times 10^5$ atoms for Kr (a diameter of 140 to 240 Å). Based on the experiment noise level we can put an approximate upper bound on the size of Ne or He clusters of <3000 atoms per cluster, if clusters of these atoms exist at all in the gas jet. Estimates of the Hagena parameter for helium, for example, suggest that no clusters should form for the backing pressures used. The size of the possible error in these estimates is large due to a lack of detailed knowledge about the average gas density as well as possible errors in estimating the throughput of our scattered light detection set-up.

4. Soft X-Ray Emission Characteristics of the Various Gases.

With a confirmation of the presence of large clusters in our experiments we examined the nature of the x-ray emission from the various gases upon irradiation of the laser. Figure 2 compares the time integrated soft x-ray emission from Ne, N_2, Ar, and Kr. In all cases the gas jet was backed with 500 psi of pressure and the incident peak laser intensity was 2×10^{16} W/cm^2. The only signal observed in Ne comes from weak harmonics. The signal from the other gases, all gases forming large clusters, exhibit strong line emission across the spectrum. The lines in N_2 originate predominately from N^{3+}, N^{4+}, N^{5+}, the lines in Ar originate from Ar^{5+}, Ar^{6+}, and Ar^{7+}, and the lines in Kr are largely from Kr^{5+}, Kr^{6+}, and Kr^{7+} [24]. In general, the time integrated signal levels from these three gases is roughly equivalent in this range. The difference in the x-ray emission characteristics between Ne and the other gases is significant since clusters are expected in the gas flows of the Ar, Kr and N_2 but none are expected in the Ne.

The time histories of select lines from the gases exhibiting strong emission under identical conditions are compared in figure 3. Here we compare the N^{3+} 4p->2s line at 189Å, the Ar^{7+} 4d->3p line at 179 Å, and the strong feature in Kr at around 175Å which probably arises from Kr^{7+}. The time resolution of this data is approximately 280 ps. For all three gases the emission starts promptly after the passage of the laser through the plasma and the decay times are roughly the same for all three species, between 6 and 8 ns.

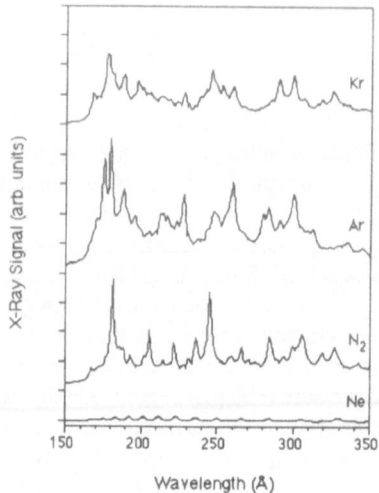

Figure 2: Time integrated soft x-ray emission from Ne, N_2, Ar, and Kr with a peak laser intensity of 2×10^{16} W/cm^2 and a gas jet backing pressure of 500 psi.

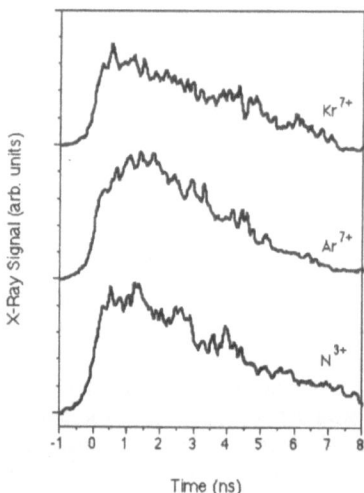

Figure 3: Time histories of select lines from around 180A from the spectral data of figure 2.

This behavior is qualitatively consistent with our supposition that these high charge states are produced during the laser pulse in the cluster. The long decay time is also consistent with the production of a reasonably hot plasma. For example, if we assume that the temperature of the plasma is given by the calculated ATI temperature ($kT_e \sim 10$ eV), the three-body recombination time for Ar^{7+} is ~100 ps. This much more rapid than is observed.

5. Helium/Cluster Mixture Emission Characteristics.

To more carefully investigate the production of heated electrons by the illumination of the clusters, we conducted a series of experiments in which a small fraction (0.01 - 0.1) of a gas known to cluster is mixed with He and passed through the gas jet. It is well known that the use of He as a carrier gas will significantly enhance the formation of clusters of heavier atoms in a gas expansion while the He does not itself undergo clustering [23]. Since the He does not itself cluster, observation of the intensity of the Lyman series transitions in He^+ allowed us to study the dynamics of the bulk plasma without concern for intra-cluster effects. The presence of the clusters in the gas mixtures, which we confirmed by Rayleigh scattering, serves to absorb laser energy and results in a thermal plasma which can then collisionally excite the He levels in the low density "warm" plasma after the clusters expand.

Figure 4 compares the emission spectra of helium for two conditions. The dashed line shows the spectrum when helium alone backs the gas jet. Here the peak intensity is about 3×10^{16} W/cm^2. The intensity required to optically ionize He to He^{2+} is

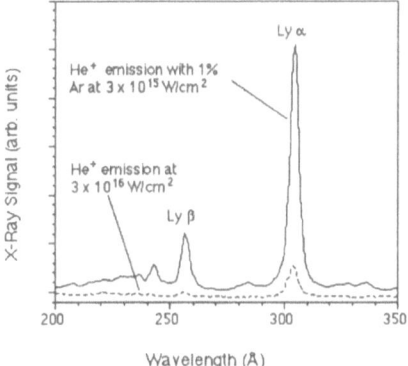

Figure 4: Emission spectrum with a peak irradiance of 3×10^{16} W/cm^2 in pure helium (dashed line) compared with the emission spectrum from a 1% Ar/ 99% He mixture irradiated with an intensity of 3×10^{15} W/cm^2.

362

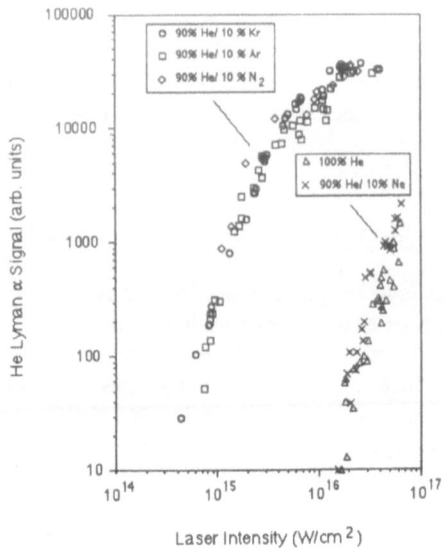

Figure 5: Integrated yield on the Ly α line in He is plotted as a function of intensity for a variety of gas mixtures.

approximately 1.5×10^{16} W/cm^2 with a 130 fs laser pulse. Above this intensity Ly α light will be emitted due to the recombination of electrons into the upper levels of the doubly ionized He. Since no clusters are present, a small amount of signal is observed on the He Ly α line resulting from this recombination. The solid line in figure 4, however, shows the spectrum of the helium when the backing gas contained 1% Ar mixed with the helium. Though the peak intensity is only 3×10^{15} W/cm^2, below the threshold for optically ionizing the helium to the required charge state for recombination, the spectrum exhibits strong emission on the n = 2->1 line as well as the n = 3->1 and n = 4->1 transitions. This represents evidence for the presence of plasma temperatures that are sufficient to collisionally ionize the helium to the doubly ionized state.

In figure 5 the integrated yield on the Ly α line in He is plotted as a function of intensity for a variety of gas mixtures. When a plasma is formed from 100% pure helium we observe a small amount of Ly α light at peak intensities above which the tunneling ionization to He^{2+} begins. Addition of a small fraction of Ne does not significantly change the observed Ly α signal. This is consistent with the fact that large clusters are absent in the He and He/Ne expansions and the small observed signal is due only to direct strong-field ionization by the laser. When a small amount (10%) of Ar, Kr, or N$_2$ is mixed with the helium, all gases with a strong propensity for forming large clusters, the magnitude of the He Ly α signal is significantly enhanced, exceeding the signal of the pure helium by nearly a factor of 100 at the highest intensities. Furthermore, the Ly α

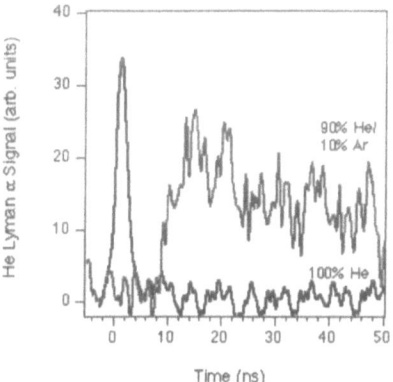

Figure 6: He Lyman α time history for an intensity of 7 x 10^{16} W/cm^2 in pure He (black line), and an intensity of 6 x 10^{15} W/cm^2 in a mixture of 10% Ar and 90% He (gray line).

signal appears at an intensity that is 20 times lower than the threshold for the production of He^{2+} predicted by tunneling ionization.

The difference in the plasma temperature between the optically ionized (pure He) case and the cluster heated case is dramatically illustrated in the comparison of the Ly α time decay dynamics. The measured time histories over the first 50 ns after passage by the pulse for both cases are shown in figure 6. These plots represent an average over ten laser shots. The time history of the helium extends out to >150 ns after the laser. The optically ionized helium (100% He I = 7 x 10^{16} W/cm^2) exhibits emission immediately (< 1 ns) after creation by the laser followed by a fast (< 2 ns) fall. The plasma heated by the presence of clusters (10% Ar I = 6 x 10^{15} W/cm^2) shows no Ly α signal immediately after the laser, due to the fact that the intensity of the laser is sufficient to ionize to He$^+$ only. The hot electrons from the clusters, however, serve to collisionally ionize the He on a long (~10 ns) time scale resulting in Ly α emission on a ~100 ns time scale. The slow turn-on of the He line is because of the slow rate of collisional ionization to He^{2+} in the relatively low density of the underdense plasma. Calculations indicate that plasma temperatures of 150 - 250 eV are required to explain this time history of the He emission [14].

6. 50 - 100Å Emission in Kr.

The strong soft x-ray emission observed resulting from the enhanced laser absorption by clusters, suggests that short wavelength (<100Å) x-rays may be produced at relatively modest laser intensity. Figure 7 shows the time integrated spectrum between 40 Å and 100 Å produced by the laser focused to an intensity of approximately 1.5 x 10^{16} W/ cm^2

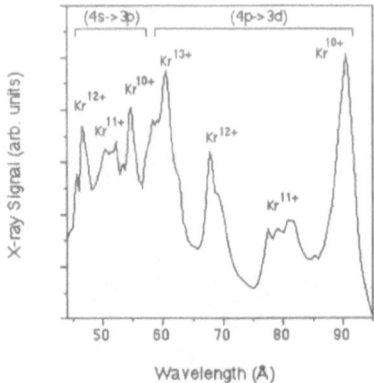

Figure 7: Time integrated spectrum in Kr with an intensity of 1.5×10^{16} W/ cm^2 into the gas jet backed by 500 psi..

into the gas jet backed by 500 psi of pure Kr. The spectrum exhibits strong emission from the 4p-3d, and 4s-3p arrays of Kr^{10+}, Kr^{11+}, Kr^{12+}, and Kr^{13+} [25]. Tunnel ionization theory predicts that intensities of 3, 4, 6 and 8 x10^{17} W/cm^2 respectively are required to produce these states by optical ionization, over an order of magnitude in excess of the actual laser intensity used here.

Similar to the soft (170 - 400 Å) x-ray emission of Kr when large clusters are present in the gas jet target, the emission from these lines is long lived. The time history of the Kr^{10+} 4p-3d line is shown in figure 8; the streak camera temporal resolution of this data is approximately 280 ps. The radiation from this line is emitted for nearly 3 ns after the laser produces the plasma. The long lifetime is consistent with the long lived emission of a hot, low density, bulk plasma. Evidence for the interaction of the laser with solid density clusters is found when the time history of the Kr emission is observed over the first 100 ps. Data showing the streaked time history of the Kr^{10+} 90 Å line for the first 100 ps after illumination by the laser is shown in figure 9. The time resolution of this data is roughly 10 ps. We observe an initial spike in the emission of the Kr^{10+} line, faster than the time resolution of the streak camera, followed by the long lived emission demonstrated in fig. 8. A similar spike is seen on all the charge states' lines as well as in the background continuum emission. This initial spike is indicative of intense x-ray emission by the dense cluster micro-plasmas during and immediately after heating by the laser. The fast (< 1 ps) expansion of the cluster is then followed by lower intensity emission by the low density bulk plasma on a long (3 ns) time scale. The fast rise time of the emission from the high charge states indicates that rapid ionization (< 1 ps) occurs to Kr^{10+} - Kr^{13+} within the cluster. Though the number of photons contained within this spike is much less than the total number of photons that are emitted by the plasma, the relative brightness is higher than the long lived emission.

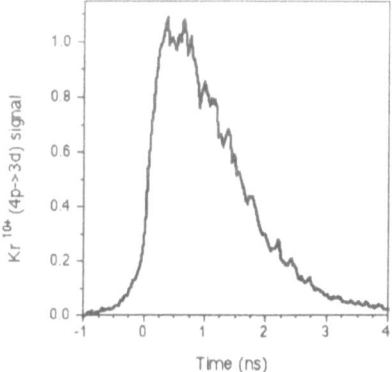

Figure 8: Time history of the Kr^{10+} 4p->3d line. Streak camera resolution is 280 ps.

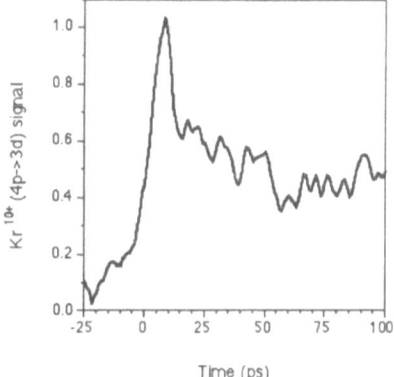

Figure 9: Time history of the Kr^{10+} 90Å line for the first 100 ps (with ~10 ps time resolution).

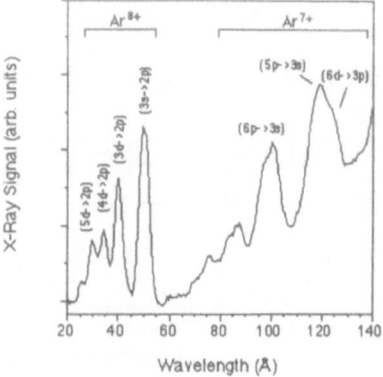

Figure 10: Time integrated spectrum of Ar emission from 20 to 140 Å with a gas jet backing pressure of 450 psi and an intensity of 8 x 10^{15} W/cm^2.

If the emission is < 1 ps in duration, then the relative brightness of the spike would be two orders of magnitude above that of the underdense plasma emission.

7. Short Wavelength Emission in Ar.

Short wavelength emission is observed under similar conditions in Ar as well. A time integrated spectrum of Ar emission from 20 to 140 Å is shown in figure 10. The gas jet backing pressure was 450 psi and the laser was focused to 8 x 10^{15} W/cm^2. The most dramatic aspect of this spectrum is the observation of strong emission from lines at wavelengths below 50Å in neon-like Ar (Ar^{8+}). To produce recombination radiation in this charge state through tunnel ionization would require a focused intensity of >1.5 x 10^{18} W/cm^2, more than two orders

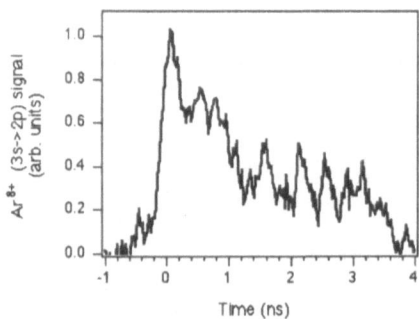

Figure 11: Measured emission history of the Ar^{8+} 3s->2p line at 49Å.

of magnitude higher than that used in the experiment. This emission is strong; we estimate that the detected photon yield on these lines indicates that up to 1 - 5% of the laser energy is converted to x-rays from in this wavelength range.

The time history of the neon-like Ar emission is of a duration that is comparable to the Kr emission. Figure 11 illustrates the characteristic emission history of the Ar^{8+} 3s->2p line at 49 Å. The signal exhibits a ~ 1ns drop followed by a longer (~ 4 ns) tail. Calculations suggest that the strong initial signal over the first ~ 1 ns results from collisional excitation, and the longer tail comes from recombination of Ar^{9+} to Ar^{8+}. Modeling of this emission time history in the Ar plasma suggests that the initial plasma temperature may be as high as 1 keV [14].

8. Prepulse Experiments

Since the clusters expand rapidly after heating by the laser, the presence of a laser prepulse of sufficient energy to ionize the cluster should be sufficient to destroy the clusters. To explore this effect we examined the yields and time history of a cluster plasma with and without a laser prepulse. To accurately determine the intensity at which the clusters disassemble, we conducted an experiment in which a small prepulse 14 ns before the main pulse is generated by allowing some pulse energy to leak out of the regenerative amplifier cavity in the LiSAF laser. Changing the timing of the Pockels cell in the cavity permitted control over the amplitude of the prepulse which was monitored on the fast photodiode and a sampling scope.

Figure 12 illustrates the dramatic difference in the x-ray spectrum between the case when a mixture of He and 10% Ar is irradiated with and without the prepulse. This figure shows the helium spectrum with a peak intensity of 7×10^{15} W/cm^2 in the main pulse. Addition of a prepulse of 8×10^{14} W/cm^2 effectively lowers the helium emission to unobservable levels. This indicates that at an intensity of ~ 10^{15} W/cm^2 the cluster sees sufficient intensity to undergo some ionization to cause it to disassemble, however, it does not acquire any significant amount of thermal energy, and does not excite any emission in the surrounding underdense He plasma. The threshold for the disassembly of the clusters begins abruptly with increasing prepulse intensity. The He Ly α yield for a fixed main pulse intensity of 7×10^{15} W/cm^2 is shown as a function of prepulse intensity in figure 13. The helium signal drops rapidly for prepulse intensities above about 4×10^{14} W/cm^2. This sharp drop is due to the nonlinear nature of the tunnel ionization rate for the Ar atoms in the cluster.

9. Conclusion

In conclusion, we have shown that large clusters produced in expanding gas jets can be used to produce hot, low density plasmas with intense, short pulse lasers. We find that the thermal plasmas produced by the illumination of clusters by femtosecond pulses of

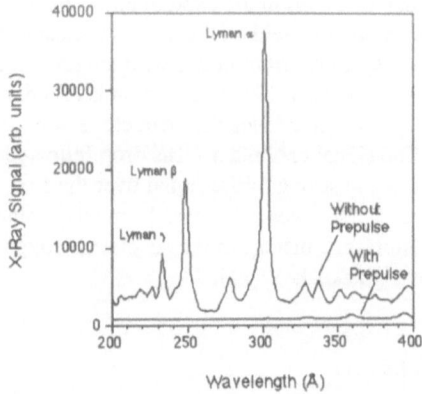

Figure 12: Helium emission spectrum with a 10% Ar mixture with a peak intensity of 7 x 10^{15} W/cm^2 in the main pulse with and without a 14 ns prepulse of 1x 10^{15} W/cm^2.

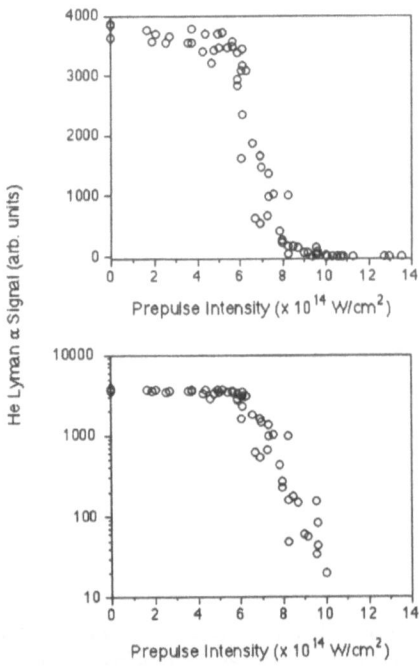

Figure 13: He Ly α yield with a 10% Ar mixture for a fixed main pulse intensity of 7 x 10^{15} W/cm^2 as a function of 14 ns prepulse intensity.

10^{16} to 10^{17} W/cm^2 dominate the plasma kinetics, producing emission from high charge states that can last for many nanoseconds. These novel plasmas exhibit dramatically enhanced absorption of the laser light relative to pure gases, with plasma temperatures well in excess of that expected from the illumination of a low density gas alone. Furthermore, the x-ray yields are comparable to those that can be achieved with solid targets. These cluster heated plasmas have the potential for providing a source of strong, x-ray radiation with the modest irradiance (10^{15} to 10^{17} W/cm^2) produced by small-scale short-pulse lasers through a unique combination of the advantages inherent to both solid and gas targets.

We would like to acknowledge the help of R. A. Smith and the technical assistance of R. Jones. This work was supported by the AFOSR and conducted under the auspices of the DOE under contract W-7405-Eng-48.

10. References

1. Perry, M.D. and Mourou, G. (1994) Terawatt to Petawatt Subpicosecond Lasers, *Science* **264**, 917.
2. L'Huillier, A., Lompré, L., Mainfray, G., and Manus, C. (1992) High-Order Harmonic Generation in Rare Gases, in M. Gavrila (ed.) *Atoms in Intense Laser Fields,* Academic Press, Boston, pp. 139-202.
3. Nagata, Y., Midorikawa, K., Kubodera, S., Obara, M., Tashiro, H., and Toyoda, K. (1993) Soft X-Ray Amplification of the Lyman-α Transition by Optical-Field-Ionization, *Phys. Rev. Lett.* **71**, 3774.
4. Lemoff, B.E., Yin, G.Y., III, C.L.G., Barty, C.P.J., and Harris, S.E. (1995) Demonstration of a 10-Hz Femtosecond-Pulse-Driven XUV Laser at 41.8nm in Xe IX, *Phys. Rev. Lett.* **74**, 1574.
5. Kmetec, J.D., III, C.L.G., Macklin, J.J., Lemoff, B.E., Brown, G.S., and Harris, S.E. (1992) MeV X-Ray Generation with a Femtosecond Laser, *Phys. Rev. Lett* **68**, 1527.
6. Fews, A.P., Norreys, P.A., Beg, F.N., Bell, A.R., Dangor, A.E., Danson, C.N., Lee, P., and Rose, S.J. (1994) Plasma Ion Emission from High Intensity Picosecond Laser Pulse Interactions with Solid Targets, *Phys. Rev. Lett.* **73**, 1801.
7. Murnane, M.M., Kapteyn, H.C., and Falcone, R.W. (1989) High-Density Plasmas Produced by Ultrafast Laser Pulses, *Phys. Rev. Lett.* **62**, 155.
8. McPherson, A., Thompson, B.D., Borisov, A.B., Boyer, K., and Rhodes, C.K. (1994) Multiphoton-induced X-ray Emission at 4-5 keV from Xe Atoms with Multiple Core Vacancies, *Nature* **370**, 631.
9. Hagena, O.F. and Obert, W. (1972) Cluster Formation in Expandion Supersonic Jets: Effect of Pressure, Temperature, Nozzle Size, and Test Gas, *J. Chem. Phys.* **56**, 1793.
10. Murnane, M.M., Kapteyn, H.C., Gordon, S.P., Bokor, J., Glytsis, E.N., and Falcone, R.W. (1993) Efficient Coupling of High-Intensity Subpicosecond Laser Pulses into Solids, *Appl. Phys. Lett.* **62**, 1068.
11. Burnett, N.H. and Corkum, P.B. (1989) Cold-Plasma Production for Recombination Extreme-Ultraviolet Lasers by Optical-Field-Induced Ionization, *J. Opt. Soc. Am. B* **6**, 1195.
12. Glover, T.E., Donnelly, T.D., Lipman, E.A., Sullivan, A., and Falcone, R.W. (1994) Subpicosecond Thomson Scattering Measurements of Optically Ionized Helium Plasmas, *Phys. Rev. Lett.* **73**, 78.
13. Blyth, W.J., Preston, S.G., Offenberger, A.A., Key, M.H., Wark, J.S., Najmudin, Z., Modena, A., Djaoui, A., and Dangor, A.E. (1995) Plasma Temperature in Optical Field Ionization of Gases by Intense Ultrashort Pulses of Ultraviolet Radiation, *Phys. Rev. Lett.* **74**, 554.
14. Ditmire, T., Donnelly, T., Rubenchik, A.M., Falcone, R.W., and Perry, M.D.,(1995) The Interaction of Intense Laser Pulses with Atomic Clusters, *Phys. Rev. A* submitted.
15. Ditmire, T., Nguyen, H., and Perry, M.D. (1994) Design and Performance of a Multiterawatt Cr:LiSrAlF$_6$ Laser System, *J. Opt. Soc. Am. B* **11**, 580.
16. Ditmire, T., Nguyen, H., and Perry, M.D. (1995) Amplification of Femtosecond Pulses to 1J in Cr:LiSrAlF$_6$, *Opt. Lett.* **20**, 1142.

17. Perry, M.D., Darrow, C., Coverdale, C., and Crane, J.K. (1992) Measurement of the Local Electron Density by Means of Stimulated Raman Scattering in a Laser-Produced Gas Jet Plasma, *Opt. Lett.* **17**, 523.
18. Hagena, O.F. (1981) Nucleation and Growth of Clusters in Expanding Nozzle Flows, *Sur. Sci.* **106**, 101.
19. Wörmer, J., Guzielski, V., Stapelfeldt, J., and Möller, T. (1989) Fluorescence Excitation Spectroscopy of Xenon Clusters in the VUV, *Chem. Phys. Lett.* **159**, 321.
20. Abraham, O., Kim, S.-S., and Stein, G.D. (1981) Homogeneous Nucleation of Sulfur Hexaflouride Clusters in Laval Nozzle Molecular Beams, *J. Chem. Phys.* **75**, 402.
21. Farges, J., Feraudy, M.F.d., Raoult, B., and Torchet, G. (1986) Noncrystalline Structure of Argon Clusters II: Multilayer Icosahedral Structure of Ar_N Clusters 50<N<750, *J. Chem. Phys.* **84**, 3491.
22. Birkhofer, H.P., Haberland, H., Winterer, M., and Worsnop, D.R. (1984) Penning, Photo and Electron Impact Ionization of Argon Clusters, *Ber. Bunsenges. Phys. Chem.* **88**, 207.
23. Wu, B.J.C., Wegener, P.P., and Stein, G.D. (1978) Homogeneous Nucleation of Argon Carried in Helium in Supersonic Nozzle Flow, *J. Chem. Phys.* **69**, 1776.
24. Kelly, R.L. (1987) *Atomic and Ionic Spectrum Lines Below 2000 Angstroms: Hydrogen through Krypton*, Vol. 1, American Chemical Society, Washington, D. C.
25. Bleach, R.D. (1980) XUV Spectra from Kr XI-XIV, *J. Opt. Soc. Am.* **70**, 861.

A COLLINEAR SETUP FOR OPTICAL/X-RAY CROSS CORRELATION MEASUREMENTS BASED ON LASER-ASSISTED AUGER DECAY

R.C. CONSTANTINESCU AND H.G. MULLER
FOM-Institute for Atomic and Molecular Physics
Kruislaan 407, 1098 SJ Amsterdam, The Netherlands

J.M. SCHINS*, P. BREGER AND P. AGOSTINI
Service des Photons, Atomes et Molécules
CE Saclay, 91191 Gif Sur Yvette, France
* present address: Department of Applied Optics
Twente University PO box 217, 7500 AE Enschede, The Netherlands

H.-J. VOORMA, E. LOUIS AND F. BIJKERK
FOM-Institute for Plasma Physics Rijnhuizen
P.O. Box 1207, 3430 BE Nieuwegein, the Netherlands

AND

A. BOUHAL, G. GRILLON, A. ANTONETTI AND A. MYSYROWICZ
Laboratoire d'Optique Appliquée, ENSTA-Polytechnique
91120 Palaiseau, France

1. Introduction

With the advent of high-intensity, subpicosecond laser systems, great progress has been made in the development of ultrafast x-ray sources, ranging from the ultraviolet (XUV) to the hard x-ray spectral region. High-order harmonic generation (HHG), using high-energy, femtosecond laser pulses, has proved to be a convenient way of generating coherent radiation in the soft-x-ray and ultraviolet spectral region. Recently, x-ray sources based on emission from high-density, laser-produced plasma's have been reported [1]. These x-rays are incoherent but have high brightness as a result of the small size, short lifetime and high temperature of the radiating plasma.

The pulse duration of these x-ray sources has been demonstrated to be a picosecond [1] or, in the case of a gallium plasma, even less [2]. Such short pulse durations defy accurate measurement by electronic means. For-

H. G. Muller and M. V. Fedorov (eds.), Super-Intense Laser-Atom Physics IV, 371–380.

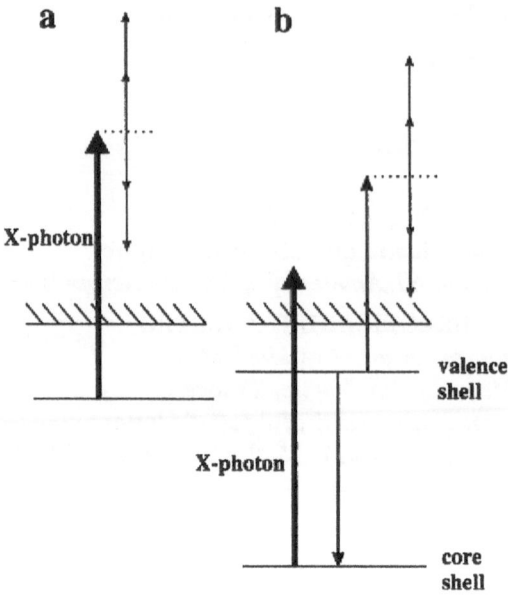

Figure 1.   Schemes of mixed two-photon processes. a: One x-ray photon (thick arrow) ionizes a valence shell of an atom. The photoelectron can absorb or emit a number of optical photons (thin arrows) when the ionization process takes place in the presence of a second, optical 'dressing' field. The laser-assisted photoionization in He has already been demonstrated [3]. b: Laser-assisted Auger decay is another example of a mixed two-photon process. One x-ray photon ionizes a core shell. This will be filled by an electron out of a valence shell. The energy released in this process will be carried away by a second electron, leaving the valence shell doubly ionized. This process, involving the L core-shell and the M valence-shell in Ar is discussed in [2]. A similar process is the NOO Auger decay in Xe, discussed here.

tunately, both the HHG and the x-rays emitted by laser-produced plasma's can be very useful light sources for driving atomic physics processes (Fig. 1).

The first two-color experiments using one x-ray photon and one or more optical photons have already been demonstrated [2], [3], [4]. Besides being a very powerful tool in investigating free-free transitions [3], any mixed-color two-photon process in the perturbative regime opens the possibility for measuring a cross-correlate of the temporal profile of the pulses involved.

Measurements on the basis of cross-correlation are limited only by the duration of the probe pulse. Such measurements using optical and XUV photons have been reported by van Woerkom et al. [5], Sher et al. [6] and Schins et al. [2] and have been used to determine the temporal profile of laser-generated x-ray pulses near 90 eV (the 3d-5p transition in Kr) and 250 eV (the Ar L-edge), respectively. In a similar experiment, using the cross-correlate of the pulses involved, the temporal profile of high order harmonics near 30 eV has been determined and was found to be of a few tens of femtoseconds [4].

In order to be able to produce a strong-enough correlation signal, high intensities of the ionizing fields are required. By focusing the harmonics or the laser-generated x-rays with suitable x-ray optics, good spatial overlap with a focused optical laser beam is possible. In addition, combined with the demonstrated subpicosecond duration of these sources, the x-ray intensities achieved can be of the order of 10^{11}W/cm^2 or more, allowing for non-linear processes with x-ray photons.

Laser-assisted Auger decay has been experimentally demonstrated for the first time in a recent work [2] using a laser-generated gallium-plasma as a broadband x-ray source. The main drawback of the gallium source is the production of debris, which makes the use of nearby focusing mirrors impractical. A second problem with that experiment is, that although the measurement on the basis of cross-correlation is in principle only limited by the duration of the probe pulse, the duration of the photons in the range of 250-400 eV could not be measured with the resolution of 150 fs (the duration of the dressing pulse). Due to the geometry of the experiment, with the pump beam and dressing beam crossing at right angles, the limiting factor for the time resolution of the method is the transit time through the area of sensitivity of the electron spectrometer.

We now present a new experimental setup for collinearly measuring cross correlates between optical pulses and x-ray pulses refocused by a multilayer mirror. This setup allows measurement of the x-ray pulse duration with a resolution equal to the duration of the dressing pulse. In order to avoid the mirror contamination problem, we performed the measurements on a plasma generated on a liquid mercury target.

2. Experimental

In the present experiment, we use the NOO Auger decay in xenon induced by 90 eV radiation from a pulsed, broadband mercury plasma as the basic process. This process results in Auger electrons, with an energy of about 32 eV. Sidebands at 1.55 eV (one photon energy) from the Auger peak are created by having this decay occur in the presence of the infrared radiation (800 nm), at intensities around $1.5 \times 10^{11}\text{W/cm}^2$, through above-threshold ionization (ATI). At this intensity, according to the Simpleman's theory [2], only one sideband is expected on both sides of the Auger peak.

Fig. 2 shows a schematic of the experimental set-up. The output pulse of a self-mode-locked Ti:S laser is stretched to 400 ps and amplified in two stages. After recompression one obtains 40 mJ pulses of 150 femtoseconds at 800 nm (infrared) with a repetition rate of 10 Hz and a near-gaussian

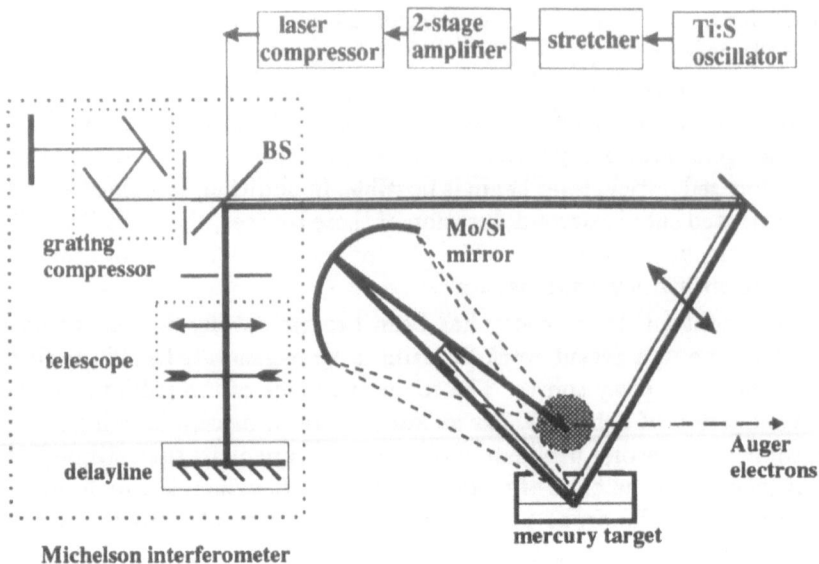

Figure 2. Schematic of the experimental setup. An amplified Ti:S laser delivers 40 mJ energy per pulse at 800 nm. The laser compressor can be set such, that the pulse duration varies between 150 fs and 3 ps. The laser beam is sent into a Michelson interferometer and split into two branches. The beam in one of the branches (thin line) passes through an aperture and a single-pass grating compressor, is then focused on the mercury target and pumps a plasma. The x-rays emitted (dashed cone) are collected in the specular direction by a Mo/Si multilayer mirror and refocused into the sensitivity region (shaded circle) of an electron-spectrometer. The IR beam reflected from the mercury surface is blocked in front of the Mo/Si mirror. The beam in the other arm (of variable length) of the interferometer (thick line) is the dressing beam. This is passed through an aperture and a telescope, and hits the mercury target in the same point as the pump beam, is specularly reflected from the mercury surface and reflected and refocused by the Mo/Si mirror, such that it envelopes the x-ray focus in the sensitivity region.

spatial profile of 12 mm FWHM [7].

In a Michelson interferometer, the Ti:S laserbeam is split into two co-propagating pulses. One of the pulses is focused onto a mercury target and pumps a plasma. The x-rays emitted from this plasma are collected and refocused by a multilayer mirror, and induce Auger processes in Xe. The second pulse from the interferometer reflects from the mercury surface and x-ray mirror, and is used to dress these Auger electrons.

The pulse used to produce the x-rays by pumping the plasma is passed through a single-pass grating compressor in one arm of the interferometer. This allows for individually adjustable pulse duration of the x-ray pump beam with respect to the dressing beam. The pump beam is focused onto a mercury target with a lens of 200 mm focal length, creating a plasma burst.

The emitted pulsed, broadband x-radiation is collected in the specular direction in a 0.12 sr solid angle by means of a concave multilayer mirror, 70 mm radius of curvature and 30 mm diameter. This mirror is coated for 90 eV radiation and is made of alternating layers of silicon and molybdenum [8]. The reflectivity of this Mo/Si mirror is estimated to be 40%-50%, and the reflectivity FWHM is about 5 eV, measured over the full active area. The selected 90 eV radiation is refocused into the sensitivity region of a magnetic-bottle electron spectrometer [9]. With the multilayer mirror used near 1 to 1 imaging, the x-radiation can in principle be refocused to a spot as small as a few tens of micrometers, depending on the focusing conditions of the pump beam on the mercury target. Such a small focal spot size of the x-radiation yields very satisfactory rates when used in Auger processes.

The second pulse from the interferometer is used as a dressing beam on the Auger electrons emitted. This pulse is passed through a slightly defocused telescope (+200 mm, -100 mm focal length lenses respectively) in the other arm of the interferometer, which gives the dressing pulse individually adjustable divergence. In addition, the length of this arm can be varied to provide an adjustable time delay between the pulses. By observing the interference fringes between the outgoing beams at zero delay, the dressing beam is interferometrically overlapped with the plasma-pumping beam. The energy of the two pulses is controlled by placing apertures in the respective arms of the interferometer. The dressing beam hits the mercury target at the same point as the pump beam, and it is specularly reflected from the mercury surface (reflectivity for IR is about 90 %). This reflected dressing beam is then also reflected and refocused by the central part of the Mo/Si multilayer mirror.

The Auger electrons are analyzed in a time-of-flight (TOF) electron spectrometer of the magnetic-bottle type, which collects the electrons over 2π sr solid angle. The detection volume, sketched in Fig. 3, is a cylinder of 2 mm length along the spectrometer axis and 0.25 mm diameter. Due to the high magnetic field directed along the spectrometer axis, the electrons spiral within this cylinder until they reach the low magnetic field region. After traversing a distance of 1.5 m in a flight tube, the electrons are detected on a microchannel plate.

The target gas, xenon, is leaked into the spectrometer through a pulsed valve, mounted in the magnetic pole piece opposite to the TOF section. It allows for local pressures of approximately 10^{-1} Torr, synchronized with the laser pulses, while maintaining an acceptable mean pressure (10^{-6} Torr). The small hole on the right-hand pole piece acts as a pump resistance; the pressure in the time-of-flight part could be kept at 10^{-8} Torr during the experiment.

The plasma is created 20 mm below the spectrometer axis. The liquid

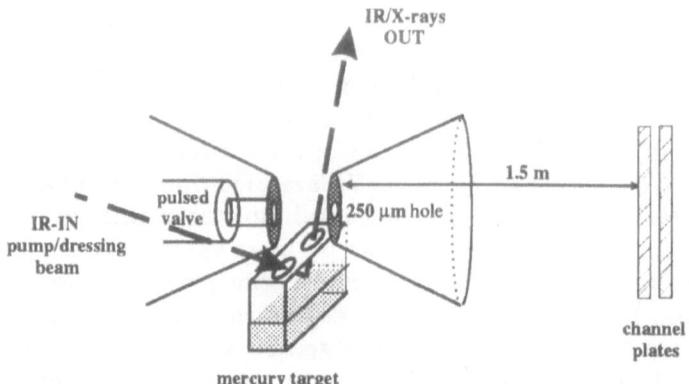

Figure 3.   Schematic of the interaction region. The sensitivity region is located in between the two magnetic polepieces (cones). The 250 μm hole in the right-hand pole-piece determines the diameter of the sensitivity volume. The mercury target is placed 20 mm below this volume. The target gas, xenon, is leaked into the interaction region through a pulsed valve, positioned in the left-hand polepiece. The emitted Auger electrons traverse a 1.5 m time-of-flight region and are detected on a microchannel plate.

mercury is kept in a copper box, uncooled. In the lid of the box, there are two elliptical holes, such that the pumping beam enters through the right hole and exits through the left one (shown in Fig. 2). The size of the holes could be kept small. This helps maintaining an acceptable mean pressure in the spectrometer despite the high vapor pressure of mercury of 1.2×10^{-3} Torr, and screens the polepiece from being directly hit by the x-rays.

When focusing the pump beam by means of a 200 mm focal length lens on the mercury target, the size of the plasma is of the order of 15 μm. This source, as stated above, is placed 20 mm below the sensitivity region. For the Mo/Si mirror used near 1 to 1 imaging, the object lies 10 mm off the optical axis, meaning that the incident cone of rays will strike the mirror at an angle. This results in a line-image with a size of 600 μm due to astigmatism. (This situation could in principle be improved by positioning the plasma closer to the sensitivity region.)

The crucial requirement of the experiment is that all the Auger electrons produced by the imaged 90 eV x-rays should be enveloped by the dressing beam. This imposes the requirement on the dressing beam size that should be larger than the imaged x-ray beam size of 600 μm. By defocusing the telescope through which the dressing beam passes, the divergence of this beam can be chosen such, that the focus created by the Mo/Si mirror for the dressing beam is larger than 600 μm.

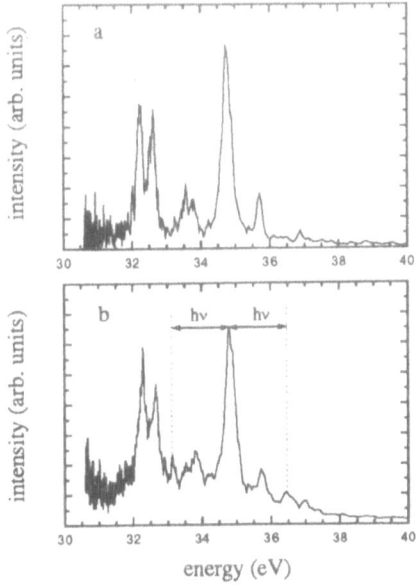

Figure 4. a: Experimental spectrum of the $N_{4,5}$ $O_{2,3}$ $O_{2,3}$ Auger transition in Xe. b: Spectrum of the same Auger transition in the presence of the dressing field. The intensity of the dressing field is about 1.9×10^{11} W/cm^2. Two-sided arrows indicate the positions of the sidebands around the peak at 34.8 eV. The sidebands are separated from this Auger peak by exactly 1.55 eV, the optical-photon energy.

3. Results and Discussion

Fig. 4a shows the $N_{4,5}$ $O_{2,3}$ $O_{2,3}$ Auger spectrum of Xe. This spectrum is taken in absence of the dressing field. The energy of the plasma pumping beam is 500 μJ, the pulse duration is 2 picoseconds and the position of the 200 mm focal length is such, that the pump beam is focused on the mercury target. The collinear setup described in section 2, sets an upper limit on the energy used for pumping the plasma: the liquid-mercury surface might be agitated strongly by the pump beam, if the pump energy gets larger than 3 mJ. This might also influence the dressing beam, which reflects from the same liquid surface.

An undesirable aspect of collecting the x-rays in the specular direction is that in principle, besides refocusing the x-rays, the multilayer mirror could also refocus the IR plasma pumping beam. This IR beam will be specularly reflected from the mercury surface and would in principle be collected and focused in the sensitivity region by the Mo/Si mirror, just as

378

the dressing beam. The occurrence of this undesired process can be easily detected by looking at ATI electrons produced by the infrared light from xenon at the position of the x-ray focus. A huge ATI signal is present under these conditions. This aspect is undesired, because (apart from producing large space charge) this focused IR beam itself can dress the Auger electrons in an uncontrollable way.

In order to circumvent this problem, the pump beam in the interferometer is passed through an aperture, cutting the tails of the spatial profile. A stripe of variable width is finally the beam profile used to pump the plasma with. In front of the multilayer mirror a needle of 2 mm thickness was placed to block the IR pump beam reflecting from the mercury target. This was placed on a rotating stage and the position of complete blocking could be easily found by minimizing the ATI signal. The dressing beam is so much wider that most of the energy passes this needle unhindered.

With a stripe of 20 mm by 5 mm as pumping beam profile, the intensity on the target is about $6 \times 10^{13} \text{W/cm}^2$. Under these pumping conditions, despite of the much larger x-ray focus in the sensitivity region due to astigmatic aberrations of the Mo/Si mirror, Auger electrons from the $N_{4,5} O_{2,3} O_{2,3}$ decay in xenon could be detected. The Auger electron energy spectrum presented in Fig. 4a is a result of averaging over 1000 laser shots. The resolution of our spectrometer permits us to resolve seven Auger lines. This result compares very well with Auger spectra of Xe investigated using synchrotron radiation [10].

The dressed Auger spectrum is shown in Fig. 4b. This spectrum is taken at zero delay between the pump and the probe pulse. The estimated intensity of the dressing field is $1.9 \times 10^{11} \text{W/cm}^2$. At this value, according to the Simpleman's theory, one sideband should appear at each side of each Auger peak. The two arrows in the figure indicate the two sidebands at 1.55 eV around the Auger peak at 34.8 eV. The finestructure of the Auger line obscures the sidebands present around the other peaks in the spectrum.

In order to get efficient overlap, the duration of the dressing pulse, 3 ps, was chosen to be longer than the duration of the pump pulse. The size of the dressing beam in the sensitivity region is estimated to be 900 μm, and is larger than the astigmatic x-ray focus. The energy per pulse in the dressing beam is limited by two factors: the fluence on the multilayer mirror and the intensity of the dressing beam on the mercury target.

The first requirement, that the fluence of the dressing beam on the Mo/Si mirror does not damage the coating, imposes a restriction on both the energy and the size of the beam. The size of the dressing beam on the mirror should be as large as possible. To this end, the divergence of the dressing beam in the Michelson interferometer has to be chosen such that the beam slightly converges. The converging dressing beam will be focused

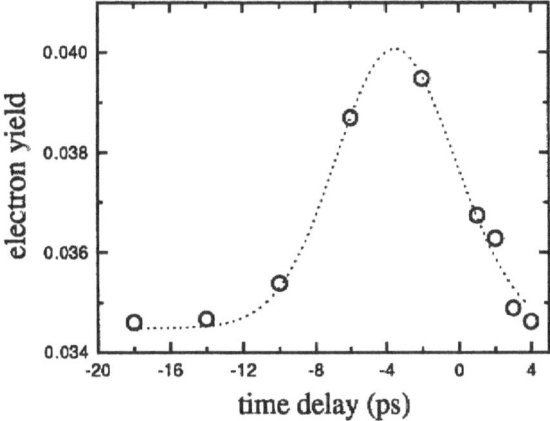

Figure 5. Cross correlation of the optical and the 90 eV x-ray pulses. The datapoints (open circles) are obtained by integrating the electron counts in an energy window around the sideband at 36.4 eV for different values of the time-delay between the pump and the dressing pulse.

by the 200 mm focal length lens in front of the mercury target, not on the mercury surface, resulting in a larger beam size on the Mo/Si mirror itself. This choice of the divergence of the dressing beam is also consistent with the requirement that the beam size in the sensitivity region has to be larger than 600 μm, the x-ray focus.

Furthermore, when the dressing beam is not focused on the mercury target, also the second requirement is met, namely that the intensity on the mercury surface is low enough such that no x-rays are produced by the dressing beam. This is easily checked by monitoring the Auger electrons when using the dressing beam alone with the pump beam blocked.

As discussed in section 1, each type of two-color photoionization process can be used to provide a cross-correlation measurement of the pulses involved. This is also the case for the laser-assisted Auger decay in Xe. By delaying the pump and dressing beam in the Michelson interferometer with respect to each other and monitoring the magnitude of the sideband in the dressed spectrum (Fig. 4b), a cross-correlate of the IR and the 90 eV x-ray pulse can be measured. This measurement has been obtained with the time resolution of the dressing beam, namely 3 ps and is shown in Fig. 5. The FWHM of this cross-correlate is 9 ps.

From the cross-correlation measurement, assuming a gaussian profile, it can be seen that the 90 eV x-ray pulse is long, on the order of 8.5 ps FWHM. This explains the impossibility to improve on the magnitude of the sidebands by changing the energy of the dressing beam. It is the poor temporal overlap of the x-ray pulse with the dressing pulse that limits the magnitude of the sidebands.

4. Conclusions

In principle, the new experimental setup allows measurement of the x-ray pulse duration with a resolution equal to the duration of the dressing pulse. In this experiment, it was not useful to go below 3 ps, because the x-ray pulse itself was much longer (although still shorter than the duration reported by Sher *et al.* [6], who used a much higher pump intensity). Scaling laws for the cooling of the laser-produced plasma [11] predict that shorter x-ray pulses should be obtained for shorter pump pulses, when the energy is lowered to keep the peak temperature constant. However, in our experiment the count rate became too low to perform useful measurements when we attempted this.

For shorter wavelengths the emissivity of the plasma is much higher, so much more energy would be radiated from a plasma with the same cooling time. In future experiments, a cobalt-carbon multilayer mirror will therefore be used to collect 250 eV x-ray photons from the plasma, exploiting the high time resolution of the collinear setup to the fullest.

5. References

1. M.M. Murnane, H.C. Kapteyn, R.W. Falcone, Phys. Rev. Lett. **62**, 155 (1989).
2. J.M. Schins, P. Breger, P. Agostini, R.C. Constantinescu, H.G. Muller,
   G. Grillon, A. Antonetti, and A. Mysyrowicz, Phys. Rev. Lett. **73**, 2180 (1994);
   Phys. Rev. A. **52**, 1272 (1995).
3. J.M. Schins, P. Breger, P. Agostini, R.C. Constantinescu, H.G. Muller, A. Bouhal, G. Grillon, A. Antonetti, and A. Mysyrowicz, JOSA B, to be published.
4. L.D. van Woerkom, R.R. Freeman, S. Davey, W.E. Cooke, and T.J. McIlrath,
   in *High-Energy Density Physics with Subpicosecond Laser Pulses*, Vol.17 of 1989
   Technical Digest Series (Optical Society of America, Washington, D.C., 1989).
5. M.H. Sher, U. Mohideen, H.W.K. Tom, O.R. Wood II, G.D. Aumiller,
   R.R. Freeman, and T.J. McIlrath, Opt. Lett. **18** (1993) 646.
6. C. Le Blanc, G. Grillon, J.P. Chambaret, A. Migus, and A. Antonetti,
   Opt. Lett. **18**, 140 (1993).
7. E. Louis, H.-J. Voorma, N.B. Koster, F. Bijkerk, Yu.Ya. Platonov, S.Yu. Zuev,
   S.S. Andreev, E.A. Shamov and N.N. Salaschenko, Microel. Engineering **27**,
   235 (1995) and in E. Louis, H.-J. Voorma, N.B. Koster, F. Bijkerk, Yu.Ya.
   Platonov, G.E. van Dorssen and H.A. Padmore, Proc. Physics of X-Ray Multilayer Structures, Technical Digest Series vol 6, 35-38 (1994).
8. H. Aksela, S. Aksela, G.M. Bancroft, and K.H. Tan and H. Pulkkinen,
   Phys. Rev. A. **33**, 3867 (1986).
9. M.D. Rosen, in *Femtosecond to Nanosecond High-Intensity Lasers and Applications* M. Campbell, Ed. [Society of Photo-Optical Instrumentation
   Engineers, Bellingham, WA, 1990],Vol.1229, pp.160-169

TWO-COLOR PHOTOIONIZATION USING HIGH-ORDER HARMONIC RADIATION

V. VENIARD, R. TAIEB and A. MAQUET
Laboratoire de Chimie Physique - Matière et Rayonnement
Université Pierre et Marie Curie and CNRS
11, Rue Pierre et Marie Curie 75231 Paris Cedex 05 FRANCE

1. Introduction

The recently observed generation, by samples of atomic rare gas, of high-order harmonics of the fundamental frequency of an infrared laser opens the possibility to have at one's disposal a table-top source of coherent and pulsed VUV, (or even soft X-ray) radiation. The characteristics of such sources, as compared to conventional UV and soft X-ray sources, has recently attracted much attention [1]-[3].

We shall discuss here the main characteristics of a new class of two-color ionization processes, dubbed Laser-Assisted-Single-Photon-Ionization (LASPI). The idea is to expose simultaneously an atomic target to the fields of a high order harmonic radiation with frequency ω_H and the fundamental of the laser, with frequency ω_L, which has been used to generate the harmonic field. Then, if the harmonic is high enough, i.e. is such that : $\omega_H > E_I$, where E_I is the ionization energy of the atom, LASPI can be observed. As we shall show below, our simulations indicate that one can find a regime for the respective intensities of the fields, in which the ionization of the target results of the simultaneous absorption of one single photon from the harmonic field and of the exchange of a number of laser photons. The first experimental observation of this process has been reported at this conference (see R. C. Constantinescu's contribution). Note also that the observation of a similar process, namely Laser-Assisted Auger transitions, has been also reported recently [4].

A simplified picture of LASPI is that of an atom being first brought in the continuum via the absorption of the single high-frequency photon, followed by the exchange of one or several photons from the laser field, as shown in Fig.1

Several general features of this class of processes have already been discussed [5]. Here, we shall present the results of a set of computations, based on the numerical resolution of the time-dependent Schrödinger equation for a (3D-) hydrogen atom, in the presence of two radiation fields.

H. G. Muller and M. V. Fedorov (eds.), Super-Intense Laser-Atom Physics IV, 381–390.
© 1996 *Kluwer Academic Publishers.*

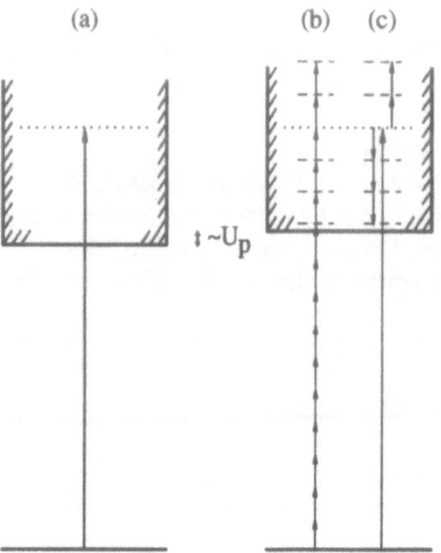

Fig. 1 Energy diagrams schematically representing photoionization processes involving respectively: (a) single-photon ionization through the absorption of one-high frequency photon; (b) multiphoton ATI; (c) laser-assisted single-photon ionization in the simultaneous presence of the IR laser and of its high-order harmonic. The shift ($\sim U_p$) of the ionization threshold in the presence of the (strong) laser is indicated.

Then, we shall compare the photoelectron spectra obtained through this "exact" analysis with those deduced from simplified approaches. Our motivations were twofolds: one was to investigate the validity ranges of simple models commonly used to discuss atom-laser interactions. The other was that, given the relative sophistication of the "exact" calculation, which is quite computer-time consuming, it is of interest to have at one's disposal simplified formulations which can be most useful as a guide for future discussions.

Within this perspective, the organisation of our paper is as follows: in Sec. 2 and 3 we shall briefly present the computation and the main features of the 2-color photoionization spectra. Sec. 4.1 will be dedicated to the comparison with a S-matrix formulation using Coulomb-Volkov waves, proposed some time ago by Leone *et al.* In Sec. 4.2 we shall discuss the "essential states" approach introduced by Rzazewski *et al.*

2. Model and theory

The model we have chosen in order to illustrate the general properties of this class of processes is the one of a hydrogen atom, initially in its ground state, submitted to a pulse of radiation containing two fields, the fundamental of an infrared laser and one of its high-order harmonic. The pulse shape is trapezoidal, its duration is eight laser cycles with one-cycle linear turn-on and turn-off. The polarization for both fields is assumed to be linear and parallel to the quantization axis. We have solved the time-dependent Schrödinger equation for this model:

$$i \frac{\partial \psi(r,t)}{\partial t} = [\, H_{at} + H_L(t) + H_H(t) \,] \, \psi(r,t)$$

where H_{at} is the atomic hamiltonian for hydrogen and $H_L = -z \, F_L(t) \sin(\omega_L t)$ and $H_H = -z \, F_H(t) \sin(\omega_H t + \varphi)$ are the interaction hamiltonians for the two fields (written in the length gauge) with time-dependent amplitudes $F_L(t)$ and $F_H(t)$, the indexes L and H labelling the laser and its harmonic respectively. The angle φ denotes the phase difference between the two fields at time t=0. Note that we have also used the velocity gauge in our calculations. The calculation is conducted along a line similar to the one developed by Kulander *et al* [6]. The computations have been performed on a radial grid from r=0 to r=1250 a.u. with steps δr=0.25 a.u. and we have included up to l_{max}=20 angular momenta in the partial wave expansion when the computation was performed in the velocity gauge. The photoelectron spectra have been determined with the help of a spectral analysis of the atomic wave function, as obtained directly after the pulse [7].

3. Results.

For the sake of illustration, we have considered here the case of a laser frequency equal to that of a Ti:Sapphire operated at ω_L=1.55 eV. The high frequency field consists of its 13th harmonic, i.e. ω_H=20.15 eV. We illustrate in Fig. 2 the influence of the laser intensity I_L on the angle-integrated photoelectron spectra. To this end, we have kept the (13th-) harmonic intensity fixed at the value I_H=3.10^8 W/cm^2. Note that such an intensity seems to be attainable, after focalisation, by currently developed harmonic sources. On the other hand, we have varied the intensity of the fundamental radiation I_L between 0, see Fig. 2a and 2.10^{13} W/cm^2, see Fig. 2f. In the following, we consider only the case of fields with the same phase at time t=0, i.e. φ=0.

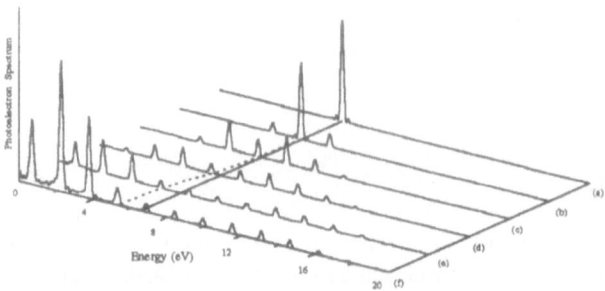

Fig. 2 Effect of the laser intensity on the two-color photoelectron spectra, for radiation pulses containing the fundamental frequency of a Ti:Sapphire laser, $\omega_L = 1.55$ eV and its 13th harmonic with a fixed intensity $I_H = 3.\ 10^8$ W/cm^2. (a) : $I_L=0$ (single-photon ionization); (b): $I_L = 5.\ 10^{11}$ W/cm^2 ; (c): $I_L=3.\ 10^{12}$ W/cm^2 ; (d): $I_L= 8.\ 10^{12}$ W/cm^2 ; (e): $I_L = 1.75\ 10^{13}$ W/cm^2 ; (f): $I_L= 2.\ 10^{13}$ W/cm^2. Note that the peaks are shifted towards the lower energies due to the ponderomotive shift.

The curve (a) represents the single-photon ionization spectrum obtained with the harmonic radiation alone. The maximum of the photoelectron peak is located at the position given by energy conservation $E_0 = \omega_H + E_{1s} = 6.55$ eV. LASPI begins to show up at relatively moderate intensity, see Fig. 2b corresponding to $I_L=5.10^{11}$ W/cm^2. Two satellite peaks are clearly visible on each side of the main peak, located at E_0. At such a low laser intensity, the two satellites can be unambiguously associated to two-photon transitions, implying the absorption of a high-frequency photon ω_H and the absorption (or emission) of a low frequency one ω_L[8].

At higher laser field intensity, the number of satellite peaks increases, as can be seen in Figs. 2c and 2d, corresponding to $I_L=3.\ 10^{12}$ W/cm^2 and $8.\ 10^{12}$ W/cm^2 respectively. We are clearly in a nonperturbative regime as large numbers of photons are exchanged with the atomic system. However, for the short pulse considered here, the laser itself does not contribute to the ionization process. This can be checked by verifying that the total area under the photoelectron peaks is the same as in the laser-free case (Fig. 2a). In these cases, the role of the laser is to redistribute the photoeletrons into continuum dressed states.

Interferences between ATI and LASPI are observed at higher laser field intensity, as shown in Figs. 2e and 2f, corresponding to $I_L=1.75\ 10^{13}$ W/cm^2 and $2.\ 10^{13}$ W/cm^2 respectively. Two important features can be noted in these cases. First, one can see an important growth of the low-energy peaks, which can be ascribed to the opening of new channels, associated to ATI, leading directly to the ionization of the target. Here, the laser radiation contributes to the ionization yield as the relative importance of ATI increases with the laser intensity. Then, one can note also the extension of the spectra towards high energies, corresponding to an increase of the number of satellites.

A remarkable result of the present simulation is that the change in the spectra between the regime in which ATI is almost not visible and the one in which ATI is dominant in the low photoelectron energy range takes place in a narrow laser intensity range : here it occurs between $I_L = 10^{13}$ W/cm^2 and $I_L = 2. \, 10^{13}$ W/cm^2. This is precisely in this range that quantum interferences can be identified. In this situation, the relative phase φ between the two fields plays an important role and governs the relative heights of the photoelectron peaks when both processes (ATI and LASPI) have comparable probabilities. In spite of the fact that the question of the control of the relative phase between the laser field and one of its higher harmonic is still an open problem for the currently developed harmonic sources [1]-[3], we have illustrated this point in Fig. 3. In the figure are shown the variations of the magnitudes of two different photoelectron peaks, which display a typical dependence on the relative phase between the laser field and its harmonic. One observes important changes when the phase is changed from 0 to $\pi/2$ and to π. Note that these changes are notable in spite of the fact that, in contrast with most previous studies on phase effects, the field strengths are quite different, since here the ratio $I_H/I_L = 2. \, 10^{-5}$. They open the possibility to realize a, at least partial, coherent control of the photoelectron current [9], [10].

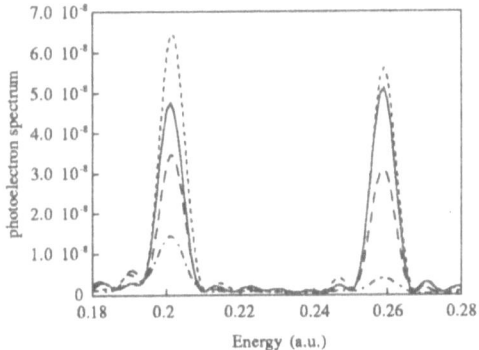

Fig. 3 Typical dependence of two peaks of the two-color photoelectron spectrum on the initial relative phase φ of the laser and its 13th harmonic, with $I_H = 3.10^8$ W/cm^2 and $I_L = 1.75 \, 10^{13}$ W/cm^2. The peaks have energies $E_0 = E_{1s} + \omega_H - U_p$ and $E_1 = E_{1s} + \omega_L + \omega_H - U_p$. The dot-dashed line represents the ATI spectrum for the laser alone. Long-dashed line, $\varphi = 0$; full line $\varphi = \pi/2$; dashed line, $\varphi = \pi$.

4. Comparison with simplified models.

At moderate laser intensities, see for instance Fig. 2b, only a few laser photons are exchanged and a perturbative approach is adequate, see Ref. [8] and A. Cionga's contribution in this volume for a more detailed discussion. On the other hand, in the non-perturbative regime, simple classical arguments [4] can provide correct estimate of the number of sidebands. It is however of interest to have at one's disposal more accurate data and it is in this spirit that we have investigated other approaches, which have been proposed to deal with the non-perturbative regime.

4.1 S-MATRIX METHOD

We have compared our results with the ones derived previously by means of a S-matrix formalism, using Coulomb-Volkov waves, for general two-color IR-UV process with no constraint on the frequencies ratio [11]. As expected for such a model, which does not use exact solutions of the Schrödinger equation, no gauge invariance can be obtained and the comparison has to be made for both the $E.r$ and $A.p$ forms of the interaction.

We show in Fig.4 the results of our simulation for a high frequency photon $\omega_H = 50$ eV and an intensity $I_H = 3.\ 10^9$ W/cm^2 in the presence of an infrared laser field with frequency $\omega_L = 1.17$ eV (Nd:Yag laser) and intensity $I_L = 5.\ 10^{12}$ W/cm^2. The pulse characteristics are the same as before. These values of the frequencies and intensities are the ones chosen by Leone $et\ al$ in Ref [11]. In this case, the frequencies are not multiple from each other, but, as shown in our simulation, the laser intensity is too low for ATI to occur. Accordingly, interference effects cannot affect significantly the magnitude of the photoelectron peaks.

Also shown in the figure are the results of the simplified calculation [11]. As the S-matrix formalism used here does not take into account the finite duration of the pulse, we have to scale the results in order to compare with our values, obtained by means of a fully time-dependent approach. This is done by imposing that the areas under both curves (i.e. the total ionization yield) are the same. More details on the procedure used to perform the comparison will be presented elsewhere. It appears that, in the case presented here, the length form of the interaction hamiltonian compares better with the "exact" calculation than the velocity form. Another interesting result is that, besides this discrepancy, the model gives a correct value of the width of the bunch of satellites lines. This indicates that, for this set of parameters and for such a moderate laser intensity, the Coulomb-Volkov waves provide a fair representation of the dressed continuum states.

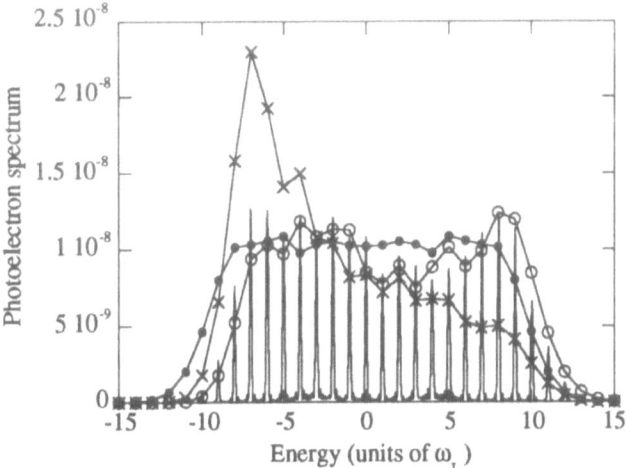

Fig. 4 Two-color photoionization spectra as obtained for a high frequency field with $\omega_H =$ 50 eV and $I_H = 3. \ 10^9$ W/cm^2. The frequency of the laser field is $\omega_L = 1.17$ ev and the intensity is $I_H = 5. \ 10^{12}$ W/cm^2. Full line: "exact " calculation; $\times$: S-matrix formalism, A.p gauge; o : S-matrix formalism, E.r gauge; $\bullet$: "Essential states" model.

In Fig. 5, we compare the results of our simulation to the ones of the simplified model for the case of a Ti:Sapphire laser operated at $\omega_L = 1.55$ eV as discussed earlier. The laser intensity is $I_L = 8. \ 10^{12}$ W/cm^2 and the high frequency field is the 13th harmonic of the fundamental radiation with an intensity $I_H = 3. \ 10^8$ W/cm^2. Again it appears that in the situation in which ATI is negligible, the Coulomb-Volkov waves used in the S-matrix formalism provide a fair description of the process. One can note the discrepancy between the two representations of the interaction hamiltonian (E.r and A.p forms) although it is not obvious in this case which gauge is better.

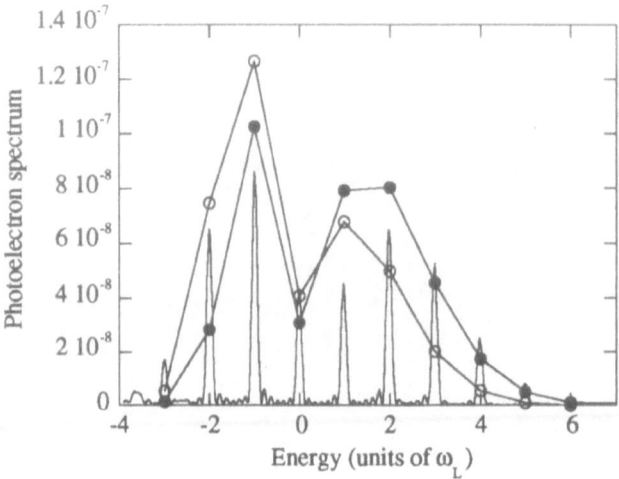

Fig. 5 Two-color photoionization spectra as obtained for a laser field with $\omega_L = 1.55$ eV and $I_L = 8. \ 10^{12}$ W/cm^2. The high frequency field is the 13th harmonic of the laser field and the intensity is $I_H = 3. \ 10^8$ W/cm^2. Full line: "exact " calculation; o : S-matrix formalism, A.p gauge; ● : S-matrix formalism, E.r gauge.

4.2 ESSENTIAL STATES APPROACH

We finally turn to the comparison with a method developed by Rzazewski *et al* [12], related to the essential states model [13]. The model includes one bound state and N continua, each having specific angular momentum, coupled by two fields of different frequencies. The high frequency field ionizes the atom, while the low frequency one redistributes the electron between the continuum states. The electron cannot be ionized directly by the low frequency photons. The continuum-continuum matrix elements depend only on the energy difference between the continuum states and exhibit a singularity as expected for dipole matrix elements. The intensity parameter in the model is F_L/ω_L^2, where F_L is the amplitude of the laser field. Analytic expressions can be obtained for the time evolution of the probability amplitudes without recourse to the rotating frame approximation. On the other hand, the energy range of the continuum is not restricted and thus threshold effect cannot be accounted for.

In Fig. 4 are also shown the results obtained from this simplified model for the same set of parameters as in the previous subsection. As expected for such a high frequency ω_H, where threshold effects can be neglected and for such a low laser intensity, where ATI does not play a role, the agreement between the "exact" calculation and the simplified model is fairly good. The strength of the continuum-continuum coupling, which is an adjustable parameter in the model, has been chosen in such a way that in the perturbative regime ($I_L = 5.10^{11}$ W/cm^2, few photon exchanged) both calculations give the same results. Note that in this model, the duration and the shape of the radiation pulses are taken into account.

The situation is different when the frequency ω_H is lower. We show in Fig. 6 the comparison between our calculation and this simplified model for the 13th harmonic of the Ti:Sapphire laser. The laser intensity is $I_L = 3.10^{12}$ W/cm^2 and the harmonic intensity is $I_H = 3.10^8$ W/cm^2 . The full line is the result of our time-dependent calculation and the open circles are the result of the simplified model obtained by changing the intensity parameter according to F_L/ω_L^2, while keeping the value of the continuum-continuum coupling the same as before. It appears that the agreement for the heights of the photoelectron peaks together with the width of the spectrum is poor. The full circles are obtained by adjusting also the value of the continuum-continuum coupling for this set of parameters. Note that the dashed line indicates the position of the threshold, which is not taken into account in the model.

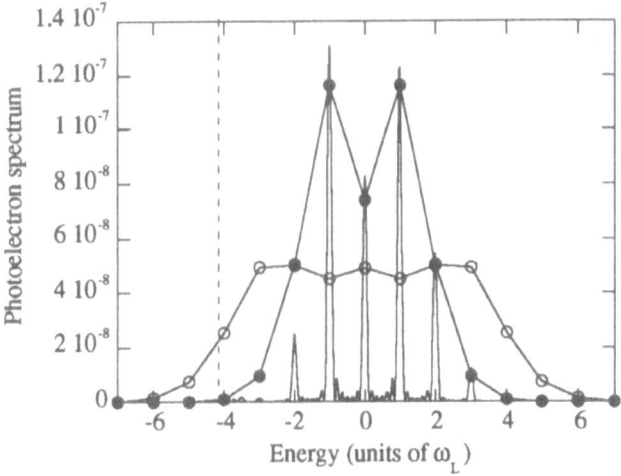

Fig. 6 Two-color photoionization spectra as obtained for a laser field with $\omega_L = 1.55$ eV and $I_L = 3. 10^{12}$ W/cm^2. The high frequency field is the 13th harmonic of the laser field and the intensity is $I_H = 3. 10^8$ W/cm^2. Full line: "exact " calculation; o : simplified model of Rzazewski et al[12]; • : same as before with a different continuum-continuum coupling (see text).

This set of results shows that the "S-matrix+Coulomb Volkov waves" and "essential sates" formalisms lead to predictions which are in fair agreement with the "exact" calculation, when ATI is negligible. The number of satellite lines is in particular correctly reproduced. One of the weakness of the essential states approach however is the need for an adjustable parameter, necessary to approximately represent the continuum-continuum coupling. On the other hand, the problem of the gauge consistency of the S-matrix formalism is not solved.

Acknowledgments
The Laboratoire de Chimie Physique-Matière et Rayonnement is a Unité de Recherche Associée au CNRS, URA 176. This work was partially supported by the EC contracts ERB CHRXCT 920013 and 940470. Parts of the computations have been performed at the Centre de Calcul pour la Recherche (CCR, Jussieu, Paris) and at the Institut du Développement et des Ressources en Informatique Scientifique (IDRIS).

References

[1] L'Huillier, A., Lompré, L. A., Mainfray, G. and Manus, C. (1992) in *Atoms in Strong Fields*, edited by M. Gavrila, Adv. Atom. Molec. Opt. Phys. Suppl. 1, Acad. Press, San Diego, p 139.

[2] Haight, R. and Peale, D. R. (1993) *Phys. Rev. Lett.* **70**, 3979; Haight, R. and Seidler, P. F. (1994) *Appl. Phys. Lett.* **65**, 517.

[3] Larsson, J., Mevel, E., Zerne, R., L'Huillier, A., Wahlström, C-G. and Svanberg, S. (1995). *J. Phys. B* **28** L51.

[4] Schins, J. M., Breger, P., Agostini, P., Constantinescu, R. C., Muller, H. G., Grillon, G., Antonetti, A. and Mysyrowicz, A. (1994) *Phys. Rev. Lett.* **73**, 2180; ibid. (1995) *Phys. Rev. A* **52**, 1272.

[5] Véniard, V., Taïeb, R. and Maquet, A. (1995) *Phys. Rev. Lett.* **74**, 4161.

[6] Kulander, K. C., Schafer, K. J. and Krause, J. L. (1992) in *Atoms in Strong Fields*, edited by M. Gavrila, Adv. Atom. Molec. Opt. Phys. Suppl. 1, Acad. Press, San Diego, p 247.

[7] Schafer, K. J. (1991) *Comp. Phys. Comm.* **63**, 427.

[8] Cionga, A., Florescu, V., Maquet, A. and Taïeb, R. (1993) *Phys. Rev. A* **47**, 1830.

[9] Shapiro, M. and Brumer, P. (1994) *Int. Rev. Phys. Chem.* **13**, 187 and references therein.

[10] Schafer, K. J. and Kulander, K. C. (1992) *Phys. Rev. A* **45**, 8026.

[11] Leone, C., Bivona, S., Burlon, R. and Ferrante, G. (1988) *Phys. Rev A* **38**, 5642; Bivona, S., Burlon, R. and Leone, C. (1992) *Phys. Rev A* **45**, 3268; ibid. (1993) **48**, R3441.

[12] Rzazewski, K., Li Wang and Haus, J. W. (1989) *Phys. Rev A* **40**, 3453; Li Wang, Haus, J. W. and Rzazewski, K. (1990) *Phys. Rev A* **42**, 6784.

[13] Deng, Z. and Eberly, J. H. (1984) *Phys. Rev. Lett.* 53, 1810; ibid (1985) *J. Opt. Soc. Am.* **2**, 486.

LASER ASSISTED PHOTOEFFECT
AT MODERATE INTENSITIES

A. CIONGA
Institute for Atomic Physics,
Institute for Gravitation and Space Sciences,
Bucharest-Măgurele, R-76900 Romania

1. Introduction

Two-color ionization designs a particular case of multiphoton ionization, in which the atom interacts simultaneously with two fields of different frequencies. This topic has recently attracted both experimental [1]-[4] and theoretical [5]-[10] interest. The increasing attention for two-color ionization is generated by the fact it provides additional information [1], over that available from multiphoton ionization by a single frequency field.

By Laser Assisted Photoeffect (LAP) one denotes a distinct case of two-color ionization: the case when one of the fields, of high frequency, is capable to ionize the atom by one photon absorption while, the second one is a laser of low frequency that modifies the ionization process caused by the first one. High-order harmonic generation experiments gave a new impetus to this topic. Schins *et al* [4] reported recently the ionization of He atoms by a combination of two pulses: the fundamental of a Ti:S laser and its 21st harmonic. For more details on this experiment, see the contribution of R. Constantinescu in this volume. V. Véniard presents here also a theoretical study of LAP, based on the numerical integration of the time dependent Schrödinger equation [10].

The aim of this paper is to review an approach, we shall call it *radiative dressing approach*, and its application to the study of LAP in the domain of moderate intensities, where this treatment is suitable. The approach is based on an approximate description of the ejected electron subject simultaneously to the Coulomb field of the nucleus and to the laser field [11] and it has been used in the context of LAP in Refs.[8, 9]. We begin by recalling in Sec. 2 the basic formulae that allow the study of LAP. Only the key features of the radiative dressing approach are outlined in Sec. 3.

H. G. Muller and M. V. Fedorov (eds.), Super-Intense Laser-Atom Physics IV, 391–400.
© *1996 Kluwer Academic Publishers.*

Some theoretical results predicted for the LAP in hydrogen are discussed in Sec. 4. Special attention is paid to the case in which the two fields are described by a set of parameters that are related to the experimental ones [4], for which results based on our approach are presented for the first time.

2. Basic Formulae

We shall focus our attention on the case of hydrogen, initially in the ground state. The process we are interested in is formally represented as:

$$H(1s) + \gamma(\vec{\epsilon}_H, \omega_H) + N\gamma(\vec{\epsilon}, \omega) \rightarrow H^+ + e^- \left(E_f, \vec{k}_f\right) + (N + \nu)\gamma(\vec{\epsilon}, \omega). \quad (1)$$

$\vec{\epsilon}$ and ω denote the polarization and frequency of the photons, the subscript H refers to the high frequency one. The ejected electron has the energy E_f and the asymptotic momentum $\hbar \vec{k}_f$. Positive values of ν correspond to the stimulated emission and the negative ones to the absorption of laser photons. We consider here the case in which the low frequency field is a relatively intense laser, but the high frequency one is weak.

The total hamiltonian of this system is

$$H = H_{atom} + V_{laser} + V_H, \quad (2)$$

where V_{laser} and V_H denote the interaction with the laser and the high frequency field, respectively. Because the high frequency field is weak, we shall treat the corresponding interaction as a perturbation.

We adopt here the semiclassical treatment; the electromagnetic fields are assumed to be monochromatic, spatially homogeneous plane waves. Choosing the Coulomb gauge, the corresponding vector potentials are given by

$$\vec{A}_H(t) = A_{0H}\vec{\epsilon}_H \exp(-i\omega_H t + i\varphi_H) + \text{c.c.}, \quad \vec{A}(t) = A_0\vec{\epsilon} \exp(\omega t) + \text{c.c.}, \quad (3)$$

where c.c. denotes the complex conjugate of the first term. The relative phase between the fields, φ_H is a relevant parameter only for the case of commensurable frequencies [1]. For both interactions, we shall use the dipole approximation in the velocity form.

The time evolution of the system is governed by the following evolution operator:

$$\begin{aligned} U(t, t_0) &= U_0(t, t_0) + \frac{1}{i\hbar} \int_{t_0}^{t} U_0(t, t') V_H(t') U(t', t_0) dt' \\ &\simeq U_0(t, t_0) + \frac{1}{i\hbar} \int_{t_0}^{t} U_0(t, t') V_H(t') U_0(t', t_0) dt', \quad (4) \end{aligned}$$

where $U_0(t, t_0)$ is the evolution operator connected to the unperturbed hamiltonian, taken as $H_0 \equiv H_{atom} + V_{laser}$.

As a consequence, the transition amplitude to find the ejected electron at the time t in a state $\mid f\rangle$, if it was initially in a state $\mid i\rangle$, is given by

$$\mathcal{M}_{i \to f} \simeq \langle f \mid \Phi_i(t)\rangle + \frac{1}{i\hbar} \int_{t_0}^{t} \langle \Phi_f(t') \mid V_H(t') \mid \Phi_i(t')\rangle dt'. \qquad (5)$$

Here

$$\mid \Phi_i(t)\rangle = U_0(t, t_0) \mid i\rangle, \quad t_0 \to -\infty, \qquad (6)$$
$$\mid \Phi_f(t')\rangle = U_0(t', t) \mid f\rangle, \quad t \to \infty, \qquad (7)$$

determine the evolution of the eigenstates of H_{atom} ($\mid i\rangle$ and $\mid f\rangle$) in the laser field only, and have the appropriate asymptotic behavior.

The form of the transition amplitude in Eq. (5) is the starting point of this analysis. The main task is therefore to find the states $\mid \Phi_i(t)\rangle$ and $\mid \Phi_f(t)\rangle$, describing the atomic system in the laser field. They are solutions of the Schrödinger equation for a charged particle which experiences simultaneously the Coulomb field of the nucleus and the laser field

$$\frac{\hbar}{i} \frac{\partial}{\partial t} \Phi(\vec{r}, t) + \left[\frac{1}{2m} \vec{P}^2 + \frac{e}{m} \vec{A}(t) \cdot \vec{P} + \frac{e^2}{2m} \vec{A}^2(t) - \frac{e^2}{4\pi\varepsilon_0 r}\right] \Phi(\vec{r}, t) = 0, \quad (8)$$

where m is the electron mass, $-e$ ($e > 0$) is its electric charge and ε_0 is the dielectric permittivity of the vacuum. In what follows, the $\vec{A}^2$ term will be eliminated through a unitary transformation.

There is no exact solution of this equation. For $\vec{A}(t) = 0$, Eq.(8) describes the hydrogen atom; when the Coulomb potential of the nucleus is neglected, one get the Volkov solution [12] that describes a "free" electron in the presence of a monochromatic electromagnetic field

$$\chi_V(\vec{r}, t) = e^{-i\vec{k} \cdot \vec{\alpha}(t)} \chi_0(\vec{r}, t). \qquad (9)$$

Here $\chi_0(\vec{r}, t)$ denotes the plane wave and $\vec{\alpha}$, defined by

$$\vec{\alpha}(t) = \frac{e}{m} \int^{t} \vec{A}(\tau) d\tau, , \qquad (10)$$

corresponds to the "quiver" motion of a classical electron in the laser field; the parameter $\alpha_0 = eA_0/m\omega$ represents the amplitude of this motion.

We mention here, in order of increasing sophistication, three nonperturbative models for the study of LAP. Using Eq.(8), they approximate in different ways the dressing of the bound and continuum states:

- KFR-like calculations [13, 14] are based on the Volkov solution (9) to describe the ejected electron. They do not take into account the dressing of the initial state and the Coulomb field of the nucleus in the final one.
- Coulomb-Volkov approach [15, 16] is an attempt to remove the later deficiency, using a Coulomb-Volkov solution

$$\phi(\vec{r}, t) = e^{-i\vec{k}\cdot\vec{a}(t)} \Psi_{\vec{k}}^{-}(\vec{r}, t), \tag{11}$$

instead of (9) as a final state. $\Psi_{\vec{k}}^{-}(\vec{r}, t)$ is a Coulomb wave with incoming behavior. Note that, in this model, the ground state is still "undressed".

- The *radiative dressing approach* has been used to treat LAP in the moderate intensity regime [8, 9]. It is based on the radiative dressing of both the ground and the Coulomb-Volkov states. Some results have been also presented in a review lecture devoted to two-color processes [17].

All these models have the merit to allow analytic calculations and are of interest in gaining useful qualitative understanding of the physical process. A critical review of the first two models is given in Ref.[8]. The next section is devoted to the third approach.

3. Radiative Dressing Approach

As long as the laser field intensity remains moderate, the dressing of the bound state can be safely described [8] to first order in the perturbation theory:

$$\Phi_{1s}(\vec{r}, t) = e^{\left(-\frac{i}{\hbar}E_1 t\right)} \left\{ \psi_{1s}(\vec{r}) - \frac{\alpha_0\omega}{2} \left[e^{-i\omega t} \vec{\epsilon} \cdot \vec{w}_1(\Omega_1; \vec{r}) + e^{i\omega t} \vec{\epsilon}^* \cdot \vec{w}_1(\Omega_2; \vec{r}) \right] \right\}, \tag{12}$$

where ψ_{1s} represents the stationary ground state wave function. We adopt here a notation that displays "the linear response" of the hydrogen atom, studied in Ref.[18]

$$\vec{w}(\Omega; \vec{r}) = G_C(\Omega) \vec{P} u(\vec{r}), \tag{13}$$

where G_C is the Coulomb Green function associated to $H_C = \vec{P}^2/2m - e^2/4\pi\varepsilon_0 r$ and $u(\vec{r})$ is an eigenfunction of H_C. The parameters Ω entering Eq.(12) have the following values:

$$\Omega_{1,2} = E_1 \pm \hbar\omega + i0. \tag{14}$$

Note also the existence of an equivalent formalism, based on the Sturmian expansion of the Coulomb Green function [19].

The radiative dressing approach for LAP is based on the following approximate solution:

$$\Phi_{\vec{k}}(\vec{r}, t) = e^{\left(-\frac{i}{\hbar}E_k t\right)} e^{-i\vec{k}\cdot\vec{a}(t)} \left\{ \left[1 - \frac{\alpha_0}{2}\vec{k}\cdot\left(e^{-i\omega t}\vec{\epsilon} - e^{i\omega t}\vec{\epsilon}^*\right)\right] \psi_{\vec{k}}^-(\vec{r}) \right.$$

$$\left. - \frac{\alpha_0\omega}{2}\left[e^{-i\omega t}\vec{\epsilon}\cdot\vec{w}^-(\Omega_1';\vec{r}) + e^{i\omega t}\vec{\epsilon}^*\cdot\vec{w}^-(\Omega_2';\vec{r})\right]\right\}, \tag{15}$$

to describe the ejected electron in the laser field. This is an improvement of the ansatz proposed by Banerji and Mittleman [21], improvement that takes into account first order radiative corrections to a Coulomb-Volkov wave [11]. $\psi_{\vec{k}}^-$ represents a Coulomb state with incoming behavior and $\vec{w}^-$ is the linear response in Eq.(13) for $u(\vec{r}) \equiv \psi_{\vec{k}}^-(\vec{r})$, see Ref.[20]. In this case the parameters Ω' are given by

$$\Omega_{1,2}' = E_k \pm \hbar\omega + i0. \tag{16}$$

The functions given by Eqs.(12) and (15) are those we use as approximate expressions for $| \Phi_i(t) \rangle$ and $| \Phi_f(t) \rangle$ in the transition amplitude (5). Supposing the fields are switched on and off adiabatically, the general structure of the S-matrix related to this amplitude becomes

$$S_{i,f} = \sum_{\nu=-\infty}^{\infty} S^{(\nu)} \delta\left(E_{k_\nu} - E_1 - \hbar\omega_H + \nu\hbar\omega\right), \tag{17}$$

where the magnitude of the final momentum $\vec{k}_\nu$ satisfies the energy conservation relation:

$$\frac{\hbar^2 k_\nu^2}{2m} = E_1 + \hbar\omega_H - \nu\hbar\omega. \tag{18}$$

$S^{(\nu)}$ is associated to the exchange of ν laser photons, in addition to the absorption of one high-frequency one, and has the expression [8, 9]

$$S^{(\nu)} = -2\pi i\left(f_I + f_{II} + f_{III}\right)\left(f_{pol}\right)^\nu, \tag{19}$$

where

$$f_I = \frac{eA_{0X}}{2m}\langle\psi_{\vec{k}_\nu}^- | \vec{\epsilon}_H \cdot \vec{P} | \psi_{1s}\rangle J_\nu(\zeta), \tag{20}$$

$$\begin{aligned} f_{II} = \quad & \frac{eA_{0X}}{4m}\left\{\zeta\langle\psi_{\vec{k}_\nu}^- | \vec{\epsilon}_H \cdot \vec{P} | \psi_{1s}\rangle\right. \\ & - \alpha_0\omega\left[\langle\vec{\epsilon}^* \cdot \vec{w}^-(\Omega_2') | \vec{\epsilon}_H \cdot \vec{P} | \psi_{1s}\rangle\right. \\ & \left.\left. + \langle\psi_{\vec{k}_\nu}^- | \vec{\epsilon}_H \cdot \vec{P} | \vec{\epsilon}\cdot\vec{w}_1(\Omega_1)\rangle\right] f_{pol}\right\} J_{\nu+1}(\zeta), \end{aligned} \tag{21}$$

and

$$
\begin{aligned}
f_{III} = \quad - \quad & \frac{eA_0 X}{4m} \Big\{ \zeta \langle \psi_{\vec{k}_\nu}^- \mid \vec{\epsilon}_H \cdot \vec{P} \mid \psi_{1s} \rangle \\
+ \quad & \alpha_0 \omega \left[\langle \vec{\epsilon} \cdot \vec{w}^- (\Omega_1') \mid \vec{\epsilon}_H \cdot \vec{P} \mid \psi_{1s} \rangle \right. \\
+ \quad & \left. \langle \psi_{\vec{k}_\nu}^- \mid \vec{\epsilon}_H \cdot \vec{P} \mid \vec{\epsilon}^* \cdot \vec{w}_1 (\Omega_2) \right] f_{pol}^* \Big\} J_{\nu-1} (\zeta).
\end{aligned}
\tag{22}
$$

Due to the polarization dependence of the Volkov exponential $e^{-i\vec{k}\cdot\vec{a}(t)}$ in Eq.(15), the factor f_{pol} and the argument ζ of the Bessel function J_ν are polarization dependent, see Ref.[9].

Let us make some remarks on Eqs.(20-22). While f_I contains only the matrix element for one-photon bound-free transition, the other two amplitudes contain also matrix elements for two-photon bound-free transitions. f_{II} contains the matrix elements for two-photon ionization (which multiply f_{pol}), and f_{III} contains the matrix elements for stimulated Compton (which multiply f_{pol}^*). This structure is entirely determined by the *radiative dressing approach* we have adopted.

For fixed energy of the ejected electron, which implies fixed value of ν, the differential cross-section is obtained from $S^{(\nu)}$ as

$$
\frac{d\sigma^{(\nu)}}{d\Omega} = \frac{1}{\pi} \frac{\omega_H}{I_H} \mid S^{(\nu)} \mid^2 .
\tag{23}
$$

The structure (17) of the S-matrix is related to the sidebands exhibited by the experimental energy spectrum of the ejected electrons [4]. We stress that, although our Eq.(19) predicts the exchange of any number of laser photons, due to the hybrid character of the dressing scheme for the ejected electron, only the cases with $\nu \pm 1$ are consistently described. Nevertheless, the analysis in Sec.4 suggests that the major contribution to $\nu = 0$ comes from the terms in Eqs.(19-22). An analysis, including the next term in the radiative dressing is in progress [22]; it will allow a consistent study of the next pair of sidebands ($\nu = \pm 2$). Therefore, we choose to discuss here only the cases $\nu = 0, \pm 1$.

4. Results and Discussion

The radiative dressing scheme leads to important contributions near the ionization threshold. This statement is illustrated in Figure 1, where we present the differential cross-section of the ejected electron, Eq.(23), for the first upper sideband ($E_f = E_1 + \hbar\omega_H + \hbar\omega$ that means $\nu = -1$), but a similar behavior has been observed for the first lower one ($\nu = 1$). The dressing field is a Nd:YAG laser ($\hbar\omega = 1.17$ eV) and its polarization is parallel to that of the high-frequency field ($\vec{\epsilon} \parallel \vec{\epsilon}_H$), chosen as quantization axis. θ and

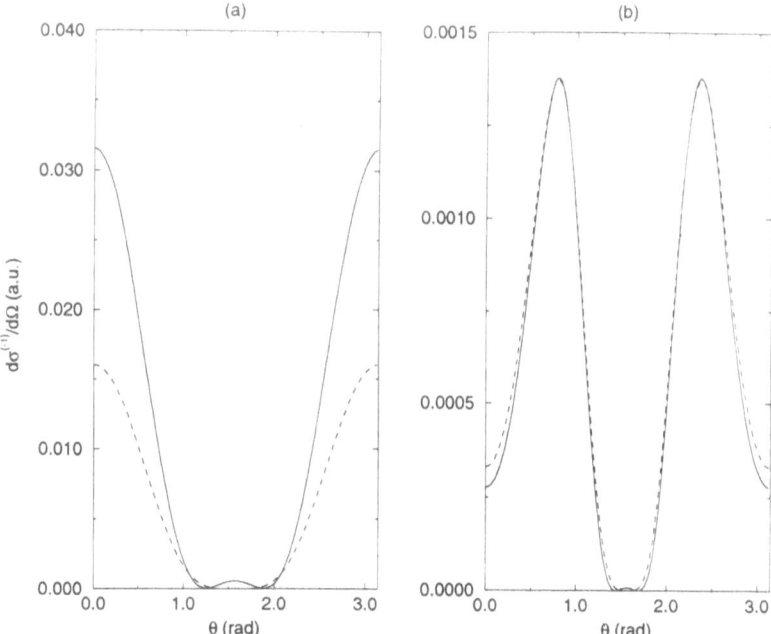

Figure 1. (a) The differential cross section of ejected electrons, Eq.(23), as a function of the scattering angle θ for $\hbar\omega = 1.17$ eV and $\hbar\omega_H = 16$ eV. The laser intensity is $I = 5 \times 10^{11}\,\mathrm{W/cm}^2$ and $\vec{\epsilon}\|\vec{\epsilon}_H$. The full line corresponds to the S−matrix in Eqs.(19-22), the dashed one to the case when only the amplitude (20) is computed. (b) Idem, but $\hbar\omega_H = 50$ eV.

ϕ are then the polar angles of $\vec{k}_\nu$. Fig.1(a) corresponds to $\hbar\omega_H = 16$ eV and Fig.1(b) to $\hbar\omega_H = 50$ eV. Close to the threshold, for 16 eV, the differential cross-section computed with $S^{(-1)}$ given by Eqs.(19-22) (full line), differs significantly from that (dashed line) in which only the first term, f_I, is taken into account in Eq.(19). The entire dynamics is simpler when the electron is ejected far into the continua, therefore less sophisticated models give adequate results, too. See a detailed analysis in the contribution of V. Véniard.

The radiative dressing approach is appropriate to study LAP in the domain of moderate field intensities. In the case of a Nd:YAG laser, we estimated [8] this domain below $10^{13}\mathrm{W/cm}^2$. Recent comparisons with results based on the numerical integration of the TDSE [10] emphasize that, in the case of Ti:S laser ($\hbar\omega = 1.55$ eV) and its 13th harmonic ($\hbar\omega_H = 20.15$ eV), our dressing scheme gives correct predictions below $10^{12}\mathrm{W/cm}$. In Figure 2 we show the total cross section, implying an angle integration of Eq.(23), for the main peak ($\nu = 0$) and the first pair of sidebands ($\nu = \pm1$). This graph

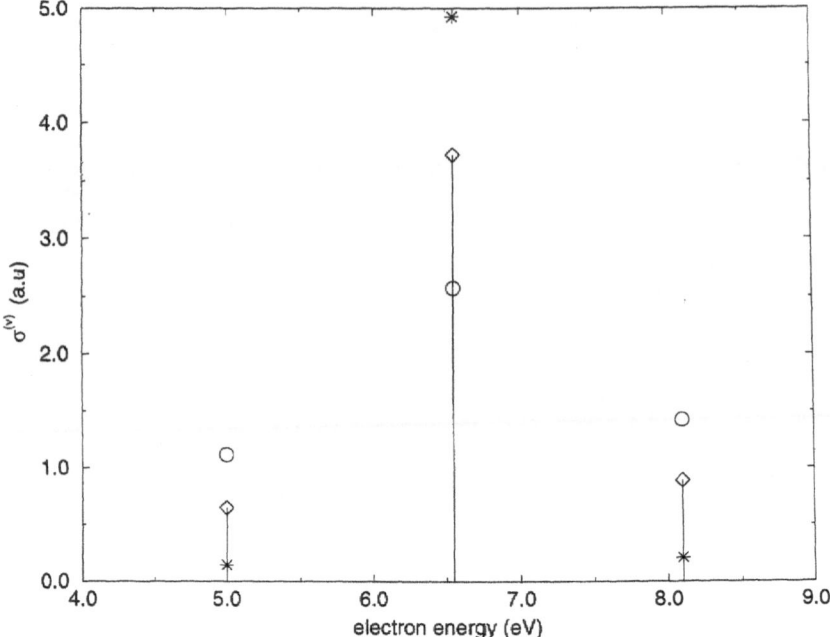

Figure 2. Total cross section for the main peak ($E_f \simeq 6.55$ eV, $\nu = 0$) and the the first pair of satellites ($E_f \simeq 5$ eV, $\nu = 1$; $E_f \simeq 8.10$ eV, $\nu = 1$) in the case of a Ti:S laser and its 13th harmonic. The laser intensity is 10^{11} W/cm^2 (star), 5×10^{11} W/cm^2 (diamond), and 10^{12} W/cm^2 (circle)

emphasizes the intensity dependence of the corresponding "electron spectrum". We choose to illustrate the following values of the laser intensity: 10^{11} W/cm (star), 5×10^{11} W/cm (diamond), and 10^{12} W/cm (circle).

In the validity domain of the radiative dressing approach, not only the total cross section, but also the angular distributions are in agreement with the "exact" ones [23]. As we expected, only qualitative agreement could be obtained for higher intensities, where higher-order corrections in the dressing procedure become important and the number of sidebands increases.

We have also investigated the differential cross sections of ejected electrons for a set of parameters which are related to the experiment of Schins *et al*, [4] which used a Ti:S laser and its 21st harmonic. In Figure 4 we show the differential cross sections corresponding to the main peak ($\nu = 0$) and its nearest sidebands ($\nu = \pm 1$). The intensity of the field is 10^{11} W/cm^2 in Fig.4(a), 5×10^{11} W/cm^2 in Fig.4(b), and 10^{12} W/cm^2 in Fig.4(c). Important deviations of the angular distribution from the perturbative shape [8], are observed even at low laser intensities. We estimate that for the last intensity, the contribution of higher order corrections in the dressing

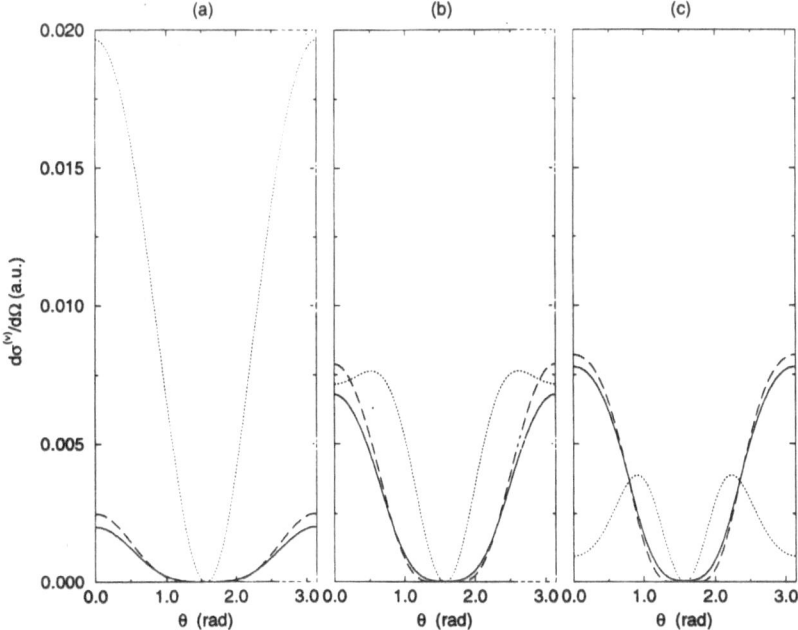

Figure 3. (a)The differential cross sections in Eq.(23) for Ti:S laser and its 21st harmonic, corresponding to $\nu = 0$ (dotted line), $\nu = 1$ (full line), and $\nu = -1$ (dashed line). The laser intensity is $I = 10^{11}\,\mathrm{W/cm^2}$. (b) same as (a) for $I = 5 \times 10^{11}\,\mathrm{W/cm^2}$. (c) same as (a) for $I = 10^{12}\,\mathrm{W/cm^2}$.

scheme should be investigated, too. Due to the higher dipole polarizability of hydrogen, the sidebands appear in its electron energy spectrum at lower intensities than those necessary for helium ($10^{12}\mathrm{W/cm^2}$) [4].

The compact forms (19-22) of the S-matrix element allow to study in detail the effects due to the polarization of the fields. For linear polarizations, the θ dependence of the differential cross-section was investigated in Ref.[8] at low and moderate intensities of the laser field.

We have also investigated [9] the azimuthal angular distribution of the ejected electrons. The axial symmetry, valid for parallel polarizations, leads to a ϕ-independent angular distribution. This symmetry is broken for orthogonal polarizations and the angular distribution in the azimuthal plane behaves like $\cos^2\phi$; the θ-dependence of the differential cross-section modulates the radius of the circle and the magnitude of the "butterfly".

The interest for the study of circular polarization was generated by the experiment of Bashkansky *et al* [24], in which asymmetric angular distributions of the ejected electrons have been detected for multiphoton ionization of noble gases by elliptically polarized light. In a *case study* [9] that mimics the experiment, we have predicted asymmetric angular distributions, too.

Acknowledgments

It is a pleasure to thank V. Florescu, A. Maquet, and R.Taïeb, who contributed to this project. Part of the computations have been performed in the Computer Center of Quantum and Statistical Physics Group (Bucharest-Măgurele), supported by SOROS Foundation. The author is indebted to P. Agostini for interesting discussions. A critical reading of the manuscript by V. Florescu is warmly acknowledged.

References

1. Muller, H.G., van Linden van der Heuvell, H.B. and van der Wiel, M.J. (1986) *J. Phys B.* **19**, L733; Muller, H.G., Bucksbaum, P.H., Schumacher, D.W. and Zavriyev, A. (1990) *J. Phys B.* **23**, 2761; Muller, H.G., (1990) *Comments At. Mol. Phys.* **24**, 355
2. Schumacher,D.W., Weihe, F., Muller, H.G. and Bucksbaum, P.H. (1994) *Phys. Rev. Lett.* **73**, 1344
3. Watanabe,S., Kondo,K., Nabekawa,Y., Sagisaka,A. and Kobayashi,Y. (1994) *Phys. Rev. Lett.* **73**, 2692
4. Schins, J.M., Breger, P., Agostini, P., Constantinescu, R.C., Muller, H.G., Bouhal, A., Grillon, G., Antonetti, A. and Mysyrowicz A. (1995) to be published
5. Dörr, M., Potvliege, R.M., Proulx, D. and Shakeshaft, R. (1991) *Phys. Rev. A* **44**, 547
6. Potvliege, R. M. and Smith, P.H.G. (1991) *J. Phys. B* **24**, L641; Potvliege, R. M. and Smith, P.H.G. (1992) *J. Phys. B* **25**, 2501
7. Schafer, K.J. and Kulander, K.C. (1992) *Phys. Rev. A* **45**, 8062
8. Cionga, A., Florescu, V., Maquet, A. and Taïeb, R. (1993) *Phys. Rev. A* **47**, 1830
9. Cionga, A., (1993) *Rom.J.Phys* **38**, 483
10. Véniard, V., Taïeb, R. and Maquet, A., (1995) *Phys. Rev. Lett.* **74**, 4161
11. Joachain, C.J., Francken, P., Maquet, A., Martin, P. and Véniard, V. (1988) *Phys. Rev. Lett.* **61**, 165 ; see also Martin, P., Véniard, V., Maquet, A., Francken, P. and Joachain, C.J. (1989) *Phys. Rev. A* **39**, 6178
12. Faisal, F.H.M., (1987) *Theory of Multiphoton processes*, Plenum Press, New York
13. Jain, M. and Tzoar, N. (1977) *Phys. Rev. A* **15**, 147
14. Fonseca, L.A. and Nunes, A. C. (1988) *Phys. Rev. A* **37**, 400
15. Leone, C., Bivona, S., Burlon, R. and Ferrante, G. (1988) *Phys. Rev. A* **38**, 5642; Leone, C., Bivona, S., Burlon, R. and Ferrante, G. (1989) *J. Mod. Opt.* **36**, 909
16. Bivona, S., Burlon, R., Leone, C. and Ferrante, G. (1989) *Nuovo Cimento D* **11**, 1751
17. Florescu, V. (1994) in *Topics in Atomic and Nuclear Collisions*, eds. Remaud, B., Calboreanu, A. and Zoran A. Plenum Press, New York, p. 343
18. Florescu, V. and Marian, T. (1986) *Phys. Rev. A* **34**, 4641
19. Maquet, A., Martin, P. and Véniard, V. (1988) *Phys. Lett. A* **129**, 26
20. Florescu, V. (1986) *Phys. Lett. A* **115**, 147
21. Banerji, J. and Mittleman, M.H. (1981) *J. Phys. B* **13**, 3717
22. Fifirig, M., Cionga, A., Florescu V. (1995) to be published
23. Taïeb, R., Véniard, V. and Maquet, A. (1995) to be published
24. Bashkansky, M., Bucksbaum, P.H., and Schumacher, D.W., (1988) *Phys. Rev. Lett.* **60** 2458

LASER-ELECTRON SCATTERING AT RELATIVISTIC INTENSITIES

D.D. MEYERHOFER
Dept. of Mechanical Engineering
and
Laboratory for Laser Energetics

250 E. River Rd.
Rochester, NY 14623

Tel: 716-275-7769
email: ddm@lle.rochester.edu

Abstract

Two recent experiments have observed laser-electron scattering at relativistic intensities. In both experiments the average quiver energy of an electron in the field approaches its rest mass. In one experiment, low energy electrons are born into the laser field and subsequently interact with the intense laser field. The energy and angular distribution relative the laser wave vector show the effects of the relativistic quiver motion, including the mass shift in the field. In the second experiment the laser interacts with 47 GeV electrons. Multiphoton Compton scattering has been observed with up to four laser photons have been simultaneously scattered. In this paper, the two experiments are described briefly and relationship between the two experiments is discussed.

1. Introduction

In two recent experiments,[1, 2] laser-electron scattering has been measured at laser intensities where the average quiver energy of an electron in the field approaches its rest mass. At these intensities, multiphoton-electron scattering becomes probable. Classically this can be viewed as a consequence of the anharmonic motion of an electron in the laser field. The anharmonicity is due both to the magnetic component of the electromagnetic field and due to the relativistic mass shift of the electron. Multiphoton scattering and the mass shift also appear in the quantum mechanical

H. G. Muller and M. V. Fedorov (eds.), Super-Intense Laser-Atom Physics IV, 401–410.
© 1996 *Kluwer Academic Publishers.*

402

calculation. One of the experiments involves the interaction of the laser with electrons which are born with essentially zero energy in the field and are subsequently accelerated out of the laser focus.[2] The multiphoton or anharmonic nature of the interaction is manifest in the observed transfer of laser momentum to the electrons. The photon energy is much less than the electron rest energy. In the second experiment, the laser interacts with counter-propagating 47 GeV electrons.[1] In the electron rest frame, the photon energy is comparable to the electron rest energy and a quantum mechanical description of the interaction is required. The scattered electrons have energies below the Klein-Nishina limit indicating a multiphoton interaction with the laser field.

This paper will discuss the relationship between the two experiments and between the classical and quantum mechanical interpretation. The multiphoton, or relativistic, laser electron interaction is described in Sec. 2 and the two experiments are described briefly in Sec. 3. The relationship between the two experiments is discussed in Sec. 4.

2. Relativistic laser-electron interactions

The multiphoton interaction is characterized both quantum mechanically[3-10] and classically[11, 12] by

$$\eta^2 = \left\langle \frac{e^2 \vec{A}^2}{m_0^2 c^2} \right\rangle,$$ (1)

where m_0 is the electron rest mass, $\vec{A}$ is the vector potential of the electromagnetic wave and $\langle \rangle$ represents an average over the laser cycle. It is clear from Eq. 1 that η is relativistically invariant. η^2 represents the average of the square of the transverse momentum divide by $m_0 c$ and is related to the ponderomotive potential (average quiver energy) of the laser by $\Phi_{pond} = \eta^2 mc^2/2$.[12]

Quantum mechanically, two processes occur during the interaction of the intense laser field with free electrons. The electron is dressed by the laser field so that its effective mass becomes $m^* = m_0\sqrt{1+\eta^2}$ and multiphoton scattering becomes probable.[3, 4, 10] The probability of an n-photon scattering event,

$$e^- + n\omega \rightarrow e'^- + \gamma_{sc},$$ (2)

is related to η^2 by $P_n \propto \eta^{2n}$.

In the Compton regime, where the photon energy is comparable to or greater than the electron rest energy, the scattered photon energy can be related to the incident by for photons colliding with electrons with energy $\gamma m_0 c^2$. The maximum backscattered photon energy for head-on collisions is[10]

$$\gamma_{sc} = \frac{4n\omega\gamma^2}{1 + \dfrac{4n\gamma\omega}{mc^2} + \eta^2}. \tag{3}$$

This corresponds to a minimum scattered electron energy of

$$E_{min} = \frac{\gamma m_0 c^2}{1 + \dfrac{4n\gamma\omega}{m^* c^2}}. \tag{4}$$

Both of these formulae reduce to the well-known Klein-Nishina formulae as $\eta \to 1$ and $n \to 1$. Thus, η^2 enters both in the rate of multiphoton emission and in causing a shift in the frequency. In the Thomson regime, where the photon energy is much less than the electron rest energy, harmonic emission and a mass shift of the emitted radiation are also governed by η^2, though the electron recoil can be ignored.[5-9] Under these conditions the mass shift can be thought of as a Doppler shift associated with the forward momentum of the electrons in the field.[8] The forward momentum arises from conservation of energy and momentum in the dressing of the electron.

The interaction can also be considered classically in the Thomson regime. In this case, the laser pulse is considered as a classical electromagnetic field, and the electron trajectory can be solved.[12-14] The trajectory includes anharmonic motion of the electron in the field and a forward drift. The anharmonic motion is a figure-8 motion in the plane made by the laser polarization and k-vector for linear polarization.[12-14] The nonlinear motion leads to the radiation of harmonics from the laser,[11, 12, 15] while the forward motion leads to a wavelength shift of the harmonics as shown in the quantum mechanical calculations.[8]

In an intense field, the interaction of the laser pulse with free electrons yields shifted harmonics of the laser frequency. This is true in both the Compton and Thomson regimes, though in the Compton regime, the scattered photon energy can be much higher in the laboratory frame than in the rest frame of the electron.[10]

The high intensity laser-electron interaction is also characterized by

$$\Psi = \frac{E}{E_{crit}}. \tag{5}$$

When the laser field, E, approaches the QED critical field[16], E_{crit}, photon-photon scattering becomes possible,[3, 4, 10, 17]

$$\gamma + n\omega \rightarrow e^- + e^+. \qquad (6)$$

The QED critical field corresponds to a laser intensity of 4×10^{29} W/cm^2, well beyond the reach of current laser technology. This intensity can be reached, however, in the rest frame of an energetic electron because the laser intensity transforms as $I_{rest} = 4\gamma^2 I_{lab}$. Thus photon-photon scattering may be observed during the scattering of an intense laser with energetic electrons in two-step process.[17, 18] The high energy gammas produced by Compton scattering subsequently interact with the laser field, leading to the production of electron positron pairs.

3. Description of the Experiments

In two recent experiments,[1, 2] laser-electron scattering has been measured at laser intensities where the average quiver energy of an electron in the field approaches its rest mass. In one experiment, the electron is born at rest so that the laser photon energy is much less than the electron rest energy, while in the second, the interaction is with 47 GeV electrons so that the photon energy is comparable to the electron rest energy in the rest frame. In both experiments, multiphoton-electron scattering has been observed. They are described briefly below.

3.1 LOW ELECTRON ENERGY EXPERIMENT

In the low electron energy experiment, electrons are born at rest during the ionization of Ne and Kr atoms and ions by a high-intensity laser.[2, 19] The electrons gain both energy and longitudinal momentum from the field. They are subsequently accelerated out of the focus by the ponderomotive force, retaining their longitudinal momentum. The longitudinal momentum is equivalent to the multiphoton recoil energy shown in Eq. 4. A particularly simple formulation of the forward momentum in terms of conservation of energy and momentum from the field was shown by Reiss[20] and by Corkum $et\ al.$[21]The longitudinal momentum p_z, which accompanies the energy absorption from multiple scattering in the focused laser field, is related to the perpendicular momentum by[20, 21]

$$p_z = \frac{p_\perp^2}{2m_0 c}. \qquad (7)$$

The angle of the ejected electron relative to the wave vector of the laser (θ) depends on the final electron energy γmc^2 as

$$\tan\theta = \sqrt{\frac{2}{\gamma-1}}. \qquad (8)$$

A magnetic spectrometer has been constructed to measure the energy and angular (relative to $\vec{k}$) distributions of electrons emitted from a high-intensity laser focus. The spectrometer consists of an energy-resolving magnet and a detector consisting of a scintillator coupled to a photo-multiplier tube (PMT). The magnet lies in a plane above the focus with an energy-angular resolving gap. The central axis of the spectrometer is perpendicular to the $\vec{k}$ direction of the laser and passes through the laser focus. The angular distribution of electrons relative to the laser $\vec{k}(\theta)$ is measured by rotating the spectrometer about its central axis (ϕ). The energy is resolved by varying the magnetic field in the gap.

The energy window of the spectrometer is varied by changing the magnetic field in the gap of the steering magnet and was calibrated using an electron gun placed at the laser focus and aimed toward the gap in the steering magnet. The calibration showed an energy window of $\Delta E/E \sim 0.3$ FWHM.

The angular distribution of electrons in θ (relative to $\vec{k}$) is measured by rotating the entire spectrometer about the cylindrical axis that passes through the laser focus at 90° to the laser axis. The gap in the magnet is offset from the central axis of the spectrometer and is always aligned so that a clear line of sight can be traced from anywhere on this central axis through the gap in the magnet. The angular resolution is ±1.5°. The experiments were conducted with a 1.053-μm, 1-ps laser using chirped-pulse amplification (CPA).[22] The laser is focused with $f/3$ optics producing a 5-μm ($1/e^{-2}$ radius) focal spot and a peak laser intensity of approximately 10^{18} W/cm^2 into neon and krypton at a pressure of 10^{-3} Torr with circular polarization.

The angular distribution of electrons from Ne^{3+} through Ne^{8+} and Kr^{10+} and Kr^{11+} were measured. Ne^{8+} emerged from the circularity polarized focus with an energy of 80±5 keV and at an angle of 75±1.5° from the laser axis. The observed electron energies were 85±3 keV for Kr^{10+} and 130±10 keV for Kr^{11+}. The energy and angular spectra are in good agreement with the predictions of calculations which include the relativistic mass shift.[2, 19] The observed energy and angular distribution for $Ne^{3+} - Ne^{8+}$, Kr^{10+}, and Kr^{11+} were in good agreement with the relativistic calculations and with the energies predicted by classical field ionization[23] and subsequent ejection from the focus by ponderomotive and canonical momentum effects

(Eq 7).[20, 21] These drifts are consistent with the classical drifts and mass shifts described above.

3.2 HIGH ENERGY ELECTRON EXPERIMENTS

Nonlinear Quantum Electrodynamics(QED) is studied during the interaction of a high intensity, $I > 10^{18}$ W/cm^2, laser with 47 GeV electrons at the Stanford Linear Accelerator Center.[18] When the transverse energy of the electron in the laser field becomes comparable to the electron rest mass (eA/mc ~ 1) multiphoton Compton scattering becomes probable. Scattering events,

$$e^- + n\omega \rightarrow e'^- + \gamma,$$ (9)

in which up to 4 photons participate with both 1 μm and 0.5 μm, picosecond laser pulses. The observed rates are in good agreement with theoretical predictions.[1]

The exploration of nonlinear Quantum Electrodynamics during laser-electron collisions puts severe constraints on a laser system.[18] Multiphoton Compton scattering requires that the quiver velocity of the electron in the laser field approach the speed of light ($I\lambda^2 \sim 10^{18}$ W μm^2/cm^2),[10, 12] while the production of electron positron pairs in by multiphoton - gamma scattering requires intensities in excess of 10^{28} W/cm^2.[3, 4, 10, 17] In addition, the laser must be synchronized to the electron beam with a time jitter less than the convolved pulse widths of the laser and electron beam. The laser must have a high repetition rate to take advantage of repetition rate of the electron accelerator and to obtain good statistics. Finally, longer wavelengths increase the ratio of the quiver velocity to the speed of light, while shorter wavelengths reduce the number of photons required for pair production. Thus, frequency conversion extends the experimental capabilities.

We have developed and demonstrated[24]a 0.5 Hz repetition rate, 1 μm, 1 ps, terawatt, chirped pulse amplification (CPA) laser system using a flashlamp-pumped, Nd:glass, zig-zag slab amplifier. The laser produces energies greater than 2 Joules with an average power in excess of 1 W and 1.4 times diffraction limited focusing. Frequency doubling of these pulses is accomplished with up to 55% efficiency. The laser has been synchronized with a 47 GeV electron beam at the Stanford Linear Accelerator. with a temporal jitter of $\sigma = 2$ ps between the two beams.[25]

In the experiment the scattered electrons are observed by passing them through a magnetic spectrometer and detecting them with silicon calorimeters.[1, 18] The energy of the electrons undergoing multiphoton Compton scattering can be lower than those allowed for linear Compton scattering as shown in Eq. 4. Thus, the existence of low energy scattered electrons is evidence for multiphoton Compton scattering. For multiphoton orders higher than 2, multiple linear scattering is significantly less probable than multiphoton scattering. The experiment has observed nonlinear Compton scattering with up to 4 photons participating in a single scattering event. There is good agreement with theory for laser wavelengths of 1.053 and 0.527 μm.[1]

As mentioned above, the laser electric field approaches the QED critical field in this experiment. This experiment is currently studying pair production in the laser field,[17, 18]

$$\gamma + n\omega \rightarrow e^- + e^+.$$ (10)

4. Discussion and Conclusion

As described above, two experiments have observed multiphoton-electron scattering at relativistic laser intensities, $\eta^2 \sim 1$.[1, 2, 19] While there are many similarities between the two experiments, there are significant differences. In particular, in the experiment with low energy electrons,[2, 19] the laser photon energy is always significantly less than the electron rest energy. The highest electron energies observed are in excess of 100 keV, indicating that in excess of 10^5 laser photons participated in the scattering. In contrast in the experiment at SLAC with 47 GeV electrons,[1] the photon energy in the electron rest frame is comparable to the electron rest energy. In this case, up to 4 laser photons participated in the scattering. Thus, the numbers of photons participating in the scattering is different even though the values of η^2 were comparable. Part of the explanation for this is that the experiment at SLAC must be described quantum mechanically, while the experiment with low energy electrons can be described classically.

Another distinction between the two experiments is the relationship between the average quiver energy in the field to the electron energy. In both cases, the quiver energy can be as high as ~100 keV, and is less in general. This is large compared to either the electron kinetic energy or laser photon energy in the low energy experiments, while it is small compared to the initial electron kinetic energy or laser photon energy in the electron rest frame in the SLAC experiment. This suggests that the effects of the dressing field are different. In the SLAC experiment, the dressing of the electron mass plays a small role in the interaction, causing a small change in the kinematics of the scattering. In the experiment with low energy electrons, the final electron energy is comparable to the dressing energy suggesting that stimulated processes[26] may play an important role in the interaction.

Finally one can consider the quiver velocity and amplitude in the field. At low intensities and for low energy electrons, η can be written as the ratio of the quiver velocity to the speed of light,

$$\eta = \frac{v_{osc}}{c}. \tag{11}$$

This is a reasonable definition for the low energy electron experiment, but it must be considered carefully in the SLAC experiment. The definition of η in terms of the vector potential, Eq. 1 is always valid, but in terms of the quiver velocity, one must consider the fact that the electron mass is relativistically boosted by the γ ($\sim 10^5$) of the electron beam and so that while η^2 is comparable in the two experiments, the quiver velocity, and by integration, the amplitude, are many orders of magnitude smaller in the SLAC experiments. The high energy electron's oscillation in the field is negligible.

All of these distinctions are essentially due the fact that the SLAC multiphoton Compton scattering experiment is inherently a quantum mechanical interaction, while the low energy electron experiment is better described classically. A full quantum mechanical description of the low electron energy experiment is not yet available.

It is a pleasure to acknowledge conversations with J.H. Eberly, J.P. Knauer, K.T. McDonald, A.C. Melissinos, and N.B. Nahrozhny. This work was supported by the National Science Foundation and the United States Department of Energy.

5. References

1.   C. Bula, K.T. McDonald, E.J. Prebys, C. Bamber, S. Boege, T. Kotseroglou, A.C. Melissinos, D.D. Meyerhofer, W. Ragg, D.L. Burke, R.C. Field, G. Horton-Smith, A.C. Odian, J.E. Spencer, D. Walz, S.C. Berridge, W.M. Bugg, K. Shmakov, and A.W. Weidemann, (1996) "Observation of Nonlinear Effects in Compton Scattering," Phys. Rev. Lett. , submitted .

2.   C.I. Moore, J.P. Knauer, and D.D. Meyerhofer, (1995) "Observation of the transition from Thomson to Compton scattering in multiphoton interactions with low energy electrons," Phys. Rev. Lett. 74, 2439 .

3.   A.I. Nikishov and V.I. Ritus, (1964) "Quantum processes in the field of a plane electromagnetic wave and in a constant field. I," Sov. Phys. Jetp 19, 529-541 .

4.   N.B Narozhny, A.I. Nikishov, and V.I. Ritus, (1965) "Quantum processes in the field of a circularly polarized electromagnetic wave," Sov. Phys. JETP 20, 622-629 .

5.   L.S. Brown and T.W.B. Kibble, (1964) Phys. Rev. **133**, A705 .

6.   O. von Roos, (1964) "Interaction of Very Intense Radiation Fields with Atomic Systems," Phys. Rev. **135**, A43-A50 .

7.   Z. Fried and J.H. Eberly, (1964) "Scattering of a high-intensity, low-frequency electromagnetic wave by an unbound electron," Phys. Rev. **136**, B871 .

8.   T.W.B. Kibble, (1966) "Frequency shift in high intensity Compton scattering," Phys. Rev. **150**, 1060 .

9.   J.H. Eberly, (1969) "Interaction of very intense light with free electrons," in *Progress in Optics*, , edited by E. Wolf North-Holland, Amsterdam, Vol. 7.

10.   V.B. Berestetskii, E.M. Lifshitz, and L.P. Pitaevski, (1982)*Quantum Electrodynamics* Pergamon Press, Oxford, pp. sec. 101.

11.   J.H. Eberly and A. Sleeper, (1968) "Trajectory and Mass Shift of a Classical Electron in a Radiation Pulse," Phys. Rev. **176**, 1570-1573 .

12.   E.S. Sarachik and G.T. Schappert, (1970) "Classical theory of scattering of intense laser ratdiation by free electrons," Phys. Rev. D **1**, 2738-2753 .

13.   Vachaspati, (1962) "Exact solution of relativistic equations of motion of an electron in an external radiation field," Proc. Nat. Inst. Sci. (India) **29**, 138-142 .

14.   L.D. Landau and E.M. Lifshitz, (1971)*The Classical Theory of Fields*, Course of Theoretical Physics, Vol. 2 Pergamon Press, Oxford.

15.   Vachaspati, (1962) "Harmonics in the scattering of light by free electrons," Phys. Rev. **128**, 664 .

16.   J. Schwinger, (1954) "The Quantum Correction in the Radiation by Energetic Accelerated Electrons," Proc. Nat. Acad. Sci. **40**, 132 .

17.   H.R. Reiss, (1971) "Production of pairs from a zero-mass state," Phys. Rev. Lett. **26**, 1072-1075 .

18.   C. Bula et al., "Study of QED at Critical Field Strength at SLAC," Princeton University, University of Rochester, Stanford Linear Accelerator Center, University of Tennessee, SLAC proposal E-144 (1992).

19.   D.D. Meyerhofer, J.P. Knauer, S.J. McNaught, and C.I. Moore, (1996) "Observation of Relativistic Mass Shift Effects during High-Intensity, Laser-Electron Interactions," J. Opt. Soc. Am. B , to be published .

20.   H.R. Reiss, (1990) "Relativistic strong-field photoionization," J. Opt. Soc. Am. B 7, 574-586 .

21.   P.B. Corkum, N.H. Burnett, and F. Brunel, (1992) "Multiphoton Ionization in Large Ponderomotive Potentials," in *Atoms in Intense Fields,* , edited by M. Gavrila Academic Press, New York, pp. 109-137.

22.   Y.-H. Chuang, D.D. Meyerhofer, S. Augst, H. Chen, J. Peatross, and S. Uchida, (1991) "Pedestal Suppression in a Chirped Pulse Amplification Laser," J. Opt. Soc. Am. B 8, 1226-1235 .

23.   S. Augst, D. Strickland, D. D. Meyerhofer, S. L. Chin, and J. H. Eberly, (1989) "Tunneling Ionization of Noble Gases in a High-Intensity Laser-Fields," Phys. Rev. Lett. 63, 2212 .

24.   C. Bamber, T. Blalock, S. Boege, J. Kelly, T. Kotseroglou, A.C. Melissinos, D.D. Meyerhofer, W. Ragg, and M. Shoup III, (1996) "0.5 Hz, Phase-stabilized Terawatt Laser System with a Nd:Glass Slab Amplifier for Nonlinear QED Experiments," Opt. Lett. , submitted .

25.   T. Kotseroglou, et al., (1996) "Picosecond synchronization of a terawatt laser pulse with a 47 GeV electron beam," Nucl. Instrum. & Methods A , submitted .

26.   N.M. Kroll and K.M Watson, (1973) "Charged particle scattering in the presence of a strong electromagnetic field," Phys. Rev. A 8, 804-809 .

PHOTON EMISSION BY AN ELECTRON COLLIDING WITH A SHORT FOCUSED LASER PULSE

N.B. NAROZHNY, M.S. FOFANOV
Moscow Engineering Physics Institute, 31
Kashirskoe Shosse, Moscow 115409, Russia
International Institute of Physics, 17
Leningradsky prospekt, Moscow 125040, Russia

Probability for photon emission by an electron colliding with a short circularly polarized focused laser pulse is derived. Spectral-angular distributions and spectra of photons are calculated. Applicability of plane wave model for description of a short laser pulse is discussed.

1. Introduction

Photon emission by an electron moving in a strong laser field was studied in the middle 60th under assumption that the laser field is a plane monochromatic wave with frequency ω (see the detailed review of the most important results in Ritus paper [1]). The experimental study of the process became possible only recently due to the development of "table-top" lasers with intensities $I \propto 10^{18} \div 10^{19} W / cm^2$. Such intensities correspond to the field strength $\propto 10^{11} V / cm$ when the dimensionless parameter [1]

$$\eta^2 = \frac{e^2 \langle E^2 \rangle}{m^2 \omega^2},$$ (1)

which is responsible for nonlinear multiphoton effects in the interaction of an electron with an intense wave field becomes of order of unity. If another parameter

$$\chi \propto \eta \frac{\gamma \omega}{m},$$ (2)

where $\gamma = \varepsilon/m$, ε-electron energy is also of order of unity the field strength in the rest frame of the electron is $\propto 10^{16} V / cm$. This is close to the characteristic QED field strength $E_c = m^2 / e = 1.3 \times 10^{16} V / cm$ and therefore quantum corrections become important in the radiation problem. The experiment with $50 Gev$ electrons and a laser with $I \propto 10^{18} W / cm^2$ is now in process at SLAC [2].

---

[1] We use units in which $c = \hbar = 1$

H. G. Muller and M. V. Fedorov (eds.), Super-Intense Laser-Atom Physics IV, 411–420.
© 1996 *Kluwer Academic Publishers.*

412

Nevertheless the adequate description of the experiment by the formulas derived in the scheme with monochromatic model of the laser field is doubtful since high intensities of the laser field are achieved by means of very short pulses focused almost to the diffraction limit. The amplitude of such fields demonstrate strong spatial and temporal inhomogeneity. This results in the so-called ponderomotive effect which at least in classical limit strongly affects the electrons trajectory [3] as well as Thompson scattering spectra [4].

In present paper we derive the probability for photon emission by an electron colliding with a short circularly polarized laser pulse. As a model of the pulse field a plane but nonmonochromatic wave is used. It is reasonable for an ultrarelativistic electron transverse displacement of which during the time of interaction τ is small compared to the focusing radius R.

Under condition $R \gg \lambda$ where λ is characteristic wave length of the laser pulse our results are true for electron wave packet with transverse size $\propto \lambda$ moving before collision with the pulse with impact factor ρ, if one understands the parameter η in our formulas as a function of ρ. Then to describe collision of the laser pulse with electron beam of radius $Re > R$ one should average the probability of emission by a single electron over impact factors.

We present the results of calculations of spectral-angular and spectral distributions of photons emitted by a single electron as well as by a broad beam of electrons. The distinction of our results from those obtained in the monochromatic scheme is discussed.

All the calculations are made under assumption that parameter $\omega\tau$ where $\omega = 2\pi/\lambda$ is characteristic frequency of the laser pulse is large $\omega\tau \gg 1$.

2. Probability of photon emission

To describe the field of a circularly polarized laser pulse we use 4-potential

$$A_\mu = g\left(\frac{\varphi}{\omega\tau}\right)\left\{a_{1\mu}\cos\varphi + a_{2\mu}\sin\varphi\right\} \qquad (3)$$

where $\varphi = kx, k^\mu = (\omega, \mathbf{k})$- wave 4-vector, $a_{1\mu}, a_{2\mu}$ -amplitudes of the potential[2] $k^2 = 0, ka_1 = ka_2 = a_1 a_2 = 0, a_1^2 = a_2^2$, and $g(\varphi/\omega\tau)$ is an envelope of the potential such that $g(0)=1$ and g is exponentially small when $|\varphi| > \omega\tau$. τ can be considered as duration time of the pulse.

It is well known that there is exact Volkov solution of Dirac equation in an arbitrary plane wave field

$$\Psi_{pr} = \left\{1 + \frac{e(\gamma k)(\gamma A)}{2(pk)}\right\} \frac{u_{pr}}{\sqrt{2p_0}} \exp\left(iS_p\right), \qquad (4)$$

where p^μ is 4-momentum of electron outside the pulse, u_{pr}-constant Dirac spinor and S_p is classical action of electron, which for the potential (3) in approximation $\omega\tau \gg 1$

---

[2] We use a metric such that $AB = A_\mu B^\mu = A_0 B^0 - \mathbf{AB}$

can be written as

$$S_p \cong -\int q^\mu dx_\mu - g\left(\frac{\varphi}{\omega\tau}\right)\left[\frac{e(a_1q)}{(kq)}\sin\varphi - \frac{e(a_2q)}{(kq)}\cos\varphi\right] \quad (5)$$

with

$$q^\mu = p^\mu + g^2\left(\frac{\varphi}{\omega\tau}\right)\frac{m^2\eta^2}{2pk}k^\mu \quad (6)$$

$$\eta^2 = -\frac{e^2a_1^2}{m^2} = -\frac{e^2a_2^2}{m^2} \quad (7)$$

Hence classical action (5) of electron in the field (3) in approximation $\omega\tau\gg1$ coincides with the corresponding expression for the monochromatic wave with that difference that 4-vector q^μ and amplitudes of the potential in our case depend on variable φ. Such structure of classical action corresponds to separation of classical motion of electron on systematic translational movement along some smooth trajectory and fast oscillations with frequency ω around it (compare paragraph 30 in reference [5]). 4-vector q^μ is classical kinetic momentum of electron averaged over fast oscillations. Its dependence on φ is manifestation of the effect of ponderomotive scattering. It's easy to see that

$$q^2 = m_*^2(\varphi) = m^2\left[1 + g^2\left(\frac{\varphi}{\omega\tau}\right)\eta^2\right] \quad (8)$$

This equation is the definition of effective mass of electron m_* which also depends on φ in inhomogeneous plane wave.

The separation of fast oscillations and " translational " movement of electron in our problem allows one to expand in Fourier series on the interval $[0,2\pi]$ oscillating functions of φ in S-matrix element for photon emission in the same manner as in the case of monochromatic wave (see for example reference [1]).

After this procedure the S-matrix element is represented as a sum over harmonics each of them describing emission of a photon by the electron due to absorbing of s=1,2,3... wave photons. Its structure differs from that in the case of monochromatic wave by the absence of conservation law for all 4 components of momenta. Because of slow dependence on φ of coefficients in mentioned above Fourier series δ–function with + components of momenta[3] is substituted by an integral over φ of the type

$$I = \int_{-\infty}^{+\infty} d\xi B(\xi)\exp(i\omega\tau G(\xi)) \quad (9)$$

where $\xi=\varphi/\omega\tau$,

$$G(\xi) = \left(\frac{1_+ + p'_+ - p_+}{2\omega} - s\right)\xi + \frac{m^2\eta^2(kl)}{2(pk)(p'k)}\int_{-\infty}^{\xi}d\xi g^2(\xi) \quad (10)$$

and $B(\xi)$ is a slow function of ξ.

---

[3] We remind that + and - components of 4-vector A^μ are defined as $A_\pm = A^0 \pm A^3$

The integral (9) can be calculated under assumption $\omega\tau \gg 1$ by stationary phase method. The equation for stationary phase points can be written in the form

$$G'(\xi_*) = \frac{1_+ + q'_+(\xi_*) - q_+(\xi_*)}{2\omega} - s = 0 \qquad (11)$$

If $s>0$ this equation has two solutions on the real axis[4] and the integral (9) can be calculated as a sum of two integrals each of them represents independent contribution of one of two stationary phase points. The length of formation $\Delta\varphi$ of the integral (9) which we will occasionally call "coherence interval" in standard stationary phase method is determined by relation

$$\Delta\varphi \propto \sqrt{\omega\tau / |G''(\xi_*)|}$$

and for $\eta \propto \chi \propto 1$ and angles of photon emission with respect to direction of initial electron momentum $\theta \leq 1/\gamma$ is $\propto (\omega\tau)^{\frac{1}{2}}$.

For stationary phase points situated near the center of the pulse $\varphi=0$ the distance between them may be $\leq (\omega\tau)^{\frac{1}{2}}$. Therefore such points can not be treated independently since interference between them must be taken into account. The latter is achieved by keeping the term with third derivative in expansion of phase $G(\xi)$ near the point $\xi=0$. Such modified stationary phase method leads to the following result for the integral (9)

$$I = 2\left[|g''(0)|\frac{m^2\eta^2(lk)}{2(pk)(p'k)}\right]^{-\frac{1}{3}} B(0)\Phi(y)\exp(i\omega\tau G(0)), \qquad (12)$$

where

$$\Phi(y) = \int_0^\infty dx \cos\left(yx + \frac{x^3}{3}\right)$$

is Airy function with the argument

$$y = (\omega\tau)^{\frac{2}{3}}\left[|g''(0)|\frac{m^2\eta^2(lk)}{(pk)(p'k)}\right]^{-\frac{2}{3}}\left(s - \frac{1_+ + p'_+ - p_+}{2\omega} - \frac{m^2\eta^2(lk)}{2(pk)(p'k)}\right). \quad (13)$$

The length of formation of Airy functions is of order $(\omega\tau)^{\frac{2}{3}}$ if $\eta \propto \chi \propto 1$ and $\theta \leq 1/\gamma$. Therefore formula (12) must be used for integral (9) if the distance between two stationary phase points is $\leq (\omega\tau)^{\frac{2}{3}}$.

The expression for probability of photon emission averaged over polarizations of initial electron and summarized over polarizations of particles in finite state can be found by a straightforward but rather cumbersome procedure and we have

---

[4] Solutions of equation (11) for $s \leq 0$ are complex and the integral (9) then is exponentially small. That was the reason why we took into consideration harmonics only with $s>0$ and not $s = 0,\pm1,\pm2,\dots$ as they appeared in Fourier series.

$$\frac{dW}{d\omega'd\Omega} = \frac{e^2\eta^2(\omega\tau)^2\omega'}{32\pi^3 m^2\chi^2} \sum_{s=1}^{\infty} \left\{ \left(2 + 2u + u^2\right)\left(\left|F_{s,1}^{(1)}\right|^2 + \left|F_{s,-1}^{(1)}\right|^2\right)\right.$$

$$\left. -\left(2 + 2u - u^2 + 2su\frac{\chi}{\eta} - \frac{z^2\chi^2}{\eta^4}\right)\left|F_{s,0}^{(0)}\right|^2 - 2\eta^2(1+u)\,\mathrm{Re}\left(F_{s,0}^{(2)}F_{s,0}^{(0)*}\right)\right\}, \quad (14)$$

where ω' is the frequency of the emitted photon $\chi = \dfrac{\eta(\mathrm{pk})}{m^2}, u = \dfrac{(\mathrm{kl})}{(\mathrm{kp}) - (\mathrm{kl})}$,

$$F_{s,i}^{(k)} = \int_{-\infty}^{+\infty} d\xi\, g^k(\xi) J_{s+i}\big(z(\xi)\big)\exp\{i\omega\tau G_s(\xi)\},$$

$$z = (1+u)\frac{g(\xi)\eta}{m^2\chi}\left[e^2\big(a_1l\big)^2 + e^2\big(a_2l\big)^2\right]^{\frac{1}{2}}.$$

3. Spectral-angular and spectral distributions of emitted photons

We present here the results of calculations of spectral-angular and spectral distributions of emitted photons described by formula (14). The form of the envelope $g(\varphi/\omega\tau)$ used in calculations is

$$g\left(\frac{\varphi}{\omega\tau}\right) = \frac{1}{\mathrm{ch}\left(\dfrac{\varphi}{\omega\tau}\right)}.$$

Spectral-angular distribution consists of harmonics which in contrast to the distributions for the case of monochromatic wave are characterized by finite bandwidth (which doesn't depend on parameter $\omega\tau$) and existence of fine structure, Fig. 1.

The right border of harmonics $\omega_c^s(\theta)$ corresponds to radiation emitted from the periphery of the pulse and is equal to the frequency of Compton photon in reaction $s\omega + e = \omega' + e'$. The left border $\omega_\eta^s(\theta)$ is the frequency of radiation of electron in a plane monochromatic wave with intensity parameter η. The photons with $\omega_\eta^s(\theta)$ are emitted from the center of the pulse. The bandwidths of harmonics depend on h and when η is large enough the adjacent harmonics overlap, Fig. 2.

The fine structure of harmonics is explained by interference of radiation emitted from two points with equal field amplitude and situated so close to the center of the pulse that corresponding coherence intervals overlap.

The fact that bandwidths of harmonics doesn't depend on $\omega\tau$ means that the spectral-angular distribution in the pulse of finite duration can not be transformed to one for the case of monochromatic wave at any $\omega\tau$. These two models of laser field refer to two different experimental schemes. The monochromatic wave model is good for the scheme in which spectrometer is switched on only for limited time not exceeding the coherence time or $\leq \sqrt{\omega\tau}\,/\,\omega$ in our problem. If we speak about a pulse the spectrometer detects radiation during all the time of interaction τ. The spectrum then is formed by all coherence intervals inside the pulse. Each of these intervals corresponds to different frequencies in the region from ω_η^s to ω_c^s and this is the reason of broadening of harmonics.

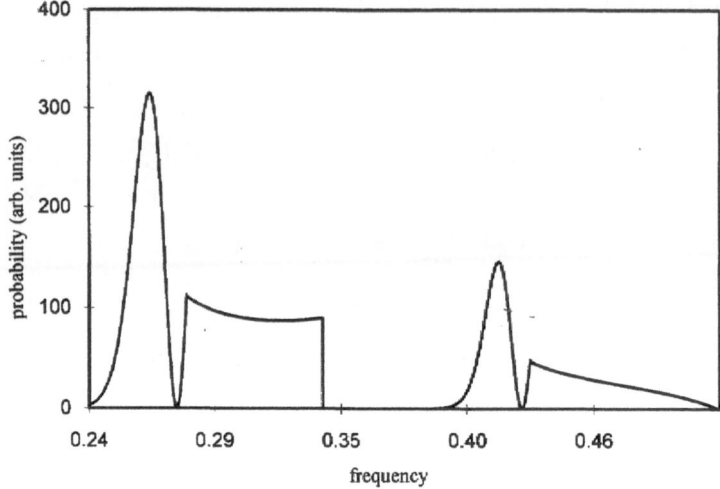

Figure 1. 1st and 2nd harmonics in spectral-angular distribution of photon emission probability. Parameters are: $\eta=1$, $\gamma = 10^5$, $\theta = 10^{-5}$, $\omega\tau = 50$. (Frequency is scaled in units of $4\gamma^2\omega$).

The spectral distribution of emitted photons can be found by integration of expression (14) over angles. The result is presented in Fig.3. The new elements compared to the monochromatic model are appearance of asymmetric maximum instead of sharp edge at frequency $\omega_\eta^s(0)$ and shift of the right border of harmonics from $\omega_\eta^s(0)$ to $\omega_c^s(0)$. That means in particular that criterion of experimental observation of the second harmonic is detection of radiation with $\omega' \geq \omega_c^s(0)$ and not $\omega_\eta^s(0)$ as it appears in the monochromatic model (Fig. 4).

The origin of existence of radiation with $\omega' \geq \omega_\eta^s(0)$ is the pondermotive effect or in other words dependence of effective mass m_* on φ (see eq. (8)). Nonmonotonious behaviour of the spectrum in this frequency region is a rather weak manifistation of interference effect discussed in connection with the fine structure of spectral-angular distribution.

Our results can be applied to the focused pulse with focusing radius R if one describes the initial electron as a wave packet moving towards the pulse with an impact factor ρ. Intensity parameter η must be understood then as a function of ρ.

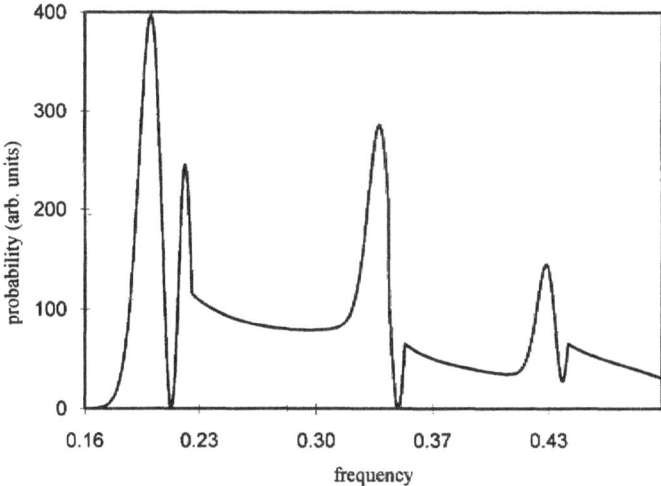

Figure 2. Three first harmonics in spectral-angular distribution.
Parameters are: $\eta=1.5$, $\gamma = 10^5$, $\theta = 10^{-5}$, $\omega\tau = 50$.

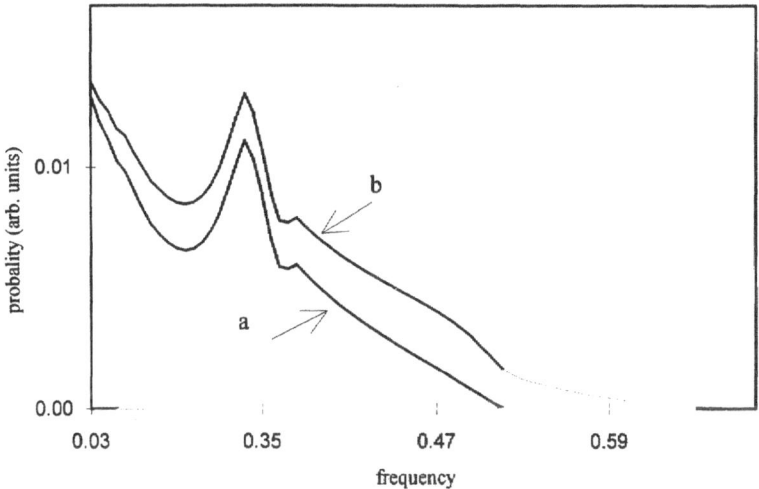

Figure 3. Spectral density of probability of photon emission by electron
over a passage through a short laser pulse. a - 1st harmonic; b - 1st and
2nd harmonics. Parameters are: $\eta = 1$, $\gamma = 10^5$, $\omega\tau = 50$

For applicability of formula (14) to the case of a focused pulse some conditions must
be satisfied. First, the transverse size b of the packet must be at least of order of
transverse length of formation of S-matrix element or δ-function responsible for

418

conservation of transverse components of 4-momentum. If we accept ω as a natural for our problem measure of accuracy for the conservation law the transverse length of formation is of order of λ. Hence

$$b \propto \lambda \ll R .$$ (15)

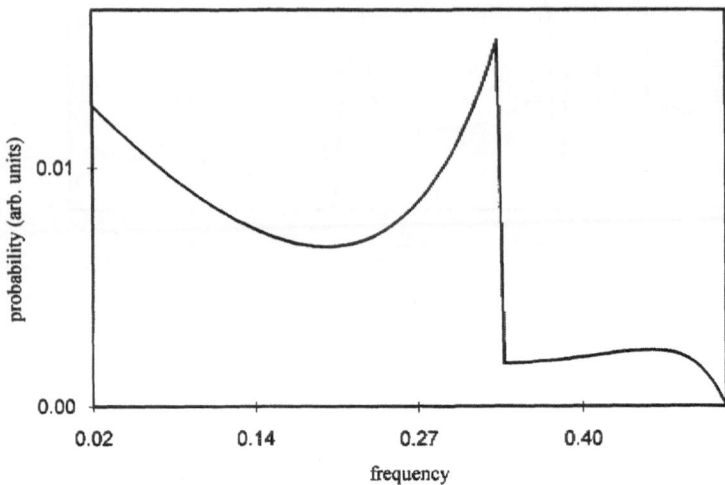

Figure 4. Spectral density of probability of photon emission by electron in a monochromatic wave. 1st and 2nd harmonics. Parameters are: $\eta = 1, \gamma = 10^5$.

The transverse displacement of the packet $\Delta r_\perp$ during the time of interaction τ also must be less than R. $\Delta r_\perp$ can be easily estimated using equations of motion of an electron in a short focused laser pulse [3] and the condition $\Delta r_\perp \ll R$ leads then to

$$\frac{R}{\lambda} \gg \omega \tau \frac{\eta}{\gamma} .$$ (16)

At last the transverse spreading of the packet during τ must be small compared to R. For the packet with transverse size b it means [6]

$$\omega \tau \ll \frac{m\gamma}{\omega} .$$ (17)

Note that conditions (16),(17) afford parameters R, τ such that $\omega \tau \gg 1, \tau \gg R$ if $\gamma \gg 1, \eta \propto 1$.

To describe emission of photons by a broad beam of electrons colliding with a short focused laser pulse one should average probability (14) over impact factors. The result of calculations for

$$\eta(\rho) = \eta \exp\left(-\frac{\rho^2}{R^2}\right)$$

are presented on Fig. 5,6.

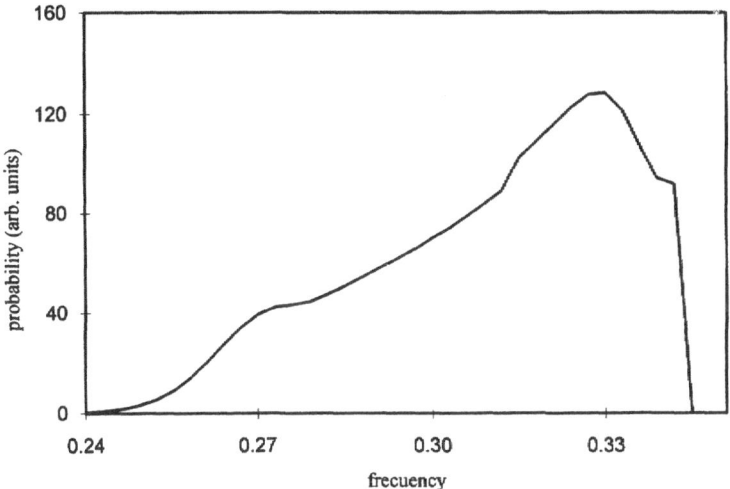

Figure 5. 1st harmonic of spectral-angular distribution of photon emission probability averaged over impact factors. Parameters are: $\eta = 1, \gamma = 10^5, \theta = 10^{-5}, \omega\tau = 50$.

The spectral-angular distribution is crucially affected by such averaging , Fig. 5. Though the bandwidth of harmonics remains the same, the maximum is shifted to the right. This is explained by relatively large weight of peripherical electrons in averaged probability compared to electrons with impact factors close to 0. The fine structure of harmonics is smashed by averaging proceedure. It can be traced only in slightly nonuniform growth of radiation in the left side of the harmonic, Fig. 5.

The changes in the spectrum (Fig. 6) are not so dramatic. The averaged spectrum has a smoother, in comparison with Fig. 3 , maximum which is slightly shifted to the right.

420

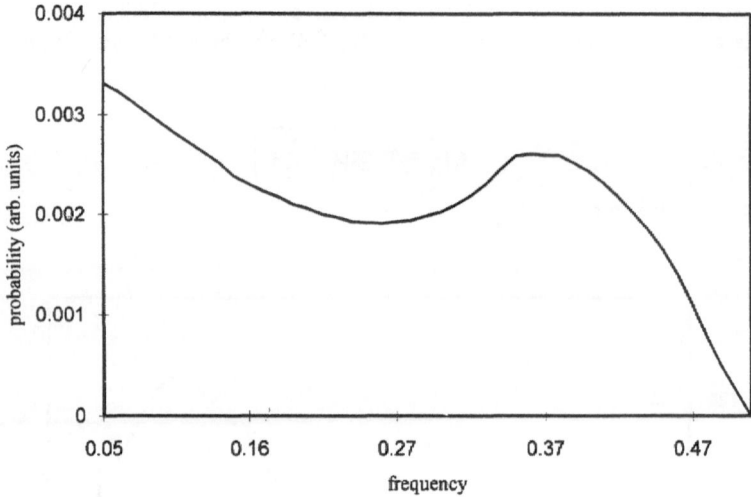

Figure 6. Spectral density of photon emission probability averaged over
impact factors. 1st harmonic. Parameters are: $\eta = 1, \gamma = 10^5, \omega\tau = 50$.

This research was supported by the Russian Foundation for Fundmental Research, Grant No. 95-02-06056-a.

4. References

1. Ritus, V.I. (1979) Quantum Effects of Interaction of Elementary Particles with Intensive Electromagnetic Fields, *Trudy FIAN* **111**, 5-151, Nauka, Moscow.
2. McDonald, K.T. et.al. (1986) Proposal for Experimental Studies and Non-linear Quantum Electrodynamics, DOE/ER/3072-38. (See also D.D.Meyerhofer paper in this issue).
3. Goreslavsky, S.P., Narozhny, N.B., Yakovlev, V.P. (1991) Relativistic Electron Collision with Focused Short Laser Pulse, *Laser Physics* **1**, 670-677.
4. Goreslavsky, S.P., Narozhny, N.B., Shcherbachev, O.V. (1995) Determination of the Effective Electron Mass in an Optical Field from the Measured Emission Spectrum of Ultrarelativistic Electrons at a Laser Focus, *Pis'ma Zh. Eksp. Teor. Fiz.* **61**, 251-253.
5. Landau, L.D., Lifshitz, E.M. (1988) *Mechanics*, Nauka, Moscow.
6. Goldberger, M.L., Watson, K.M. (1964) *Collision Theory*, John Wiley & Sons, Inc, New York-London-Sydney.

HIGH HARMONIC GENERATION IN ATOMS AND IONS

A. SANPERA, S.G. PRESTON, M. ZEPF, J.B. WATSON,
J.S. WARK AND K. BURNETT
Clarendon Laboratory. Department of Physics
University of Oxford, Parks Road
Oxford OX1 3PU, United Kingdom.

1. Introduction

In the last few years the availability of very intense and very short laser pulses has created a new experimental domain for laser atom interactions. Among many new phenomena, the generation of high order harmonics has opened the possibility of production of coherent radiation in the extreme ultraviolet and the X-ray region. In this paper we will first review the current state in generating harmonics from atoms and ions. We shall then present experimental and theoretical results that allow us to determine the contribution of ions in the harmonic response unambiguously and show the shortest harmonic wavelength generated to date. Finally, we shall discuss how to generate high order harmonics at moderate intensities by taking advantage of the higher ionisation potential of ions compared to neutral atoms.

Harmonic generation refers to the conversion of the fundamental wavelength to high odd harmonic wavelengths when an atom is irradiated by a strong laser field. A complete description of the process should include both the response of the single atom to the laser radiation as well as the collective response of all atoms. The collective response determines, to a great extent, the experimental efficiencies in the harmonic generation since the ionisation of the medium modifies the refractive index and therefore the phase matching between the harmonics and the fundamental. Nevertheless is the strongly nonlinear response of the single atom that produces a large conversion efficiency from the fundamental wavelength to high order harmonic wavelength. This is commonly referred to as the harmonic generation plateau. There is an upper limit to the harmonic plateau determined by the maximum kinetic energy that an electron (that has been previously

H. G. Muller and M. V. Fedorov (eds.), Super-Intense Laser-Atom Physics IV, 421–431.

promoted to the continuum) could have if it is to be driven back to the core by the action of the field. This maximum kinetic energy is approximately $3.2U_p$ ($U_p = E^2/4\omega^2$ in atomic units) and depends only on the intensity and frequency of the incident radiation, because the interaction of the electron in the continuum with the external field dominates over the Coulomb interaction at such intensities [1, 2]. This is the origin of the well known cut-off rule

$$h\nu_{max} = U_i + 3.2U_p, \tag{1}$$

to the harmonic generation where U_i refers to the ionization potential. Harmonics corresponding to higher photon energies are produced with an exponentially diminishing efficiency thus becoming rapidly indistinguishable from the background.

From a theoretical point of view the interaction of a high-intensity field with an atom has to be calculated without introducing the usual perturbation theories; the study of this interaction entails numerical simulation that needs massive computations. In the strong regime, a much better understanding of the process have been achieved by using quasi classical models. Those models pioneered by Keldysh [3, 4, 5] and valid in the regime of tunnelling ionisation, have been very successfully applied to understand the cut-off rule in the harmonic generation. The quasi classical models assume two completely separate steps in the harmonic generation. In the first step the electron, initially in the ground state, is promoted to the continuum either by tunnelling through the atomic barrier (tunnelling ionisation) or by absorbing several photons (multiphoton ionisation) from the field. The second step deals with the evolution of the electron after it has been released. If we neglect the Coulomb interaction the subsequent evolution could be described as a first approximation, classically. Once in the continuum and depending on the initial velocity and the phase of the field, there is some probability that electron will be driven back to the atomic core by the field emitting a photon in the transition back to the ground state. Since the atom has central symmetry, only odd harmonics will be produced. Finally, the first step determines the efficiency on the harmonic signal which is found to be directly proportional to the ionisation rate, with a system-dependent constant. Therefore this suggests that at best, for a given ionisation potential, the highest harmonic order achievable would be the one produced just below the saturation intensity for each wavelength.

2. NEUTRAL VERSUS IONS ?

Most harmonic generation research to date has been performed using 1.053 nm Nd:YAG or 800 nm Ti:Sapphire lasers [6, 7], partly because these are the most commonly available laser systems, but also because it was considered that the best way to generate short wavelength harmonics was to maximise the ponderomotive term in the plateau cut-off equation which is proportional to the square of the laser wavelength. However, recent work using a 248 nm KrF laser [8] indicates that shorter wavelength driver lasers can produce short wavelength harmonics not just from the ponderomotive energy term in the cut-off equation, but by using the higher ionisation potentials of ions rather than neutrals. He^+, for example, has an ionisation potential of 54.4 eV compared to 24.6 eV for neutral Helium. This leads to a much higher ionisation (and saturation) intensity for the ion compared to that for the neutral atom which allows us to work at correspondingly higher ponderomotive potentials. However, at long wavelengths, harmonic signal from ions, has shown to be, at best, very week in comparison to the neutral response[9]. For 248 nm nevertheless ionic efficiencies can be of the same order as those from neutral atoms. The generation of high harmonics from ions using short wavelengths can be attributed to different effects. The first one is related to the collective response; the phase mismatch between fundamental and harmonic is less when the refractive index is dominated by free electrons, as is the case when the medium is ionised. The single atom response is also greater for short wavelengths. Using the quasi classical models this could be attributed to:

(a) the larger interaction time between the recolliding electron and the core. The velocity of the electron driven back to the core scales as E/ω. Diminishing the velocity of the particle (shorter wavelengths) when it "crosses" the nucleus increases the interaction time and therefore the cross section of scattering.

(b) the shorter available time for diffusion of the electron wavefunction between tunnelling through the Coulomb barrier and the first recollison with the core: this time being half a laser cycle [5]. The transverse spreading of the wavefunction is wavelength dependent rather than intensity dependent. In this simple model, with significant harmonic generation occurring only during the first recollision, the reduced diffusion of the wavepacket increases the cross section for the scattering, and therefore the harmonic signal.

Furthermore, numerical simulations using the full pulse response, rather than the steady state response, have also shown that the harmonic signal in the plateau disappears when ions are driven by long wavelengths[10]. This effect is related to the intensity of the laser changing with time: the variations of intensity during the trajectory back to the core will introduce a phase shift relative to the field. During the turn on of the pulse, this shift will be different for each cycle. This results in a structureless plateau, where the harmonics are not differentiated from the background and they are only clearly present again in the cut-off region. Since the classical velocity of the

424

electron is proportional to E/ω this phase shift effect is more pronounced for high intensities and/or low frequency driving fields.

With harmonic generation from ions already demonstrated, it is timely to investigate whether higher harmonic photon energies can be produced from ions using short wavelength lasers rather than from neutrals with longer wavelength lasers. In the next section we will present single atom/ion simulations and experimental data that show that 248 nm light can produce much shorter harmonic wavelengths than those presently achieved using 800 nm or 1.053 nm light, and can produce higher generation efficiencies at the longer harmonic wavelengths [11].

3. EXPERIMENTAL RESULTS

We present now experimental harmonic generation results from He and Ne irradiated with 248 nm KrF laser at intensities I= 10^{17} W/cm^2. The experiments were performed on the Sprite KrF CPA laser system at the Rutherford Appleton Laboratory, U.K [12]. The laser delivered up to 250 mJ on target in a 380 fs FWHM pulse at 248.6 nm. The beam was focused using a 100 cm OAP which produced, on average, a 6 times diffraction limited focal spot of 18 μm FWHM with a mean focused intensity of just over 10^{17} W/cm^2 over a length of approximately 700 μm. A solenoid-valve gas jet provided a gas target of Helium or Neon at the laser focus, with atomic densities ranging from 10^{16} to 10^{19} cm^{-3} and gas lengths between 0.5 and 2 mm. Densities produced at the focus were determined by Thomson scattering [12] . The harmonic emission was detected with a slitless flat-field grazing incidence XUV spectrometer (1200 line/mm Hitachi grating at 3.77°) used with an additional gold-coated grazing incidence toroidal mirror. The grating and mirror were oriented so that their grazing incidence reflections gave perpendicular astigmatic line images of the focus of the laser beam at the detector plane. The acceptance solid angle of the grating and gold mirror was 1.5 x 10^{-5} steradians. Filters of 20 - 200 μg/cm^2 of Carbon or 27- 108μg/cm^2 of Aluminium were used to discriminate against the fundamental laser light and to attenuate the bright lower order harmonics. In addition, a beam block was inserted in the zero order position to eliminate scattered XUV radiation. Figure 1 shows a lineout from a typical laser shot into low pressure Helium. The first and second order diffracted signals are labelled in the lineout. This data is not corrected for the spectrometer, detector, or filter responses. The first order diffracted signal from the 13th to the 37th harmonics can clearly be seen. To quantitatively analyse the harmonic response we must integrate the total energy of each harmonic. In order to compare harmonic spectra produced by different laser intensities (caused by unavoidable shot-to-shot variations in the laser energy and focal spot profile) we will present the conversion efficiencies for each harmonic. Conversion efficiencies are calculated assuming that the total solid angle of emission is equal to that of the laser beam[13]. Further calibration factors come from the absolute response of the MCP, the grating reflectivity, the filter transmission, the collection solid angle of the spectrometer, the focusing mirror reflectivity and its collection solid angle. Using an f/10 OAP we generate over 30 nJ at 355Å (7th harmonic), i.e. a conversion efficiency of 10^{-7}. If we assume that this energy is generated over the entire length of the laser pulse (380 fs) this corresponds to a power of over 80 kW. In reality, He^{+} is fully ionised long before the peak of the pulse which implies that harmonic powers are much higher than this and that the pulse length of the harmonic is less than the laser pulse length [14].

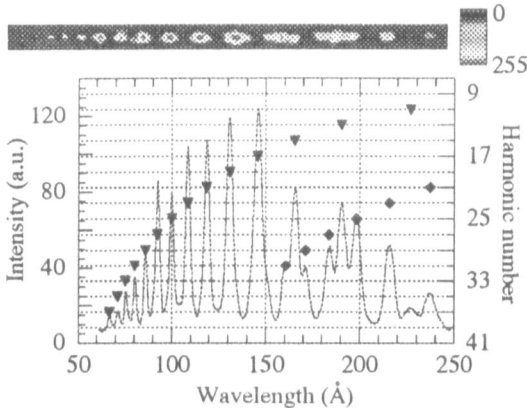

Figure 1. Lineout of SN20118 (Helium). The laser was focused 1 mm above and 0.5 mm beyond the 0.5 mm x 1 mm gas jet nozzle (with the 0.5 mm dimension along the laser focus. The gas jet backing pressure was 0.25 bar gauge.

4. THEORETICAL RESULTS

The relationship between single atom calculations and the data recorded in an experiment is not straightforward because the net generation depends on the time and phase dependent integration over the whole focal region. Although single atom calculations can take into account the temporal variations in laser intensity, they ignore spatial variations of intensity, variations of phase and propagation effects. When making a full simulation of the harmonic experimental data one should consider not only the single atom response but also all the issues of beam propagation which may introduce effects such as: blue-shifting of the fundamental laser frequency; filamentation of the laser resulting in large intensity variations over a small radial scale in the focus; ponderomotive expulsion of electrons and ions from the focus which can significantly change the plasma density; and phase matching effects which can greatly enhance the harmonic generation efficiencies, especially of the lower harmonics. The simulations presented here do not consider beam propagation, they model the single atom/ion response to the laser field. However, these numerical simulations of the individual atoms/ions can match the experiments reasonably well by combining the response of the neutral single atom response with that of their ions for a full pulse.

Our numerical approach is based on a N-step procedure where, as a first step, the ionisation and harmonic generation for the neutral N-electron atom is calculated in the Single Active Electron (SAE) approximation with an absorbing boundary condition. From the decrease of the norm of the neutrals' wavefunction, a time-dependent ionisation function

$$P_{ion}(t) = 1 - \langle \Psi(r,t) | \Psi(r,t) \rangle, \qquad (2)$$

is obtained as the pulse proceeds. In the next step, we solve the time-dependent Schrödinger equation for the ion, and then the ion wavefunction is weighted by the time-dependent ionisation function [15]. The procedure is repeated for every ionisation stage we consider to be contributing to the harmonic generation. Therefore, using this scheme, to determine the time dependent evolution of He is reduced from the time propagation of a two-electrons wavefunction to the propagation of two one-electron wavefunctions. The validity of this approximation relies on the assumption that ionisation is a stepwise process. This model is generally too computationally expensive to use with Neon, due to the greater number of ionisation stages and bound electrons involved, we have therefore used the same procedure but with a one dimensional soft core potential, $V(x) = -Z_{eff}/\sqrt{a^2 + x^2}$ to calculate the response of Neon and its ions, where a is chosen for each ionisation stage to give the correct ionisation potential. We solve the Schrödinger equation in the length gauge, with the laser field linearly polarised. The numerical time-evolution of the one-electron wavefunction is performed using standard techniques. We use a partial wave decomposition of the wavefunction, together with a split operator and a Crank-Nicholson algorithm as described in detail elsewhere [15]. Our simulations are performed for a number of laser intensities in the range of 10^{15} - 10^{17} W/cm^2 at a fundamental wavelength of 248 nm for Helium and its ion and for the Neon atom and its charge states up to Ne^{3+}. The simulated laser pulse length has a sine-squared temporal profile with a FWHM of 350 fs. These intensities range from close to well above the saturation intensity of the neutral and so it is necessary to study the contributions both from neutral atoms and from ions. The saturation intensities of Ne, Ne^+, Ne^{2+} and Ne^{3+} are well below 10^{17} W/cm^2 and therefore they are mostly ionised before the experimental peak laser intensity. Although higher ionised states of the atom, i.e. Ne^{4+} and above, have higher saturation intensities and could therefore experience larger intensities before being completely ionised, their contribution to harmonic generation is negligible [11]. The simulations include temporal laser variations in the laser intensity, but ignore spatial variations in the laser intensity and phase. In a real laser focus, atoms do not all experience a single intensity, the laser intensity will vary along and radially outwards from the centre of the focus. To more realistically simulate the laser focus we can make a superposition of the atom/ion response for varying laser intensities, weighted for the relative volume in the focus at which these intensities occur (assuming a Gaussian, TEM00, beam profile). The result of such a superposition is to enhance the generation efficiency for the lower harmonics with respect to the higher harmonics, as they will be produced over a much larger volume. To allow for phase matching effects, we can also weight the response by a q^{-2} factor (where q is the harmonic order) to allow for the coherence length dependence in the case when phase matching is dominated by the free electron refractive index [14]. Our simulations are compared with a selection of experimental data for (a) Helium, and (b) Neon in Figure 2. We see in that figure that both, the full Helium and the one dimensional Neon simulations provide a good fit with experiment. The experimental data shown represents a range of gas densities and gas jet nozzle configurations. The neutral Helium simulation models the form of the data very well, but appears to be about an order of magnitude too low. This can be explained by the increased phase matching during generation from a neutral when the coherence length is dominated by Gaussian focusing and by dispersion in the gas compared to free electrons for the ions. If we were to multiply the neutral Helium simulation by a factor of about ten the changeover from neutral to ionic response would occur near the 15th to 17th harmonic rather than the 9th or 11th which is currently predicted. The one dimensional Neon simulation provides a remarkably good fit to experiment. Not only does it exactly predict the cut-off at the 35th harmonic, but it models the changes in the slope of the response at the 9th, 17th, 21st and 27th harmonics. This excellent match implies that the harmonic signal we are seeing is truly from Neon and its ionisation stages up to the doubly ionised stage. This means that harmonics above the 9th are primarily due to Ne^+, and harmonics above the 19th are primarily due to Ne^{2+}. Harmonic efficiencies above the 27th do appear to be approximately a factor of two or three higher than the simulation predicts. This is not surprising given the shot-to-shot variation in signal. However, it is interesting that above about the 31st harmonic the response of Ne^{3+} becomes approximately the same as Ne^{2+} until the 35th harmonic when the Ne^{2+} response falls off. The results as they stand can all be explained without having to invoke a third ionisation stage of Neon, but had any harmonic been detected above the 35th we would have had a very strong case for attributing it to the response of Ne^{3+}.

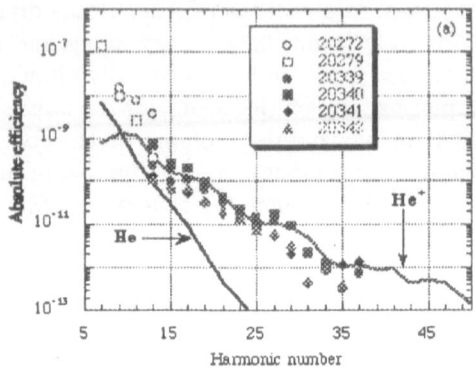

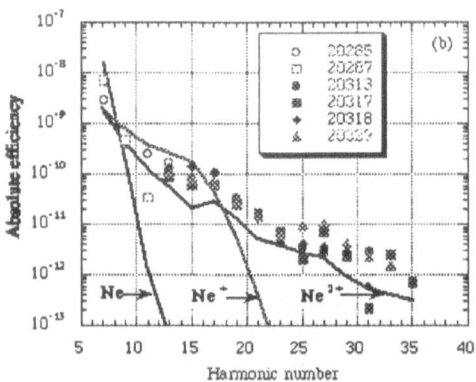

Figure 2. A comparison of experimental data with the weighted simulated response for (a) Helium, and (b) Neon. The simulation has been multiplied by an arbitrary constant

5. HARMONIC GENERATION FROM A COHERENT SUPERPOSITION

The cut-off rule seems to exhaust all possibilities in generating harmonics for a given ionisation potential, intensity and frequency. Despite the convenience of using long wavelengths in ions to generate very high harmonics we have seen that in fact, only ions driven by short wavelengths can produce high harmonic photons with high efficiency. However, preparing the initial state as a coherent superposition of ground and excited state we can induce dipole transitions between the continuum and the ground state via the intermediate (excited) state responsible for the ionisation. In this way it is possible to generate very high-order harmonics with high conversion efficiencies using moderate laser intensities and long wavelengths in ions [16]. By moderate intensities we mean that the intensity is high enough to ionise the excited state, but the field is too weak to directly ionise the ground state. In this case we found that the harmonic spectrum is composed of two distinct sets of harmonics. The first (lowest energy) plateau corresponds to transitions back to the initial state, and as one might expect, has a cut-off at $U_e + 3U_p$. The second plateau begins at an energy $U_g - U_e$ and has a cut-off at energy $U_g + 3U_p$ (U_g, U_e refers to the ionisation potential of ground and excited states respectively). We shall show that this set of harmonics is generated like "transitions" from the excited state into the continuum and then back to the ground state. However, these harmonics cannot be obtained by computing the responses starting from the ground and excited states separately, and therefore it is essential to have coherence between the ground and excited state populations. Our results are based on numerical solutions of the time dependent Schrödinger equation for a superposition of ground and excited states of He^+. We use ions to take advantatge of their higher ionisation energy but the process we wish to present here is very general, and is not restricted to ions. We prepare the initial state as a coherent superposition of the ground state 1s (54.4 eV) and the excited (metastable) state 2s (13.6 eV)

$$\Psi(\vec{r}, t = 0) = \alpha|1s\rangle + e^{i\phi}\beta|2s\rangle, \tag{3}$$

where α and β are the amplitudes of ground and excited states, and $|\alpha|^2 + |\beta|^2 = 1$. We have varied the initial relative phase between the states ϕ, and found that the results are not modified, so we shall take $\phi = 0$ for the remainder of this section. We present results for He^+ starting from a coherent superposition of 1s and 2s states ($\alpha = \beta = 1/\sqrt{2}$) irradiated by a laser frequency of $\omega = 0.060$ a.u.($\lambda = 746nm$) at an intensity of $I = 8.8 \times 10^{13}$ W/cm^2. For this choice of intensity and wavelength the ionisation of the ground state is negligible, and there is only transference of population from

430

the excited state to the continuum. The harmonic spectrum corresponding to this case is shown in Figure 3a. The spectrum clearly shows two different set of harmonics, the second one starting after the two strong peaks corresponding to the transition energy between 1s and 2s states. For the sake of comparison we also display in Figure 3b the harmonic spectra corresponding to an initial state of the ground ($\alpha = 1$, $\beta = 0$) and excited ($\alpha = 0$, $\beta = 1$) states alone.

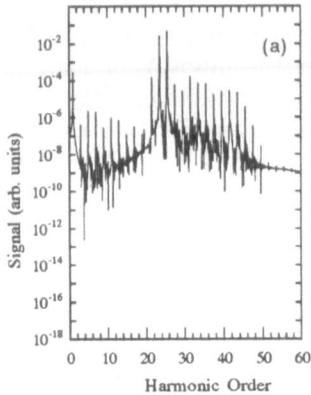 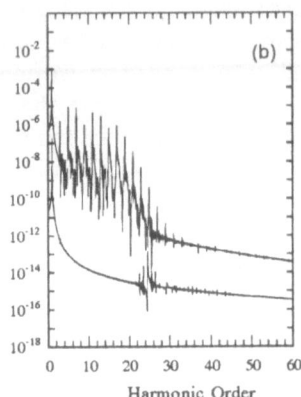

Figure 3. Harmonic spectra of He^+ at $I = 8.8 \times 10^{13}$ W/cm^2 and $\lambda = 746$ nm. Fig. 3a corresponds to HG spectrum starting from a coherent superpostion of ground and excited states with equally weighted populations. Fig 3b: the upper curve corresponds to the HG spectrum starting from the excited state alone, and lower curve shows the HG spectrum starting from ground state alone, otherwise identical parameters that Fig 3a.

This method allows us to extend the harmonic plateau to higher energies with higher efficiencies than it is possible to obtain starting the process from the ground or excited state. Since both cut-offs are separated by the transition energy between the ground and the excited state, this method of generating high order harmonics is particularly suitable for ions driven by long wavelengths. (To avoid phase matching problems we can choose to use a resonant laser to match the transition energy between ground and excited state.) In this way we get the advantage of the high ionisation energy of (ground state of) the ion, while still using moderate intensities and avoiding the pulse shape related difficulties mentioned previously.

6. ACKNOWLEDGMENTS

This work is financially supported by the U.K. Engineering and Physical Science Research Council, as well as the Human Capital and Mobility program of the European Community. A.S also thanks and the Fleming/MEC program for financial support and DGICYT (contract number PB92-0600)

References

1. Krause, J.L., Schafer K.J and Kulander K.C. (1992) High-order harmonic generation from atoms and ions in the high intensity regime, *Phys. Rev. Lett.* **64**: *pp. 3535.*
2. Corkum P.B (1993)*Phys. Rev. Lett.* **71**, *pp. 1994-1997.*
3. Keldysh L.V. (1964) *Zh. Ekps. Teor. Fiz* **47**, *pp. 1945.*
4. Reiss H.R (1980) Effect of an intense electromagnetic field on a wealky bound system, *Phys. Rev. A* **22**, *pp. 1768.*
5. Lewenstein M, Balcou Ph, Ivanov M.Y, L'Huillier A, Corkum P.B (1994) Theory of high-harmonic generation by low-frquency laser fields, *Phys. Rev A* **49**, *pp. 2117-2132.*
6. Perry, M.D. and G. Mourou (1994) Terawatt to petawatt subpicosecond lasers, *Science*, **264**, *pp. 917-924.*
7. Macklin, J.J., J.D. Kmetec, and C.L. Gordon III (1993) High order harmonic generation using intense femtosecond pulses, *Phys. Rev. Lett.* **70**, *pp. 766-769.*
8. Sarukura, N., et al., (1991) Coherent soft x-ray generation by the harmonics of an ultrahigh-power KrF laser, *Phys. Rev. A* **43**, *pp. 1669 - 1674.*
9. Wahlstrom, C.-G, Larsson J., Perrson A., Starczewski T., Svanberg S., Salieres P., Balcou P, and L'Huillier A, (1993) High order harmonic generation in rare-gases with an intense short-pulse laser, *Phys. Rev. A*, **48**, *pp. 4709-4720.*
10. Watson, J.B, A. Sanpera, and K. Burnett, (1995) Pulse-shape effects and blueshifting in the single-atom harmonic-generation from neutral species and ions, *Phys. Rev. A*, **51**, *pp. 1458-1463.*
11. Preston S.G, Sanpera A, Zepf M, Blyth W.J, Smith C.G, Wark J.S, Key M.H, Burnett K, Nakai M, Neely D, Offenberger A.A, (1995) High-order harmonics of 248.6 nm KrF laser from Helium and Neon ions, *Submitted to Physical Review Letters.*
12. Lister, J.M.D, Divall E.J, Downes S.J, Edwards C.B, Hirst C.J, Hooker C.J, Key M.H, Ross I.N, Shaw M.J, and Toner W.T (1994) Sprite - A very high brightness, ultraviolet laser system. *Journal of Modern Optics*, **41**, *pp 1203-1215.*
13. Zepf, M., et al., (1995) Comparison of X-ray detectors,*Daresbury Rutherford Laboratory Annual Report.*
14. Ditmire, T, Crane J.K, Nguyen H, DaSilva L.B, and Perry M.B, (1995) Energy-yield and conversion-efficiency measurements of high-order harmonic radiation, *Phys. Rev. A*, **51**, *pp. R902-R905.*
15. Sanpera, A. Jönson, P, Watson J.B, and Burnett K ,(1995) Harmonic generation beyond the saturation intensity in Helium, *Phys. Rev. A*, **51**, *pp. 3148-3153.*
16. Watson J.B, Sanpera A, Chen X, Bunett K (1995) Harmonic generation from a superposition of coherent states *Submitted to Phy. Rev. Lett.*

OPTIMISATION AND APPLICATIONS OF HARMONIC GENERATION

C.-G. Wahlström[1], C. Altucci[1*], S. Borgström[1], B. Carré[2],
M.B. Gaarde[1†], J. Larsson[1], A. L'Huillier[1,2], C. Lyngå[1], E. Mevel[1],
A. Persson[1], T. Starczewski[1], S. Svanberg[1], and R. Zerne[1]

[1]*Division of Atomic Physics, Lund Institute of Technology*
Lund, Sweden
[2]*Service des Photons, Atomes et Molécules, Centre d'Etudes de Saclay,*
France

High-order harmonic generation has been studied rather intensively over the past eight years. It has been appreciated as a very rewarding and fascinating phenomena to investigate. It involves, on the one hand, the highly nonlinear response of an atom exposed to an intense laser field of ultra-short duration. On the other hand, the macroscopic field observed experimentally, requires phase matching to take place in the nonlinear medium, which might simultaneously undergo rapid ionisation. A fairly good understanding has now been achieved with respect to both these effects. However, there are continuously new questions arising regarding fundamental physical processes involved, and harmonic generation will certainly continue to be the subject of many investigations in the future as well. At the same time as it is an interesting phenomenon from a fundamental point of view, harmonic generation is also a way of producing ultra-short pulses of coherent radiation, suitable for applications in the extreme ultraviolet (XUV) and soft x-ray spectral range. Indeed, a number of experiments utilising this radiation in various applications have already been reported. These experiments have demonstrated that it is feasible to use high-order harmonic radiation for applied as well as fundamental applications. In this paper, we will discuss the potentials of this source for applications, and different ways of optimising its characteristics.

1. Applications of Harmonic Radiation

In order to identify the applications where high-order harmonics can be expected to be most valuable, we present in Table 1 a comparison between harmonic radiation and radiation from an undulator at a synchrotron storage ring. The numbers given in the two columns represent typical values, at 100 eV, obtained with harmonics and with a state-of-the-art undulator. The harmonics are generated in Ne, using 150 mJ of 800 nm

433

H. G. Muller and M. V. Fedorov (eds.), Super-Intense Laser-Atom Physics IV, 433–443.
© 1996 *Kluwer Academic Publishers.*

radiation from the 150 fs Ti:S terawatt laser at the Lund-High Power Laser Facility [1], and the undulator data [2] are for an undulator at the new 1.5 GeV storage ring in Lund, MAX II.

TABLE 1. A comparison at 100 eV between the characteristic data for radiation generated as high-order harmonics, using a short-pulse laser and with a state-of-the art undulator.

	Harmonics 150 mJ @ 1.5eV 150 fs, $h = 10^{-8}$	Undulator Max II 200 mA, 1.5 GeV
Photons /pulse	10^8	10^5
Repetition rate	10 Hz	500 Mhz
P_{ave}	15 nW	1 mW
Pulse width	0.1 ps	60 ps
P_{peak}	15 kW	40 mW

One notices that the number of photons per pulse is several orders of magnitude higher in the case of harmonics. The average power, however, is still significantly higher for the undulator, because of its extremely high repetition rate. Applications relying on, e.g., absorption measurements, the time integrated flux or single-photon-counting coincidence measurements will therefore likely benefit from being performed at an undulator. The pulse duration is considerably shorter for the harmonics. This short pulse duration, together with the larger number of photons per pulse, results in a peak power in the harmonic pulses, many orders of magnitude higher than the corresponding peak power from the undulator. Applications requiring ps or sub-ps resolution, or very high peak powers, will benefit from these special features of the harmonic source. In addition, estimates based on the spatial coherence of the harmonic radiation, and the effective source size, indicates that it should be possible to focus the XUV radiation from harmonic generation to a smaller focal spot than the corresponding radiation from the undulator. We therefore expect that the ratio between the peak intensities will differ with an even larger factor than for the peak powers.

If the above comparison would have been made at a lower photon energy, the ratio between the number of photons per pulse, obtained through harmonic generation and with an undulator, would be even larger. On the other hand, at much higher energies, this ratio would be lower, since the efficiency for the harmonic generation is lower. Above about 180 eV, harmonics are no longer produced, but the efficiency of the undulator extends to very high energies, into the hard x-ray range.

The short pulse duration was utilised in a very nice experiment by Haight and Peale [3]. They used relatively low-order harmonics (up to the 11th) of a 1 ps dye laser to study the dynamics of surface states in solids. They utilised the convenient synchronisation between the harmonic pulse and an additional laser pulse and could study the dynamics with ps resolution in the XUV. The high temporal resolution required for this study ruled out the use of a synchrotron source.

Balcou *et al.* [4] showed that high-order harmonics could be used also in atomic spectroscopy. They measured the relative photoionisation cross-sections of rare gases. Coarse tunablity was obtained in that study by using a fixed frequency short-pulse laser and selecting successive harmonic orders. The range from a few eV up about 104 eV was covered using different gases for generation of the harmonics. This represents the type of experiments that frequently are performed at synchrotron facilities. However, the use of harmonics as the source for this study demonstrated that this type of studies can now be performed also in small scale laboratories.

In a more specific application, requiring finer tunability and a narrower linewidth, we recently utilised harmonics of a 80 ps tunable dye laser to measure the radiative lifetime of the 1s2p ^{1}P [5] and 1s3p ^{1}P states [6] of helium in a pump/probe configuration. A selected harmonic resonantly excited the atomic state, which was subsequently ionised by another laser pulse. The experimental setup is shown schematically in Figure 1. A distributed feedback dye laser oscillator was used to generate 80-ps pulses tunable between 715 and 900 nm. These pulses were amplified in two dye cells and in a Ti:S crystal up to an energy of 50 mJ. Harmonics were generated in a pulsed jet of rare gas and separated by a normal-incidence spherical grating.

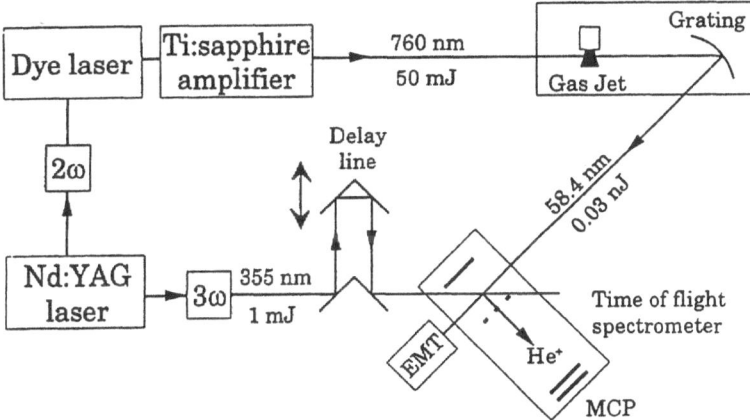

Figure 1. Experimental setup used for the measurement of atomic radiative lifetimes using a pump-probe technique, with the pump pulse in the XUV.

For the 2p state, we tuned the laser wavelength around 760 nm and selected the 13th harmonic (58 nm, 21 eV) as the pump pulse. The excited atoms were subsequently ionised by the third harmonic (355 nm, 3.5 eV) of a fraction of the 80 ps Nd-YAG laser whose second harmonic was used to pump the dye laser (see Fig. 1). For the 3p state, we tuned the laser around 752 nm. We generated the second harmonic in a KDP doubling crystal and selected the 7th harmonic of the doubled frequency (54 nm). The excited atoms were ionised by the second harmonic (532 nm, 2.2 eV) of the YAG laser. In both cases, the photon energy of the probe laser was chosen in order to barely reach the threshold, thus optimising the ionisation cross-section. The two beams crossed at 45 degrees inside a time-of-flight spectrometer. A variable delay line placed in the beam path of the probe beam allowed us to adjust the relative time between the two light pulses. The ions generated were separated in mass in the time-of-flight tube and detected by a microchannel plate (MCP).

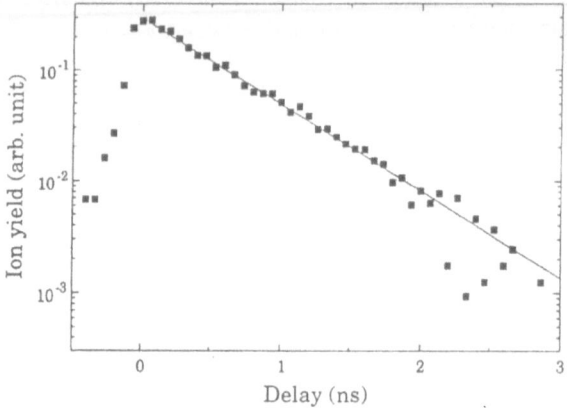

Figure. 2. Number of ions produced as a function of the delay between the pump and the probe pulse. The slope of the curve yields the radiative lifetime of the 1s2p ^{1}P state in neutral He.

Figure 2 shows, on a logarithmic scale, the number of He$^+$ ions, created via selective excitation to the 2p state, as a function of the delay time between the pump and the probe. Each point is an average of 200 shots. The atomic density for this measurement was 7×10^{-6} mbar. The probe photon flux was sufficiently high to ionise all the excited atoms in the interaction region.

With this experiment, we have demonstrated that the harmonics can be used for two-colour pump-probe experiments. It enables cascade-free studies of radiative properties of atomic and ionic systems in the XUV spectral range. Note that it is possible to go to much shorter wavelength and shorter pulse duration than those used in the present experiment, by using a pump laser with a shorter pulse duration.

Another type of application where the characteristic of high-order harmonic radiation appears to be very attractive, is high-intensity experiments. Intensities in the range of 10^{12} W/cm^2 should be feasible within the next few years using the latest developments in x-ray optics, such as normal incidence multilayer mirrors. This will open doors to new fields of physics, such as multiphoton processes in the XUV, which are not possible to explore using conventional sources of XUV radiation. Values for the third-order nonlinear susceptibilities of He, for photon energies in the range between 25-50 eV have been calculated by van Enk et al. [7]. Their estimates indicates that it should be possible to observe nontrivial amounts of third-harmonic generation with photons in this energy range when the intensity exceeds about 10^9 W/cm^2. This intensity should be within reach in a near future.

2. Optimisation

2.1 PHOTON ENERGY

As soon as a new source of radiation is beginning to be used in applications, the need for optimisation arises. A great deal of work has been devoted lately to optimise the harmonic generation process with respect to the highest order observed, or equivalently, to the shortest wavelength generated. There are today a fairly good understanding of the mechanisms determining the cutoff of the characteristic harmonic plateau. The width of the plateau varies as $I_p + 3U_p$, where I_p is the field-free ionisation potential of the atom and U_p the ponderomotive energy. $U_p(eV) = 9.33 \times 10^{-14} \times I(W/cm^2) \times \lambda^2(\mu m^2)$, where I is the laser intensity and λ the laser wavelength. This scaling law is valid up to the intensity at which the atom becomes ionised (saturation intensity). It shows that one should preferentially use light rare gases (with a high ionisation energy), low-frequency lasers, and short pulses (such that neutral atomic species can survive to a higher intensity), in order to produce high-energy photons.

The extent of the harmonic plateau obtained experimentally in rare gases [8,9] is found to be shorter than in the single-atom response. It varies with the intensity as approximately Ip+α×Up with α ≈ 2-2.5. This can be understood by considering the effect of propagation in the nonlinear medium, and more precisely by including the geometrical phase mismatch introduced by focusing [9].

Ions have a high ionisation energy and require a high intensity to be further ionised. According to the prediction of the cutoff law, these systems should produce harmonics of extremely high orders. However, there are several problems involved in generating harmonics from ions [10,11]. The shortest wavelengths to date, about 6.5 nm, generated in ions [12,13], are only marginally shorter than the shortest wavelengths generated in neutral atoms using long-wavelength lasers. This subject is discussed more in details in the paper by Sanpera et al. in this volume.

2.2 PEAK POWER

When one turns to applications, one usually wants to optimise the characteristics of the radiation at a particular wavelength, corresponding to one harmonic order. If a particular application requires high peak power, one wants to generate a large number of photons per pulse, while retaining as short pulse duration as possible and a good spatial coherence. The conditions required to optimise these parameters might not coincide with the conditions required to generate the highest possible harmonic orders. The pulse duration of the harmonics follows the duration of the laser, and is in general shorter than the driving pulse [14]. The use of a short-pulse laser will therefore result in short-pulse harmonic radiation. However, special attention must be given to the techniques used to separate one particular harmonic order from the rest and from the fundamental field, and those used to focus the harmonics. The use of gratings and various types of optics can introduce temporal spreading of the harmonic pulse. Since the harmonic pulses are ultrashort, less than 100 fs for harmonics created by ultrashort laser pulses, temporal spreadings of the order of picoseconds will seriously reduce the peak power and peak intensity. The spatial coherence is another factor that must be given proper attention in order to be able to focus the radiation to a small spot for high-intensity applications. It was recently shown by Salières *et al.* [15] how the spatial coherence, and in particular the angular distribution of the harmonic radiation critically depends on the focusing of the laser relative to the nonlinear medium (the gas jet). That work also shows that when the laser is focused at the optimum position, which is before the nonlinear medium, the harmonic radiation can be described by a nearly Gaussian spatial profile. We will not discuss the temporal and angular properties more in this paper, but instead devote the rest to the optimisation of the number of photons, at a given harmonic order.

2.3   NUMBER OF HARMONIC PHOTONS OF A GIVEN ORDER

With a given laser, there are a number of parameters that should be optimised in order to obtain the maximum number of photons of a given harmonic order. One is the focusing of the laser radiation, another is the laser wavelength (fundamental or frequency doubled laser radiation). The type of gas used as the nonlinear medium and the pressure of this gas are other parameters which must be optimised for each energy range of interest.

2.3.1 *Focusing*
Focusing conditions have been shown to considerably affect the conversion efficiency [16]. In order to optimise phase matching, the focusing should be as weak as possible. The optimum conversion efficiency depends, of course, on the laser power available, since the intensity at the focus must be high enough (to reach the plateau level). Actually, it frequently turns out that the efficiency is not limited by the available laser power but by the damage threshold of the optics used to analyse the radiation.

2.3.2 *Laser Wavelength*

The ponderomotive energy is proportional to the square of the laser wavelength. According to the I_p+3U_p rule discussed above, the energy of the highest harmonic in the plateau will therefore be much reduced if the laser radiation is frequency doubled before the generation of the harmonics in the gas. However, low photon energies in the XUV are generated with higher efficiency using frequency doubled laser radiation compared with the fundamental laser radiation [16,17]. This is illustrated in Figure 3. Harmonics are generated in Ne, using 800 nm radiation directly from a Ti:S laser, and

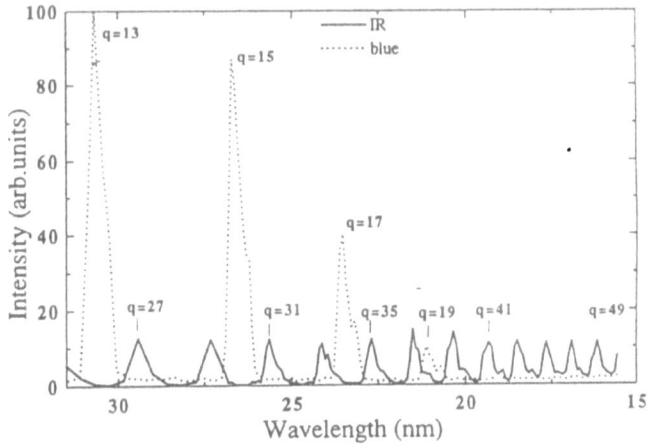

Figure 3. The harmonic signal obtained in Ne, using 800 nm and 400 nm laser radiation.

with 400 nm radiation obtained by frequency doubling the laser radiation in a thin (1.5 mm) KDP crystal. The laser energy was kept constant, and the figure therefore includes the approximately 60% energy reduction inherent in the doubling process. For applications requiring photon energies above about 50 eV, one must use the fundamental laser frequency in order to generate the harmonic radiation. To generate photons of energy less than about 50 eV, it is clearly advantageous to first frequency double the laser radiation. For still lower photon energies, it is possible to generate the harmonics in a heavier gas as discussed below. The choice of gas and the choice of fundamental or frequency doubled laser radiation obviously depend on the energy range of interest.

2.3.3. *Non-Linear Medium*

Most investigations regarding high-order harmonic generation have been performed with rare gases as the nonlinear medium. It is well known that among the rare gases the conversion efficiency increases with the atomic number (decreasing ionisation potential). Considering the relatively large values of the ionisation potentials and the saturation intensities for ionisation, they are likely to be the best neutral medium for generation of very high harmonic orders. However, for relatively low-order harmonics, say below 30 eV, molecular gases can also be used. It is not obvious how the relative conversion efficiencies of such gases will be dependent of the species and molecular structure. There has, e.g., been arguments [18] that the conversion efficiencies should be correlated to the static polarisabilities, and those can be much larger for certain molecules compared to the rare gases. We are presently studying harmonic generation in a number of different molecules, O_2, N_2, H_2, CO_2, N_2O, SF_6, CH_4, C_3H_8 and regular air. The conversion efficiencies as a function of wavelength are recorded and compared, using both 800 and 400 nm laser radiation. The result of this study is currently being analysed and will be published elsewhere [19]. However, the preliminary results indicates that with 800 nm radiation and harmonic photon energies up to about 35 eV, all the molecular gases investigated have efficiencies higher than Ne, but lower than Ar.

2.3.3. *Pressure*

Harmonic generation is a coherent process, and the total number of photons generated should be proportional to the square of the number of (emitting) atoms in the interaction region, or for a given length of the medium, to the square of the pressure in the medium. However, such a simple square dependency cannot be expected to hold for arbitrary high pressures. There are a number of reasons for deviations from a quadratic behavior. High-order harmonic generation in the low-frequency high-intensity regime can be understood as a two-step process [20,21]: electrons tunnel out through the atomic potential, oscillate in the laser field and can come back towards the nucleus with a high kinetic energy, recombining with the core and producing high-order harmonics. The quiver amplitude of an electron exposed to an intensity of 10^{15} W/cm^2 at 800 nm is 2.8 nm. At a typical "high pressure" of, say 80 mbar ($\sim 2\times10^{18}$ atoms/cm^3), the average distance between two atoms is 8 nm, i.e. only a factor of three higher than the quiver amplitude. One could therefore expect the return of the electron towards the nucleus to be disturbed by the presence of other atoms, a disturbance that would increase with the pressure. The pressure will also influence the propagation through the medium, of the fundamental as well as of the generated radiation. There will be, to some extent, reabsorption of the XUV radiation. This is, of course, pressure dependent. There will also be pressure dependent phase matching due to the dispersion in the neutral medium, and due to the density of free electrons created through ionisation of the medium during the laser pulse. This free-electron density can also influence the intensity in the medium. The electron density has a radial gradient, since the degree of ionisation depends on the radial intensity profile of the laser beam. This gradient will act as a negative lens for the fundamental beam and hence lead to defocusing and a corresponding reduction in the laser intensity obtained in the centre of the nonlinear medium. The peak intensity will consequently be pressure dependent and add to the effects that reduces the harmonic efficiency as a function of the pressure.

In a recent investigation [22], we studied the number of harmonic photons produced as a function of the pressure for different process orders, laser intensities, etc. The atomic

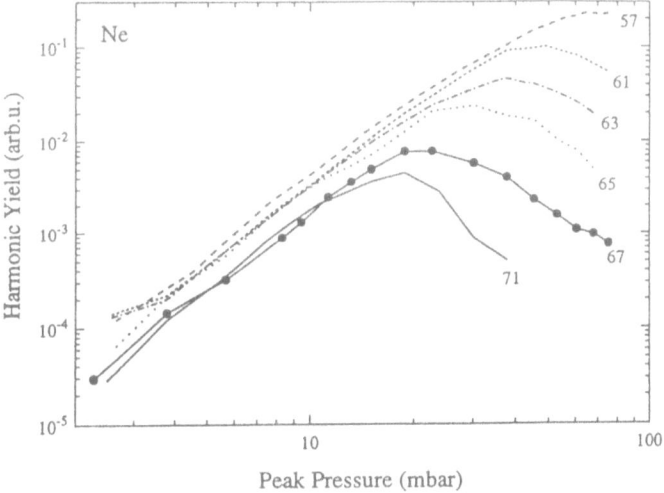

Figure. 4. The number of harmonic photons as a function of the pressure in the nonlinear medium. Results are shown for different harmonic orders, from the 57th in the plateau to the 71st, the last visible harmonic in the cutoff.

density below the nozzle of the gas jet was measured using an interferometric technique. In general, the number of photons increased as the square of the atomic density over the range investigated (1-80 mbar). However, an interesting effect was observed in the cutoff region, which is shown in Figure 4. The harmonic yield goes through a maximum and then decreases. The pressure corresponding to this maximum decreases as the process order increases. The figure shows the evolution of several harmonics from the 57th, in the plateau region, to the 71st, the last visible harmonic in the cutoff. The intensity is estimated to be 5×10^{14} W/cm^2. Comparing these experimental results with results from detailed theoretical calculations, including the different propagation effects discussed above, but ignoring the disturbance from neighbouring atoms, a good agreement was obtained. These calculations showed that the saturation effect obtained as the pressure increased to a large extent results from electron dispersion. The decrease in harmonic intensity, after the optimum pressure, results from the defocusing of the fundamental laser radiation. The effective medium, over which the intensity remains significant, is considerably reduced. This allows us to understand why the optimum pressure observed experimentally depends on the process order and is higher for the harmonics in the plateau. The harmonics in the cutoff region are produced at the highest intensity, and are strongly reduced as the intensity in the medium and the effective interaction length decrease. The harmonics in the plateau, however, are produced also at lower intensities, which means over a large volume and at the beginning of the pulse, for which the defocusing of the fundamental is not as important. Therefore, the conversion efficiency for these harmonics is not affected so much by the defocusing of the fundamental. The fact that good agreement was obtained between experiment and theory, without including the possible disturbance from neighbouring atoms, means that this effect plays a minor role, if any, in the range of pressures investigated.

3. Conclusion

High-order harmonic generation has matured to the point where it has started to be used in applications. Harmonic radiation has recently been applied as a source of XUV radiation in experiments in atomic, molecular and solid-state physics. The short pulse duration and the high peak power inherent in the harmonic pulses makes this source particularly promising for applications requiring high temporal resolution or high intensity, e.g., multiphoton processes in the XUV. For many applications, it is important to be able to produce as many photons as possible per pulse, of a particular harmonic order, using a given laser. This requires knowledge about the influence of the focusing of the laser and the laser wavelength (fundamental or frequency doubled radiation) as well as of the species and the pressure in the nonlinear medium. We have discussed these different parameters, and how the photon flux can be optimised. In particular, we have shown that the choices of these parameters depend on the energy range of interest.

4. Acknowledgement

We acknowledge the support of the Swedish Natural Science Research Council, the Direction des Recherches, Etudes et Techniques under contract 92-439 and the EC Human Capital and Mobility Programme under contract ERBCHX920028

† Also at Niels Bohr Institute, Copenhagen University, Denmark

* Present address: Dipartimento Scienze Fisiche, Università di Napoli, Naples, Italy

References

1. S. Svanberg, J. Larsson, A. Persson, and C.-G. Wahlström, Physica Scripta 49, 187 (1994).
2. S. Werin (1994), Private Communications.
3. R. Haight and D. R. Peale, Phys. Rev. Lett. 70, 3979 (1993).
4. Ph. Balcou, P. Salières, K. S. Budil, T. Ditmire, M. D. Perry and A. L'Huillier, Z. Physik D, 34, 107 (1995).
5. J. Larsson, E. Mevel, R. Zerne, A. L'Huillier, C.-G. Wahlström and S. Svanberg, J. Phys. B. Letters, 28, L53-(1995).
6. A. L'Huillier, T. Auguste, Ph. Balcou, B. Carré, P. Monot, P. Salières, C. Altucci, M.B. Gaarde, J. Larsson, E. Mevel, T. Starczewski, S. Svanberg, C.-G. Wahlström R. Zerne, K.S. Budil, T. Ditmire and M.D. Perry. Journal of Nonlinear Optical Physics and Materials, 4, no 3, ... (1995).
7. S.J. van Enk, J. Zhang and P. Lambropoulos, Preprint.

8. C.-G. Wahlström, J. Larsson, A. Persson, T. Starcszewski, S. Svanberg, P. Salières, Ph. Balcou and A. L'Huillier, Phys. Rev. A **48**, 4709 (1993).
9. A. L'Huillier, M. Lewenstein, Ph. Balcou, P. Salières, M. Y. Ivanov, J. Larsson and C.-G. Wahlström, Phys. Rev. A **48**, R3433 (1993).
10. C.-G. Wahlström, S. Borgström, J. Larsson and S.-G. Pettersson, Phys. Rev. A **51**, 585 (1995)
11. S. Kubodera, Y. Nagata, Y. Akiyama, K. Midorikawa, M. Obara, H. Tashiro, and K. Toyoda, Phys. Rev. A **48**, 4576 (1993).
12. Y. Nagata, K. Midorikawa, M. Obara and K. Toyoda, Preprint.
13. S.G. Preston, A. Sanpera, M. Zepf, W.J. Blyth, C.G. Smith, J.S. Wark, H.M. Key, K. Burnett, M. Nakai, D. Neely and A.A. Offenberger. Preprint.
14. T. Starczewski, J. Larsson, C.-G. Wahlström, M.H.R. Hutchinson, J.E. Muffett, R.A. Smith, and J.W.G. Tisch,. J.Phys. B. **27**, 3291 (1994). See also the paper by R.C. Constantinescu *et al.* in this volume.
15. P. Salières, A. L'Huillier and M. Lewenstein, Phys. Rev. Lett **75**, 3776 (1995).
16. Ph. Balcou, A. S. L. Gomes, C. Cornaggia, L. A. Lompré, A. L'Huillier, J. Phys. B **25**, 4467 (1992).
17. T. Ditmire, J.K. Crane, H. Nguyen, L.B. DaSilva, M.D. Perry, Phys. Rev. A **51**, R902 (1995).
18. S.L. Chin and P.A. Golovinski, J.Phys. B **28**, 55 (1995).
19. C. Lyngå, A. L'Huillier and C.-G. Wahlström, to be published.
20. P.B. Corkum, Phys. Rev. Lett. **73**, 1995 (1993).
21. K.C. Kulander, K.J. Schafer and J.L. Krause, in "Super-Intense Laser-Atom Physics", ed. B. Piraux, A. L'Huillier and K. Rzazewski, NATO ASI Series B, Vol 316, p. 95 (Plenum Press, New York, 1993).
22. C. Altucci, T. Starcszewski, C.-G. Wahlström, E. Mevel, B. Carré and A. L'Huillier, *J. Opt. Soc. Am B*, **13**, no 1 (1996), In press.

TIME DEPENDENCE OF HARMONIC GENERATION BY A SINGLE ATOM

R. TAIEB, A. MAQUET
Laboratoire de Chimie Physique-Matière et Rayonnement
Université Pierre et Marie Curie,
11 rue Pierre et Marie Curie
75231 Paris Cedex 05, France.

P. ANTOINE* AND B. PIRAUX
Institut de Physique, Université Catholique de Louvain,
2, Chemin du Cyclotron, B1348 Louvain-La-Neuve Belgium.

1. Introduction

Since the observation of high-order harmonic generation by gaseous atomic targets submitted to intense laser radiation [1], the question of the actual time dependence of the harmonic emission has attracted a lot of attention [2-4]. In fact, this question is of importance in view of the current investigations regarding the possible applications of harmonics as an ultra-intense source of pulsed and coherent radiation in the X-UV spectral range. The first pump-probe type experiments using harmonics have been reported [5] and the possibility of generating, via an ingenious two-color scheme, sub-femtosecond pulses of harmonics has been envisioned, [6]. Within this context, it has been already shown that a time-frequency (Gabor or/and wavelet) analysis of the atomic dipole acceleration can be a valuable tool in order to analyse the time profile of harmonic generation by a single atom, submitted to an intense radiation pulse, [2]. The main purposes of the present paper are to discuss in general terms, some typical features of the harmonic emission rate, depending on the characteristics (time duration, shape of the envelope, peak intensity, frequency, etc...) of the laser pulse used to generate the harmonics. Then, we shall show that such a time frequency analysis sheds some light on several intriguing features of the basic mechanisms governing high-order harmonic generation by an atom in the presence of an infrared laser. The method used is based on the analysis of the atomic dipole acceleration, as computed via the numerical resolution of the Time-Dependent Schrödinger Equation (TDSE) for a model atom.

The paper is organized as follows : the second Section will be dedicated to a brief exposition of the model used and of the time-frequency (Gabor) analysis developed here. The general features of the time profiles of harmonic emission by an atom submitted to a low frequency radiation pulse with a sine-square envelope are presented in

445

H. G. Muller and M. V. Fedorov (eds.), Super-Intense Laser-Atom Physics IV, 445–454.
© *1996 Kluwer Academic Publishers.*

Section 3. Then, considering an idealised pulse shape with a linear ramp extending over several laser periods, we shall present, in Section 4, the corresponding time variation of the Gabor expansion coefficients associated to an harmonic frequency present in the spectra. Our results permit to define an "Appearence Intensity" for a given harmonic, above which it will rejoin the plateau observed in the power spectra. Also shown in this Section is that this simple analysis brings interesting views on the width of the plateau, depending on the laser field strength intensity and indirectly on the ionization regime (multiphoton or tunnel) prevailing at a given intensity. Still regarding the time dependence of the emission of a given harmonic, our results provide a direct confirmation of the validity of the currently accepted picture of an emission process taking place in the course of a recollision of the active electron with the nucleus (or ionic core) [7-9]. Another outcome of the presented time-frequency analysis is the possibility to discuss the recently predicted occurrence of frequency shifts of the harmonic lines [10]. This latter point will be briefly addressed in Section 5.

2. Theory

We use in our calculation a (1-D) model potential introduced by Eberly et al. [11]. One has to solve numerically, using a Crank-Nicholson scheme [12], the (1-D) TDSE which reads:

$$i \frac{\partial}{\partial t} \psi(x,t) = \left(\frac{p^2}{2} - \frac{1}{\sqrt{\delta + x^2}} - x \, F(t) \sin(\omega_L t) \right) \psi(x,t) \tag{1}$$

We took $\delta=2$ in order to get a ground-state energy close to 0.5 a.u.(Hydrogen 1s energy), ω_L is the laser frequency. We also used the velocity gauge expression of the laser-atom interaction. F(t) represents the field envelop and was taken either as a sine square or as a linear turn-on followed by a constant value. The radiated spectrum of a single atom is then obtained by computing the expectation value of the dipole acceleration. Its Fourier transform gives the power spectrum:

$$P(\omega) \propto \left| \int_0^T a(t) \, e^{-i\omega t} dt \right|^2 \tag{2}$$

where T is the pulse duration and a(t) the dipole acceleration given by the Ehrenfest's theorem [13] :

$$a(t) = -\left\langle \psi(x,t) \left| \frac{\partial V}{\partial x} + F(t)\sin(\omega_L t) \right| \psi(x,t) \right\rangle \tag{3}$$

Such standard Fourier analysis does not allow to determine at which time the emission of a given harmonic takes place. As shown in Ref. [2], a time frequency analysis can provide the information .The idea is to introduce a window depending on two parameters α, τ where α is related to the given frequency and τ to the time localization. We have found convenient to use in our case the Gabor transform [14] :

$$C\,(\alpha,\tau) = |C\,(\alpha,\tau)|\,e^{\,i\phi\,(\alpha,\tau)} \propto \int_{-\infty}^{\infty} \left(a(t)\,e^{-i\alpha t}\right)g_{\sigma_0}(t-\tau)\,dt \qquad (4)$$

with
$$g_{\sigma_0}(t) = \frac{1}{\sigma_0\,\sqrt{2\pi}}\,e^{-\frac{t^2}{2\sigma_0^2}} \qquad (5)$$

α defines the frequency of the modulation whereas the width of the window function is fixed by σ_0. The parameter σ_0 can be related to the resolution in time Δt (FWHM of the window function g) and in frequency $\Delta \omega$ (FWHM of the Fourier transform of g). These two are given by:

$$\Delta t = 2\sqrt{2\ln 2}\,\sigma_0 \quad,\quad \Delta\omega = 2\sqrt{2\ln 2}\,\sigma_0^{-1} \qquad (6)$$

If σ_0 is set to $0.6T_L$, one finds $\Delta t{\approx}1.4\ T_L$ and $\Delta\omega{\approx}0.6\omega_L$. By dividing σ_0 by ten, the time resolution is increased and the overlap for neighboring frequencies becomes larger. This is illustrated in Fig.1

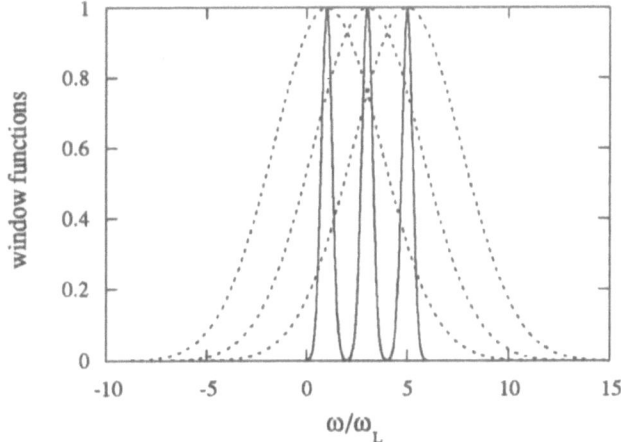

Fig.1 : Fourier transform of the chosen window functions for different values of α ($\sigma_0 = 0.06\ T_L$ dashed lines, $\sigma_0 = 0.6T_L$ solid lines) centered at $\alpha = 1,\ 3,\ 5\omega_L$.

3. Time duration of harmonic emission in a laser pulse.

Let us now consider the interaction of an atom in its ground state with an intense sine-square shape laser pulse of duration $32\ T_L$ ($T_L{\sim}146$ a.u.), whose frequency is 0.043 a.u. (1.17 eV) corresponding to a Nd:YAG laser. We first take a moderate peak intensity $I_0=3.5\ 10^{12}\ W/cm^2$ ($F_0{\sim}10^{-2}$ a.u.) well within the perturbative regime with only a few harmonics. In this regime one expexts a simple "power" law for the intensity (and time) dependence of the harmonic [15]. In Fig 2 we show the usual (Fourier) power spectrum for this pulse and in Fig 3 the time profile of the harmonic emission given by the Gabor coefficients for the harmonics 1, 3, and 5. The curves in Fig.3 are almost indistiguishable from the corresponding power of the sine-square envelop function (not plotted here).

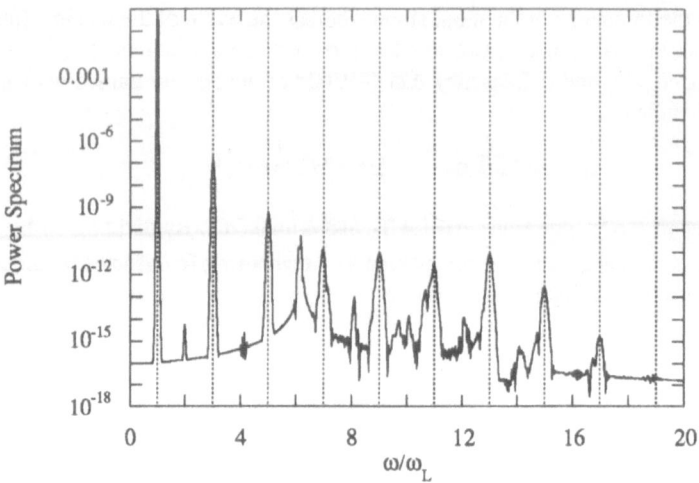

Fig. 2.Harmonic spectrum in H (1-D) at 1.064 μm and a peak intensity I_0=3.5 10^{12} W/cm^2.The pulse has a sine-square shape and is 32T_L long

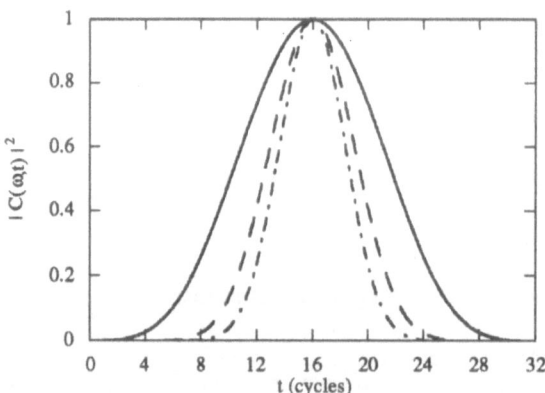

Fig. 3. Time profile given by the Gabor coefficientsfor α=1, 3, 5ω_L (solid , dashed, dotted lines respectively). Same conditions as Fig. 2.

For a higher peak intensity (I_0 = 3.5 10^{14} W/cm^2, F_0 = 0.1 a.u.) ionization occurs, which entails a significant depletion of the ground state population. Therefore, [2] the duration of emission is much shorter than the pulse duration, as seen in Fig 3. For example we find that the duration time of the 7th harmonic is only of the order of five laser cycles (which correspond to approximatively 17.2 fs). We also note that the emission time interval decreases with the harmonic order which should lead to a broadening of the lines with the order. Fig 3 also shows that the harmonic profiles seem to appear successively as a function of the order. We will address this point in the next section with the use of a different pulse shape, namely a linear turn-on.

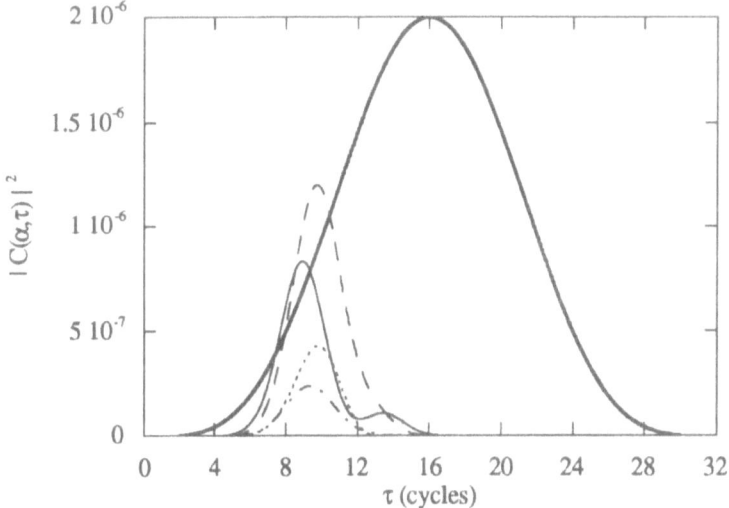

Fig. 4. Time profile given by the Gabor coefficients for, α=5, 7, 13, 17ω_L (thin solid dashed, dotted, and dot-dashed lines respectively). The envelop of the pulse is drawn in thick solid line.Same conditions as in Fig. 2 except that I_0 = 3.5 10^{14} W/cm^2.

4. Time dependent analysis with a linear ramp.

We now consider the case of a linear turn-on We have chosen a linear turn-on for the field extending over 20T_L up to an intensity of 10^{14} W/cm^2. Note that we have checked that a linear turn-on for the intensity provides essentially the same results. Making a Fourier analysis between t=16T_L and 24T_L, gives the spectrum of Fig.5, with rather well defined harmonics and some "hyper-Raman" peaks corresponding to atomic transitions [16]. The end of the plateau is around the 41st harmonic which agrees with the so-called "3Up rule" [7-9]:

$$n_{max} \, \omega_L = I_p + 3.17 \, U_p \qquad (7)$$

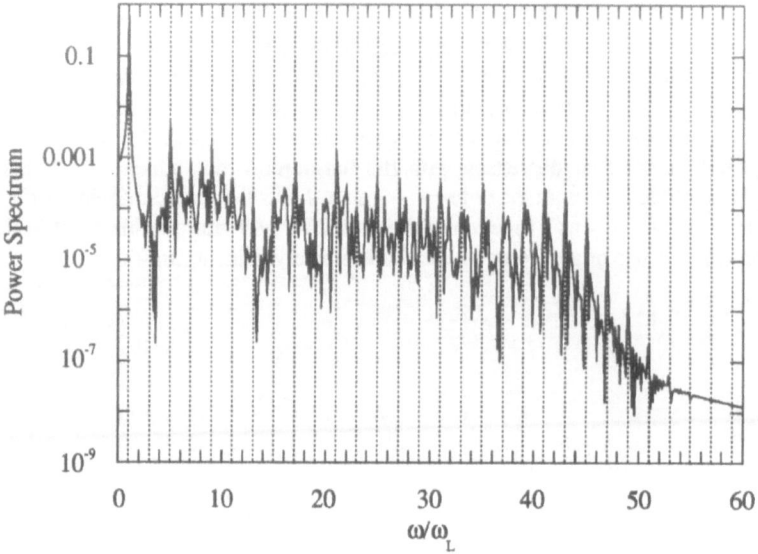

Fig. 5.Harmonic spectrum in H (1-D) at 1.064 μm and a peak intensity
$I_0 = 10^{14}$ W/cm². The linear turn-on is $20T_L$ long.

If we analyse now the dipole acceleration using the Gabor transform we obtain
the results of Fig.6. Almost all the harmonics have the same behavior as a function of
the intensity. The Gabor coefficient modulus first grows rapidly in time (~Intensity) and
then reaches a plateau with oscillations. We can attribute to each harmonic an
"Appearance Intensity" which is defined as the intensity where a given harmonic
coefficient is equal to the previous one (crossing of the two curves). We can then assign
this "Appearance Intensity" to the one at which a harmonic leaves the cut-off region and
rejoins the plateau of harmonics. These curves are reminiscent of the ones obtained by
Lewenstein et al. [9]. By plotting the "Appearance Intensity" as a function of the
harmonic order, we find that it follows a linear law given by:

$$n \, \omega_L = .58 + 3.1 \, U_p \, (I_{AI}) \tag{8}$$

where U_p (I_{AI}) is the ponderomotive energy computed at the "Appearance Intensity",
which is very close to expression (7).It is intersting to note that this law is valid not
only in the tunneling regime, where the Keldysh parameter γ is lower than 1, but also
in the multiphoton regime.

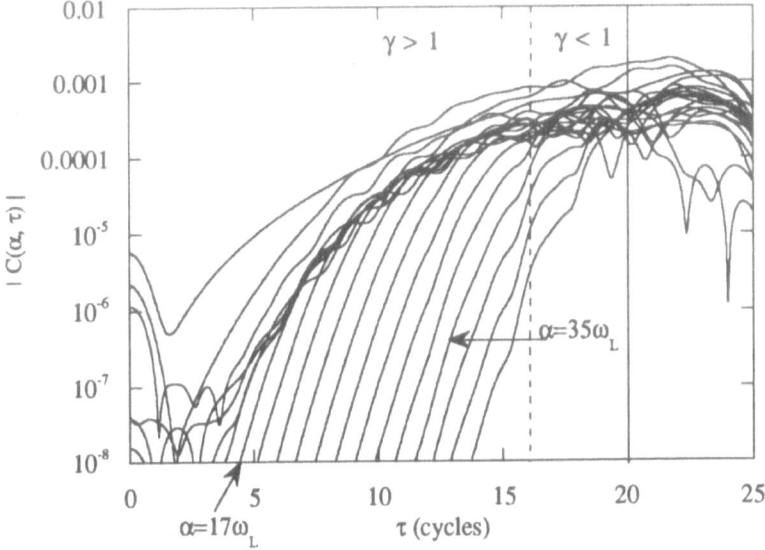

Fig. 6. Time profile given by the Gabor coefficients for $\alpha=(2n+1)\omega_L$, n=1 to 20 and σ_0 is set to $0.6T_L$. Same conditions as in Fig. 5. The vertical solid line indicates the end of the turn-on, whereas the dashed line shows the transition time from the multiphoton to the tunneling regime.

The other question we want to address is the possibility to observe the $2\ \omega_L$ periodicity in the emission of the harmonics [2]. This is predicted using either the semi-classical models [7,8] or the model developed by Lewenstein et al. [9] . Using the former formalism one should see that the harmonics in the cut-off are all emitted at the same time (namely the one when classical trajectories return close to the nucleus with a maximum kinetic energy) and is defined as:

$$\frac{[p - A(t-\tau)]^2}{2} + I_p = 0 \qquad (9)$$

The real part of τ, τ', can be interpreted as the duration of the trajectory and its imaginary part as the tunneling time. This real part, in the case of a harmonic in the cut-off region maximises the function $C(\tau')$ defined as:

$$C(\tau') = \sin(\omega_L\tau') - 4\ \sin^2(\omega_L\tau'/2)/(\omega_L\tau') \qquad (10)$$

The numerical value obtained for τ' is $\approx 0.65T_L$. As the phase of the electrical field, when the electron is born close to the nucleus, is $\approx 0.30T_L$, the harmonic amplitude should have maxima at $\approx 0.95T_L$. This can be seen if we increase the time resolution in our time-frequency analysis. By decreasing σ_0 by a factor of ten we obtain the Fig.7:

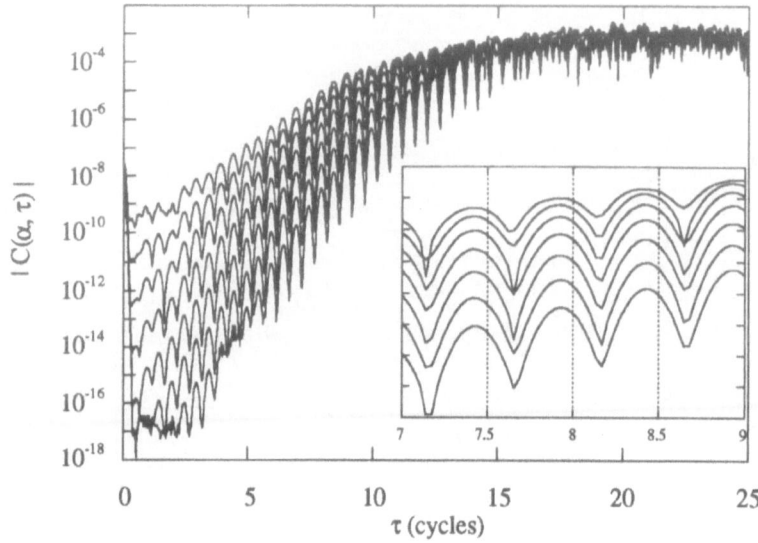

Fig. 7 Same as in Fig.6 for $\alpha=(2n+1)\omega_L$, n=9 to 15 with $\sigma_0 = 0.06T_L$.
The inset represents an enlargement between 7 and 9 T_L.

     The general behavior of the curves are the same as Fig.6 except that on top of that we see oscillations at the frequency $2\omega_L$. We can see in the inset that the curves oscillates in phase showing maxima of emission of harmonics at the same times. The maxima of these curves are at $t\approx(n/2+0.95)T_L$, which were predicted above. In the plateau region, after the appearance intensity, no clear oscillations can be detected, because the harmonics are emitted out of phase (for a different return time) and the overlap between the different harmonics is much too large to differenciate the frequencies.

5. Frequency Analysis with a Linear Ramp

In this section we study the same dipole as in the Sec. 5 except that we are looking at the phase of the Gabor coefficient instead of its modulus. Our motivation is to address the question of the blue shift of the harmonics, due to a change of intensity of the laser field, recently found in numerical simulations by Watson et al. [10]. The phase of the Gabor coefficient, more precisely its time derivative, can be directly related to an instantaneous frequency [17]. By centering the Gabor window around a given harmonic one could see if the atom is emitting this frequency or a slightly shifted one. To illustrate this, we study two different frequencies ($\alpha=23\omega_L$, $\alpha=31\omega_L$) and take $\sigma_0=0.6T_L$. The results are presented in Fig 8:

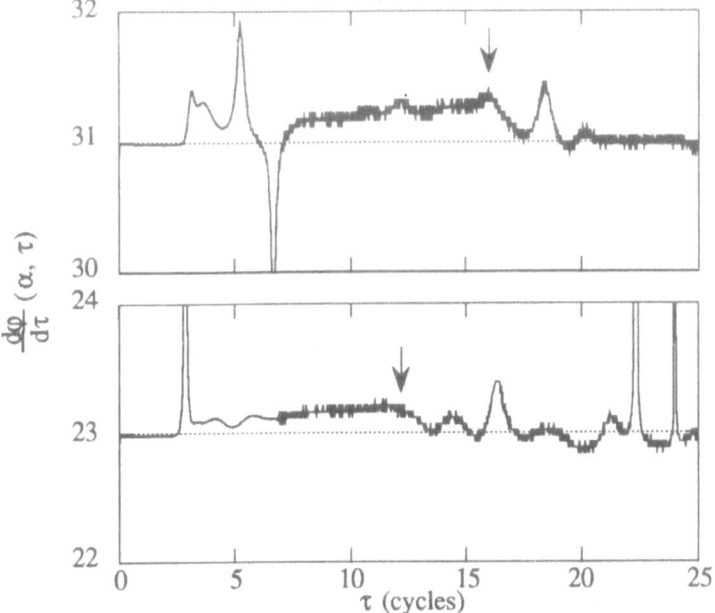

Fig. 8 Time derivative of the phase of the Gabor coefficients for $\alpha=23\omega_L$ (lower part) and $31\omega_L$ (upper part). Same conditions as Fig.6.
The arrows indicate the times corresponding to the "Appearance Intensities", i.e. when the harmonic reaches the plateau.

The variations of the instantaneous frequencies present two different domains: a first one for an intensity lower to the appearance intensity where the frequencies is higher (blueshifted) than studied one; a second one where the frequencies oscillates around the right value. These oscillations are probably due to appearances and disappearances of hyper-Raman frequency around the harmonics. The sharp peaks also observed may be due to resonances crossed as the field is increasing. It seems then that when a given harmonic reaches the plateau its frequency is not anymore blueshifted.

Acknowledgments:

The Laboratoire de Chimie Physique-Matière et Rayonnement is a Unité de Recherche Associée au CNRS, URA 176. This work was partially supported by the EC contract ERB CHRXCT 920013 and 940470. Parts of the computations have been performed at the Centre de Calcul pour la Recherche (CCR, Jussieu, Paris) and at the Institut du Développement et des Ressources en Informatique Scientifique (IDRIS).

454

References:

* Present address: Services des Photons, Atomes et Molécules, Centre d'Etudes de Saclay, 91191 Gif sur Yvette, France.
1. L'Huillier, A., Schafer, K.J., and Kulander, K.C. (1991), *J. Phys. B* **24**, 3315.
2. Antoine, P., Piraux, B., and Maquet,A. (1995), *Phys. Rev A* **51**, R1750.
3. Rae, S., Burnett, K., and Cooper, J. (1994), *Phys. Rev A* **50**, 1946.
4. Starczewski, T., Walhström, C.-G., Tisch, J.W.G., Smith, R.A., Muffett, J.E., and Hutchinson, M.R.H. (1994), *J. Phys. B* **27**, 3291.
5. Larsson, J., Mevel, E., Zerne, R., L'Huillier, A., Walhström, C.-G., and Svanberg, S. (1995) *J. Phys. B* **28**, L53.
6. Ivanov, M., Corkum, P.B., Zuo, T., and Bandrauk, A. (1995), *Phys. Rev. Lett.* **74**, 2933.
7. Corkum, P.B. (1993), *Phys. Rev. Lett.* **71**, 1994.
8. Krause, J.L., Schafer, K.J., and Kulander, K.C. (1992), *Phys. Rev. Lett.* **68**, 3535; Kulander, K.C., Schafer, K.J., and Krause, J.L. (1993), in *Super-Intense Laser-Atom Physics*, Edited by Piraux, B., L'Huillier, A., and Rzazewski, K., Plenum Press, New York.
9. Lewenstein, M., Balcou, P., Ivanov, M.Y., L'Huillier, A., and Corkum, P.B. (1994), *Phys. Rev A* **49**, 2117.
10. Watson, J.B., Sanpera, A., and Burnett, K. (1995), *Phys. Rev A* **51**, 1458.
11. Eberly, J.H., Grobe, R., Law, C.K., and Su, Q. (1992) in *Atoms in Strong Fields*, edited by M. Gavrila, Academic Press, San Diego, pp 301.
12. Press, W.H., Vettering, W.T., Teukolsky, S.A., and Flannery, B.P. (1988) *Numerical Recipes* Cambridge University Press, Cambridge.
13. Burnett, K., Reed, V.C., Cooper, J., and Knight, P.L. (1992) *Phys. Rev A* **45**, 3347.
14. Chui, C.K. (1992), *An introduction to wavelets*, Academic Press, San Diego.
15. Reintjes, J.F. (1984), *Nonlinear Optical Parametric Processes in Liquids and Gases*, Academic Press, Orlando.
16. Millack, T., and Maquet, A. (1993), *J. Mod. Opt.* **40**, 2161; LaGattuta, K.J. (1993), *Phys. Rev A* **48**, 666.
17. Combes, J.M., Grossmann, A., and Tchamitchian, P. (1989), in *Wavelets*, Springer-Verlag, Berlin.

HIGH-ORDER HARMONIC GENERATION WITH A 25 FEMTOSECOND LASER PULSE

J. PEATROSS, J. ZHOU, A. RUNDQUIST, M. M. MURNANE, AND
H. C. KAPTEYN
Department of Physics
Washington State University
Pullman, WA 99164-2814

I. P. CHRISTOV
Department of Physics
Sofia University
1126 Sofia, Bulgaria

Abstract

We have studied harmonic generation in noble gases using a 25fs Ti:Sapphire laser system. For a variety of noble gases we observe significantly higher harmonic orders than previously observed using longer laser pulses. When we increase our pulse duration with fixed peak intensity, the number of harmonic orders produced decreases markedly. We discuss the widths of the individual harmonic peaks in the context of the wide bandwidth of our laser. We also describe how macroscopic phase matching in the interaction is not an important element in determining the spectral widths of individual harmonic peaks.

1. Introduction

Recent advances in the development of high-intensity ultrashort laser pulses provide a new regime for the study of high-order harmonic generation. We have investigated high harmonics generated by a 25fs, 806nm Ti:Sapphire laser [1] in various noble gases. The pulse duration is roughly an order of magnitude shorter than typical pulses previously applied to the study of high harmonics. [2-4] The extreme shortness of our pulses, about ten optical cycles, means that the laser intensity cannot be assumed

455

H. G. Muller and M. V. Fedorov (eds.), Super-Intense Laser-Atom Physics IV, 455–466.
© *1996 Kluwer Academic Publishers.*

to vary adiabatically in time as in the traditional approach. For example, at the half maximum, the intensity of our laser pulse varies by more than 25% in a single optical cycle. Thus, not only will the instantaneous intensity that an atom experiences during the pulse be important, but also the history of the laser pulse leading up to that intensity.

In our investigations [5] using a 25fs laser pulse, we have seen from the heavier noble gases harmonics with photon energies up to 35% higher than those reported previously. We observed up to the 29th harmonic (44eV) in xenon, the 41st (64eV) in krypton, the 61st (94eV) harmonic in argon, and the 135th (209eV) in Ne. In comparison, L'Huillier *et al.* [2] using a 1ps, 1054nm laser pulse reported the observation of harmonics up to the 27th (32eV) in Xe, the 55th (65eV) in Ar, and the 135th (159eV) in Ne. Macklin *et al.* [3,4] using a 125fs, 806nm laser pulse, conditions closest to ours, saw harmonics up to the 53rd (82eV) in Ar and the 109th (169eV) in Ne. It should be mentioned that Wahlstrom *et al.* [6] reported seeing comparatively high harmonic orders generated in ions, up to the 45th in Xe+ and up to the 65th in Ar+ using a 200fs, 800nm laser pulse. However, the emission from the ions was weaker than the emission from neutral atoms by four orders of magnitude; they saw up to the 19th harmonic (29eV) in neutral xenon and the 39th (60eV) in neutral argon.

Observations previous to the present work are in reasonable agreement with the rule that the highest harmonic photon energy cannot exceed $I_p + 3.2 U_p$, where I_p is the atomic ionization potential and U_p is the ponderomotive potential that an electron experiences at the time of detaching from its parent atom. [7-9] This rule predicting the cutoff of the harmonic plateau is consistent with the behavior exhibited by a variety of theoretical models. [10,11]

Since the ponderomotive potential is proportional to the wavelength squared, the 1054nm laser used in the experiments by L'Huillier et al. [2] has a 42% higher ponderomotive potential than our 806nm laser at a given intensity. However, the ponderomotive potential must be calculated for the intensity at which the atom ionizes, and this can be higher for shorter pulses owing to a rapid turn-on time. Thus, in spite of a shorter laser wavelength, extremely short pulses may be expected to generate harmonics with higher photon energies because atoms are able to experience higher intensities before ionizing.

Another possible enhancement to the number of harmonic orders generated by very short pulses arises from the non-adiabatic nature of the pulse envelope. As will be discussed in the next section, the $I_p + 3.2 U_p$ rule is derived for a plane wave. This is a reasonable approximation for longer pulses for which the intensity varies only slightly from cycle to cycle, but the assumption is not valid for our pulses which vary significantly during an optical cycle.

2. Enhancement to the Plateau Cutoff for Short Pulses

To study the effect of pulse duration on ionization, we investigated numerically the ionization rates of a model argon atom [12] for different pulse durations, holding the peak intensity fixed. The calculations were performed by integrating the time-dependent Schrodinger equation in one dimension using the potential $-(1.185^2+x^2)^{-1/2}$ which has a ground state binding energy the same as the outer electron of argon, 0.58 atomic units. While this calculation cannot be expected to provide an exact description of ionization, it is useful to compare the ionization rates within this frame work for different applied pulse durations. The calculation indicates that for an applied Gaussian pulse of peak intensity 5×10^{14} W/cm^2 (our experimental conditions) an electron detaches at an intensity ~25% higher for a 25fs pulse than for 100fs pulse. The ionization rate was characterized in time by computing the probability of the electron being found in the field-free ground state. The higher ionization intensity (i.e., higher U_p) associated with the 25fs pulse translates into a higher maximum harmonic order according to the $I_p+3.2U_p$ rule, which may explain in part the unusually high harmonic orders that we observed.

The $I_p+3.2U_p$ rule can be understood in terms of a simple classical picture wherein the electron detaches from the atom and releases energy when it is recaptured by the atom after approximately a laser cycle. [8,9] The rule is derived by considering the maximum energy that an electron can have when it collides with its parent ion. In the plane-wave case, the classical equation of motion for a suddenly freed electron is solved:

$$\ddot{x} = \frac{eE_o}{m}\cos(\omega t + \phi) \qquad \dot{x}(0) = 0 \qquad x(0) = 0. \qquad (1)$$

The return energy as a function of phase ϕ is

$$V_{return}(\phi) = \frac{1}{2}m\dot{x}^2\left(t_{return}(\phi)\right), \qquad (2)$$

where t_{return} is the time at which the electron returns to the nucleus. The maximum value of V_{return} is $3.2U_p$ which occurs when $\phi = 0.05\pi$ radians. The ionization potential I_p plus the maximum return energy V_{return} provides an estimate for the maximum energy available to be converted into a harmonic photon.

Since this mechanism requires a time on the scale of a laser cycle, the fact that the ponderomotive potential grows during this time will influence the process. For our laser pulses, E_o can no longer be treated as a constant , but must be treated as a function of time, i.e., $E_o(t)$. Depending on how rapidly the field envelope varies in time, this can cause the maximum electron return energy to change significantly. However, rather than alter the $I_p+3.2U_p$ rule for a rapidly changing laser field it is convenient to enter into the formula an effective ponderomotive potential which takes into account the nature of the pulse field envelope. That is, rather than express the return energy in terms of the ponderomotive potential at the moment that the electron breaks away from the atom, it is better to choose an effective ponderomotive potential (or equivalently, an effective intensity) such that the factor 3.2 remains appropriate.

Figure 2 (a) shows the laser intensity and the effective intensity calculated within this frame work for a 25fs, 806nm laser pulse. During the rising edge of the pulse, the relevant intensity for calculating the ponderomotive potential is higher than the laser intensity at the moment when the electron breaks away from the atom since the field strength is stronger when the electron is pushed back toward the atom. On the trailing edge of the pulse, the effective intensity is lower. Figure 2 (b) shows the intensity and the effective intensity for a 100fs laser pulse. Because for a 100fs or a longer pulse, the laser intensity varies only slightly during a cycle, there is little difference between the effective intensity and the electron break-away intensity.

In the 25fs case, there is over a 20% enhancement of the laser intensity midway up on the rising edge of the pulse, and this may account in part for the high harmonic orders that we observed. This is in addition to the enhancement discusses above caused by atoms surviving to higher intensities for shorter pulses. These results are in agreement with recent calculations by I. P. Christov *et al.* that show an increase in the number of harmonic orders for faster rise times of the excitation pulse. [13]

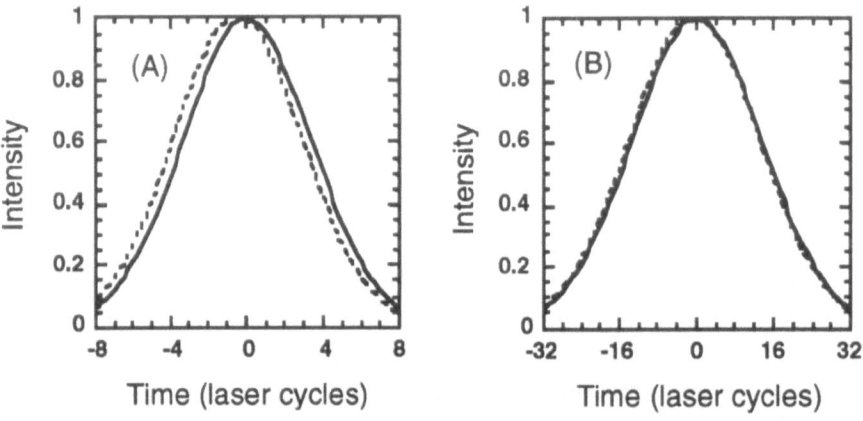

Figure 1. The intensity and effective intensity for an electron liberated by (a) a 25fs and (b) a 100fs laser pulse.

3. Experimental Setup

The experiments were performed using a 10 Hz Ti:sapphire laser system [1] based on chirped pulse amplification. The bandwidth of the pulses is 35nm centered at a wavelength of 806 nm. The laser system can provide up to 70 mJ of energy per pulse, although we used only a fraction of this in the experiments. The final compression of the pulses takes place inside the vacuum. The pulse duration was measured to be 26fs (FWHM) using a single-shot autocorrelator, also inside of the vacuum. The laser beam is linearly polarized and focused by a curved mirror of focal length 1m. The beam diameter ($1/e^2$ intensity) at the mirror is ~12mm resulting in f/80 focusing. The focal spot diameter was measured with a CCD camera to be ~100µm, about 1.2 times the diffraction limit.

A gas target ~1mm thick was placed at the focus. Because the laser confocal parameter (b=15mm) is much larger than the gas target thickness , the harmonics can be thought of as emerging from a plane so that geometric phase matching does not critically depend on the axial position of the target. The gas target [14] consists of two thin aluminum plates with a 500µm hole drilled through them. The laser goes through the hole and interacts with the gas which enters the hole from between the plates. The device was backed continuously with a pressure of typically 5-25 Torr, and the pressure inside the hole was estimated to be about one fifth of that amount.

The harmonic signals were resolved with HIREF-SXR-1.75 Monochromator (Hettrick Scientific) and measured with microchannel plate detector. The resolution was limited because we removed the entrance slit of the spectrometer to avoid damaging it by the laser light. Thus, the effective entrance slit was determined by the size of the actual interaction region wherein the harmonics were generated (~100µm). The exit slit was set at 100 µm which gave a spectrometer resolution of 1.0 Å. The spectrometer was scanned through wavelengths at a rate ~0.02 Å per laser shot (every 0.1 seconds) and the harmonic signal from each laser shot was recorded so that approximately 12,000 points typically make up a given spectrum (ranging through tens of harmonic peaks). For the spectra shown in this paper every 25 points have been binned and averaged together.

4. Experimental Results

Figure 2 shows high harmonic spectra generated in 5 Torr of argon (target backing pressure 25 Torr) with a peak laser intensity of 5×10^{14} W/cm^2. The two lowest harmonics seen in each spectrum are artificially damped by the spectrometer grating efficiency. The laser pulse durations for the spectra are (a) 100fs, (b) 50fs, and (a) 25fs. The laser pulse energy was adjusted to maintain the same peak laser intensity for each case. The pulse duration was controlled by cutting the amount of spectrum injected into the laser amplifying system. The duration of the final pulse was determined by autocorrelation, and a check of the final laser spectral profile revealed that the amplifier system reshaped the cut spectra into smooth peaks.

As is immediately evident from figure 2, many more harmonic peaks are produced with the 25fs pulse than with the 100fs pulse. In the context of the $I_p+3.2U_p$ rule, this indicates about a 60% higher ponderomotive potential for the 25fs case in comparison with the 100fs case. This is in qualitative agreement with the 25% higher ionization intensity along with the additional 20% intensity enhancement described in section 2 for a 25fs pulse.

As a reference, it is interesting to note that the ionization intensity for Ar predicted by Barrier Suppression Ionization (BSI) [15] is 2.5×10^{14} W/cm^2, half of the peak laser intensity used for figure 2. While BSI is a good estimate of the ionization intensity for longer pulses, it does not account for the effect of pulse duration. This is

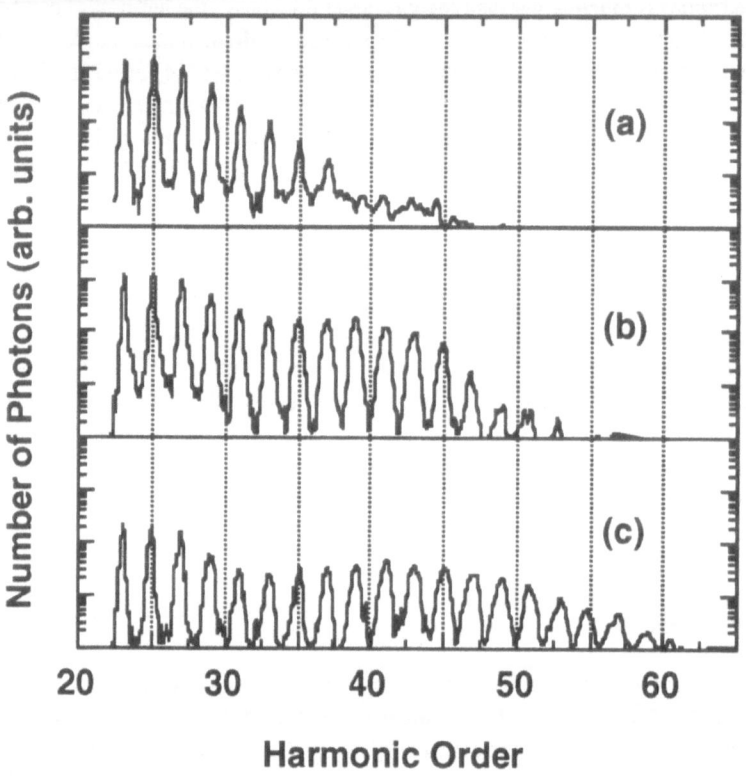

Figure 2. Harmonic spectra produced in 5 Torr of Ar with a peak intensity of 5×10^{14} W/cm^2. The pulse duration is 100fs (a), 50fs (b), and 25fs (c).

in contrast with the numerical calculation described in section 2. Under the assumption that argon atoms ionize at the BSI intensity, the $I_p+3.2U_p$ rule predicts the 41st to be the maximum harmonic order observed. This agrees with the 100fs case of figure 2 (a), but it is in sharp contrast with the 25fs case of figure 2 (c). For the other noble gases, the BSI assumption similarly predicts a much lower maximum harmonic order than we observed with a 25fs pulse.

It should be noted that the spectra shown in figure 2 display the harmonic emission that is accumulated during the entire pulse and that emerges from the interaction region within the cone angle of the laser. Because the laser intensity varies radially within the interaction region, the emitting atoms experience different peak laser intensities depending on their radial locations, and a higher number of atoms are available at the larger radii. Also, electrons which become free through ionization can have a serious impact on the phase matching [16] of harmonic emission from remaining neutral atoms. For example, if an electron from each of 20% of the atoms in 5 Torr of gas are liberated, the coherence length for the 51st harmonic generated in that environment is ~0.6mm, which is shorter than the gas target thickness.

The spectra seen in figure 2 were produced with a peak laser intensity approximately a factor of two above that at which ionization should readily occur. In fact, recombination light from the interaction region was visible to the unaided eye. The possibility therefore exists that the highest observed harmonics arise from ions, in which case they would not be considered to be of unusually high orders. To check for this, we observed harmonic generation in argon as a function of gas pressure (laser intensity held constant). We observed that all of the harmonic peaks decreased in strength together as the pressure was gradually reduced from 5 Torr to 1 Torr. If the higher-order harmonic peaks were produced by ions while the lower ones by neutral atoms, one would expect the brightness of the different harmonics to scale differently with pressure. This is because with a decrease in the pressure, the coherence length for harmonics emitted from ions should increase because of a reduced number of free electrons present. Our observations suggest that all of the harmonics that we observe come from neutral atoms.

The spectrum in Fig. 2(a) appears to contradict the result of Macklin *et al.* [4] who observed up to the 53rd harmonic using a 125fs pulse with the same wavelength. The explanation may lie in the fact that they used three times higher laser intensity than we did. We were unable to duplicate their intensity at 100fs because of the damage threshold of our amplifier. However, when we increased our intensity by 50% at 100fs, the highest harmonic peak that we observed increased by two orders. Additional harmonic peaks may occur with higher peak laser intensities because the percentage change in laser intensity from cycle to cycle is highest far away from the peak of the pulse (assuming, for example, a Gaussian temporal profile). For the laser pulses of Macklin *et al.*, the argon atoms ionized at approximately a sixth of the peak laser intensity; for our pulses, the argon atoms ionized at approximately half of the peak intensity. The percentage increase in the laser intensity over a cycle on the rising edge of a 100fs pulse at one sixth of the peak intensity is about the same as that of a 60fs pulse at half of the peak intensity.

5. Bandwidth of Individual Harmonics

The spectral widths of the high harmonics are individually broad enough to support pulse durations which are much shorter than the temporal profile of the laser. For example, the full-width at half maximum of the 41st harmonic in figure 2 (c) is about $0.6\langle\omega_L\rangle$, where $\langle\omega_L\rangle$ is the mean frequency of the laser. This corresponds to a width of 3Å centered at 197Å, which translates into a pulse possibly as short in duration as 2fs. However, it is unlikely that the necessary phase conditions are met to sustain such a short pulse since the harmonic emission arises from an ensemble of atoms inhomogeneously illuminated by the laser. [17]

The laser wavelength changes by approximately 5% across its bandwidth. The 39th and 41st harmonics are also separated in wavelength by about 5%, so it might be expected that harmonics orders higher than this will have merging spectra. This is not the case, however, as is plainly evident in figure 2. To see why this is so, it is helpful to make the simple assumption that a photon of the qth harmonic is comprised of q photons randomly chosen from the spectral distribution of the laser. This assumption allows us to apply some simple statistics to the problem.

We can write the frequency of the qth harmonic as the sum of a random assortment of frequencies from the laser spectral profile:

$$\omega_q \equiv \sum_{i=1}^{q} \omega_L^{(i)}. \tag{3}$$

The expectation value of the harmonic frequency is seen to be

$$\langle\omega_q\rangle = \sum_{i=1}^{q} \langle\omega_L^{(i)}\rangle = q\langle\omega_L\rangle \tag{4}$$

The width of the laser spectral profile $\Delta\omega_L$ can be described by

$$\Delta\omega_L^2 \equiv \mathbf{D}[\omega_L] = \langle(\omega_L - \langle\omega_L\rangle)^2\rangle, \tag{5}$$

where the operator $\mathbf{D}$ denotes the deviation. Then the spectral width $\Delta\omega_q$ of the qth harmonic is related to the laser spectrum by

$$\Delta\omega_q^2 = \mathbf{D}\left[\sum_{i=1}^{q} \omega_L^{(i)}\right] = \sum_{i=1}^{q} \mathbf{D}\left[\omega_L^{(i)}\right] = q\Delta\omega_L^2. \tag{6}$$

The operator $\mathbf{D}$ may be taken inside of the summation [18] since it is assumed that the q laser photons are independently selected.

Eq. (6) implies that each harmonic spectral width is broader in frequency than the laser spectral width by a factor of $\sqrt{q}$. Since each harmonic peak is separated from its neighbors by $2\langle\omega_L\rangle$, the harmonic peaks will overlap when $q > 4\langle\omega_L\rangle^2/\Delta\omega_L^2$. Given the 5% spectral width of our laser, this estimate gives overlapping harmonic peaks for orders above 1600. In this context, it is not surprising that our harmonic peaks can be resolved for orders well above the 41st.

It should be noted, however, that the nonlinear process of harmonic generation may broaden the harmonic spectral peaks beyond that implied by this estimate so that an overlap might occur at a much lower harmonic order. For example, this estimate gives $0.3\langle\omega_L\rangle$ for the width of the 41st harmonic, but, as mentioned, the measured width is $0.6\langle\omega_L\rangle$.

6. Non-Effect of Phase Matching on Individual Harmonic Bandwidths

It might be conjectured that macroscopic phase matching can cause harmonic spectra to become more discrete than the spectra from individual atoms. However, while phase-matching can indeed influence the shape of the overall emission spectra, it does not in general preferentially select harmonic frequencies over other arbitrary frequencies emitted by the laser-stimulated atoms. This can be understood from a simple one-dimensional analysis of phase matching.

Consider an atom positioned at z which is stimulated by the laser field $E_L(t - n_L z/c)$, where n_L is the optical index for the laser frequency. The Fourier transform of the dipole acceleration depends on the atom's position in the following way:

$$\ddot{d}(z,\omega) = \int_{-\infty}^{\infty} \ddot{d}\left[E_L(t - n_L z/c)\right]e^{-i\omega t}dt = e^{-i\omega n_L z/c}\int_{-\infty}^{\infty} \ddot{d}\left[E_L(t)\right]e^{-i\omega t}dt. \tag{7}$$

Figure 3 depicts an atom being stimulated by the laser field and emitting to a point in the far field (i.e., $L \gg z$). In the far field, the electric field emitted from the single atom may be written as

$$E_{\text{Emis.}}^{(S.A.)}(z,\omega) = \frac{\ddot{d}(z,\omega)e^{-i\omega n_\omega (L-z)/c}}{4\pi\varepsilon_o c^2 L},$$

(8)

where n_ω is the optical index at frequency ω.

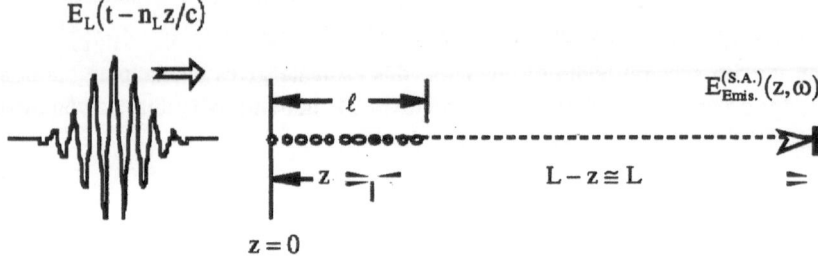

Figure 3. The emission from an atom stimulated by the laser propagates to a point at a far distance L.

The electric field which arises from an ensemble atoms distributed over $0 \le z \le \ell$ with uniform density N_o is found by summing up the emission from all of the atoms:

$$E_{\text{Emis.}}^{(\text{total})}(\omega) = N_o \int_0^\ell E_{\text{Emis.}}^{(S.A.)}(z,\omega)dz = \frac{N_o e^{-i\omega n_\omega L/c}}{4\pi\varepsilon_o c^2 L}\left[\frac{e^{i\omega\Delta n(\omega)\ell/c}-1}{i\omega\,\Delta n(\omega)/c}\right]\int_{-\infty}^{\infty}\ddot{d}[E_L(t)]e^{-i\omega t}dt$$

(9)

This summation over the individual fields takes place under the assumption that the laser field is not affected by its interaction with the atoms and that the response of each atom is not affected by emission from neighboring atoms. $\Delta n(\omega) \equiv n_\omega - n_L$ is a frequency-dependent index mismatch which in general shows no preference for harmonic frequencies; nor does the phase-matching factor in the central brackets of Eq. (9) exhibit a form which discriminates *a priori* for or against harmonic frequencies. Other than what might be contained in the single-atom dipole response to the laser temporal profile,

Eq. (9) shows no preference for harmonic frequencies over other frequencies. Similarly, more complicated multi-dimensional calculations which include geometrical as well as time-dependent phase mismatches do not in general enhance harmonic frequencies over other arbitrary frequencies. While phase matching is likely to influence the over spectral envelope of the harmonic plateau, the widths of individual harmonic peaks are determined primarily by the single-atom response to the laser temporal profile. Nevertheless, the spectra seen in figure 2 must be viewed as averaged spectra arising from an ensemble of atoms experiencing different laser intensities according to their radial positions in the focus.

7. Summary

We have investigated harmonic generation using laser pulses of much shorter duration than has been used previously. In many of the noble gases, we see an extended plateau in the harmonic spectra. The additional harmonic peaks arise because of the non-adiabatic nature of our short pulses. Besides having the obvious advantage of producing higher harmonic orders, short laser pulses can be generated with higher repetition rates since less energy per pulse is required to reach a given intensity. This makes short laser pulses an attractive source for the production of high-harmonic radiation.

This project was supported by the National Science Foundation, by the Air force Office of Scientific Research, and by the US Department of Energy Division of Advanced Energy Projects. H. C. Kapteyn acknowledges support from a Sloan Foundation fellowship.

References

1. J. Zhou, C.-P. Huang, M. M. Murnane, and H. C. Kapteyn, Amplification of 26-fs, 2-TW Pulses Near the Gain-Narrowing Limit in Ti:Sapphire, *Opt. Lett.* **20**, 64 (1995).
2. Anne L'Huillier and Ph. Balcou, High-Order harmonic Generation in Rare Gases with a 1-ps 1053nm Laser, *Phys. Rev. Lett.* **70**, 774 (1993).
3. J. J. Macklin, J. D. Kmetec, and C. L. Gordon III, High-Order Harmonic Generation Using Intense Femtosecond Pulses, *Phys. Rev. Lett.* **70**, 766 (1993).
4. J. J. Macklin, High-Order Harmonic Generation in Gases Using Intense Femtosecond Laser Pulses, Ph.D. Dissertation, Stanford University (December 1993).

5. J. Zhou, J. Peatross, M. M. Murnane, H. C. Kapteyn, and I. P. Christov, Enhanced High-Harmonic Generation Using 25 Femtosecond Laser Pulses, Submitted to *Phys. Rev. Lett.* (1995).

6. C.-G. Wahlstrom, J. Larsson, A. Persson, T. Starczewski, and S. Svanberg, High-Order Harmonic Generation in Rare Gases with an Intense Short-Pulse Laser, *Phys. Rev. A* **48**, 4709 (1993).

7. J. L. Krause, K. J. Schafer, and K. C. Kulander, High-Order Harmonic Generation from Atoms and Ions in the High Intensity Regime, *Phys. Rev. Lett.* **68**, 3535 (1992).

8. K. C. Kulander, K. J. Schafer, and J. L. Krause, Dynamics of Short-Pulse Excitation, Ionization and Harmonic Conversion, in *Super Intense Laser-atom Physics*, B. Piraux, A. L'Huillier, and K. Rzazewski, eds., Vol. 316 of NATO ASI Series (Plenum, New York, 1993), pp. 95-110.

9. P. B. Corkum, Plasma Perspective on Strong-Field Multiphoton Ionization, *Phys. Rev. Lett.* **71**, 1994 (1993).

10. W. Becker, S. Long, and J. K. McIver, Higher-Harmonic Production in a Model Atom with Short-Range Potential, *Phys. Rev. A* **41**, 4112 (1990).

11. M. Lewenstein, Ph. Balcou, M. Yu. Ivanov, A. L'Huillier, and P. B. Corkum, Theory of High Harmonic Generation by Low Frequency Laser Fields, *Phys. Rev. A* **49**, 2117 (1994).

12. J. B. Watson, A. Sanpera, and K. Burnett, Pulse-Shape Effects and Blueshifting in the Single-Atom Harmonic Generation from Neutral Species and Ions, *Phys. Rev. A* **51**, 1458 (1995).

13. I. P. Christov, J. Zhou, J. Peatross, A. Rundquist, M. M. Murnane, and H. C. Kapteyn, Non-Adiabatic Effects in High Harmonic Generation with Short Laser Pulses, to be published (1995).

14. J. Peatross and D. D. Meyerhofer, Novel Gas Target for Use in Laser Harmonic Generation, *Rev. Sci. Instrum.* **64**, 3066 (1992).

15. S. Augst, D. Strickland, D. D. Meyerhofer, S. L. Chin, and J. H. Eberly, Tunneling Ionization of Noble Gases in a High-Intensity Laser-Field, *Phys. Rev. Lett.* **63**, 2212 (1989).

16. A. L'Huillier, K. J. Schafer, and K. C. Kulander, Theoretical Aspects of Intense Field Harmonic Generatation, *J. Phys. B: At. Mol. Opt. Phys.* **24**, 3315 (1991).

17. J. Peatross, M. V. Fedorov, and K. C. Kulander, Intensity-Dependent Phase-Matching Effects in Harmonic Generation, *J. Opt. Soc. of Am. B* **12**, 863 (1995).

18. Y. A. Rozanov, *Probability Theory: a Concise Course*, Richard A. Silverman, ed., (Dover, New York, 1969), pp. 48-49.

HARMONIC EMISSION WITH ELLIPTICAL POLARIZATION

P. ANTOINE, B. CARRÉ, M. LEWENSTEIN AND A. L'HUILLIER[1]
Commissariat à l'énergie atomique, DSM/DRECAM/SPAM
Centre d'Etudes de Saclay, 91191 Gif-sur-Yvette, France.
1. also at Department of Physics, Lund Institute of Technology
P.O. Box 118, 221 00 Lund, Sweden.

AND

B. PIRAUX AND M. GAJDA
Institut de Physique, Université catholique de Louvain
chemin du cyclotron, 2 B-1348 Louvain-la-Neuve, Belgium.

1. Introduction

High-order harmonic generation (HG) is one of the most rapidly developing topics in the field of intense laser–atom interactions. It does not only provide an effective mechanism for the production of short wavelength but it is also a probe for investigating the dynamics of an atom exposed to a strong laser field.

In the high-intensity (10^{13}-10^{14} W/cm^2), low frequency regime, the harmonic spectra are characterized by a broad plateau followed by an abrupt cutoff. The understanding of the process, in particular of the location of the cutoff, has been improved by the development of the so-called two-step model [1, 2]. After tunneling out through the atomic potential, the electron oscillates back and forth in the presence of the laser field. If it returns close to the nucleus, the electron may recombine back to the ground state and emit an harmonic photon. The energy of the emitted photon is equal to $E_{kin} + I_p$ where I_p and E_{kin} stand for the ionization potential and the kinetic energy of the returning electron, respectively. According to this model, the extent of the plateau is determined by the maximum kinetic energy of the electron close to the nucleus $E_{kin} = 3.2U_p$ where U_p, the ponderomotive potential, is the mean kinetic energy of the quivering electron. The cutoff appears at an energy equal to $I_p + 3.2U_p$. This classical interpretation shows that in order to control harmonic generation process, one has to control the

H. G. Muller and M. V. Fedorov (eds.), Super-Intense Laser-Atom Physics IV, 467–476.

motion of the free electron in the laser field. Shaping appropriately electron trajectories allows for various fascinating applications, including for example, the generation of very short, high frequency pulses [3]. This is illustrated in the last part of this work.

One way of controlling the electron trajectory is to alter the polarization of the driving field, making it elliptical rather than linear. In the linear case, the classical trajectories of the electron pass the nucleus periodically, thus allowing for recombination and harmonic generation. There are, strictly speaking, no such trajectories for elliptical polarizations. HG is, in that case, possible only thanks to the finite extent of the electronic wavepacket and to quantum diffusion effect. HG is expected to decrease very rapidly with an increase of the ellipticity of the laser. This has been observed by several groups [4-7].

In this paper, we investigate high order harmonic generation with elliptical polarization by means of two theoretical approaches : the first one is the generalisation of the recently formulated theory of high-order harmonic generation by low-frequency laser fields [8] to the case of elliptically polarized light. Contrary to previous theoretical descriptions [7, 9], our model includes both the single-atom response and propagation. We also consider polarization properties of harmonics. The results of our calculations compare very well with recent experimental observations. The second theoretical approach is the numerical integration of the time-dependent Schrödinger equation. In order to illustrate the method we present results concerning laser pulse with a polarization varying in time.

2. Model 1 : Strong field approximation

2.1. SINGLE ATOM RESPONSE

We consider an atom in a single-electron approximation under the influence of the laser field $\vec{\mathcal{E}}(t)$ of arbitrary polarization. We skip all the details of the model which are thoroughly discussed in Ref. [10]. Briefly, we neglect the contribution of all bound states except the ground state, as well as the effects of the atomic potential on continuum electronic states. Our approach is valid in the tunneling regime for ionization, $i.e.$ when U_p is comparable or larger than I_p. With the help of these assumptions, we can derive a very simple expression for the atomic dipole.

In Fig. 1, we present typical results for the intensity dependence of the x-component of the induced atomic dipole at the 43-rd harmonic, for three values of the ellipticity (the y-component is very similar). The intensity dependences of both dipole strengths (Fig. 1(a)) show the characteristic transition from the cutoff region (where the dipole strengths increase rapidly) to the plateau region (where the dipole strengths saturate and are

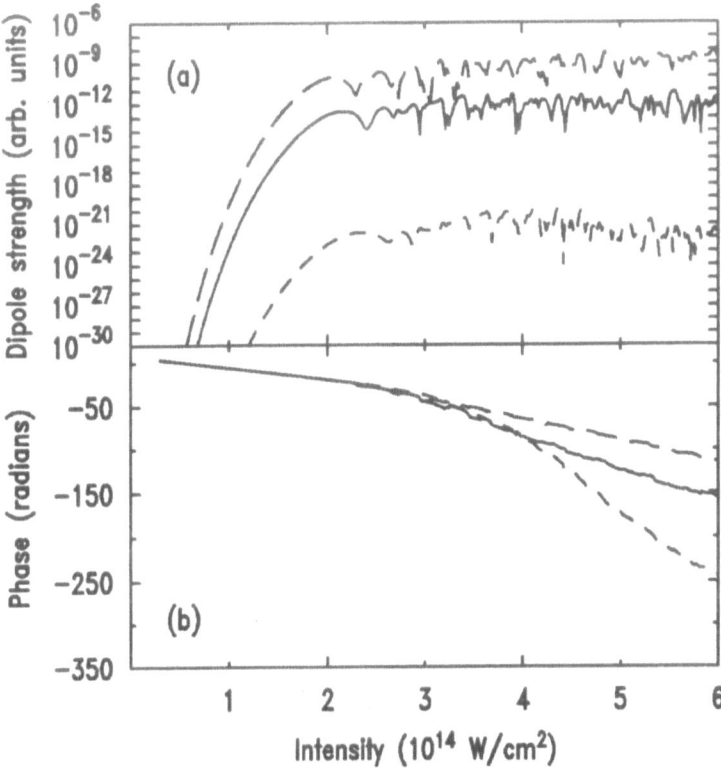

Figure 1. Strength (a) and phase (b) of the x-component of the dipole for the 43-rd harmonic frequency in Neon as a function of the laser intensity, for three values of the ellipticity: $\epsilon = 0$ (long-dashed line), $\epsilon = 0.3$ (solid line) and $\epsilon = 0.6$ (short-dashed line).

dominated by quantum interference effects). With increasing ellipticity, the dipole strength decreases whereas the cutoff position shifts slightly toward higher intensities (hence, for a given intensity, towards lower harmonic orders). In Fig. 1(b), we show the intensity dependences of the phase of the dipole. This phase, as we stressed in Refs. [11, 12], determines, to a great extent, the coherence properties of the propagated signal and can be interpreted in quasi-classical terms. It exhibits a piece-wise linear behavior as a function of the laser intensity. The slope of the phase for intensities below the cutoff-plateau transition point practically does not change with ellipticity. In the plateau region, the phase exhibits oscillations due to quantum interferences. The average slope is larger than in the cutoff, and increases with ellipticity. This increase of the slope with the laser ellipticity takes place over a limited range of intensities (from ~ 2.7 to 6×10^{14} W/cm^2) in

Fig. 1(b). It strongly depends on the process order, being more and more pronounced as the harmonic order increases (it is very significant for the 63-rd harmonic). As we have shown in Ref. [12], the slope is related to the return time of the electron for the most relevant electronic trajectories. The large slopes obtained for large ellipticities seem to imply that the trajectories corresponding to long return times (*i.e.* longer than one period with, possibly, multiple returns) play a dominant role in this case, especially for high harmonic orders. Numerical analysis confirms this interpretation.

2.2. MACROSCOPIC RESPONSE

The second step of the theoretical description consists of solving the propagation equations in the paraxial and slowly-varying envelope approximations, using the dipole moments discussed previously as source terms. The method for solving the propagation equations has been discussed previously for linearly-polarized fundamental (and harmonic) fields [13]. We have generalized it for the elliptically-polarized fields. Here, we restrict the discussion to few results only, the complete discussion can be found in reference [10].

In Fig. 2, we present the comparison of our calculation (including propagation) with experimental data for the 43-rd harmonic in Neon. Three theoretical curves corresponding to the laser peak intensities 2, 4 and 6 × 10^{14} W/cm^2 are drawn. The full circles denote the result of the experiment performed by Budil *et al.* [4]. The open squares are the results of recent experiments carried out at Saclay with a Ti:Sapphire laser at a slightly different wavelength 790 nm, but otherwise in very similar conditions. The agreement between theory and experiment is very good, irrespectively of the laser intensity used in the calculations (which does not influence much the results). In particular, the theory reproduces extremely well the significant narrowing of the ellipticity dependence with increasing harmonic number. The deviation observed at large ellipticities is simply due to the fact that, in the experiments, the signal was barely above the noise level, and measured with poor accuracy in this region.

2.3. POLARIZATION OF HARMONICS

The polarization of the harmonics is more sensitive to the dynamics of the interaction than the conversion efficiency, so that it study might lead to a deeper understanding of the process. In a non perturbative regime, the polarization of the harmonics is not equal to the polarization of the incident laser. For example, Weihe and coworkers have observed that the polarization ellipse of low-order harmonics is rotated by some angle with respect to the polarization of the laser [14] .

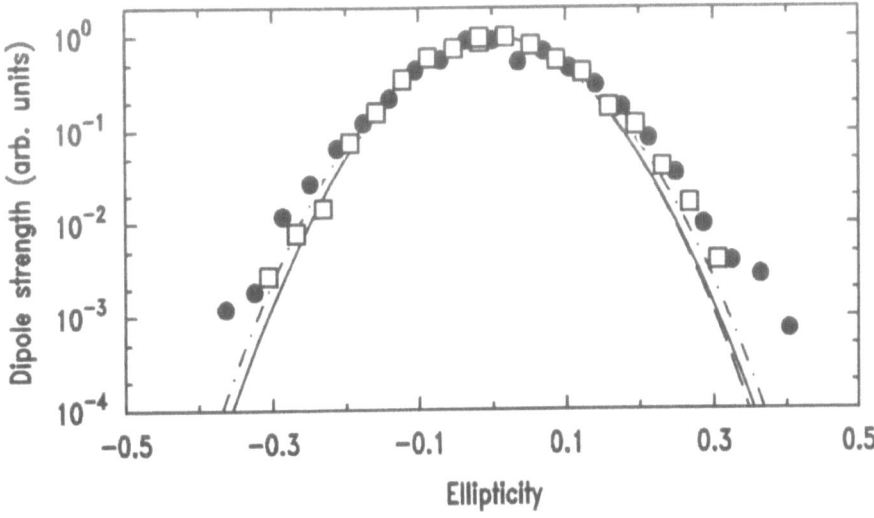

Figure 2. Comparison of the simulated harmonic strengths with experimental data for 43-rd harmonic in Neon. Theoretical curves correspond to three different peak intensities: 2×10^{14} W/cm² (long-dashed line), 4×10^{14} W/cm² (dot-dashed line) and 6×10^{14} W/cm² (solid line). Full circles denote experimental data from Ref. [4]. Open squares denote recent results obtained at Saclay with a Ti:S laser

Following Born and Wolff [15], we define the polarization properties of the harmonic field with the help of the Stokes parameters. These parameters are very useful since they can be defined as the result of simple experiments. For example, the third parameter is a measure of the excess in intensity of light transmitted by a device which accepts right-handed circular polarization, over that transmitted by a device which accepts left-handed circular polarization. The calculations show that harmonic radiation generated by high-order conversion of an elliptically-polarized laser field is only *partially*-polarized, because the phase difference $\phi(\vec{r}, t)$ between the x and y components varies, in *space*, over the beam profile, and in *time*, over the pulse duration. In these conditions, all the Stokes parameters are requested to determine the polarization of the light.

Fig. 3 shows the variation of the ellipticity of the 23-rd harmonic after propagation through the medium, as a function of the laser ellipticity and for different values of laser intensity. The ellipticity of the harmonics is always smaller than the ellipticity of the laser. All the curves present some oscillations. We attribute these variations to quantum interference effects. It has to be noted that these structures are much more pronounced in the single atom response. Propagation smoothes out the quantum interference, but only partly. In Fig. 4, we present the macroscopic result concerning

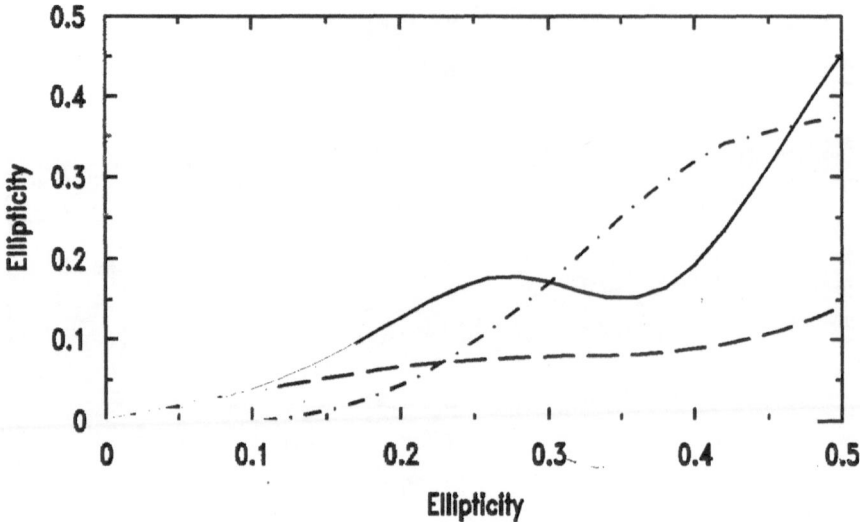

Figure 3. Ellipticity of the 23–rd harmonic after propagation as a function of the laser ellipticity, and for three values of the laser intensity: 2×10^{14} W/cm^2 (long-dashed line), 4×10^{14} W/cm^2 (dot-dashed line) and 6×10^{14} W/cm^2 (solid line). Full circles denote recent experiment obtained at Saclay

the rotation angle of the harmonic ellipse with respect to the fundamental, as a function of the laser ellipticity. The rotation angle corresponding to the single atom response is a rapidly varying function of both laser ellipticity and intensity, and exhibits strong quantum interference effects. Propagation, however, smoothes out interference features quite efficiently. The theoretical result is consistent with the experiment done recently at Saclay. For higher harmonic order, the rotation angle remains practically equal to zero.

3. Model 2 : time dependent Schrödinger equation

In this third part, we study the time profile of the harmonics emitted by atomic hydrogen exposed to a strong laser pulse of an arbitrary polarization by solving numerically the Schrödinger equation :

$$i\frac{\partial}{\partial t}\Psi(\vec{r},t) = \left(H_0 + \vec{A}(t)\cdot\vec{p}\right)\Psi(\vec{r},t), \qquad (1)$$

where H_0 is the atomic Hamiltonian and $\mathbf{A}(t)$ the vector potential associated to the field:

$$\vec{A}(t) = A_1 f(t)\sin(\omega_1 t + \phi_1)\hat{e}_x + A_2 f(t)\sin(\omega_2 t + \phi_2)\hat{e}_y; \qquad (2)$$

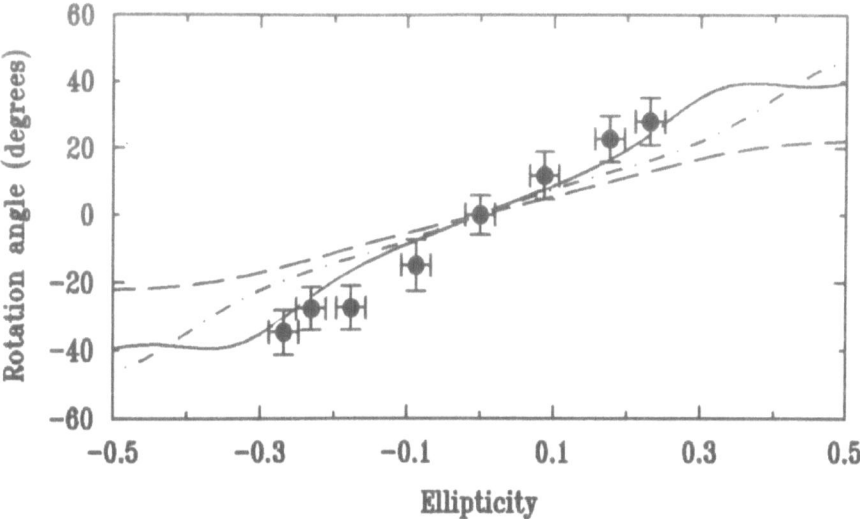

Figure 4. Rotation angle of the harmonic ellipse with respect to the fundamental for the (a) 23–rd harmonic after propagation as a function of the laser ellipticity, and for three values of the laser intensity: 2×10^{14} W/cm^2 (long-dashed line), 4×10^{14} W/cm^2 (dot-dashed line) and 6×10^{14} W/cm^2 (solid line). Open squares denote recent results obtained at Saclay with a Ti:S laser.

A_1 and A_2 are the amplitudes of the two components of the potential, $\hat{e}_x$ and $\hat{e}_y$ are the unit vectors along the x-axis and the y-axis, ω_1 and ω_2 are the frequencies of the laser field along the x- and the y-axis and $f(t)$ is a slowly varying envelope. Depending on the amplitudes, the frequencies and the difference of phase $\phi_2 - \phi_1$, we can study linear, elliptical, circular or time-dependent polarization. The numerical integration of the time-dependent Schrödinger equation is a hard task, but it is very useful since it can test simple models, such as the strong field approximation model presented in the first section, but also since its validity range is much larger.

The numerical procedure is the following: we first expand the wave function $\Psi(\vec{r}, t)$ in Coulomb-Sturmian functions [16] of the radial coordinate, and in spherical harmonics of the angular coordinates. This function is then propagated numerically by means of a high order fully implicit Runge-Kutta method. Knowing the wave function $\Psi(\vec{r}, t)$, it is then a simple matter to evaluate the acceleration $\vec{a}(t)$ of the atomic dipole by means of Ehrenfest's theorem. The power spectrum is obtained by calculating the modulus square of the Fourier transform of $\vec{a}(t)$. This standard Fourier method however, does not provide any information on the time profile of the emission; this can be obtained only through a time-frequency analysis [17].

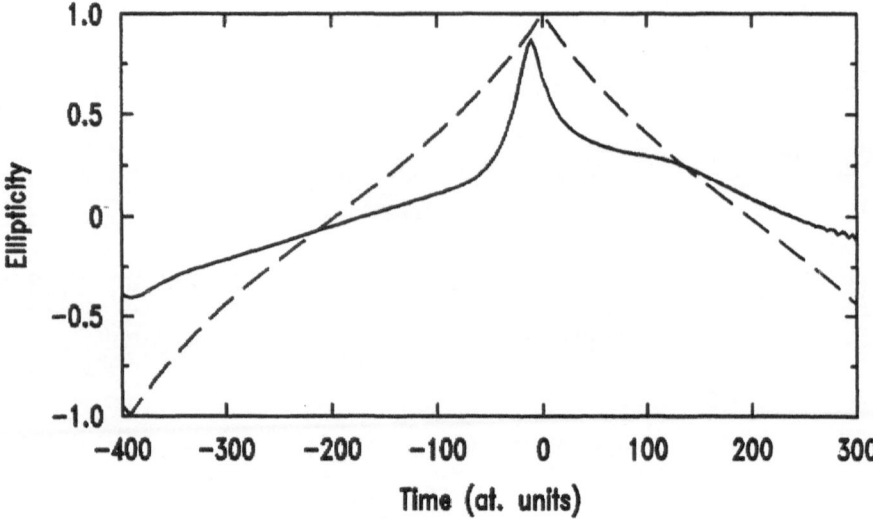

Figure 5. Ellipticity of respectively the laser (dashed line) and the 11th harmonic (solid line) as a function of time

In the present work, we consider a laser pulse with a polarization which varies in time. The aim is to produce short harmonic pulse by taking advantage of the rapidly decreasing harmonic efficiency with an increase of the ellipticity [3]: we expect to observe harmonic generation only when the polarization is linear (or at least when the ellipticity is very small). So, we may produce short harmonic pulses, if the polarization remains linear during a very short time. Such a situation could be obtained if the two components of the potential vector along x-axis and y-axis oscillate at two slightly different frequencies ω_1 and ω_2 [3]. Using a time-dependent polarization allows one to control temporally harmonic generation.

Let us consider the interaction of atomic hydrogen initially in its ground state, with an intense laser pulse (peak intensity=10^{14}W/cm^2), whose frequencies are respectively 0.110 and 0.118 in atomic units for the x- and the y-component. The pulse has a flat top shape and its duration is 20 optical cycles. We expect to observe two harmonic pulses centered around t=±196 (in atomic units of times) since the polarization is linear twice during the laser pulse (Fig. 5).

We first calculate the harmonic spectrum. We observe up to the 13-th harmonic, but the spectrum does not exhibit a real plateau because the intensity is too low at the considered frequency. The harmonic peaks correspond to odd multiples of the mean frequency 0.114. We then concentrate on the time profile of the 11-th harmonic. The amplitudes of both compo-

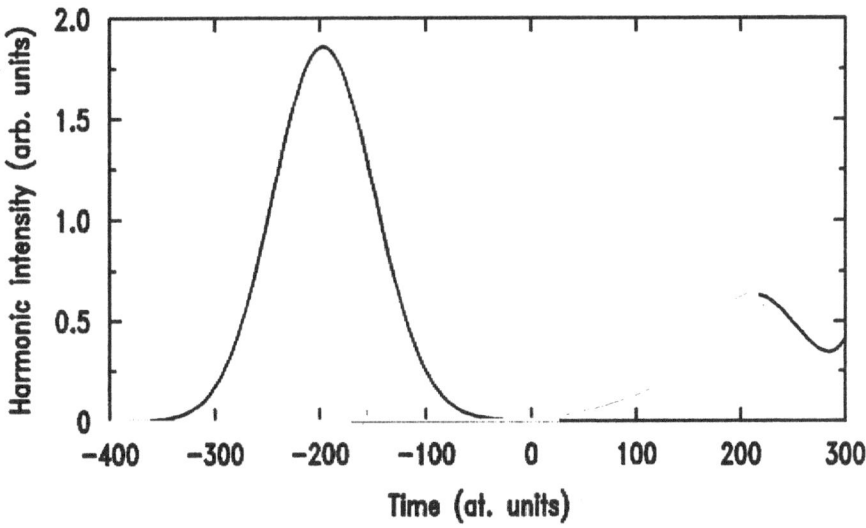

Figure 6. Time dependence of the 11th harmonic intensity

nents of the acceleration $\vec{a}(t)$ correponding to the choosen harmonic, as well as the difference of phase between the x- and y-axis components are determined by means of a Gabor analysis [17]. Fig. 6 shows the time dependence of the 11-th harmonic.

The emissions occur during short time interval, and their duration (full width at half maximum) are equal to 100 atomic units, this corresponds to 2.5 fs, less than two optical periods. It has to be noted that the maxima of the curves do not appear exactly when the polarization is linear, but a little later. This shift is related to the return time of the electron *i.e.* the length of the time interval between the moments of its ionization and its recombination back to the ground state. The pulse duration here is longer than the value obtained in Ref. [3] (<1 fs). The main difference between the results of [3] and our analysis is the bandwidth of the emission. In our case, the selected bandwidth is less than 2ω, the energy difference between two adjacents harmonics. Therefore, we select in fact only one harmonic. In such conditions, the pulse duration is defined by the interval of time required by the ellipticity to change significantly. According to the semi-classical model, the smallest pulse duration should be longer than half of an optical period. On the contrary, if we accept several harmonics by increasing the bandwidth of the analysis, the time profile of the harmonics is modulated at 2ω due to the oscillations of the ionized electron in the driving field, and consists of a train of very short pulses [17]. Their durations are less than 1/4 period. The aim of the time-dependent polarization is to select one, or

476

a few of these very short pulses. Therefore, the time profile of the emission, the pulse duration in particular, depends on how we look at it, in other words, on how we analyse it both experimentally or theoretically.

We have also computed the polarization of the harmonic radiation as a function of time. Fig. 5 shows the ellipticity of the 11-th harmonic. The harmonic polarization (the ellipticity as well the orientation of the polarization) follows the polarization of the driving field. However, the ellipticity of the harmonic remains smaller.

4. Conclusion

In this paper, we have first described a model which characterizes completely the harmonic emission with elliptical polarization. The single atom response as well as the macroscopic response including propagation effects were considered. Our results compare very well with recent experimental findings. Secondly, we have demonstrated by solving numerically the time-dependent Schrödinger equation, how one could control temporally harmonic generation by means of a time-dependent polarization.

References

1. K. C. Kulander, K. J. Schafer and J. L. Krause, in *Super-Intense Laser-Atom Physics*, ed. B. Piraux, Anne L'Huillier, and K. Rząźewski, NATO ASI Series B, vol. 316, p. 95 (Plenum Press, New York,1993).
2. P. B. Corkum, Phys. Rev. Lett. **73**, 1995 (1993).
3. M. Ivanov, P. B. Corkum, T. Zuo and A. Bandrauk, Phys. Rev. Lett. **74**, 2933 (1995).
4. K. S. Budil, P. Salières, Anne L'Huillier, T. Ditmire, and M. D. Perry, Phys. Rev. A **48**, R3437 (1993).
5. P. Dietrich, N. H. Burnett, M. Ivanov and P. B. Corkum, Phys. Rev. A**50**, R3585 (1995).
6. Y. Liang, M. V. Ammosov, and S. L. Chin, J. Phys. B **27**, 1296 (1994).
7. N. H. Burnett, C. Kan, and P. B. Corkum, Phys. Rev. A**51**, R3418 (1995).
8. M. Lewenstein, Ph. Balcou, M. Yu. Ivanov, Anne L'Huillier, and P. Corkum, Phys. Rev. A **49**, 2117 (1994).
9. W. Becker, S. Long and J.K. McIver, Phys. Rev A**50**, 1540 (1995).
10. Ph. Antoine, A. L'Huillier, M. Lewenstein, P. Salières and B. Carré, submitted to Phys. Rev. A (1995).
11. P. Salières, A. L'Huillier, and M. Lewenstein, Phys. Rev. Lett. **75**, 3376 (1995).
12. M. Lewenstein, P. Salières, and A. L'Huillier, Phys. Rev. A in press.
13. Anne L'Huillier, Ph. Balcou, S. Candel, K. J. Schafer, and K. C. Kulander, Phys. Rev. A **46**, 2778 (1992).
14. F. Weihe, S. K. Dutta, G. Korn, D. Du, P. H. Bucksbaum, P. L. Shkolnikov, Phys. Rév. A**51**, R3433 (1995).
15. Born and E. Wolf, *Principles of Optics*, (Pergamon Press, New-York),1964).
16. see *e.g.* M. Rotenberg, Adv. At. Mol. Phys. **63**, 233 (1970).
17. Ph. Antoine, B. Piraux and A. Maquet, Phys. Rev. A **51**, R1750 (1995).

POLARIZATION-DEPENDENT ONE- AND TWO-COLOR HIGH-HARMONIC GENERATION

A. LOHR[1], S. LONG[2], W. BECKER[1,2], and J. K. McIVER[2]

[1] *Physik-Department T30, Technische Universität München,*
85747 Garching, Germany

[2] *Center for Advanced Studies, Department of Physics and Astronomy,*
University of New Mexico, Albuquerque, NM 87131, USA

1. Introduction

The generation of high harmonics of a laser field irradiating a gaseous target has been extensively investigated in recent years; for a review see Ref. [1]. For a linearly polarized monochromatic laser field, there is by now good agreement between numerical simulations and experiments, with respect to both the single-atom and the collective effects. The physical understanding of the single-atom aspects has been particularly promoted by the semiclassical model of the electron revisiting the ionic core [2, 3]. The situation is much less clear when the driving field is bi- or multichromatic and/or its polarization is no longer linear. First, the extension of the semiclassical model to this case is far from obvious. In order to generate high harmonics, does the electron have to revisit the core very closely or can it pass by at some distance taking advantage of quantum mechanical wave-packet spreading to ensure overlap with the ionic core [4]? Second, the classical kinematics of the electron in the laser field is much richer for a two-color field, particularly so for commensurate frequencies, where the relative phase plays a vital role. The rate of injection of electrons in the continuum depends on the electric field while their subsequent kinematics is a function of the vector potential. For a one-color field, it so happens that the electrons that have gained the highest energy when they return to the ionic core have been injected at about the time where the electric field was at its peak. This is not necessarily so for a two-color field, and this may have very noticeable consequences. The tremendous parameter space of two-color high-harmonic generation (intensities, frequencies, polarizations, relative phase) raises the prospect that it may be possible to tailor a harmonic spectrum to some extent to one's liking. Schemes are under investigation which exploit these possibilities, for example, to generate subfemtosecond harmonic pulses [5, 6].

In this paper we will report on results for one- and two-color high-harmonic generation by driving fields of various polarizations. They are based on a radically simplified model atom, idealized by a zero-range potential . The model has been

H. G. Muller and M. V. Fedorov (eds.), Super-Intense Laser-Atom Physics IV, 477–488.
© *1996 Kluwer Academic Publishers.*

shown to describe correctly many features of harmonic generation in a real atom [7, 8, 9]. It is closely related to the effective-dipole model of Lewenstein et al. [10].

2. High-Harmonic Generation by an Elliptically Polarized One-Color Field

Harmonics generated by elliptically polarized driving fields are now being investigated. Besides providing much more stringent tests for theory they are of great interest for the possible generation of ultrashort pulses [5]. Several experiments have been carried out thus far [11]-[14]. Here we will investigate the polarization properties of harmonics generated by the elliptically polarized driving field

$$\mathbf{A}(t) = \frac{a}{2} \left((\cos \epsilon \, \hat{\mathbf{x}} + i \sin \epsilon \, \hat{\mathbf{y}}) e^{-i\omega t} + \text{c.c.} \right) \tag{1}$$

with frequency ω and ellipticity $\xi = \tan \epsilon$.

The dipole moment generated by this field in the delta-function atom has been written down in Eq. (3.15) of Ref. [8]. It is

$$
\begin{aligned}
\mathbf{d}(\Omega) &= a \sum_k \delta(\Omega - (2k+1)\omega) \left[(\cos \epsilon \, \hat{\mathbf{x}} - i \sin \epsilon \, \hat{\mathbf{y}}) A_k - (\cos \epsilon \, \hat{\mathbf{x}} + i \sin \epsilon \, \hat{\mathbf{y}}) B_k \right] \\
&\equiv \sum_k \mathbf{d}_k \delta(\Omega - (2k+1)\omega)
\end{aligned} \tag{2}
$$

where A_k and B_k are closed-form one-dimensional integrals involving a Bessel function with oscillating argument, exponentials, and powers. The explicit forms can be found in Ref. [8]. Here it only matters that in lowest-order perturbation theory A_k and B_k are of order $(a^2 \cos 2\epsilon)^{2k+2}$ and $(a^2 \cos 2\epsilon)^{2k}$, respectively, in the amplitude a and the ellipticity ξ of the driving field (1). The field of the harmonic with frequency Ω that is generated by the dipole moment (2) is, in general, elliptically polarized again. Its polarization ellipse is rotated versus the axis of the driving field (1) by an angle ϕ given by (below we will suppress the subscript k of A_k, B_k, and $\mathbf{d}_k$)

$$\tan 2\phi = i \sin 2\epsilon \frac{BA^* - B^*A}{AB^* + BA^* - \cos 2\epsilon(|A|^2 + |B|^2)}, \tag{3}$$

and its ellipticity $\xi' = \tan \epsilon'$ derives from

$$\sin 2\epsilon' = \sin 2\epsilon \frac{|B|^2 - |A|^2}{|A|^2 + |B|^2 - \cos 2\epsilon(A^*B + AB^*)}. \tag{4}$$

The total intensity of the respective harmonic is related to

$$|\mathbf{d}|^2 \sim a^2[|A|^2 + |B|^2 - \cos 2\epsilon(AB^* + A^*B)]. \tag{5}$$

The power laws given above for the integrals A_k and B_k imply that in lowest-order perturbation theory (LOPT) with respect to the driving field (1) the emitted harmonics have the same polarization as the former. This changes as soon as one goes beyond LOPT. The origin of this effect is closely related to the mechanism of phase conjugation in standard $\chi^{(3)}$-nonlinear optics. Consider the polarization $\mathbf{P}(t)$ of the atom induced by the driving field (not to be confused with the polarization of the emitted field) to order $2k+1$,

$$
\begin{aligned}
P_i^{(2k+1)}(t) \;=\; & \sum_{i_1,i_2,\cdots,i_{2k+1}} \int dt_1\, dt_2 \cdots dt_{2k+1} \\
& \chi_{i,i_1,\cdots,i_{2k+1}}^{(2k+1)}(t-t_1, t-t_2, \cdots, t-t_{2k+1}) \\
& (E+E^*)_{i_1}(t_1)(E+E^*)_{i_2}(t_2)\cdots(E+E^*)_{i_{2k+1}}(t_{2k+1}),
\end{aligned} \quad (6)
$$

where E and E^* are the positive and negative frequency parts, respectively, of the driving field whose vector potential is given by Eq.(1). For an isotropic medium, the tensor χ must be a sum of products of Kronecker symbols, viz.

$$
\begin{aligned}
\chi_{i,i_1,\cdots,i_{2k+1}}^{(2k+1)} \quad & (t-t_1, t-t_2, \cdots, t-t_{2k+1}) \\
& = \sum_P \delta_{i,P1}\delta_{i_1,P2}\cdots\delta_{i_{2k+1},P(2k+2)}\,\chi(t-t_1, ...t-t_{2k+1}),
\end{aligned} \quad (7)
$$

where P denotes all permutations of the indices $(1,2,...2k+2)$. In LOPT, at the frequency $\Omega = (2k+1)\omega$, just the product of the $2k+1$ E's contributes. Therefore, in LOPT the polarization $\mathbf{P}$ has the same polarization as the driving field. Beyond LOPT we need to consider $P_i^{(2k+3)}$ (and higher orders). A term with exactly one E^* and $2k+2$ E's yields the same frequency as before. However, such a term if contracted with an appropriate Kronecker symbol δ_{i,i_i} now generates a contribution to the polarization $\mathbf{P}$ having the complex conjugate polarization.

The offset angle ϕ goes through zero when A and B are either in phase or out of phase by $180°$. In particular, consider the case where a certain harmonic is completely suppressed for a linearly polarized driving field. For $\xi \ll 1$, we then have $A = a + \xi^2 a_1 + ...$ and $B = b + \xi^2 b_1 + ...$ with $a = b$. Under these conditions Eq.(3) shows that $\tan 2\phi = O(\xi^3)$. Hence, whenever a certain harmonic is suppressed for linear polarization the offset angle ϕ for elliptic polarization will stay close to zero up to comparatively large ellipticities.

A detailed discussion of the polarization properties of the emitted harmonics based on Eqs. (3,4) will be given elsewhere [15]. Here we will present some preliminary results. For the harmonics near the beginning of the plateau we observe a fairly erratic behavior of the offset angle (3). The sign varies without a recognizable pattern, and the absolute magnitude is typically much smaller than found experimentally [13]. This suggests that the actual behavior in a real atom does depend sensitively on the atomic structure. As a function of the scaled intensity

$\eta = U_p/\hbar\omega$ we find a quasi-periodic behavior of the offset angle with a period of $\Delta\eta \sim 1$. Most likely, this is related to the effect of the ponderomotive channel closings as discussed in Refs.[16, 8]. A regular pattern, however, emerges near the end of the plateau. Figure 1 displays the offset angles for a series of harmonics at and slightly beyond the end of the plateau. For the harmonics just below the cut-off, the axis of the polarization ellipse switches rather abruptly from $\phi \sim 0$ to $\phi \sim 90°$ for surprisingly small ellipticities. Beyond the end of the plateau, the offset angle never exceeds a few degrees within the range of the figure. It appears that this behavior is a general feature.

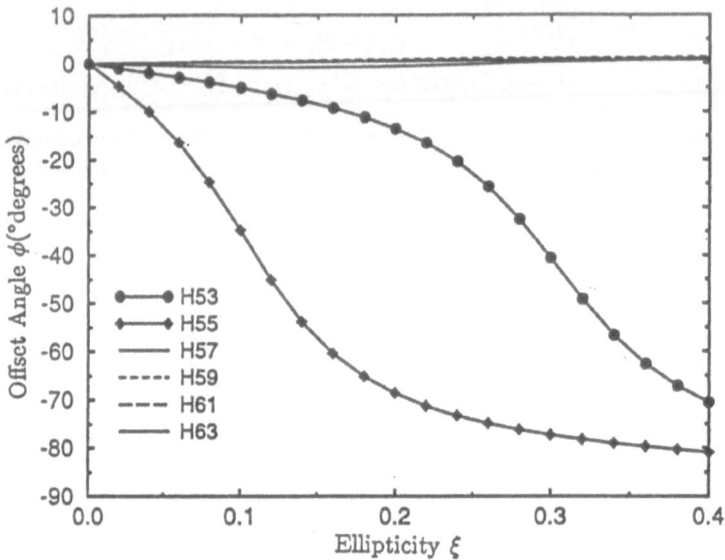

Figure 1. Ellipticity dependence of the offset angle ϕ of the polarization ellipse of the 53rd to 63rd harmonics for a binding energy of $|E_0| = 0.9$a.u., frequency $\hbar\omega = 0.043$a.u. and driving intensity corresponding to $\eta = U_p/\hbar\omega = 10.6$. The harmonics selected are on either side of the single-atom cut-off at $|E_0| + 3.17U_p$.

3. Two-Color High-Harmonic Generation

In this section we will deal with various situations out of the bewildering multitude of possibilities. All calculations are based on the three-dimensional delta-function atom. We will not write down the formulas that underly the calculations. They consist of one-dimensional quadratures and sums over Bessel functions and can be found in Ref. [17] where many more examples are given. Other calculations of two-color high-harmonic emission include Refs. [18, 19].

3.1 TWO PERPENDICULAR LINEAR POLARIZATIONS; INCOMMENSU-RATE FREQUENCIES

If the two driving frequencies ω_1 and ω_2 are incommensurate an *arbitrary* frequency can be arbitrarily closely approximated by a mixing process such that

$$\Omega = m_1\omega_1 + m_2\omega_2. \qquad (8)$$

where m_1 and m_2 are positive or negative integers such that one is even and the other one odd. Figure 2 shows a typical spectrum. It is, of course, completely

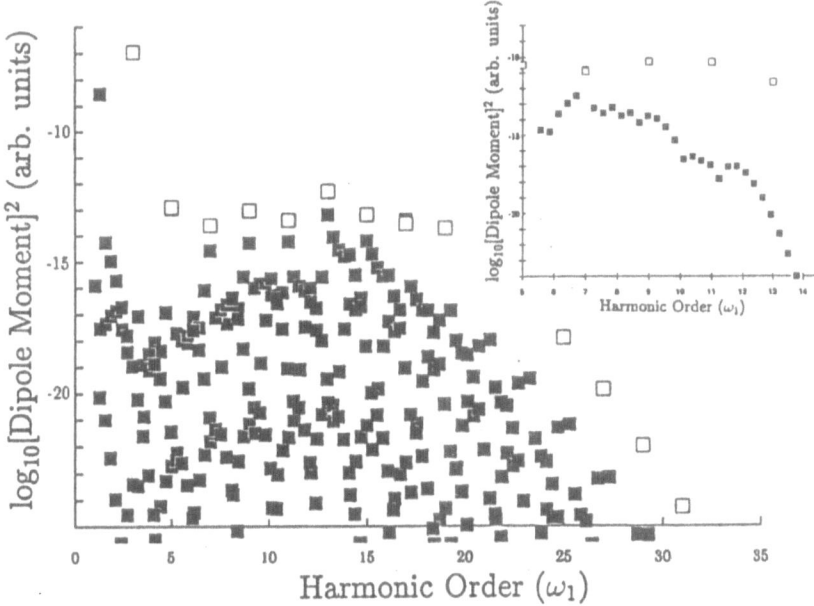

Figure 2. Harmonic spectrum (solid squares) for $|E_0|/\hbar\omega_1 = 10$ in a linearly polarized laser field with an additional low-frequency background field with $\omega_2 = 0.1414\omega_1$ polarized perpendicularly to the first field. The intensities are $\eta_1 = U_{p1}/\hbar\omega_1 = 2$, $\eta_2 = 2.828$. Only the harmonics polarized in the same direction as the first field are shown. The open squares specify the harmonics due to the first field only ($\eta_2 = 0$). Notice that whenever open and solid squares overlap open squares are given. The odd harmonics of ω_1 are unaffected by the low-frequency field upwards from the 19th. The inset shows the sidebands ($m_2 = \pm 2, \pm 4, ...$) to the 7th harmonic of the first field for $|E_0|/\hbar\omega_1 = 6.44, \eta_1 = 1, \eta_2 = 7.08$.

academic. The vast majority of the frequencies displayed are far below the limit of experimental detectability. In fact, the corresponding emission rates could

482

not possibly be calculated via a numerical solution of the Schrödinger equation, due to the limited numbers of cycles of the driving field that such a calculation can handle. Figure 2 exhibits, however, one important feature. The one-color harmonics (where the second field is turned off) are dimmed by the addition of the second field within the plateau, but neither above nor below. This is plausible: if we think in terms of the semiclassical model then a second perpendicular field will obstruct the ability of the electron to revisit its mother ion. However, the semiclassical model does not apply outside the plateau, neither above nor below, where lowest-order perturbation theory essentially holds. Hence, in this region the second field has only a minor impact on the one-color harmonics of the first field.

The apparent chaos of Fig. 2 hides some underlying order. The spectrum can be envisioned as consisting of sidebands. The inset displays one such sideband, for fixed $m_1 = 7$ and arbitrary even m_2 in the nomenclature of Eq.(8). It is interesting to notice, though again academic, that the side bands exhibit plateau structures themselves. While, as a very general feature, the one-color harmonics decrease in intensity when a second field is added the intensities of mixing frequencies may considerably exceed the intensities of either one-color harmonics (where one or the other field is turned off); see Fig.2 of Ref.[17].

3.2 TWO PERPENDICULAR LINEAR POLARIZATIONS; COMMENSURATE FREQUENCIES

In this case the spectrum is simpler. If $\omega_1 = p\omega$ and $\omega_2 = q\omega$ then the emitted spectrum consists of multiples of ω. As a new feature, the same frequency can be reached via different pathways, and, consequently, the relative phase between the two colors has an important impact on the spectrum. In the two-color high-harmonic measurements [20, 21, 22] performed thus far this phase was not under control yet. However, techniques to control it exist [13].

Here we will concentrate on the effect of the phase on selected harmonics near the end of the plateau. Figure 3 deals with the case where $\omega_1 = 2\omega_2 \equiv 2\omega$ and the high-frequency field is dominant ($U_{p1}/\hbar\omega_1 = 5$, $U_{p2}/\hbar\omega_2 = 0.5$). According to the number of photons that must be absorbed from or emitted into the second (low-frequency) field, the emitted frequencies Ω can be separated in four groups, for example,

$$
\begin{aligned}
60\omega &= 29 \times (2\omega) + 2(\omega) &= 31 \times (2\omega) - 2(\omega), \\
59\omega &= 30 \times (2\omega) - (\omega) &= 28 \times (2\omega) + 3(\omega), \\
58\omega &= 29 \times (2\omega), \\
57\omega &= 28 \times (2\omega) + (\omega) &= 30 \times (2\omega) - 3(\omega),
\end{aligned}
$$

and analogously for the other harmonics modulo four. Not surprisingly, 58ω which does not require participation of field 2 is least dependent on the relative phase.

The strongest phase dependence occurs for 60ω which requires at least two low-frequency photons either emitted or absorbed such that two paths of about equal significance are able to interfere. The emission rates of 59ω and 55ω are strongly phase dependent as well. This may come as a surprise since the path involving 3 low-frequency photons would seem to be less important. However, we have observed frequently that the low-frequency field prefers to contribute photons rather than to absorb them. This observation explains why 59ω displays a pronounced phase dependence while 57ω does not. If both fields are of comparable magnitude the dependence on the relative phase is more complicated and does no longer look sinusoidal as in Fig.3; for an example, see Fig.5 of Ref.[17].

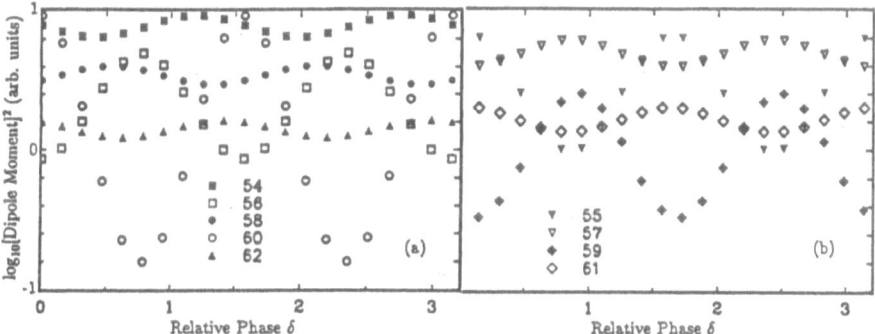

Figure 3. Two-color harmonic efficiencies versus the relative phase between the two fields for $|E_0|/\hbar\omega_1 = 10.44, \omega_1 = 2\omega_2, \eta_1 = 5, \eta_2 = 0.5$ and perpendicular polarizations. The parameters correspond to the experiment [20]; (a) even harmonics, (b) odd harmonics.

3.3 TWO PARALLEL LINEAR POLARIZATIONS

This is the case where most of the experiments have been carried out [20, 21, 22] and the one easiest to interpret since the semiclassical return-of-the-electron model should apply in a straightforward way. There is, however, one complication as compared to the one-color case and this is what we want to concentrate on in this section. For a monochromatic field, the electrons that revisit the core with maximal kinetic energy have been born at a time where the electric field was very close to its peak value. Hence, when the intensity of the driving field is raised into the tunneling regime the end of the plateau (in terms of the ponderomotive energy U_p) is largely unaffected. This is not necessarily so for a two-color field: depending on the frequencies and, if they are commensurate, the relative phase the electrons that return with maximal energy may have been born at a comparatively low intensity. If so, then the relative contribution of these electrons will decrease when the driving field approaches the tunneling regime. As a consequence, the cut-off

484

law for the plateau will be more complicated than the simple form $|E_0| + \text{const} \times U_p$ familiar from the one-color case.

We can study these effects with the help of a simple graphical method [23]. The condition that an electron born at the origin at time t_0 return there at a later time t_1 is

$$F(t_1) = F(t_0) + (t_1 - t_0)F'(t_0), \qquad (9)$$

where $F(t)$ is the indefinite integral of the vector potential, $F(t) = \int dt A(t)$. The electron returns with the kinetic energy

$$E_{\text{kin,ret}} = \frac{e^2}{2m}(A(t_1) - A(t_0))^2. \qquad (10)$$

Demanding that this energy be extremal yields a second equation,

$$A(t_0) = A(t_1) + (t_0 - t_1)A'(t_1) \qquad (11)$$

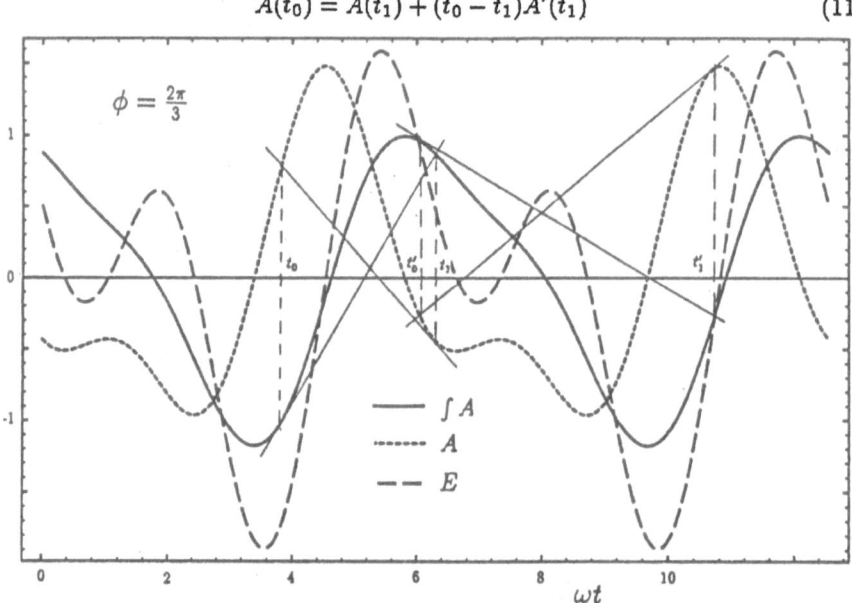

Figure 4. The field $E/\omega A$ (long dashes), the vector potential (short dashes), and its integral F/A (solid line) for the field (12) with relative phase $\phi = 2\pi/3$. The initial and final times t_0, t_0' and t_1, t_1', resp., that yield maximal return energies are marked and the tangent construction according to Eqs.(9) and (11) is given.

Equations (9) and (11) can easily be solved graphically and this provides a simple method to gain some idea of what to expect.

An example is given in Fig.4, for the field

$$E = \omega A(\cos(\omega t) + \cos(2\omega t + \phi)). \tag{12}$$

The figure displays $E/\omega A$, the associated vector potential as well as F/A for a relative phase of $\phi = 2\pi/3$. There are two initial times $\omega t_0 = 3.81$ and $\omega t_0' = 6.06$ that lead to maximal return energies (10). They are marked in the figure along with the corresponding return times $\omega t_1 = 6.29$ and $\omega t_1' = 10.70$. The kinetic energies upon return are $E_{\mathrm{kin,ret}}(t_1) = 2.3 U_p$ and $E_{\mathrm{kin,ret}}(t_1') = 4.8 U_p$, respectively, where U_p is the total ponderomotive potential of the field (12). The electric fields at the initial times are very different, $E(t_0) = 0.92 E_{\mathrm{peak}}$ and $E(t_0') = 0.5 E_{\mathrm{peak}}$, with E_{peak} the peak value of the electric field (12). In the multiphoton regime, the initial times are equally distributed and both return energies will show in the harmonic spectrum. In the tunneling regime, however, emission at time t_0 will be strongly favored and, consequently, the higher return energy will be severely suppressed. It can readily be checked along the same lines that for other phases, e.g. for $\phi = \pi/3$, this is not so and both return energies are similar and so are the fields at the initial times. High-harmonic spectra of, for example, xenon and neon are usually recorded in the multiphoton and in the tunneling regime, respectively. We may then expect that the two-color spectra of these two atoms, each scaled in terms of U_p, will be qualitatively different for this reason already. Notice, that this phenomenon is particular to the two-color environment. Effects like this can, in principle, be exploited in order to achieve coherent control of the harmonic cut-off. The simple graphical method sketched above can be used for a first orientation.

3.4 TWO COPLANAR CIRCULAR POLARIZATIONS

While owing to angular momentum conservation one circularly polarized driving field does not generate any harmonics via dipole transitions, two such fields can do so. However, if the two fields rotate in the same plane angular momentum still restricts the emitted spectrum considerably. In place of Eq. (8) we now have

$$\Omega = m\omega_1 \pm (m+1)\omega_2 \quad \text{or} \quad \Omega = m\omega_1 \pm (m-1)\omega_2 \tag{13}$$

where the upper (lower) sign holds for polarizations which rotate in the opposite (same) direction and m is any positive integer. The frequencies (13) define a one-parameter set in contrast to the two-parameter set (8). Again, explicit expressions for the emission rates can be found in Ref.[17]. Here we want to emphasize one distinguishing and potentially important property of these harmonics: if there is just one pathway to a given frequency Ω then it is circularly polarized. This may be a very feasible method to generate monochromatic circular polarization in the far-uv region. An experiment along these lines has been carried out [22] even though, for technical reasons, the circular polarization of the emitted harmonics could not be corroborated.

4. A Synopsis of Various Polarizations

A recent experiment for the commensurate frequency ratio 2:1 has recorded harmonic spectra for all of the various polarization configurations discussed above [22]. The experimental spectra have been compared to calculations based on the delta-function potential. We reproduce both in Fig.5.

The left-hand half of Fig.5 shows the experimental spectra for the two fields separately, the two fields with linear polarizations either parallel or perpendicular, and the two fields having circular polarizations which co-rotate or counterrotate in the same plane. Several features are worth emphasizing. For the two linear polarizations, the harmonics due to parallel fields are consistently more intense than those due to the perpendicular fields, in agreement with what would be expected on the basis of the semiclassical model. The low-frequency part of the spectrum is more intense than either one-color spectrum. For perpendicular polarizations, there is a characteristic difference between the even and the odd harmonics. The harmonics due to the counterrotating circular polarizations are much more intense than those due to the corotating fields. According to Eq.(13) there should be no $6\omega, 9\omega, 12\omega, ...$ for the case of the counterrotating polarizations. In the experiment, these harmonics are observed, though suppressed. This is a consequence of the less than perfect circular polarizations. The $\times$ mark in Fig.5c gives the observed intensity of the 3rd harmonic due to the circularly polarized 2ω-field only. It should be absent for pure circular polarization. On the basis of this observed intensity, the ellipticity was estimated as 0.9 in place of the ideal 1. In so much as the driving fields are perfectly circularly polarized, the harmonics in the counterrotating case must be circularly polarized as well, alternating between one and the other polarization. This is not so in the corotating case where two pathways exist to each harmonic frequency.

The theoretical spectra are evaluated with the help of the explicit formulas given in Ref.[17]. Experimentally, the intensities of the fundamental and the second harmonic driving field were estimated as comparable. For the calculation, they were adjusted so as to produce an optimal description of the one-color spectra, especially of their cut-off. This resulted in a higher intensity of the fundamental as compared to the second harmonic. These intensities were then used for the calculation of the two-color spectra. Experimentally, the relative phase of the two driving fields was not under control. For the calculation, the phase average was taken whenever the results do depend on the phase, that is, for the two linear polarizations. In this case, we notice that the spectra for parallel and perpendicular polarizations are closer together in the experiment than in the theory. This may be due to less than perfect perpendicularity in the experiment. A very pronounced sensitivity to a deviation from 90° has been observed in calculations [17]. A major discrepancy between experiment and theory occurs for the corotating circular polarizations. The theoretical intensities have been multiplied with a factor of 10^3 in order that they fit in the figure. We surmise that this is at least partially due to the deviation from perfect circular polarization. In conclusion, varying the polarization of the fundamental and the second harmonic provides another example of the potential of coherent control of the high-harmonic output.

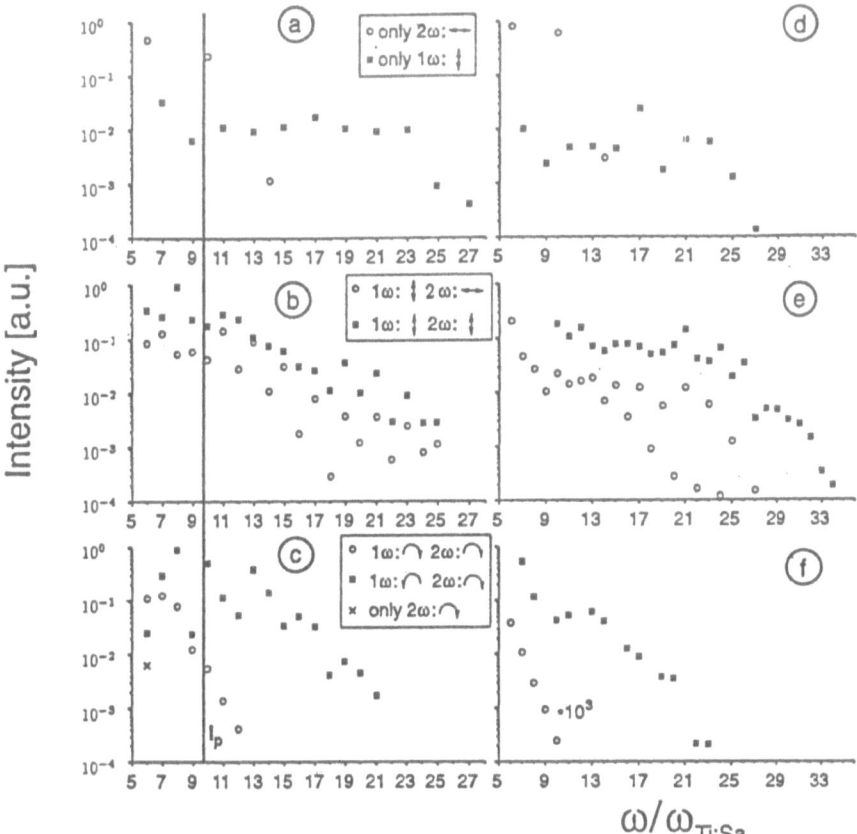

Figure 5: Two-color $1\omega + 2\omega$ harmonics for argon irradiated by a Ti-sapphire laser (1.6eV) and its second harmonic. The relative intensity scales are identical for (a)-(c), and for (d)-(f), respectively. (a)-(c): Experimental results; I_p indicates the ionization potential of argon (15.76eV). (d)-(f): calculated intensities for $I(\omega) = 1.33 \times 10^{14} \text{W/cm}^2$, $I(2\omega) = 0.58 \times 10^{14} \text{W/cm}^2$. (e) is averaged over the relative phase. The intensities for the corotating fields in (f) have been multiplied by a factor of 10^3. The ground state energy $|E_0|$ of the zero-range potential has been adjusted to the energy difference between the ground state and the first excited state of argon rather than to its binding energy [8].

488

References

[1] L'Huillier, A., Lompré, L.-A., Mainfray, G., and Manus, C., in: *Atoms in Intense Fields,* ed. by M. Gavrila (1992), Advances in Atomic, Molecular, and Optical Physics, Supplement, Academic, London, pp. 139-206.

[2] Kulander, K. C., Schafer, K. J., and Krause, J. L. (1993), in: *Super-Intense Laser-Atom Physics,* ed. by B. Piraux et al. (1993), NATO ASI Series B316, Plenum, New York, pp. 95-110.

[3] Corkum, P. B. (1993), Phys. Rev. Lett. 71, 1994-7.

[4] Dietrich, P., Burnett, N. H., Ivanov, M., and Corkum, P. B. (1994), Phys. Rev. A50, R3585-8.

[5] Corkum, P. B., Burnett, N. H., and Ivanov, M. Y. (1994), Opt. Lett. 19, 1870-2.

[6] Ivanov, M., Corkum, P. B., Zuo, T., and Bandrauk, A. (1995), Phys. Rev. Lett. 74, 2933-6.

[7] Becker, W., Long, S., and McIver, J. K. (1990), Phys. Rev. A41, 4112-5.

[8] Becker, W., Long, S., and McIver, J. K. (1994), Phys. Rev. A50, 1540-60.

[9] Becker, W., Lohr, A., and Kleber, M. (1995), Quantum Semiclass. Opt. 7, 423-48.

[10] Lewenstein, M., Balcou, Ph., Ivanov, M. Yu., L'Huillier, A., and Corkum, P. B. (1994), Phys. Rev. A49, 2117-32.

[11] Budil, K. S., Saliéres, P., L'Huillier, A., Ditmire, T., and Perry, M. D. (1993), Phys. Rev. A48, R3437-40.

[12] Burnett, N. H., Kan, C., and Corkum, P. B. (1995), Phys. Rev. A51, R3418-21.

[13] Weihe, F. A., Dutta, S. K., Korn, G., Du, D., Bucksbaum, P. H., and Shkolnikov, P. L. (1995), Phys. Rev. A51, R3433-6.

[14] Antoine, Ph., Carré, B., Saliéres, P., L'Huillier, A., Lewenstein, M., Piraux, B., and Gajda, M. (1995), this volume.

[15] Lohr, A., Becker, W., and Kleber, M., to be published.

[16] Becker, W., Long, S., and McIver, J. K. (1992), Phys. Rev. A46, R5334-7.

[17] Long, S., Becker, W., and McIver, J. K. (1995), Phys. Rev. A52, 2262-78.

[18] Protopapas, M., Knight, P. L., and Burnett, K. (1994), Phys. Rev. A49, 1945-9.

[19] Protopapas, M., Sanpera, A., Knight, P. L., and Burnett, K. (1995), Phys. Rev. A52, xxxx.

[20] Perry, M. D. and Crane, J. K. (1993), Phys. Rev. A48, R4051-4.

[21] Watanabe, S., Kondo, K., Nabekawa, Y., Sagisaka, A., and Kobayashi, Y. (1994), Phys. Rev. Lett. 73, 2692-5.

[22] Eichmann, H., Egbert, A., Nolte, S., Momma, C., Wellegehausen, B., Becker, W., Long, S., and McIver, J. K. (1995), Phys. Rev. A51, R3414-7.

[23] Paulus, G. G., Becker, W., and Walther, H. (1995), Phys. Rev. A52, xxxx.

DYNAMICS OF THE STIMULATED RAMAN SCATTERING OF SUBPICOSECOND LASER PULSE IN THE SELF-PRODUCED PLASMA

O.G. KOSAREVA, V.P. KANDIDOV, and S.A. SHLENOV

Physics Department, Moscow State University,

Vorobyovi Gori, Moscow, 119899, Russia

A model is developed to calculate the dynamics of the stimulated Raman scattering (SRS) of an intense $(10^{14} - 10^{16} \text{ W}/\text{cm}^2)$ subpicosecond ionizing laser pulse in atmospheric-density gas. The plasma inhomogeneity is shown to crucially affect the SRS dynamics: for electron density change of 15% the Stokes pulse amplification decreases by a factor of 10^3. The spatio-temporal instability and breakup of the ionizing pulse in dense gas is considered. The contributions of the spatio-temporal instability and the SRS to the laser pulse spectrum are compared.

1. Introduction

An intense subpicosecond laser pulse focused into an atmospheric-density gas with peak intensity of about $10^{13} - 10^{16} \text{ W}/\text{cm}^2$ undergoes severe transformation of its spatio-temporal and spectral characteristics. The study of this phenomenon is of great importance for many applications including creation of the source of continuum radiation, long channels of high-intense radiation in air and production of dense cold plasmas for X-ray lasing. There could be several reasons for the pulse shape and spectrum transformation, in particular: 1) Kerr nonlinearity of neutral atoms and ions,

H. G. Muller and M. V. Fedorov (eds.), Super-Intense Laser-Atom Physics IV, 489–502.
© 1996 *Kluwer Academic Publishers.*

which leads to the increase in the index of refraction, self-focusing and spectral broadening of the incoming pulse to both blue and red sides; 2) creation of the plasma by strong-field ionization, which leads to the decrease in the index of refraction, frequency upshift and defocusing of the incoming radiation; 3) stimulated Raman scattering of the pulse in the self-produced inhomogeneous plasma which may lead to the red-side spectral broadening up to $\omega_L - \omega_{pmax}$, where ω_L is the laser frequency and ω_{pmax} is the plasma frequency, corresponding to the maximum plasma density in the medium; 4) spatio-temporal instability, which arises when high-power ionizing pulse propagates in dense gas.

The first and the second mechanisms of the pulse transformation have been thoroughly studied experimentally and theoretically [1-6]. To our knowledge the role of the SRS in the spectral and spatial changes of subpicosecond pulses at nonrelativistic intensities has not been reported in the literature. At the same time the contribution of Raman scattering often remains unclear in the experiments where high-intense ultrashort laser pulse propagates in dense gas. The possibility of spatio-temporal instability in defocusing medium has been discussed in [7], but the conditions of the propagation were quite different from those discussed here: the pulse breakup occured due to the joint effect of the group velocity dispersion and negative third-order nonlinear susceptibility.

Both the SRS in plasma and spatio-temporal instability develop if the region of the radiation-plasma interaction is long enough. The effectivness of the spatio-temporal instability grows simultaneously with the effectiveness of the SRS with increasing laser power. In the first part of this paper we calculate dynamics of the SRS in inhomegeneous plasma and contribution of the Stokes scattering to the pulse spectrum transformation. In the second part we present the simulations of the spatio-temporal instability of the ionizing pulse in dense gas and consider the effect of this instability on the pulse spectrum.

2. Stimulated Raman Scattering

The SRS mainly develops in the focus region of the propagation path, where the intensity and plasma density reach their maximum values. We will investigate the dynamics of the SRS in the focus region using a plane wave approximation. The electric field in the medium is given by:

$$\mathbf{E} = \frac{1}{2}\mathbf{e}E(z,t)\exp\left(i\left(\omega_L t - k_L z\right)\right) + \frac{1}{2}\mathbf{e}E_s(z,t)\exp\left(i\left(\omega_s t - k_s z\right)\right) + c.c., \quad (1)$$

where $E(z,t)$, $E_s(z,t)$ are complex amplitudes and ω_L, ω_s are frequencies of the laser pulse and the forward scattered Stokes pulse respectively. The backward scattering will not be considered in this paper since we are mainly interested in the contribution of the SRS to the transmitted laser pulse spectrum. The equation for the complex amplitude of the Stokes pulse $E_s(z,t)$ in the slowly varuing envelope approximation is given by [8]:

$$\frac{\partial E_s}{\partial z} + \frac{1}{c}\frac{\partial E_s}{\partial t} = i\frac{\pi e^2}{mc\left(\omega_L - \omega_{p0}\right)}\left(n_e^* E \exp\left(i\left(\omega_{p0} - \omega_p(z)\right)t\right) + N_e E\right). \quad (2)$$

The terms on the right-hand side of the Eq.(2) represent the response of the plasma, whose density can be written as the sum of the ambient value $N_e(z)$ and a small perturbation δn:

$$\delta n = \frac{1}{2}n_e(z,t)\exp\left(i\left(\omega_p(z)t - k_p z\right)\right) + c.c., \quad (3)$$

where $\omega_p(z) = \left(4\pi e^2 N_e(z)/m\right)^{1/2}$ is plasma frequency. At the entrance to the medium (in the plane $z{=}0$) the plasma frequency satisfies the equation:

$$\omega_L = \omega_{p0} + \omega_s. \quad (4)$$

Here $\omega_{p0} = \omega_p(z = 0)$.

Assuming that the plasma is a cold fluid and that the complex amplitude $n_e(z,t)$ changes little during the plasma wave period, we write the equation for electron density perturbation in the following form:

$$\frac{\partial n_e}{\partial t} = i\frac{N_e e^2}{4m^2 c^2}\frac{\omega_{p0}}{\omega_L\left(\omega_L - \omega_{p0}\right)}\frac{\omega_{p0}}{\omega_p}E\left(E_s^* + \mu^*\right)\exp\left(i\left(\omega_{p0} - \omega_p(z)\right)t\right). \quad (5)$$

The Stokes pulse builds up from the noise background of plasma radiation $\mu(z,t)$.

492

In inhomogeneous plasma the Stokes pulse undergoes phase modulation. This phase modulation results in broadening of the Stokes pulse spectrum. Under the assumption $|E_s| << |E|$ the spectrum of the radiation transmitted through gas is the superposition of the laser and the Stokes pulse spectra.

In numerical simulations of the system of equations (2,5) the complex amplitude of the laser pulse has the form:

$$E(z,t) = E_0 \exp\left(-(t - z/c)^2 / 2(\tau_0/2)^2\right), \qquad (6)$$

where z changes from 0 to L and τ_0 is the pulse duration. The parameters were taken from the experiment [4], where the 1 ps 1053 nm pulse was focused in 1 atm. argon with peak vacuum intensity 10^{16} W/cm^2. With regard to the defocusing in plasma we take $I_{L0} = (c/8\pi)|E_0|^2 \approx 10^{15}$ W/cm^2 and $L=1$ cm.

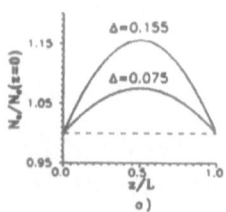

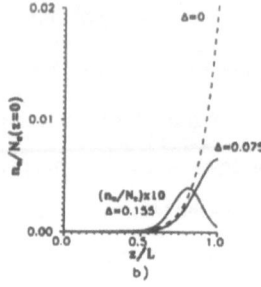

Figure 1. a) Plasma density profile (7); b) Growth of the maximum (in time) plasmon amplitude n_e with z for different Δ.

Let the plasma density change along the propagation direction following the equation (Fig. 1a):

$$N_e(z) = N_e(z = 0) \cdot (1 + \Delta) \cdot \exp\left(-4(z - L/2)^2 \ln(1+\Delta)/L^2\right), \qquad (7)$$

where Δ is inhomogeneity parameter, defined as

$$\Delta = \left(N_e(z = L/2) - N_e(z = 0)\right)/N_e(z = 0). \qquad (8)$$

The electron density $N_e(z = 0) = 2.68 \cdot 10^{19}$ cm^{-3} corresponds to the complete single ionization of atmospheric density gas. Boundary condition for the complex amplitude of the Stokes and electron plasma wave are of the form:

$$n_e(z = 0, t) = 0 \qquad (9)$$

$$E_s(z = 0, t) = 0 \qquad (10)$$

The estimate of the noise intensity in the conditions of [4] yields 360 W/cm^2.

In the case of homogeneous plasma (i. e. $\omega_p(z) = \omega_{p0}$, $\Delta=0$) the amplitude of the Stokes pulse and the electron plasma wave grow exponentially at large z [9] (Fig. 1b and Fig. 2). The intensity of the Stokes pulse reaches 10^{14} W/cm^2 that is $9.4 \cdot 10^{11}$ times larger than the noise intensity. At the same time the assumption $|E_s| << |E|$ is valid: the maximum Stokes pulse intensity is ten times smaller than the laser pulse intensity.

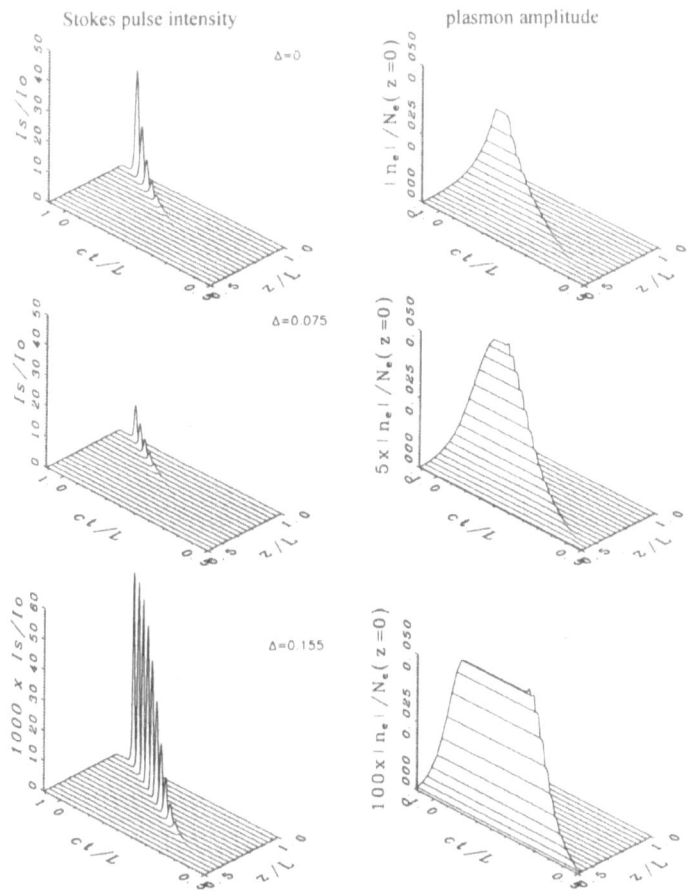

Figure 2. The growth of the Stokes pulse intensity I_s (on the left) and plasmon amplitude $|n_e|$ (on the right) with z for different Δ. $I_0 = 10^{13}$ W / cm^2, $N_e(z = 0) = 2.68 \cdot 10^{19}$ cm^{-3}.

The plasma inhomogeneity has a crucial effect on the SRS dynamics. Already at Δ =0.075 the growth of the Stokes pulse intensity and plasmon amplitude slows down. At larger inhomogeneity Δ =0.155 the growth of the Stokes pulse intensity saturates and plasmon amplitude starts to decrease when z>0.7L (Fig. 1b and Fig. 2). Frequency mismatch acts as a damping factor for the electron plasma wave.

494

The dependence of the Stokes pulse amplification on inhomogeneity parameter Δ is shown in Fig. 3. As parameter Δ increases from $\Delta=0$ to $\Delta=0.5$ the amplification decreases by a factor of $3 \cdot 10^6$.

The frequency mismatch arising due to plasma inhomogeneity determines the phase growth during the pulse. At the end of the propagation path $z=L$ the phase difference $\Delta\varphi$ between the trailing and the leading edges of the pulse may be approximated by the equation:

$$\Delta\varphi \approx \tau_0 \omega_{p0} \left(1 - \sqrt{1+\Delta}\right) \qquad (11)$$

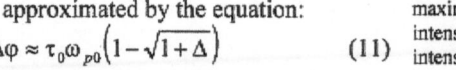

Figure 3. The ratio of the maximum Stokes pulse intensity at $z=L$ to the intensity of the noise radiation $I_\mu = 360 \ W / cm^2$

This equation yields $\Delta\varphi \approx -10$ for $\Delta=0.075$ and $\Delta\varphi \approx -20$ for $\Delta=0.155$. Thus the derivative $\partial\varphi / \partial t$ is negative and the central wavelength of the scattered radiation shifts to the red as the pulse propagates through the inhomogeneous plasma.

Figure 4 shows a formation of the Stokes pulse spectrum from the noise radiation. In the homogeneous plasma (Fig. 4, $\Delta=0$) the central wavelength λ_s remains unchanged and equal to $\lambda_{s0} = 2\pi c / \left(\omega_L - \omega_{p0}\right) = 1258$ nm for all $0<z<L$. As inhomogeneity parameter increases the central wavelength changes with z according to the plasma density profile. At $z<L/2$ the wavelength λ_s shifts to the red while at $z>L/2$ the spectrum broadens acquiring new frequencies from $\omega_L - \omega_{p0}$ to $\omega_L - \omega_{p\max}$ during the propagation (Fig. 4, $\Delta=0.075$ and $\Delta=0.155$). The central wavelength at $z=L$ is 1267 nm for $\Delta=0.075$ and 1276 nm for $\Delta=0.155$. The shifts $(1267-\lambda_{s0})$ and $(1276-\lambda_{s0})$ correspond to the maximum along z plasma frequencies (Fig. 1a). This obviously follows from the fact that the optimum conditions for the amplification are satisfied in the vicinity of $z=L/2$, where the plasma density N_e takes its maximum value and the derivative dN_e / dz takes its minimum value.

At small values of inhomogeneity parameter $(0<\Delta<0.155)$ the intensity of the Stokes pulse is as large as $10^{-1} \div 10^{-4}$ of the laser pulse maximum intensity I_{L0}. As it follows from the experiments [1,2], the large spectral broadening of the order of 100 - 300 nm is observed at $10^{-2} \div 10^{-5}$ intensity level, while FWHM of spectra is usually not larger than 10-20 nm. Hence the SRS may contribute to the pedestal of the spectral continuum (see also Fig. 7c).

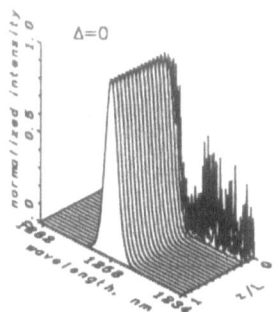

3. Spatio-temporal Instability

Another important phenomenon, arising when a high power $(10^9 - 10^{12}$ W) subpicosecond laser pulse is tightly focused in dense (1-5 atm) gas, is spatio-temporal instability. The propagation of the pulse is described by the following system of equations [6]:

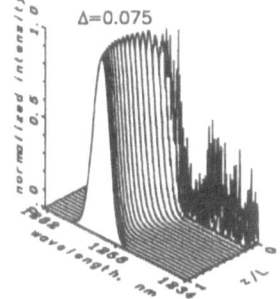

$$2ik\left(\frac{\partial}{\partial z} + \frac{1}{v_g}\frac{\partial}{\partial t}\right)E =$$

$$\Delta_\perp E + \frac{2k^2}{n_0}\left(\frac{1}{2}n_2|E|^2 - \frac{2\pi e^2 N_e}{m\omega_L^2 n_0}\right)E - ik\alpha E , \quad (12)$$

$$N_e = \sum_l l N_l , \quad \frac{dN_l}{dt} = -R_l\left(|E|^2\right)N_l + R_{l-1}\left(|E|^2\right)N_{l-1} ,$$

$$n_2 = \frac{3\pi}{n_0}\sum_k \chi_l^{(3)} N_l , \quad \alpha = I^{-1}\sum_l \varepsilon_l N_l R_l\left(|E|^2\right), \quad (13)$$

where E is a complex amplitude of the electric field, $\Delta_\perp = \frac{1}{r}\frac{\partial}{\partial r}\left(r\frac{\partial}{\partial r}\right)$ is transverse Laplacian, n_0 is linear refractive index of a gas ($n_0 \approx 1$), N_l is the population, ε_l is the ionization potential and $\chi_l^{(3)}$ is the third order nonlinear susceptibility of the lth ionization stage of a gas. R_l is optical-field-induced ionization rate,

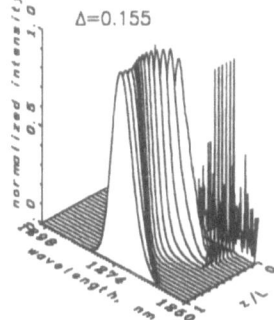

Figure 4. Spectra of the Stokes pulse. Each spectrum is normalized to its maximum value.

calculated according to [10]. The numerical simulations of the system of equations (12-13) have been done in [6] for the propagation of 100 fs $3 \div 30$ μJ laser pulse in 1-5 atm noble gases. The characteristic features of the propagation were formation of the ring in the trailing edge of the pulse due to the nonstationary defocusing in plasma. The spectrum of the initial pulse was blueshifted by 10-50 nm depending on the gas pressure and energy.

496

As the pulse energy increases up to $1 \div 10$ mJ the pulse undergoes structural

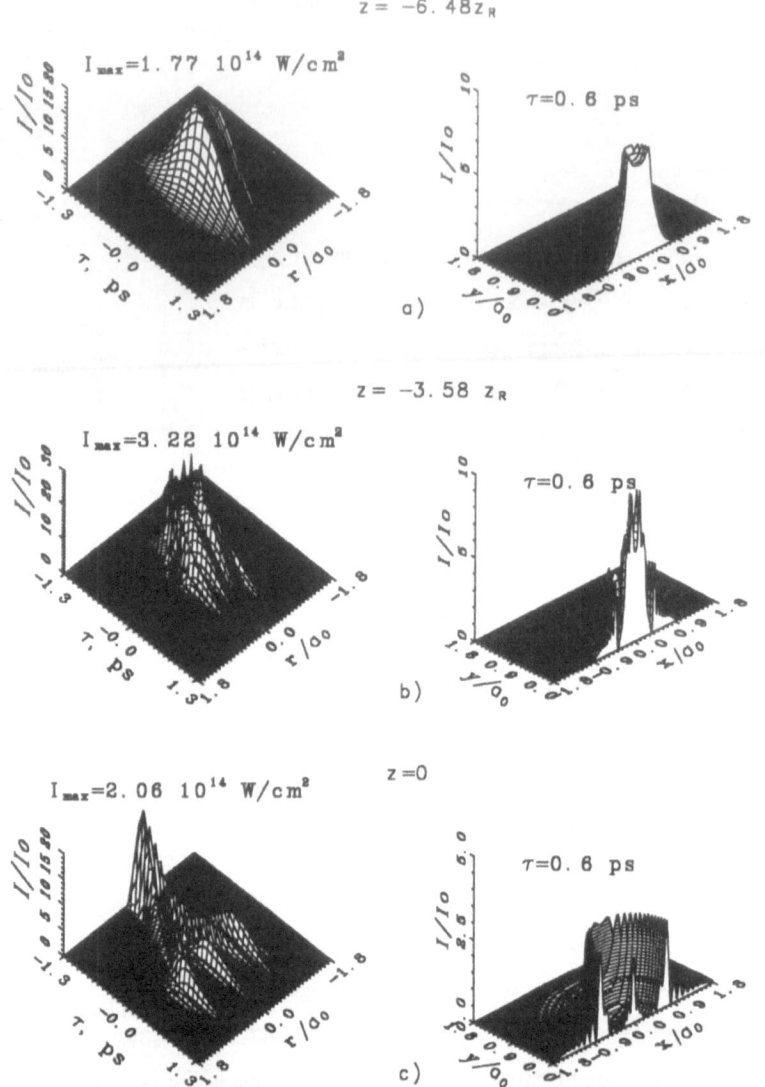

Figure 5. Development of the spatio-temporal instability in 1 atm argon. P=7.5GW, $a_0 = 89$ μm, $I_0 = 10^{13}$ W / cm^2, z=0 corresponds to the position of the focus in vacuum.

instability. This instability originated from the inertial interaction of the electromagnetic radiation with gas. Indeed, the electron concentration accumulates with time and the contribution of the ionization nonlinearity to the self-phase modulation at a given space-time point (r, z, τ) (here $\tau = t - z/v_g$ is time in comoving coordinate system) depends on the evolution of the pulse at this point from $-\infty$ to τ. Consequently, each part of the pulse behaves according to the number of electrons accumulated by this time moment in the medium. For the subpicosecond time scale and tightly focused laser pulse the transverse and longitudinal sizes are of the same order of magnitude. For example, in the conditions of the experiment [4] the longitudinal size of 1 ps pulse is 300 μ m and the minimum transverse size is 80 μ m. This means that the formation of the structural instability in the transverse direction is immediately reflected in the temporal shape of the pulse.

The mechanism of the instability development is the nonlinear defocusing of the near-axis part of the pulse while the periphery continues to focus. The role of initial perturbation belongs to the ring created in a gas long before (about 6-10 Rayleigh lengths) the vacuum focus of the pulse (Fig. 5a). This is in contrast to the case of comparatively low energies of the order of $3 \div 30$ μ J [6], when the first ring is formed at the point of vacuum focus or after it.

Numerical simulations of the system of equations (12-13) are performed in the conditions of experiment [4]: $\tau_0 = 1$ ps, $\lambda = 1053$ nm, $P = 1 \div 10$ GW, Rayleigh length $z_R = 162 \mu$ m, argon pressure 1 atm. The incoming pulse has Gaussian temporal and radial profiles. The evolution of the spatio-temporal instability with z is shown in Fig. 5. Local minimum in the trailing edge of the pulse (Fig. 5a) transforms to three rings at $z = -3.58z_R$ (Fig. 5b) and to eight rings at $z=0$ (Fig.5c). The distribution of the laser intensity in (r, τ) plane demonstrates that the number and radius of rings depend on time τ. This is the obvious consequence of the fact that the trailing edge of the pulse propagates in the plasma with a higher density than the leading edge. Therefore the distortions in the trailing edge are more dramatic.

Due to the structural instability different parts of the pulse experience different degree of the self-phase modulation. For each space point (r, z) we may calculate frequency spectrum of the pulse by performing Fourier transformation of the complex function $E(r, z, \tau)$ with fixed r and z. As a result we obtain spatially resolved spectra demonstrated in Fig.6. Similar procedure in actual experimental conditions has been recently done in [11].

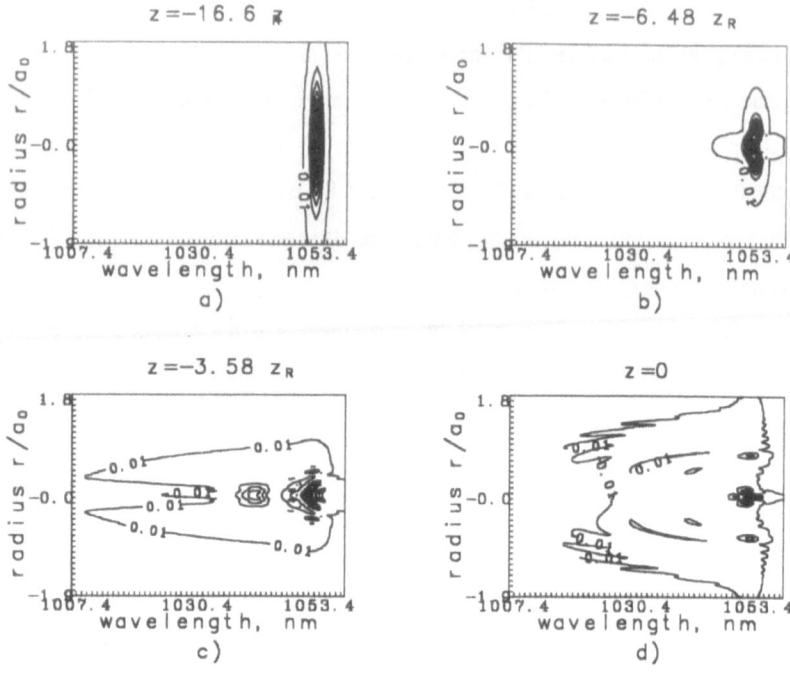

Figure 6. Spatially resolved spectra of 1 ps 7.5 GW pulse in 1 atm. argon in different z-planes. $a_0 = 89$ μm, $z=0$ corresponds to the position of the focus in vacuum.

At the entrance to the medium all the energy is concentrated near the fundamental frequency $\lambda = 1053$ nm (Fig. 6a). As the pulse propagates through the gas the electron density builds up first on laser axis. Therefore far from the focus spectral changes are observed on the axis (Fig. 6b). After the first ring is created (Fig.5a), the electron density starts to grow at the periphery, where the intensity is larger than on the axis. The next plots (Fig.6c,d) show the effect of this process on the pulse spectrum. The maximum spectral broadening is observed in the ring. Multiple ring formation leads to the production of 'color' rings in the spectrum (Fig. 6.d). Up to $r \approx a_0$ (this is roughly the boundary of the region, inside which the largest part of the pulse energy is concentrated) the outer rings correspond to the shorter wavelength. At $r > a_0$ the nonlinear interaction becomes weaker and spectral broadening gradually disappears

(Fig. 6d). The 'color' ring formation has been observed in the experiment on long self-channeling of subpicosecond pulse in air [12], where the main mechanisms of the pulse transformation were Kerr and ionization nonlinearity.

In order to find an averaged spectrum, which is usually registered in the experiments, we integrated spatially resolved spectra over the aperture (Fig.7). The intensity-axis scale is logarithmic in order to show the broadening of the spectral wings. The symmetric broadening at large distances from the focus is due to the noninertial Kerr nonlinearity (Fig.7a). As the pulse propagates through the gas and ionizes it, the ionization nonlinearity suppresses Kerr nonlinearity and blue spectral broadening becomes the main feature of spectral transformation (Fig. 7b,c).

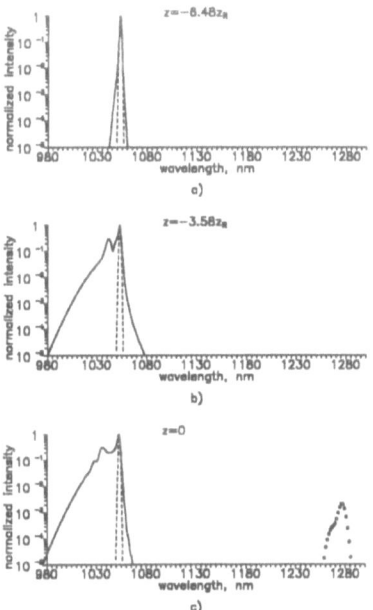

Figure 7. Averaged over the aperture frequency spectrum of 1 ps 7.5GW pulse in 1 atm. argon. The dashed curve in Figs. (a) and (b) and (c) corresponds to the spectrum of the incoming pulse. The curve marked by circles shows the possible contribution of the SRS in plasma to the pulse spectrum.

Figure 7c shows the possible contribution of the SRS to the pulse spectrum (dashed curve). Note that the contribution of the SRS has not been obtained as a solution of the self-consistent system of equations (12-13). The spectrum of the Stokes pulse, shown in Fig.7c, is the result of the simulations presented in the first part of this paper. The gap between 1080 and 1240 nm is explained by the suppression of the Kerr nonlinearity in the conditions of tight focusing geometry. In this case the contribution of the SRS is overestimated, because the actual plasma profile is not so smooth as that shown in Fig.1a. Nevertheless, even if the maximum intensity of the scattered radiation is by a factor of 10 or 10^2 smaller than that shown in Fig 7c (dashed curve), the SRS may substantially contribute to the pedestal of the pulse spectrum.

4. Conclusions

In this paper we considered two mechanisms of the transformation of the subpicosecond laser pulse tightly focused in atmospheric density gas: the stimulated Raman scattering in inhomogeneous plasma and the spatio-temporal instability. The study of the SRS dynamics showed a crucial effect of plasma inhomogeneity on the Stokes pulse amplification: as inhomogeneity parameter Δ increases from $\Delta = 0$ to $\Delta = 0.5$ the amplification decreases by a factor of $3 \cdot 10^6$. At the same time the SRS does not erode the laser pulse itself, since the maximum intensity of the scattered radiation is as large as $10^{-1} \div 10^{-4}$ of the maximum intensity in gas.

The spatio-temporal instability leads to the pulse breakup in both transverse and longitudinal directions. Simultaneously large (up to 100 nm) blue spectral broadening occurs. The spectrum of the pulse in the conditions of spatio-temporal instability may be spatially resolved: the outer rings correspond to the shorter wavelength.

The red side of the spectrum is not continuous since the ionization nonlinearity suppresses Kerr nonlinearity and SRS contribution represents a remote 30 nm wide maximum shifted by 200 nm from the fundamental wavelength. The situation may change in the conditions of the recent experiment on the long self-channelling of the subpicosecond laser pulse in air [12], where the propagation is governed by the same mechanisms discussed here with the only difference that the incoming beam is collimated. In this case the red-side broadening most likely will not be suppressed by Kerr nonlinearity. At the same time the long channel with ionized air may create the optimum conditions for the SRS amplification. Spectrum could be continuously broadened to both blue and red sides.

The authors are grateful to Prof. N.B.Delone, Prof. V.T.Platonenko and Dr. F.A.Ilkov for many helpful discussions.

References

1. Corkum, P.B. and Rolland, C.(1989) Femtosecond continua produced in gases, *IEEE J. QE*, **25**, 2634-2639.

2. Ilkov, F.A., Francois, V. and Chin, S.L. (1993) Supercontinuum generation in a CO_2 gas in the presence of ionization, *SPIE Proceedings*, **2041**, 127-139.

3. Wood, W.M., Siders, C.W. and Downer, M.C. (1993) Femtosecond growth dynamics of an underdense ionization front measured by spectral blueshifting, *IEEE Trans. Plasma. Sci.*, **21**, 20-33.

4. Auguste, T., Monot, P., Mainfray, G. and Manus, C. (1994) Interaction of an ultrastrong laser field with a gaseous target", *Laser Physics*, **4**, 52-57.

5. Rae, S.C. (1994) Spectral blueshifting and spatial defocusing of intense laser pulses in dense gases, *Opt. Comm.*, **104**, 330-335.

6. Kandidov, V.P., Kosareva, O.G and Shlenov, S.A. (1994) Influence of transient self-defocusing on the propagation of high-power femtosecond laser pulses in gases under ionization conditions, *Quantum Electronics*, **24**, 971-977.

7. Liou, L.W., Cao, X.D. et al. (1992) Spatio-temporal instabilities in dispersive nonlinear media, *Phys. Rev. A*, **46**, 4202-4207.

8. Rosenbluth, M.N. (1972) Parametric instabilities in inhomogeneous media, *Phys. Rev. Lett.*, **29**, 565-567.

9. Shen, Y.R. (1984) *The Principles of Nonlinear Optics*, John Wiley & Sons, New York.

10. Ammosov, M.V., Delone, N.B and Krainov, V.P. (1986) Tunnel ionization of complex atoms and of atomic ions in an alternating electromagnetic field, *Sov. Phys. JETP*, **64**, 1191-1194.

11. Siders, C.W., Turner III, N.C., Downer, M.C., Babine, A. and Stepanov, A. (in press) Blueshifted third harmonic generation during ultrafast barrier suppression ionization of subatmospheric density noble gases.

12. Braun, A., Korn, G., Liu, X., Du, D., Squier, J. and Mourou, G. (1995) Self-channeling of high-peak-power femtosecond laser pulses in air, *Optics Letters*, **20**, 73-75.

CHANGE OF THE FREQUENCY SPECTRUM
OF RELATIVISTICALLY STRONG LASER PULSES
DURING THEIR INTERACTION WITH A PLASMA

S.V. BULANOV AND N.M. NAUMOVA
General Physics Institute, Russian Academy of Sciences
Vavilov Street 38, 117942 Moscow, Russia

T.ZH. ESIRKEPOV
Moscow Institute of Physics and Technology
Dolgoprudny, 141730 Moscow region, Russia

AND

F. PEGORARO
Department of Theoretical Physics, University of Turin
via P. Giuria 1, 10125 Turin, Italy

Abstract. The evolution of laser pulses during their interaction with underdense and overdense plasmas is investigated. The spectra of both the reflected and the transmitted radiation are examined. The structure of these spectra is due to the stimulated Raman scattering, nonlinear phase self modulation and to the Doppler effect.

1. Introduction

Recent progress in technology has made available to investigators laser pulses with duration shorter than a picosecond that reach intensities exceeding 10^{18} W/cm^2 at the focus [1]. At such intensities, the pulse electric field exceeds the intra-atomic field and matter turns out to be ionized instantaneously. Electrons in the laser field oscillate at relativistic velocities and the pulse radiation can exert pressures in the *gigabar* range. The laser-plasma dynamics in this range of parameters is determined by collective effects. Among the different applications of such intense, short laser pulses

H. G. Muller and M. V. Fedorov (eds.), Super-Intense Laser-Atom Physics IV, 503–512.
© *1996 Kluwer Academic Publishers.*

are the acceleration of charged particles, photon acceleration, and the production of x-ray sources [2, 3, 4].

The interaction of a laser pulse with a plasma depends critically on the ratio of the laser pulse carrier frequency ω to the Langmuir frequency $\omega_p = \sqrt{4\pi n e^2/m}$. During the interaction in a transparent plasma $(\omega \gg \omega_p)$ wake field generation takes place, together with acceleration of charged particles and photons, excitation of solitary waves and generation of quasistatic magnetic fields. The most important nonlinear processes for the pulse evolution are the backward and forward stimulated Raman scattering (BSRS and FSRS), relativistic self focusing and phase self modulation. In overdense plasmas $(\omega \ll \omega_p)$ the interaction takes place near the boundary and depends on the nonuniformity of the plasma boundary, on the pulse polarization and angle of incidence. The results of the investigation of the interaction of a short pulse with an overdense plasma have a wide range of applications to inertial fusion, plasma diagnostics and to the development of power sources of electromagnetic radiation etc. These regimes are also related to the investigation of charged particle acceleration in a narrow channel. In this scheme, the pulse is transferred over large distances inside a thin channel in matter, either hollow or filled by a low density plasma [5, 6]. The laser radiation is absorbed due to different nonlinear processes that occur in the plasma resonance region, and/or in the course of the electron vacuum heating [7, 8].

Particle-in-cell (PIC) simulations provide great help in the study of the non-linear processes that occur during the interaction of super-intense ultra-short laser pulses with plasmas and give an insight into the dynamics of the pulse and a detailed information on that of the particles. PIC simulations describe this interaction self-consistently, including wave-breaking and self-intersection of electron trajectories, heating and acceleration of particles.

The aim of this report is to outline the mechanisms that occur during the interaction of laser pulses of finite width and length with underdense and overdense plasmas and their influence on the pulse spectrum. In this paper the underdense plasma limit is investigated in Secs. 2–5, while the change of the radiation spectrum during the interaction with overdense plasmas is considered in Sec. 6.

2. Change of the electromagnetic radiation frequency in the course of the pulse depletion

When a short laser pulse with length much shorter than the Langmuir wave-length (or a pulse with a sharp leading front $\tau \leq 1/\omega_p|a|$) propagates in an underdense plasmas, it can excite a wake wave. The corresponding maximum electric field in the wake wave is approximately

$$E_p = \frac{mc\omega_p}{e}\frac{|a|}{2^{1/2}},$$ (1)

where $|a| = eE/m\omega c$ is the dimensionless amplitude of the laser pulse. In this case the group velocity of the pulse

$$v_g \approx c\left(1 - \frac{\omega_p^2}{2\omega^2}\right) \approx c$$ (2)

is close to the speed of light. The wake wave has a phase velocity $v_{ph} = \beta_{ph}c$ equal to (2). During the excitation of the wake wave the laser pulse loses its energy. The characteristic time of the non–linear pulse depletion is [9, 10, 11]

$$t_{nl} \approx \omega_p^{-1}\,(\omega/\omega_p)^2\,|a|^{-1}.$$ (3)

Conservation of the number of electromagnetic quanta leads to the down-shift of the average frequency value according to the formula

$$< \omega(t) >= (1 - t/t_{nl})^{1/3} < \omega(0) > .$$ (4)

For longer pulses, the role of plasma instabilities becomes more impor-tant and they can deteriorate the regular pattern of the wake field. In its initial stage, a smooth laser pulse with length greater than the Langmuir wave-length does not excite a wake. As a result of the variation in its group velocity due to amplitude non-uniformity and to relativistic mass growth, the leading front of the pulse can become steeper [11, 12]. This shortens the scale of the pulse amplitude non-uniformity and, when this scale-length becomes comparable with the Langmuir wave-length, a wake field is gener-ated, accompanied by changes in the frequency spectrum.

The other physical mechanisms that can affect the laser field evolution are the forward and the backward stimulated Raman scattering. The growth rates of these instabilities are [11, 13]

$$\gamma_{BSRS} \approx \omega(\omega_p/\omega)^{2/3}, \ \gamma_{FSRS} \approx \omega(\omega_p/\omega)^2,$$ (5)

and their development can cause respectively a non-uniformity in the laser pulse profile, the formation of shock fronts and ultra-fast pulse depletion due to strong excitation of wake waves [11, 12].

In our PIC simulations we have observed the formation of solitary waves [11, 14] together with the pulse distortion. These localized electro-magnetic waves, with carrier frequency smaller than plasma frequency, show quite a stable behaviour for a time interval comparable with the dispersion time of short wave packets.

3. Foton accelerator

In a wake plasma wave, modulations in electron density propagate with relativistic phase velocity. Since the local value of the plasma frequency is a function of the phase of the wake field, a short wave packet of electromagnetic radiation with carrier frequency ω_1 placed in a wake plasma wave may increase or decrease its energy depending on its phase and, since in the limit $\omega_1 \gg \omega_p$ the photon number is conserved, the frequency of the wave packet may up-shift or down-shift. This mechanism of up-shift of the frequency of the laser radiation is called "photon accelerator" [3].

In the "photon accelerator" scheme, the wave packet frequency changes as in the case when a wave gains energy interacting with a moving mirror. The wave packet in the photon accelerator moves in the same direction as the plasma wave. Taking into account the photon number conservation, we obtain for the final electromagnetic radiation frequency the value [15]

$$\omega_2 \approx \frac{\Omega_1^2}{2\omega_1(1 - \beta_{ph})}, \tag{6}$$

where Ω_1 is the local Langmuir frequency in the plasma wave. In [16] it has been shown that the maximum frequency up-shift is obtained when the wave packet moves together with a non-linear plasma wave near its wave-break limit. In this case the maximum frequency up-shift is

$$\omega_{max} = \omega_1 \frac{1 + \beta_{ph}}{1 - \beta_{ph}} = \omega_1 \gamma_{ph}^2, \tag{7}$$

corresponding to Einstein's solution of the problem of the reflection of an electromagnetic wave from a mirror moving at a relativistic speed.

4. Induced focusing

For the one dimensional approximation to be valid, the transverse scale of the pulse must be larger than the Langmuir wave-length. On the other hand, in order to obtain high radiation intensity, we need to enter the regime where the laser pulse undergoes relativistic self-focusing down to a region with transverse size $l_\perp$ of order of several wave lengths. In this case the size of the field localization in the longitudinal direction prescribed by the Rayleigh length $(l_\perp^2/\lambda)$ may be shorter than the depletion length and the acceleration length.

One possibility in order to avoid pulse spreading is to use self-focusing [17]. The threshold of relativistic self-focusing corresponds to the power $P > P_{cr} = 15(\omega/\omega_p)^2 GW$ [18].

In [19] it has been shown that a short laser pulse in a wake wave may be focused or de-focused depending on the phase of its location in the plasma wave. 2D PIC simulations of a semi-infinite, ultrarelativistic laser pulse with a sharp leading front and transverse length shorter than the plasma wave-length in an underdense plasma, show the development of a "horse-shoe" structure [20]. The leading edge of the pulse acts as a driver for the plasma wave and later as a focusing and de-focusing driver for the rest of the pulse.

5. Phase self modulation

Recent experiments [21, 22] on the interaction of subpicosecond $1\mu m$ wave-length laser pulses, with intensity $\approx 10^{18} W/cm^2$ in plasmas with density $10^{17} \div 10^{18} cm^{-3}$, revealed a quite complicated dependence of the backscat-tered radiation spectrum on the pulse amplitude. 1D PIC simulation of this interaction [22, 23] has shown similar features (fig.1). Detailed analysis re-vealed backward and forward stimulated Raman scattering and phase self modulation [23]. According to [24] the pulse spectrum can be expressed in the form

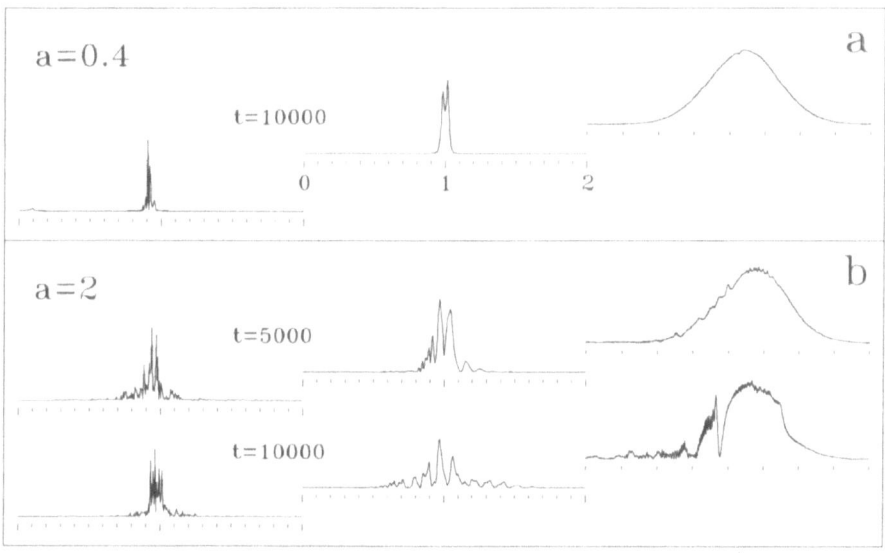

Figure 1. Frequency spectra of the reflected (left) and transmitted radiation and the pulse envelope for pulses with amplitudes (a) $a = 0.4$, (b) $a = 2$ at time $t = 5000\omega_o^{-1}$ and t = $10000\omega_o^{-1}$ in a plasma with density $n = 0.01n_{cr}$.

508

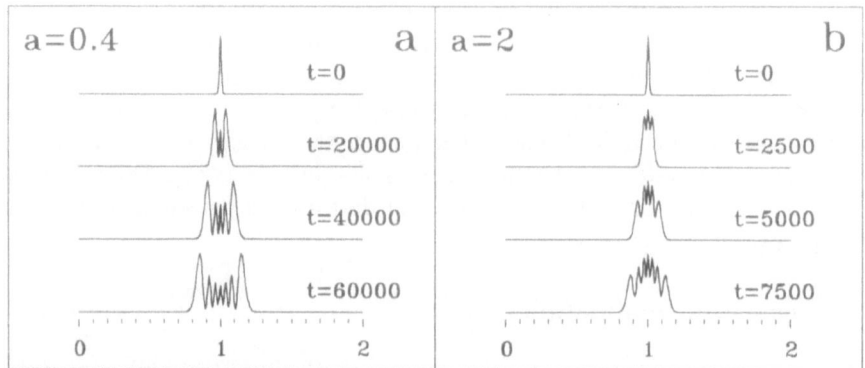

Figure 2. Evolution of the frequency spectra of smooth symmetrical pulses in an underdense plasma ($n = 0.01n_{cr}$) due to the phase self modulation for (a) $a = 0.4$ and (b) $a = 2$.

$$|E(\omega)|^2 = \left| \int_0^\infty \mathcal{E}(t) e^{i\omega t + i\Delta\phi(t)} dt \right|, \tag{8}$$

where

$$\Delta\omega(t) = -\partial(\Delta\phi)/\partial t \tag{9}$$

describes frequency self modulation. When a relativistically strong laser pulse propagates in a plasma the phase self modulation is given by

$$\Delta\phi(t) \approx \frac{t\omega_p^2}{2\omega_0} \left(1 - \frac{1}{\sqrt{1 + a^2(t)}} \right). \tag{10}$$

This expression is obtained by describing the electromagnetic wave packet in the envelope approximation. Eq. (10) accounts for the relativistic increase of the electron mass and for the change of the electron density inside the wave packet (see also [25]).

In the case of a symmetrical pulse, the frequency self modulation $\Delta\omega$ has a minimum in the leading part of the pulse and has a maximum at the rear. The frequency spectra calculated with the model dependence of $\Delta\phi$ given by (10) have a minimum in the center in a low-relativistic case ($a < 1$) (fig.2a) and a maximum for an ultrarelativistic case ($a > 1$) (fig.2b). A pulse asymmetry results in an asymmetry of the frequency spectrum as is shown in fig.3 for a sharp front edge and for a sharp rear edge.

These novel features could be used for determining the amplitude of the pulse after it has passed the plasma slab. When the pulse is self-focused in

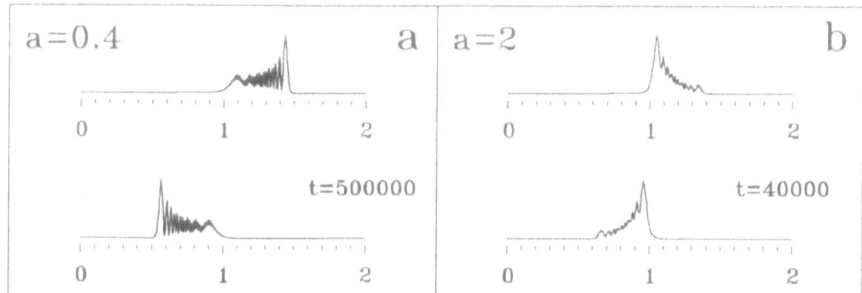

Figure 3. Frequency spectra of pulses with sharp front edge (upper) and sharp rear edge affected by phase self modulation for (a) $a = 0.4$ and (b) $a = 2$ in an underdense plasma ($n = 0.01n_{cr}$).

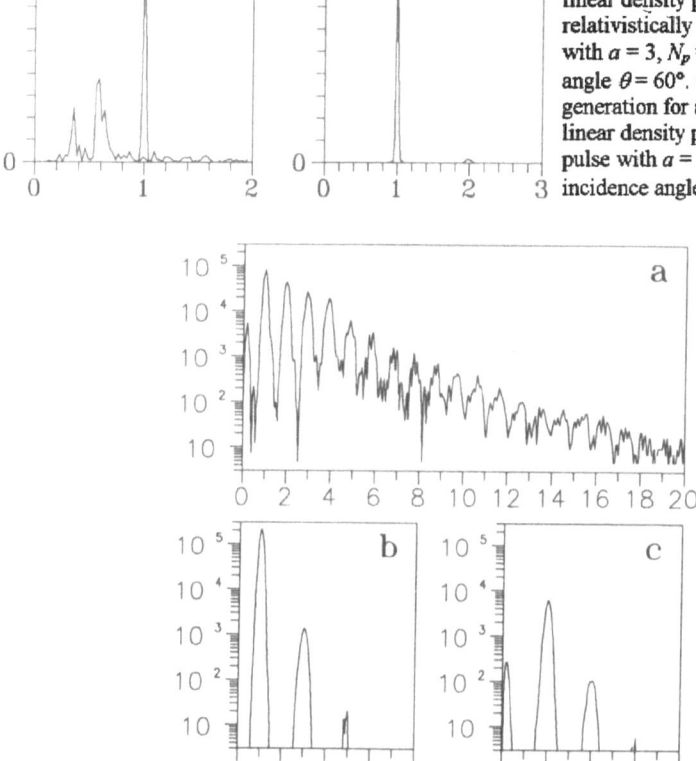

Figure 4. (a) Low-frequency generation for a plasma with a linear density profile and a relativistically strong laser pulse with $a = 3$, $N_p = 30$ and incidence angle $\theta = 60°$. (b) Second harmonic generation for a plasma with a linear density profile and a laser pulse with $a = 0.1$, $N_p = 30$ and incidence angle $\theta = 15°$.

Figure 5. High harmonic and low-frequency generation for a sharp boundary plasma and a relativistically strong laser pulse with $a = 3$, $N_p = 8$, incidence angle $\theta = 60°$ and ion density $n_o = 6n_{cr}$. The incident wave is p-polarized in (a) and s-polarized in (b). An additional p-polarized wave is reflected in the case of s-polarization (c).

the plasma, its amplitude is increased and this in turns affects the pulse spectrum.

6. High harmonic generation

If the plasma is overdense $\omega/\omega_p < 1$, i.e., non-transparent for the laser pulse, the interaction takes place near the plasma boundary and depends critically on the ratio $\eta = r_E/L$, where $r_E = eE/m\omega^2$ is electron oscillation amplitude and $L = (\partial \ln n/\partial x)^{-1}|_{n=n_{cr}}$ is the scale length of the plasma non-uniformity.

If $\eta \ll 1$, the main absorption mechanisms occur in the plasma resonance region for a p-polarized electromagnetic wave [26]. Similarly to the case of an underdense plasma, there is an effective plasma wave excitation in an underdense plasma region and the development of stimulated Raman scattering when $|a| > 1$.

The results of the simulations presented below have been performed with a 1D PIC code using a Lorentz transformation to a reference frame moving along the plasma boundary [27]. A similar approach has been used in [8]. This procedure reduces the problem of oblique incidence to that of normal incidence on the plasma [28].

The spectrum of the reflected radiation during oblique incidence of a p-polarized wave at the optimal angle for $|a| > 1$ exhibits peaks in the low-frequency region and is broaden in the high-frequency region (fig.4a). In the low-relativistic case ($|a| < 1$) we can see the second harmonic generation (fig.4b).

The interaction with a sharp-boundary plasma ($\eta \gg 1$) depends on the pulse polarization (fig.5). In case of p-polarized waves we can see an effective generation of odd and even harmonics (a). The low frequency component is due to the laser pulse envelope. In case of s-polarized waves high harmonics generation is less effective. Only odd harmonics are produced with s-polarization (b) while an additional p-polarized wave appears with even harmonics (c).

The laser pulse interacting with plasma forces the boundary electrons to oscillate in the y-direction and, due to its relativistic amplitude, in the x-direction. Afterwards the boundary acts as a moving mirror, generating harmonics.

In these simulations the electromagnetic radiation reflected from the plasma is produced by a source near the plasma boundary which has the form of a flat charge sheet. From the one-dimensional Lienard-Wiechert potential we obtain for the y component of the electric field

$$E_y = \frac{2\pi n' e l v_y(t^*) \, \text{sign}\,[x - x_0 - \xi(t^*)]}{c - (\partial \xi/\partial t^*)}, \qquad (11)$$

where v_y is the quiver velocity of the electrons in the y direction, $l \approx c/\omega'$ is the thickness of the reflecting charge sheet, and the retarded time t^* obeys the equation

$$c(t' - t^*) = x - x_0 - \xi(x_0, t^*). \tag{12}$$

By expanding $v(t^*)$ and the denominator of (11) into a series of the ratio $(\partial \xi / \partial \tau)(1/c)$, we see that the reflected electromagnetic radiation contains all harmonics of the wave frequency ω'.

7. Conclusions

The interaction of high intensity, ultrashort laser pulses with plasmas is accompanied by various phenomena known in nonlinear physics. A host of instabilities can take place, transformation of radiation energy into fast particle energy, soliton excitation, shock wave formation, self-focusing and radiation channeling. These nonlinear processes change the pulse spectrum and shape drastically. Some processes are related to specific instabilities such as BSRS, while other processes appear when the total energy of the pulse decreases or increases. A third class is related to the nonlinear motion of particles in the vicinity of sharp plasma boundaries. The results we have obtained may be used for identifying the nonlinear mechanisms occurring during the laser-plasma interaction and are useful for diagnostics and for the production of sources of hard electromagnetic radiation.

Financial support from the International Science Foundation and from the Russian Foundation for the Promotion of Basic Research (Grant No. 93-02-169220) is acknowledged.

References

1. Mourou, G., and Umstadter, D. (1992) Development and applications of compact high-intensity lasers, *Phys Fluids B* **4**, 2315–2325.
2. Tajima, T., Dawson, J.M. (1979) Laser electron accelerator, *Phys. Rev. Lett.* **34**, 267–270.
3. Wilks, S.C., Dawson, J.M., Mori, W.B., Katsouleas, T., and Jones, M.E. (1989) Photon accelerator, *Phys. Rev. Lett.* **62**, 2600–2603.
4. Murnane, M.M., Kapteyn, H.C., and Falcone, R.W. (1991) Generation of efficient ultrafast laser-plasma x-ray sources, *Phys. Fluids B* **3**, 2409–2413.
5. Bulanov, S.V., Pegoraro, F., Pukhov, A.M. (1994) Short, relativistically strong laser pulse in a narrow channel, *Phys. Lett. A* **195**, 84–89.
6. Chiou, T.C., Katsouleas, T., Decker, C., Mori, W.B., Wurtele, J.S., Shvets, G., and Su, J.J. (1995) Laser wake-field acceleration and optical guiding in a hollow plasma channel, *Phys. Plasmas* **2**, 310–318.
7. Bulanov, S.V., Kovrizhnykh, L.M., and Sakharov, A.S. (1990) Regular mechanisms of electron and ion acceleration in the interaction of strong electromagnetic waves with a plasma, *Phys. Rep.* **186**, 1–51.
8. Gibbon, P., Bell, A.R. (1992) Collisionless absorption in sharp-edged plasmas, *Phys. Rev. Lett.* **68**, 1535–1538.

512

9. Bulanov, S.V., Kirsanov, V.I., and Sakharov, A.S. (1989) Excitation of ultrarelativistic plasma waves by pulse of electromagnetic radiation, *JETP Lett.* **50**, 198–201.
10. Bulanov, S.V., Kirsanov, V.I., and Sakharov, A.S. (1990) Ultrarelativistic theory of a laser plasma wake accelerator, *Sov. J. Plasma Phys.* **16**, 543–548.
11. Bulanov, S.V., Inovenkov, I.N., Kirsanov, V.I., Naumova, N.M., and Sakharov, A.S. (1992) Nonlinear depletion of ultrashort and relativistically strong laser pulses in an underdense plasma, *Phys. Fluids B* **4**, 1935–1942.
12. Bulanov, S.V., Inovenkov, I.N., Kirsanov, V.I., Naumova, N.M., Sakharov, A.S., and Shah, H.A. (1993) Stationary shock-front of a relativistically strong electromagnetic radiation in an underdense plasma, *Phys. Scripta* **47**, 209–213.
13. Sakharov, A.S., and Kirsanov, V.I. (1994) Theory of Raman scattering for a short ultrastrong laser pulse in a rarefied plasma, *Phys. Rev. E* **49**, 3274–3282.
14. Bulanov, S.V., Esirkepov, T.Zh., Kamenets, F.F., and Naumova, N.M. (1995) Electromagnetic solitons formation during the interaction of relativistically strong laser pulses with plasmas, *Plasma Physics Reports* **21**, 600–610.
15. Bulanov, S.V., Kirsanov, V.I., Pegoraro, F., and Sakharov, A.S. (1993) Charged particle and photon acceleration by wake field plasma waves in nonuniform plasmas, *Laser Physics* **3**, 1078–1087.
16. Bulanov, S.V., Inovenkov, I.N., Kirsanov, V.I., Naumova, N.M., and Sakharov, A.S. (1991) Up-shifting of frequency of electromagnetic radiation during the interaction with non-linear plasma waves, *Soviet Physics - Lebedev Institute Reports*, No.6, 9-11.
17. Askar'an, G.A. (1962) Action of intensity field gradient of electromagnetic ray on electrons and atoms, *JETP* **42**, 1567–1571.
18. Litvak, A.G. (1969) Finite amplitude wave beams in a magnetoactive plasma, *Sov. Phys. JETP* **57**, 629–647.
19. Bulanov, S.V., and Sakharov, A.S. (1991) Induced focusing of electromagnetic wave in a wake plasma wave, *JETP Lett.* **54**, 203–207.
20. Bulanov, S.V., Pegoraro, F., and Pukhov, A.M. (1994) Two-dimensional regimes of self-focusing, wake field generation, and induced focusing of a short intense laser pulse in an underdense plasma, *Phys. Rev. Lett.* **74**, 710–713.
21. Darrow, C.B., Coverdale, C.A., Crane, J.K., Perry, M.D., Mori, W.B., Decker, C., Joshi, C., and Clayton, C. (1994) Spectrally Modulated Stimulated Raman and anomalous backscatter in high-intensity, sub-picosecond laser, underdense gas-target experiments, *Bulletin of the American Physical Society* **39**, 1519-1528.
22. Coverdale, C.A., Darrow, C.B., Decker, C.D., Naumova, N.M., Bulanov, S.V., Mori, W.B., Tzeng, K.C. Spectral anomalies and reflectivity saturation of stimulated Raman backscatter from a sub-picosecond laser in an underdense plasma, *Phys. Rev. Lett.*, submitted for publication.
23. Bulanov, S.V., Esirkepov, T.Zh., and Naumova, N.M. (1995) Properties of spectra of the reflected and transmitted radiation during propagation of relativiatically strong laser pulses in underdense plasmas, *Proceedings of the XII International Conference on Laser Interaction and Related Plasma Phenomena*, Osaka, Japan, April 24-29, AIP.
24. Shen, Y. R. (1984) *The principles of nonlinear optics*, John Wiley and Sons, New York.
25. Borovskiy, A.M., Zhileikin, Ya.M., Korobkin, V.V., Makarova, E.A., Osipik, Yu.I., and Prokhorov, A.M. (1994) Spatial modulation and giant broadening of spectra of high-power ultrashort laser pulses subjected to self-channeling in nonlinear media, *Laser Physics* **4**, 1173-1184.
26. Ginzburg, V. L. (1964) *The Propagation of Electromagnetic Waves in Plasmas*, Pergamon, New York.
27. Bulanov, S.V., Naumova, N.M., and Pegoraro F. (1994) Interaction of an ultrashort, relativistically strong laser pulse with an overdense plasma, *Phys. Plasmas* **1**, 745–757.
28. Bourdier, A. (1983) *Phys. Fluids* **26**, 1804-1807.

NONLINEAR OPTICS IN PLASMA WAVEGUIDES

T.J. MCILRATH, H.M. MILCHBERG and C.G. DURFEE III
Institute for Physical Science and Technology
University of Maryland, College Park, MD 20742

ABSTRACT

It is shown that a recently demonstrated plasma waveguide [C.G. Durfee III and H.M. Milchberg, Phys. Rev. Lett.. **71**, 2407 (1993)] can provide phase matching over extended interaction lengths for the generation of high-order harmonics and difference frequencies. The factors affecting phase matching in plasma waveguides are evaluated in the perturbation limit, including the effect of residual ions. It is shown that long coherence lengths should be attainable for higher order parametric processes.

1. INTRODUCTION

We have recently developed and characterized a method for optically guiding high-intensity laser pulses in a plasma waveguide [1,2,3]. The extremely large product of interaction length and intensity made possible by this waveguide suggests its application to short wavelength generation by very high-order frequency conversion, which has received considerable recent attention [4]. In this paper, the properties of the plasma fibers are summarized. The properties of the optical modes supported by the fibers are discussed along with their dispersion characteristics and the effects of ions on the mode properties. Several waveguide-based schemes are proposed for phase matching the conversion process. Essential to these schemes is that the waveguide eliminates the focusing phase shift. Since the plasma waveguide evolves dynamically in time, its dispersion relation is tunable and offers a degree of freedom for phase matching not previously available.

In our guiding technique, an axicon lens brings a laser pulse to a line focus in a gas. The shock expansion of the resulting spark forms a favourable refractive index profile into which a second laser pulse is injected after an adjustable delay. In experiments to date, pulses have been guided distances of 2.2 cm ($\sim$ 70 Rayleigh lengths) at intensities greater than 10^{14} W/cm^2, sufficient to be in the regime of high harmonic generation. In the plasma waveguide, the nonlinear polarization is induced by the input field(s) in ions and high ionization potential neutrals which may be present. Th goal of this paper is to demonstrate phase matching in such plasma waveguides. Phase matching will allow effective utilization of the long confinement

513

H. G. Muller and M. V. Fedorov (eds.), Super-Intense Laser-Atom Physics IV, 513–521.
© *1996 Kluwer Academic Publishers.*

length of the waveguide for efficient generation of high intensity, short wavelength radiation.

2. WAVEGUIDE CHARACTERISTICS

2.1. FREE ELECTRON WAVEGUIDE

We first consider the case in which the free electron contribution to the refractive index dominates over the bound electron contribution. The channel electron density profile is modeled to be monotonically increasing with radius $|r_\perp|$ from the channel axis out to some boundary, outside of which it remains constant. Such a density profile, which we have called the "finite" profile [2], can support bound modes, and possesses a cutoff, In the real experimental profile, the electron density decreases beyond its peak at the position of the shock wave, leading to tunneling or leaking of field energy for waves which are near cutoff [2]. Here, however, we are concerned mainly with bound modes, so that the finite profile can be employed to good approximation. The transverse eigenmodes $u(r_\perp)$ of the plasma waveguide are found from

$$[\nabla_\perp^2 - 4\pi r_e N_e(r_\perp)]u = -\xi^2 u \tag{1}$$

where $E_\beta(r_\perp, z, \omega) = u(r_\perp)exp(i\,\beta(\omega)z)$ is a pure propagating mode, $\nabla_\perp^2$ is the ransverse Laplacian, $r_o = 2.82\times10^{-13}$ cm is the classical electron radius, $\xi^2 = k^2 - \beta^2$ is the eigenvalue, and k and β are the vacuum and waveguide wave numbers, respectively. Some general conclusions may be drawn without specifying the exact plasma density profile $N_e(r_\perp)$. The left side of Eqn. (1) is independent of k so that the eigenvalues ξ^2 and eigenmodes $u(r_\perp)$ are *wavelength independent*. Therefore, any light generated in the waveguide may be decomposed into the same set of modes as the driving wave(s). As an example, if only the lowest order transverse mode of the driving wave is bound by the finite guide, then only the lowest order mode of the generated wave is bound, and these transverse modes are spatially the same.

For the useful special case of an infinite parabolic profile $N_{eo}(r) = N_{eo}(1 + \delta^2 r^2)$ (with no limit in r), it can be shown [5] that $u(r_\perp) = u_{pl}(r,\phi) = e^{il\phi}\, s^l\, L_p^l(2s^2)exp(-s^2)$ (Laguerre-Gaussian functions), where $s^2 = r^2(\pi r_o N_{eo})^{1/2}\delta$ and p and l are the radial and angular integral mode indices. Here, the lowest order mode u_{oo} is a Gaussian with $1/e$ radius $r = (\pi r_o \delta^2 N_{eo})^{1/4}$. For general density profiles, the sets of u and p,l must be calculated numerically [3]. We will use the parabolic profile to illustrate the physics associated with the plasma fibers.

The waveguide propagation wavenumber is given by $\beta^2 = k^2 - k_p^2 - 2k_p(|l| + 2p + 1)\delta = k^2(1 - A^2/k^2)$, where $A^2 = k_p^2 + 2k_p\delta(|l| + 2p + 1)$ and $k_p = 4\pi r_o N_{eo} = \omega_p^2/c^2$ with ω_p the plasma frequency at $r = 0$. The A^2/k^2 term is small (typical values of these terms for a guided optical driving wave are 10^{-2}), so that

$$\beta \approx k - A^2/2k = k - (k_p^2/2k) - (k_p/k)\delta\ (|l|+2p+1) \qquad (2)$$

The propagation phase βz has two contributions: the plasma dispersion term $(k - k_p^2/2k)z$ and the waveguide geometric contribution $-(k_p/k)\delta(|l|+2p+1)z$. The latter replaces the free space focusing phase (which for Gaussian beams is $tan^{-1}[\lambda(2p+l+1)z/\pi w_0^2]$, where w_0 is the minimum $1/e$ beam radius).

In summary, the wave equation in a plasma waveguide with a parabolic density profile composed entirely of free electrons is characterized by an eigenvalue equation which is independent of wavelength. The number of bound modes supported and their scaled spatial distribution is constant from the wavelength determined by the plasma critical density down to a wavelength determined by relativistic effects. The propagation wavenumber has normal dispersion.

2.2 FREE ELECTRONS PLUS IONS

The effect of ions is easily included if one considers that the propagation of waves through any medium is determined by the complex susceptibility $\chi = \chi' + i\chi''$. We will only concern ourselves with the real part of the susceptibility. For a mixture of ions and free electrons, the susceptibility can be written with the help of the Sellmeir relationship as,

$$\chi' = \sum_i (N_i 4\pi r_0 c^2 \sum_j \frac{f_{ij}}{\omega_{ij}^2 - \omega^2}) - \frac{N_e 4\pi r_0 c^2}{\omega^2} \qquad (3)$$

where N_i is the number density of the ith ionization species (characterized by both a stage of ionization and internal state of the ion), ω_{ij} is the transition frequency from the occupied ion state to an excited state j, f_{ij} is the oscillator strength of the transition and N_e is the free electron density. It is easy to see that we can take our previous results for a free electron waveguide and simply modify them by using an effective electron density,

$$N_e' = N_e[1 - \sum_i (\frac{N_i}{N_e} \sum_j \frac{f_{ij}\omega^2}{\omega_{ij}^2 - \omega^2})] \qquad (4)$$

This effective electron density is used to calculate the "effective plasma frequency", $(\omega_p')^2 = 4\pi r_0 c^2 N_e'$. Because of the ion resonances, the effective electron density and the associated effective plasma frequency are now wavelength dependent. By considering equation (4) it is readily seen that at high frequencies, $\omega >> \omega_{ij}$ for all i,j, the ion electrons are essentially free and contribute to the negative susceptibility. For low frequencies, $\omega << \omega_{ij}$, the tightly bound ion electrons become unresponsive and contribute little to the susceptibility. Thus the effect of the ions is to conteract the normal dispersion. The $1/\omega^2$ decrease in susceptibility at higher frequencies may be compensated by the increase in the number of effective electrons contributing to χ'. This has a major effect on phase matching in plasma fibers. Obviously, the effect of the ions when ω is near resonances is complex and requires detailed knowledge of the atomic physics, but it equally clearly has a major effect on phase matching for such frequencies.

3. PHASE MATCHING

We now want to consider nonlinear mixing processes, $m\omega_1 \pm n\omega_2 = q\omega_3$ where ω_1 is the driving frequency, ω_2 is either a second input frequency, or a spontaneously generated frequency, and ω_3 is the generated short wavelength frequency. For $n=0$ and $q=1$ we have harmonic generation. Other processes correspond to higher order parametric processes of various types.

We can expand the bound portion of the nonlinearly generated field in waveguide eigenmodes[6] according to $\underline{E}(r,t) = \frac{1}{2}[\Sigma_j a_j(z,t)u_j(r_\perp)\exp i(\beta_{j_0}z - \omega t) + c.c.]$, where the mode amplitudes a_j are slowly varying in z and t, and c.c. is the complex conjugate of the previous term. (We ignore the contribution of leaky and free waves.). Inserting this into the wave propagation equation, projecting onto the jth bound channel mode, neglecting group velocity dispersion (GVD), and returning to the time domain gives

$$\frac{\partial}{\partial z}a_j(z,\tau) = \frac{2\pi i \; \omega_0^2}{\beta_{j_0} \; c^2} \; e^{-i \; \beta_{j_0}z} \int d^2r_\perp \underline{P}_{NL} \; u_j^*(r_\perp) \tag{5}$$

for the growth of the amplitude a_j in the nonlinearly generated field of frequency ω. Here $v_{g_0}=(\partial\omega/\partial\beta_j)_0$ is the group velocity of the jth mode at the center frequency $\omega=\omega_0$, $\beta_{j_0}=\beta_j(\omega_0)$, $\tau = t - z/v_{g_0}$ is a time coordinate local to the pulse ($\tau = 0$ corresponds to the pulse peak), $\underline{P}_{NL}$ is the nonlinear polarization (slowly varying in time but not in space), and the integration is over the channel cross section. To evaluate the growth of the mode amplitude for a fixed position on the pulse (constant τ), we put $\partial/\partial\tau = 0$ in Eqn. (5) and integrate with respect to z.

If two interacting *modal* fields $\underline{E}_1$ and $\underline{E}_2$ of frequency ω_1 and ω_2 are quasi-monochromatic and have envelopes varying slowly in time compared to the medium response, the resulting nonlinear polarization at frequency ω can be written

$$\underline{P}_{NL} = P_{NL}(E_1,E_2)e^{i \; \Psi(E_1,E_2,\varphi_1,\varphi_2)} \tag{6}$$

where $P_{NL}(E_1,E_2)$ and Ψ are real; P_{NL}, $E_1(r_\perp,z,\tau_1,\omega_1)$, and $E_2(r_\perp,z,\tau_2,\omega_2)$ are slowly varying amplitudes in time and space, $\tau_{1,2}=t-z/v_g(\omega_{1,2})$ are local time coordinates of the interacting fields (with group velocities $v_g(\omega_{1,2})$), and $\varphi_1(z)$ and $\varphi_2(z)$ are their modal propagation phases. If ω_1 and ω_2 are commensurate, then P_{NL} may also depend on φ_1 and φ_2, but we do not consider that case here. In the low intensity limit, where lowest order perturbation theory applies, $\Psi(E_1,E_2,\varphi_1,\varphi_2)=\psi_0+m\varphi_1\pm n\varphi_2$ and ψ_0 is constant. For sum or difference frequency generation $q\omega = m\omega_1 \pm n\omega_2$ ($m+n$ odd, $\omega_3>0$). At higher intensities beyond the perturbation limit, Shkolnikov et al. have shown [7] that this can be generalized by writing $\Psi(E_1,E_2,\varphi_1,\varphi_2)=\psi(E_1,E_2)+m\varphi_1\pm n\varphi_2$, where $\psi(E_1,E_2)$ allows for the possibility of an intensity dependent phase. An intensity-dependent phase of the field-induced dipole moment has been predicted in calculations [8], with some supporting experimental observations [9]. Equation (5) then becomes, for constant τ,

$$\frac{\partial}{\partial z}a_j(z,\tau) = \frac{2\pi i\ \omega_0^2}{\beta_{j0}\ c^2}\ e^{i\ \Delta kz}\ \int d^2 r_\perp P_{NL}(E_1,E_2)\ u_j^*(r_\perp)\ e^{i\ \psi(E_1,E_2)} \tag{7}$$

where $\Delta k = m\beta_1(\omega_1) \pm n\beta_2(\omega_2) - q\beta_j(\omega_3)$ for the case where the interacting modal fields are channel eigenmodes 1 and 2 with propagation phases $\varphi_1= \beta_1(\omega_1)z$ and $\varphi_2=\beta_2(\omega_2)z$.

The intensity-dependent phase $\psi(E_1,E_2)$ contributes to the modal overlap of the nonlinear polarization through the transverse integration in Eqn. (6). It may also contribute to the phase mismatch if there is a significant difference in the group velocities among the generated and driving waves. For sufficiently small GVD however, $\tau \approx \tau_1 \approx \tau_2$, and $\psi(E_1, E_2)$ does not contribute to the z-integration of Eqn. (7), and that case is assumed here. We therefore identify ΔkL as the phase mismatch in a plasma waveguide of length L, with optimum phase matching for $|\Delta kL| << 2\pi$. Light is generated in any channel mode u_j to which P_{NL} couples; the efficiency of generation depends on Δk and on the overlap of P_{NL} with u_j.

Recalling that $\beta \approx k -(k_p^2/2k) - (k_p/k)\delta(|l| + 2p + 1) = k(1-A^2/2k^2)$ where $k_p^2 \approx 4\pi r_0 N_{eo}$ we can write

$$\Delta k = (\frac{qA^2(\omega_3)}{2k_3} - \frac{mA^2(\omega_1)}{2k_1} \mp \frac{nA^2(\omega_2)}{2k_2}) \tag{8}$$

subject to the boundary condition $mk_1 \pm nk_2 - qk_3 = 0$ and q, m, n integral. It is easy to show that if we consider only the free electron dispersion, with all beams in the same mode simple harmonic generation cannot be phase matched if A^2 is independent of ω. The mixing of spatial modes does allow phase matching under limited onditions, but we find much greater flexibility if we consider the effect of ions and consider more complicated parametric processes which combine absorption of m photons at ω_1 with absorption or emission of n photons at ω_2 to generate q photons at ω_3. These processes include, in the lowest order case, the very useful process of parametric down-conversion where both ω_2 and ω_3 represent lower frequencies than the driving frequency ω_1. The process can be made more reliable if ω_2 is injected along with ω_1. If m > 1 then we can generate short wavelength radiation at ω_3. The existence of high harmonic generation confirms that such high order process can be effective.

3.1 PHASE MATCHING EXAMPLES

We now start with $A^2(\omega) = 4\pi r_o N_e'(\omega) + 2\delta(4\pi r_o N_e'(\omega))^{1/2}(|l|+2p+1)$ with

$$N_e' = N_e[1 - \sum_i (\frac{N_i}{N_e} \sum_j \frac{f_{ij}\omega^2}{\omega_{ij}^2 - \omega^2})] \tag{4}$$

and calculate Δk. We recognize that phase matching can be achieved by varying the ions stage as well as the combination of frequencies ω_1, ω_2, and ω_3. For the purposes of illustration we will assume that the effect of the ions can be included with a single resonance frequency ω_{ij} and an associated oscillator strength f. This should be sufficient when all the frequencies are far from resonance.

In the tables below we show the variation of the coherence length ($1/\Delta k$) as different parameters of the parametric process or the plasma fiber are changed. Note that the coherence length is in millimeters while all other dimensions are in microns, and also note that the sign of the coherence length is not significant experimentally. Table 1 shows the variation with the order of the parametric process. We input ω_1 and ω_2 and generate ω_3 by $q\omega_3 = m\omega_1 - n\omega_2$. In the table we vary n. Because q, m and n are necessarily integers, we cannot fine tune to phase match.

$\lambda 1$ μ	m	$\lambda 2$ μ	n	q	$\lambda 3$ μ	res μ	f	Ni/Ne	$1/\Delta k$ mm
0.4	9	0.80	8	1	0.08	0.089	1	1	0.0099
0.4	9	0.80	6	1	0.067	0.089	1	1	0.02177
0.4	9	0.80	4	1	0.057	0.089	1	1	-0.3302
0.4	9	0.80	2	1	0.05	0.089	1	1	-0.0196

TABLE 1: Effect of varying order of parametric process (m) on coherence length ($1/\Delta k$).

In table 2 we vary the wavelength of the imput driving beam ω_2. It is seen that the coherence length is very sensitive to the wavelength. If a broad band input is used, or if the parametric process is driven by spontaneous process, then it is expected that the output will reflect the position of optimum phase matching and gain.

$\lambda 1$ μ	m	$\lambda 2$ μ	n	q	$\lambda 3$ μ	res	f	Ni/N e	$1/\Delta k$ mm
0.4	9	0.80	4	1	0.057	0.089	1	1	-0.33021
0.39	9	0.80	4	1	0.055	0.089	1	1	-2.22178
0.388	9	0.80	4	1	0.055	0.089	1	1	14.38578
0.387	9	0.80	4	1	0.055	0.089	1	1	3.03226

TABLE 2: Effect of varying input wavelength for parametric process on coherence length $(1/\Delta k)$.

In table 3 we vary the ratio of ions to free electrons while keeping the number of free electrons fixed. This could be accomplished by varying the temperature of the plasma by altering the laser power. Clearly the coherence length is very senssitive to the ion/free electron ratio. This will probably make any parametric process quite sensitive to the conditions in the plasma. If the process is driven by input beams at ω_1 and ω_2, then the efficiency of the parametric process will be very sensitive to plasma conditions. If the wavelengths are allowed to arise from spontaneous processes, the the output wavelength will be very sensitive to plasma conditions.

$\lambda 1$ μ	m	$\lambda 2$ μ	n	q	$\lambda 3$ μ	res	f	Ni/N e	$1/\Delta k$ mm
0.4	9	0.80	4	1	0.057	0.089	1	1	-0.33021
0.4	9	0.80	4	1	0.057	0.089	1	1.2	-0.61866
0.4	9	0.80	4	1	0.057	0.089	1	1.4	-4.88331
0.4	9	0.80	4	1	0.057	0.089	1	1.6	0.828795

TABLE 3: Effect of varying ion/free electron ratio on coherencd length $(1/\Delta k)$.

In table 4 we see that varying the total density of the plasma has roughly a linear effect on the coherence length. This simple density effect will not be useful for achieving phase matching except under exceptional circumstances.

$\lambda 1$ μ	m	$\lambda 2$ μ	n	q	$\lambda 3$ μ	Neo μ^{-3}	$1/\Delta k$ mm
0.4	9	0.80	4	1	0.057	1.0e+07	-0.33021
0.4	9	0.80	4	1	0.057	5.0e+06	-0.61926
0.4	9	0.80	4	1	0.057	1.0e+06	-2.55808

TABLE 4: Effect of varying total density on phase matching length ($1/\Delta k$).

The four tables shown above demonstrate that the phase mismatch associated with parametric upconversion in plasma fibers is a sensitive function of several variables which are amenable to experimental control, in particular the driving wavelengths and the plasma temperature. The exact conditions which lead to long coherence lengths will depend on the atomic properties of the ions involved, but the can be estimated and finally determined by experimental measurement. Naturally the output of such processes will reflect a convolution of the effects of phase matching with the gain derived from the nonlinear polarization of the medium, as well as the incorporation of higher order, intensity dependent effects. A primary concern associated with these processes will be that of robustness. The high sensitivity to ion/free electron ratio indicates that systematic experiments will require a highly reproducible laser driver so that the plasma conditions are as stable as possible. However, it is clear that conditions do exist to achieve long coherence lengths in the newly developed plasma fibers and they should prove to be exciting and fruitful environments in which to study high intensity, non-linear processes.

This work is supported by the NSF (ECS-9224520) and the AFOSR (F49620-92-J-0059).

References

1. C. G. Durfee III and H.M. Milchberg, Phys. Rev. Lett. **71**, 2409 (1993).

2. C.G. Durfee III, J. Lynch, and H.M. Milchberg, Opt. Lett. **19**, 1937 (1994).

3. C.G. Durfee III, J. Lynch, and H.M. Milchberg, Phys. Rev. E **51**, 2368 (1995).

4. *for example*, X.F. Li, A. L'Huillier, M. Feray, L.A. Lompre', and G. Mainfray, Phys. Rev. A **39**, 5751 (1989); J.J. Macklin, J.D. Kmetec, and C.L. Gordon III, Phys. Rev. Lett. **70**, 766 (1993).

5. A.W. Snyder and J.D. Love, *Optical Waveguide Theory* (Chapman and Hall,

London, 1983)

6.   H.M. Milchberg, C.G. Durfee III, and T.J. McIlrath, Phys. Rev. Lett. In press.

7.   P.L. Shkolnikov, A.E. Kaplan and A. Lago, Opt. Comm. **111**, 93 (1994).

8.   P.B. Corkum, Phys. Rev. Lett. **71**, 1994 (1993); M. Lewenstein, Ph. Balcou, M. Yu Ivanov, A. L'Huillier, and P.B. Corkum, Phys. Rev. A **49**, 2117 (1994).

9.   Ph. Balcou, P. Salieres, and A. L'Huillier, in *Super Intense Laser-Atom Physics*, ed. B. Piraux, A. L'Huillier, and K. Rzazewski, NATO ASI Series B, vol. 316, p. 9 (Plenum Press, NY, 1993), D.D. Meyerhofer and J. Peatross, *ibid*, p. 19.

10.   P.L. Shkolnikov, A.E. Kaplan, and A. Lago, Opt. Lett. **18**, 1700 (1993).

OBSERVATION OF RELATIVISTIC AND CHARGE-DISPLACEMENT SELF-CHANNELING USING X-RAY FLUORESCENCE

A.B. BORISOV, B.D. THOMPSON, A. MCPHERSON, K. BOYER
AND C.K. RHODES
Laboratory for Atomic, Molecular and Radiation Physics
Department of Physics
University of Illinois at Chicago
Chicago, IL 60607, USA

Abstract

The propagation of ultrashort, super-intense ($\sim 10^{19}$ W/cm^2) laser pulses through underdense plasmas has been studied through measurements of X-ray fluorescence. By comparing the spatial intensity distribution of the X-ray signals produced in the plasmas with the predicted behavior obtained with a numerical simulation of the propagation, it is concluded that the pulses were undergoing relativistic and charge-displacement-induced self-channeling. The observed channel lengths were ~ 25 times the Rayleigh range of the optical system used for focusing and the peak intensity in the channels was estimated to be $\sim 10^{20}$ W/cm^2.

1. Introduction

There has been considerable recent experimental interest [1-9] concerning the nonlinear propagation of ultrashort, super-intense laser pulses through underdense plasmas. This activity has been facilitated with the increase in the number of laser systems capable of the generation of terawatt laser pulses with subpicosecond widths. Theoretical results [10,11] indicate that plasmas produced by the initial part of such high power pulses can develop properties which will strongly modify the propagation of the high intensity core of the pulse. In the regime reached with subpicosecond 248 nm terawatt pulses focused to intensities above $\sim 10^{19}$ W/cm^2, the electron quiver velocity becomes relativistic and the ponderomotive force is sufficiently strong to expel electrons from the central high-intensity zone of the beam. Meanwhile, the ions are inertially confined in this central core region due to their much higher mass. These effects, relativistic motion and charge-displacement, lead to modes of nonlinear propagation of the radiation which are capable of compressing the propagating power into channels smaller than the unperturbed focal spot size over lengths much longer than the corresponding Rayleigh range. Furthermore, theoretical results indicate that there is a threshold power for this process to occur. This critical power [10,11] for self-

H. G. Muller and M. V. Fedorov (eds.), Super-Intense Laser-Atom Physics IV, 523–533.
© 1996 *Kluwer Academic Publishers.*

channelling is given by $P_{cr} = 1.6 \times 10^{10}$ (n_{cr}/n_e) W where n_e is the free electron density in the plasma and n_{cr} is the critical electron density given by $n_{cr} = m_e\omega^2/(4\pi e^2)$ with m_e, ω, and e denoting the electron mass, the angular frequency of the laser radiation, and the electronic charge, respectively. Applications for these long pathlength, high intensity channels include X-ray lasers [12] and high harmonic production [13].

There are several effects which serve to alter the propagation of an intense pulse as it propagates through a plasma. These effects include (1) the Kerr effect, a modulation of the electron orbitals in atoms and ions by the laser field which creates a nonlinear polarization increasing the index of refraction, (2) a radial variation in the refractive index of the media arising from the nonlinear ionization of the atoms and the radial intensity dependence of the incident radiation, (3) stimulated Raman scattering of the pulse causing a loss of energy and heating of the plasma, and (4) diffraction of the leading edge of the pulse which is locally below the self-focusing power threshold.

In order to perform quantitative experimental studies of the self-channeling process, it is necessary to control the spatial density distribution of the material and use a target with an abrupt interface between the high density zone of the target and the adjacent vacuum. This is required in order to reduce the amount of defocusing and erosion of energy that can occur before the electron density increases sufficiently so that the critical power for self-channeling drops below the incident laser power. Suitably configured pulsed gas-jets provide a convenient method for producing such a target.

2. Experiment

The present paper describes experimental and computational studies of the propagation of high power 248 nm subpicosecond pulses focused to intensities of ~ 8 x 10^{18} W/cm^2 into gas-jet targets. In particular, these studies examined the process of self-channeling in the relativistic and charge-displacement regime. Linearly polarized laser pulses (λ = 248 nm, pulse duration ~ 270 fs, laser energy ~ 180 mJ) from a Ti:Al$_2$O$_3$/KrF* laser system [14] were focused with an f/3 off-axis parabolic mirror to a spot of ~ 3 μm in diameter into gas jets of xenon produced from a pulsed valve. The backing pressures of the gas ranged from 65 psia to 90 psia, conditions suitable for the formation of xenon clusters. In addition, a wall was attached to the nozzle to block expansion of the gas jet in the direction of the incoming laser pulse. The focal plane of the optical system was positioned on a 500 μm diameter aperture placed ~ 0.8 mm below the orifice of the nozzle. The average atom density was estimated to be ~ 10^{19}/cm^3 and the Rayleigh range of the optical system was ~ 28.5 μm.

In this experiment, the X-ray fluorescence was used to study the propagation of the 248 nm pulses through the target. The detector consisted of a pinhole camera with a CCD chip and a 25 μm diameter pinhole. The radiation passing through the pinhole was filtered with a membrane whose composition was 2000 Å Aluminum + 1

μm polycarbonate + 10 μm Beryllium, a filter which effectively blocks all radiation at wavelengths longer than ~ 20 Å. The spatial resolution of the camera, as determined by the pinhole diameter and the pixel size of the CCD chip, was ~ 30 μm. The images obtained with the camera were all made with a single-shot exposure.

Previous experimental research [15-17] on X-ray generation has examined the production of X-rays in multiphoton-excited clusters produced in pulsed gas jets. These studies were performed under conditions similar to those of the present experiment. It was determined [15] from these studies that the threshold intensity for the production of Xe(M) radiation (8 Å < λ < 19 Å) is ~ 3 x 10^{15} W/cm^2.

In order to interpret the X-ray images, we compared them with corresponding results obtained from a numerical model which simulates the propagation of the pulses through the plasma. This model [10,11] is based on the nonlinear Schrödinger equation in the Coulomb gauge and describes the propagation of sufficiently short pulses propagating through cold, underdense, and spatially inhomogeneous media. The model incorporates several processes affecting the propagation including (1) the diffraction due to the finite aperture of the laser pulse, (2) the refraction due to the spatial variations in the electron density and, hence, the index of refraction of the plasma, (3) the electron mass increase due to the relativistic motion, and (4) the charge displacement arising from the ponderomotive force.

3. Results

Figure 1(a) shows an X-ray image of the propagation taken in the transverse direction under conditions for which the nozzle had a stagnation pressure of the plenum of 65 psia and a temperature of 233 K. The focal plane of the optical system was located at z = 940 μm. The electron density was estimated to be ~ 6.5 x 10^{20}/cm^3, a value which corresponds to 1.32 critical powers (P_{laser}/P_{cr} = 1.32). It is evident from the observed morphology that there is considerable spatial variation of the X-ray emission, both in terms of the radial structure, which is seen to fluctuate in size, and longitudinally on the axis, where two regions of strong emission are visible. The observed axial (r = 0) longitudinal variations of the X-ray emission occurring in Fig. 1(a) are shown graphically in Fig. 1(b). The abscissa corresponds to the same scale as that used in Fig. 1(a). The two peaks correspond to the two regions of strong emission and their measured separation is 200 ± 30 μm.

The numerical model was used to simulate the propagation for conditions corresponding to those of the experiment. This result, which is shown in Fig. 1(c), shows the intensity (I) of the pulse as a function of both the radial distance (r) from the optical axis (r = 0) and the longitudinal distance (z) from the focal plane (z = 940 μm). The constants I_o and r_o represent the peak intensity in the focal plane and the radius of the initial focal spot, respectively. The simulation predicts that the high intensity core of the pulse undergoes oscillations between I_o and ~ 12 I_o as it propagates through the plasma. This behavior stems from the competing effects of self-focusing and diffraction which cause the beam to expand and contract as it propagates. The separation between the high intensity zones is 214 μm, a value which falls within the measured range shown in Fig. 1(b).

Previous experimental and theoretical work [1] on self-channeling has demonstrated that the inclusion of the charge-displacement mechanism in the theoretical model is essential in order to obtain results that agree with experiment. Figure 1(d) illustrates the prediction of the simulation for the spatial distribution of the normalized electron density in the plasma for conditions corresponding to those of Fig. 1(c). The unperturbed electron density in the plasma is given by $N_{e,o} = ZN_{ion}$ in which Z is the average level of ionization and N_{ion} is the average atom density. As expected, the electron density is greatly perturbed in the regions of high intensity. In particular, the calculation predicts complete expulsion of the electrons due to the ponderomotive force from the core of the channel, a region in which the intensity significantly exceeds I_o.

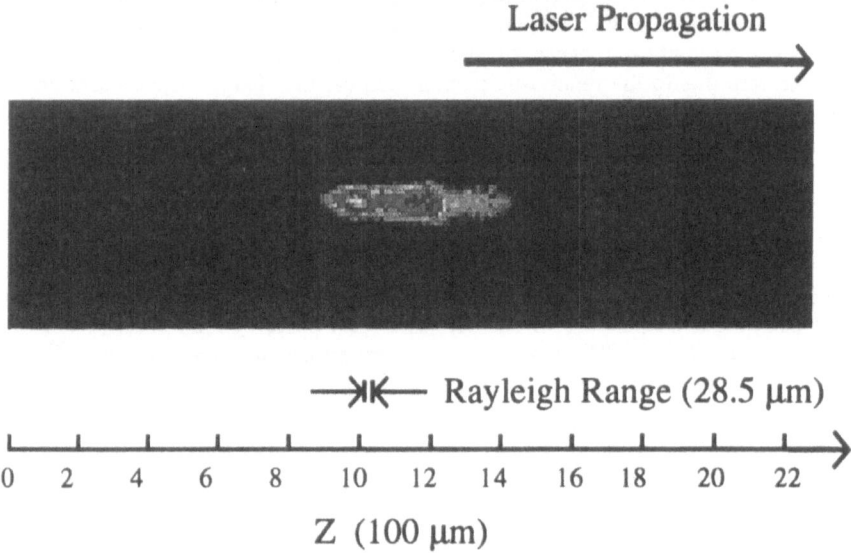

Fig. 1(a): Spatially resolved X-ray fluorescence emission ($\lambda < 20$ Å) from xenon plasma produced by irradiating a xenon gas jet with a 248 nm laser pulse focused to an intensity of $\sim 8 \times 10^{18}$ W/cm^2. The gas pressure and temperature in the nozzle were 65 psia and 233 K, respectively. The focal plane of the laser beam is in the region of z = 940 μm on the scale and the Rayleigh range ($\sim$ 28.5 μm) is shown.

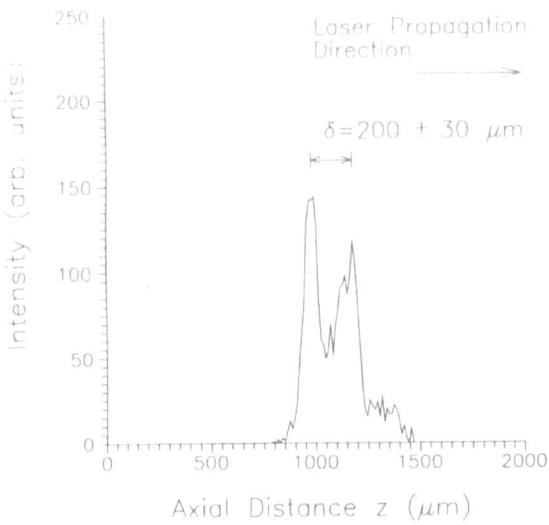

Fig. 1(b): Axial intensity distribution of the X-ray pinhole photograph shown in Fig. 1(a).

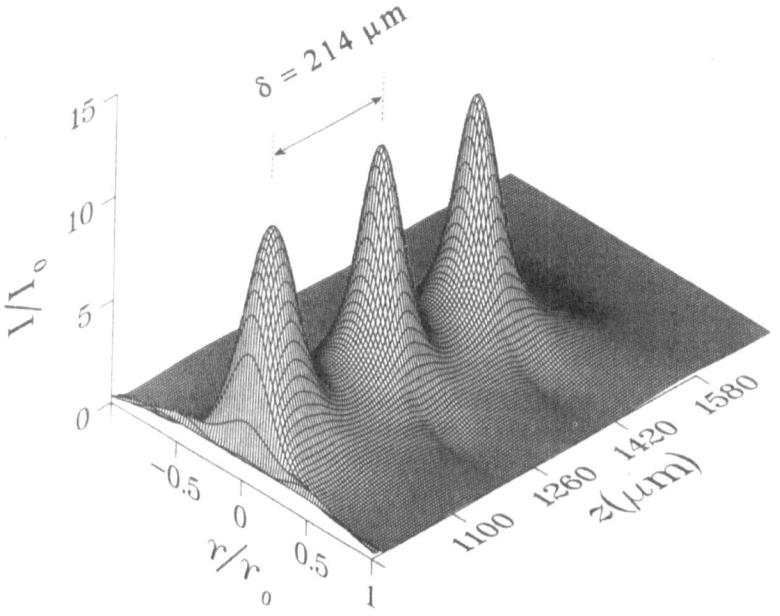

Fig. 1(c): Computer simulation of a high intensity 248 nm laser pulse propagating through an underdense and inhomogeneous plasma corresponding to the same conditions as in Fig. 1(a). The laser pulse power is equal to 1.32 critical powers ($P_{laser}/P_{cr} = 1.32$).

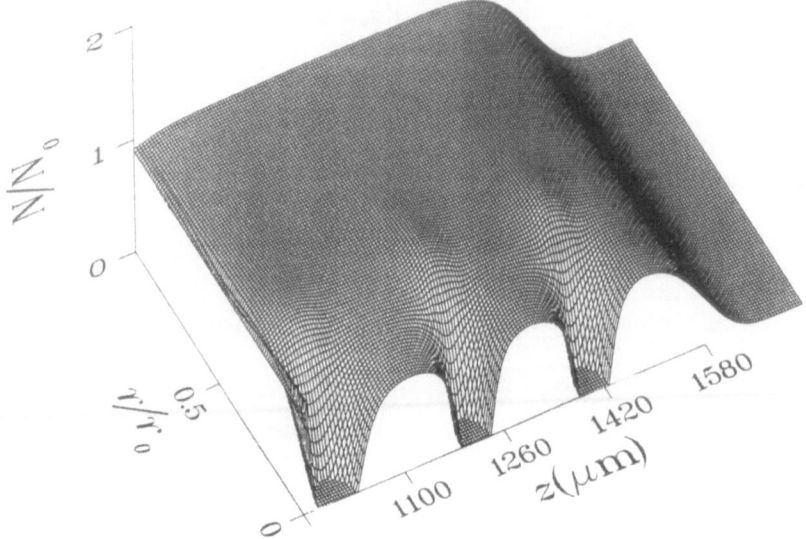

Fig. 1(d): Spatial distribution of the normalized electron density at the temporal peak of the laser pulse as predicted by the numerical model for the same conditions as in Fig. 1(c).

Strong transverse modulation can also be a characteristic of the propagation. In Fig. 1(e) the radius of the image in Fig. 1(a) is shown as a function of the longitudinal position. Since previous measurements have shown that the threshold intensity for X-ray production is $\sim 3 \times 10^{15}$ W/cm^2, this curve also represents the isophote corresponding to this intensity. This result can be compared with the isophote for the same intensity obtained using the simulation. This comparison shows that these two isophotes exhibit the same qualitative behavior. The difference in the absolute values can be attributed in part to the resolution of the camera, which for this measurement would be ~ 30 μm, and in part to an approximation incorporated in the formulation of the model. This approximation, which involves the dynamics of the electron distribution, assumes that the distribution is <u>instantaneously</u> formed as a result of the balance between the outward ponderomotive force and the inward electrostatic attraction to the ions. This approximation results in an underestimation of the electron density on the axis of the channel and, hence, an overestimation of the plasma index of refraction.

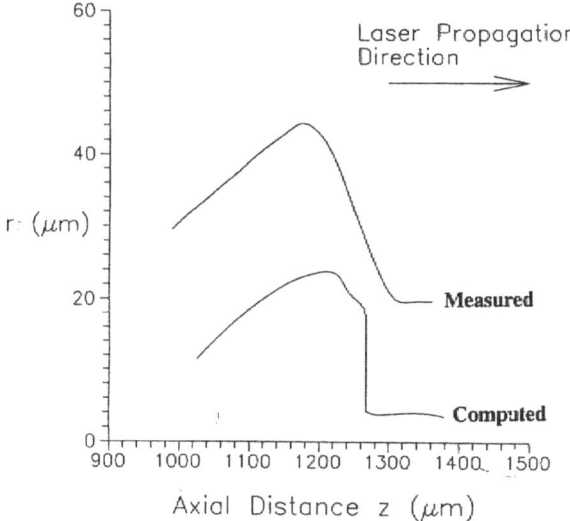

Fig. 1(e): Computed and measured isophotes corresponding to Xe(M) emission. The measured curve was obtained by measuring the radius of the pinhole image in Fig. 1(a). The computed curve utilized the fact that the observed threshold intensity for producing X-rays is 3×10^{15} W/cm^2.

Figure 2(a) shows a CCD image taken for the same experimental conditions as in Fig. 1(a), but with the stagnation pressure of the plenum increased to 90 psia. The focal plane of the optical system in this case is located at $z = 655$ μm. The increase in density of the target represents a 50% increase in the electron density and a corresponding decrease in the critical power. The ratio of laser power to critical power in this case is $P_{laser}/P_{cr} \cong 1.94$. This change in the experimental conditions accounts for the obvious difference in spatial structure of the image. Figure 2(b) shows the axial X-ray intensity distribution for the image shown in Fig. 2(a). Rather than a modulated profile, the distribution is now uniform along the entire length of the X-ray emission. This behavior implies that the pulse can undergo minimal modulation along the longitudinal axis for appropriately selected conditions. With the increase in the ratio P_{laser}/P_{cr}, the simulation now predicts that the core of the pulse will undergo much more rapid and smaller fluctuations and that it quickly stabilizes into a narrow uniform channel having a radius ~ 0.25 r_0 and a peak intensity of ~ 20 I_0, as indicated in Fig. 2(c). Figure 2(d) shows the distribution of the normalized electron density for this case. It shows that the electrons have been completely expelled from the high intensity core of a highly uniform channel. Figure 2(e) shows the corresponding measured and computed isophotes, results which should be compared to the data shown in Fig. 1(e). As in the previous case, the two curves qualitatively reflect the same behavior. The difference in the absolute values can again be attributed to the resolution of the camera and to the approximation used in the simulation for the electron dynamics described above.

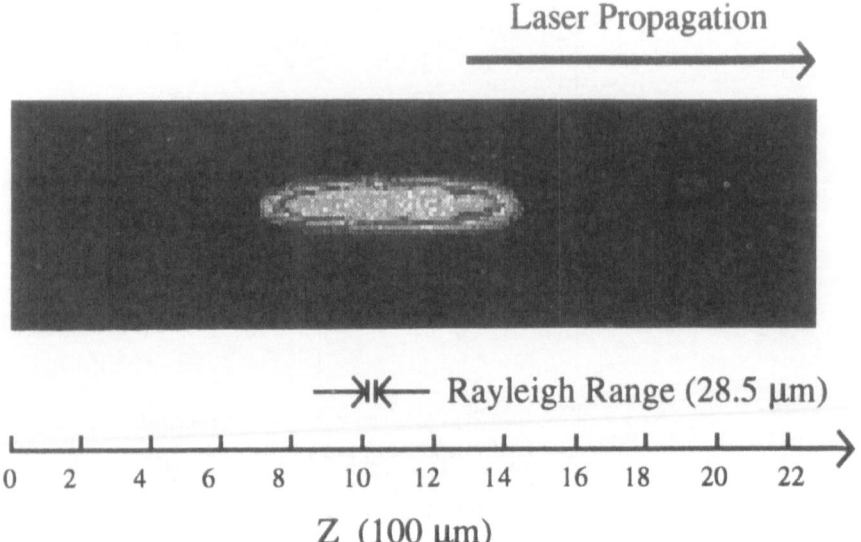

Fig. 2(a): Spatially resolved X-ray fluorescence emission ($\lambda < 20$ Å) from xenon plasma produced by irradiating a xenon gas jet with a 248 nm laser pulse focused to an intensity of $\sim 8 \times 10^{18}$ W/cm^2. The gas pressure and temperature in the nozzle were 90 psia and 233 K, respectively. The focal plane of the laser pulse is in the region of $z = 655$ μm on the scale.

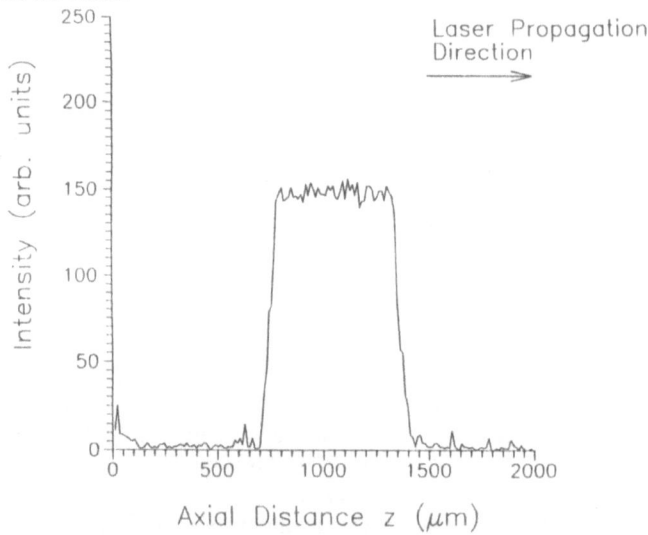

Fig. 2(b): Axial intensity distribution of the X-ray pinhole photograph shown in Fig. 2(a).

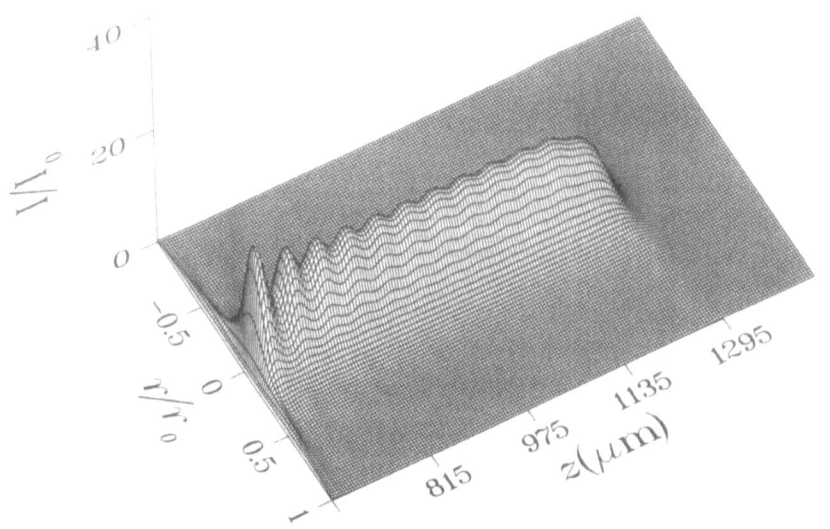

Fig. 2(c): Computer simulation of a high intensity 248 nm laser pulse propagating through an underdense and inhomogeneous plasma corresponding to the same conditions as in Fig. 2(a). The laser pulse power is equal to 1.94 critical powers ($P_{laser}/P_{cr} = 1.94$

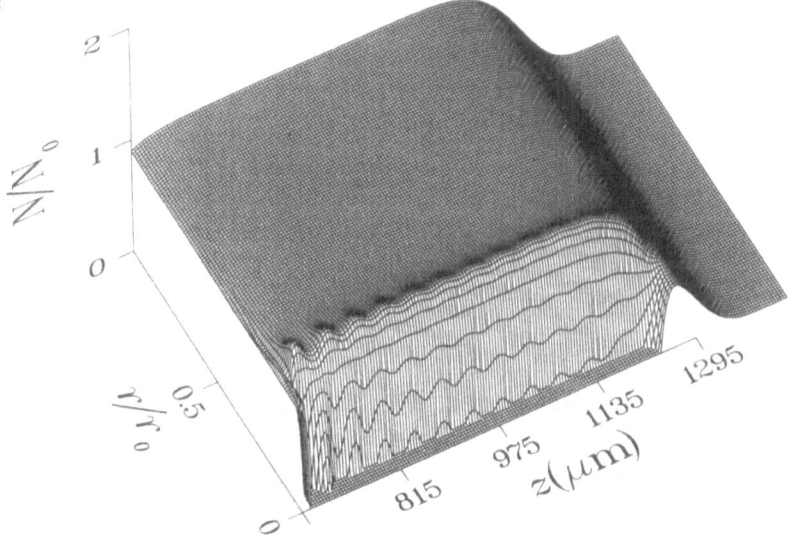

Fig. 2(d): Spatial distribution of the normalized electron density at the temporal peak of the laser pulse as predicted by a numerical model for the same conditions as in Fig. 2(c).

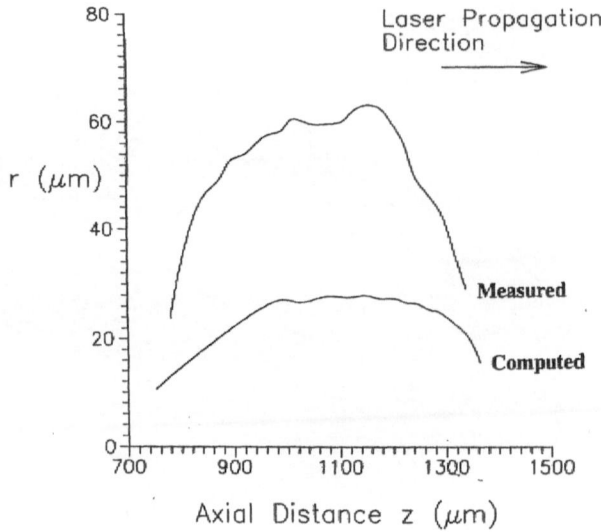

Fig. 2(e): Computed and measured isophotes corresponding to an intensity of Xe(M) emission. The measured curve was obtained by measuring the radius of the pinhole image in Fig. 2(a). The computed curve utilized the fact that the observed threshold intensity for X-ray production is 3×10^{15} W/cm^2.

4. Conclusion

The correspondence of the experimental findings with the theoretical results generated from the simulation leads to the conclusion that the relativistic charge-displacement mechanism can be effective in compressing the propagating power over extended lengths of propagation. In particular, in the second case discussed above, the radiation propagated in a stable channel for over 700 μm ($\sim$ 25 Rayleigh lengths) and reached a peak intensity of $\sim 10^{20}$ W/cm^2.

5. Acknowledgements

The authors respectfully acknowledge the expert technical assistance of J. Wright and P. Noel. Support for this research was provided under contracts with SDI/NRL (N00014-93-K-2004), ARO (DAAH04-94-G-0089) and the University of California/Lawrence Livermore National Laboratory (W-7405-eng-48).

6. References

1. Borisov, A.B., Borovskiy, A.V., Korobkin, V.V., Prokhorov, A.M., Shiryaev, O.B., Shi, X.M., Luk, T.S., McPherson, A., Solem, J.C., Boyer, K. and Rhodes, C.K. (1992) Observation of Relativistic and Charge-Displacement Self-Channeling of Intense Subpicosecond Ultraviolet (248 nm) Radiation in Plasmas, *Phys. Rev. Lett.* **68**, 2309-2312.

2. Borisov, A.B., Shi, X., Karpov, V.B., Korobkin, V.V., Solem, J.C., Shiryaev, O.B., McPherson, A., Boyer, K., and Rhodes, C.K. (1994) Stable Self-Channeling of Intense Ultraviolet Pulses in Underdense Plasma, Producing Channels Exceeding 100 Rayleigh Lengths, *J. Opt. Soc. Am. B* **11**, 1941-1947.

3. Sullivan, A., Hamster, H., Gordon, S.P., Falcone, R.W. and Nathel, H. (1994) Propagation of Intense, Ultrashort Laser Pulses in Plasmas, *Opt. Lett.* **19**, 1544-1546.

4. Borisov, A.B., McPherson, A., Boyer, K., and Rhodes, C.K. (1994) Controlled Power Compression in Materials for X-Ray Amplification, in *X-Ray Laser 1994*, Eder, D.C. and Matthews, D.L., eds., Vol. 332, AIP Conference Proceedings (American Institute of Physics, N.Y., 1994) 134-136.

5. Borisov, A.B., McPherson, A., Thompson, B.D., Boyer, K. and Rhodes, C.K. (1995) Ultrahigh Power Compression for X-Ray Amplification: Multiphoton Cluster Excitation Combined with Non-Linear Channeled Propagation, *J. Phys. B* **28**, 2143-2158.

6. Coverdale, C.A., Darrow, C.B., Decker, C.D., Tzeng, K-C, Marsh, K.A., Clayton, C.E. and Joshi, C. (1995) Propagation of Intense Subpicosecond Laser Pulses through Underdense Plasmas, *Phys. Rev. Lett.* **74**, 4659-4662.

7. Monot, P., Auguste, T., Gibbon, P., Jakober, F., Mainfray, G., Dulieu, A., Louis-Jacquet, M., Malka, G., and Miquel, J.L. (1995) Experimental Demonstration of Relativistic Self-Channeling of a Multiterawatt Laser Pulse in an Underdense Plasma, *Phys. Rev. Lett.* **74**, 2953-2956.

8. Borisov, A.B., McPherson, A., Boyer, K. and Rhodes, C.K. (in press) Z-λ Imaging of Xe(M) and Xe(L) Emissions from Channeled Propagation of Intense Femtosecond 248 nm Pulses in a Xe Cluster Target, *J. Phys B*.

9. Borisov, A.B., McPherson, A., Boyer, K. and Rhodes, C.K. (in press) Intensity Dependence of the Multiphoton-Induced Xe(L) Spectrum Produced by Subpicosecond 248 nm Excitation of Xe Clusters, *J. Phys B*.

10. Borisov, A.B., Borovskiy, A.V., Shiryaev, O.B., Korobkin, V.V., Prokhorov, A.M., Solem, J.C., Luk, T.S., Boyer, K. and Rhodes, C.K. (1992) Relativistic and Charge-Displacement Self-Channeling of Intense Ultrashort Laser Pulses in Plasmas, *Phys. Rev. A* **45**, 5830-5845.

11. Borisov, A.B., Shiryaev, O.B., McPherson, A., Boyer, K. and Rhodes, C.K. (1995) Stability Analysis of Relativistic and Charge-Displacement Self-Channeling of Intense Laser Pulses in Underdense Plasmas, *Plasma Phys. Control. Fusion* **37**, 569-597.

12. Amendt, P., Eder, D.C. and Wilks, S.C. (1991) X-ray Lasing by Optical-Field-Induced Ionization, *Phys. Rev. Lett.* **66**, 2589-2592.

13. Keitel, C.H., Knight, P.L. and Burnett, K. (1993) Relativistic High-Harmonic Generation, *Europhys. Lett.* **24**, 539-544.

14. Bouma, B., Luk, T.S., Boyer, K., and Rhodes, C.K. (1993) High-Brightness Subpicosecond Terawatt KrF* System Driven with a Frequency-Converted Self-Mode-Locked Pulse-Compressed Ti:Al$_2$O$_3$ Laser, *J. Opt. Soc. Am. B* **10**, 1180-1184.

15. McPherson, A., Luk, T.S., Thompson, B.D., Borisov, A.B., Shiryaev, O.B., Chen, X., Boyer, K. and Rhodes, C.K. (1994) Multiphoton Induced X-ray Emission from Kr Clusters on M-shell (~100Å) and L-shell(~6Å) Transitions, *Phys. Rev. Lett.* **72**, 1810-1813.

16. McPherson, A., Thompson, B.D., Borisov, A.B., Boyer, K. and Rhodes, C.K. (1994) Multiphoton-Induced X-Ray Emission at 4 - 5 keV from Xe Atoms with Multiple Core Vacancies, *Nature* **370**, 631-634.

17. Boyer, K., Thompson, B.D., McPherson, A. and Rhodes, C.K. (1994) Evidence for Coherent Electron Motions in Multiphoton X-ray Production from Kr and Xe Clusters, *J. Phys. B* **27**, 4373-4389.

QUASI-STEADY STATES OF IONISED MEDIA IN INTENSE LASER FIELDS

G.FERRANTE[+] and P.I.PORSHNEV[#]

[+]*INFM and Dipartimento di Energetica ed Applicazioni di Fisica, Viale delle Scienze, 90128 Palermo, Italy*
[#]*Physics Department, Bielorussian State University, F.Skorina av.4, 220020, Minsk, Bielorussia*

1. Introduction

Experiments with strong laser fields are presenting, with increasing frequency, physical situations, where a fully or partially ionised medium is created by the action of the radiation field, and the charged particles of such a medium keeps to interact with it. If the density of the ionised medium particles is not small, in the interaction with the radiation field the collective behaviour of the medium particles too is expected to play a role. It broadens significantly the frontiers of strong field laser-matter interactions, which now are requested to include also the interactions of strong lasers with many-particle systems, like plasmas, for instance. Below, for brevity, we will refer to the wealth of processes in which one has interaction of strong lasers with ionised media as to strong laser-plasma interactions.

Actually, within the plasma physics community, the topics of laser-plasma interactions have been of large interest since many years in different contexts [1], mainly within nuclear fusion oriented projects and plasma heating schemes. Theoretically, as a rule, the laser-plasma interaction processes have been investigated: a) within essentially perturbative approaches; or b) numerically for specifically designed cases, aimed at particular applications. The development of very powerful laser systems with unique properties (with extremely short-pulses, for instance), and the numerous strong laser-plasma experiments in several research laboratories are opening a new stage in this area of physics, and call for a renewed effort on the subject, with the aim of removing previous limitations and improving the knowledge of the most general features of strong laser-plasma interactions.

A beneficial aspect of the present-day situation is the awareness that the previously widespread among plasma physicists belief that the several plasma instabilities will make very difficult the interpretation of plasma behaviour has revealed partially ungrounded. Experiments confirm the leading, dominating role that collisions have in controlling the plasma characteristics and in shaping the electron velocity distribution. It, in its turn, emphasises the importance in this new context too of the Boltzmann equation as a theoretical tool, and of its methods of solution. Thus, wishing to address the topics of strong laser-plasma interactions, among the first, preliminary

535

H. G. Muller and M. V. Fedorov (eds.), Super-Intense Laser-Atom Physics IV, 535–546.
© 1996 *Kluwer Academic Publishers.*

536

items to deal with, there is the answer to the question: how the electron velocities are altered in a plasma interacting with a strong laser field? In other words, which kind of electron velocity distribution function (EDF) characterises a plasma interacting with a strong laser field? Implicit in such a question is the expectation that the EDF of a plasma embedded in a laser field will result, in general, different as compared with a maxwellian.

In this paper we investigate theoretically how a given EDF the plasma electrons have when they start to interact with the radiation field, is altered by this interaction, and how these alterations evolve with time. Numerical solutions of the appropriate kinetic equations have shown that the shape of the velocity distribution undergoes fast changes in relatively few radiation cycles starting from the beginning and, practically, looses any relation with the initial form. When the initial time interval of fast changes is elapsed, the electron velocity distribution continues to evolve slowly due to heating, but maintaining approximately the same shape. Thus, the electrons of the ionised medium have entered a quasi-steady state, characterised by some quasi-stationary distribution function. In terms of solutions of the pertinent kinetic equation, it amounts to conclude that possibly some self-similar (SS) solution is established. We investigate this possibility, considering different physical situations. Particular emphasis is placed in the cases when the radiation field is expected to strongly modify the shape of the EDF. To deal with highly anisotropic EDF, we develop a new version of the Legendre polynomial expansion (LPE) for the unknown solution of the kinetic equation. In practice, we apply the LPE only after the kinetic equation has been transformed into an oscillating and appropriately contracted reference system. It amounts to a renormalization of the unknown laser-embedded velocity distribution in such a way to make its bulk isotropic on the average. Quasi-steady state analytical solutions covering a broad range of situations when the laser field forms EDF's strongly departing from the conventional ones will be reported. The relevance of the reported results for applications of atomic and single particle data will be briefly discussed.

All the results reported below will be obtained within the widely used model of a fully ionised, homogeneous collisional plasma. Next, we take a linearly polarised laser field, and consider the physical situation when in this kind of plasma nonlinear inverse bremsstrahlung is the main absorption mechanism. In other words, we do not take into account processes like resonance absorption, for which the laser field inhomogeneity plays a crucial role, nor nonlinear effects like parametric instabilities.

While it is rather evident the need, in general, to understand the laser-plasma interaction regularities for this kind of plasma, to get a closer contact with laser-plasma experiments a model of an inhomogeneous plasma would be better suited. For such a case, the theoretically analysis is more difficult, but not prohibitive. We hope to report soon on such a case as well.

2. The Kinetic Equation

The evolution of the EDF in a uniform, collisional, fully-ionised plasma can be described by the equation [2]

$$\frac{\partial f}{\partial t} + \frac{e\,E_0}{m}\cdot\cos\omega t\,\frac{\partial f}{\partial v} = \frac{1}{2}\frac{\partial}{\partial v_n}\left(v_{ei}(v)\cdot\left(v^2\delta_{nm} - v_n v_m\right)\frac{\partial f}{\partial v_m}\right),\qquad(1)$$

where e and m are the electron charge and mass, v_n is the n-th component of the electron velocity, $v_{ei}(v) = 4\pi e^4 n_e Z\ln\varLambda/(m^2 v^3)$, n_e is the number of electrons per unit volume, Z is the ion charge, $\ln\varLambda$ is the Coulomb logarithm, in the modified form suggested by Silin [3] for interactions between plasmas and fast oscillating fields, and δ_{nm} is the Kroneker delta symbol.

In Eq.(1) the e-e collision term has not been included. In general, for sufficiently intense fields and/or high ionic charge Z, the e-e collision term is negligible. In such a case, the inequality $\left(v_e/v_T\right)^2 \gg 3/Z$ must hold, where $v_e = eE_0/m\omega$ is the peak quiver velocity of the plasma electrons and $v_T = \sqrt{T_e/m}$ is the electron thermal velocity. It is satisfied in the cases considered throughout this work. Eq. (1), based on general kinetic equations of particles ensembles, is derived, e.g., in [2], where the conditions of its validity are discussed in detail. Generally, Eq. (1) is not solved directly. Provided the anisotropy is small, simplified equations are derived by expanding the unknown solution in spherical harmonics. For the solutions of such well-known simplified equations, see, e.g., [2].

To solve Eq. (1), it is useful to change to a velocity reference frame oscillating with the same frequency as the external field [4]. For the electron velocity one has

$$\mathbf{u} = \mathbf{v} - v_e\sin\omega t.$$

In the coordinates $\left(u_\perp, u_z\right)$, with u_z and $u_\perp$, respectively, the velocity component parallel and perpendicular to the direction of the external field polarisation, Eq. (1) has the form

$$u_\perp\frac{\partial\varphi}{\partial t'} = \frac{\partial}{\partial u_\perp}\left[\frac{\delta\,u_\perp u_t}{\left(u_\perp^2 + u_t^2\right)^{3/2}}\left(u_t\frac{\partial\varphi}{\partial u_\perp} - u_\perp\frac{\partial\varphi}{\partial u_z}\right)\right] +$$
$$\frac{\partial}{\partial u_z}\left[\frac{\delta\,u_\perp^2}{\left(u_\perp^2 + u_t^2\right)^{3/2}}\left(u_\perp\frac{\partial\varphi}{\partial u_z} - u_t\frac{\partial\varphi}{\partial u_\perp}\right)\right],\qquad(2)$$

where $\varphi(\mathbf{u}, t') = f(\mathbf{u} + \mathbf{v}\sin t, t)$, $t' = t$. Eq. (2) is dimensionless: the velocity components are in units of v_e, time is in units of ω^{-1}, $u_t = u_z + \sin t'$, and

$$\delta = \frac{v_{ei}(v_e)}{2\omega}.\qquad(3)$$

Explicitly Eq. (2) contains only the small parameter δ, which expresses the strength of the laser-plasma interaction. Another important parameter is the ratio v_e/v_T, stemming from the initial conditions. In [5], performing a two dimensional (2D)

538

calculation, we solved Eq. (2) and investigated numerically the time evolution of the resulting EDF. In the present work, taking advantage of the information of Ref. 5, we aim to develop a systematic procedure able, in principle, to provide accurate analytical EDF in the domain of high anisotropy.

3. Small Anisotropy

To clarify the connection of our approach to the usual procedure, we first briefly review the well-known method of expanding the unknown EDF into Legendre polynomials.

This method is most effective in the case of small anisotropy. We outline the basic steps, writing Eq. (1) in spherical coordinates in a fixed velocity frame:

$$v_x = v \sin\theta \cos\phi; \qquad v_y = v \sin\theta \sin\phi; \qquad v_z = v \cos\theta,$$

$$\frac{\partial f}{\partial t} + \cos t \left(\cos\theta \frac{\partial f}{\partial v} - \frac{\sin\theta}{v} \frac{\partial f}{\partial \theta} \right) = \frac{\delta}{v^3 \sin\theta} \frac{\partial}{\partial \theta} \left(\sin\theta \frac{\partial f}{\partial \theta} \right). \qquad (4)$$

In Eq. (4) dimensionless variables are also used. In a spherical frame, the nature of the collision operator (the right side of the equation) appears in its clearest form. It changes only the anisotropic part of the EDF, which is present because of the external field. When the field vanishes, any isotropic function can be taken as a solution. The maxwellian EDF is formed only because of the e-e collisions.

The LPE for the EDF of Eq. (4) is written as

$$f(v, \theta, t) = f_0(v, t) + f_1(v, t) P_1(\cos\theta) + f_2(v, t) P_2(\cos\theta) + \dots, \qquad (5)$$

where the coefficients f_0, f_1, f_2, ... are unknown. Substituting (5) into (4) and leaving only the first two terms under the assumption of small anisotropy, we have

$$\frac{\partial f_0}{\partial t} + \cos t \left(\frac{1}{3} \frac{\partial f_1}{\partial v} + \frac{2}{3v} f_1 \right) = 0, \qquad (6a)$$

$$\frac{\partial f_1}{\partial t} + \cos t \frac{\partial f_0}{\partial v} + \frac{2\delta}{v^3} f_1 = 0. \qquad (6b)$$

This is the system of two equations for the isotropic and the leading anisotropic part of the EDF, f_0 and f_1. The next step is connected with some basic assumption about the properties of f_0 and f_1. If f_0 is a slowly oscillating function of time (within a field period), while f_1 is a rapidly oscillating one, it is possible to solve (6b) to get

$$f_1 = -\frac{\partial f_0}{\partial v} \left(\frac{\sin t}{1 + \frac{4\delta^2}{v^6}} + \frac{2\delta}{v^3} \frac{\cos t}{1 + \frac{4\delta^2}{v^6}} \right). \qquad (7)$$

Now, substituting (7) into (6a), using the smallness of δ and removing fast oscillations by averaging over the field period, we get the equation for f_0, e.g. [2],

$$\frac{\partial f_0}{\partial t} = \frac{\delta}{3v^3} \frac{\partial^2 f_0}{\partial v^2} - \frac{\delta}{3v^4} \frac{\partial f_0}{\partial v}. \tag{8}$$

The fast convergence of the series (5) is crucial for the derivation of Eq. (8), which implies the inequality $f_0 \gg f_1$. The latter is equivalent to the condition $v_T \gg v_e$. Thus the domain of validity of Eq. (8) is restricted to moderately intense laser fields. The entire procedure is useful insofar as it can be truncated after the first few terms. In the present version, the procedure is of little use when $v_T \approx v_e$ or $v_T < v_e$.

4. Large Anisotropy

Here we show how the LPE can be extended to the large anisotropy domain by exploiting information from existing numerical calculations. The numerical solution of Eq. (1) in Ref. 5 has shown that, after a relatively short time interval of fast changes, the EDF averaged over the external field period acquires some regular shape in the oscillating frame. Specifically, the shape is stretched out along the field polarisation direction and, of course, is far from that of a spherically symmetric function (see, for instance, Fig. 8 of Ref. 5).

An appreciation that a quasistationary, anisotropic EDF has been established is given by Fig. 7 of Ref. 5, which shows the time evolution up to the first 100 field periods of the ratio $E_\perp(t) / E_\parallel(t)$, with $E_\perp$ and $E_\parallel$ being, respectively, the ensemble-averaged perpendicular and parallel kinetic energy. This ratio, which is rigorously equal to two for an undistorted, initial maxwellian EDF, undergoes significant changes during (approximately) the first 20 field cycles to become almost constant afterwards (but numerically smaller than two).

The idea at the basis of the extension of the LPE into the large anisotropy domain is to perform a time-dependent transformation of the parallel (with respect to the field polarisation) velocity scale in the oscillating frame, amounting to a contraction. An appropriate time-dependent transformation coefficient a(t) is introduced. As a result, we work in a scaled oscillating frame, where the originally anisotropic, quasistationary EDF is expected to be squeezed into a distribution function with an isotropic bulk. The small residual anisotropy left after the transformation then may be treated following the standard procedure of LPE.

We point out that : i) the transformation coefficient a(t) needs to be a function of time, because the anisotropy of the EDF, as a rule, changes with absorption of energy from the external field; ii) the deviation of a(t) from unity is a measure of the departure of the EDF from an isotropic shape; iii) as in the usual LPE, to achieve fast convergence it is necessary that the terms accounting for the EDF anisotropy be much smaller than the isotropic term. However, it is not necessary that the electron oscillatory velocity v_e be smaller than the thermal velocity v_T. The release of this constraint for fast convergence of the LPE derives from the circumstance of working in a scaled oscillating frame.

To develop the procedure outlined above, we define a scaled, oscillating frame:

$$s_\perp = u_\perp = s\sin\vartheta, \qquad a(\tau)\cdot s_z = u_z = a(t)\cdot s\cdot\cos\vartheta, \qquad \tau = t, \qquad (9)$$

where $a(\tau)$ is an unknown time-dependent factor altering the length of u_z. Thus s_z is the contracted parallel velocity component, while u_z is the uncontracted counterpart. Considering the relations (9), or the similar relations between scaled moving and fixed coordinates

$$s_\perp = \tilde{v}\sin\theta = s\sin\vartheta, \qquad a(\tau)\cdot s_z = \tilde{v}\cos\theta - \sin\tau = a(t)\cdot s\cdot\cos\vartheta,$$

$$\tilde{v} = v/v_e, \qquad (10)$$

Eq. (2), or Eq. (1), is transformed into

$$s\cdot\sin\vartheta\,\frac{\partial F}{\partial\tau} - \left(\frac{s^2\cos^2\vartheta\,\sin\vartheta}{a}\frac{da}{dt}\right)\frac{\partial F}{\partial s} + \left(\frac{s\cdot\sin^2\vartheta\,\cos\vartheta}{a}\frac{da}{dt}\right)\frac{\partial F}{\partial\vartheta} = I_{coll}(F), \qquad (11)$$

where $F(s, t) = \varphi(u, t')$ and $I_{coll}(F)$ is the collision integral. In analogy with Eq. (5), we write the EDF as a series of Legendre polynomials

$$F(s, \vartheta, \tau) = F_0(s, \tau) + F_1(s, \tau)\,P_1(\cos\vartheta) + F_2(s, \tau)\,P_2(\cos\vartheta) + \ldots \qquad (12)$$

Substituting (12) into (11) and integrating over the angle ϑ, we get the equation for the isotropic part of the EDF in the scaled oscillating frame

$$2s\frac{\partial F_0}{\partial\tau} - \frac{2}{3}\frac{s^2}{a}\frac{da}{dt}\frac{\partial F_0}{\partial s} = I_{coll}^0(F_0), \qquad (13)$$

where in the collision integral $I_{coll}^0(F_0)$ has been left the dependence on F_0 only. The explicit expression of $I_{coll}^0(F_0)$ is found in Ref. 6.

It can be demonstrated that Eq. (13) averaged over the field period turns into the well-known Eq. (8) in the limit of small anisotropy. With the restriction $a\cdot s > 1$ (the EDF is elongated in electric field direction), with the same averaging over the field period from Eq. (13) we obtain the equation for the isotropic part F_0 [6]:

$$s^2\frac{\partial F_0}{\partial\tau} = \frac{s^3}{3a}\frac{da}{dt}\frac{\partial F_0}{\partial s} + \delta\frac{\partial}{\partial s}\left(\frac{c_1}{s}\frac{\partial F_0}{\partial s} + c_2 s\frac{\partial F_0}{\partial s}\right). \qquad (14)$$

The coefficients c_1 and c_2 appearing in Eq. (14) as well as all the other coefficients c_i appearing below are reported in full in [6]. They are functions of $a(\tau)$. Under the condition $a \to 1$, Eq. (14) goes over to Eq. (8). Thus, Eq. (14) is the "final" equation we need to solve for the isotropic part of the EDF in the scaled, oscillating frame for arbitrary values of $a(\tau)$. The EDF must satisfy the normalisation condition

$$4\pi\, a(\tau)\int_0^\infty F(s)\, s^2\, ds = 1. \qquad (15)$$

The equations for the high order coefficients of the LPE F_1 and F_2 necessary to estimate the convergence of the expansion procedure are reported in [6]. If the conditions of fast convergence are met, the sole Eq. (14) will be sufficient to describe the plasma characteristics.

Under the assumption that values of $a(\tau)$ exist such that the goal of making the EDF $F(s, \tau)$ isotropic on the average is achieved, we obtain the equation giving the time evolution of $a(\tau)$ in the form

$$\frac{\langle s^2 \rangle}{4\pi} \frac{da}{dt} = \delta \, c_9(a) F_0(0) - \delta \, c_{10}(a) \int_0^\infty s F_0(s, \tau) \, ds, \qquad (16)$$

where

$$\langle s^2 \rangle = \frac{\int s^2 F(s, \tau) d^3 s}{\int F(s, \tau) d^3 s},$$

and small terms containing F_1 and F_2 are consistently omitted. Eq. (14) for F_0, and Eq. (16) for $a(\tau)$ form the system of two coupled equations, which substitutes the kinetic equation, Eq. (1), with the restriction that the averaging over the field period has been performed and $a(\tau) \geq 1$. Of course, the proposed procedure can be considered complete only when the calculated F_0 and $a(\tau)$ give $F_1 \ll F_0$. When this inequality is satisfied, all the LPE coefficients are small compared to the first one, F_0.

Using again the assumption that $a(\tau)$ makes $F(s, \tau)$ isotropic on the average, the following results is also obtained

$$a^2(\tau) = \frac{2\langle u_z^2 \rangle}{\langle u_\perp^2 \rangle}, \qquad (17)$$

where

$$\langle u_x^2 \rangle = \frac{\int u_x^2 \varphi(\mathbf{u}, t) \, d^3 u}{\int \varphi(\mathbf{u}, t) \, d^3 u}.$$

Thus, the squared scaling function $a(\tau)$ is twice the ratio of the average parallel to the average perpendicular kinetic energies (evaluated in the unscaled, oscillating frame). To some extent, Eq. (17) makes unnecessary the self-consistent solution of Eqs. (14) and (16). As done in Ref. 5 (see, especially, Fig. 7), we first find numerically $a(\tau)$ in the unscaled, oscillating frame and then concentrate on the approximate solution of Eq. (14) in an effort to find F_0 analytically for the calculated values of $a(\tau)$. Examples of this procedure are worked out in the next section for such time intervals that the establishment of the SS solutions is a possible outcome of the EDF evolution. We conclude by observing, that the collision integral has been approximately evaluated under the assumption that $a(\tau) \geq 1$. The latter assumption expresses the expectation (corroborated by numerical calculations) that a linearly polarised laser electric field along the u_z- direction will yield on the average a flattened at the poles (oblate) distribution. Thus, the Eqs.(14) and (16), in their present form, are not suited to treat the case when $a(\tau) < 1$, i.e. if the field action produces an elongated towards the poles (prolate) distribution.

5. Self Similar Model Solutions

In this Section we concentrate on SS solutions, i.e. on solutions corresponding to a relatively late stage, when the rapid variations in EDF due laser-plasma interactions are over, and the further time evolution due to heating takes place with an EDF keeping its shape largely unchanged. In the case of small anisotropy, the solution is known (see, e.g. [7-9], and [10] for the case when electron-electron collisions are included). We consider this case, too, because our equations, for a small but finite degree of anisotropy, allow the possibility of finding corrections to and improving upon the known result. Besides, the small anisotropy case serves as a check of our entire procedure. For the second case (large anisotropy), to the best our knowledge, no accurate solution is known.

5.1. SMALL BUT FINITE ANISOTROPY ($a \gtrsim 1.$)

Under this condition, we can simplify Eqs. (14) and (16) to obtain

$$s^2 \frac{\partial G_0}{\partial \tau} = \frac{\partial}{\partial s}\left(\frac{da}{dt}\frac{s^3}{3a}G_0 + \delta\frac{c_1}{s}\frac{\partial G_0}{\partial s}\right), \tag{18}$$

$$\frac{da}{dt}\int_0^\infty s^4 G_0 ds \approx \delta\, c_9(a)G_0(0), \tag{19}$$

where $G_0(s,\tau) = a(\tau)\,F_0(s,\tau)$. With the transformation $T = \int \delta\, c_1 d\tau$ and the SS transformations

$$s' = s / W(T) \qquad\qquad G' = G_0\, W^3, \tag{20}$$

the heating equation and the SS EDF are found to be

$$W^5 = W_i^5 + (25 - \frac{5K_9}{3c_1})T \tag{21a}$$

$$G'(s') = G_i \exp(-s'^5), \tag{21b}$$

where W_i is an initial value and $K_9 = c_9(1)G'(0)/\int_0^\infty s'^4 G' ds'$. The normalisation condition gives $G_i = 0.2672$. We observe that the usual SS solution of Eq. (8) [7] for $W(\tau)$ does not contain the last term in the brackets of Eq.(21a). Accordingly (21a) shows that heating proceeds slower than in Ref. 7.

5.2. LARGE ANISOTROPY (a >> 1)

This regime is described by the equations

$$s^2 \frac{\partial G_0}{\partial \tau} = \frac{\partial}{\partial s}\left(\frac{da}{dt}\frac{s^3}{3a}G_0 + \delta\, c_2 s\frac{\partial G_0}{\partial s}\right), \tag{22}$$

$$\frac{da}{dt}\int_0^\infty s^4 G_0\; ds \approx -\delta\, c_{10}(a)\int_0^\infty sG_0 ds. \tag{23}$$

Making the SS transformation of Eq. (20), proceeding as above, the SS solution to Eq.(22) is found as

$$G'(s') = G_i \exp(-s'^3), \tag{24}$$

and solution to the heating equation as

$$W(\tau) = \left(W_i^3 + \frac{135}{52}\delta\tau\right)^{1/3}, \tag{25}$$

6. Calculations and Concluding Remarks

In Fig. 1 a) and b) we report some calculations showing EDF's evaluated under different physical conditions and approximations. The relevant parameters are: a) $R = v_e / v_T^0 = 1$; $n_e = 8 \cdot 10^{20}$ cm^{-3} ; electric field strength $E_0 = 1.45 \cdot 10^9$ V/cm ; photon energy $\hbar\omega = 1.17$ ev; unbalanced ion charge $Z = 10$; $\delta = 6.4 \cdot 10^{-3}$ and ratio of parallel to perpendicular temperature $A = T_Z^0 / T_\perp^0 = 1$, b) $R = 3.0$; $n_e = 5 \cdot 10^{19}$ cm^{-3} ; $E_0 = 9.4 \cdot 10^8$ V/cm ; $Z = 2$; $\delta = 2 \cdot 10^{-4}$; $A = 20$. The basic remark, having far reaching consequences for applications of laser-plasma interactions, is that a laser-embedded EDF as a rule, exhibit two distinctive features: i) as compared to a maxwellian, a considerable smaller number of slow electrons are present. This feature has been observed also experimentally [11, 12], and has consequences on plasma lasing schemes based on recombination, which is effective when large numbers of slow electrons are available. ii) Fast electrons are slightly affected by the radiation field. Thus, if one has an initially maxwellian EDF, its tail remains essentially unchanged. It is essentially the bulk of plasma electrons velocities which undergoes a reshaping. Accordingly, when we need to evaluate plasma properties or processes depending on the bulk of plasma electron velocities, significant changes are to be expected. On the contrary, properties depending essentially on fast electrons are likely to remain unchanged.

In Fig. 2 a) and b) we report some trial calculations showing the parallel time evolution of two different initial EDF's: i) a multipeaked EDF, mimicking some initial velocity distribution resulting in a multichannel multiphoton ionisation process; and ii) a maxwellian EDF with the same mean initial energy. The relevant parameters are:

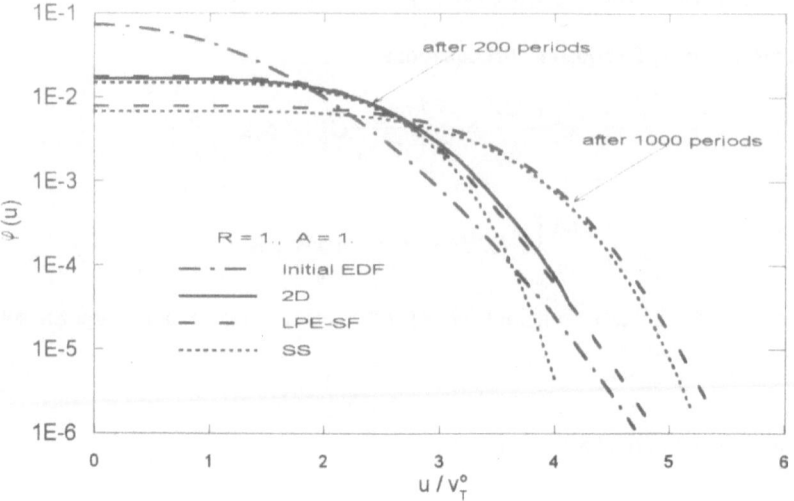

Figure 1a. Time evolution of an initial maxwellian. The EDF are given in the oscillating coordinate system. The velocity is measured in units of the initial velocity v_T^0. Dot-dashed lines: initial maxwellian EDF; continuous line: direct 2D calculations; dashed lines: LPE calculations in the scaled frame (SF); dotted lines: small anisotropy SS EDF.

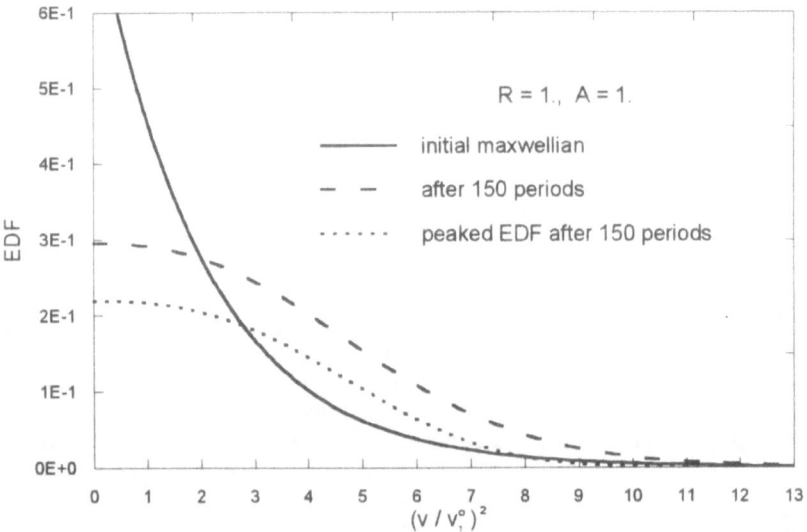

Figure 1b. One-dimensional EDF after 200 field periods in the large anisotropy regime. Continuous line: 2D calculations; dotted line: initial bi-maxwellian; dashed line: LPE-SF calculations; open circles: SS EDF.

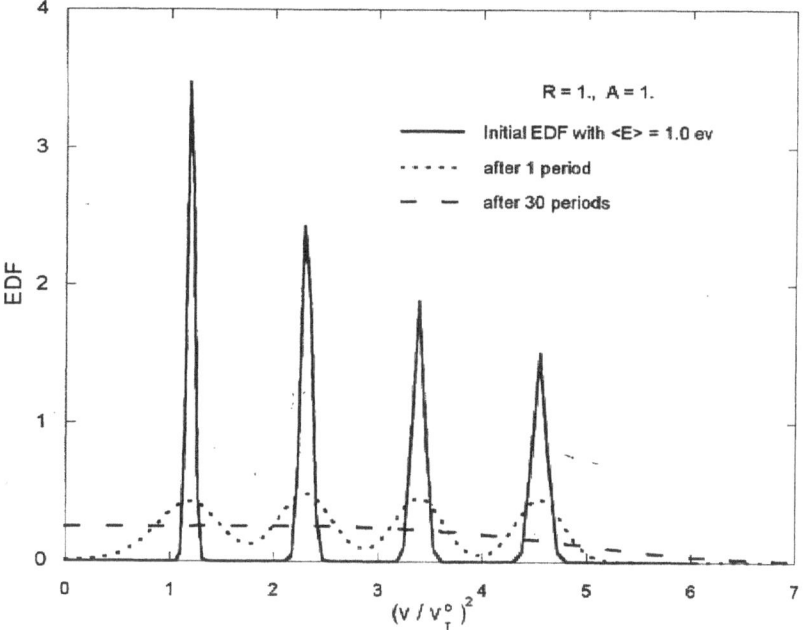

Figure 2a. Time evolution of an initial peaked EDF (continuos line). Dashed line: calculation after 30 field periods; dotted line: calculations after 1 period.

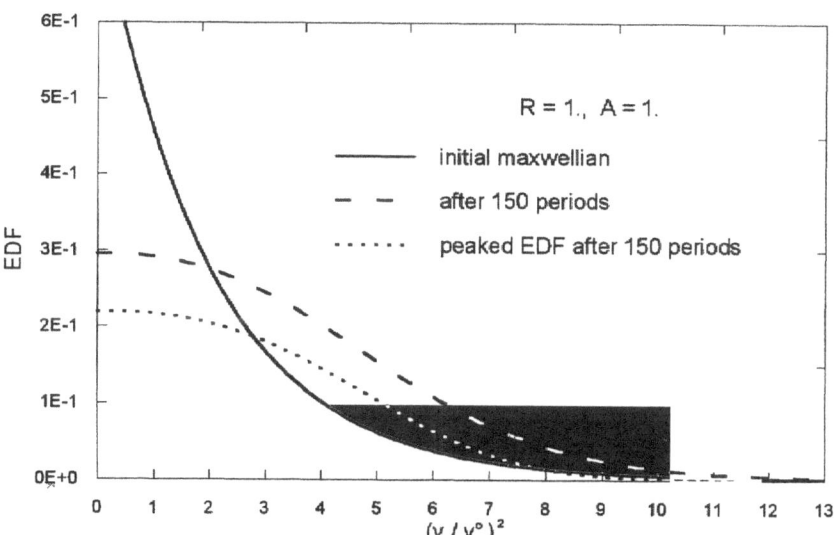

Figure 2b. One-dimensional EDF's after 150 field periods in the small anisotropy regime. Continuos line: initial maxwellian; dashed line: maxwellian; dotted line: peaked EDF.

R=2.3; $n_e = 10^{17}$ cm^{-3} ; $E_0 = 10^8$ V/cm ; photon energy $\hbar\omega = 1.17$ ev; unbalanced ion charge Z = 1; $\delta = 10^{-4}$. Note that after 150 field periods the two distributions become similar, with the maxwellian considerably heated and the multipeaked one still concentrated on a smaller portion of the velocity space.

References

1. Kruer , W.L. (1988) The physics of Laser Plasma Interactions, Addison-Wesley Publishing.
2. Shkarofsky, I.P., Johnston, T.W., and Bachynski, M.P. (1966) The Particle Kinetics of Plasma (Addison-Wesley, Reading, MA.
3. Silin, V.P. (1960) Sov.Phys.JETP 11, 1277.
4. Chichkov, B.N., Shumsky, S.A. and Uryupin, S.A. (1992) Phys.Rev.A 45, 7475.
5. Porshnev, P.I., Bivona, S. and Ferrante, G. (1994) Phys.Rev.E 50, 3943.
6. Porshnev, P.I., Khanevich, E.I., Bivona, S. and Ferrante, G. (1995) Phys.Rev.E 52. No.6 (Dec.).
7. Langdon, A.B. (1980) Phys.Rev.Lett. 44, 575.
8. Balescu, R. (1982) J.Plasma Phys. 27, 553.
9. Jones, R.D. and Lee, K. (1982) Phys.Fluids 25, 2307.
10. Porshnev, P.I., Ferrante, G. and Zarcone, M. (1993) Phys.Rev.E 48, 2081.
11. Donnelly, T.D., Lee, R.W. and Falcone, R.W. (1995) Phys. Rev.A 51, R2691.
12. Glover, T.E., Crane, J.K., Perry, M.D., Lee, R.W. and Falcone, R.W. (1995) Phys.Rev.Lett. 75, 445.

COHERENT CONTROL OF ABOVE-THRESHOLD IONIZATION

P.H. BUCKSBAUM AND D.W. SCHUMACHER[1]
Physics Department, University of Michigan
Ann Arbor, MI 48109-1120

AND

C.W.S. CONOVER[2]
Center for Ultrafast Optical Science, University of Michigan
Ann Arbor, MI 48109-2099

Abstract. Above-threshold ionization (ATI) of atoms occurs when laser fields are comparable to the atomic binding fields. This suggests that coherent control of the optical field should produce dramatic effects in the ATI spectrum, and even lead to control of atomic ionization. Coherent control for this problem can occur in two different ways: First, the the electric field during each optical cycle can be sculpted to control the ionization rate and ATI spectrum; alternatively, the intensity envelope may be adjusted to control ponderomotive shifts, multiphoton resonances, and ultimately, the ATI rate.

1. Introduction

Above-threshold ionization (ATI) of atoms by intense lasers occurs for laser fields comparable to the Coulomb fields experienced by valence electrons, 1-10 v/Å. For visible or near infrared lasers, a field this strong produces an electronic ponderomotive energy U_P, or "quiver energy" equal to or larger than the photon energy $h\nu$.

Since the laser field is a sizable fraction of the Coulomb field, its direct manipulation might be a good way to control the ionization spectrum. In order to explore this possibility the laser field must depart significantly

---

[1]Present Address: Physics Department, University of Virginia, Charlottesville, Virginia
[2]Permanent Address: Physics Department, Colby College, Waterville, Maine

H. G. Muller and M. V. Fedorov (eds.), Super-Intense Laser-Atom Physics IV, 547–556.

from a pure sine wave. This can be accomplished by mixing together laser beams with very different colors, maintaining their phase coherence. Such coherent control of the carrier (i.e. *carrier control*) should be particularly effective if the peak laser field F is high enough for an electron to tunnel out of the atom during one optical cycle.

Instead of directly altering each optical cycle, one might control ATI by manipulating the ponderomotive potential U_P during the pulse, to shift excited states of the atom into multiphoton resonance with the initial state. U_P depends on the laser field and frequency:

$$U_P = \frac{e^2 F^2}{4m_e \omega^2},$$ (1)

where F is the peak electric field over a cycle, and ω is the angular frequency of the laser. Since this quantity is linearly proportional to intensity, amplitude modulation of a nearly-monochromatic laser pulse may be useful for control. Ponderomotively shifted bound-state resonance are known to dominate the ionization probability in ATI experiments where the ponderomotive shift is approximately equal to $h\nu$ [1]. These resonances appear as avoided crossings in a dressed-state energy diagram for the atom. Several recent experiments have shown how photoionization and photoexcitation can be manipulated by *controlling the intensity envelope* to adjust which states come into resonance, and to control the duration of the resonances [2] [3] [4].

Whether envelope control or carrier control is more useful depends on the details of the laser focus, pulse duration, and pulse spectrum. Envelope control demands a well-defined ionization rate vs. cycle-averaged intensity, whereas carrier control dominates if the ionization probability during a single optical cycle is very large. The "Keldysh parameter" γ is the time it takes an electron to tunnel through the atom-laser potential barrier in units of the optical period [5]:

$$\gamma = \sqrt{I_p/2U_P} = \sqrt{2I_p}\omega/F$$ (2)

Here I_p is the ionization potential. Carrier control is more important for $\gamma < 1$, and envelope control if $\gamma > 1$. ATI occurs in the transition regime.

We shall review two ATI experiments that test each of these control strategies in the intensity regime $\gamma \approx 1$. The first experiment tests carrier control of ATI [6][7]. It uses very long (100 psec) pulses, so that the amplitude may be assumed nearly constant as the shape of the individual field cycle is varied. The second experiment tests envelope control using a pair of 100 fsec pulses, where the intensity modulation is large but the frequency is nearly constant. [8]

2. Carrier Control

Our carrier control experiment employed a two-color laser pulse in which one of the colors was the second harmonic of the other, so there was a well-defined controllable phase between the colors. In weak fields this technique has been exploited to study a different kind of coherent control suggested by Brumer and Shapiro, where two distinct pathways between the initial and final state interfere according to the adjustable phase [9][10][11]. In the Brumer-Shapiro form of coherent control, the *transition amplitudes* of the two interfering channels must be carefully equalized to maximize the control. This usually means that the laser intensities of the two colors are very different, by as much as a factor of a million in the case where the second color is the third harmonic of the first. In carrier control of ATI, the intensity is high enough that the ionization per cycle is significant, and phase control takes a different form: We must adjust the *laser intensities* to be equal, so that we can maximize control over the shape of the superposed fields.

2.1. TWO-STEP MODEL

Numerical integrations of Schrodinger's equation for a 2-color carier-control experiment in hydrogen have been carried out by Schafer and Kulander [12]. Most of their results are in good qualitative agreement with those of a very simple semiclassical theory, which we describe here. [13] The 2-color electric field of the laser is

$$F(t) = F_1 \cos \omega t + F_2 \cos 2\omega t + \phi. \tag{3}$$

If the optical field frequency ω is slow enough compared to the ionization rate Γ, then ionization can be treated as a quantum tunneling process, where the electron tunnels through the saddle-point in the combined coulomb-laser quasi-electrostatic potential. This rate is exponential in field amplitude [14]

$$\Gamma(t) = 4I_p^{5/2} \frac{1}{F(t)} exp \left[\frac{-2}{3} I_p^{3/2} \frac{1}{F(t)} \right] \tag{4}$$

Following ionization, the electron moves in the field as a free classical particle starting at rest. (The ion field can be neglected here.) The cycle-averaged kinetic energy of the electron, then, is given by:

$$\begin{aligned} E_k = \quad & U_{p1} + 2U_{p1} \sin^2 \omega t_0 + U_{p2} + 2U_{p2} \sin^2(2\omega t_0 + \phi) \\ & + 4(U_{p1} U_{p2})^{1/2} \sin \omega t_0 \sin(2\omega t_0 + \phi) \end{aligned} \tag{5}$$

where U_{pi} is the ponderomotive potential for field i, and e and m are the electron charge and mass. The first two terms represent the cycle-averaged wiggle energy of the electron, and the remaining terms describe the net drift energy just after ionization. Since the drift energy is phase-dependent, the energy spectrum is as well.

Since the laser pulse duration is long compared to the electron travel time out of the focus, the wiggle energy component is converted to drift energy, and the sum is seen at the detector. Not only the energy, but also the phase-dependent emission direction can be measured by placing the detector so that it only sees emission current in one direction. The minimum energy electron, one that only wiggles and does not drift in the laser field, leaves the focus due to the ponderomotive force and is detected with ponderomotive energy $U_{p1} + U_{p2}$. Conversely, the maximum energy possible in this model, for a monochromatic laser field, is is $3U_P$. With two colors, the maximum depends on the relative phase.

2.2. APPARATUS

Two-color phase-coherent laser pulses are derived from 100 ps, 70 mJ 1.06 μm laser pulses, by passing them through a half-wave plate, a KD^*P doubling crystal cut for type II phase matching, and a dielectric polarizer. The field amplitude of the two colors is set equal by means of an attenuator and polarization rotator (half-wave plate) positioned before the doubling crystal. The relative phase ϕ is controlled using phase-velocity dispersion in a 1 cm thick fused silica plate. We can select any ϕ by tilting the angle of incidence.

The light is focused into a field-free region of a vacuum chamber chamber (base pressure $2x10^{-9}$ torr), which is backfilled with up to 10^{-4} torr of krypton or xenon gas. The kinetic energy of ATI photoelectrons is measured by the time-of-flight of electrons striking a multichannel plate subtending a small solid angle along the laser polarization.

The phase ϕ between the two fields in the interaction region is measured with a type-I phase-matched KD^*P crystal $\approx 15cm$ beyond the focus. The two colors combine in the crystal to produce difference- and sum-frequency radiation at ω and 2ω which interfere with the incident light depending on: the phase ϕ, phase shift of $\frac{\pi}{2}$ that develops between the fields as they leave the focus, and the phase shift between the 1.06 μm light and the newly generated second harmonic in the second crystal. The amount of 532nm light that exits the chamber then tells us the value of ϕ up to a sign ambiguity. This final uncertainty is removed by detecting optical rectification in the same crystal. We have adopted a phase convention where $\phi = 0$ means that there is maximum constructive interference of the two

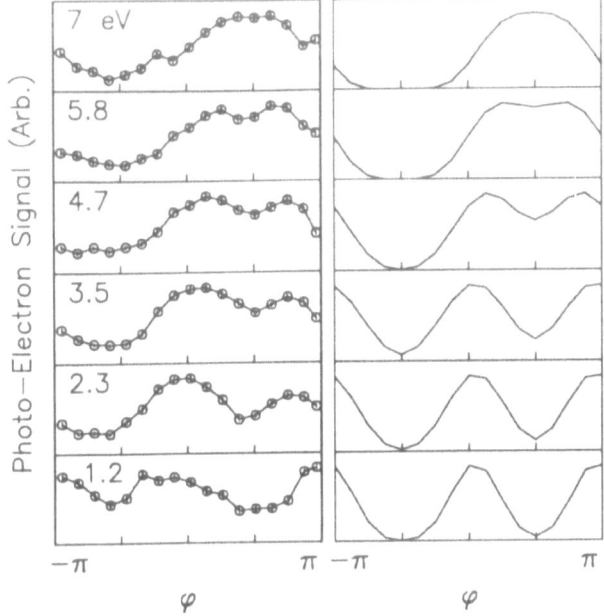

Figure 1. Left: Phase dependence of ATI in Kr, for $I_1 = I_2 = 2x10^{13}W/cm^2$, for the first through sixth $^2P_{3/2}$ ATI peaks (bottom to top). Right: Model using $I_1 = I_2 = 4x10^{13}W/cm^2$.

fields, in a direction away from the detector.

2.3. DATA AND ANALYSIS

The tunneling character of ATI becomes evident in the phase-dependence. Figure 1 shows the relative ϕ-dependence for the first through sixth $^2P_{3/2}$ peak in Kr, and calculations using the semiclassical model. Most electrons arrive when $\phi = 0$ and $\phi = \pi$, where constructive interference makes the largest total field. On the other hand, electrons produced at this point in the cycle have zero net drift, so they contribute only to the low energy part of the spectrum. The phases that produce the largest rates do not produce the highest energy electrons.

Although the detector only measures forward rates, we determine how many electrons go backward from symmetry: The electric field is unchanged under $\phi \rightarrow \phi + \pi$, $z \rightarrow -z$, and $\omega t \rightarrow \omega t + \pi$. The forward/backward (towards/away from the detector) symmetry of the electron current is therefore

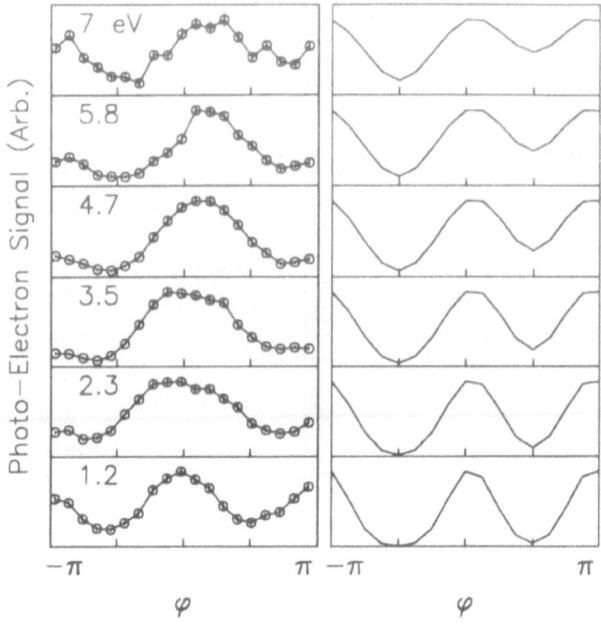

Figure 2. Left: Phase dependence of ATI in Kr, for $I_1 = I_2 = 4x10^{13}W/cm^2$, for the first through sixth $^2P_{3/2}$ ATI peaks (bottom to top). Right: Model using $I_1 = I_2 = 8x10^{13}W/cm^2$.

the balance between signal at ϕ and $\phi + \pi$. Low energy electrons have a fairly symmetric distribution, but higher energy electrons show a strong forward/backward asymmetry: the peak signal appears at two places π radians in phase apart for the low energy electrons, but the two peaks move closer in phase for the higher energy electrons.

If the intensity is doubled, to $8x10^{13}W/cm^2$, the forward/backward asymmetry diminishes (figure 2), but a new feature appears: There are still two peaks in the signal vs. phase for the high energy electrons, but they are no longer equal height.

The model calculations displayed to the right of the data in figures 1 and 2 show most of the same features as the data. One significant difference occurs for high energy electrons at high intensity. According to the model, the *total* rates into any energy channel must be symmetric about $\phi = 0$. The high intensity data show a small but clear asymmetry about $\phi = 0$.

A likely mechanism to break the symmetry is collisions between the outgoing electron and the ion core[15]. To test this we repeated the experiment with slightly elliptical polarization (about 10% of an orthogonal electric field component). In the semiclassical model, this diverts the electron trajectory away from the atom, thereby restoring the symmetry. This was observed.

Similar experiments were also carried out in xenon, with similar results for the most part. Again, an asymmetry occurred in the data for high energy electrons emitted at high intensity. Now, however, introduction of a small ellipticity in the polarization did not remove the asymmetry. Perhaps the xenon core is sufficiently large that the electron wave is substantially scattered even if it returns to the core with a small offset.

3. Envelope Control

The experiments on carrier control have shown that ATI can be effectively altered by changing the shape of the driving field. These conclusions support recent semiclassical models of ATI. Yet we know that multiphoton resonances, which are totally absent from such models, can have an important influence on the ATI rate [1]. Thus to complete the picture of coherent control of ATI, we must also study the influence of these resonances for control.

In envelope control, the intensity envelope is shaped in order to affect the ATI rate or spectrum by controlling Stark-shifted resonances. Several recent experiments at relatively low intensities have shown that this can be very effective [2][3][4]. At high intensities new problems complicate these control experiments: The states shift more rapidly and further, usually by more than one photon. In addition, ATI complicates the coupling to the continuum. Finally, experiments must be performed in a tight focus, where spatial averaging dilutes all resonant intensity-dependent phenomena.

We have attempted envelope control experiments in ATI in xenon, krypton, and H_2. We find that in some cases phase-sensitive enhancements in ATI may be observed, although the underlying physics is still unclear. The control experiments that show the biggest effects are those where a two-peak pulse is formed by superposing two temporally separated and phase-coherent laser pulses, which then pass through a second-harmonic crystal to produce a double-peaked ultraviolet pulse. When the peaks are separated by one to two single-pulse-widths, the ATI yield increases by nearly an order of magnitude compared to the yield when the single pulses are overlapped, despite the decrease in peak field. Other experiments employing higher order ATI with near infrared pulses show no such shape-dependence.

In these experiments, we applied two identical intense laser pulses with a variable time delay between them. In some runs the two fields were linearly

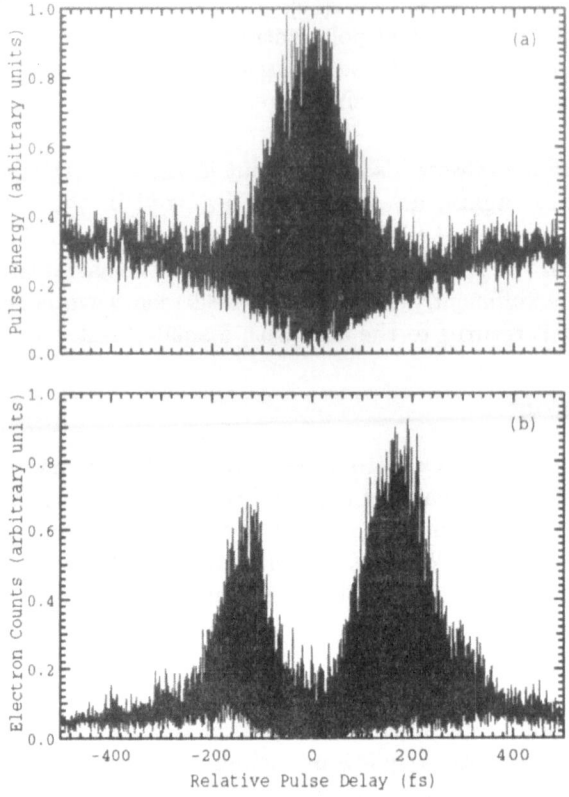

Figure 3. (a) Frequency-doubled energy in the double pulse, vs. time delay between the two legs of the Michelson interferometer. (b) ATI signal in the $5p^5(1/2)6s$ resonant peak, vs. time delay.

superposed, so that the combined field was simply

$$F(t) = F_o(t)cos(\omega t) + F_o(t - \tau)cos(\omega(t - \tau)),\qquad(6)$$

where F_o is the envelope function of the electric field pulse, ω is its frequency and τ is the relative time delay between the two pulses. Such a field has a maximum amplitude of $2F_o$ when $\tau = 0$, and a peak amplitude of F_o when the pulses are well separated. The fluence is also twice as great when $\tau = 0$. In other runs, the pulse-pair was frequency-doubled before entering the chamber. Since this is a nonlinear process, the fluence may be up to eight times greater when $\tau = 0$, than when the pulses are well-separated.

In either case, the light was focused to a maximum intensity of $\approx 2 \times 10^{14}$ W/cm^2 per pulse, or a peak field of about 4 V/Å. The Keldysh parameter γ was approximately equal to 1.

By gating on a single peak in the ATI photoelectron spectrum, it is possible to measure the relative rate of ionization through a given state, as well as the total ionization rate. However, in our apparatus the data are necessarily spatially averaged over a laser focus, so it is not possible to observe a single intensity profile. The data were taken by sweeping the delay of the two pulses through zero delay and monitoring the energy of the light and the ionization through the various multiphoton resonances.

Fig. 3a shows the total frequency-doubled light energy versus relative pulse delay after it passes through the vacuum chamber. There is four times the energy at zero delay than when the pulses are well separated. This is consistent with a depletion-limited conversion efficiency of about 14% for a single pulse. The electron signal in Fig. 3b has a relative minimum at $\tau = 0$, and maxima when $\tau \approx \pm 150 fs$. Data for all of the ATI peaks show the same behavior.

A simple numerical integration of Schrödinger's equation over the temporal intensity profile of the pulse, with several Rydberg states interacting with the dressed ground state, displays the essential features seen in the data; however, when several intensity profiles are added together with weights appropriate for a focused Gaussian laser beam, the model ultimately fails to reproduce the experiment. Therefore we still cannot conclude that we have observed envelope control in an ATI experiment.

4. Conclusions

Both envelope control and carrier control of ATI have been studied. We find that carrier control works well. Results where two colors are mixed together generally follow the predictions of the simple semi-classical models of ATI. Furthermore, departures from the model predictions appear consistent with the presence of core scattering, which is a known additional complication that can be included in refinements of the simplest two-step model.

Envelope control of ionization, which has been demonstrated in low-intensity near-resonance multiphoton experiments, is made difficult in ATI by the inhomogeneous focal volume. We see some effect on the pulse envelope in simple two-pulse experiments, but have not succeeded in modeling our results. Therefore envelope control seems possible, but is not yet established.

The work reviewed in this paper was supported by the National Science Foundation. One of us (CWSC) participated as a Sabbatical Fellow at the Center for Ultrafast Optical Physics at the University of Michigan.

References

1. R. R. Freeman, P. H. Bucksbaum, H. Milchberg, S. Darack, D.W. Schumacher, and M. E. Geusic (1987) Phys. Rev. Lett. **59**, 1092.
2. R.R. Jones, Phys. Rev. Letters (1995), **74**, 1091.
3. J.G. Story, D.I. Duncan, and T.F. Gallagher (1993), Phys. Rev. Lett. **70**, 3012.
4. R. B. Vrijen, J. H. Hoogenraad, H.G. Muller, and L.D. Noordam (1993), Phys. Rev. Lett. **70**, 3016.
5. Keldysh, L. V. (1965) Sov. Phys. JETP **20**, 1307.
6. D. W. Schumacher, F. Weihe, H. G. Muller, and P. H. Bucksbaum, (1994) Phys. Rev. Lett. **73**, 1344, (1994).
7. D. W. Schumacher and P. H. Bucksbaum, (1994) submitted to Phys. Rev. A.
8. C. W. S. Conover and P. H. Bucksbaum, submitted for publication (1995).
9. Moshe Shapiro, John W. Hepburn, and Paul Brumer (1988), Chem. Phys. Lett **149**, 451; P. Brumer and M. Shapiro (1989), Acc. Chem. Res. **22**, 407.
10. Ce Chen and D. S. Elliot, (1990) Phys. Rev. Lett. **65**, 1737; Ce Chen, Yi-Yian Yin, and D. S. Elliot, (1990) Phys. Rev. Lett. **64**, 507 (1990); Yi-Yian Yin, Ce Chen, D. S. Elliot, and A. V. Smith, (1992) Phys. Rev. Lett. **69**, 2353.
11. Seung Min Park, Shao-Ping Lu, and Robert J. Gordon (1991), J. Chem. Phys. **94**, 8622; Shao-Ping Lu, Seung Min Park, Yongjin Xie, and Robert J. Gordon (1992), J. Chem. Phys. **96**, 6613.
12. Kenneth J. Schafer and Kenneth C. Kulander (1992), Phys. Rev. A **45**, 8026.
13. H. B. van Linden van den Heuvell and H. G. Muller (1988), *Multiphoton Processes*, S. J. Smith and P. L. Knight Eds., (Cambridge University Press, Cambridge).
14. P. B. Corkum, N. H. Burnett, and F. B. Brunel (1989), Phys. Rev. Lett. **62**, 1259.
15. P. B. Corkum (1993), Phys. Rev. Lett. **71**, 1994. K. C. Kulander, J. Cooper, and K. J. Schafer (1995), Phys. Rev. A **51**, 561.

LASER-INDUCED ATOMIC STRUCTURE

D. Charalambidis[1,2], N. E. Karapanagioti[1], O. Faucher[3] and Y.L. Shao[4]
1. Foundation for Research and Technology-Hellas, Institute of Electronic Structure and Laser, P.O. Box 1527, 711 10 Heraklion, Crete, Greece
2. University of Crete, Physics Department, P.O. Box 2208, 714 09, Heraklion, Crete, Greece
3. Université de Bourgogne, Laboratoire de Physique, U.R.A. CNRS 1796, Faculté des Sciences, 6 Boulevard Gabriel, 21000 Dijon, France
4. Imperial College of Science, Technology and Medicine, Physics Department, The Blackett Laboratory, Prince Consort Road, London SW7 2BZ, U.K.

1. Introduction

Even at moderate power densities of the order of 10^9 - 10^{10}W/cm^2 in the pulsed laser mode, an atomic or molecular system can be electromagnetically strongly driven in case of resonant couplings. Strong modification or induction of structure may then occur in both the bound and continuum part of the spectrum of the system. Coherent interactions of such systems with electromagnetic fields in the so called Λ, V and ladder coupling schemes show very interesting properties which have been extensively studied theoretically for many years now. In the last ten years, the subject has received back widespread attention due to a number of experimental demonstrations and investigations of related phenomena. Examples of those are laser - induced continuum structure LICS [1] in smooth and structured continua, electromagnetically induced transparency (EIT) [2] coherent population trapping (CPT) [3], adiabatic population transfer [4] and resonance stabilization. In the present work we review our experimental results on LICS obtained recently, i.e. on the induction of structure in the continuum of an atom by means of a strong electromagnetic field. Examples include smooth but also structured continua in which the structure is induced in the vicinity of autoionizing resonances. Inducing structure on an autoionizing state and consequently strongly modifying the decay dynamics of the state is considered as a special case. A variety of different Λ and ladder coupling schemes of bound and autoionizing states is presented. Depending on the atomic, field and coupling scheme parameters, they show a controllable enhancement or suppression (partial or complete stabilization) of ionization, autoionization or third harmonic generation. The relation of all these effects with a.c. Stark splitting is discussed.

557

H. G. Muller and M. V. Fedorov (eds.), Super-Intense Laser-Atom Physics IV, 557–567.
© 1996 *Kluwer Academic Publishers.*

558

We would like to point out that since at higher laser intensities dynamical resonances may occur due to the a.c. Stark shifting of discrete levels, the effects under consideration in the present work may affect in a transient manner strong field interactions with atomic or molecular systems, as discussed in section 3. Thus the present studies contribute towards a more complete understanding of short and strong laser pulse interactions with atoms.

2. Results and discussion

2.1 GENERAL REMARKS

Before presenting the results, the general idea of what is called laser-induced continuum structure (LICS), first predicted by Heller and Popov [5], will be illustrated since this effect will be the basis of presenting and discussing all investigated schemes.

An excited discrete state embedded in a continuum provides a system that, upon decay, may result in the observation of a quantum-mechanical interference. Autoionization may be the best known phenomenon of this nature in atomic and molecular physics. However, the embedded state does not need to be a state of the atomic system (e.g. an autoionizing state), it can also be induced by means of a strong electromagnetic wave that embeds a bound state in the continuum.

Such a system is illustrated in Fig. 1. State $|2\rangle$ is embedded via ω_2, which we will refer to as the coupling laser, in the continuum and can be probed from state $|1\rangle$ via ω_1, by measuring e.g. absorption of ω_1, ionization or polarization rotation. State $|2\rangle$ embedded in the continuum via ω_2 acts now as an autoionizing state, which probed via ω_1, will exhibit a Fano-Beutler profile.

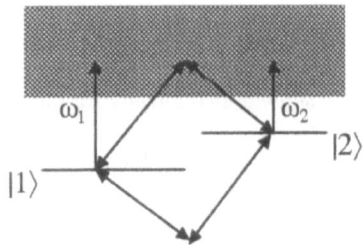

Figure 1. The LICS scheme

The full analogy of such a scheme to autoionization has been shown by Dai et al. [6] and relies on the two bound states being coupled to each other via two Raman processes and to the same continuum (Fig. 1), so that the energy balance condition $E_{|1\rangle} + \hbar\omega_1 = E_{|2\rangle} + \hbar\omega_2$ is satisfied. The interference process manifests itself as an asymmetric autoionizing-like resonance, the shape of which is determined, under low intensity and single rate approximation conditions by a q parameter equivalent to the Fano parameter in autoionization:

$$q = \frac{\text{Re}\,\Omega_{12}}{\text{Im}\,\Omega_{12}} = \frac{M_{12}}{\mu_{1c}\mu_{2c}}, \quad M_{12} = \sum \frac{\mu_{1c}\mu_{2c}}{-\omega_2 + \omega_{1c}} + P\int \frac{\mu_{1c}\mu_{2c}}{\omega_2 + \omega_{2c}}d\omega_c \quad (1)$$

where M_{12} is the two-photon transition moment between states 1 and 2, μ_{1c}, μ_{2c} are the bound-free transition moments and Ω_{12} is the two photon Rabi frequency.

It should be noted that due to the incoherent channels generally involved when measuring ionization, such as ionization of the system from state |1> via multiphoton absorption of the field ω_2 or ionization of state |2> via ω_1, a practically constant ionization background will contribute to the spectrum. Thus the ionization rate will be [7]

$$\partial P_{ion}/\partial t \propto [1 + kf(q,\delta)] \qquad (2)$$

where k is a laser intensity dependent parameter involving the widths of states $|1\rangle$, $|2\rangle$ and laser bandwidths and δ is the detuning $(E_1+\hbar\omega_1) - (E_2+\hbar\omega_2)$ normalized to the half of the sum of all the ionization widths and laser bandwidths.

It is worth noting that when LICS is probed through ionization, the degree of the asymmetry is given by the q parameter and the observability of the structure depends on the contrast parameter k. This contrast does not appear when probing occurs through rotation of polarization measurements in which the experimental conditions for observing LICS become less restrictive. At higher laser intensities and conditions in which the single rate approximation breaks down, the shape of the LICS resonance will strongly depend on the intensity.

2.2 Λ-SYSTEMS

2.2.1 *Smooth continuum (Na)*

In the earlier experiments the ionization enhancement in the form of an asymmetric LICS in the smooth continuum of Na (one-valence-electron atom) has been demonstrated in both low atomic density (effusive beam) [1(a)] and high atomic density (heat-pipe) [1(b)] environments. In our experiment (effusive beam), state $|1\rangle$ has been the Na ground state and state $|2\rangle$ the $4s\ ^2S_{1/2}$ excited state. The small asymmetry of the observed structure could be understood through the relatively large value of the q-parameter due to the moderate overall bound-bound coupling as discussed in ref. 1(a). Another main feature of this earlier work has been the demonstration of the role of a.c. Stark and saturation effects as the coupling intensity I_2 is increased. As confirmed through density matrix calculations of Lambropoulos and Zhang , such effects tend to smear out the asymmetry of the induced structure.

2.2.2 *Structured continuum (Ca)*

The picture described above is altered significantly when dealing with a structured continuum, that is when LICS is in the vicinity of autoionizing states (AIS). The AIS is now a third discrete state $|3\rangle$ lying near or at the position of the virtual state of the Raman processes that goes through the continuum (Fig. 2). The presence of the AIS plays a dominant role and hence controls the modification characteristics. Experiments have been performed in a Ca effusive beam as well as in Ca vapor in a

560

heat pipe. The effects under consideration have been probed by observing both the total ionization as well as the third harmonic generation (THG) signal [8] [9] [10]. State $|3\rangle$ has been among others the 3d4f ^{1}P AIS, state $|2\rangle$ one of the bound states 4s6s ^{1}S$_0$ or 4p^2 ^{1}D$_2$ while the 4s4d ^{1}D$_2$ state happens to be near two-photon resonant $(2\hbar\omega_1)$ with the ground state, thus playing effectively the role of state $|1\rangle$.

The main result of these experiments is that when scanning ω_2 an induced structure

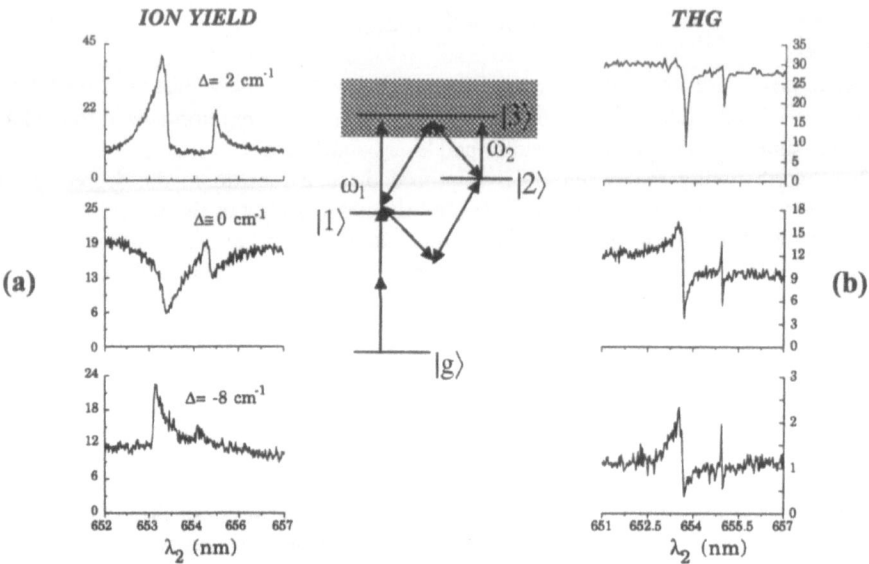

Figure 2. The coupling scheme involving an AIS. LICS shape dependence on the detuning from the AIS (a) and laser power density (b).

can be observed in both ionization and THG, the shape of which can be controlled either through the power density of the lasers or the detuning $3\hbar\omega_1$ - E$_{|3\rangle}$) from the AIS. Examples for the two coupled bound states $|2\rangle$ are depicted in Fig. 2 showing the dependence of the ion yield on the detuning and of the THG on the power density of the field ω_1. The shape of the induced structure changes dramatically from a dip to a symmetric peak, or an asymmetric interference structure that reverses asymmetry upon changing sign of the detuning. These experimental results have been verified through density matrix calculations of Lambropoulos and Zhang [8]. Although for at least the on-resonance case with the AIS the rate approximation may not be valid due to an induced strong coherence, we will apply for the sake of an intuitive picture, the Fano q parameter argumentation for the interpretation of the experimental observations. The dip, the asymmetry and its reversal can then be understood through the modification of the real part of the Rabi frequency ReΩ_{12} in the numerator of the q parameter due to the presence of the AIS which adds to it a dispersive term:

$$\mathrm{Re}\,\Omega_{12} = \mathrm{Re}\!\left(\frac{\mu_{13}\mu_{32}}{\omega_2 + \omega_{23} + i\Gamma}\right) + \sum \frac{\mu_{1c}\mu_{c2}}{-\omega_2 + \omega_{1c}} + P\!\int d\omega_c \frac{\mu_{1c}\mu_{c2}}{\omega_2 + \omega_{2c}} \qquad (3)$$

where $|3\rangle$ is the AIS and Γ its width. The value and sign of the dispersive term changes rapidly as we tune through the AIS, thus giving the possibility that at some detuning this term compensates the other two. The resulting q value is then close to zero. Above and below this point of detuning the sign of q changes, thus causing the observed asymmetry reversal.

The shape behavior of the induced structure can be further interpreted equivalently by means of an a.c. Stark splitting caused through the strong coupling of the AIS (state $|3\rangle$) with the bound state $|2\rangle$. In this picture, which is also strictly valid for the on resonance case, the shape of the induced structure is defined by the detuning from and thus by the complete or partial destructive interference of the two Autler-Townes components. The laser power density dependence of the observed induced structure may then be attributed to the dependence of the widths, splittings and positions (a.c. Stark shifting) of the states involved on the strength of the laser field. In fact, due to the large number of states (not all of them depicted in Fig.2) non- or near-resonant couplings cause dynamical shiftings of the states involved in the scheme, thus dynamicaly altering the detuning during the rise and fall of the pulse.

Note further that LICS can be equivalently described in an a.c. Stark splitting picture in any coupling scheme. The virtual state of the Raman processes can be considered as the sum of all a.c Stark split components of all parity and angular momentum allowed states of the system, due to their non resonant coupling with state $|2\rangle$ via ω_2 (Fig. 1). Whenever a near resonance with state $|3\rangle$ occurs (Fig. 2), the contribution of this state to the virtual state may be dominating the contribution of all other allowed states and thus only this state has to be taken into account. Furthermore the Raman process which is now near-resonant with state $|3\rangle$ may dominate the second one.

Such an a.c. Stark splitting $\delta = (\Delta^2 + 4\Omega^2)^{1/2}$ has been observed when the ^1P AIS was coupled with the 4s6s ^1S bound state [11]. $\Delta = \hbar\omega_2 - (E_{|3\rangle} - E_{|2\rangle})$ is the detuning from the AIS state and Ω the Rabi frequency.

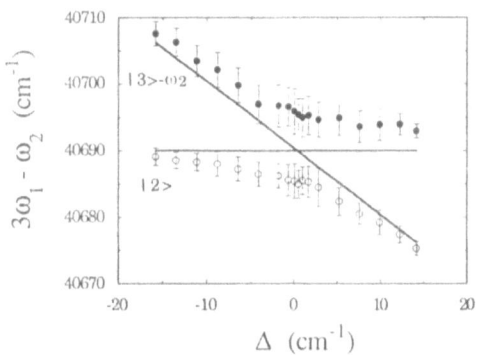

Fig. 7 depicts the observed laser avoided crossing for the two dressed (atom + laser field) states. The energy position of the AIS $|3\rangle$ or the bound state $|2\rangle$ is shown as probed by the $3\omega_1 - \omega_2$ frequency versus the detuning of the dressing laser ω_2. The data (circles) are compared to the spectroscopic energy of the states (full line). From a fit resulting from a two level system model a minimum splitting (2Ω) of 10.4 cm^{-1} has been obtained [11].

Figure 3. The laser avoided crossing

When the $4p^2$ ^{1}D bound state is coupled though, no splitting was detectable for coupling power densities up to ~5GW/cm^2 due to the lower coupling strength. In the later case the argumentation above applies for the now a.c. Stark broadened state.

2.3 LADDER-SYSTEMS

Starting from the LICS schemes described in the previous sections, a ladder system results by moving state $|2\rangle$ from the bound part of the spectrum to the continuum above state $|3\rangle$ or the virtual sate of the Raman process of the Λ scheme (Fig. 4).

This scheme is expected to have behavior similar to the Λ-system, the energy balance condition now becoming $\hbar\omega_1 + \hbar\omega_2 = E_{|2\rangle} - E_{|1\rangle}$ for a two-color single photon case. If a third state $|3\rangle$ is involved, this can be an AIS, as before but it can also be a bound state of the system. Since this work is dealing with continuum interactions we do not consider the case of states $|2\rangle$ and $|3\rangle$ being bound states.

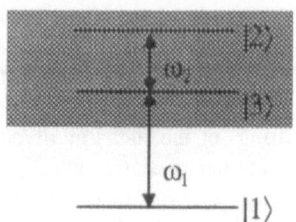

Figure 4. The ladder coupling scheme

2.3.1. *The coupling of two autoionizing states*
Very recently [12] [13] we have obtained experimental results on a ladder scheme which involves two autoionizing states (state $|3\rangle$ and $|2\rangle$) coupled to each other by a single photon ($\hbar\omega_2$). The lower lying of those was two-photon ($2\hbar\omega_1$) probed from the ground state in Mg. The corresponding states were the $3p^2$ ^{1}S (state $|3\rangle$) and 3p3d ^{1}P (state $|2\rangle$) doubly excited states. The modification of autoionizing states due to a coherent coupling with each other has been the subject of several theoretical studies in the last ten years [14] [15]. However, despite the promising features predicted by the theory there have been no experimental studies up to now to our knowledge. In this first experiment involving two electromagnetically coupled AIS, total ionization as well as energy-resolved photoelectron spectroscopy has been employed. Photoelectron spectra allow for the investigation of the decay at the two different positions in the continuum (at $2\hbar\omega_1$ and at $2\hbar\omega_1 + \hbar\omega_2$) accessible in the present schemes. The results of the ionization spectra can be summarized as follows: When $\hbar\omega_2$ is resonantly coupling the two AIS the total ionization spectra show an a.c. Stark splitting of the AIS resonance when ω_1 is scanned while keeping ω_2 fixed, and a dip in the ionization yield when ω_2 is scanned keeping ω_1 fixed. The splitting is observable although the separation of the two components is less than the width of the $3p^2$ ^{1}S state. Note that the $3p^2$ ^{1}S AIS is very broad (~280 cm^{-1}) in comparison to the 3p3d ^{1}P. For this reason

the splitting can be observed even when the Rabi frequency of the two level coupling is less then the width of the $3p^2$ ^{1}S state. For the same reason variable a.c. Stark shifts as a function of laser intensity do not play a significant role like in the case of Ca, as their magnitude is smaller than the width of the $3p^2$ ^{1}S AIS. Repetition of these measurements for slight positive or negative detunings from the center of the $3p^2$ ^{1}S state lead to the observation of asymmetric induced ionization structures when ω_2 is scanned or asymmetric structures induced on the $3p^2$ ^{1}S profile when ω_1 is scanned (see figure 5 (a),(b),(c)). An asymmetry reversal occurs for detunings of opposite sign. When detuning further away from the $3p^2$ ^{1}S the induced structure becomes a more symmetric peak.

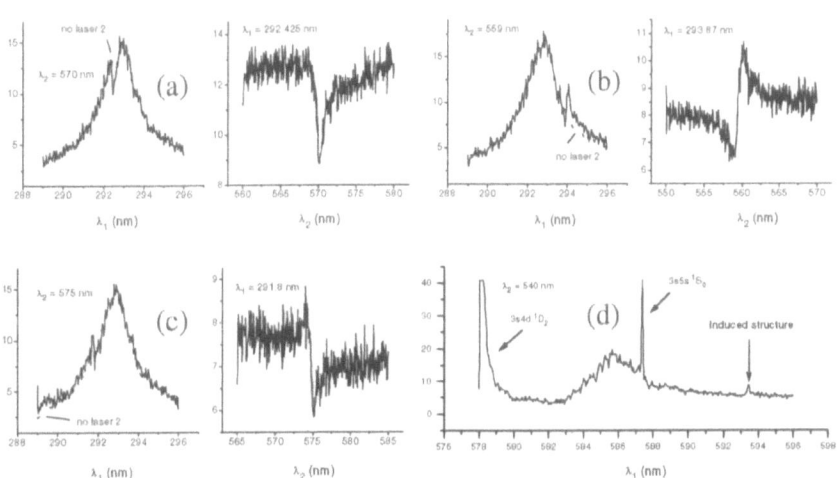

Figure 5. Total ionization spectra scanning either the probe laser (λ_1) or the coupling laser (λ_2) on resonance with the AIS (a); negatively detuned from the AIS (b); positively detuned from the AIS (c); and detuned far beyond the width of the AIS (d) .

The above observations can be interpreted using the same argumentation as in the Λ-systems in Ca. The observed suppression of autoionization and partial stabilization of the system is due to destructive interference effects at resonance. As the calculations of Bachau and Cormier show [13], the total amount of population in the two AIS decreases dramatically and the atomic population is trapped in the ground state. Complete stabilization could be achieved in the absence of incoherent channels contributing to the background by using properly chosen AIS and laser polarizations.

As the detuning increases beyond the width of the $3p^2$ state, the structure observed becomes an almost symmetric and separate ionization peak on the side of the $3p^2$ resonance (Fig. 5(d)). The precise nature of this feature however cannot be determined with only the total ionization data at hand, since the origin of the ionization

564

mechanism is not known. The peak may correspond to an increased ionization rate near the $3p^2$ ^1S state, or it might be due to ionization from the $3p3d$ ^1P state, which is now three-photon resonant ($2\hbar\omega_1 + \hbar\omega_2$).

To clarify the above question, the energy-resolved photoelectron spectra were obtained for the two AIS energy regions. Upon performing this part of the experiment, the main realization was that the spectra of the photoelectrons originating from decay in the region of the $3p^2$ ^1S state were practically identical to the total ionization spectra. They displayed the effects already outlined above in relation to total ionization. As the laser detuning increases a separate photoelectron peak appeared in the region of the $3p^2$ ^1S AIS (see figure 6).

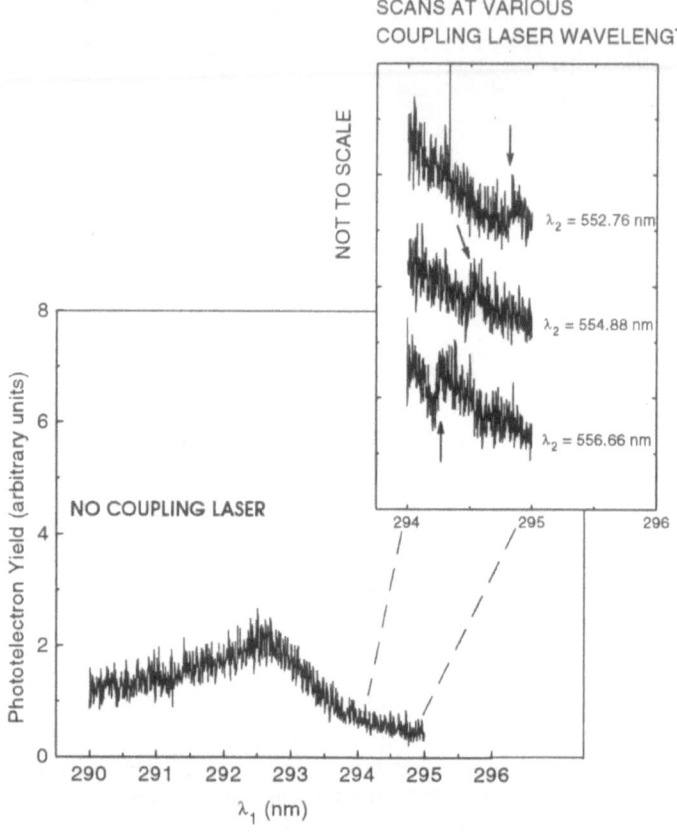

Figure 6. Energy-resolved photoelectron spectra, without and with coupling laser (inserted figure) depicting the structure induced slightly below the $3p^2$ ^1S AIS.

In addition to these observations, it also emerged that the signal obtained from the region of the $3p3d$ state was extremely weak. Very faint spectra on this vicinity were obtained through pulse counting techniques, indicating the appearance of the resonance as the coupling laser was scanned through it. This made it clear that almost all of the ionization recorded originated form decay of or around the $3p^2\,{}^1S$ state, with only a minimal contribution from the 3p3d ${}^1P$ state. Thus, the observed small resonance is mainly induced structure near the $3p^2\,{}^1S$ state and less $2\hbar\omega_1 + \hbar\omega_2$ ionization through the 3p3d ${}^1P$ state.

This non-resonant case is in close relation to the LICS scheme in a Λ-system through a smooth continuum that involves two bound states. A full analogy of the present ladder to a LICS scheme involving a Λ-system and a smooth continuum could be established if the $3p^2\,{}^1S$ AIS were not present. The bound state of the Λ-system is replaced by an AIS (3p3d ${}^1P$) which is embedded by $\hbar\omega_2$ at a lower position in the continuum. In the presence of the $3p^2\,{}^1S$, in the non resonant case, it is this state that may have the dominating contribution to the virtual state of the three-photon coupling. The scheme can be described as an "above-threshold" LICS scheme which could also be observed in a one color experiment ($\hbar\omega$) provided that $E_2 - \hbar\omega$ is above threshold.

2.3.2 Self-induced transparency

The on-resonance case of the previous section can be realized by employing only one electromagnetic wave ω if accidentally a double resonance condition $(E_{|3\rangle} - E_{|1\rangle})/n = E_{|3\rangle} - E_{|2\rangle} = \hbar\omega$ is satisfied, in which state $|3\rangle$ and $|2\rangle$ are single photon and state $|3\rangle$ and $|1\rangle$ are n-photon coupled. In such a case, the coupling of state $|2\rangle$ and $|3\rangle$ may result in an a.c. Stark splitting or broadening of the dressed states (electromagnetically mixed states $|2\rangle$ and $|3\rangle$) so that excitation of the mixed state from the ground state may be diminished. A self-induced transparency is occurring. The coherent mixing of states $|2\rangle$ and $|3\rangle$ results into an non absorbing system due to two destructively interfering components on the mixed state. The system cannot absorb the driving frequency or resonantly ionize and the population in trapped is the ground state. The field propagates without losses. Although not probable, the described situation can occur. We have demonstrated such an effect in Xe [16] where state $|3\rangle$ is a bound state and $|2\rangle$ is an autoionizing state . For $\lambda = 375$ nm, three-photon absorption from the ground state is resonant with the 5d[1/2]$_1$ state and a fourth photon happens to be resonant with the 8f autoionizing state lying between the two spin-orbit components of the ground state of the Xe$^+$ ion. In the REMPI spectrum of Xe in a low atomic density experiment (atomic beam) all other three photon allowed 5d states ([3/2]$_1$, [5/2]$_3$ and [7/2]$_3$) appear in the spectrum but not the 5d[1/2]$_1$. By contrast, in a high atomic density experiment (static cell) a strong resonance appears in the spectrum at the wavelength that the double resonance occurs. The results of the beam experiments can be understood through a self induced transparency. The results of the cell experiment have been shown to be due to THG and reabsorption of it. TH, in contrast to three-photon absorption, can be absorbed all the way along the laser beam and not only at the focus of the strong focused geome   ployed. 1 +1 ionization (the interaction volume of which is much longer than the volume of the focal area where 3+1 ionization is possible) occurs with the first photon being a TH and the second a

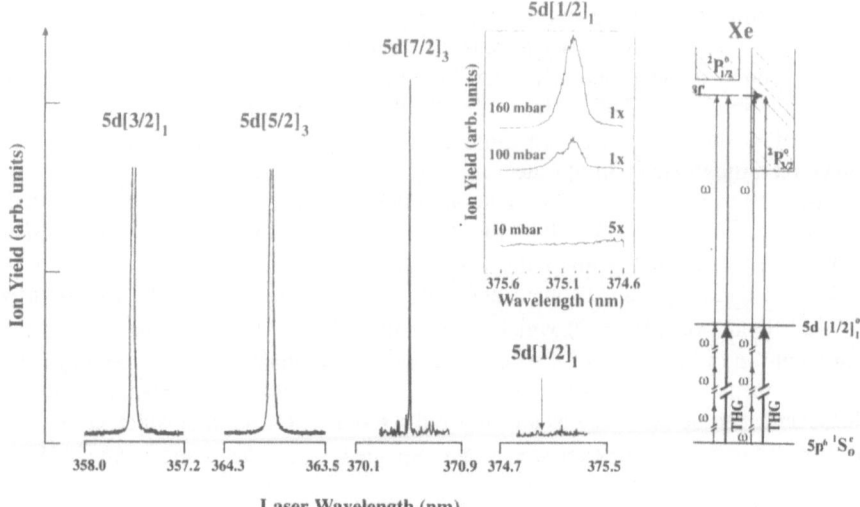

Figure 7. 3+1 REMPI spectra in Xe atomic beam and in a static cell at 10, 100 and 160mbar (inserted graph) showing the absence of the 5d[1/2], resonance in the low pressure spectra. On the right is depicted the self-induced transparency coupling scheme showing the accidental double resonance in Xe.

fundamental beam photon. In the unfocused part of the beam a.c. Stark effects become negligible and the medium is not transparent at 3ω any more. This has been verified [16] in a double cell experiment in which the beam was focused in the first cell while ionization was measured in the second one. The 5d[1/2], resonance appears in the spectrum only when both cells are filled with Xe. This proved that the resonance occurs via reabsorbing of the TH generated in the first cell in the vicinity of the focus.

3. Conclusions

Coherent interactions within atomic continua have demonstrated a variety of phenomena related to the induction of structure, modification and control of the properties of the continua. Enhancement of ionization or harmonic generation but also destructive inferences leading to suppression of ionization, autoionization or harmonic generation and hence stabilization of the system are the main features presented here. These phenomena contribute to the understanding of above threshold interactions of an atom with an electromagnetic field as well as to its optimum modification towards an effective use of these at any energy range above the threshold. The later can in particular be achieved by employing the coupling of AIS as in the Mg case. In particular the ladder schemes described above are strongly relevant to the interaction of atoms with short and intense laser pulses. It is well known that in such experiments, discrete states can be a.c. Stark shifted into resonance with n photons of the field. The case of the above threshold LICS in Mg (far detuned from the $3p^2\,{}^1S$ state) or the double resonance condition of the Xe example can then occur in a transient manner in non resonant strong field multiphoton ionization. This would then affect strongly the ionization rate, the ATI as well as the HOHG spectrum. Thus, a transient double resonance may lead to partial stabilization of the system. An "above threshold" LICS effect could lead to the observation of structure in high intensity above threshold processes due to the a.c. Stark shifting as AIS into resonance. This induced structure is expected to be present for instance in all the ATI peaks above and below the shifted AIS. The same effect may also cause a transient enhancement of all the harmonics in a HOHG experiment

ACKNOWLEDGMENTS

The work presented was conducted in the laboratories of the Ultraviolet Laser Facility operating at FO.R.T.H. -I.E.S.L. (Human Capital and Mobility, Access to Large Scale Facilities E.U. program, Contract No. CHGE-CT92-0007)

References

1. (a) Shao, Y.L., Charalambidis, D., Fotakis, C., Zhang, Jian, and Lambropoulos P. (1991) Observation of laser-induced continuum structure in ionization of sodium, *Phys. Rev. Lett.* **67**, 3669, (b) Cavalieri, S., Pavone, F.S., and Matera, C. (1991) Observation of a laser-induced resonance in the photoionization spectrum of sodium, *Phys. Rev. Lett.* **67**, 3673, (c) Knight, P.L., Lander, M.A., and Dalton, B.J. (1990) Laser induced continuum structure, *Phys. Rep.* **190**, 1 and references therein.
2. Boller, K.-J., Imamoglu, A., and Harris, S.E. (1991) Observation of electromagnetically induced transparency, *Phys. Rev. Lett.* **66**, 2593.
3. Aspect, A., Arimondo, E., Kaiser, R., Vansteenkiste, N., Cohen-Tannoudji, C. (1988) Laser cooling below the one-photon recoil energy by velocity-selective coherent population trapping, *Phys. Rev. Lett.* **61**, 826 and references therein.
4. Kuklinski, J.R., Gaubatz, U., Hioe, F.T., and Bergmann, K. (1989) Adiabatic population transfer in a three-level system driven by delayed laser pulses, *Phys. Rev. A* **40**, 6741.
5. Heller, Yu.I., Lukinykh, V.F., Popov, A.K., and Slabko, V.V. (1981) Experimental evidence for a laser-induced autoionizing-like resonance in the continuum, *Phys. Lett. A* **82**, 4.
6. Dai, Bo-nian, and Lambropoulos, P. (1987) Laser-induced autoionizing like behavior, population trapping, and stimulated Raman processes in real atoms, *Phys. Rev. A* **36**, 5205.
7. Zhang, Jian, (Thesis).
8. Faucher, O., Charalambidis, D., Fotakis, C., Zhang, Jian, and Lambropoulos, P. (1993) Control of laser induced continuum structure in the vicinity of autoionizing states, *Phys. Rev. Lett.* **70**, 3004.
9. Faucher, O., Shao, Y.L. and Charalambidis, D. (1993) Modification of a structured continuum through coherent interactions observed in third harmonic generation, *J. Phys. B* **26**, L309.
10. Faucher, O., Shao, Y.L., Charalambidis, D., and Fotakis, C. (1994) Laser-induced modification of a structured continuum observed in ionization and harmonic generation, *Phys. Rev. A* **50**, 641.
11. Shao, Y.L., Faucher, O., Zhang, Jian, and Charalambidis, D. (1995) Laser avoided crossing in the ionization continuum, *J. Phys. B* **28**, 755.
12. Karapanagioti, N.E., Faucher, O., Shao, Y.L., Charalambidis, D., Bachau, H., and Cormier, E. (1995) Observation of autoionization suppression through coherent population trapping, *Phys. Rev. Lett.* **74**, 2431.
13. Karapanagioti, N.E., Charalambidis, D., Uiterwaal, C.J.G.J., Fotakis, C., Bachau, H., Sánchez, I., and Cormier, E. (1995) Effects of coherent coupling of autoionizing states on photoionization, submitted to *Phys. Rev. A*.
14. Lambropoulos, P., and Zoller, P. (1981) Autoionizing states in strong laser fields *Phys. Rev. A* **24**, 379.
15. Bachau, H., Lambropoulos, P., and Shakeshaft, Robin (1986) Theory of laser-induced transitions between autoionizing states of He, *Phys. Rev. A* **34**, 4785.
16. Charalambidis, D., Stockdale, J.A.D., and Fotakis, C. (1994) Modification of multiphoton ionization spectra via third harmonic generation in focused laser beams, *Z. Phys. D.* **32**, 191.

IONIZATION OF RYDBERG HYDROGEN BY A HALF-CYCLE PULSE

ALEJANDRO BUGACOV, MARCEL PONT AND
ROBIN SHAKESHAFT
Physics Department, University of Southern California,
Los Angeles, CA 90089-0484

AND

BERNARD PIRAUX
Institut de Physique, Universite Catholique de Louvain,
B-1348 Louvain-la-Neuve, Belgium.

1. Introduction

Recently Jones, Bucksbaum, *et al*[1, 2] experimentally investigated the ionization of a sodium atom in a high Rydberg state by a half-cycle (hc) linearly polarized pulse of light. In this paper we present results of quantum calculations [3, 4] for ionization of hydrogen by a hc pulse, using the same parameters as in the experiments, i.e. we have used the same pulse (shape and duration) and the same principal quantum number for the initial state of the atom.

Ionization of a Rydberg atom by a powerful hc pulse of light differs significantly from photoionization by a long pulse. In particular, photoionization by a long pulse can occur only if the electron scatters from the nucleus during the passage of the pulse; in contrast, the electron-nucleus interaction may hinder ionization by a powerful hc pulse. A dramatic illustration of this was provided by Jones *et al* [2], in their experiment where sodium atoms were prepared in parabolic downhill or uphill Rydberg states, depending on whether the hc electric field was oriented either parallel or antiparallel with the dipole moment of the atom. They found that the probability for ionization was enhanced if the electron was directly pushed away, rather than scattered, from the nucleus by the force of the hc electric field. We find that the behavior of the electron is qualitatively very different de-

569

H. G. Muller and M. V. Fedorov (eds.), Super-Intense Laser-Atom Physics IV, 569–582.

pending on whether ionization occurs from an uphill or downhill state, and our results support the conclusions of Jones *et al* [2].

The electric field, $\mathbf{F}(t)$, can be treated within the dipole approximation. Choosing the z-axis to be both the electric field polarization axis and the quantization axis, and writing $\mathbf{F}(t) = F(t)\hat{z}$, we choose the pulse profile to be similar to the one used in the experiments, namely

$$F(t) \;=\; 0, \; t < 0, \tag{1}$$

$$F(t) \;=\; 29.56 F_0 [17.75(t/\tau)^3 e^{-8.87t/\tau} - 0.412(t/\tau)^5 e^{-4.73t/\tau}], \; t \geq 0, \tag{2}$$

where $\tau = 1$ psec. The pulse given by Eq. (2) consists of a large half-cycle lobe of short duration, whose peak value is F_0 and whose full-width at half-maximum is roughly $t_{\text{hc}} = 0.44\tau$. This main lobe is followed by a long but shallow tail that has opposite (negative) polarity — see Ref. [1]. The impulse communicated to the electron by the electric field over the time interval $[0, t]$ is (we use atomic units)

$$Q(t) = -\int_0^t dt' \, F(t'). \tag{3}$$

The momentum transferred to the electron by the main hc lobe is

$$\Delta p_{\text{hc}} = Q(t_0), \tag{4}$$

where 0 and t_0 are the end points of the main lobe, i.e. t_0 is the duration of the main lobe.

Let $p_n \equiv 1/n$ and $T_n \equiv 2\pi n^3$ be the characteristic momentum and orbital period, respectively, of the electron in the initial Rydberg orbit whose principal quantum number is n. Following Reinhold *et al* [5, 6] we introduce the scaled momentum transferred by the main hc lobe:

$$\Delta p_0 = \Delta p_{\text{hc}}/p_n, \tag{5}$$

and the scaled duration of the main hc lobe:

$$T_0 = t_{\text{hc}}/T_n. \tag{6}$$

The experiment of Ref. [2] was conducted in the presence of a very weak static (dc) electric field F_{dc} aligned along the positive z-axis. This dc field splits the Rydberg manifold into Stark states.[9] For a pure hydrogenic system (whose atomic number is Z) the energy shifts of the Stark states are $(3/2Z)n(n_1 - n_2)F_{\text{dc}}$, where n_1 and n_2 are the parabolic quantum numbers, with $n_1 + n_2 = n - 1 - |m|$, where m is the magnetic quantum number which we choose to be zero, as in the experiment. If $n_1 > n_2$ the Stark shift is positive and the electron is predominantly on the positive side of the z-axis, i.e. on the uphill side of the potential energy $F_{dc}z$; a state with $n_1 > n_2$ is therefore called an uphill state. On the other hand, if $n_1 < n_2$ the Stark shift is negative and the electron is predominantly on the negative side of the z-axis, i.e. on the downhill side of the potential energy $F_{dc}z$; a state with $n_1 < n_2$ is therefore called a downhill state. In our calculations with oriented states, we consider only the most uphill (i.e. $n_1 = n - 1, n_2 = 0$) and downhill (i.e. $n_1 = 0, n_2 = n-1$) states, and we ignore the dc field since it is so weak that its only effect is to cause a slight Stark splitting which allows the most uphill and downhill states to be selected in the experiment.

2. Impulse approximation

If the main lobe of the pulse delivers a sudden ($T_0 << 1$) and powerful ($\Delta p_0 >> 1$) kick to the electron, the impulse approximation applies. This approximation amounts to neglecting the influence of the nucleus on the electron while the electron interacts with the electric field. Hence, during the time the electric field acts the Hamiltonian can be approximated as $H_{imp}(t) = Q^2(t)/2 + F(t)z$, and the evolution operator becomes simply $\exp[-i \int_0^t dt'\, H_{imp}(t')]$. Since $\int_0^t dt'\, H_{imp}(t')$ is just a spatial constant plus $Q(t)z$, the ionization probability, within the impulse approximation, is[6]

$$P_{imp} = \int d^3k \, |\langle \psi_{\mathbf{k}}^- | e^{iQz} | \psi_{nlm} \rangle|^2, \tag{7}$$

where $|\psi_{nlm}\rangle$ and $|\psi_{\mathbf{k}}^-\rangle$ represent the initial and final states of the electron, and where Q is the appropriate momentum transfer, i.e. we insert $Q = \Delta p_{hc}$ to yield the probability for ionization by the main lobe.

We can evaluate P_{imp} rather easily by first rewriting Eq. (7) as

$$P_{imp} = \int_0^\infty dE \, P_{imp}(E), \tag{8}$$

where, with $G(E)$ the Coulomb Green's function,

$$
\begin{aligned}
P_{imp}(E) &= \int d^3k' \, \delta(k'^2/2 - E) |\langle \psi_{\mathbf{k}'}^- | e^{iQz} | \psi_{nlm} \rangle|^2, \\
&= -(1/\pi) \mathrm{Im} \langle \psi_{nlm} | e^{-iQz} G(E) e^{iQz} | \psi_{nlm} \rangle. \tag{9}
\end{aligned}
$$

Using a (complex) Sturmian representation of $G(E)$ in parabolic coordinates, we were able to evaluate $P_{imp}(E)$ without much computational effort.

The kick delivered to the electron by the main lobe is partially offset by the opposite kick delivered by the negative tail [i.e. Q is reduced by about

17% for the pulse form of Eq. (2)], and so within the impulse approximation the full pulse is not as effective as the main lobe in ionizing the atom. Note that in making the dipole approximation, we treat the pulse as an infinite plane wave. Hence, if the impulse approximation were applicable during the entire passage of the pulse, and if the kick delivered by the main lobe were completely offset by the reverse kick of the negative tail, no ionization would occur; from Eq. (7) we have $P_{imp} = 0$ when $Q = 0$. However, in the case of a plane-wave pulse, electron scattering from the nucleus cannot be neglected during the long negative tail; indeed, the net momentum imparted to the electron must ultimately be supplied by electron scattering from the nucleus, even though the electron may not have time to scatter from the nucleus during the main lobe. Nevertheless, if the negative tail is very long and shallow it will hardly influence the ionization probability.

In reality, of course, the pulse is not an infinite plane-wave, but rather a focussed pulse. Therefore, the electron will leave the focal region, along the electric field axis, long before it can experience the entire negative tail. As the electron leaves the focal region it will be accelerated by the ponderomotive force, and thereby acquire momentum in addition to that imparted by the main lobe of the pulse. Consequently, if the main lobe delivers a sudden powerful impulse to the electron, the ionization probability can be accurately calculated from applying the sudden approximation to the main lobe, even though the energy and angular distributions will not be given corrrectly.

Suppose that the main lobe is indeed sudden ($T_0 << 1$) and powerful ($\Delta p_0 >> 1$). If, just before being hit by the pulse, the electron has a momentum $\mathbf{p}$ and is at a distance r from the nucleus, its energy after being hit by the main lobe, i.e. after receiving an impulse $\Delta\mathbf{p}_{hc}$, is $(\mathbf{p} + \Delta\mathbf{p}_{hc})^2/2 - 1/r$ (the electron does not have time to move during the main lobe). Averaging over all directions of $\mathbf{p}$, and noting that $\mathbf{p}^2/2 - 1/r = -p_n^2/2$, the energy distribution of the emergent electron, immediately after the main lobe has passed, is centered at roughly $(\Delta p_{hc})^2/2 - (p_n^2/2)$, i.e. at

$$E_{peak} = \frac{1}{2}[(\Delta p_0)^2 - 1]p_n^2, \qquad (10)$$

with a width proportional to $p_n \Delta p_{hc}$. (If $\Delta p_0 < 1$ the kick is too weak for the electron to be impulsively knocked out of the atom.) The angular momentum distribution of the emergent electron, immediately after the main lobe has passed, is centered at about $r_n \Delta p_{hc}$, where $r_n \equiv n^2$ is the characteristic radius of the atom in its initial Rydberg state. The electron can emerge with a linear momentum of magnitude k that is much larger than Δp_{hc} only if it is ejected from a high momentum (i.e. $|\mathbf{p}| >> p_n$) component of the initial bound state momentum distribution $|\langle \mathbf{p}|\psi_{nlm}\rangle|^2$.

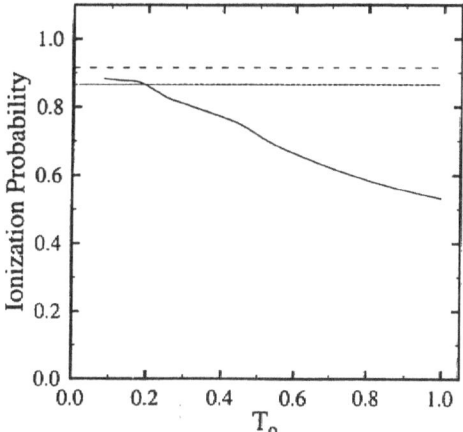

Figure 1. Probability for ionization from the 9d state *vs* the scaled (dimensionless) pulse duration T_0 when the scaled momentum imparted to the electron is held fixed. The long-broken and short-broken horizontal lines are the probabilities calculated within the sudden approximati for ionization by the main lobe and the full pulse, respectively.

However, although the angular momentum $l = kr$ increases with increasing k, the electron can have a high momentum component in the initial bound state only if it is initially at a distance r from the nucleus that is very small ($r << r_n$). As a result, l decreases rapidly in the high-energy tail of the energy distribution.

Of course, as already noted, both the negative tail of the pulse (even if it is weak) and ponderomotive scattering may significantly influence the energy and angular distributions, while not effecting the ionization probability.

We proved in Ref.[4] within the impulse approximation, the probability for ionization by a specified pulse, polarized along the z-axis, is invariant under the interchange of n_1 and n_2; in other words, the ionization probability is the same for an uphill state as it is for its mirror downhill state.

3. Results: Spherical States

In Fig. 1 we show the probability for the atom to be ionized if it is initially in the 9d $(m = 0)$ state. We show the ionization probability for different durations of the pulse, holding the scaled momentum fixed, at $\Delta p_0 = 1.9$. The solid curve is the "exact" probability for ionization by the full pulse, calculated by solving the time-dependent Schrödinger equation.

The horizontal long-broken and short-broken lines in Fig. 1 are the ionization probabilities, calculated within the impulse approximation, for the atom to be ionized by the main lobe and by the full pulse, respectively.

We see that the exact ionization probability begins to level off as T_0 approaches zero, rising above the impulse approximation result for the full pulse (since electron scattering from the nucleus mitigates the effect of the reverse kick by the negative tail), but remaining below the impulse approximation result for the main lobe (since the reverse kick by the negative tail is not eliminated entirely). In the present case, where Δp_0 is larger than unity but not much larger, the impulse approximation begins to break down as T_0 increases beyond about 0.2. If Δp_0 were much larger than unity we would expect the impulse approximation to apply up to significantly larger T_0. Incidentally, note that, since we hold Δp_{hc} fixed, F_0 decreases as T_0 increases, and for large T_0 ionization occurs through tunneling.

In Fig. 2 we show the ionization probability density in the energy-angular momentum plane, at the end of the main lobe. These results were calculated by solving the time-dependent Schrödinger equation for a pulse whose peak field and duration are fixed, for the cases where the initial bound state is the 7d, 8d, or 9d state, for which $T_0 = 0.35, 0.24$, and 0.17, respectively, and $\Delta p_0 = 1.51, 1.73, 1.94$, respectively. We have indicated, by a vertical arrow, the point E_{peak}, i.e. the center of the energy distribution as predicted in the impulse approximation — recall Eq. (10). A horizontal arrow indicates the approximate cutoff in angular momentum quantum number l (i.e. the maximum l, beyond which the population falls off rapidly). Since a discrete basis was used, the allowed energies of the electron are discrete, so the plot of the probability density is not continuous. We see a significant population of high l-states, as did LaGattuta and Lerner.[8] We also see that the most probable l decreases rapidly in the high-energy tail of the energy distribution, as discussed in section II. Finally, recall that for $T_0 << 1$ and $\Delta p_0 >> 1$ (impulse approximation) the angular momentum distribution should be centered at roughly $r_n \Delta p_{hc}$, which grows with increasing n as n^2; however, the growth in the cutoff seen in going from Fig. 2 (a) to (b) to (c), while more rapid than linear, is not as rapid as n^2.

We have calculated, by solving the time-dependent Schrödinger equation, the probability for ionization from the 15d ($m = 0$) state by a full pulse whose main lobe has a width 0.44 psec, the same as in the first series of experiments of Jones et al [1]. In Fig. 3 we compare the ionization probability with the experimental data, and with the results of the classical calculations of Jones et al [1], for various peak field strengths. Our quantum results agree well with the classical results, and the agreement with the experimental data is also fairly good up to the highest field strength we could handle. (Our calculations are limited primarily by the basis size required for convergence; as the peak field strength F_0 increases, the required basis size increases even more rapidly.) We also show in Fig. 3 the results calcu-

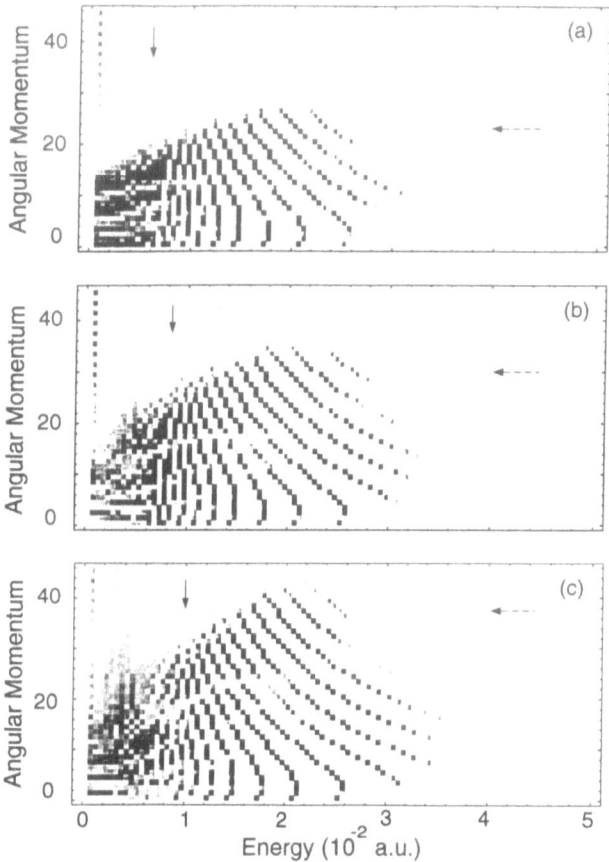

Figure 2. Ionization probability distribution over the angular momentum quantum number and the energy of the emergent electron. Darker areas are regions of higher probability The peak field of the pulse is fixed and the width of the main lobe is 0.018 psec. The initial bound state is (a) 7d, (b) 8d, and (c) 9d. The vertical arrows mark the points at which the energy distribution should peak according to the impulse approximation — see Eq. (10) of text. The horizontal arrows mark the approximate cutoff in angular momentum.

lated using the impulse approximation, for the probability of ionization *by the main lobe*, but the agreement with the impulse approximation is poor, which is perhaps expected since T_0 is not small, i.e. $T_0 = 0.9$. However, Jones et al [1] also measured the probability for ionization from the 20d and 35d ($m = 0$) states, for which $T_0 = 0.4$ and 0.07, respectively. The impulse approximation should be more accurate for these higher n states, and that is confirmed by Fig. 4, where we compare our impulse approximation results (for ionization by the main lobe) with the experimental data and

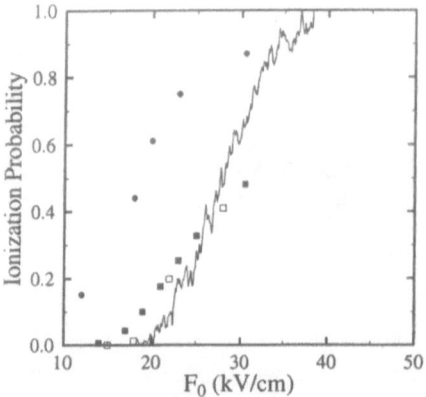

Figure 3. Probability for ionization from the 15d state *vs* the peak field F_0. Solid line: Data from experiment of Ref. [1], with F_0 shifted by factor of 2.5, as recommended in Ref. [1]. Solid squares: Present results, calculated by solving the time-dependent Schrödinger equation. Open squares: Classical results from Ref. [1]. Solid circles: Results calculated within the impulse approximation.

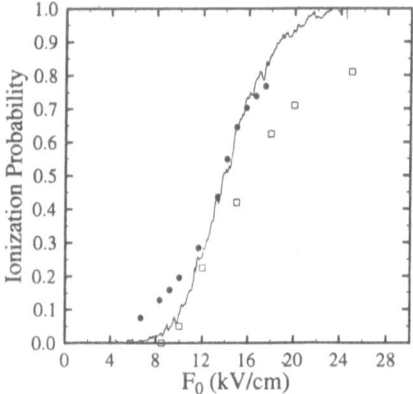

Figure 4. Probability for ionization from the 20d state, for the same pulse as in Fig. 3. Solid line: Data from experiment of Ref. [1]. Open squares: Results calculated from classical calculations of Ref. [1]. Solid circles: Results calculated within the impulse approximation.

with the classical calculation results of Jones *et al* for ionization from the 20d state.

We have not carried out full calculations for the 20d state since they would be expensive (a very large basis would be required) and we would not expect to obtain results much different from those found using the impulse approximation, in contrast to the case of the 15d state. In Fig. 5 we show results (obtained from the impulse approximation) for the energy distribu-

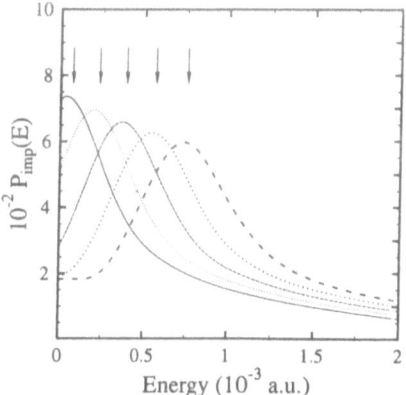

Figure 5. Energy distribution of the electron, ejected from the 20d state, according to the impulse approximation. The curves correspond, from left to right, to the peak field strengths 17, 18, 19, 20, and 21 kV/cm. The arrows mark the points at which the energy distribution is centered according to Eq. (10) of the text.

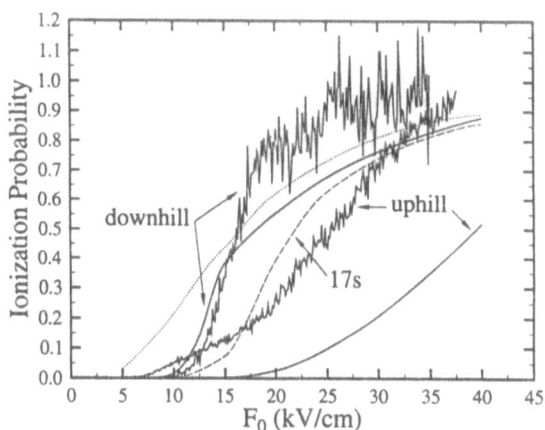

Figure 6. Probability for ionization of hydrogen from the $n = 17$ most downhill and uphill states (solid lines), and from the 17s state (dashed line), by a half-cycle pulse with peak field strength F_0. The dotted line is the probability for ionization from the most downhill and uphill states when calculated within the impulse approximation.

tion $P_{\mathrm{imp}}(E)$ of the electron at the end of the main lobe, for different peak field strengths, when the atom is initially in the 20d state. The arrows mark the energies E_{peak}, at which the energy distribution should peak according to Eq. (10).

4. Results: Oriented States

In **Fig.** 6 we show the probability for a hydrogen atom to be ionized by the main (hc) lobe of the pulse, when the atom is initially in the most downhill or most uphill $n = 17$ ($m = 0$) state. We choose the electric field of the main lobe to be aligned with the dipole moment of the atom in the downhill state, the state localized along the negative z-axis. Thus the hc electric field points along the positive z-axis and pushes the electron in the direction of the negative z-axis. In the case of a pure hydrogenic system, subject to a very weak dc field, the uphill and downhill states would be interchanged under the reflection $z \to -z$; as a consequence of this symmetry we would not obtain essentially different results by reversing the direction of the hc electric field of the main lobe. We show both results obtained using the impulse approximation (dotted line) and "exact" results obtained by solving the time-dependent Schrödinger equation (solid lines). (The solid lines would be simply interchanged were we to reverse the direction of the hc electric field.) We also show the "exact" probability for ionization from the 17s state (dashed line). The scaled pulse duration is $T_0 = 0.6$, and at the highest field strength we consider, i.e. $F_0 = 40$ kV/cm, the scaled momentum is $\Delta p_0 = 2.3$. The impulse approximation becomes more accurate as the field strength increases, but, as discussed in the preceding paragraph, it does not discriminate between uphill and downhill states. On the other hand, at small to moderate field strengths there is a very significant difference in the "exact" probabilities for ionization from the uphill and downhill states.

This asymmetry between uphill and downhill states was explained by Jones *et al* [2]; their argument goes as follows: The energy pumped into the atom by the main lobe of the pulse is $- \int_0^{t_0} dt \, \mathbf{F}(t) \cdot \mathbf{v}(t)$, where $\mathbf{v}(t)$ is the instantaneous velocity of the electron. Averaging over the velocity distribution of the electron in the initial Rydberg state, $\mathbf{v}(t)$ would be simply the velocity induced by the hc field $\mathbf{F}(t)$, and would be in the opposite direction to $\mathbf{F}(t)$, if scattering from the nucleus could be ignored. Now, in the downhill state the electron is primarily located on the left ($z < 0$) side of the nucleus, and since the force exerted on the electron by $\mathbf{F}(t)$ points towards this side the electron is pushed away from the nucleus. Hence, in the downhill state, the electron has little chance of scattering from the nucleus; consequently, $\mathbf{v}(t)$ has little chance of changing sign, and since $\mathbf{F}(t)$ does not change sign over the interval $0 \leq t \leq t_0$ the energy pumped into the atom, and therefore the probability for ionization, are relatively large. However, in the uphill state the electron is primarily located on the right ($z > 0$) side of the nucleus, and therefore the force exerted by $\mathbf{F}(t)$ pushes the electron towards the nucleus, from which it is likely to scatter. Hence, in

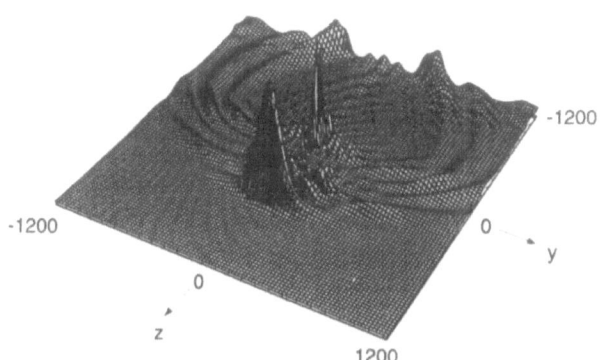

Figure 7. Spatial probability distribution for an electron, initially in the 17s state, after the the passage of the main (hc) lobe of the pulse.

the uphill state, $\mathbf{v}(t)$ is likely to change sign, and so the energy pumped into the atom, and therefore the probability for ionization, are relatively small. On the basis of this argument we can easily understand why the impulse approximation does not discriminate between uphill and downhill states: Within the impulse approximation scattering from the nucleus is neglected, and (averaging over the initial velocity distribution) $\mathbf{v}(t)$ has *no* chance of changing sign in either the uphill or downhill state; this also explains why the ionization probability calculated within the impulse approximation is larger than the "exact" probability (for ionization from either state).

In the 17s state the electron is as likely to be on one side of the nucleus as the other, but in this state the spatial probability distribution is spherically symmetric, and so the electron will more likely than not miss scattering from the nucleus as it is pushed in the direction of the negative z-axis by the hc field; this explains why the "exact" probability for ionization from the 17s state lies closer to the "exact" probability for ionization from the downhill state than that for ionization from the uphill state.[10] However, that portion of the electron probability distribution that initially lies within a narrow cylinder around the positive z-axis can be pushed into the nucleus and is therefore "stable" against ionization.

In Fig. 7 we show the electron probability distribution in the yz-plane at the end of the main lobe, i.e. at the time $t_0 = 3.88 \times 10^4$ a.u., when a pulse with peak field strength $F_0 = 35$ kV/cm is incident on hydrogen in the 17s state. The ionization probability is 80%, and so most of the probability distribution has moved far away, off the sale of the figure. The bound-state probability distribution remaining behind is largely a wavepacket located along the positive z-axis. Of course, after some time this wavepacket will orbit around the nucleus.

580

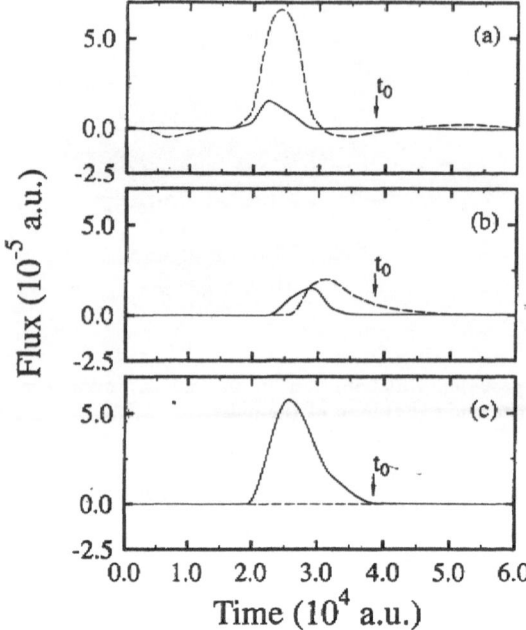

Figure 8. Flux through the left (solid line) or right (dashed line) hemisphere of radius R *vs* time, due to ionization from the $n = 17$ most uphill (right localized) or downhill (left localized) state by a hc pulse which has duration t_0, peak strength F_0, and which pushes the electron to the left. The cases are: (a) uphill state, $R = 600$ a.u., $F_0 = 30$ kV/cm; (b) uphill state, $R = 1000$ a.u., $F_0 = 30$ kV/cm; and (c) downhill state, $R = 1000$ a.u., $F_0 = 13$ kV/cm. In all cases the ionizaton probability is 20%.

Returning to the results of Fig. 6, it is of interest to ask into what direction the electron is emitted after a hc pulse impinges on a hydrogen atom in an uphill or downhill state? The answer is given by evaluating the current density $\mathbf{j} = \mathrm{Im}\Psi^*(\mathbf{x}, t)\nabla_{\mathbf{x}}\Psi(\mathbf{x}, t)$, where $\Psi(\mathbf{x}, t)$ is the electron wavefunction at time t. We considered a sphere of radius R centered at the nucleus. We divided this sphere into left ($z < 0$) and right ($z > 0$) hemispheres, separated by the xy-plane, and we integrated the current density over the surface of each of the hemispheres. Some results are shown in Figs. 8 (a) and (b) for the $n = 17$ most uphill state and in Fig. 8 (c) for the $n = 17$ most downhill state; the solid and dashed lines represent the fluxes through the left and right hemispheres, respectively, *versus* the time. In Fig. 8 (c) the radius of the sphere is $R = 600$ a.u. while in Figs. 8 (a) and (b) we have $R = 1000$ a.u. Let us focus first on Fig. 8(a). A sphere of radius $R = 600$ a.u. is not quite large enough to contain the initial probability distribution for the uphill state; although most of the distribution is contained within

591

the right hemisphere, the tail extends beyond the surface. Hence, in Fig.
8(a) we see that the flux through the right hemisphere (the dashed line)
first becomes negative as t increases; the tail of the electron distribution is
pushed into the right hemisphere from outside by the hc field. As t increases
further, the electron distribution is pushed towards the nucleus, from which
it partially rebounds, giving rise to a large positive flux through the right
hemisphere. Part of the electron distribution is pushed past the nucleus,
and gives rise to a postive flux through the left hemisphere (the solid line).
At still later times, but before t_0, some portion of the scattered distribution (that rebounded) is pushed back into the right hemisphere by the hc
field, given rise again to a negative flux through this hemisphere. Turning now to Fig. 8(b), where $R = 1000$ a.u., the sphere is large enough to
contain essentially all of the initial probability distribution for the uphill
state (the electron initially resides in the right hemisphere). We see that the
flux through the right hemisphere is larger than the flux through the left
hemisphere. In other words, *although the hc field pushes the uphill electron
to the left, most of the electron distribution rebounds from the nucleus and
emerges to the right*. Figure 8(c) pertains to ionization from the downhill
state. The electron initially resides in the left hemisphere, and since the hc
field pushes the electron to the left all of the flux passes through the left
hemisphere.

Finally, we note that the results of Fig. 6 (the solid lines) are in good
qualitative agreement with the data measured in the second series of experiments by Jones *et al* [2] for the probabilities for ionization from the
$n = 17$ most uphill and downhill states of sodium. However, there is an
important quantitative difference: Due to its nonCoulombic core, sodium,
unlike hydrogen, does not possess the symmetry that the uphill and downhill states are interchanged under the reflection $z \rightarrow -z$; consequently, the
probabilities for ionization of the uphill and downhill states are not interchanged when the direction of the hc field is reversed, and this discrepancy
is undoubtedly accentuated by the role played by the core if it scatters the
electron. Therefore quantitative agreement between the probabilities for
ionization from hydrogen and sodium cannot be expected when the initial
states are Stark states.

Acknowledgements

This work was supported by the National Science Foundation under Grant
No. PHY9315704, by NATO under grant CRG 931418, and by the National
Center for Supercomputing Applications, under Grant No. PHY940011N;
we utilized the Connection Machine Model-5 at the National Center for
Supercomputing Applications, University of Illinois at Urbana-Champaign.

One of us (B. P.) was supported in Belgium by the Fonds National de la Recherche Scientifique.

References

1. R. R. Jones, D. You, and P. H. Bucksbaum, Phys. Rev. Lett. **70**, 1236 (1993).
2. R. R. Jones, N. E. Tielking, D. You, C. Raman, and P. H. Bucksbaum, Phys. Rev. A **51**, R2687 (1995).
3. A. Bugacov, B. Piraux, M. Pont, and R. Shakeshaft, Phys. Rev. A **51**, 1490 (1995).
4. A. Bugacov, B. Piraux, M. Pont, and R. Shakeshaft, Phys. Rev. A **51**, 4877 (1995).
5. C. O. Reinhold, M. Melles, and J. Burgdörfer, Phys. Rev. Lett. **70**, 4026 (1993).
6. C. O. Reinhold, M. Melles, H. Shao, and J. Burgdörfer, J. Phys. B **26**, L659 (1993).
7. H. A. Bethe and E. E. Salpeter, *Quantum Mechanics of One- and Two-Electron Atoms*, (Springer-Verlag, New York, 1957).
8. K. J. LaGattuta and P. B. Lerner, Phys. Rev. A **49**, R1547 (1994).
9. H. A. Bethe and E. E. Salpeter, *Quantum Mechanics of One- and Two-Electron Atoms*, (Springer-Verlag, New York, 1957).
10. Incidentally, the angular momentum of the emergent electron is higher when the initial state is an s-state rather than a parabolic downhill or uphill state. Within the impulse approximation the angular momentum of the emergent electron is proportional to the charactenistic distance of the electron, in its initial state, from the z-axis; this distance is relatively small for uphill and downhill states that are elongated along the z-axis.

ATOM-ATOM CORRELATIONS AND NON-LOCALITY

Y. GONTIER and M. TRAHIN

Commissariat à l'Energie Atomique
Service des photons et des Molécules
Centre d' Etudes de Saclay
91191 Gif-sur-Yvette Cedex - France

1-Introduction

The correlations between distant particles has stimulated many theoretical and experimental works [1,2]. An important subject of controversy concerned the question of whether local deterministic hidden variables theories could reproduce all the predictions of quantum mechanics. Most of papers have shown that the answer is "no". The conclusion is that the concept of locality is to be dropped with the allowance for two systems to be correlated in the absence of any mutual interaction.

In this account, consider two atoms interacting separately with an external radiation field. We assume that the atoms do not interact with each other and that they are located at positions where the field has the same intensity, the same phase and the same polarisation, whatever the distance between the atoms be.

It is worthwhile to examine how the problems we encounter have been already approached in other studies dealing with particle-particle correlations.

In this respect, the first example concerns the Bohm's "gedankenexperiment" [3] which illustrates the EPR paradox [4]. In this kind of experiment,a measurement on the subsystem N°1 gives with certainty the state of the system N°2 without interacting with it. By doing as many measurements as desired, one can know with certainty all the sates of twe two subsystems. Since quantum theory cannot provide this knowledge, it was assumed that additional parameters (hidden variables) had to be introduced in the theory to complete the specification of the states. By considering local deterministic theories including such additional parameters, Bell [5] has set up the relations which must be verified by the joint probability of measuring well-defined values taken by two different observables in the two subsystems. These inqualities are strongly violated in many cases which make unrealistic such models and leave open the problem of non-locality. The other chapter heading which is in favor of our model deal with cooperative effects. In the work of Dicke [6], the cooperative emission can be interpreted in terms of the correlations generated by processes where the atom N°1, initialy in the upper state $|a_1\rangle$ makes a transition toward the lower state $|b_1\rangle$, the emitted photon of frequency ω is reabsorbed by the atom N°2 which is excited in the upper state $|a_2\rangle$ etc ... It is important to notice that each atom is

H. G. Muller and M. V. Fedorov (eds.), Super-Intense Laser-Atom Physics IV, 583–592.

584

to its own field and to that of the other atoms within interaction volume which is assumed to be much smaller than the wavelength of the radiation. this last condition prevents from any important variation of the field over the dimensions of the interaction volume. As a consequence, if we look at the mechanism of absorption-emission which is responsible of cooperative emission, we remark that for any pair of atoms, the correlation can be due to the absorption of photons produced in every parts of the interaction volume instead of coming exclusively from exchange of photons between the two components of the pair. This can be interpreted in terms of the non-locality of atom-field ineraction due to the uniformity of the field over the interaction volume, which makes undistinguishable the photons of the field. The restrictions concerning the size of the interaction volume (coherence length) are necessary when the correlations are due to the radiation field generated by the atoms themselves but one can imagine that such correlations could be generated by the resonant coupling of these atoms with an external field whose characteristics are the same in all parts of the interaction region (of limited size). The next step of our discussion consists of considering extended domain where the field duplicates periodically the ensemble of its characteristics. According to the preceding arguments, it is not senseless to expect that particular regions of space be correlated via this radiation field displaying identical features at that places. Our idea is to put a detector at any one of such reciprocal places (places where the system atom-plus-field shows the same characteristics), and to observe the perturbation of another system lying at a remote reciprocal place. It is this expectation which has motivated the present work. Such an effect, called STOC (Signal Transmission by Optical Correlations) would enable to observe distant perturbations. If exists, it could be a way to transmit the information between two distant points without resorting to intensity modulation of the radiation field (Laser) connecting these points.

2-Atom-Atom correlations

2.1. SPIN-1/2 FORMALISM

To begin with, we consider the simple case of two two-level atoms interacting with a radiation field whose characteristics are the same at the positions of the atoms. We assume that the interaction between the jth -atom and the field proceeds via the exchange of a photon labelled k . This label k accounts for all the features of the field in a well defined region of space. Any change of the field parameters will correspond to another value of k. Since two-level atoms are involved, it is convenient to relate operator commutations and correlations within the formalism of spin-flip operators.

The hamiltonian of this two-atom system is (in atomic units)

$$H = \sum_{j=1}^{2} \omega_{0j} S_3^j + \sum_{j,k} \left(\alpha_k^j S_+^j a_k + \alpha_k^{j^*} S_-^j a_k^+ \right) + \sum_k \omega_k a_k^+ a_k, \qquad (2.1)$$

where the rotating wave approximation is made and $S_\pm$ are the spin-flip operators [7] obeying the following commutation relations

$$\left[S_\pm^j , S_\mp^{j'} \right] = 2 S_3^j \, \delta_{jj'} \qquad (2.2)$$

and

$$\left[S_3^j , S_\pm^{j'} \right] = \pm S_\pm^j \, \delta_{jj'}. \qquad (2.3)$$

In Eq. (2.1), the coefficient α_j^k is expressed in terms of the single-photon flux F'/F_0 (F' = flux / photon number, $F_0 = 3.22 \times 10^{34} \, cm^{-2} \, s^{-1}$) and the dipole matrix element corresponding to jth -atom and to the photon polarisation $\bar{\varepsilon}_k$

$$\alpha_k^j = -i \left(\frac{F'}{F_0} \right)^{\frac{1}{2}} \omega_k^{\frac{1}{2}} \,\, {}_j\langle \pm | \bar{r}_j . \bar{\varepsilon}_k | \mp \rangle_j, \qquad (2.4)$$

where for clarity the upper and the lower states of the jth-atom are denoted by $|+\rangle_j$ and $|-\rangle_j$ respectively. The Hamiltonian of Eq.(2.1) can be rewritten into the following more convenient form

$$H = H_0^A + H_F + \sum_{j=1}^{2} \sum_k \left(V_{j,k}^+ + V_{j,k}^- \right), \qquad (2.5)$$

where

$$H_0^A = \omega_{01} S_3^1 + \omega_{02} S_3^2, \qquad (2.6)$$

$$H_F = \sum_k \omega_k a_k^+ a_k, \qquad (2.7)$$

and

$$V_{j,k}^- = \left(V_{j,k}^+ \right)^* = \alpha_k^j S_+^j a_k \quad . \quad (j = 1, 2) \qquad (2.8)$$

From Eq.(2.2) and (2.3) one obtains the commutation relations for the operators $V^{\pm}$

$$\left[V_{j,k}^{\pm}, V_{j,k}^{\pm} \right] = 0, \tag{2.9}$$

and

$$\left[V_{j,k}^{\pm}, V_{j',k'}^{\mp} \right] = \mp\, a_k^j\, a_{k'}^{j'} \left\{ 2\, S_3^j\, a_k\, a_{k'}^+\, \delta_{jj'} + S_{\pm}^j\, S_{\mp}^{j'}\, \delta_{kk'} \right\}. \tag{2.10}$$

Eq.(2.10) shows that the necessary condition for the atoms to be correlated is that the commutator must not vanish. Such a condition is fulfilled if $j = j'$ and / or $k = k'$ i.e, if the atoms are idendical and / or if they see the same field. Since our model is based on the undistinguishability of the two atom-field systems, the operators are independent of space. Therefore, Eq.(2.10) holds for atoms separated by arbitrarily large distances. It may be viewed as a necessary condition for a generalization to the macrophysical world the non-locality of the microphysical world predicted by the quantum theory. In this respect, we note that Eq.(2.9) is unconclusive because the commutator allways vanishes because it contains field operators which commute in every cases.

Finally, we note that the possible naked states for two correlated atoms are

$$|Correl.\rangle = \begin{cases} |a_1\, a_2\rangle \\ \dfrac{1}{\sqrt{2}} \left(|a_1\, b_2\rangle + |b_1\, a_2\rangle \right) \\ |b_1\, b_2\rangle \end{cases}. \tag{2.11}$$

2.2. RE-SUMMED THEORY

Since the effect we study requires high intensity radiation fields, we must resort to a nonperturbative model. To this end, we make use of a theory making extensive use of exact resummation of perturbation series. In the case of two non-commuting operators (absorption and emission operators of a single atom) , the technique has been discussed in details in preceding accounts [8]. Here we are faced with a much more complicated situation since one has to handle four noncommuting operators (absorption and emission operators of two atoms). Such a problem has been solved in the general case [9]. For the sake of completeness, only the salient results will be recalled.

In general, the behaviour of any system can be predicted once the time evolution operator is known. This operator can be calculated from the resolvent operator by means of the inversion integral

$$U(t) = \frac{1}{2\pi i} \oint e^{-izt} \, G(z) \, dz, \tag{2.12}$$

where

$$G(z) = \frac{1}{z - H}. \tag{2.13}$$

We refrain from presenting in more details the resolvent theory which is well-known and widely utilized [10]. According to Eqs.(2.5) and (2.13), $G(z)$ can be expressed as

$$G(z) = G_0(z) + G_0(z) \, H_I \, G(z), \tag{2.14}$$

where

$$H_I = \sum_{j=1}^{2} \left(V_j^+ + V_j^- \right), \tag{2.15}$$

and

$$G_0(z) = \frac{1}{z - \left(H_0^{AT} + H^F \right)} \tag{2.16}$$

In Eq.(2.15), the subscript k refering to the field state is dropped because it is assumed that the field is the same everywhere. The solution of Eq.(2.14) is obtained by iteration techniques which provide infinite series of increasing powers of the interaction H_I.

The problem under consideration consists of calculating the resonant emission probability of the atom N°1 in the presence of the atom N°2 (which also can resonantly absorb and emit photons of the field). Initially, the two atoms are assumed to be in their upper states. Thus, it is not necessary to resort to symetrization and the initial and the final states of the system which are $|a_1, a_2, n\rangle$ and $|b_1, a_2, n+1\rangle$ respectively i.e, the atom N°1 emits a photon by making a transition from the upper state a_1 toward the lower state b_1, the atom N°2 remains into its initial state a_2 while the photon number n is increased by one unit . Notice that the influence of atom N°2 on atom N°1 is independent of the choice of the states.

The matrix element one has to calculate is

$$^{(k+1)}G_{b_1 a_2, a_1 a_2}(z) = \langle b_1 a_2, n+1 | \, ^{(k+1)}G(z) | a_1 a_2, n\rangle, \tag{2.17}$$

where $^{(k+1)}G(z)$ is the operator describing the resonant emission of a photon by the atom N°1. As a result of the resummation, this operator can be expressed as

$$^{(k+1)}G(z) = G(z)\, B\, G^+(z),\tag{2.18}$$

where

$$G(z) = \frac{1}{1 - \rho^+(z) - \rho^-(z)},\tag{2.19}$$

$$G^\pm(z) = \frac{1}{1 - \rho^\pm(z)},\tag{2.20}$$

and

$$\rho^+(z) = B\, G^+(z)\, A + Y\, G^+(z)\, X,\tag{2.21}$$

$$\rho^-(z) = A\, G^-(z)\, B + X\, G^-(z)\, Y.\tag{2.22}$$

The absorption-emission operators A, B, X and Y appearing in Eqs.(2.18), (2.21) and (2.22) are defined by $A = G_0 V_1^-$, $B = G_0 V_1^+$, $X = G_0 V_2^-$ and $Y = G_0 V_2^+$. To get the computational formulas, we replace $G^\pm(z)$ in Eqs.(3.21) and (3.22) by their values obtained from Eq.(2.20) by iteration, then one substitute the expressions thus obtained into Eq.(2.19).

3. Numerical analysis

Eqs. (2.17)-(2.22) enable to compute self-consistently to all orders the probability for the resonant net emission of a photon by atom 1. To make tractable the calculation without resorting to exagerated computation time, we have retained, to each order, the most significant contributions. In addition, the number of iterations in the continued fractions has been limited to that ensuring the stability of the results. According to what it was done in previous accounts [10], we put the matrix element of Eq.(2.17) into a form characteristic of two-level problems. For brevity, we do not display the details of the algebra involved in the derivation of the final formula which reads

$$^{(k+1)}G_{b_1a_2,a_1a_2}(z) = \frac{\alpha_1}{\left[z - \frac{\omega_{01}}{2} - \frac{\omega_{02}}{2} - R_{a_1a_1}(z)\right]\left[z - \omega + \frac{\omega_{01}}{2} - \frac{\omega_{02}}{2} - R_{a_2a_2}(z)\right] - |\alpha_1|^2}$$

$$(3.1)$$

In Eq.(3.1), we have changed the origin of the energies by substracting the quantity $n\omega$ everywhere. On the other hand, for each atom this origin is half the distance of the (naked) levels i.e, $\omega_{0(1,2)} = \left|\omega_{a(1,2)} - \omega_{b(1,2)}\right|$. The atom-field parameters α_1 and α_2 are those of Eq.(2.4) where the subscripts and the superscrits are replaced by a single subscript which enables to distinguish the two systems through both their atomic and field parameters. The $\alpha's$ are related to the intensity by the relation

$$I_{W/cm^2} = \frac{14.038 \times 10^{16}}{\left(\ |\bar{\varepsilon}.\bar{r}|\ \right)^2_{a.u}} \alpha^2_{a.u},$$

$$(3.2)$$

which in the case of 1S - 2P transition in hydrogen reduces to $I_{W/cm^2} = 2.53 \times 10^{17} \times \alpha^2_{a.u}$.

The matrix elements of R(z), the effective operator (called also the shift operator because it provides the diagonal contributions to the shifts of the levels a_1 and b_1) are given by

$$R_{a_1a_1}(z) = \frac{|\alpha_1|^2}{z + \omega + \frac{\omega_{01}}{2} - \frac{\omega_{02}}{2} - \frac{|\alpha_1|^2}{z + 2\omega - \frac{\omega_{01}}{2} - \frac{\omega_{02}}{2} - \cdots} - \frac{|\alpha_2|^2}{z + 2\omega + \frac{\omega_{01}}{2} + \frac{\omega_{02}}{2} \cdots}},$$

$$(3.3)$$

and

$$R_{b_1b_1}(z) =$$

$$\frac{|\alpha_1|^2}{z - 2\omega - \frac{\omega_{01}}{2} - \frac{\omega_{02}}{2} - \frac{|\alpha_1|^2}{z - 3\omega + \frac{\omega_{01}}{2} - \frac{\omega_{02}}{2} - \frac{|\alpha_1|^2}{z - 4\omega - \frac{\omega_{01}}{2} - \frac{\omega_{02}}{2} - \cdots} - \frac{|\alpha_2|^2}{z - 4\omega + \frac{\omega_{01}}{2} + \frac{\omega_{02}}{2} + \cdots}}}$$

$$+\cfrac{|\alpha_1|^2}{z-2\omega+\dfrac{\omega_{01}}{2}+\dfrac{\omega_{02}}{2}-\cfrac{|\alpha_1|^2}{z-3\omega-\dfrac{\omega_{01}}{2}+\dfrac{\omega_{02}}{2}-\cfrac{|\alpha_1|^2}{z-4\omega+\dfrac{\omega_{01}}{2}+\dfrac{\omega_{02}}{2}-\cdots}-\cfrac{|\alpha_2|^2}{z-4\omega-\dfrac{\omega_{01}}{2}-\dfrac{\omega_{02}}{2}+\cdots}}}$$

$$+\cfrac{|\alpha_2|^2}{z+\dfrac{\omega_{01}}{2}+\dfrac{\omega_{02}}{2}-\cfrac{|\alpha_2|^2}{z+\omega+\dfrac{\omega_{01}}{2}-\dfrac{\omega_{02}}{2}-\cfrac{|\alpha_1|^2}{z+2\omega-\dfrac{\omega_{01}}{2}-\dfrac{\omega_{02}}{2}-\cdots}-\cfrac{|\alpha_2|^2}{z+2\omega+\dfrac{\omega_{01}}{2}+\dfrac{\omega_{02}}{2}+\cdots}}}$$

$$(3.4)$$

One must note that these matrix elements of R(z) contain the whole physics of the process we are concerned with. Thus, the correlations come from crossed processes where one or several photon emissions or absorptions by one of the two atoms is followed by an equivalent number of absorptions or emissions by the other atom. More precisely, the atoms are correlated by the mixing of the two species of atom-field quantities in the denominators of the continued fractions.

To calculate the integral of Eq.(2.12) by the residues, one needs the poles of $^{(k+1)}G_{b_1 a_2, a_1 a_2}(z)$. They are determined to any desired accuracy by a technique which consists of searching the position of the divergences which these poles induce.

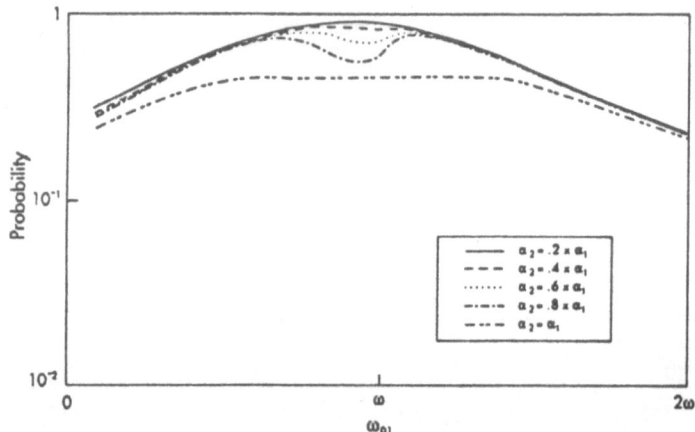

Figure 1. All-order resonant single-photon emission probability for a two-level atom (atom 1) in the presence of another identical atom (atom 2), as a function of the energy level separation ω_{01}. The atom-field parameter of atom 1 is $\alpha_1 = .3$ a.u. and that of atom 2 varies from .2 to 1 a.u.

591

In Fig.(1) we have plot the probability of resonant single photon emission for atom N°1 as a finction of ω_{01} for $\alpha_2 / \alpha_1 = .2, .4, .6, .8, 1$. The field frequency whose value is chosen to be unity in the calculations is used as scaling parameter. For values less than .2, the resonance curves are similar the corresponding curves of a single atom. Their amplitudes tend to unity and their maxima are shifted with respect to the resonant frequency in the absence of Bloch-Siegert shift. If the ratio increases, The maxima decrease and the greatest flattening does occur when the two systems are identical. In this case, the amplitude is lowered by a factor greater than two. This variation of amplitude is not negligible even in the case of two atoms . Its detection can be favoured by a modulation of the atom-field parameters of system N°2. By this way, even if the effect of this "intelligent" modulation on system N°1 is small, it is possible to detect it by resorting to techniques used to extract a signal from a noise. On the other hand, by generalizing the calculation to a more complicated system of several atoms, one can expect an enhancement of the effect. This is due to the presence of additional terms appearing in the denominators of the continued fractions which reinforce the correlations and thus, contribute to magnify the amplitude variations observed with two atoms.

4- Concluding remarks

The effect we have discussed in the preceding lines can be considered as the starting point of the mechanism, called STOC in the introduction, which enable the transmission of informations from the system N°2 to system N°1 without amplitude modulation of the laser beam connecting the two systems. From Eqs. (3.4) we see that its origin lies in the possibility for a photon to be indifferently absorbed or emitted in two undistinguishable regions of space. A break in the symetry produced by a slight change (modulation) of any parameter characterizing one of the two systems can be observed from a measurement performed on the other system. Our calculation does not provide how the information is transmitted, but it seems that the undeterminacy of the photon path, which holds in single photon interferences, can be invoked here in order to preserve the Einstein 's causality principle. In contrast, a controversy can arise if one considers that the photon involved in the exchange can be found indifferently in two regions of space. In this case, the instantaneous transmission of the information can be interpreted in terms of non-locality of quantum theory and becomes open to discussion with respect to causality. At the present time, we limit ourselves to the detection on system N°1 the effect of the perturbation suffered by system N°2 without worry about the instant at which the perturbation was initiated and the time spent by the signal to reach the measurement device. In this respect, we adopt the point of view according to which only correlated measurements on the two systems require the presence of two observers exchanging information at a velocity less than that of the light [11]. Anyway, only experimental tests of such an effect, if observable, will provide reliable arguments concerning the way chosen by the signal to propagate.

5- References

1. Kafatos, M. (1989) *Bell's theorem, quantum theory and conception of the universe*, Kluwer Academic Publishers, Dordrecht.

2. Omnès, R. (1992) Consistent interpretation of quantum mechanics, *Rev. Mod. Phys.* **64** , 339-82.

3. Bohm, D. (1951) *Quantum Theory,* Prentice-hall, Inc. Englewood Cliffs, New Jersey.

4. Einstein, A., Rosen, N. and Podolsky, B. (1935) Can quantum-mechanical description of physical reality be considered complete ? *Phys. Rev* **47**, 777-80.

5. Bell, J. S. (1965) On the Einstein Podolsky Rosen paradox, *Physics* **1**, 195-200.

6. Dicke, R. H. (1954) Coherence in spontaneous radiation processes, *Phys. Rev.* **93**, 99-110.

7. Buley, E. R. and Cummings, F. W. (1964) Dynamics of a system of N atoms interacting with a radiation field, *Phys. Rev.* **A134**, 1454-60.

8. Gontier, Y. Rahman, N. K. and Trahin, M. (1976) Exact summation of higher-order terms in multiphoton processes, *Phys. Rev.* **A14**, 2109-25.

9. Gontier, Y. (1985) Summation of four-operator perturbation series: Application to atom-atom intensity-dependent correlations, *Phys. Rev.* **A31**, 279-96.

10. Gontier, Y. and Trahin, M. (1989) Effect of an intense radiation field onb the discrete and continuum spectra of atomic hydrogen, *Phys. Rev.* **A40**, 1351-61.

11. Biswas, A. K. Compagno, G., Palma, G. M., Passante, R., Persico, F. (1990) Virtual photons and causality in the dynamics of a pair of two-level atoms, *Phys. Rev,.* **A42,** 4291-301.

CLASSIFICATION OF LASERS WITHOUT INVERSION

F.B. DE JONG, R.J.C. SPREEUW AND A. DÖNSZELMANN
Van der Waals-Zeeman Institute, University of Amsterdam
Valckenierstraat 65, 1065 XE Amsterdam

H.G. MULLER
FOM-Institute for Atomic and Molecular Physics,
Kruislaan 407, 1098 SJ Amsterdam The Netherlands

AND

H.B VAN LINDEN VAN DEN HEUVELL
Van der Waals-Zeeman Institute,
FOM-Institute for Atomic and Molecular Physics

1. Introduction

The last few years there has been an enormous interest in the subject of lasing without inversion. Many theoretical contributions have been made but until very recently [1] only 'amplification without inversion' (AWI) was demonstrated experimentally. Reviews were given by Kocharovskaya [2], Scully [3] and Mandel [4]. The main interest in LWI is the hope to build a laser at high frequencies, entering the VUV, x-ray or even γ-ray regime. Since the rate for spontaneous emission between two levels separated by frequency ω scales with ω^3, it is much harder to create inversion in this regime. Moreover the gain cross section decreases proportional to ω^{-2}. If the inversion condition can somehow be relaxed, lasing can be achieved with less stringent pumping requirements.

The phrase lasing without inversion can be confusing. From a strict point of view, lasing without inversion is actually a paradox because the S and the E in L.A.S.E.R. stand for a net stimulated emission, (more stimulated emission than absorption) which requires inversion. Consider an *isolated* system of thermally excited levels. There is no inversion in thermal equilibrium; for every choice of two levels, the lower level is more populated than the upper. Laser action in this system is impossible, it violates the second law of thermodynamics: if it would be possible, we could have la-

H. G. Muller and M. V. Fedorov (eds.), Super-Intense Laser-Atom Physics IV, 593–602.
© 1996 Kluwer Academic Publishers.

ser action, wait until thermal equilibrium is restored and have laser action again thus cooling our device. In thermal contact with an external bath we would get an energy source for free. The argument can be extended to include incoherent pumping mechanisms: Not only in a thermally excited medium but also in an incoherently pumped medium LWI is impossible. This is equivalent with the notion that any system that can be described by the Einstein rate equations requires inversion in order to show gain.

A different situation arises in the presence of a pumping mechanism that creates a non-zero atomic coherence. Energy transfer from pump field to medium and subsequently from medium to probe field is possible depending on the relative phases between the induced dipole and the probe field. In the next section we discuss this 'phase-driven gain' versus normal 'inversion-driven gain' for pulsed pumps and probes. In the third section the extension to cw phase-driven systems is made. In between the extreme situations of phase-driven gain and inversion-driven gain there is an interesting area of gain processes in which both coherences and populations contribute to gain: the population-driven LWI, described in section four. Here phase-driven gain and population-driven gain interplay, resulting in schemes that might prove useful in the pursuit of high-frequency lasers.

2. Phase-driven versus inversion-driven systems; pulsed regime

In the interaction between a single atom and a probe field, the relative phase between atomic dipole and the optical field determines whether the atom absorbs energy from the field or emits energy into the field. In an ensemble of atoms also phase-independent gain can exist. Consider an ensemble of two-level atoms. These levels are embedded in a multi-level system. The influence of the other levels is incorporated in incoherent pump processes and level decays as indicated in Fig.1. The λ's denote the incoherent pumping rates from a large ground-state reservoir into the levels $|a\rangle$ and $|b\rangle$. The well-known Bloch equations for this system interacting with a field $\vec{E}(t)$ are in the rotating frame [6]:

$$\dot{\rho}_{ab} = -\gamma\rho_{ab} + \frac{i}{\hbar}V_{ab}(t)(\rho_{aa} - \rho_{bb}) \tag{1}$$

$$\dot{\rho}_{aa} = \lambda_a - \gamma_a\rho_{aa} + \frac{2}{\hbar}\text{Im}(V_{ab}(t)\rho_{ab}^*) \tag{2}$$

$$\dot{\rho}_{bb} = \lambda_b - \gamma_b\rho_{bb} - \frac{2}{\hbar}\text{Im}(V_{ab}(t)\rho_{ab}^*) \tag{3}$$

Where $V_{ab}(t) = -\frac{1}{2}pE(t)e^{i\omega_0 t}$, with p the component of the atomic transition dipole moment along $\vec{E}(t)$, $\gamma = \gamma_{ab} + \gamma_{ph}$ with γ_{ph} the contribution from dephasing processes and $\gamma_{ab} = (\gamma_a + \gamma_b)/2$.

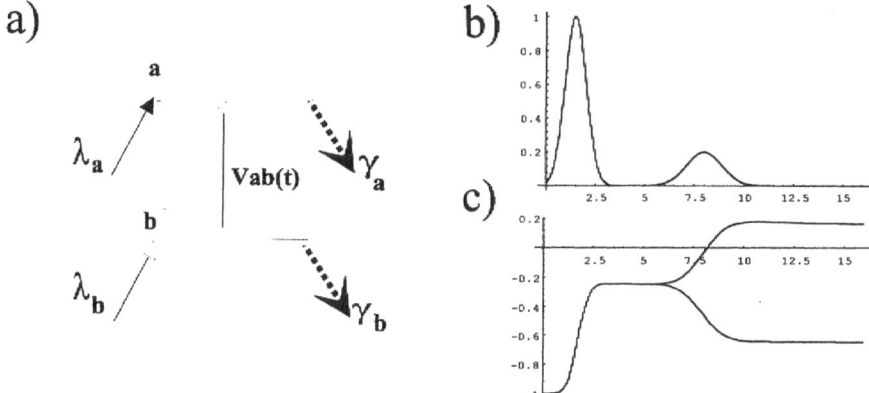

Figure 1.   a) A two-level system embedded in a multi-level scheme. Both levels represent excited states. b) Sequence of pump and probe pulse. c) Population difference $\rho_{aa} - \rho_{bb}$ for pump-probe phase difference equal to zero (upper curve, probe absorption) and π (lower curve, probe amplification) as a function of time.

To illustrate phase-driven gain we discuss the solution of Eqn.(1-3) for a particular case. At ($t = 0$) we assume the steady-state values for the density-matrix elements as they would be obtained without interaction with the field $E(t)$: $\rho_{ab}(0) = 0$, $\rho_{aa}(0) = \lambda_a/\gamma_a$ and $\rho_{bb}(0) = \lambda_b/\gamma_b$. For a resonant field we may write in the rotating wave approximation:

$$E(t) = E_0(t)e^{-i\omega_0 t + i\phi} \quad \rightarrow \quad V_{ab}(t) = -\frac{1}{2}pE_0(t)e^{i\phi} \qquad (4)$$

where $E_0(t)$ is a pulse-envelope function. Its time dependence has to be slow compared to $e^{-i\omega_0 t}$ in order to make the separation between envelope and carrier useful. There is gain for the field if, due to its interaction with the atoms, ρ_{aa} decreases. The gain is thus proportional to $-\text{Im}(V_{ab}\rho_{ab}^*)$.

If the incoherent processes provide population inversion in the absence of the atom-field interaction the field will experience gain, regardless of the phase ϕ: in Eqn.(1) $V_{ab}(t)$ couples to the population difference to generate a non-zero ρ_{ab} with the same phase. In the gain expression the phase cancels. This is 'regular' *inversion-driven* gain.

If the incoherent processes provide a negative inversion, *phase-driven* gain is still possible: We will follow the system as it is pumped with a short coherent pulse and probed after a time τ by a weak pulse with the same frequency but a different phase:

$$V_{ab}(t) = -\frac{1}{2}pE_0(t) - \epsilon\frac{1}{2}pE_0(t-\tau)e^{i\phi}, \qquad \epsilon \ll 1 \qquad (5)$$

The pulse durations of pump and probe are chosen so short that populations and coherences are not effected by the incoherent processes during the pump-probe interaction. At t=0 the coherent pump is turned on and the population starts its Rabi cycle. While ρ_{aa} is still smaller than ρ_{bb} the pump is turned off. Afterwards a probe pulse of the same phase as the pump will excite ρ_{aa} even further, thus losing intensity. A probe pulse with opposite phase, however, can lower ρ_{aa} thus extracting energy from the medium and gaining intensity, as is shown in Fig.1. This is phase-driven gain: not the inversion between the levels but the phase difference between atomic dipole and probe field determines the gain. The phase difference between the dipole moment of the atom and probe field is the argument of $V_{ab}\rho_{ab}^*$.

Summarising the distinction between the two gain mechanisms:
Phase-driven gain occurs when the probe couples directly to the atomic dipole as created by the pumping mechanism.
Population-driven gain occurs when the probe couples to a positive inversion, creating a dipole moment to which it can couple for gain.

3. Continuous-wave, phase-driven gain

3.1. INTRODUCTION

Until now we discussed phase-driven and inversion-driven gain in a two-level system interacting with one pulsed field. The extension to cw operation for the inversion-driven gain is well known: The steady state solutions of equations (1-3) in the rate equation approximation form its basis.

Phase-driven gain as described in the previous section can also occur cw. Consider two fields with the same frequency, ω_0.

$$V_{ab}(z) = -\frac{1}{2}pE_0e^{ik_1z} - \epsilon\frac{1}{2}pE_0e^{ik_2z+i\phi}, \qquad \epsilon \ll 1 \qquad (6)$$

Where k_1 is the wavevector of the pump field (propagating in the z direction) and k_2 is the projection of the probe wavevector on the z axis. As long as the angle between the two beams is small (that is $(k_2 - k_1)l \ll 1$, with l the length of the medium) ϕ dependent gain is possible in analogy with Eqn.5.

Other cw phase-driven systems include Raman lasers and wave-mixing processes. The phase dependence of the gain process is not always obvious because the gain does not start from a probe pulse but from vacuum fluctuations. However the existence of the phase-matching condition is a remnant of the phase dependence in the gain process.

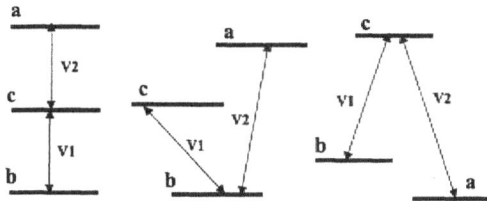

Figure 2.   Three level systems; Cascade scheme, V scheme and Λ scheme. V1 and V2 denote the coupling fields.

3.2. MOLLOW GAIN

Consider the same two-level system coupled to two coherent fields of *different* frequency:

$$V_{ab}(t) = -\frac{1}{2}pE_0 - \epsilon\frac{1}{2}pE_0 e^{-i\delta\nu t+i\phi}, \qquad \epsilon \ll 1 \qquad (7)$$

with the pump on resonance and $\delta\nu$ the pump-probe detuning. When substituted in Eqn.(1-3) we see that the the time dependence is not eliminated in the rotating frame. The detuning between the two coherent fields coupling to the same transition prevents the populations to reach a steady state. This is typical for excitation with a bichromatic field. Note that in the present time-resolved description the gain is both time dependent and phase dependent.

In the approximation that the probe intensity is much smaller than the pump intensity and therefore the effect of the probe on the populations and coherences can be neglected in first order, there *is* a steady state. In this case an analytical expression for the gain can be found in refs.[5] [6]. In the bare-state basis $|a\rangle, |b\rangle$ (the eigen states of the Hamiltonian unperturbed by the coupling field $E(t)$) there is no inversion, irrespective of the pump strength. It can be shown that in the dressed state basis (the eigen values of the Hamiltonian describing atom plus interacting field) inversion is necessary for probe amplification [7]. In this dressed state description Mollow gain can be seen as inversion driven gain where the phase of the probe plays no role. This is a well known feature of the dressed-state description. The ratio of the population amplitudes in the dressed description is a measure of the phase difference between the amplitudes in the bare-state description [8].

4. Population-driven LWI

4.1. INTRODUCTION

In the system we discussed in section 2, the coherent pump and the probe coupled the same levels. In three-level systems coherences can be pumped between two states other than the upper and lower state that are probed. The various possible combinations of probe and coherent pump are depicted in Fig.2. If there is a coherent pump between two states, and no further incoherent pumping, then the probe pulse can never experience gain on transitions other than the driven transition. However it can lead to a reduction or even cancellation of probe absorption, known as electromagnetically induced transparency (EIT) [9].

4.2. PULSED POPULATION DRIVEN LWI

In 1993 three LWI experiments were done in the pulsed regime [10] [11] [12]. In all experiments a Λ scheme was used where coherence was created between the two lower levels. The coherence (the phase relation between the lower states) was created in different ways: The experiment by Fry et al. [11] used the fact that two Rabi oscillations involving one common upper level tend to synchronise the population of the two lower levels. Switching off the driving fields quickly preserves this phase relation also without any light. The experiment by Nottelmann [10] uses Raman transitions. The down pumping step in the Raman scheme redistributes the cycled wave function over all involved lower states. This process also tends to synchronise the population of the lower levels. In contrast to the other two experiments, the experiment by Van der Veer [12] starts with empty lower states. Therefore one short pulse suffices to populate the lower levels with a well defined phase relation. In all three experiments the coherent superposition of the lower states forms a time dependent wavefunction. Consequently the coupling of the probe field also oscillates while the population of the atomic eigenstates is constant. The reduction of probe absorption due to destructive interference between the excitation of the two eigenstates is known as EIT. In the three experiments mentioned coherent population trapping (CPT) occurs periodically with a frequency of the energy separation of the lower states. At the moment that the lower-state population is not coupled to the probe field a small amount of population in the upper state can result in gain.

In view of our earlier discussion, the schemes that were discussed are not strictly phase driven nor driven by population inversion between bare states. It can be seen as gain driven by a *hidden* population inversion or as phase-driven gain, where not the phase of the carrier level of the probe light determines the gain but the relative phase of the envelope of the probe light with respect to the lower-level coherence.

Just as in section 3.2 there is freedom of description: An explanation in terms of an inversion in the dressed model (coupled model would be a more appropriate name in this case) is equivalent to a phase dependent description in the bare states (or again more appropriate eigen states).

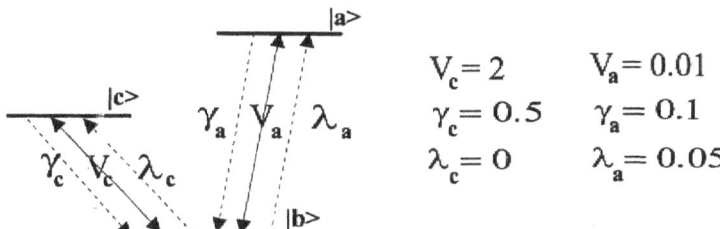

Figure 3. Driven V-scheme. The dashed lines indicate the incoherent processes, the solid lines the coherent processes. For the listed parameter values AWI occurs on the coherent probe V_a.

4.3. CONTINUOUS WAVE POPULATION DRIVEN LWI

In the case of a population-driven, pulsed LWI the presence or absence of inversion was a function of the particular choice of basis functions and the particular choice of the lower state. For the situation that we would like to discuss in the present section, a continuous-wave, population-driven LWI, no inversion is found in the atomic system, irrespective of the choice of basis. The most clearcut example of a cw population-driven LWI is the three-level system. We will discuss the V-scheme, where the choice of parameters is as indicated in Fig.3. A strong field V_c is driving Rabi oscillations in the $|b\rangle$-$|c\rangle$ transition. The strengths $\lambda_{a,c}$ denote incoherent pumping into levels $|a\rangle$ and $|c\rangle$. With the right choice of parameters the probe V_a can experience gain, independent of its phase and without any inversion between the levels $|a\rangle$, $|b\rangle$ and $|c\rangle$ as was shown experimentally [1].

We have performed a Monte-Carlo wavefunction simulation of this scheme [14]. The population of level $|a\rangle$ as a function of time is shown in Fig.4. We see a coherent evolution of the wavefunction which is interrupted by the incoherent pump and decay processes. In the calculation the incoherent processes are mimicked by jumps of the wavefunction to states $|a\rangle$ or $|b\rangle$ at random times. The statistics of these jumps are determined by the incoherent rates. The coherent evolution of $|b\rangle$ is frequently interrupted due to the large spontaneous decay from level $|c\rangle$ to $|b\rangle$. In fact, the collapse out of $|a\rangle$ is caused by the small admixture of $|b\rangle$. This small admixture will Rabi-oscillate and finally stop the coherent evolution due to spontaneous emission from $|c\rangle$ to $|b\rangle$. This difference in evolution time of $|a\rangle$ and $|b\rangle$ is crucial because the build-up of admixtures in $|a\rangle$ and $|b\rangle$ (and therefore the occurrence of gain and absorption) is *quadratic* in time. In other words, $|b\rangle$ is populated more often than $|a\rangle$ (since there is no inversion). But since the build-up of admixture starting from $|b\rangle$ is interrupted much more often than the build-up starting from $|a\rangle$, the population change of $|b\rangle$ is much slower. This can be interpreted as the quantum Zeno effect [15],

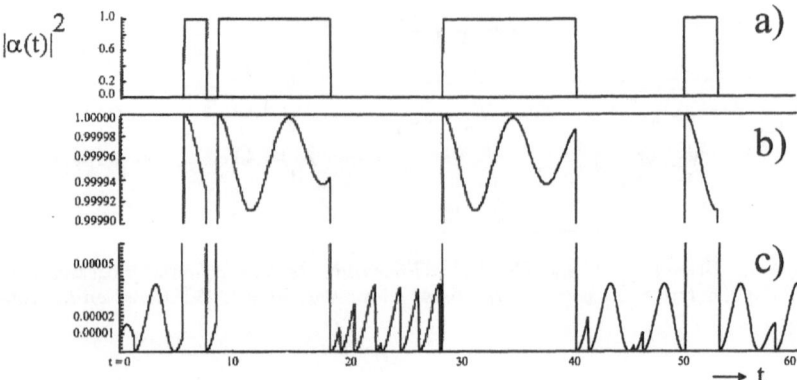

Figure 4. a) The population of $|a\rangle$ as function of time as created by a Monte-Carlo simulation of the V scheme. The quantum jumps into $|a\rangle$ are caused by the incoherent pump on the $|a\rangle$-$|b\rangle$ transition, b) upper part of the curve enlarged, c) lower part of the curve enlarged.

where spontaneous emission (and therefore an implicit measurement that the atom is in its lower state) prevents the effective build-up of an admixture of the upper state. Therefore this type of LWI could be called equally well a quantum Zeno laser. Figure 4 does not show why this net stimulated emission is population-driven. This becomes more obvious from the power spectrum of the population amplitude of the three involved states. For $|a\rangle$ and $|b\rangle$ these power spectra are given in Fig.5. In terms of the spectrally integrated population, there is of course still no inversion. Relevant, however, are those parts of the spectrum that are coupled by the probe laser, in this particular case the central peaks, here given by zero detuning. The sidebands in the powerspectrum of $|b\rangle$ do not contribute to absorption into $|a\rangle$. The wavefunction of $|b\rangle$ changes sign twice every period of the Rabi oscillation between $|b\rangle$ and $|c\rangle$. The Rabi oscillation between $|b\rangle$ and $|a\rangle$ is much slower, so during the build-up of $|a\rangle$ out of $|b\rangle$ the wavefunction of $|b\rangle$ changes sign many times, leading to a cancellation of the contributions to $|a\rangle$. This cancellation of absorptions has also been explained in terms of Fano-type interferences [13]. As in the previous cases it is apparently possible to isolate a sub-system that actually drives the gain of the probe and for which real inversion exists.

There is an alternative explanation for the observed gain in the V scheme. When we do not limit our description to the atomic state, but also include the radiation field, spontaneous emission does not lead to an incoherent phase jump, but to population of a new quantum state (describing both atom and field). Because $|b\rangle$ is much more frequently involved in a spontaneous process than $|a\rangle$, the occupation probability is spread out over

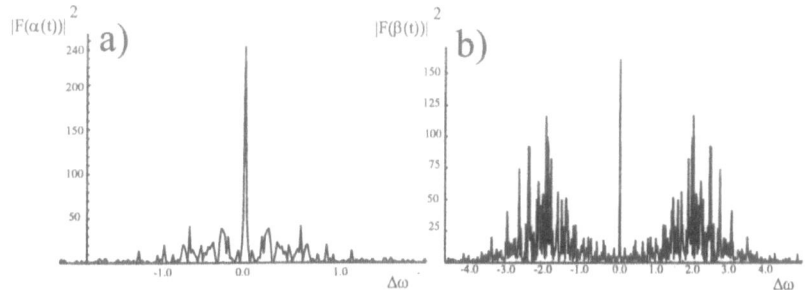

Figure 5. a) *Power spectrum of the amplitude of level* $|a\rangle$, b) *Power spectrum of the amplitude of level* $|b\rangle$.

much more states than the occupation probability of $|a\rangle$ after a few lifetimes of $|c\rangle$. In the originally populated sub-systems this will lead to inversion (and therefore gain). All the newly populated sub-systems are populated due to spontaneous emission and therefore fed via the lower state. All these sub-systems are therefore in the minimum of a Rabi cycle, where there is no absorption. Again, this reduced absorption is due to the quadratic onset of the coupling of light in a two-level system with an empty upper state. And again, the gain is driven by a sub-system with inversion.

The concept of dividing a lower state in sub-systems with inversion and other sub-systems that do *not* couple to the light, is encountered more often. The earlier discussed population trapping can be interpreted in these terms. Another example is the well-known excimer laser: when the lower state is defined as the lower electronic state, irrespective of the internuclear distance, also this is a LWI. However, when the lower state is split up in parts that couple with light (leading via the Franck-Condon principle to a reduction of allowed internuclear distances) and other parts of the lower state (that describe the already dissociated molecule) then, again, no inversion is found.

5. Conclusion

There are two kinds of lasers without inversion, phase driven and population driven:

- Phase-driven LWI can be truly inversionless. However there is no conversion of energy from an incoherent pump to a coherent field.
- All schemes for population-driven LWI have hidden inversion in a subset of the lower state. In this paper we identified various realizations of this mechanisms like CPT and spectral decomposition of the lower state population. An overview of our classification is given in table 1.

Class:	Phase driven LWI		Population driven LWI		Inversion laser
nature of pumping:	Coherent pumping		Coherent and incoherent pumping		Incoherent pumping
output:	pulsed	cw	pulsed	cw	pulsed / cw
example:	3π/4 pulse	Mollow ('72) Wu ('90) Wave mixing Raman lasers	Fry, Nottelmann, Veer ('93)	Zibrov ('95) V,Λ scheme	regular lasers
nature of inversion:	No inversion	No inversion (hidden in dressed state)	hidden inversion (CPT)	hidden inversion in sub-set	inversion

Table 1: *Classification of various Lasing Without Inversion schemes.*

On the left the completely phase driven schemes where the gain is determined by the phase difference between the atomic dipole (induced by a coherent pump) and probe field. On the right the normal inversion schemes where the (mostly incoherent) pump mechanism creates population inversion leaving the coherences zero. The light to be amplified couples to the inversion and creates the atomic dipoles through which it can be amplified. The population driven LWI schemes in the middle are the most interesting in the context of this contribution. In these systems the coherence introduced by the first pump reduces the amount of inversion that has to be created by the second pump, to allow for gain.

References

1. A.S. Zibrov, M.D. Lukin, D.E. Nikonov, L.W. Hollberg, M.O. Scully, V.L. Velichansky, and H.G. Robinson, Phys. Rev. Lett. **75**, 1499 (1995).
2. O. Kocharovskaya, Physics Reports **219**, 175-190 (1992).
3. M.O. Scully, Phys. Rep. **219**, 191-201 (1992).
4. P. Mandel, 'Lasing without inversion a useful concept?', Contemporary Physics **34** 235 (1993).
5. B.R. Mollow, Phys. Rev. A. **5**, 2217 (1972).
6. P. Meystre and M. Sargent III, 'Elements of Quantum Optics', Springer-Verlag 1990.
7. P.L. Knight and P.W. Milonni, 'The Rabi frequency in optical spectra', Physics Reports **66** 21-107 (1980).
8. Qilin Wu, D.J. Gauthier, and T.W. Mossberg, Phys. Rev. A. **49**, R1519 (1994).
9. K.-J. Boller, A. Imamoglu, and S.E. Harris, Phys. Rev. Lett. **66**, 2593 (1991).
10. A. Nottelmann, C. Peters, and W. Lange, Phys. Rev. Lett. **70**, 1783 (1993).
11. E.S. Fry, X. Li, D. Nikonov, G.G. Padmabandu, M.O. Scully, A.V. Smith, F.K. Tittel, C. Wang, S.R. Wilkinson, and S.Y. Zhu, Phys. Rev. Lett. **70**, 3235 (1993).
12. W.E. van der Veer, R.J.J. van Diest, A. Dönszelmann, and H.B. van Linden van den Heuvell, Phys. Rev. Lett. **70**, 3243 (1993).
13. M. Fleischhauer, T. McIllrath, and M.O. Scully, Appl. Phys. B. **60**, S123 (1995).
14. J. Dalibard, Y. Castin, and K. Mølmer, Phys. Rev. Lett. **68**, 580 (1992).
15. B. Misra and E.C.G. Sudarshan, J. Math. Phys. **18**, 756 (1977).
16. A. Lezama, Y. Zhu, M. Kanskar, and T.W. Mossberg, Phys. Rev. A. **41**, 1576 (1990).

Subject index

Author Index